Engineering Optimization

Engineering Optimization
Theory and Practice

Fourth Edition

Singiresu S. Rao

WILEY

JOHN WILEY & SONS, INC.

Library of Congress Cataloging-in-Publication Data:

Rao, S. S.
  Engineering optimization : theory and practice / Singiresu S. Rao.–4th ed.
      p. cm.
  Includes index.
  ISBN 978-0-470-18352-6 (cloth)
  1.   Engineering—Mathematical models. 2.   Mathematical optimization. I. Title.
  TA342.R36 2009
  620.001′5196—dc22

                                                                              2009018559

Printed in the United States of America

10 9 8 7 6 5 4 3 2

Contents

10 Integer Programming   588

11 Stochastic Programming   632

Preface

The ever-increasing demand on engineers to lower production costs to withstand global competition has prompted engineers to look for rigorous methods of decision making, such as optimization methods, to design and produce products and systems both economically and efficiently. Optimization techniques, having reached a degree of maturity in recent years, are being used in a wide spectrum of industries, including aerospace, automotive, chemical, electrical, construction, and manufacturing industries. With rapidly advancing computer technology, computers are becoming more powerful, and correspondingly, the size and the complexity of the problems that can be solved using optimization techniques are also increasing. Optimization methods, coupled with modern tools of computer-aided design, are also being used to enhance the creative process of conceptual and detailed design of engineering systems.

The purpose of this textbook is to present the techniques and applications of engineering optimization in a comprehensive manner. The style of the prior editions has been retained, with the theory, computational aspects, and applications of engineering optimization presented with detailed explanations. As in previous editions, essential proofs and developments of the various techniques are given in a simple manner without sacrificing accuracy. New concepts are illustrated with the help of numerical examples. Although most engineering design problems can be solved using nonlinear programming techniques, there are a variety of engineering applications for which other optimization methods, such as linear, geometric, dynamic, integer, and stochastic programming techniques, are most suitable. The theory and applications of all these techniques are also presented in the book. Some of the recently developed methods of optimization, such as genetic algorithms, simulated annealing, particle swarm optimization, ant colony optimization, neural-network-based methods, and fuzzy optimization, are also discussed. Favorable reactions and encouragement from professors, students, and other users of the book have provided me with the impetus to prepare this fourth edition of the book. The following changes have been made from the previous edition:

- Some less-important sections were condensed or deleted.
- Some sections were rewritten for better clarity.
- Some sections were expanded.
- A new chapter on modern methods of optimization is added.
- Several examples to illustrate the use of Matlab for the solution of different types of optimization problems are given.

Features

Each topic in *Engineering Optimization: Theory and Practice* is self-contained, with all concepts explained fully and the derivations presented with complete details. The computational aspects are emphasized throughout with design examples and problems taken from several fields of engineering to make the subject appealing to all branches of engineering. A large number of solved examples, review questions, problems, project-type problems, figures, and references are included to enhance the presentation of the material.

Specific features of the book include:

- More than 130 illustrative examples accompanying most topics.
- More than 480 references to the literature of engineering optimization theory and applications.
- More than 460 review questions to help students in reviewing and testing their understanding of the text material.
- More than 510 problems, with solutions to most problems in the instructor's manual.
- More than 10 examples to illustrate the use of Matlab for the numerical solution of optimization problems.
- Answers to review questions at the web site of the book, www.wiley.com/rao.

I used different parts of the book to teach optimum design and engineering optimization courses at the junior/senior level as well as first-year-graduate-level at Indian Institute of Technology, Kanpur, India; Purdue University, West Lafayette, Indiana; and University of Miami, Coral Gables, Florida. At University of Miami, I cover Chapters 1, 2, 3, 5, 6, and 7 and parts of Chapters 8, 10, 12, and 13 in a dual-level course entitled *Mechanical System Optimization*. In this course, a design project is also assigned to each student in which the student identifies, formulates, and solves a practical engineering problem of his/her interest by applying or modifying an optimization technique. This design project gives the student a feeling for ways that optimization methods work in practice. The book can also be used, with some supplementary material, for a second course on engineering optimization or optimum design or structural optimization. The relative simplicity with which the various topics are presented makes the book useful both to students and to practicing engineers for purposes of self-study. The book also serves as a reference source for different engineering optimization applications. Although the emphasis of the book is on engineering applications, it would also be useful to other areas, such as operations research and economics. A knowledge of matrix theory and differential calculus is assumed on the part of the reader.

Contents

The book consists of fourteen chapters and three appendixes. Chapter 1 provides an introduction to engineering optimization and optimum design and an overview of optimization methods. The concepts of design space, constraint surfaces, and contours of objective function are introduced here. In addition, the formulation of various types of optimization problems is illustrated through a variety of examples taken from various fields of engineering. Chapter 2 reviews the essentials of differential calculus useful in finding the maxima and minima of functions of several variables. The methods of constrained variation and Lagrange multipliers are presented for solving problems with equality constraints. The Kuhn–Tucker conditions for inequality-constrained problems are given along with a discussion of convex programming problems.

Chapters 3 and 4 deal with the solution of linear programming problems. The characteristics of a general linear programming problem and the development of the simplex method of solution are given in Chapter 3. Some advanced topics in linear programming, such as the revised simplex method, duality theory, the decomposition principle, and post-optimality analysis, are discussed in Chapter 4. The extension of linear programming to solve quadratic programming problems is also considered in Chapter 4.

Chapters 5–7 deal with the solution of nonlinear programming problems. In Chapter 5, numerical methods of finding the optimum solution of a function of a single variable are given. Chapter 6 deals with the methods of unconstrained optimization. The algorithms for various zeroth-, first-, and second-order techniques are discussed along with their computational aspects. Chapter 7 is concerned with the solution of nonlinear optimization problems in the presence of inequality and equality constraints. Both the direct and indirect methods of optimization are discussed. The methods presented in this chapter can be treated as the most general techniques for the solution of any optimization problem.

Chapter 8 presents the techniques of geometric programming. The solution techniques for problems of mixed inequality constraints and complementary geometric programming are also considered. In Chapter 9, computational procedures for solving discrete and continuous dynamic programming problems are presented. The problem of dimensionality is also discussed. Chapter 10 introduces integer programming and gives several algorithms for solving integer and discrete linear and nonlinear optimization problems. Chapter 11 reviews the basic probability theory and presents techniques of stochastic linear, nonlinear, and geometric programming. The theory and applications of calculus of variations, optimal control theory, and optimality criteria methods are discussed briefly in Chapter 12. Chapter 13 presents several modern methods of optimization including genetic algorithms, simulated annealing, particle swarm optimization, ant colony optimization, neural-network-based methods, and fuzzy system optimization. Several of the approximation techniques used to speed up the convergence of practical mechanical and structural optimization problems, as well as parallel computation and multiobjective optimization techniques are outlined in Chapter 14. Appendix A presents the definitions and properties of convex and concave functions. A brief discussion of the computational aspects and some of the commercial optimization programs is given in Appendix B. Finally, Appendix C presents a brief introduction to Matlab, optimization toolbox, and use of Matlab programs for the solution of optimization problems.

Acknowledgment

I wish to thank my wife, Kamala, for her patience, understanding, encouragement, and support in preparing the manuscript.

S. S. Rao

srao@miami.edu
January 2009

1

Introduction to Optimization

1.1 INTRODUCTION

Optimization is the act of obtaining the best result under given circumstances. In design, construction, and maintenance of any engineering system, engineers have to take many technological and managerial decisions at several stages. The ultimate goal of all such decisions is either to minimize the effort required or to maximize the desired benefit. Since the effort required or the benefit desired in any practical situation can be expressed as a function of certain decision variables, *optimization* can be defined as the process of finding the conditions that give the maximum or minimum value of a function. It can be seen from Fig. 1.1 that if a point x^* corresponds to the minimum value of function $f(x)$, the same point also corresponds to the maximum value of the negative of the function, $-f(x)$. Thus without loss of generality, optimization can be taken to mean minimization since the maximum of a function can be found by seeking the minimum of the negative of the same function.

In addition, the following operations on the objective function will not change the optimum solution x^* (see Fig. 1.2):

1. Multiplication (or division) of $f(x)$ by a positive constant c.
2. Addition (or subtraction) of a positive constant c to (or from) $f(x)$.

There is no single method available for solving all optimization problems efficiently. Hence a number of optimization methods have been developed for solving different types of optimization problems. The optimum seeking methods are also known as *mathematical programming techniques* and are generally studied as a part of operations research. *Operations research* is a branch of mathematics concerned with the application of scientific methods and techniques to decision making problems and with establishing the best or optimal solutions. The beginnings of the subject of operations research can be traced to the early period of World War II. During the war, the British military faced the problem of allocating very scarce and limited resources (such as fighter airplanes, radars, and submarines) to several activities (deployment to numerous targets and destinations). Because there were no systematic methods available to solve resource allocation problems, the military called upon a team of mathematicians to develop methods for solving the problem in a scientific manner. The methods developed by the team were instrumental in the winning of the Air Battle by Britain. These methods, such as linear programming, which were developed as a result of research on (military) operations, subsequently became known as the methods of operations research.

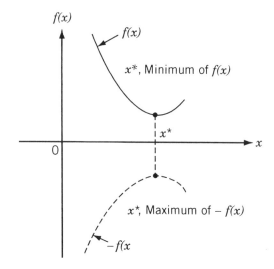

Figure 1.1   Minimum of $f(x)$ is same as maximum of $-f(x)$.

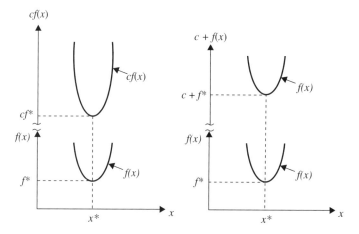

Figure 1.2   Optimum solution of $cf(x)$ or $c + f(x)$ same as that of $f(x)$.

Table 1.1 lists various mathematical programming techniques together with other well-defined areas of operations research. The classification given in Table 1.1 is not unique; it is given mainly for convenience.

Mathematical programming techniques are useful in finding the minimum of a function of several variables under a prescribed set of constraints. Stochastic process techniques can be used to analyze problems described by a set of random variables having known probability distributions. Statistical methods enable one to analyze the experimental data and build empirical models to obtain the most accurate representation of the physical situation. This book deals with the theory and application of mathematical programming techniques suitable for the solution of engineering design problems.

Table 1.1   Methods of Operations Research

Mathematical programming or optimization techniques	Stochastic process techniques	Statistical methods
Calculus methods	Statistical decision theory	Regression analysis
Calculus of variations	Markov processes	Cluster analysis, pattern recognition
Nonlinear programming	Queueing theory	
Geometric programming	Renewal theory	Design of experiments
Quadratic programming	Simulation methods	Discriminate analysis (factor analysis)
Linear programming	Reliability theory	
Dynamic programming		
Integer programming		
Stochastic programming		
Separable programming		
Multiobjective programming		
Network methods: CPM and PERT		
Game theory		

Modern or nontraditional optimization techniques

Genetic algorithms
Simulated annealing
Ant colony optimization
Particle swarm optimization
Neural networks
Fuzzy optimization

1.2   HISTORICAL DEVELOPMENT

The existence of optimization methods can be traced to the days of Newton, Lagrange, and Cauchy. The development of differential calculus methods of optimization was possible because of the contributions of Newton and Leibnitz to calculus. The foundations of calculus of variations, which deals with the minimization of functionals, were laid by Bernoulli, Euler, Lagrange, and Weirstrass. The method of optimization for constrained problems, which involves the addition of unknown multipliers, became known by the name of its inventor, Lagrange. Cauchy made the first application of the steepest descent method to solve unconstrained minimization problems. Despite these early contributions, very little progress was made until the middle of the twentieth century, when high-speed digital computers made implementation of the optimization procedures possible and stimulated further research on new methods. Spectacular advances followed, producing a massive literature on optimization techniques. This advancement also resulted in the emergence of several well-defined new areas in optimization theory.

It is interesting to note that the major developments in the area of numerical methods of unconstrained optimization have been made in the United Kingdom only in the 1960s. The development of the simplex method by Dantzig in 1947 for linear programming problems and the annunciation of the principle of optimality in 1957 by Bellman for dynamic programming problems paved the way for development of the methods of constrained optimization. Work by Kuhn and Tucker in 1951 on the necessary and sufficiency conditions for the optimal solution of programming problems laid the foundations for a great deal of later research in nonlinear programming. The contributions of Zoutendijk and Rosen to nonlinear programming during the early 1960s have been significant. Although no single technique has been found to be universally applicable for nonlinear programming problems, work of Carroll and Fiacco and McCormick allowed many difficult problems to be solved by using the well-known techniques of unconstrained optimization. Geometric programming was developed in the 1960s by Duffin, Zener, and Peterson. Gomory did pioneering work in integer programming, one of the most exciting and rapidly developing areas of optimization. The reason for this is that most real-world applications fall under this category of problems. Dantzig and Charnes and Cooper developed stochastic programming techniques and solved problems by assuming design parameters to be independent and normally distributed.

The desire to optimize more than one objective or goal while satisfying the physical limitations led to the development of multiobjective programming methods. Goal programming is a well-known technique for solving specific types of multiobjective optimization problems. The goal programming was originally proposed for linear problems by Charnes and Cooper in 1961. The foundations of game theory were laid by von Neumann in 1928 and since then the technique has been applied to solve several mathematical economics and military problems. Only during the last few years has game theory been applied to solve engineering design problems.

Modern Methods of Optimization. The modern optimization methods, also sometimes called nontraditional optimization methods, have emerged as powerful and popular methods for solving complex engineering optimization problems in recent years. These methods include genetic algorithms, simulated annealing, particle swarm optimization, ant colony optimization, neural network-based optimization, and fuzzy optimization. The genetic algorithms are computerized search and optimization algorithms based on the mechanics of natural genetics and natural selection. The genetic algorithms were originally proposed by John Holland in 1975. The simulated annealing method is based on the mechanics of the cooling process of molten metals through annealing. The method was originally developed by Kirkpatrick, Gelatt, and Vecchi.

The particle swarm optimization algorithm mimics the behavior of social organisms such as a colony or swarm of insects (for example, ants, termites, bees, and wasps), a flock of birds, and a school of fish. The algorithm was originally proposed by Kennedy and Eberhart in 1995. The ant colony optimization is based on the cooperative behavior of ant colonies, which are able to find the shortest path from their nest to a food source. The method was first developed by Marco Dorigo in 1992. The neural network methods are based on the immense computational power of the nervous system to solve perceptional problems in the presence of massive amount of sensory data through its parallel processing capability. The method was originally used for optimization by Hopfield and Tank in 1985. The fuzzy optimization methods were developed to solve optimization problems involving design data, objective function, and constraints stated in imprecise form involving vague and linguistic descriptions. The fuzzy approaches for single and multiobjective optimization in engineering design were first presented by Rao in 1986.

1.3 ENGINEERING APPLICATIONS OF OPTIMIZATION

Optimization, in its broadest sense, can be applied to solve any engineering problem. Some typical applications from different engineering disciplines indicate the wide scope of the subject:

1. Design of aircraft and aerospace structures for minimum weight
2. Finding the optimal trajectories of space vehicles
3. Design of civil engineering structures such as frames, foundations, bridges, towers, chimneys, and dams for minimum cost
4. Minimum-weight design of structures for earthquake, wind, and other types of random loading
5. Design of water resources systems for maximum benefit
6. Optimal plastic design of structures
7. Optimum design of linkages, cams, gears, machine tools, and other mechanical components
8. Selection of machining conditions in metal-cutting processes for minimum production cost
9. Design of material handling equipment, such as conveyors, trucks, and cranes, for minimum cost
10. Design of pumps, turbines, and heat transfer equipment for maximum efficiency
11. Optimum design of electrical machinery such as motors, generators, and transformers
12. Optimum design of electrical networks
13. Shortest route taken by a salesperson visiting various cities during one tour
14. Optimal production planning, controlling, and scheduling
15. Analysis of statistical data and building empirical models from experimental results to obtain the most accurate representation of the physical phenomenon
16. Optimum design of chemical processing equipment and plants
17. Design of optimum pipeline networks for process industries
18. Selection of a site for an industry
19. Planning of maintenance and replacement of equipment to reduce operating costs
20. Inventory control
21. Allocation of resources or services among several activities to maximize the benefit
22. Controlling the waiting and idle times and queueing in production lines to reduce the costs
23. Planning the best strategy to obtain maximum profit in the presence of a competitor
24. Optimum design of control systems

1.4   STATEMENT OF AN OPTIMIZATION PROBLEM

An optimization or a mathematical programming problem can be stated as follows.

$$\text{Find } \mathbf{X} = \begin{Bmatrix} x_1 \\ x_2 \\ \vdots \\ x_n \end{Bmatrix} \text{ which minimizes } f(\mathbf{X})$$

subject to the constraints

$$
\begin{aligned}
g_j(\mathbf{X}) &\leq 0, & j = 1, 2, \ldots, m \\
l_j(\mathbf{X}) &= 0, & j = 1, 2, \ldots, p
\end{aligned}
\tag{1.1}
$$

where $\mathbf{X}$ is an n-dimensional vector called the *design vector*, $f(\mathbf{X})$ is termed the *objective function*, and $g_j(\mathbf{X})$ and $l_j(\mathbf{X})$ are known as *inequality* and *equality* constraints, respectively. The number of variables n and the number of constraints m and/or p need not be related in any way. The problem stated in Eq. (1.1) is called a *constrained optimization problem*.[†] Some optimization problems do not involve any constraints and can be stated as

$$\text{Find } \mathbf{X} = \begin{Bmatrix} x_1 \\ x_2 \\ \vdots \\ x_n \end{Bmatrix} \text{ which minimizes } f(\mathbf{X}) \tag{1.2}$$

Such problems are called *unconstrained optimization problems*.

1.4.1   Design Vector

Any engineering system or component is defined by a set of quantities some of which are viewed as variables during the design process. In general, certain quantities are usually fixed at the outset and these are called *preassigned parameters*. All the other quantities are treated as variables in the design process and are called *design* or *decision variables* x_i, $i = 1, 2, \ldots, n$. The design variables are collectively represented as a design vector $\mathbf{X} = \{x_1, x_2, \ldots, x_n\}^{\mathrm{T}}$. As an example, consider the design of the gear pair shown in Fig. 1.3, characterized by its face width b, number of teeth T_1 and T_2, center distance d, pressure angle ψ, tooth profile, and material. If center distance d, pressure angle ψ, tooth profile, and material of the gears are fixed in advance, these quantities can be called *preassigned parameters*. The remaining quantities can be collectively represented by a design vector $\mathbf{X} = \{x_1, x_2, x_3\}^{\mathrm{T}} = \{b, T_1, T_2\}^{\mathrm{T}}$. If there are no restrictions on the choice of b, T_1, and T_2, any set of three numbers will constitute a design for the gear pair. If an n-dimensional Cartesian space with each coordinate axis representing a design variable x_i $(i = 1, 2, \ldots, n)$ is considered, the space is called

---

[†]In the mathematical programming literature, the equality constraints $l_j(\mathbf{X}) = 0$, $j = 1, 2, \ldots, p$ are often neglected, for simplicity, in the statement of a constrained optimization problem, although several methods are available for handling problems with equality constraints.

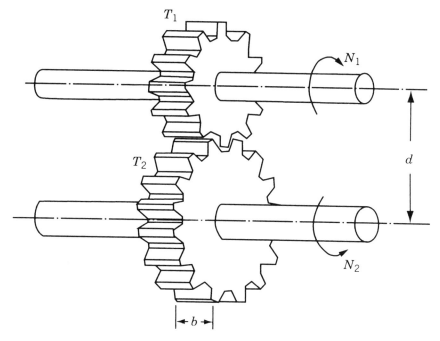

Figure 1.3   Gear pair in mesh.

the *design variable space* or simply *design space*. Each point in the n-dimensional design space is called a *design point* and represents either a possible or an impossible solution to the design problem. In the case of the design of a gear pair, the design point $\{1.0, 20, 40\}^{\mathrm{T}}$, for example, represents a possible solution, whereas the design point $\{1.0, -20, 40.5\}^{\mathrm{T}}$ represents an impossible solution since it is not possible to have either a negative value or a fractional value for the number of teeth.

1.4.2   Design Constraints

In many practical problems, the design variables cannot be chosen arbitrarily; rather, they have to satisfy certain specified functional and other requirements. The restrictions that must be satisfied to produce an acceptable design are collectively called *design constraints*. Constraints that represent limitations on the behavior or performance of the system are termed *behavior* or *functional constraints*. Constraints that represent physical limitations on design variables, such as availability, fabricability, and transportability, are known as *geometric* or *side constraints*. For example, for the gear pair shown in Fig. 1.3, the face width b cannot be taken smaller than a certain value, due to strength requirements. Similarly, the ratio of the numbers of teeth, T_1/T_2, is dictated by the speeds of the input and output shafts, N_1 and N_2. Since these constraints depend on the performance of the gear pair, they are called behavior constraints. The values of T_1 and T_2 cannot be any real numbers but can only be integers. Further, there can be upper and lower bounds on T_1 and T_2 due to manufacturing limitations. Since these constraints depend on the physical limitations, they are called side *constraints*.

1.4.3   Constraint Surface

For illustration, consider an optimization problem with only inequality constraints $g_j(\mathbf{X}) \leq 0$. The set of values of $\mathbf{X}$ that satisfy the equation $g_j(\mathbf{X}) = 0$ forms a hypersurface in the design space and is called a *constraint surface*. Note that this is an $(n-1)$-dimensional subspace, where n is the number of design variables. The constraint surface divides the design space into two regions: one in which $g_j(\mathbf{X}) < 0$ and the other in which $g_j(\mathbf{X}) > 0$. Thus the points lying on the hypersurface will satisfy the constraint $g_j(\mathbf{X})$ critically, whereas the points lying in the region where $g_j(\mathbf{X}) > 0$ are infeasible or unacceptable, and the points lying in the region where $g_j(\mathbf{X}) < 0$ are feasible or acceptable. The collection of all the constraint surfaces $g_j(\mathbf{X}) = 0$, $j = 1, 2, \ldots, m$, which separates the acceptable region is called the *composite constraint surface*.

Figure 1.4 shows a hypothetical two-dimensional design space where the infeasible region is indicated by hatched lines. A design point that lies on one or more than one constraint surface is called a *bound point*, and the associated constraint is called an *active constraint*. Design points that do not lie on any constraint surface are known as *free points*. Depending on whether a particular design point belongs to the acceptable or unacceptable region, it can be identified as one of the following four types:

1. Free and acceptable point
2. Free and unacceptable point
3. Bound and acceptable point
4. Bound and unacceptable point

All four types of points are shown in Fig. 1.4.

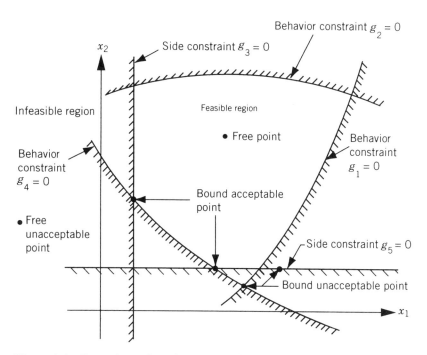

Figure 1.4   Constraint surfaces in a hypothetical two-dimensional design space.

1.4.4   Objective Function

The conventional design procedures aim at finding an acceptable or adequate design that merely satisfies the functional and other requirements of the problem. In general, there will be more than one acceptable design, and the purpose of optimization is to choose the best one of the many acceptable designs available. Thus a criterion has to be chosen for comparing the different alternative acceptable designs and for selecting the best one. The criterion with respect to which the design is optimized, when expressed as a function of the design variables, is known as the *criterion* or *merit* or *objective function*. The choice of objective function is governed by the nature of problem. The objective function for minimization is generally taken as weight in aircraft and aerospace structural design problems. In civil engineering structural designs, the objective is usually taken as the minimization of cost. The maximization of mechanical efficiency is the obvious choice of an objective in mechanical engineering systems design. Thus the choice of the objective function appears to be straightforward in most design problems. However, there may be cases where the optimization with respect to a particular criterion may lead to results that may not be satisfactory with respect to another criterion. For example, in mechanical design, a gearbox transmitting the maximum power may not have the minimum weight. Similarly, in structural design, the minimum weight design may not correspond to minimum stress design, and the minimum stress design, again, may not correspond to maximum frequency design. Thus the selection of the objective function can be one of the most important decisions in the whole optimum design process.

In some situations, there may be more than one criterion to be satisfied simultaneously. For example, a gear pair may have to be designed for minimum weight and maximum efficiency while transmitting a specified horsepower. An optimization problem involving multiple objective functions is known as a *multiobjective programming problem*. With multiple objectives there arises a possibility of conflict, and one simple way to handle the problem is to construct an overall objective function as a linear combination of the conflicting multiple objective functions. Thus if $f_1(\mathbf{X})$ and $f_2(\mathbf{X})$ denote two objective functions, construct a new (overall) objective function for optimization as

$$f(\mathbf{X}) = \alpha_1 f_1(\mathbf{X}) + \alpha_2 f_2(\mathbf{X}) \tag{1.3}$$

where α_1 and α_2 are constants whose values indicate the relative importance of one objective function relative to the other.

1.4.5   Objective Function Surfaces

The locus of all points satisfying $f(\mathbf{X}) = C = $ constant forms a hypersurface in the design space, and each value of C corresponds to a different member of a family of surfaces. These surfaces, called *objective function surfaces*, are shown in a hypothetical two-dimensional design space in Fig. 1.5.

Once the objective function surfaces are drawn along with the constraint surfaces, the optimum point can be determined without much difficulty. But the main problem is that as the number of design variables exceeds two or three, the constraint and objective function surfaces become complex even for visualization and the problem

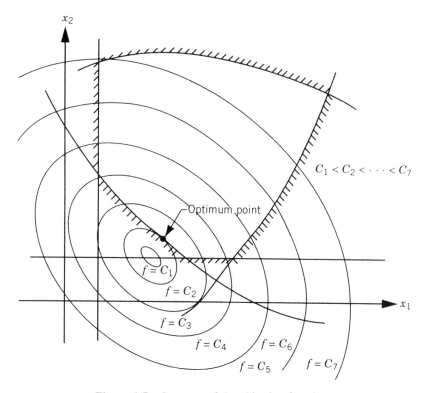

Figure 1.5   Contours of the objective function.

has to be solved purely as a mathematical problem. The following example illustrates the graphical optimization procedure.

Example 1.1   Design a uniform column of tubular section, with hinge joints at both ends, (Fig. 1.6) to carry a compressive load $P = 2500\,\text{kg}_\text{f}$ for minimum cost. The column is made up of a material that has a yield stress (σ_y) of $500\,\text{kg}_\text{f}/\text{cm}^2$, modulus of elasticity (E) of $0.85 \times 10^6\,\text{kg}_\text{f}/\text{cm}^2$, and weight density ($\rho$) of $0.0025\,\text{kg}_\text{f}/\text{cm}^3$. The length of the column is $250\,\text{cm}$. The stress induced in the column should be less than the buckling stress as well as the yield stress. The mean diameter of the column is restricted to lie between 2 and $14\,\text{cm}$, and columns with thicknesses outside the range 0.2 to $0.8\,\text{cm}$ are not available in the market. The cost of the column includes material and construction costs and can be taken as $5W + 2d$, where W is the weight in kilograms force and d is the mean diameter of the column in centimeters.

SOLUTION   The design variables are the mean diameter (d) and tube thickness (t):

$$\mathbf{X} = \begin{Bmatrix} x_1 \\ x_2 \end{Bmatrix} = \begin{Bmatrix} d \\ t \end{Bmatrix} \tag{E$_1$}$$

The objective function to be minimized is given by

$$f(\mathbf{X}) = 5W + 2d = 5\rho l \pi\, dt + 2d = 9.82x_1 x_2 + 2x_1 \tag{E$_2$}$$

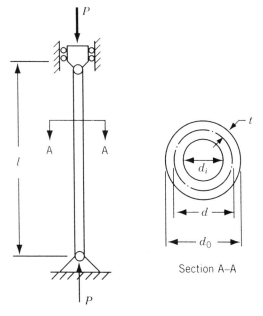

Figure 1.6 Tubular column under compression.

The behavior constraints can be expressed as

$$\text{stress induced} \le \text{yield stress}$$

$$\text{stress induced} \le \text{buckling stress}$$

The induced stress is given by

$$\text{induced stress} = \sigma_i = \frac{P}{\pi \, dt} = \frac{2500}{\pi x_1 x_2} \tag{E_3}$$

The buckling stress for a pin-connected column is given by

$$\text{buckling stress} = \sigma_b = \frac{\text{Euler buckling load}}{\text{cross-sectional area}} = \frac{\pi^2 E I}{l^2} \frac{1}{\pi \, dt} \tag{E_4}$$

where

$$I = \text{second moment of area of the cross section of the column}$$

$$= \frac{\pi}{64}(d_o^4 - d_i^4)$$

$$= \frac{\pi}{64}(d_o^2 + d_i^2)(d_o + d_i)(d_o - d_i) = \frac{\pi}{64}[(d+t)^2 + (d-t)^2]$$

$$\times [(d+t) + (d-t)][(d+t) - (d-t)]$$

$$= \frac{\pi}{8} dt(d^2 + t^2) = \frac{\pi}{8} x_1 x_2 (x_1^2 + x_2^2) \tag{E_5}$$

Thus the behavior constraints can be restated as

$$g_1(\mathbf{X}) = \frac{2500}{\pi x_1 x_2} - 500 \leq 0 \qquad (E_6)$$

$$g_2(\mathbf{X}) = \frac{2500}{\pi x_1 x_2} - \frac{\pi^2(0.85 \times 10^6)(x_1^2 + x_2^2)}{8(250)^2} \leq 0 \qquad (E_7)$$

The side constraints are given by

$$2 \leq d \leq 14$$

$$0.2 \leq t \leq 0.8$$

which can be expressed in standard form as

$$g_3(\mathbf{X}) = -x_1 + 2.0 \leq 0 \qquad (E_8)$$

$$g_4(\mathbf{X}) = x_1 - 14.0 \leq 0 \qquad (E_9)$$

$$g_5(\mathbf{X}) = -x_2 + 0.2 \leq 0 \qquad (E_{10})$$

$$g_6(\mathbf{X}) = x_2 - 0.8 \leq 0 \qquad (E_{11})$$

Since there are only two design variables, the problem can be solved graphically as shown below.

First, the constraint surfaces are to be plotted in a two-dimensional design space where the two axes represent the two design variables x_1 and x_2. To plot the first constraint surface, we have

$$g_1(\mathbf{X}) = \frac{2500}{\pi x_1 x_2} - 500 \leq 0$$

that is,

$$x_1 x_2 \geq 1.593$$

Thus the curve $x_1 x_2 = 1.593$ represents the constraint surface $g_1(\mathbf{X}) = 0$. This curve can be plotted by finding several points on the curve. The points on the curve can be found by giving a series of values to x_1 and finding the corresponding values of x_2 that satisfy the relation $x_1 x_2 = 1.593$:

x_1	2.0	4.0	6.0	8.0	10.0	12.0	14.0
x_2	0.7965	0.3983	0.2655	0.1990	0.1593	0.1328	0.1140

These points are plotted and a curve $P_1 Q_1$ passing through all these points is drawn as shown in Fig. 1.7, and the infeasible region, represented by $g_1(\mathbf{X}) > 0$ or $x_1 x_2 < 1.593$, is shown by hatched lines.[†] Similarly, the second constraint $g_2(\mathbf{X}) \leq 0$ can be expressed as $x_1 x_2(x_1^2 + x_2^2) \geq 47.3$ and the points lying on the constraint surface $g_2(\mathbf{X}) = 0$ can be obtained as follows for $x_1 x_2(x_1^2 + x_2^2) = 47.3$:

[†]The infeasible region can be identified by testing whether the origin lies in the feasible or infeasible region.

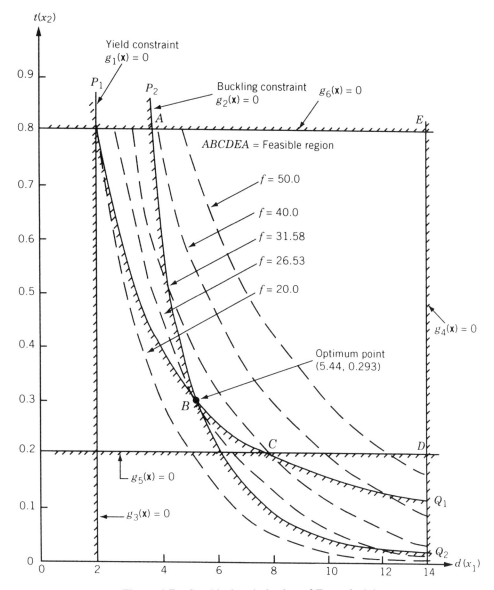

Figure 1.7   Graphical optimization of Example 1.1.

x_1	2	4	6	8	10	12	14
x_2	2.41	0.716	0.219	0.0926	0.0473	0.0274	0.0172

These points are plotted as curve $P_2 Q_2$, the feasible region is identified, and the infeasible region is shown by hatched lines as in Fig. 1.7. The plotting of side constraints is very simple since they represent straight lines. After plotting all the six constraints, the feasible region can be seen to be given by the bounded area *ABCDEA*.

Next, the contours of the objective function are to be plotted before finding the optimum point. For this, we plot the curves given by

$$f(\mathbf{X}) = 9.82x_1x_2 + 2x_1 = c = \text{constant}$$

for a series of values of c. By giving different values to c, the contours of f can be plotted with the help of the following points.

For $9.82x_1x_2 + 2x_1 = 50.0$:

x_2	0.1	0.2	0.3	0.4	0.5	0.6	0.7	0.8
x_1	16.77	12.62	10.10	8.44	7.24	6.33	5.64	5.07

For $9.82x_1x_2 + 2x_1 = 40.0$:

x_2	0.1	0.2	0.3	0.4	0.5	0.6	0.7	0.8
x_1	13.40	10.10	8.08	6.75	5.79	5.06	4.51	4.05

For $9.82x_1x_2 + 2x_1 = 31.58$ (passing through the corner point C):

x_2	0.1	0.2	0.3	0.4	0.5	0.6	0.7	0.8
x_1	10.57	7.96	6.38	5.33	4.57	4.00	3.56	3.20

For $9.82x_1x_2 + 2x_1 = 26.53$ (passing through the corner point B):

x_2	0.1	0.2	0.3	0.4	0.5	0.6	0.7	0.8
x_1	8.88	6.69	5.36	4.48	3.84	3.36	2.99	2.69

For $9.82x_1x_2 + 2x_1 = 20.0$:

x_2	0.1	0.2	0.3	0.4	0.5	0.6	0.7	0.8
x_1	6.70	5.05	4.04	3.38	2.90	2.53	2.26	2.02

These contours are shown in Fig. 1.7 and it can be seen that the objective function cannot be reduced below a value of 26.53 (corresponding to point B) without violating some of the constraints. Thus the optimum solution is given by point B with $d^* = x_1^* = 5.44$ cm and $t^* = x_2^* = 0.293$ cm with $f_{\min} = 26.53$.

1.5 CLASSIFICATION OF OPTIMIZATION PROBLEMS

Optimization problems can be classified in several ways, as described below.

1.5.1 Classification Based on the Existence of Constraints

As indicated earlier, any optimization problem can be classified as constrained or unconstrained, depending on whether constraints exist in the problem.

1.5.2   Classification Based on the Nature of the Design Variables

Based on the nature of design variables encountered, optimization problems can be classified into two broad categories. In the first category, the problem is to find values to a set of design parameters that make some prescribed function of these parameters minimum subject to certain constraints. For example, the problem of minimum-weight design of a prismatic beam shown in Fig. 1.8a subject to a limitation on the maximum deflection can be stated as follows:

$$\text{Find } \mathbf{X} = \begin{Bmatrix} b \\ d \end{Bmatrix} \text{ which minimizes}$$

$$f(\mathbf{X}) = \rho l b d \tag{1.4}$$

subject to the constraints

$$\delta_{\text{tip}}(\mathbf{X}) \leq \delta_{\text{max}}$$

$$b \geq 0$$

$$d \geq 0$$

where ρ is the density and δ_{tip} is the tip deflection of the beam. Such problems are called *parameter* or *static optimization problems*. In the second category of problems, the objective is to find a set of design parameters, which are all continuous functions of some other parameter, that minimizes an objective function subject to a set of constraints. If the cross-sectional dimensions of the rectangular beam are allowed to vary along its length as shown in Fig. 1.8b, the optimization problem can be stated as

$$\text{Find } \mathbf{X}(t) = \begin{Bmatrix} b(t) \\ d(t) \end{Bmatrix} \text{ which minimizes}$$

$$f[\mathbf{X}(t)] = \rho \int_0^l b(t) \, d(t) \, dt \tag{1.5}$$

subject to the constraints

$$\delta_{\text{tip}}[\mathbf{X}(t)] \leq \delta_{\text{max}}, \qquad 0 \leq t \leq l$$

$$b(t) \geq 0, \qquad 0 \leq t \leq l$$

$$d(t) \geq 0, \qquad 0 \leq t \leq l$$

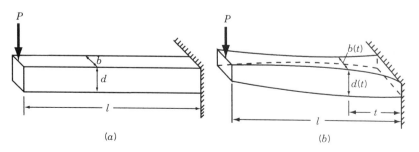

Figure 1.8   Cantilever beam under concentrated load.

Here the design variables are functions of the length parameter t. This type of problem, where each design variable is a function of one or more parameters, is known as a *trajectory* or *dynamic optimization problem* [1.55].

1.5.3 Classification Based on the Physical Structure of the Problem

Depending on the physical structure of the problem, optimization problems can be classified as optimal control and nonoptimal control problems.

Optimal Control Problem. An *optimal control* (OC) *problem* is a mathematical programming problem involving a number of stages, where each stage evolves from the preceding stage in a prescribed manner. It is usually described by two types of variables: the control (design) and the state variables. The *control variables* define the system and govern the evolution of the system from one stage to the next, and the *state variables* describe the behavior or status of the system in any stage. The problem is to find a set of control or design variables such that the total objective function (also known as the *performance index*, PI) over all the stages is minimized subject to a set of constraints on the control and state variables. An OC problem can be stated as follows [1.55]:

$$\text{Find } \mathbf{X} \text{ which minimizes } f(\mathbf{X}) = \sum_{i=1}^{l} f_i(x_i, y_i) \qquad (1.6)$$

subject to the constraints

$$q_i(x_i, y_i) + y_i = y_{i+1}, \quad i = 1, 2, \ldots, l$$
$$\mathbf{g}_j(x_j) \leq 0, \qquad j = 1, 2, \ldots, l$$
$$\mathbf{h}_k(y_k) \leq 0, \qquad k = 1, 2, \ldots, l$$

where x_i is the ith control variable, y_i the ith state variable, and f_i the contribution of the ith stage to the total objective function; $\mathbf{g}_j, \mathbf{h}_k$, and q_i are functions of x_j, y_k, and x_i and y_i, respectively, and l is the total number of stages. The control and state variables x_i and y_i can be vectors in some cases. The following example serves to illustrate the nature of an optimal control problem.

Example 1.2 A rocket is designed to travel a distance of $12s$ in a vertically upward direction [1.39]. The thrust of the rocket can be changed only at the discrete points located at distances of $0, s, 2s, 3s, \ldots, 12s$. If the maximum thrust that can be developed at point i either in the positive or negative direction is restricted to a value of F_i, formulate the problem of minimizing the total time of travel under the following assumptions:

1. The rocket travels against the gravitational force.
2. The mass of the rocket reduces in proportion to the distance traveled.
3. The air resistance is proportional to the velocity of the rocket.

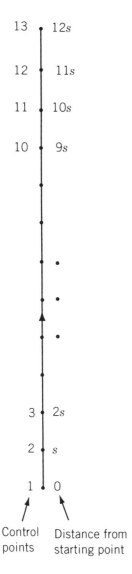

Control points   Distance from starting point   **Figure 1.9**   Control points in the path of the rocket.

SOLUTION   Let points (or control points) on the path at which the thrusts of the rocket are changed be numbered as $1, 2, 3, \ldots, 13$ (Fig. 1.9). Denoting x_i as the thrust, v_i the velocity, a_i the acceleration, and m_i the mass of the rocket at point i, Newton's second law of motion can be applied as

$$\text{net force on the rocket} = \text{mass} \times \text{acceleration}$$

This can be written as

$$\text{thrust} - \text{gravitational force} - \text{air resistance} = \text{mass} \times \text{acceleration}$$

or

$$x_i - m_i g - k_1 v_i = m_i a_i \tag{E_1}$$

where the mass m_i can be expressed as

$$m_i = m_{i-1} - k_2 s \tag{E_2}$$

and k_1 and k_2 are constants. Equation (E_1) can be used to express the acceleration, a_i, as

$$a_i = \frac{x_i}{m_i} - g - \frac{k_1 v_i}{m_i} \tag{E_3}$$

If t_i denotes the time taken by the rocket to travel from point i to point $i + 1$, the distance traveled between the points i and $i + 1$ can be expressed as

$$s = v_i t_i + \tfrac{1}{2} a_i t_i^2$$

or

$$\frac{1}{2} t_i^2 \left(\frac{x_i}{m_i} - g - \frac{k_1 v_i}{m_i} \right) + t_i v_i - s = 0 \tag{E_4}$$

from which t_i can be determined as

$$t_i = \frac{-v_i \pm \sqrt{v_i^2 + 2s \left(\dfrac{x_i}{m_i} - g - \dfrac{k_1 v_i}{m_i} \right)}}{\dfrac{x_i}{m_i} - g - \dfrac{k_1 v_i}{m_i}} \tag{E_5}$$

Of the two values given by Eq. (E_5), the positive value has to be chosen for t_i. The velocity of the rocket at point $i + 1$, v_{i+1}, can be expressed in terms of v_i as (by assuming the acceleration between points i and $i + 1$ to be constant for simplicity)

$$v_{i+1} = v_i + a_i t_i \tag{E_6}$$

The substitution of Eqs. (E_3) and (E_5) into Eq. (E_6) leads to

$$v_{i+1} = \sqrt{v_i^2 + 2s \left(\frac{x_i}{m_i} - g - \frac{k_1 v_i}{m_i} \right)} \tag{E_7}$$

From an analysis of the problem, the control variables can be identified as the thrusts, x_i, and the state variables as the velocities, v_i. Since the rocket starts at point 1 and stops at point 13,

$$v_1 = v_{13} = 0 \tag{E_8}$$

Thus the problem can be stated as an OC problem as

$$\text{Find } \mathbf{X} = \begin{Bmatrix} x_1 \\ x_2 \\ \vdots \\ x_{12} \end{Bmatrix} \text{ which minimizes}$$

$$f(\mathbf{X}) = \sum_{i=1}^{12} t_i = \sum_{i=1}^{12} \left\{ \frac{-v_i + \sqrt{v_i^2 + 2s\left(\dfrac{x_i}{m_i} - g - \dfrac{k_1 v_i}{m_i}\right)}}{\dfrac{x_i}{m_i} - g - \dfrac{k_1 v_i}{m_i}} \right\}$$

subject to

$$m_{i+1} = m_i - k_2 s, \qquad i = 1, 2, \ldots, 12$$

$$v_{i+1} = \sqrt{v_i^2 + 2s\left(\frac{x_i}{m_i} - g - \frac{k_1 v_i}{m_i}\right)}, \qquad i = 1, 2, \ldots, 12$$

$$|x_i| \leq F_i, \qquad i = 1, 2, \ldots, 12$$

$$v_1 = v_{13} = 0$$

1.5.4   Classification Based on the Nature of the Equations Involved

Another important classification of optimization problems is based on the nature of expressions for the objective function and the constraints. According to this classification, optimization problems can be classified as linear, nonlinear, geometric, and quadratic programming problems. This classification is extremely useful from the computational point of view since there are many special methods available for the efficient solution of a particular class of problems. Thus the first task of a designer would be to investigate the class of problem encountered. This will, in many cases, dictate the types of solution procedures to be adopted in solving the problem.

Nonlinear Programming Problem.   If any of the functions among the objective and constraint functions in Eq. (1.1) is nonlinear, the problem is called a *nonlinear programming* (NLP) *problem*. This is the most general programming problem and all other problems can be considered as special cases of the NLP problem.

Example 1.3   The step-cone pulley shown in Fig. 1.10 is to be designed for transmitting a power of at least 0.75 hp. The speed of the input shaft is 350 rpm and the output speed requirements are 750, 450, 250, and 150 rpm for a fixed center distance of a between the input and output shafts. The tension on the tight side of the belt is to be kept more than twice that on the slack side. The thickness of the belt is t and the coefficient of friction between the belt and the pulleys is μ. The stress induced in the belt due to tension on the tight side is s. Formulate the problem of finding the width and diameters of the steps for minimum weight.

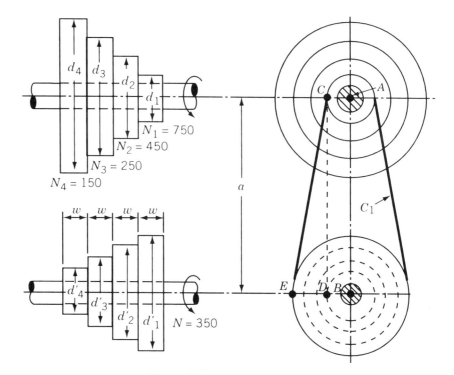

Figure 1.10 Step-cone pulley.

SOLUTION   The design vector can be taken as

$$\mathbf{X} = \begin{Bmatrix} d_1 \\ d_2 \\ d_3 \\ d_4 \\ w \end{Bmatrix}$$

where d_i is the diameter of the ith step on the output pulley and w is the width of the belt and the steps. The objective function is the weight of the step-cone pulley system:

$$f(\mathbf{X}) = \rho w \frac{\pi}{4} (d_1^2 + d_2^2 + d_3^2 + d_4^2 + d_1'^2 + d_2'^2 + d_3'^2 + d_4'^2)$$

$$= \rho w \frac{\pi}{4} \left\{ d_1^2 \left[1 + \left(\frac{750}{350} \right)^2 \right] + d_2^2 \left[1 + \left(\frac{450}{350} \right)^2 \right] \right.$$

$$\left. + d_3^2 \left[1 + \left(\frac{250}{350} \right)^2 \right] + d_4^2 \left[1 + \left(\frac{150}{350} \right)^2 \right] \right\} \tag{E_1}$$

where ρ is the density of the pulleys and d_i' is the diameter of the ith step on the input pulley.

To have the belt equally tight on each pair of opposite steps, the total length of the belt must be kept constant for all the output speeds. This can be ensured by satisfying the following equality constraints:

$$C_1 - C_2 = 0 \tag{E_2}$$

$$C_1 - C_3 = 0 \tag{E_3}$$

$$C_1 - C_4 = 0 \tag{E_4}$$

where C_i denotes length of the belt needed to obtain output speed N_i ($i = 1, 2, 3, 4$) and is given by [1.116, 1.117]:

$$C_i \simeq \frac{\pi d_i}{2} \left(1 + \frac{N_i}{N} \right) + \frac{\left(\frac{N_i}{N} - 1 \right)^2 d_i^2}{4a} + 2a$$

where N is the speed of the input shaft and a is the center distance between the shafts. The ratio of tensions in the belt can be expressed as [1.116, 1.117]

$$\frac{T_1^i}{T_2^i} = e^{\mu \theta_i}$$

where T_1^i and T_2^i are the tensions on the tight and slack sides of the ith step, μ the coefficient of friction, and θ_i the angle of lap of the belt over the ith pulley step. The angle of lap is given by

$$\theta_i = \pi - 2 \sin^{-1} \left[\frac{\left(\frac{N_i}{N} - 1 \right) d_i}{2a} \right]$$

and hence the constraint on the ratio of tensions becomes

$$\exp \left\{ \mu \left[\pi - 2 \sin^{-1} \left\{ \left(\frac{N_i}{N} - 1 \right) \frac{d_i}{2a} \right\} \right] \right\} \geq 2, \qquad i = 1, 2, 3, 4 \tag{E_5}$$

The limitation on the maximum tension can be expressed as

$$T_1^i = stw, \qquad i = 1, 2, 3, 4 \tag{E_6}$$

where s is the maximum allowable stress in the belt and t is the thickness of the belt. The constraint on the power transmitted can be stated as (using lb_f for force and ft for linear dimensions)

$$\frac{(T_1^i - T_2^i)\pi d_i'(350)}{33{,}000} \geq 0.75$$

which can be rewritten, using $T_1^i = stw$ from Eq. (E_6), as

$$stw \left(1 - \exp \left[-\mu \left(\pi - 2 \sin^{-1} \left\{ \left(\frac{N_i}{N} - 1 \right) \frac{d_i}{2a} \right\} \right) \right] \right) \pi d_i'$$

$$\times \left(\frac{350}{33{,}000} \right) \geq 0.75, \qquad i = 1, 2, 3, 4 \tag{E_7}$$

Finally, the lower bounds on the design variables can be taken as

$$w \geq 0 \tag{E_8}$$

$$d_i \geq 0, \quad i = 1, 2, 3, 4 \tag{E_9}$$

As the objective function, (E_1), and most of the constraints, (E_2) to (E_9), are nonlinear functions of the design variables d_1, d_2, d_3, d_4, and w, this problem is a nonlinear programming problem.

Geometric Programming Problem.

Definition   A function $h(\mathbf{X})$ is called a *posynomial* if h can be expressed as the sum of power terms each of the form

$$c_i x_1^{ai1} x_2^{ai2} \cdots x_n^{ain}$$

where c_i and a_{ij} are constants with $c_i > 0$ and $x_j > 0$. Thus a posynomial with N terms can be expressed as

$$h(\mathbf{X}) = c_1 x_1^{a11} x_2^{a12} \cdots x_n^{a1n} + \cdots + c_N x_1^{aN1} x_2^{aN2} \cdots x_n^{aNn} \tag{1.7}$$

A *geometric programming* (GMP) *problem* is one in which the objective function and constraints are expressed as posynomials in $\mathbf{X}$. Thus GMP problem can be posed as follows [1.59]:

Find $\mathbf{X}$ which minimizes

$$f(\mathbf{X}) = \sum_{i=1}^{N_0} c_i \left(\prod_{j=1}^{n} x_j^{p_{ij}} \right), \quad c_i > 0, \quad x_j > 0 \tag{1.8}$$

subject to

$$g_k(\mathbf{X}) = \sum_{i=1}^{N_k} a_{ik} \left(\prod_{j=1}^{n} x_j^{q_{ijk}} \right) > 0, \quad a_{ik} > 0, \ x_j > 0, \ k = 1, 2, \ldots, m$$

where N_0 and N_k denote the number of posynomial terms in the objective and kth constraint function, respectively.

Example 1.4   Four identical helical springs are used to support a milling machine weighing 5000 lb. Formulate the problem of finding the wire diameter (d), coil diameter (D), and the number of turns (N) of each spring (Fig. 1.11) for minimum weight by limiting the deflection to 0.1 in. and the shear stress to 10,000 psi in the spring. In addition, the natural frequency of vibration of the spring is to be greater than 100 Hz. The stiffness of the spring (k), the shear stress in the spring (τ), and the natural frequency of vibration of the spring (f_n) are given by

$$k = \frac{d^4 G}{8 D^3 N}$$

$$\tau = K_s \frac{8FD}{\pi d^3}$$

$$f_n = \frac{1}{2} \sqrt{\frac{kg}{w}} = \frac{1}{2} \sqrt{\frac{d^4 G}{8 D^3 N} \frac{g}{\rho (\pi d^2 / 4) \pi D N}} = \frac{\sqrt{Gg}\, d}{2 \sqrt{2 \rho} \pi D^2 N}$$

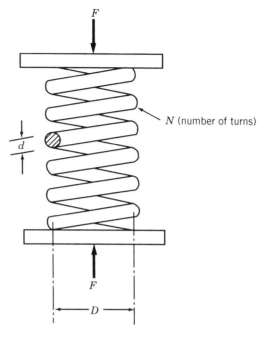

Figure 1.11   Helical spring.

where G is the shear modulus, F the compressive load on the spring, w the weight of the spring, ρ the weight density of the spring, and K_s the shear stress correction factor. Assume that the material is spring steel with $G = 12 \times 10^6$ psi and $\rho = 0.3\,\text{lb/in}^3$, and the shear stress correction factor is $K_s \approx 1.05$.

SOLUTION   The design vector is given by

$$\mathbf{X} = \begin{Bmatrix} x_1 \\ x_2 \\ x_3 \end{Bmatrix} = \begin{Bmatrix} d \\ D \\ N \end{Bmatrix}$$

and the objective function by

$$f(\mathbf{X}) = \text{weight} = \frac{\pi d^2}{4}\pi D N \rho \tag{E_1}$$

The constraints can be expressed as

$$\text{deflection} = \frac{F}{k} = \frac{8FD^3 N}{d^4 G} \leq 0.1$$

that is,

$$g_1(\mathbf{X}) = \frac{d^4 G}{80FD^3 N} > 1 \tag{E_2}$$

$$\text{shear stress} = K_s \frac{8FD}{\pi d^3} \leq 10{,}000$$

that is,

$$g_2(\mathbf{X}) = \frac{1250\pi d^3}{K_s FD} > 1 \tag{E_3}$$

$$\text{natural frequency} = \frac{\sqrt{Gg}}{2\sqrt{2}\rho\pi} \frac{d}{D^2 N} \geq 100$$

that is,

$$g_3(\mathbf{X}) = \frac{\sqrt{Gg}\, d}{200\sqrt{2}\rho\pi\, D^2 N} > 1 \tag{E_4}$$

Since the equality sign is not included (along with the inequality symbol, $>$) in the constraints of Eqs. (E$_2$) to (E$_4$), the design variables are to be restricted to positive values as

$$d > 0, \quad D > 0, \quad N > 0 \tag{E_5}$$

By substituting the known data, F = weight of the milling machine/4 = 1250 lb, $\rho = 0.3$ lb/in^3, $G = 12 \times 10^6$ psi, and $K_s = 1.05$, Eqs. (E$_1$) to (E$_4$) become

$$f(\mathbf{X}) = \tfrac{1}{4}\pi^2(0.3)d^2 DN = 0.7402 x_1^2 x_2 x_3 \tag{E_6}$$

$$g_1(\mathbf{X}) = \frac{d^4(12 \times 10^6)}{80(1250)D^3 N} = 120 x_1^4 x_2^{-3} x_3^{-1} > 1 \tag{E_7}$$

$$g_2(\mathbf{X}) = \frac{1250\pi d^3}{1.05(1250)D} = 2.992 x_1^3 x_2^{-1} > 1 \tag{E_8}$$

$$g_3(\mathbf{X}) = \frac{\sqrt{Gg}\, d}{200\sqrt{2}\rho\pi\, D^2 N} = 139.8388 x_1 x_2^{-2} x_3^{-1} > 1 \tag{E_9}$$

It can be seen that the objective function, $f(\mathbf{X})$, and the constraint functions, $g_1(\mathbf{X})$ to $g_3(\mathbf{X})$, are posynomials and hence the problem is a GMP problem.

Quadratic Programming Problem.   A quadratic programming problem is a nonlinear programming problem with a quadratic objective function and linear constraints. It is usually formulated as follows:

$$F(\mathbf{X}) = c + \sum_{i=1}^{n} q_i x_i + \sum_{i=1}^{n}\sum_{j=1}^{n} Q_{ij} x_i x_j \tag{1.9}$$

subject to

$$\sum_{i=1}^{n} a_{ij} x_i = b_j, \quad j = 1, 2, \ldots, m$$

$$x_i \geq 0, \quad i = 1, 2, \ldots, n$$

where c, q_i, Q_{ij}, a_{ij}, and b_j are constants.

Example 1.5 A manufacturing firm produces two products, A and B, using two limited resources. The maximum amounts of resources 1 and 2 available per day are 1000 and 250 units, respectively. The production of 1 unit of product A requires 1 unit of resource 1 and 0.2 unit of resource 2, and the production of 1 unit of product B requires 0.5 unit of resource 1 and 0.5 unit of resource 2. The unit costs of resources 1 and 2 are given by the relations $(0.375 - 0.00005u_1)$ and $(0.75 - 0.0001u_2)$, respectively, where u_i denotes the number of units of resource i used $(i = 1, 2)$. The selling prices per unit of products A and B, p_A and p_B, are given by

$$p_A = 2.00 - 0.0005x_A - 0.00015x_B$$

$$p_B = 3.50 - 0.0002x_A - 0.0015x_B$$

where x_A and x_B indicate, respectively, the number of units of products A and B sold. Formulate the problem of maximizing the profit assuming that the firm can sell all the units it manufactures.

SOLUTION Let the design variables be the number of units of products A and B manufactured per day:

$$\mathbf{X} = \begin{Bmatrix} x_A \\ x_B \end{Bmatrix}$$

The requirement of resource 1 per day is $(x_A + 0.5x_B)$ and that of resource 2 is $(0.2x_A + 0.5x_B)$ and the constraints on the resources are

$$x_A + 0.5x_B \leq 1000 \tag{E_1}$$

$$0.2x_A + 0.5x_B \leq 250 \tag{E_2}$$

The lower bounds on the design variables can be taken as

$$x_A \geq 0 \tag{E_3}$$

$$x_B \geq 0 \tag{E_4}$$

The total cost of resources 1 and 2 per day is

$$(x_A + 0.5x_B)[0.375 - 0.00005(x_A + 0.5x_B)]$$
$$+ (0.2x_A + 0.5x_B)[0.750 - 0.0001(0.2x_A + 0.5x_B)]$$

and the return per day from the sale of products A and B is

$$x_A(2.00 - 0.0005x_A - 0.00015x_B) + x_B(3.50 - 0.0002x_A - 0.0015x_B)$$

The total profit is given by the total return minus the total cost. Since the objective function to be minimized is the negative of the profit per day, $f(\mathbf{X})$ is given by

$$f(\mathbf{X}) = (x_A + 0.5x_B)[0.375 - 0.00005(x_A + 0.5x_B)]$$
$$+ (0.2x_A + 0.5x_B)[0.750 - 0.0001(0.2x_A + 0.5x_B)]$$
$$- x_A(2.00 - 0.0005x_A - 0.00015x_B)$$
$$- x_B(3.50 - 0.0002x_A - 0.0015x_B) \tag{E_5}$$

As the objective function [Eq. (E$_5$)] is a quadratic and the constraints [Eqs. (E$_1$) to (E$_4$)] are linear, the problem is a quadratic programming problem.

Linear Programming Problem.   If the objective function and all the constraints in Eq. (1.1) are linear functions of the design variables, the mathematical programming problem is called a *linear programming* (LP) *problem*. A linear programming problem is often stated in the following standard form:

$$\text{Find } \mathbf{X} = \begin{Bmatrix} x_1 \\ x_2 \\ \vdots \\ x_n \end{Bmatrix}$$

$$\text{which minimizes } f(\mathbf{X}) = \sum_{i=1}^{n} c_i x_i$$

subject to the constraints   (1.10)

$$\sum_{i=1}^{n} a_{ij} x_i = b_j, \quad j = 1, 2, \ldots, m$$

$$x_i \geq 0, \quad i = 1, 2, \ldots, n$$

where c_i, a_{ij}, and b_j are constants.

Example 1.6   A scaffolding system consists of three beams and six ropes as shown in Fig. 1.12. Each of the top ropes A and B can carry a load of W_1, each of the middle ropes C and D can carry a load of W_2, and each of the bottom ropes E and F can carry a load of W_3. If the loads acting on beams 1, 2, and 3 are x_1, x_2, and x_3, respectively, as shown in Fig. 1.12, formulate the problem of finding the maximum

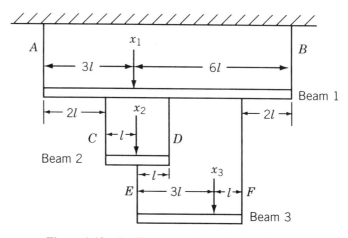

Figure 1.12   Scaffolding system with three beams.

load $(x_1 + x_2 + x_3)$ that can be supported by the system. Assume that the weights of the beams 1, 2, and 3 are w_1, w_2, and w_3, respectively, and the weights of the ropes are negligible.

SOLUTION   Assuming that the weights of the beams act through their respective middle points, the equations of equilibrium for vertical forces and moments for each of the three beams can be written as

For beam 3:

$$T_E + T_F = x_3 + w_3$$

$$x_3(3l) + w_3(2l) - T_F(4l) = 0$$

For beam 2:

$$T_C + T_D - T_E = x_2 + w_2$$

$$x_2(l) + w_2(l) + T_E(l) - T_D(2l) = 0$$

For beam 1:

$$T_A + T_B - T_C - T_D - T_F = x_1 + w_1$$

$$x_1(3l) + w_1(\tfrac{9}{2}l) - T_B(9l) + T_C(2l) + T_D(4l) + T_F(7l) = 0$$

where T_i denotes the tension in rope i. The solution of these equations gives

$$T_F = \tfrac{3}{4}x_3 + \tfrac{1}{2}w_3$$

$$T_E = \tfrac{1}{4}x_3 + \tfrac{1}{2}w_3$$

$$T_D = \tfrac{1}{2}x_2 + \tfrac{1}{8}x_3 + \tfrac{1}{2}w_2 + \tfrac{1}{4}w_3$$

$$T_C = \tfrac{1}{2}x_2 + \tfrac{1}{8}x_3 + \tfrac{1}{2}w_2 + \tfrac{1}{4}w_3$$

$$T_B = \tfrac{1}{3}x_1 + \tfrac{1}{3}x_2 + \tfrac{2}{3}x_3 + \tfrac{1}{2}w_1 + \tfrac{1}{3}w_2 + \tfrac{5}{9}w_3$$

$$T_A = \tfrac{2}{3}x_1 + \tfrac{2}{3}x_2 + \tfrac{1}{3}x_3 + \tfrac{1}{2}w_1 + \tfrac{2}{3}w_2 + \tfrac{4}{9}w_3$$

The optimization problem can be formulated by choosing the design vector as

$$\mathbf{X} = \begin{Bmatrix} x_1 \\ x_2 \\ x_3 \end{Bmatrix}$$

Since the objective is to maximize the total load

$$f(\mathbf{X}) = -(x_1 + x_2 + x_3) \tag{E_1}$$

The constraints on the forces in the ropes can be stated as

$$T_A \le W_1 \tag{E_2}$$

$$T_B \le W_1 \tag{E_3}$$

$$T_C \le W_2 \tag{E_4}$$

$$T_D \leq W_2 \tag{E$_5$}$$

$$T_E \leq W_3 \tag{E$_6$}$$

$$T_F \leq W_3 \tag{E$_7$}$$

Finally, the nonnegativity requirement of the design variables can be expressed as

$$x_1 \geq 0$$

$$x_2 \geq 0$$

$$x_3 \geq 0 \tag{E$_8$}$$

Since all the equations of the problem (E$_1$) to (E$_8$), are linear functions of x_1, x_2, and x_3, the problem is a linear programming problem.

1.5.5 Classification Based on the Permissible Values of the Design Variables

Depending on the values permitted for the design variables, optimization problems can be classified as integer and real-valued programming problems.

Integer Programming Problem. If some or all of the design variables $x_1, x_2, \ldots, x_n$ of an optimization problem are restricted to take on only integer (or discrete) values, the problem is called an *integer programming problem*. On the other hand, if all the design variables are permitted to take any real value, the optimization problem is called a *real-valued programming problem*. According to this definition, the problems considered in Examples 1.1 to 1.6 are real-valued programming problems.

Example 1.7 A cargo load is to be prepared from five types of articles. The weight w_i, volume v_i, and monetary value c_i of different articles are given below.

Article type	w_i	v_i	c_i
1	4	9	5
2	8	7	6
3	2	4	3
4	5	3	2
5	3	8	8

Find the number of articles x_i selected from the ith type ($i = 1, 2, 3, 4, 5$), so that the total monetary value of the cargo load is a maximum. The total weight and volume of the cargo cannot exceed the limits of 2000 and 2500 units, respectively.

SOLUTION Let x_i be the number of articles of type i ($i = 1$ to 5) selected. Since it is not possible to load a fraction of an article, the variables x_i can take only integer values.

The objective function to be maximized is given by

$$f(\mathbf{X}) = 5x_1 + 6x_2 + 3x_3 + 2x_4 + 8x_5 \tag{E$_1$}$$

and the constraints by

$$4x_1 + 8x_2 + 2x_3 + 5x_4 + 3x_5 \leq 2000 \tag{E_2}$$

$$9x_1 + 7x_2 + 4x_3 + 3x_4 + 8x_5 \leq 2500 \tag{E_3}$$

$$x_i \geq 0 \text{ and integral,} \quad i = 1, 2, \ldots, 5 \tag{E_4}$$

Since x_i are constrained to be integers, the problem is an integer programming problem.

1.5.6   Classification Based on the Deterministic Nature of the Variables

Based on the deterministic nature of the variables involved, optimization problems can be classified as deterministic and stochastic programming problems.

Stochastic Programming Problem.   A stochastic programming problem is an optimization problem in which some or all of the parameters (design variables and/or preassigned parameters) are probabilistic (nondeterministic or stochastic). According to this definition, the problems considered in Examples 1.1 to 1.7 are deterministic programming problems.

Example 1.8   Formulate the problem of designing a minimum-cost rectangular under-reinforced concrete beam that can carry a bending moment M with a probability of at least 0.95. The costs of concrete, steel, and formwork are given by $C_c = \$200/\text{m}^3$, $C_s = \$5000/\text{m}^3$, and $C_f = \$40/\text{m}^2$ of surface area. The bending moment M is a probabilistic quantity and varies between 1×10^5 and 2×10^5 N-m with a uniform probability. The strengths of concrete and steel are also uniformly distributed probabilistic quantities whose lower and upper limits are given by

$$f_c = 25 \text{ and } 35 \text{ MPa}$$

$$f_s = 500 \text{ and } 550 \text{ MPa}$$

Assume that the area of the reinforcing steel and the cross-sectional dimensions of the beam are deterministic quantities.

SOLUTION   The breadth b in meters, the depth d in meters, and the area of reinforcing steel A_s in square meters are taken as the design variables x_1, x_2, and x_3, respectively (Fig. 1.13). The cost of the beam per meter length is given by

$$f(\mathbf{X}) = \text{cost of steel} + \text{cost of concrete} + \text{cost of formwork}$$

$$= A_s C_s + (bd - A_s)C_c + 2(b + d)C_f \tag{E_1}$$

The resisting moment of the beam section is given by [1.119]

$$M_R = A_s f_s \left(d - 0.59 \frac{A_s f_s}{f_c b} \right)$$

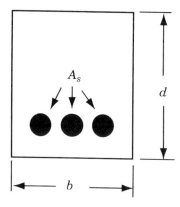

Figure 1.13 Cross section of a reinforced concrete beam.

and the constraint on the bending moment can be expressed as [1.120]

$$P[M_R - M \geq 0] = P\left[A_s f_s \left(d - 0.59\frac{A_s f_s}{f_c b}\right) - M \geq 0\right] \geq 0.95 \qquad (E_2)$$

where $P[\cdots]$ indicates the probability of occurrence of the event $[\cdots]$.

To ensure that the beam remains underreinforced,[†] the area of steel is bounded by the balanced steel area $A_s^{(b)}$ as

$$A_s \leq A_s^{(b)} \qquad (E_3)$$

where

$$A_s^{(b)} = (0.542)\frac{f_c}{f_s}\, bd \, \frac{600}{600 + f_s}$$

Since the design variables cannot be negative, we have

$$d \geq 0$$

$$b \geq 0$$

$$A_s \geq 0 \qquad (E_4)$$

Since the quantities M, f_c, and f_s are nondeterministic, the problem is a stochastic programming problem.

1.5.7 Classification Based on the Separability of the Functions

Optimization problems can be classified as separable and nonseparable programming problems based on the separability of the objective and constraint functions.

[†]If steel area is larger than $A_s^{(b)}$, the beam becomes overreinforced and failure occurs all of a sudden due to lack of concrete strength. If the beam is underreinforced, failure occurs due to lack of steel strength and hence it will be gradual.

Separable Programming Problem.

Definition   A function $f(\mathbf{X})$ is said to be *separable* if it can be expressed as the sum of n single-variable functions, $f_1(x_1), f_2(x_2), \ldots, f_n(x_n)$, that is,

$$f(\mathbf{X}) = \sum_{i=1}^{n} f_i(x_i) \tag{1.11}$$

A separable programming problem is one in which the objective function and the constraints are separable and can be expressed in standard form as

$$\text{Find } \mathbf{X} \text{ which minimizes } f(\mathbf{X}) = \sum_{i=1}^{n} f_i(x_i) \tag{1.12}$$

subject to

$$g_j(\mathbf{X}) = \sum_{i=1}^{n} g_{ij}(x_i) \le b_j, \quad j = 1, 2, \ldots, m$$

where b_j is a constant.

Example 1.9   A retail store stocks and sells three different models of TV sets. The store cannot afford to have an inventory worth more than \$45,000 at any time. The TV sets are ordered in lots. It costs \$$a_j$ for the store whenever a lot of TV model j is ordered. The cost of one TV set of model j is c_j. The demand rate of TV model j is d_j units per year. The rate at which the inventory costs accumulate is known to be proportional to the investment in inventory at any time, with $q_j = 0.5$, denoting the constant of proportionality for TV model j. Each TV set occupies an area of $s_j = 0.40\,\text{m}^2$ and the maximum storage space available is $90\,\text{m}^2$. The data known from the past experience are given below.

	TV model j		
	1	2	3
Ordering cost, a_j (\$)	50	80	100
Unit cost, c_j (\$)	40	120	80
Demand rate, d_j	800	400	1200

Formulate the problem of minimizing the average annual cost of ordering and storing the TV sets.

SOLUTION   Let x_j denote the number of TV sets of model j ordered in each lot ($j = 1, 2, 3$). Since the demand rate per year of model j is d_j, the number of times the TV model j needs to be ordered is d_j/x_j. The cost of ordering TV model j per year is thus $a_j d_j/x_j$, $j = 1, 2, 3$. The cost of storing TV sets of model j per year is $q_j c_j x_j/2$ since the average level of inventory at any time during the year is equal to

$c_j x_j / 2$. Thus the objective function (cost of ordering plus storing) can be expressed as

$$f(\mathbf{X}) = \left(\frac{a_1 d_1}{x_1} + \frac{q_1 c_1 x_1}{2} \right) + \left(\frac{a_2 d_2}{x_2} + \frac{q_2 c_2 x_2}{2} \right) + \left(\frac{a_3 d_3}{x_3} + \frac{q_3 c_3 x_3}{2} \right) \qquad (E_1)$$

where the design vector $\mathbf{X}$ is given by

$$X = \begin{Bmatrix} x_1 \\ x_2 \\ x_3 \end{Bmatrix} \qquad (E_2)$$

The constraint on the worth of inventory can be stated as

$$c_1 x_1 + c_2 x_2 + c_3 x_3 \le 45,000 \qquad (E_3)$$

The limitation on the storage area is given by

$$s_1 x_1 + s_2 x_2 + s_3 x_3 \le 90 \qquad (E_4)$$

Since the design variables cannot be negative, we have

$$x_j \ge 0, \quad j = 1, 2, 3 \qquad (E_5)$$

By substituting the known data, the optimization problem can be stated as follows:

Find $\mathbf{X}$ which minimizes

$$f(\mathbf{X}) = \left(\frac{40,000}{x_1} + 10x_1 \right) + \left(\frac{32,000}{x_2} + 30x_2 \right) + \left(\frac{120,000}{x_3} + 20x_3 \right) \qquad (E_6)$$

subject to

$$g_1(\mathbf{X}) = 40x_1 + 120x_2 + 80x_3 \le 45,000 \qquad (E_7)$$

$$g_2(\mathbf{X}) = 0.40(x_1 + x_2 + x_3) \le 90 \qquad (E_8)$$

$$g_3(\mathbf{X}) = -x_1 \le 0 \qquad (E_9)$$

$$g_4(\mathbf{X}) = -x_2 \le 0 \qquad (E_{10})$$

$$g_5(\mathbf{X}) = -x_3 \le 0 \qquad (E_{11})$$

It can be observed that the optimization problem stated in Eqs. (E_6) to (E_{11}) is a separable programming problem.

1.5.8   Classification Based on the Number of Objective Functions

Depending on the number of objective functions to be minimized, optimization problems can be classified as single- and multiobjective programming problems. According to this classification, the problems considered in Examples 1.1 to 1.9 are single objective programming problems.

Multiobjective Programming Problem.   A multiobjective programming problem can be stated as follows:

$$\text{Find } \mathbf{X} \text{ which minimizes } f_1(\mathbf{X}), f_2(\mathbf{X}), \ldots, f_k(\mathbf{X})$$

subject to   (1.13)

$$g_j(\mathbf{X}) \leq 0, \quad j = 1, 2, \ldots, m$$

where $f_1, f_2, \ldots, f_k$ denote the objective functions to be minimized simultaneously.

Example 1.10   A uniform column of rectangular cross section is to be constructed for supporting a water tank of mass M (Fig. 1.14). It is required (1) to minimize the mass of the column for economy, and (2) to maximize the natural frequency of transverse vibration of the system for avoiding possible resonance due to wind. Formulate the problem of designing the column to avoid failure due to direct compression and buckling. Assume the permissible compressive stress to be $\sigma_{\max}$.

SOLUTION   Let $x_1 = b$ and $x_2 = d$ denote the cross-sectional dimensions of the column. The mass of the column (m) is given by

$$m = \rho b d l = \rho l x_1 x_2 \tag{E$_1$}$$

where ρ is the density and l is the height of the column. The natural frequency of transverse vibration of the water tank (ω), by treating it as a cantilever beam with a tip mass M, can be obtained as [1.118]

$$\omega = \left[\frac{3EI}{(M + \frac{33}{140}m)l^3} \right]^{1/2} \tag{E$_2$}$$

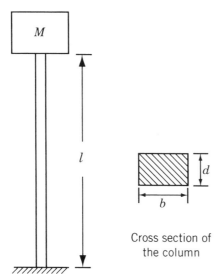

Cross section of
the column

Figure 1.14   Water tank on a column.

where E is the Young's modulus and I is the area moment of inertia of the column given by

$$I = \tfrac{1}{12} b d^3 \tag{E$_3$}$$

The natural frequency of the water tank can be maximized by minimizing $-\omega$. With the help of Eqs. (E$_1$) and (E$_3$), Eq. (E$_2$) can be rewritten as

$$\omega = \left[\frac{E x_1 x_2^3}{4 l^3 (M + \frac{33}{140} \rho l x_1 x_2)} \right]^{1/2} \tag{E$_4$}$$

The direct compressive stress (σ_c) in the column due to the weight of the water tank is given by

$$\sigma_c = \frac{Mg}{bd} = \frac{Mg}{x_1 x_2} \tag{E$_5$}$$

and the buckling stress for a fixed-free column (σ_b) is given by [1.121]

$$\sigma_b = \left(\frac{\pi^2 E I}{4 l^2} \right) \frac{1}{bd} = \frac{\pi^2 E x_2^2}{48 l^2} \tag{E$_6$}$$

To avoid failure of the column, the direct stress has to be restricted to be less than $\sigma_{\max}$ and the buckling stress has to be constrained to be greater than the direct compressive stress induced.

Finally, the design variables have to be constrained to be positive. Thus the multiobjective optimization problem can be stated as follows:

$$\text{Find } \mathbf{X} = \begin{Bmatrix} x_1 \\ x_2 \end{Bmatrix} \text{ which minimizes}$$

$$f_1(\mathbf{X}) = \rho l x_1 x_2 \tag{E$_7$}$$

$$f_2(\mathbf{X}) = - \left[\frac{E x_1 x_2^3}{4 l^2 (M + \frac{33}{140} \rho l x_1 x_2)} \right]^{1/2} \tag{E$_8$}$$

subject to

$$g_1(\mathbf{X}) = \frac{Mg}{x_1 x_2} - \sigma_{\max} \leq 0 \tag{E$_9$}$$

$$g_2(\mathbf{X}) = \frac{Mg}{x_1 x_2} - \frac{\pi^2 E x_2^2}{48 l^2} \leq 0 \tag{E$_{10}$}$$

$$g_3(\mathbf{X}) = -x_1 \leq 0 \tag{E$_{11}$}$$

$$g_4(\mathbf{X}) = -x_2 \leq 0 \tag{E$_{12}$}$$

1.6 OPTIMIZATION TECHNIQUES

The various techniques available for the solution of different types of optimization problems are given under the heading of mathematical programming techniques in Table 1.1. The classical methods of differential calculus can be used to find the unconstrained maxima and minima of a function of several variables. These methods assume that the function is differentiable twice with respect to the design variables and the derivatives are continuous. For problems with equality constraints, the Lagrange multiplier method can be used. If the problem has inequality constraints, the Kuhn–Tucker conditions can be used to identify the optimum point. But these methods lead to a set of nonlinear simultaneous equations that may be difficult to solve. The classical methods of optimization are discussed in Chapter 2.

The techniques of nonlinear, linear, geometric, quadratic, or integer programming can be used for the solution of the particular class of problems indicated by the name of the technique. Most of these methods are numerical techniques wherein an approximate solution is sought by proceeding in an iterative manner by starting from an initial solution. Linear programming techniques are described in Chapters 3 and 4. The quadratic programming technique, as an extension of the linear programming approach, is discussed in Chapter 4. Since nonlinear programming is the most general method of optimization that can be used to solve any optimization problem, it is dealt with in detail in Chapters 5–7. The geometric and integer programming methods are discussed in Chapters 8 and 10, respectively. The dynamic programming technique, presented in Chapter 9, is also a numerical procedure that is useful primarily for the solution of optimal control problems. Stochastic programming deals with the solution of optimization problems in which some of the variables are described by probability distributions. This topic is discussed in Chapter 11.

In Chapter 12 we discuss calculus of variations, optimal control theory, and optimality criteria methods. The modern methods of optimization, including genetic algorithms, simulated annealing, particle swarm optimization, ant colony optimization, neural network-based optimization, and fuzzy optimization, are presented in Chapter 13. Several practical aspects of optimization are outlined in Chapter 14. The reduction of size of optimization problems, fast reanalysis techniques, the efficient computation of the derivatives of static displacements and stresses, eigenvalues and eigenvectors, and transient response are outlined. The aspects of sensitivity of optimum solution to problem parameters, multilevel optimization, parallel processing, and multiobjective optimization are also presented in this chapter.

1.7 ENGINEERING OPTIMIZATION LITERATURE

The literature on engineering optimization is large and diverse. Several text-books are available and dozens of technical periodicals regularly publish papers related to engineering optimization. This is primarily because optimization is applicable to all areas of engineering. Researchers in many fields must be attentive to the developments in the theory and applications of optimization.

The most widely circulated journals that publish papers related to engineering optimization are *Engineering Optimization, ASME Journal of Mechanical Design, AIAA Journal, ASCE Journal of Structural Engineering, Computers and Structures, International Journal for Numerical Methods in Engineering, Structural Optimization, Journal of Optimization Theory and Applications, Computers and Operations Research, Operations Research, Management Science, Evolutionary Computation, IEEE Transactions on Evolutionary Computation, European Journal of Operations Research, IEEE Transactions on Systems, Man and Cybernetics*, and *Journal of Heuristics*. Many of these journals are cited in the chapter references.

1.8   SOLUTION OF OPTIMIZATION PROBLEMS USING MATLAB

The solution of most practical optimization problems requires the use of computers. Several commercial software systems are available to solve optimization problems that arise in different engineering areas. MATLAB is a popular software that is used for the solution of a variety of scientific and engineering problems.[†] MATLAB has several toolboxes each developed for the solution of problems from a specific scientific area. The specific toolbox of interest for solving optimization and related problems is called the *optimization toolbox*. It contains a library of programs or m-files, which can be used for the solution of minimization, equations, least squares curve fitting, and related problems. The basic information necessary for using the various programs can be found in the user's guide for the optimization toolbox [1.124]. The programs or m-files, also called functions, available in the minimization section of the optimization toolbox are given in Table 1.2. The use of the programs listed in Table 1.2 is demonstrated at the end of different chapters of the book. Basically, the solution procedure involves three steps after formulating the optimization problem in the format required by the MATLAB program (or function) to be used. In most cases, this involves stating the objective function for minimization and the constraints in "$\leq$" form with zero or constant value on the righthand side of the inequalities. After this, step 1 involves writing an m-file for the objective function. Step 2 involves writing an m-file for the constraints. Step 3 involves setting the various parameters at proper values depending on the characteristics of the problem and the desired output and creating an appropriate file to invoke the desired MATLAB program (and coupling the m-files created to define the objective and constraints functions of the problem). As an example, the use of the program, `fmincon`, for the solution of a constrained nonlinear programming problem is demonstrated in Example 1.11.

Example 1.11   Find the solution of the following nonlinear optimization problem (same as the problem in Example 1.1) using the MATLAB function `fmincon`:

$$\text{Minimize } f(x_1, x_2) = 9.82x_1x_2 + 2x_1$$

subject to

$$g_1(x_1, x_2) = \frac{2500}{\pi x_1 x_2} - 500 \leq 0$$

[†]The basic concepts and procedures of MATLAB are summarized in Appendix C.

Table 1.2 MATLAB Programs or Functions for Solving Optimization Problems

Type of optimization problem	Standard form for solution by MATLAB	Name of MATLAB program or function to solve the problem
Function of one variable or scalar minimization	Find x to minimize $f(x)$ with $x_1 < x < x_2$	`fminbnd`
Unconstrained minimization of function of several variables	Find $\mathbf{x}$ to minimize $f(\mathbf{x})$	`fminunc` or `fminsearch`
Linear programming problem	Find $\mathbf{x}$ to minimize $\mathbf{f}^T\mathbf{x}$ subject to $[A]\mathbf{x} \le \mathbf{b}, [A_{eq}]\mathbf{x} = \mathbf{b}_{eq}, \mathbf{l} \le \mathbf{x} \le \mathbf{u}$	`linprog`
Quadratic programming problem	Find $\mathbf{x}$ to minimize $\frac{1}{2}\mathbf{x}^T[H]\mathbf{x} + \mathbf{f}^T\mathbf{x}$ subject to $[A]\mathbf{x} \le \mathbf{b}, [A_{eq}]\mathbf{x} = \mathbf{b}_{eq}, \mathbf{l} \le \mathbf{x} \le \mathbf{u}$	`quadprog`
Minimization of function of several variables subject to constraints	Find $\mathbf{x}$ to minimize $f(\mathbf{x})$ subject to $\mathbf{c}(\mathbf{x}) \le \mathbf{0}, \mathbf{c}_{eq} = \mathbf{0}$ $[A]\mathbf{x} \le \mathbf{b}, [A_{eq}]\mathbf{x} = \mathbf{b}_{eq}, \mathbf{l} \le \mathbf{x} \le \mathbf{u}$	`fmincon`
Goal attainment problem	Find $\mathbf{x}$ and γ to minimize γ such that $F(\mathbf{x}) - \mathbf{w}\gamma \le \mathbf{goal}, \mathbf{c}(\mathbf{x}) \le \mathbf{0}, \mathbf{c}_{eq} = \mathbf{0}$ $[A]\mathbf{x} \le \mathbf{b}, [A_{eq}]\mathbf{x} = \mathbf{b}_{eq}, \mathbf{l} \le \mathbf{x} \le \mathbf{u}$	`fgoalattain`
Minimax problem	$\underset{\mathbf{x}}{\text{Minimize}} \ \underset{\{F_i\}}{\text{Max}} \ \{F_i(\mathbf{x})\}$ such that $\mathbf{c}(\mathbf{x}) \le \mathbf{0}, \mathbf{c}_{eq} = \mathbf{0}$ $[A]\mathbf{x} \le \mathbf{b}, [A_{eq}]\mathbf{x} = \mathbf{b}_{eq}, \mathbf{l} \le \mathbf{x} \le \mathbf{u}$	`fminimax`
Binary integer programming problem	Find $\mathbf{x}$ to minimize $\mathbf{f}^T\mathbf{x}$ subject to $[A]\mathbf{x} \le \mathbf{b}, [A_{eq}]\mathbf{x} = \mathbf{b}_{eq},$ each component of $\mathbf{x}$ is binary	`bintprog`

$$g_2(x_1, x_2) = \frac{2500}{\pi x_1 x_2} - \frac{\pi^2(x_1^2 + x_2^2)}{0.5882} \le 0$$

$$g_3(x_1, x_2) = -x_1 + 2 \le 0$$

$$g_4(x_1, x_2) = x_1 - 14 \le 0$$

$$g_5(x_1, x_2) = -x_2 + 0.2 \le 0$$

$$g_6(x_1, x_2) = x_2 - 0.8 \le 0$$

SOLUTION

Step 1: Write an M-file `probofminobj.m` for the objective function.

```
function f= probofminobj (x)
f= 9.82*x(1)*x(2)+2*x(1);
```

Step 2: Write an M-file `conprobformin.m` for the constraints.

```
function [c, ceq] = conprobformin(x)
% Nonlinear inequality constraints
c = [2500/(pi*x(1)*x(2))-500;2500/(pi*x(1)*x(2))-
(pi^2*(x(1)^2+x(2)^2))/0.5882;-x(1)+2;x(1)-14;-x(2)+0.2;
x(2)-0.8];
% Nonlinear equality constraints
ceq = [];
```

Step 3: Invoke constrained optimization program (write this in new matlab file).

```
clc
clear all
warning off
x0 = [7 0.4]; % Starting guess\
fprintf ('The values of function value and constraints
at starting point\n');
f=probofminobj (x0)
[c, ceq] = conprobformin (x0)
options = optimset ('LargeScale', 'off');
[x, fval]=fmincon (@probofminobj, x0, [], [], [], [], [],
[], @conprobformin, options)
fprintf('The values of constraints at optimum solution\n');
[c, ceq] = conprobformin(x) % Check the constraint values at x
```

This produces the solution or output as follows:

```
The values of function value and constraints at starting point
f=
 41.4960
c =
-215.7947
-540.6668
-5.0000
-7.0000
-0.2000
-0.4000
ceq =
   []
Optimization terminated: first-order optimality
measure less
```

```
than options. TolFun and maximum constraint violation
is less
than options.TolCon.
Active inequalities (to within options.TolCon = 1e-006):
lower upper ineqlin ineqnonlin
                  1
                  2
x=
 5.4510 0.2920
fval =
 26.5310
The values of constraints at optimum solution
c=
 -0.0000
 -0.0000
 -3.4510
 -8.5490
 -0.0920
 -0.5080
ceq =
   []
```

REFERENCES AND BIBLIOGRAPHY

Structural Optimization

1.1   K. I. Majid, *Optimum Design of Structures*, Wiley, New York, 1974.

1.2   D. G. Carmichael, *Structural Modelling and Optimization*, Ellis Horwood, Chichester, UK, 1981.

1.3   U. Kirsch, *Optimum Structural Design*, McGraw-Hill, New York, 1981.

1.4   A. J. Morris, *Foundations of Structural Optimization*, Wiley, New York, 1982.

1.5   J. Farkas, *Optimum Design of Metal Structures*, Ellis Horwood, Chichester, UK, 1984.

1.6   R. T. Haftka and Z. Gürdal, *Elements of Structural Optimization*, 3rd ed., Kluwer Academic Publishers, Dordrecht, The Netherlands, 1992.

1.7   M. P. Kamat, Ed., *Structural Optimization: Status and Promise*, AIAA, Washington, DC, 1993.

1.8   Z. Gurdal, R. T. Haftka, and P. Hajela, *Design and Optimization of Laminated Composite Materials*, Wiley, New York, 1998.

1.9   A. L. Kalamkarov and A. G. Kolpakov, *Analysis, Design and Optimization of Composite Structures*, 2nd ed., Wiley, New York, 1997.

Thermal System Optimization

1.10   W. F. Stoecker, *Design of Thermal Systems*, 3rd ed., McGraw-Hill, New York, 1989.

1.11   S. Stricker, *Optimizing Performance of Energy Systems*, Battelle Press, New York, 1985.

1.12   Adrian Bejan, G. Tsatsaronis, and M. Moran, *Thermal Design and Optimization*, Wiley, New York, 1995.

1.13　Y. Jaluria, *Design and Optimization of Thermal Systems*, 2nd ed., CRC Press, Boca Raton, FL, 2007.

Chemical and Metallurgical Process Optimization

1.14　W. H. Ray and J. Szekely, *Process Optimization with Applications to Metallurgy and Chemical Engineering*, Wiley, New York, 1973.

1.15　T. F. Edgar and D. M. Himmelblau, *Optimization of Chemical Processes*, McGraw-Hill, New York, 1988.

1.16　R. Aris, *The Optimal Design of Chemical Reactors, a Study in Dynamic Programming*, Academic Press, New York, 1961.

Electronics and Electrical Engineering

1.17　K. W. Cattermole and J. J. O'Reilly, *Optimization Methods in Electronics and Communications*, Wiley, New York, 1984.

1.18　T. R. Cuthbert, Jr., *Optimization Using Personal Computers with Applications to Electrical Networks*, Wiley, New York, 1987.

1.19　G. D. Micheli, *Synthesis and Optimization of Digital Circuits*, McGraw-Hill, New York, 1994.

Mechanical Design

1.20　R. C. Johnson, *Optimum Design of Mechanical Elements*, Wiley, New York, 1980.

1.21　E. J. Haug and J. S. Arora, *Applied Optimal Design: Mechanical and Structural Systems*, Wiley, New York, 1979.

1.22　E. Sevin and W. D. Pilkey, *Optimum Shock and Vibration Isolation*, Shock and Vibration Information Center, Washington, DC, 1971.

General Engineering Design

1.23　J. Arora, *Introduction to Optimum Design*, 2nd ed., Academic Press, San Diego, 2004.

1.24　P. Y. Papalambros and D. J. Wilde, *Principles of Optimal Design*, Cambridge University Press, Cambridge, UK, 1988.

1.25　J. N. Siddall, *Optimal Engineering Design: Principles and Applications*, Marcel Dekker, New York, 1982.

1.26　S. S. Rao, *Optimization: Theory and Applications*, 2nd ed., Wiley, New York, 1984.

1.27　G. N. Vanderplaats, *Numerical Optimization Techniques for Engineering Design with Applications*, McGraw-Hill, New York, 1984.

1.28　R. L. Fox, *Optimization Methods for Engineering Design*, Addison-Wesley, Reading, MA, 1972.

1.29　A. Ravindran, K. M. Ragsdell, and G. V. Reklaitis, *Engineering Optimization: Methods and Applications*, 2nd ed., Wiley, New York, 2006.

1.30　D. J. Wilde, *Globally Optimal Design*, Wiley, New York, 1978.

1.31　T. E. Shoup and F. Mistree, *Optimization Methods with Applications for Personal Computers*, Prentice-Hall, Englewood Cliffs, NJ, 1987.

1.32　A. D. Belegundu and T. R. Chandrupatla, *Optimization Concepts and Applications in Engineering*, Prentice Hall, Upper Saddle River, NJ, 1999.

General Nonlinear Programming Theory

1.33 S. L. S. Jacoby, J. S. Kowalik, and J. T. Pizzo, *Iterative Methods for Nonlinear Optimization Problems*, Prentice-Hall, Englewood Cliffs, NJ, 1972.

1.34 L. C. W. Dixon, *Nonlinear Optimization: Theory and Algorithms*, Birkhauser, Boston, 1980.

1.35 G. S. G. Beveridge and R. S. Schechter, *Optimization: Theory and Practice*, McGraw-Hill, New York, 1970.

1.36 B. S. Gottfried and J. Weisman, *Introduction to Optimization Theory*, Prentice-Hall, Englewood Cliffs, NJ, 1973.

1.37 M. A. Wolfe, *Numerical Methods for Unconstrained Optimization*, Van Nostrand Reinhold, New York, 1978.

1.38 M. S. Bazaraa and C. M. Shetty, *Nonlinear Programming*, Wiley, New York, 1979.

1.39 W. I. Zangwill, *Nonlinear Programming: A Unified Approach*, Prentice-Hall, Englewood Cliffs, NJ, 1969.

1.40 J. E. Dennis and R. B. Schnabel, *Numerical Methods for Unconstrained Optimization and Nonlinear Equations*, Prentice-Hall, Englewood Cliffs, NJ, 1983.

1.41 J. S. Kowalik, *Methods for Unconstrained Optimization Problems*, American Elsevier, New York, 1968.

1.42 A. V. Fiacco and G. P. McCormick, *Nonlinear Programming: Sequential Unconstrained Minimization Techniques*, Wiley, New York, 1968.

1.43 G. Zoutendijk, *Methods of Feasible Directions*, Elsevier, Amsterdam, 1960.

1.44 J. Nocedal and S. J. Wright, *Numerical Optimization*, Springer, New York, 2006.

1.45 R. Fletcher, *Practical Methods of Optimization*, Vols. 1 and 2, Wiley, Chichester, UK, 1981.

1.46 D. P. Bertsekas, *Nonlinear Programming*, 2nd ed., Athena Scientific, Nashua, NH, 1999.

1.47 D. G. Luenberger, *Linear and Nonlinear Programming*, 2nd ed., Kluwer Academic Publishers, Norwell, MA, 2003.

1.48 A. Antoniou and W-S. Lu, *Practical Optimization: Algorithms and Engineering Applications*, Springer, Berlin, 2007.

1.49 S. G. Nash and A. Sofer, *Linear and Nonlinear Programming*, McGraw-Hill, New York, 1996.

Computer Programs

1.50 J. L. Kuester and J. H. Mize, *Optimization Techniques with Fortran*, McGraw-Hill, New York, 1973.

1.51 H. P. Khunzi, H. G. Tzschach, and C. A. Zehnder, *Numerical Methods of Mathematical Optimization with ALGOL and FORTRAN Programs*, Academic Press, New York, 1971.

1.52 C. S. Wolfe, *Linear Programming with BASIC and FORTRAN*, Reston Publishing Co., Reston, VA, 1985.

1.53 K. R. Baker, *Optimization Modeling with Spreadsheets*, Thomson Brooks/Cole, Belmont, CA, 2006.

1.54 P. Venkataraman, *Applied Optimization with MATLAB Programming*, Wiley, New York, 2002.

Optimal Control

1.55 D. E. Kirk, *Optimal Control Theory: An Introduction*, Prentice-Hall, Englewood Cliffs, NJ, 1970.

1.56 A. P. Sage and C. C. White III, *Optimum Systems Control*, 2nd ed., Prentice-Hall, Englewood Cliffs, NJ, 1977.

1.57 B. D. O. Anderson and J. B. Moore, *Linear Optimal Control*, Prentice-Hall, Englewood Cliffs, NJ, 1971.

1.58 A. E. Bryson and Y. C. Ho, *Applied Optimal Control: Optimization, Estimation, and Control*, Blaisdell, Waltham, MA, 1969.

Geometric Programming

1.59 R. J. Duffin, E. L. Peterson, and C. Zener, *Geometric Programming: Theory and Applications*, Wiley, New York, 1967.

1.60 C. M. Zener, *Engineering Design by Geometric Programming*, Wiley, New York, 1971.

1.61 C. S. Beightler and D. T. Phillips, *Applied Geometric Programming*, Wiley, New York, 1976.

1.62 B-Y. Cao, *Fuzzy Geometric Programming*, Kluwer Academic, Dordrecht, The Netherlands, 2002.

1.63 A. Paoluzzi, *Geometric Programming for Computer-aided Design*, Wiley, New York, 2003.

Linear Programming

1.64 G. B. Dantzig, *Linear Programming and Extensions*, Princeton University Press, Princeton, NJ, 1963.

1.65 S. Vajda, *Linear Programming: Algorithms and Applications*, Methuen, New York, 1981.

1.66 S. I. Gass, *Linear Programming: Methods and Applications*, 5th ed., McGraw-Hill, New York, 1985.

1.67 C. Kim, *Introduction to Linear Programming*, Holt, Rinehart, & Winston, New York, 1971.

1.68 P. R. Thie, *An Introduction to Linear Programming and Game Theory*, Wiley, New York, 1979.

1.69 S. I. Gass, *An illustrated Guide to Linear Programming*, Dover, New York, 1990.

1.70 K. G. Murty, *Linear Programming*, Wiley, New York, 1983.

Integer Programming

1.71 T. C. Hu, *Integer Programming and Network Flows*, Addison-Wesley, Reading, MA, 1982.

1.72 A. Kaufmann and A. H. Labordaere, *Integer and Mixed Programming: Theory and Applications*, Academic Press, New York, 1976.

1.73 H. M. Salkin, *Integer Programming*, Addison-Wesley, Reading, MA, 1975.

1.74 H. A. Taha, *Integer Programming: Theory, Applications, and Computations*, Academic Press, New York, 1975.

1.75 A. Schrijver, *Theory of Linear and Integer Programming*, Wiley, New York, 1998.

1.76 J. K. Karlof (Ed.), *Integer Programming: Theory and Practice*, CRC Press, Boca Raton, FL, 2006.

1.77 L. A. Wolsey, *Integer Programming*, Wiley, New York, 1998.

Dynamic Programming

1.78 R. Bellman, *Dynamic Programming*, Princeton University Press, Princeton, NJ, 1957.

1.79 R. Bellman and S. E. Dreyfus, *Applied Dynamic Programming*, Princeton University Press, Princeton, NJ, 1962.

1.80 G. L. Nemhauser, *Introduction to Dynamic Programming*, Wiley, New York, 1966.

1.81 L. Cooper and M. W. Cooper, *Introduction to Dynamic Programming*, Pergamon Press, Oxford, UK, 1981.

1.82 W. B. Powell, *Approximate Dynamic Programming: Solving the Curses of Dimensionality*, Wiley, Hoboken, NJ, 2007.

1.83 M. L. Puterman, *Dynamic Programming and Its Applications*, Academic Press, New York, 1978.

1.84 M. Sniedovich, *Dynamic Programming*, Marcel Dekker, New York, 1992.

Stochastic Programming

1.85 J. K. Sengupta, *Stochastic Programming: Methods and Applications*, North-Holland, Amsterdam, 1972.

1.86 P. Kall, *Stochastic Linear Programming*, Springer-Verlag, Berlin, 1976.

1.87 J. R. Birge and F. Louveaux, *Introduction to Stochastic Programming*, Springer, New York, 1997.

1.88 P. Kall and S. W. Wallace, *Stochastic Programming*, Wiley, Chichester, UK, 1994.

1.89 P. Kall and J. Mayer, *Stochastic Linear Programming: Models, Theory, and Computation*, Springer, New York, 2005.

Multiobjective Programming

1.90 R. E. Steuer, *Multiple Criteria Optimization: Theory, Computation, and Application*, Wiley, New York, 1986.

1.91 C. L. Hwang and A. S. M. Masud, *Multiple Objective Decision Making: Methods and Applications*, Lecture Notices in Economics and Mathematical Systems, Vol. 164, Springer-Verlag, Berlin, 1979.

1.92 J. P. Ignizio, *Linear Programming in Single and Multi-objective Systems*, Prentice-Hall, Englewood Cliffs, NJ, 1982.

1.93 A. Goicoechea, D. R. Hansen, and L. Duckstein, *Multiobjective Decision Analysis with Engineering and Business Applications*, Wiley, New York, 1982.

1.94 Y. Collette and P. Siarry, *Multiobjective Optimization: Principles and Case Studies*, Springer, Berlin, 2004.

1.95 H. Eschenauer, J. Koski, and A. Osyczka (Eds.), *Multicriteria Design Optimization: Procedures and Applications*, Springer-Verlag, Berlin, 1990.

1.96 P. Sen and J-B. Yang, *Multiple Criteria Decision Support in Engineering Design*, Springer-Verlag, Berlin, 1998.

1.97 G. Owen, *Game Theory*, 3rd ed., Academic Press, San Diego, 1995.

Nontraditional Optimization Techniques

1.98   M. Mitchell, *An Introduction to Genetic Algorithms*, MIT Press, Cambridge, MA, 1998.

1.99   D. B. Fogel, *Evolutionary Computation: Toward a New Philosophy of Machine Intelligence*, 3rd ed., IEEE Press, Piscataway, NJ, 2006.

1.100 K. Deb, *Multi-Objective Optimization Using Evolutionary Algorithms*, Wiley, Chichester, England, 2001.

1.101 C. A. Coello Coello, D. A. van Veldhuizen and G. B. Lamont, *Evolutionary Algorithms for Solving Multi-Objective Problems*, Plenum, New York, 2002.

1.102 D. E. Goldberg, *Genetic Algorithms in Search, Optimization and Machine Learning*, Addison-Wesley, Reading, MA, 1989.

1.103 P. J. M. van Laarhoven and E. Aarts, *Simulated Annealing: Theory and Applications*, D. Reidel, Dordrecht, The Netherlands, 1987.

1.104 J. Hopfield and D. Tank, "Neural Computation of Decisions in Optimization Problems," *Biological Cybernetics*, Vol. 52, pp. 141–152, 1985.

1.105 J. J. Hopfield, Neural networks and physical systems with emergent collective computational abilities, *Proceedings of the National Academy of Sciences, USA*, Vol. 79, pp. 2554–2558, 1982.

1.106 N. Forbes, *Imitation of Life: How Biology Is Inspiring Computing*, MIT Press, Cambridge, MA, 2004.

1.107 J. Harris, *Fuzzy Logic Applications in Engineering Science*, Springer, Dordrecht, The Netherlands, 2006.

1.108 M. Hanss, *Applied Fuzzy Arithmetic: An Introduction with Engineering Applications*, Springer, Berlin, 2005.

1.109 G. Chen and T. T. Pham, *Introduction to Fussy Systems*, Chapman & Hall/CRC, Boca Raton, FL, 2006.

1.110 T. J. Ross, *Fuzzy Logic with Engineering Applications*, McGraw-Hill, New York, 1995.

1.111 M. Dorigo and T. Stutzle, *Ant Colony Optimization*, MIT Press, Cambridge, MA, 2004.

1.112 J. Kennedy, R. C. Eberhart, and Y. Shi, *Swarm Intelligence*, Morgan Kaufmann, San Francisco, CA, 2001.

1.113 J. C. Spall, *Introduction to Stochastic Search and Optimization*, Wiley Interscience, 2003.

1.114 A. P. Engelbrecht, *Fundamentals of Computational Swarm Intelligence*, Wiley, Chichester, UK, 2005.

1.115 E. Bonabeau, M. Dorigo, and G. Theraulaz, *Swarm Intelligence: From Natural to Artificial Systems*, Oxford University Press, Oxford, UK, 1999.

Additional References

1.116 R. C. Juvinall and K. M. Marshek, *Fundamentals of Machine Component Design*, 2nd ed., Wiley, New York, 1991.

1.117 J. E. Shigley and C. R. Mischke, *Mechanical Engineering Design*, 5th ed., McGraw-Hill, New York, 1989.

1.118 S. S. Rao, *Mechanical Vibrations*, 4th ed., Pearson Prentice Hall, Upper Saddle River, NJ, 2004.

1.119 J. M. MacGregor, *Reinforced Concrete: Mechanics and Design*, Prentice Hall, Englewood Cliffs, NJ, 1988.

1.120 S. S. Rao, *Reliability-Based Design*, McGraw-Hill, New York, 1992.

1.121   N. H. Cook, *Mechanics and Materials for Design*, McGraw-Hill, New York, 1984.

1.122   R. Ramarathnam and B. G. Desai, Optimization of polyphase induction motor design: a nonlinear programming approach, *IEEE Transactions on Power Apparatus and Systems*, Vol. PAS-90, No. 2, pp. 570–578, 1971.

1.123   R. M. Stark and R. L. Nicholls, *Mathematical Foundations for Design: Civil Engineering Systems*, McGraw-Hill, New York, 1972.

1.124   T. F. Coleman, M. A. Branch, and A. Grace, *Optimization Toolbox—for Use with MATLAB®*, User's Guide, Version 2 MathWorks Inc., Natick, MA, 1999.

REVIEW QUESTIONS

1.1   Match the following terms and descriptions:

(a) Free feasible point $g_j(\mathbf{X}) = 0$
(b) Free infeasible point   Some $g_j(\mathbf{X}) = 0$ and other $g_j(\mathbf{X}) < 0$
(c) Bound feasible point   Some $g_j(\mathbf{X}) = 0$ and other $g_j(\mathbf{X}) \geq 0$
(d) Bound infeasible point   Some $g_j(\mathbf{X}) > 0$ and other $g_j(\mathbf{X}) < 0$
(e) Active constraints   All $g_j(\mathbf{X}) < 0$

1.2   Answer true or false:

(a) Optimization problems are also known as mathematical programming problems.

(b) The number of equality constraints can be larger than the number of design variables.

(c) Preassigned parameters are part of design data in a design optimization problem.

(d) Side constraints are not related to the functionality of the system.

(e) A bound design point can be infeasible.

(f) It is necessary that some $g_j(\mathbf{X}) = 0$ at the optimum point.

(g) An optimal control problem can be solved using dynamic programming techniques.

(h) An integer programming problem is same as a discrete programming problem.

1.3   Define the following terms:

(a) Mathematical programming problem
(b) Trajectory optimization problem
(c) Behavior constraint
(d) Quadratic programming problem
(e) Posynomial
(f) Geometric programming problem

1.4   Match the following types of problems with their descriptions.

(a) Geometric programming problem   Classical optimization problem
(b) Quadratic programming problem   Objective and constraints are quadratic
(c) Dynamic programming problem   Objective is quadratic and constraints are linear
(d) Nonlinear programming problem   Objective and constraints arise from a serial system
(e) Calculus of variations problem Objective and constraints are polynomials with positive coefficients

1.5   How do you solve a maximization problem as a minimization problem?

1.6   State the linear programming problem in standard form.

1.7   Define an OC problem and give an engineering example.

1.8   What is the difference between linear and nonlinear programming problems?

1.9   What is the difference between design variables and preassigned parameters?

1.10 What is a design space?

1.11 What is the difference between a constraint surface and a composite constraint surface?

1.12 What is the difference between a bound point and a free point in the design space?

1.13 What is a merit function?

1.14 Suggest a simple method of handling multiple objectives in an optimization problem.

1.15 What are objective function contours?

1.16 What is operations research?

1.17 State five engineering applications of optimization.

1.18 What is an integer programming problem?

1.19 What is graphical optimization, and what are its limitations?

1.20 Under what conditions can a polynomial in n variables be called a posynomial?

1.21 Define a stochastic programming problem and give two practical examples.

1.22 What is a separable programming problem?

PROBLEMS

1.1   A fertilizer company purchases nitrates, phosphates, potash, and an inert chalk base at a cost of $1500, $500, $1000, and $100 per ton, respectively, and produces four fertilizers A, B, C, and D. The production cost, selling price, and composition of the four fertilizers are given below.

Fertilizer	Production cost ($/ton)	Selling price ($/ton)	Percentage composition by weight			
			Nitrates	Phosphates	Potash	Inert chalk base
A	100	350	5	10	5	80
B	150	550	5	15	10	70
C	200	450	10	20	10	60
D	250	700	15	5	15	65

During any week, no more than 1000 tons of nitrate, 2000 tons of phosphates, and 1500 tons of potash will be available. The company is required to supply a minimum of 5000 tons of fertilizer A and 4000 tons of fertilizer D per week to its customers; but it is otherwise free to produce the fertilizers in any quantities it pleases. Formulate the problem of finding the quantity of each fertilizer to be produced by the company to maximize its profit.

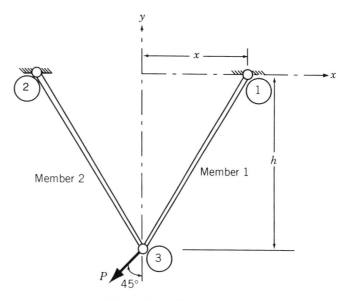

Figure 1.15 Two-bar truss.

1.2 The two-bar truss shown in Fig. 1.15 is symmetric about the y axis. The nondimensional area of cross section of the members A/A_{ref}, and the nondimensional position of joints 1 and 2, x/h, are treated as the design variables x_1 and x_2, respectively, where A_{ref} is the reference value of the area (A) and h is the height of the truss. The coordinates of joint 3 are held constant. The weight of the truss (f_1) and the total displacement of joint 3 under the given load (f_2) are to be minimized without exceeding the permissible stress, σ_0. The weight of the truss and the displacement of joint 3 can be expressed as

$$f_1(\mathbf{X}) = 2\rho h x_2 \sqrt{1 + x_1^2}\, A_{\text{ref}}$$

$$f_2(\mathbf{X}) = \frac{Ph(1 + x_1^2)^{1.5}\sqrt{1 + x_1^4}}{2\sqrt{2}Ex_1^2 x_2 A_{\text{ref}}}$$

where ρ is the weight density, P the applied load, and E the Young's modulus. The stresses induced in members 1 and 2 (σ_1 and σ_2) are given by

$$\sigma_1(\mathbf{X}) = \frac{P(1 + x_1)\sqrt{(1 + x_1^2)}}{2\sqrt{2}x_1 x_2 A_{\text{ref}}}$$

$$\sigma_2(\mathbf{X}) = \frac{P(x_1 - 1)\sqrt{(1 + x_1^2)}}{2\sqrt{2}x_1 x_2 A_{\text{ref}}}$$

In addition, upper and lower bounds are placed on design variables x_1 and x_2 as

$$x_i^{\min} \leq x_i \leq x_i^{\max}; \quad i = 1, 2$$

Find the solution of the problem using a graphical method with (**a**) f_1 as the objective, (**b**) f_2 as the objective, and (**c**) $(f_1 + f_2)$ as the objective for the following data: $E = 30 \times 10^6$ psi,

$\rho = 0.283$ lb/in^3, $P = 10,000$ lb, $\sigma_0 = 20,000$ psi, $h = 100$ in., $A_{\text{ref}} = 1$ in^2, $x_1^{\min} = 0.1$, $x_2^{\min} = 0.1$, $x_1^{\max} = 2.0$, and $x_2^{\max} = 2.5$.

1.3   Ten jobs are to be performed in an automobile assembly line as noted in the following table:

Job Number	Time required to complete the job (min)	Jobs that must be completed before starting this job
1	4	None
2	8	None
3	7	None
4	6	None
5	3	1, 3
6	5	2, 3, 4
7	1	5, 6
8	9	6
9	2	7, 8
10	8	9

It is required to set up a suitable number of workstations, with one worker assigned to each workstation, to perform certain jobs. Formulate the problem of determining the number of workstations and the particular jobs to be assigned to each workstation to minimize the idle time of the workers as an integer programming problem. *Hint:* Define variables x_{ij} such that $x_{ij} = 1$ if job i is assigned to station j, and $x_{ij} = 0$ otherwise.

1.4   A railroad track of length L is to be constructed over an uneven terrain by adding or removing dirt (Fig. 1.16). The absolute value of the slope of the track is to be restricted to a value of r_1 to avoid steep slopes. The absolute value of the rate of change of the slope is to be limited to a value r_2 to avoid rapid accelerations and decelerations. The absolute value of the second derivative of the slope is to be limited to a value of r_3

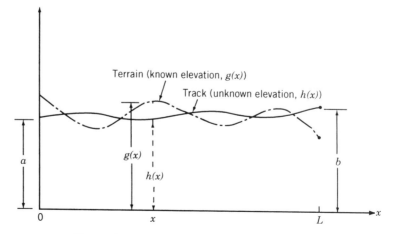

Figure 1.16   Railroad track on an uneven terrain.

to avoid severe jerks. Formulate the problem of finding the elevation of the track to minimize the construction costs as an OC problem. Assume the construction costs to be proportional to the amount of dirt added or removed. The elevation of the track is equal to a and b at $x = 0$ and $x = L$, respectively.

1.5 A manufacturer of a particular product produces x_1 units in the first week and x_2 units in the second week. The number of units produced in the first and second weeks must be at least 200 and 400, respectively, to be able to supply the regular customers. The initial inventory is zero and the manufacturer ceases to produce the product at the end of the second week. The production cost of a unit, in dollars, is given by $4x_i^2$, where x_i is the number of units produced in week $i\,(i = 1, 2)$. In addition to the production cost, there is an inventory cost of \$10 per unit for each unit produced in the first week that is not sold by the end of the first week. Formulate the problem of minimizing the total cost and find its solution using a graphical optimization method.

1.6 Consider the slider-crank mechanism shown in Fig. 1.17 with the crank rotating at a constant angular velocity ω. Use a graphical procedure to find the lengths of the crank and the connecting rod to maximize the velocity of the slider at a crank angle of $\theta = 30°$ for $\omega = 100$ rad/s. The mechanism has to satisfy Groshof's criterion $l \geq 2.5r$ to ensure $360°$ rotation of the crank. Additional constraints on the mechanism are given by $0.5 \leq r \leq 10, 2.5 \leq l \leq 25$, and $10 \leq x \leq 20$.

1.7 Solve Problem 1.6 to maximize the acceleration (instead of the velocity) of the slider at $\theta = 30°$ for $\omega = 100$ rad/s.

1.8 It is required to stamp four circular disks of radii R_1, R_2, R_3, and R_4 from a rectangular plate in a fabrication shop (Fig. 1.18). Formulate the problem as an optimization problem to minimize the scrap. Identify the design variables, objective function, and the constraints.

1.9 The torque transmitted (T) by a cone clutch, shown in Fig. 1.19, under uniform pressure condition is given by

$$T = \frac{2\pi f p}{3 \sin \alpha} (R_1^3 - R_2^3)$$

where p is the pressure between the cone and the cup, f the coefficient of friction, α the cone angle, R_1 the outer radius, and R_2 the inner radius.

(a) Find R_1 and R_2 that minimize the volume of the cone clutch with $\alpha = 30°$, $F = 30$ lb, and $f = 0.5$ under the constraints $T \geq 100$ lb-in., $R_1 \geq 2R_2$, $0 \leq R_1 \leq 15$ in., and $0 \leq R_2 \leq 10$ in.

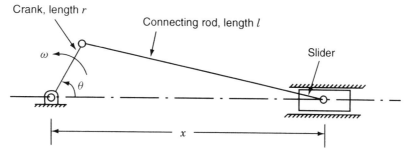

Figure 1.17   Slider-crank mechanism.

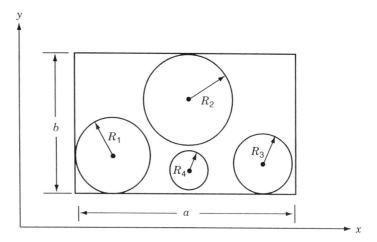

Figure 1.18   Locations of circular disks in a rectangular plate.

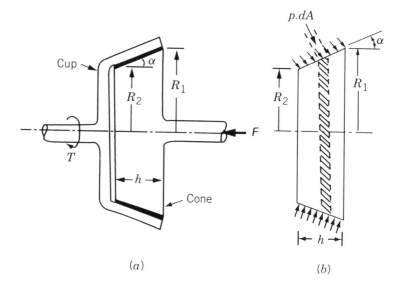

(a)   (b)

Figure 1.19   Cone clutch.

(b) What is the solution if the constraint $R_1 \geq 2R_2$ is changed to $R_1 \leq 2R_2$?

(c) Find the solution of the problem stated in part (a) by assuming a uniform wear condition between the cup and the cone. The torque transmitted (T) under uniform wear condition is given by

$$T = \frac{\pi f p R_2}{\sin \alpha}(R_1^2 - R_2^2)$$

Note: Use graphical optimization for the solutions.

1.10 A hollow circular shaft is to be designed for minimum weight to achieve a minimum reliability of 0.99 when subjected to a random torque of $(\overline{T}, \sigma_T) = (10^6, 10^4)$ lb-in., where $\overline{T}$ is the mean torque and σ_T is the standard deviation of the torque, T. The permissible shear stress, τ_0, of the material is given by $(\overline{\tau}_0, \sigma_{\tau 0}) = (50{,}000, 5000)$ psi, where $\overline{\tau}_0$ is the mean value and $\sigma_{\tau 0}$ is the standard deviation of τ_0. The maximum induced stress (τ) in the shaft is given by

$$\tau = \frac{T r_o}{J}$$

where r_o is the outer radius and J is the polar moment of inertia of the cross section of the shaft. The manufacturing tolerances on the inner and outer radii of the shaft are specified as ± 0.06 in. The length of the shaft is given by 50 ± 1 in. and the specific weight of the material by 0.3 ± 0.03 lb/in^3. Formulate the optimization problem and solve it using a graphical procedure. Assume normal distribution for all the random variables and 3σ values for the specified tolerances. *Hints:* (1) The minimum reliability requirement of 0.99 can be expressed, equivalently, as [1.120]

$$z_1 = 2.326 \le \frac{\overline{\tau} - \overline{\tau}_0}{\sqrt{\sigma_\tau^2 + \sigma_{\tau 0}^2}}$$

(2) If $f(x_1, x_2, \ldots, x_n)$ is a function of the random variables $x_1, x_2, \ldots, x_n$, the mean value of $f(\overline{f})$ and the standard deviation of $f(\sigma_f)$ are given by

$$\overline{f} = f(\overline{x}_1, \overline{x}_2, \ldots, \overline{x}_n)$$

$$\sigma_f = \left[\sum_{i=1}^{n} \left(\frac{\partial f}{\partial x_i} \bigg|_{\overline{x}_1, \overline{x}_2, \ldots, \overline{x}_n} \right)^2 \sigma_{xi}^2 \right]^{1/2}$$

where $\overline{x}_i$ is the mean value of x_i, and σ_{xi} is the standard deviation of x_i.

1.11 Certain nonseparable optimization problems can be reduced to a separable form by using suitable transformation of variables. For example, the product term $f = x_1 x_2$ can be reduced to the separable form $f = y_1^2 - y_2^2$ by introducing the transformations

$$y_1 = \tfrac{1}{2}(x_1 + x_2), \quad y_2 = \tfrac{1}{2}(x_1 - x_2)$$

Suggest suitable transformations to reduce the following terms to separable form:

(a) $f = x_1^2 x_2^3$, $x_1 > 0$, $x_2 > 0$
(b) $f = x_1^{x_2}$, $x_1 > 0$

1.12 In the design of a shell-and-tube heat exchanger (Fig. 1.20), it is decided to have the total length of tubes equal to at least α_1 [1.10]. The cost of the tube is α_2 per unit length and the cost of the shell is given by $\alpha_3 D^{2.5} L$, where D is the diameter and L is the length of the heat exchanger shell. The floor space occupied by the heat exchanger costs α_4 per unit area and the cost of pumping cold fluid is $\alpha_5 L/d^5 N^2$ per day, where d is the diameter of the tube and N is the number of tubes. The maintenance cost is given by $\alpha_6 NdL$. The thermal energy transferred to the cold fluid is given by $\alpha_7/N^{1.2}dL^{1.4} + \alpha_8/d^{0.2}L$. Formulate the mathematical programming problem of minimizing the overall cost of the heat exchanger with the constraint that the thermal energy transferred be greater than a specified amount α_9. The expected life of the heat exchanger is α_{10} years. Assume that $\alpha_i, i = 1, 2, \ldots, 10$, are known constants, and each tube occupies a cross-sectional square of width and depth equal to d.

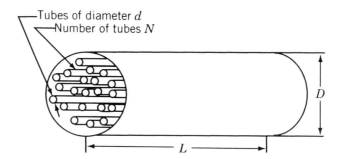

Figure 1.20   Shell-and-tube heat exchanger.

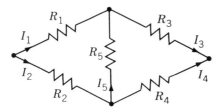

Figure 1.21   Electrical bridge network.

1.13   The bridge network shown in Fig. 1.21 consists of five resistors $R_i (i = 1, 2, \ldots, 5)$. If I_i is the current flowing through the resistance R_i, the problem is to find the resistances $R_1, R_2, \ldots, R_5$ so that the total power dissipated by the network is a minimum. The current I_i can vary between the lower and upper limits $I_{i,\min}$ and $I_{i,\max}$, and the voltage drop, $V_i = R_i I_i$, must be equal to a constant c_i for $1 \le i \le 5$. Formulate the problem as a mathematical programming problem.

1.14   A traveling saleswoman has to cover n towns. She plans to start from a particular town numbered 1, visit each of the other $n - 1$ towns, and return to the town 1. The distance between towns i and j is given by d_{ij}. Formulate the problem of selecting the sequence in which the towns are to be visited to minimize the total distance traveled.

1.15   A farmer has a choice of planting barley, oats, rice, or wheat on his 200-acre farm. The labor, water, and fertilizer requirements, yields per acre, and selling prices are given in the following table:

Type of crop	Labor cost ($)	Water required (m^3)	Fertilizer required (lb)	Yield (lb)	Selling price ($/lb)
Barley	300	10,000	100	1,500	0.5
Oats	200	7,000	120	3,000	0.2
Rice	250	6,000	160	2,500	0.3
Wheat	360	8,000	200	2,000	0.4

The farmer can also give part or all of the land for lease, in which case he gets $200 per acre. The cost of water is $0.02/m^3 and the cost of the fertilizer is $2/lb. Assume that the farmer has no money to start with and can get a maximum loan of $50,000 from the land mortgage bank at an interest of 8 %. He can repay the loan after six months. The irrigation canal cannot supply more than 4×10^5 m^3 of water. Formulate the problem of finding the planting schedule for maximizing the expected returns of the farmer.

1.16 There are two different sites, each with four possible targets (or depths) to drill an oil well. The preparation cost for each site and the cost of drilling at site i to target j are given below:

Site i	Drilling cost to target j				Preparation cost
	1	2	3	4	
1	4	1	9	7	11
2	7	9	5	2	13

Formulate the problem of determining the best site for each target so that the total cost is minimized.

1.17 A four-pole dc motor, whose cross section is shown in Fig. 1.22, is to be designed with the length of the stator and rotor x_1, the overall diameter of the motor x_2, the unnotched radius x_3, the depth of the notches x_4, and the ampere turns x_5 as design variables.

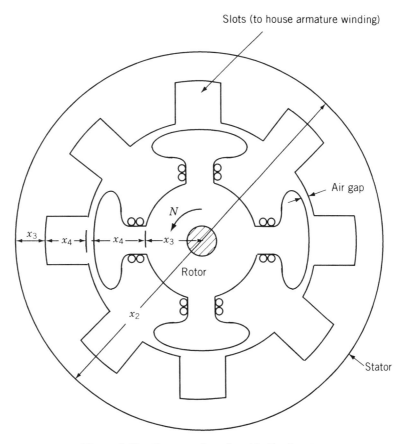

Figure 1.22 Cross section of an idealized motor.

The air gap is to be less than $k_1\sqrt{x_2 + 7.5}$ where k_1 is a constant. The temperature of the external surface of the motor cannot exceed ΔT above the ambient temperature. Assuming that the heat can be dissipated only by radiation, formulate the problem for maximizing the power of the motor [1.59]. *Hints:*

1. The heat generated due to current flow is given by $k_2 x_1 x_2^{-1} x_4^{-1} x_5^2$, where k_2 is a constant. The heat radiated from the external surface for a temperature difference of ΔT is given by $k_3 x_1 x_2 \Delta T$, where k_3 is a constant.

2. The expression for power is given by $k_4 N B x_1 x_3 x_5$, where k_4 is a constant, N is the rotational speed of the rotor, and B is the average flux density in the air gap.

3. The units of the various quantities are as follows. Lengths: centimeter, heat generated, heat dissipated; power: watt; temperature: °C; rotational speed: rpm; flux density: gauss.

1.18 A gas pipeline is to be laid between two cities A and E, making it pass through one of the four locations in each of the intermediate towns B, C, and D (Fig. 1.23). The associated costs are indicated in the following tables.

Costs for A to B and D to E

	Station i			
	1	2	3	4
From A to point i of B	30	35	25	40
From point i of D to E	50	40	35	25

Costs for B to C and C to D

From:	To:			
	1	2	3	4
1	22	18	24	18
2	35	25	15	21
3	24	20	26	20
4	22	21	23	22

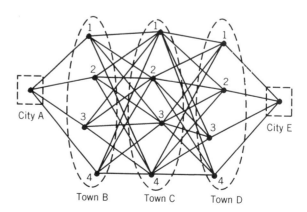

Figure 1.23   Possible paths of the pipeline between A and E.

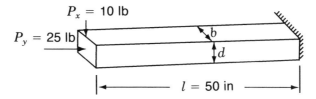

Figure 1.24   Beam-column.

Formulate the problem of minimizing the cost of the pipeline.

1.19   A beam-column of rectangular cross section is required to carry an axial load of 25 lb and a transverse load of 10 lb, as shown in Fig. 1.24. It is to be designed to avoid the possibility of yielding and buckling and for minimum weight. Formulate the optimization problem by assuming that the beam-column can bend only in the vertical (xy) plane. Assume the material to be steel with a specific weight of 0.3 lb/in^3, Young's modulus of 30×10^6 psi, and a yield stress of 30,000 psi. The width of the beam is required to be at least 0.5 in. and not greater than twice the depth. Also, find the solution of the problem graphically. *Hint:* The compressive stress in the beam-column due to P_y is P_y/bd and that due to P_x is

$$\frac{P_x l d}{2 I_{zz}} = \frac{6 P_x l}{b d^2}$$

The axial buckling load is given by

$$(P_y)_{\text{cri}} = \frac{\pi^2 E I_{zz}}{4 l^2} = \frac{\pi^2 E b d^3}{48 l^2}$$

1.20   A two-bar truss is to be designed to carry a load of $2W$ as shown in Fig. 1.25. Both bars have a tubular section with mean diameter d and wall thickness t. The material of the bars has Young's modulus E and yield stress σ_y. The design problem involves the determination of the values of d and t so that the weight of the truss is a minimum and neither yielding nor buckling occurs in any of the bars. Formulate the problem as a nonlinear programming problem.

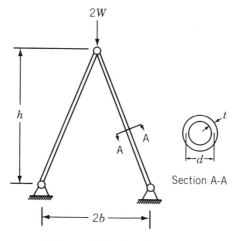

Figure 1.25   Two-bar truss.

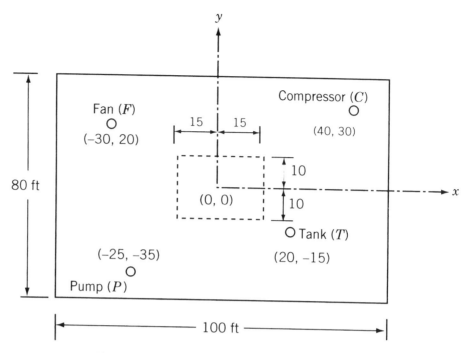

Figure 1.26 Processing plant layout (coordinates in ft).

1.21 Consider the problem of determining the economic lot sizes for four different items. Assume that the demand occurs at a constant rate over time. The stock for the ith item is replenished instantaneously upon request in lots of sizes Q_i. The total storage space available is A, whereas each unit of item i occupies an area d_i. The objective is to find the values of Q_i that optimize the per unit cost of holding the inventory and of ordering subject to the storage area constraint. The cost function is given by

$$C = \sum_{i=1}^{4} \left(\frac{a_i}{Q_i} + b_i Q_i \right), \quad Q_i > 0$$

where a_i and b_i are fixed constants. Formulate the problem as a dynamic programming (optimal control) model. Assume that Q_i is discrete.

1.22 The layout of a processing plant, consisting of a pump (P), a water tank (T), a compressor (C), and a fan (F), is shown in Fig. 1.26. The locations of the various units, in terms of their (x, y) coordinates, are also indicated in this figure. It is decided to add a new unit, a heat exchanger (H), to the plant. To avoid congestion, it is decided to locate H within a rectangular area defined by $\{-15 \le x \le 15, -10 \le y \le 10\}$. Formulate the problem of finding the location of H to minimize the sum of its x and y distances from the existing units, P, T, C, and F.

1.23 Two copper-based alloys (brasses), A and B, are mixed to produce a new alloy, C. The composition of alloys A and B and the requirements of alloy C are given in the following table:

Alloy	Composition by weight			
	Copper	Zinc	Lead	Tin
A	80	10	6	4
B	60	20	18	2
C	≥ 75	≥ 15	≥ 16	≥ 3

If alloy B costs twice as much as alloy A, formulate the problem of determining the amounts of A and B to be mixed to produce alloy C at a minimum cost.

1.24 An oil refinery produces four grades of motor oil in three process plants. The refinery incurs a penalty for not meeting the demand of any particular grade of motor oil. The capacities of the plants, the production costs, the demands of the various grades of motor oil, and the penalties are given in the following table:

Process plant	Capacity of the plant (kgal/day)	Production cost ($/day) to manufacture motor oil of grade:			
		1	2	3	4
1	100	750	900	1000	1200
2	150	800	950	1100	1400
3	200	900	1000	1200	1600
Demand (kgal/day)		50	150	100	75
Penalty (per each kilogallon shortage)		$10	$12	$16	$20

Formulate the problem of minimizing the overall cost as an LP problem.

1.25 A part-time graduate student in engineering is enrolled in a four-unit mathematics course and a three-unit design course. Since the student has to work for 20 hours a week at a local software company, he can spend a maximum of 40 hours a week to study outside the class. It is known from students who took the courses previously that the numerical grade (g) in each course is related to the study time spent outside the class as $g_m = t_m/6$ and $g_d = t_d/5$, where g indicates the numerical grade ($g = 4$ for A, 3 for B, 2 for C, 1 for D, and 0 for F), t represents the time spent in hours per week to study outside the class, and the subscripts m and d denote the courses, mathematics and design, respectively. The student enjoys design more than mathematics and hence would like to spend at least 75 minutes to study for design for every 60 minutes he spends to study mathematics. Also, as far as possible, the student does not want to spend more time on any course beyond the time required to earn a grade of A. The student wishes to maximize his grade point P, given by $P = 4g_m + 3g_d$, by suitably distributing his study time. Formulate the problem as an LP problem.

1.26 The scaffolding system, shown in Fig. 1.27, is used to carry a load of 10,000 lb. Assuming that the weights of the beams and the ropes are negligible, formulate the problem of determining the values of x_1, x_2, x_3, and x_4 to minimize the tension in ropes A and B while maintaining positive tensions in ropes C, D, E, and F.

1.27 Formulate the problem of minimum weight design of a power screw subjected to an axial load, F, as shown in Fig. 1.28 using the pitch (p), major diameter (d), nut height

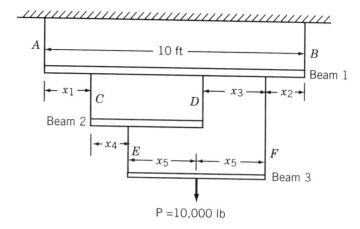

Figure 1.27 Scaffolding system.

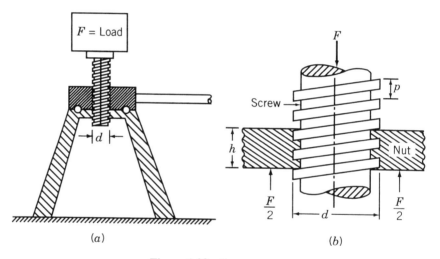

Figure 1.28 Power screw.

(h), and screw length (s) as design variables. Consider the following constraints in the formulation:

1. The screw should be self-locking [1.117].
2. The shear stress in the screw should not exceed the yield strength of the material in shear. Assume the shear strength in shear (according to distortion energy theory), to be $0.577\sigma_y$, where σ_y is the yield strength of the material.
3. The bearing stress in the threads should not exceed the yield strength of the material, σ_y.
4. The critical buckling load of the screw should be less than the applied load, F.

1.28 **(a)** A simply supported beam of hollow rectangular section is to be designed for minimum weight to carry a vertical load F_y and an axial load P as shown in Fig. 1.29. The deflection of the beam in the y direction under the self-weight and F_y should

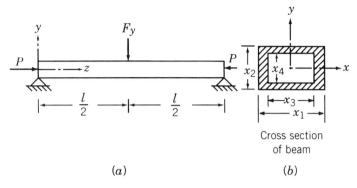

Figure 1.29 Simply supported beam under loads.

not exceed 0.5 in. The beam should not buckle either in the yz or the xz plane under the axial load. Assuming the ends of the beam to be pin ended, formulate the optimization problem using x_i, $i = 1, 2, 3, 4$ as design variables for the following data: $F_y = 300$ lb, $P = 40,000$ lb, $l = 120$ in., $E = 30 \times 10^6$ psi, $\rho = 0.284$ lb/in^3, lower bound on x_1 and $x_2 = 0.125$ in, upper bound on x_1, and $x_2 = 4$ in.

(b) Formulate the problem stated in part (a) using x_1 and x_2 as design variables, assuming the beam to have a solid rectangular cross section. Also find the solution of the problem using a graphical technique.

1.29 A cylindrical pressure vessel with hemispherical ends (Fig. 1.30) is required to hold at least 20,000 gallons of a fluid under a pressure of 2500 psia. The thicknesses of the cylindrical and hemispherical parts of the shell should be equal to at least those recommended by section VIII of the ASME pressure vessel code, which are given by

$$t_c = \frac{pR}{Se + 0.4p}$$

$$t_h = \frac{pR}{Se + 0.8p}$$

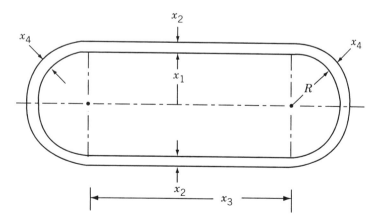

Figure 1.30 Pressure vessel.

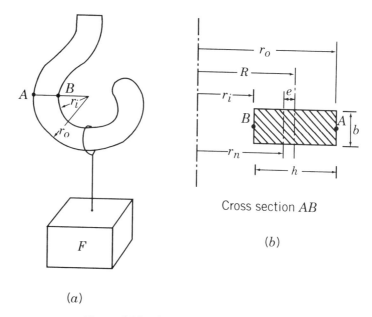

Cross section AB

(b)

(a)

Figure 1.31   Crane hook carrying a load.

where S is the yield strength, e the joint efficiency, p the pressure, and R the radius. Formulate the design problem for minimum structural volume using x_i, $i = 1, 2, 3, 4$, as design variables. Assume the following data: $S = 30{,}000$ psi and $e = 1.0$.

1.30   A crane hook is to be designed to carry a load F as shown in Fig. 1.31. The hook can be modeled as a three-quarter circular ring with a rectangular cross section. The stresses induced at the inner and outer fibers at section AB should not exceed the yield strength of the material. Formulate the problem of minimum volume design of the hook using r_o, r_i, b, and h as design variables. *Note:* The stresses induced at points A and B are given by [1.117]

$$\sigma_A = \frac{M c_o}{A e r_o}$$

$$\sigma_B = \frac{M c_i}{A e r_i}$$

where M is the bending moment due to the load ($= FR$), R the radius of the centroid, r_o the radius of the outer fiber, r_i the radius of the inner fiber, c_o the distance of the outer fiber from the neutral axis $= R_o - r_n$, c_i the distance of inner fiber from neutral axis $= r_n - r_i$, r_n the radius of neutral axis, given by

$$r_n = \frac{h}{\ln(r_o/r_i)}$$

A the cross-sectional area of the hook $= bh$, and e the distance between the centroidal and neutral axes $= R - r_n$.

1.31   Consider the four-bar truss shown in Fig. 1.32, in which members 1, 2, and 3 have the same cross-sectional area x_1 and the same length l, while member 4 has an area of

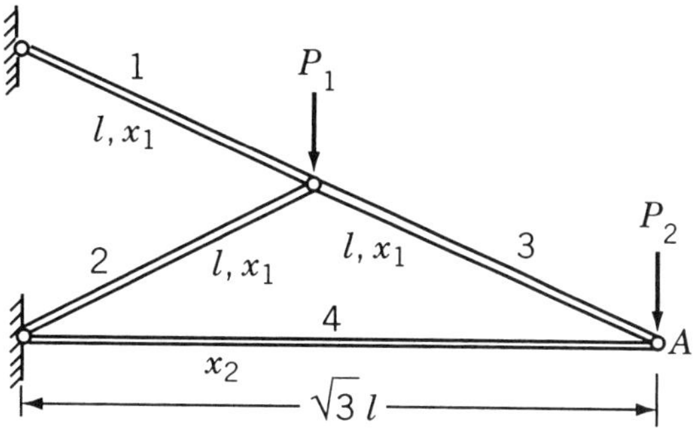

Figure 1.32 Four-bar truss.

cross section x_2 and length $\sqrt{3}\ l$. The truss is made of a lightweight material for which Young's modulus and the weight density are given by 30×10^6 psi and 0.03333 lb/in^3, respectively. The truss is subject to the loads $P_1 = 10{,}000$ lb and $P_2 = 20{,}000$ lb. The weight of the truss per unit value of l can be expressed as

$$f = 3x_1(1)(0.03333) + x_2\sqrt{3}(0.03333) = 0.1x_1 + 0.05773x_2$$

The vertical deflection of joint A can be expressed as

$$\delta_A = \frac{0.6}{x_1} + \frac{0.3464}{x_2}$$

and the stresses in members 1 and 4 can be written as

$$\sigma_1 = \frac{5(10{,}000)}{x_1} = \frac{50{,}000}{x_1}, \quad \sigma_4 = \frac{-2\sqrt{3}(10{,}000)}{x_2} = -\frac{34{,}640}{x_2}$$

The weight of the truss is to be minimized with constraints on the vertical deflection of the joint A and the stresses in members 1 and 4. The maximum permissible deflection of joint A is 0.1 in. and the permissible stresses in members are $\sigma_{\max} = 8333.3333$ psi (tension) and $\sigma_{\min} = -4948.5714$ psi (compression). The optimization problem can be stated as a separable programming problem as follows:

$$\text{Minimize } f(x_1, x_2) = 0.1x_1 + 0.05773x_2$$

subject to

$$\frac{0.6}{x_1} + \frac{0.3464}{x_2} - 0.1 \le 0, \quad 6 - x_1 \le 0, \quad 7 - x_2 \le 0$$

Determine the solution of the problem using a graphical procedure.

1.32 A simply supported beam, with a uniform rectangular cross section, is subjected to both distributed and concentrated loads as shown in Fig. 1.33. It is desired to find the cross section of the beam to minimize the weight of the beam while ensuring that the maximum stress induced in the beam does not exceed the permissible stress (σ_0) of the material and the maximum deflection of the beam does not exceed a specified limit (δ_0).
The data of the problem are $P = 10^5$ N, $p_0 = 10^6$ N/m, $L = 1$ m, $E = 207$ GPa, weight density (ρ_w) = 76.5 kN/m^3, $\sigma_0 = 220$ MPa, and $\delta_0 = 0.02$ m.

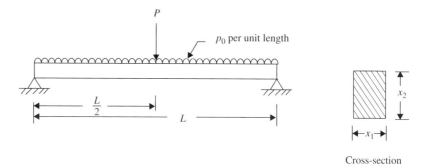

Figure 1.33 A simply supported beam subjected to concentrated and distributed loads.

(a) Formulate the problem as a mathematical programming problem assuming that the cross-sectional dimensions of the beam are restricted as $x_1 \leq x_2$, $0.04m \leq x_1 \leq 0.12m$, and $0.06m \leq x_2 \leq 0.20\ m$.

(b) Find the solution of the problem formulated in part (a) using MATLAB.

(c) Find the solution of the problem formulated in part (a) graphically.

1.33 Solve Problem 1.32, parts (a), (b), and (c), assuming the cross section of the beam to be hollow circular with inner diameter x_1 and outer diameter x_2. Assume the data and bounds on the design variables to be as given in Problem 1.32.

1.34 Find the solution of Problem 1.31 using MATLAB.

1.35 Find the solution of Problem 1.2(a) using MATLAB.

1.36 Find the solution of Problem 1.2(b) using MATLAB.

2

Classical Optimization Techniques

2.1 INTRODUCTION

The classical methods of optimization are useful in finding the optimum solution of continuous and differentiable functions. These methods are analytical and make use of the techniques of differential calculus in locating the optimum points. Since some of the practical problems involve objective functions that are not continuous and/or differentiable, the classical optimization techniques have limited scope in practical applications. However, a study of the calculus methods of optimization forms a basis for developing most of the numerical techniques of optimization presented in subsequent chapters. In this chapter we present the necessary and sufficient conditions in locating the optimum solution of a single-variable function, a multivariable function with no constraints, and a multivariable function with equality and inequality constraints.

2.2 SINGLE-VARIABLE OPTIMIZATION

A function of one variable $f(x)$ is said to have a *relative* or *local minimum* at $x = x^*$ if $f(x^*) \leq f(x^* + h)$ for all sufficiently small positive and negative values of h. Similarly, a point x^* is called a *relative* or *local maximum* if $f(x^*) \geq f(x^* + h)$ for all values of h sufficiently close to zero. A function $f(x)$ is said to have a *global* or *absolute minimum* at x^* if $f(x^*) \leq f(x)$ for all x, and not just for all x close to x^*, in the domain over which $f(x)$ is defined. Similarly, a point x^* will be a global maximum of $f(x)$ if $f(x^*) \geq f(x)$ for all x in the domain. Figure 2.1 shows the difference between the local and global optimum points.

A *single-variable optimization problem* is one in which the value of $x = x^*$ is to be found in the interval $[a, b]$ such that x^* minimizes $f(x)$. The following two theorems provide the necessary and sufficient conditions for the relative minimum of a function of a single variable.

Theorem 2.1 Necessary Condition If a function $f(x)$ is defined in the interval $a \leq x \leq b$ and has a relative minimum at $x = x^*$, where $a < x^* < b$, and if the derivative $df(x)/dx = f'(x)$ exists as a finite number at $x = x^*$, then $f'(x^*) = 0$.

Proof: It is given that

$$f'(x^*) = \lim_{h \to 0} \frac{f(x^* + h) - f(x^*)}{h} \tag{2.1}$$

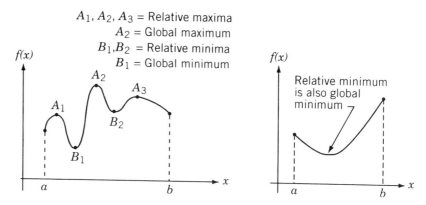

Figure 2.1 Relative and global minima.

exists as a definite number, which we want to prove to be zero. Since x^* is a relative minimum, we have

$$f(x^*) \leq f(x^* + h)$$

for all values of h sufficiently close to zero. Hence

$$\frac{f(x^* + h) - f(x^*)}{h} \geq 0 \quad \text{if } h > 0$$

$$\frac{f(x^* + h) - f(x^*)}{h} \leq 0 \quad \text{if } h < 0$$

Thus Eq. (2.1) gives the limit as h tends to zero through positive values as

$$f'(x^*) \geq 0 \tag{2.2}$$

while it gives the limit as h tends to zero through negative values as

$$f'(x^*) \leq 0 \tag{2.3}$$

The only way to satisfy both Eqs. (2.2) and (2.3) is to have

$$f'(x^*) = 0 \tag{2.4}$$

This proves the theorem.

Notes:

1. This theorem can be proved even if x^* is a relative maximum.
2. The theorem does not say what happens if a minimum or maximum occurs at a point x^* where the derivative fails to exist. For example, in Fig. 2.2,

$$\lim_{h \to 0} \frac{f(x^* + h) - f(x^*)}{h} = m^+ \text{(positive) or } m^- \text{(negative)}$$

depending on whether h approaches zero through positive or negative values, respectively. Unless the numbers m^+ and m^- are equal, the derivative $f'(x^*)$ does not exist. If $f'(x^*)$ does not exist, the theorem is not applicable.

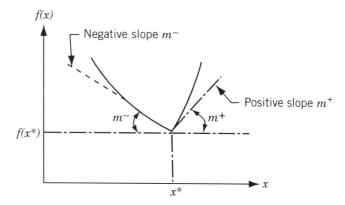

Figure 2.2   Derivative undefined at x^*.

3. The theorem does not say what happens if a minimum or maximum occurs at an endpoint of the interval of definition of the function. In this case

$$\lim_{h \to 0} \frac{f(x^* + h) - f(x^*)}{h}$$

exists for positive values of h only or for negative values of h only, and hence the derivative is not defined at the endpoints.

4. The theorem does not say that the function necessarily will have a minimum or maximum at every point where the derivative is zero. For example, the derivative $f'(x) = 0$ at $x = 0$ for the function shown in Fig. 2.3. However, this point is neither a minimum nor a maximum. In general, a point x^* at which $f'(x^*) = 0$ is called a *stationary point*.

If the function $f(x)$ possesses continuous derivatives of every order that come in question, in the neighborhood of $x = x^*$, the following theorem provides the sufficient condition for the minimum or maximum value of the function.

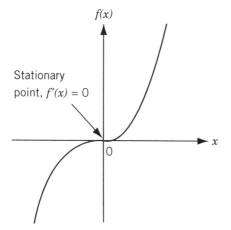

Stationary
point, $f'(x) = 0$

Figure 2.3   Stationary (inflection) point.

Theorem 2.2 Sufficient Condition Let $f'(x^*) = f''(x^*) = \cdots = f^{(n-1)}(x^*) = 0$, but $f^{(n)}(x^*) \neq 0$. Then $f(x^*)$ is (i) a minimum value of $f(x)$ if $f^{(n)}(x^*) > 0$ and n is even; (ii) a maximum value of $f(x)$ if $f^{(n)}(x^*) < 0$ and n is even; (iii) neither a maximum nor a minimum if n is odd.

Proof: Applying Taylor's theorem with remainder after n terms, we have

$$f(x^* + h) = f(x^*) + hf'(x^*) + \frac{h^2}{2!}f''(x^*) + \cdots + \frac{h^{n-1}}{(n-1)!}f^{(n-1)}(x^*)$$

$$+ \frac{h^n}{n!}f^{(n)}(x^* + \theta h) \qquad \text{for} \quad 0 < \theta < 1 \tag{2.5}$$

Since $f'(x^*) = f''(x^*) = \cdots = f^{(n-1)}(x^*) = 0$, Eq. (2.5) becomes

$$f(x^* + h) - f(x^*) = \frac{h^n}{n!}f^{(n)}(x^* + \theta h)$$

As $f^{(n)}(x^*) \neq 0$, there exists an interval around x^* for every point x of which the nth derivative $f^{(n)}(x)$ has the same sign, namely, that of $f^{(n)}(x^*)$. Thus for every point $x^* + h$ of this interval, $f^{(n)}(x^* + \theta h)$ has the sign of $f^{(n)}(x^*)$. When n is even, $h^n/n!$ is positive irrespective of whether h is positive or negative, and hence $f(x^* + h) - f(x^*)$ will have the same sign as that of $f^{(n)}(x^*)$. Thus x^* will be a relative minimum if $f^{(n)}(x^*)$ is positive and a relative maximum if $f^{(n)}(x^*)$ is negative. When n is odd, $h^n/n!$ changes sign with the change in the sign of h and hence the point x^* is neither a maximum nor a minimum. In this case the point x^* is called a *point of inflection*.

Example 2.1 Determine the maximum and minimum values of the function

$$f(x) = 12x^5 - 45x^4 + 40x^3 + 5$$

SOLUTION Since $f'(x) = 60(x^4 - 3x^3 + 2x^2) = 60x^2(x - 1)(x - 2)$, $f'(x) = 0$ at $x = 0, x = 1$, and $x = 2$. The second derivative is

$$f''(x) = 60(4x^3 - 9x^2 + 4x)$$

At $x = 1$, $f''(x) = -60$ and hence $x = 1$ is a relative maximum. Therefore,

$$f_{\max} = f(x = 1) = 12$$

At $x = 2$, $f''(x) = 240$ and hence $x = 2$ is a relative minimum. Therefore,

$$f_{\min} = f(x = 2) = -11$$

At $x = 0$, $f''(x) = 0$ and hence we must investigate the next derivative:

$$f'''(x) = 60(12x^2 - 18x + 4) = 240 \quad \text{at} \quad x = 0$$

Since $f'''(x) \neq 0$ at $x = 0$, $x = 0$ is neither a maximum nor a minimum, and it is an inflection point.

Example 2.2 In a two-stage compressor, the working gas leaving the first stage of compression is cooled (by passing it through a heat exchanger) before it enters the second stage of compression to increase the efficiency [2.13]. The total work input to a compressor (W) for an ideal gas, for isentropic compression, is given by

$$W = c_p T_1 \left[\left(\frac{p_2}{p_1} \right)^{(k-1)/k} + \left(\frac{p_3}{p_2} \right)^{(k-1)/k} - 2 \right]$$

where c_p is the specific heat of the gas at constant pressure, k is the ratio of specific heat at constant pressure to that at constant volume of the gas, and T_1 is the temperature at which the gas enters the compressor. Find the pressure, p_2, at which intercooling should be done to minimize the work input to the compressor. Also determine the minimum work done on the compressor.

SOLUTION The necessary condition for minimizing the work done on the compressor is

$$\frac{dW}{dp_2} = c_p T_1 \frac{k}{k-1} \left[\left(\frac{1}{p_1} \right)^{(k-1)/k} \frac{k-1}{k} (p_2)^{-1/k} \right.$$
$$\left. + (p_3)^{(k-1)/k} \frac{-k+1}{k} (p_2)^{(1-2k)/k} \right] = 0$$

which yields

$$p_2 = (p_1 p_3)^{1/2}$$

The second derivative of W with respect to p_2 gives

$$\frac{d^2 W}{dp_2^2} = c_p T_1 \left[-\left(\frac{1}{p_1} \right)^{(k-1)/k} \frac{1}{k} (p_2)^{-(1+k)/k} \right.$$
$$\left. - (p_3)^{(k-1)/k} \frac{1-2k}{k} (p_2)^{(1-3k)/k} \right]$$

$$\left(\frac{d^2 W}{dp_2^2} \right)_{p_2 = (p_1 p_2)^{1/2}} = \frac{2 c_p T_1 \frac{k-1}{k}}{p_1^{(3k-1)/2k} p_3^{(k+1)/2k}}$$

Since the ratio of specific heats k is greater than 1, we get

$$\frac{d^2 W}{dp_2^2} > 0 \quad \text{at} \quad p_2 = (p_1 p_3)^{1/2}$$

and hence the solution corresponds to a relative minimum. The minimum work done is given by

$$W_{\min} = 2 c_p T_1 \frac{k}{k-1} \left[\left(\frac{p_3}{p_1} \right)^{(k-1)/2k} - 1 \right]$$

2.3   MULTIVARIABLE OPTIMIZATION WITH NO CONSTRAINTS

In this section we consider the necessary and sufficient conditions for the minimum or maximum of an unconstrained function of several variables. Before seeing these conditions, we consider the Taylor's series expansion of a multivariable function.

Definition: rth Differential of f .   If all partial derivatives of the function f through order $r \geq 1$ exist and are continuous at a point $\mathbf{X}^*$, the polynomial

$$d^r f(\mathbf{X}^*) = \underbrace{\sum_{i=1}^{n} \sum_{j=1}^{n} \cdots \sum_{k=1}^{n}}_{r\text{summations}} h_i h_j \cdots h_k \frac{\partial^r f(\mathbf{X}^*)}{\partial x_i \partial x_j \cdots \partial x_k} \tag{2.6}$$

is called the *r*th differential of f at $\mathbf{X}^*$. Notice that there are r summations and one h_i is associated with each summation in Eq. (2.6).

For example, when $r = 2$ and $n = 3$, we have

$$d^2 f(\mathbf{X}^*) = d^2 f(x_1^*, x_2^*, x_3^*) = \sum_{i=1}^{3} \sum_{j=1}^{3} h_i h_j \frac{\partial^2 f(\mathbf{X}^*)}{\partial x_i \partial x_j}$$

$$= h_1^2 \frac{\partial^2 f}{\partial x_1^2}(\mathbf{X}^*) + h_2^2 \frac{\partial^2 f}{\partial x_2^2}(\mathbf{X}^*) + h_3^2 \frac{\partial^2 f}{\partial x_3^2}(\mathbf{X}^*)$$

$$+ 2h_1 h_2 \frac{\partial^2 f}{\partial x_1 \partial x_2}(\mathbf{X}^*) + 2h_2 h_3 \frac{\partial^2 f}{\partial x_2 \partial x_3}(\mathbf{X}^*) + 2h_1 h_3 \frac{\partial^2 f}{\partial x_1 \partial x_3}(\mathbf{X}^*)$$

The Taylor's series expansion of a function $f(\mathbf{X})$ about a point $\mathbf{X}^*$ is given by

$$f(\mathbf{X}) = f(\mathbf{X}^*) + df(\mathbf{X}^*) + \frac{1}{2!}d^2 f(\mathbf{X}^*) + \frac{1}{3!}d^3 f(\mathbf{X}^*)$$

$$+ \cdots + \frac{1}{N!}d^N f(\mathbf{X}^*) + R_N(\mathbf{X}^*, \mathbf{h}) \tag{2.7}$$

where the last term, called the *remainder*, is given by

$$R_N(\mathbf{X}^*, \mathbf{h}) = \frac{1}{(N+1)!}d^{N+1} f(\mathbf{X}^* + \theta \mathbf{h}) \tag{2.8}$$

where $0 < \theta < 1$ and $\mathbf{h} = \mathbf{X} - \mathbf{X}^*$.

Example 2.3   Find the second-order Taylor's series approximation of the function

$$f(x_1, x_2, x_3) = x_2^2 x_3 + x_1 e^{x_3}$$

about the point $\mathbf{X}^* = \{1, 0, -2\}^{\mathrm{T}}$.

SOLUTION   The second-order Taylor's series approximation of the function f about point $\mathbf{X}^*$ is given by

$$f(\mathbf{X}) = f \begin{pmatrix} 1 \\ 0 \\ -2 \end{pmatrix} + df \begin{pmatrix} 1 \\ 0 \\ -2 \end{pmatrix} + \frac{1}{2!}d^2 f \begin{pmatrix} 1 \\ 0 \\ -2 \end{pmatrix}$$

where

$$f \begin{pmatrix} 1 \\ 0 \\ -2 \end{pmatrix} = e^{-2}$$

$$df \begin{pmatrix} 1 \\ 0 \\ -2 \end{pmatrix} = h_1 \frac{\partial f}{\partial x_1} \begin{pmatrix} 1 \\ 0 \\ -2 \end{pmatrix} + h_2 \frac{\partial f}{\partial x_2} \begin{pmatrix} 1 \\ 0 \\ -2 \end{pmatrix} + h_3 \frac{\partial f}{\partial x_3} \begin{pmatrix} 1 \\ 0 \\ -2 \end{pmatrix}$$

$$= [h_1 e^{x_3} + h_2 (2 x_2 x_3) + h_3 x_2^2 + h_3 x_1 e^{x_3}] \begin{pmatrix} 1 \\ 0 \\ -2 \end{pmatrix} = h_1 e^{-2} + h_3 e^{-2}$$

$$d^2 f \begin{pmatrix} 1 \\ 0 \\ -2 \end{pmatrix} = \sum_{i=1}^{3} \sum_{j=1}^{3} h_i h_j \frac{\partial^2 f}{\partial x_i \partial x_j} \begin{pmatrix} 1 \\ 0 \\ -2 \end{pmatrix} = \left(h_1^2 \frac{\partial^2 f}{\partial x_1^2} + h_2^2 \frac{\partial^2 f}{\partial x_2^2} + h_3^2 \frac{\partial^2 f}{\partial x_3^2} \right.$$

$$\left. + 2 h_1 h_2 \frac{\partial^2 f}{\partial x_1 \partial x_2} + 2 h_2 h_3 \frac{\partial^2 f}{\partial x_2 \partial x_3} + 2 h_1 h_3 \frac{\partial^2 f}{\partial x_1 \partial x_3} \right) \begin{pmatrix} 1 \\ 0 \\ -2 \end{pmatrix}$$

$$= [h_1^2 (0) + h_2^2 (2 x_3) + h_3^2 (x_1 e^{x_3}) + 2 h_1 h_2 (0) + 2 h_2 h_3 (2 x_2)$$

$$+ 2 h_1 h_3 (e^{x_3})] \begin{pmatrix} 1 \\ 0 \\ -2 \end{pmatrix} = -4 h_2^2 + e^{-2} h_3^2 + 2 h_1 h_3 e^{-2}$$

Thus the Taylor's series approximation is given by

$$f(\mathbf{X}) \simeq e^{-2} + e^{-2}(h_1 + h_3) + \frac{1}{2!} (-4 h_2^2 + e^{-2} h_3^2 + 2 h_1 h_3 e^{-2})$$

where $h_1 = x_1 - 1$, $h_2 = x_2$, and $h_3 = x_3 + 2$.

Theorem 2.3 Necessary Condition   If $f(\mathbf{X})$ has an extreme point (maximum or minimum) at $\mathbf{X} = \mathbf{X}^*$ and if the first partial derivatives of $f(\mathbf{X})$ exist at $\mathbf{X}^*$, then

$$\frac{\partial f}{\partial x_1}(\mathbf{X}^*) = \frac{\partial f}{\partial x_2}(\mathbf{X}^*) = \cdots = \frac{\partial f}{\partial x_n}(\mathbf{X}^*) = 0 \qquad (2.9)$$

Proof: The proof given for Theorem 2.1 can easily be extended to prove the present theorem. However, we present a different approach to prove this theorem. Suppose that one of the first partial derivatives, say the kth one, does not vanish at $\mathbf{X}^*$. Then, by Taylor's theorem,

$$f(\mathbf{X}^* + \mathbf{h}) = f(\mathbf{X}^*) + \sum_{i=1}^{n} h_i \frac{\partial f}{\partial x_i}(\mathbf{X}^*) + R_1(\mathbf{X}^*, \mathbf{h})$$

that is,

$$f(\mathbf{X}^* + \mathbf{h}) - f(\mathbf{X}^*) = h_k \frac{\partial f}{\partial x_k}(\mathbf{X}^*) + \frac{1}{2!} d^2 f(\mathbf{X}^* + \theta \mathbf{h}), \qquad 0 < \theta < 1$$

Since $d^2 f(\mathbf{X}^* + \theta \mathbf{h})$ is of order h_i^2, the terms of order $\mathbf{h}$ will dominate the higher-order terms for small $\mathbf{h}$. Thus the sign of $f(\mathbf{X}^* + \mathbf{h}) - f(\mathbf{X}^*)$ is decided by the sign of $h_k \, \partial f(\mathbf{X}^*)/\partial x_k$. Suppose that $\partial f(\mathbf{X}^*)/\partial x_k > 0$. Then the sign of $f(\mathbf{X}^* + \mathbf{h}) - f(\mathbf{X}^*)$ will be positive for $h_k > 0$ and negative for $h_k < 0$. This means that $\mathbf{X}^*$ cannot be an extreme point. The same conclusion can be obtained even if we assume that $\partial f(\mathbf{X}^*)/\partial x_k < 0$. Since this conclusion is in contradiction with the original statement that $\mathbf{X}^*$ is an extreme point, we may say that $\partial f/\partial x_k = 0$ at $\mathbf{X} = \mathbf{X}^*$. Hence the theorem is proved.

Theorem 2.4 Sufficient Condition A sufficient condition for a stationary point $\mathbf{X}^*$ to be an extreme point is that the matrix of second partial derivatives (Hessian matrix) of $f(\mathbf{X})$ evaluated at $\mathbf{X}^*$ is (i) positive definite when $\mathbf{X}^*$ is a relative minimum point, and (ii) negative definite when $\mathbf{X}^*$ is a relative maximum point.

Proof: From Taylor's theorem we can write

$$f(\mathbf{X}^* + \mathbf{h}) = f(\mathbf{X}^*) + \sum_{i=1}^{n} h_i \frac{\partial f}{\partial x_i}(\mathbf{X}^*) + \frac{1}{2!} \sum_{i=1}^{n} \sum_{j=1}^{n} h_i h_j \frac{\partial^2 f}{\partial x_i \partial x_j}\bigg|_{\mathbf{X}=\mathbf{X}^*+\theta\mathbf{h}},$$

$$0 < \theta < 1 \tag{2.10}$$

Since $\mathbf{X}^*$ is a stationary point, the necessary conditions give (Theorem 2.3)

$$\frac{\partial f}{\partial x_i} = 0, \qquad i = 1, 2, \ldots, n$$

Thus Eq. (2.10) reduces to

$$f(\mathbf{X}^* + \mathbf{h}) - f(\mathbf{X}^*) = \frac{1}{2!} \sum_{i=1}^{n} \sum_{j=1}^{n} h_i h_j \frac{\partial^2 f}{\partial x_i \partial x_j}\bigg|_{\mathbf{X}=\mathbf{X}^*+\theta\mathbf{h}}, \qquad 0 < \theta < 1$$

Therefore, the sign of

$$f(\mathbf{X}^* + \mathbf{h}) - f(\mathbf{X}^*)$$

will be same as that of

$$\sum_{i=1}^{n} \sum_{j=1}^{n} h_i h_j \frac{\partial^2 f}{\partial x_i \partial x_j}\bigg|_{\mathbf{X}=\mathbf{X}^*+\theta\mathbf{h}}$$

Since the second partial derivative of $\partial^2 f(\mathbf{X})/\partial x_i \partial x_j$ is continuous in the neighborhood of $\mathbf{X}^*$,

$$\frac{\partial^2 f}{\partial x_i \partial x_j}\bigg|_{\mathbf{X}=\mathbf{X}^*+\theta\mathbf{h}}$$

will have the same sign as $(\partial^2 f/\partial x_i \partial x_j)|$ $\mathbf{X} = \mathbf{X}^*$ for all sufficiently small $\mathbf{h}$. Thus $f(\mathbf{X}^* + \mathbf{h}) - f(\mathbf{X}^*)$ will be positive, and hence X^* will be a relative minimum, if

$$Q = \sum_{i=1}^{n} \sum_{j=1}^{n} h_i h_j \left. \frac{\partial^2 f}{\partial x_i \partial x_j} \right|_{\mathbf{X}=\mathbf{X}^*} \tag{2.11}$$

is positive. This quantity Q is a quadratic form and can be written in matrix form as

$$Q = \mathbf{h}^{\mathrm{T}} \mathbf{J} \mathbf{h}|_{\mathbf{X}=\mathbf{X}^*} \tag{2.12}$$

where

$$\mathbf{J}|_{\mathbf{X}=\mathbf{X}^*} = \left[\left. \frac{\partial^2 f}{\partial x_i \partial x_j} \right|_{\mathbf{X}=\mathbf{X}^*} \right] \tag{2.13}$$

is the matrix of second partial derivatives and is called the *Hessian matrix* of $f(\mathbf{X})$.

It is known from matrix algebra that the quadratic form of Eq. (2.11) or (2.12) will be positive for all $\mathbf{h}$ if and only if $[\mathbf{J}]$ is positive definite at $\mathbf{X} = \mathbf{X}^*$. This means that a sufficient condition for the stationary point $\mathbf{X}^*$ to be a relative minimum is that the Hessian matrix evaluated at the same point be positive definite. This completes the proof for the minimization case. By proceeding in a similar manner, it can be proved that the Hessian matrix will be negative definite if $\mathbf{X}^*$ is a relative maximum point.

Note: A matrix $\mathbf{A}$ will be positive definite if all its eigenvalues are positive; that is, all the values of λ that satisfy the determinantal equation

$$|\mathbf{A} - \lambda \mathbf{I}| = 0 \tag{2.14}$$

should be positive. Similarly, the matrix $[\mathbf{A}]$ will be negative definite if its eigenvalues are negative.

Another test that can be used to find the positive definiteness of a matrix $\mathbf{A}$ of order n involves evaluation of the determinants

$$A = |a_{11}|,$$

$$A_2 = \begin{vmatrix} a_{11} & a_{12} \\ a_{21} & a_{22} \end{vmatrix}, \qquad A_n = \begin{vmatrix} a_{11} & a_{12} & a_{13} & \cdots & a_{1n} \\ a_{21} & a_{22} & a_{23} & \cdots & a_{2n} \\ a_{31} & a_{32} & a_{33} & \cdots & a_{3n} \\ \vdots & & & \\ a_{n1} & a_{n2} & a_{n3} & \cdots & a_{nn} \end{vmatrix} \tag{2.15}$$

$$A_3 = \begin{vmatrix} a_{11} & a_{12} & a_{13} \\ a_{21} & a_{22} & a_{23} \\ a_{31} & a_{32} & a_{32} \end{vmatrix}, \ldots,$$

The matrix $\mathbf{A}$ will be positive definite if and only if all the values $A_1, A_2, A_3, \ldots, A_n$ are positive. The matrix $\mathbf{A}$ will be negative definite if and only if the sign of A_j is $(-1)^j$ for $j = 1, 2, \ldots, n$. If some of the A_j are positive and the remaining A_j are zero, the matrix $\mathbf{A}$ will be positive semidefinite.

Example 2.4 Figure 2.4 shows two frictionless rigid bodies (carts) A and B connected by three linear elastic springs having spring constants k_1, k_2, and k_3. The springs are at their natural positions when the applied force P is zero. Find the displacements x_1 and x_2 under the force P by using the principle of minimum potential energy.

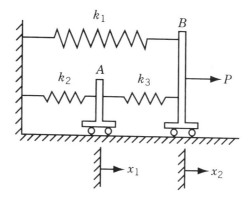

Figure 2.4   Spring–cart system.

SOLUTION   According to the principle of minimum potential energy, the system will be in equilibrium under the load P if the potential energy is a minimum. The potential energy of the system is given by

potential energy (U)

= strain energy of springs − work done by external forces

$$= [\tfrac{1}{2}k_2 x_1^2 + \tfrac{1}{2}k_3(x_2 - x_1)^2 + \tfrac{1}{2}k_1 x_2^2] - Px_2$$

The necessary conditions for the minimum of U are

$$\frac{\partial U}{\partial x_1} = k_2 x_1 - k_3(x_2 - x_1) = 0 \tag{E_1}$$

$$\frac{\partial U}{\partial x_2} = k_3(x_2 - x_1) + k_1 x_2 - P = 0 \tag{E_2}$$

The values of x_1 and x_2 corresponding to the equilibrium state, obtained by solving Eqs. (E_1) and (E_2), are given by

$$x_1^* = \frac{Pk_3}{k_1 k_2 + k_1 k_3 + k_2 k_3}$$

$$x_2^* = \frac{P(k_2 + k_3)}{k_1 k_2 + k_1 k_3 + k_2 k_3}$$

The sufficiency conditions for the minimum at (x_1^*, x_2^*) can also be verified by testing the positive definiteness of the Hessian matrix of U. The Hessian matrix of U evaluated at (x_1^*, x_2^*) is

$$\mathbf{J}\Big|_{(x_1^*,x_2^*)} = \begin{bmatrix} \dfrac{\partial^2 U}{\partial x_1^2} & \dfrac{\partial^2 U}{\partial x_1 \partial x_2} \\[2mm] \dfrac{\partial^2 U}{\partial x_1 \partial x_2} & \dfrac{\partial^2 U}{\partial x_2^2} \end{bmatrix}_{(x_1^*,x_2^*)} = \begin{bmatrix} k_2 + k_3 & -k_3 \\ -k_3 & k_1 + k_3 \end{bmatrix}$$

The determinants of the square submatrices of $\mathbf{J}$ are

$$J_1 = |k_2 + k_3| = k_2 + k_3 > 0$$

$$J_2 = \begin{vmatrix} k_2 + k_3 & -k_3 \\ -k_3 & k_1 + k_3 \end{vmatrix} = k_1 k_2 + k_1 k_3 + k_2 k_3 > 0$$

since the spring constants are always positive. Thus the matrix $\mathbf{J}$ is positive definite and hence (x_1^*, x_2^*) corresponds to the minimum of potential energy.

2.3.1 Semidefinite Case

We now consider the problem of determining the sufficient conditions for the case when the Hessian matrix of the given function is semidefinite. In the case of a function of a single variable, the problem of determining the sufficient conditions for the case when the second derivative is zero was resolved quite easily. We simply investigated the higher-order derivatives in the Taylor's series expansion. A similar procedure can be followed for functions of n variables. However, the algebra becomes quite involved, and hence we rarely investigate the stationary points for sufficiency in actual practice. The following theorem, analogous to Theorem 2.2, gives the sufficiency conditions for the extreme points of a function of several variables.

Theorem 2.5 Let the partial derivatives of f of all orders up to the order $k \geq 2$ be continuous in the neighborhood of a stationary point $\mathbf{X}^*$, and

$$d^r f|_{\mathbf{X}=\mathbf{X}^*} = 0, \qquad 1 \leq r \leq k - 1$$

$$d^k f|_{\mathbf{X}=\mathbf{X}^*} \neq 0$$

so that $d^k f|_{\mathbf{X}=\mathbf{X}^*}$ is the first nonvanishing higher-order differential of f at $\mathbf{X}^*$. If k is even, then (i) $\mathbf{X}^*$ is a relative minimum if $d^k f|_{\mathbf{X}=\mathbf{X}^*}$ is positive definite, (ii) $\mathbf{X}^*$ is a relative maximum if $d^k f|_{\mathbf{X}=\mathbf{X}^*}$ is negative definite, and (iii) if $d^k f|_{\mathbf{X}=\mathbf{X}^*}$ is semidefinite (but not definite), no general conclusion can be drawn. On the other hand, if k is odd, $\mathbf{X}^*$ is not an extreme point of $f(\mathbf{X})$.

Proof: A proof similar to that of Theorem 2.2 can be found in Ref. [2.5].

2.3.2 Saddle Point

In the case of a function of two variables, $f(x, y)$, the Hessian matrix may be neither positive nor negative definite at a point (x^*, y^*) at which

$$\frac{\partial f}{\partial x} = \frac{\partial f}{\partial y} = 0$$

In such a case, the point (x^*, y^*) is called a *saddle point*. The characteristic of a saddle point is that it corresponds to a relative minimum or maximum of $f(x, y)$ with respect to one variable, say, x (the other variable being fixed at $y = y^*$) and a relative maximum or minimum of $f(x, y)$ with respect to the second variable y (the other variable being fixed at x^*).

As an example, consider the function $f(x, y) = x^2 - y^2$. For this function,

$$\frac{\partial f}{\partial x} = 2x \quad \text{and} \quad \frac{\partial f}{\partial y} = -2y$$

These first derivatives are zero at $x^* = 0$ and $y^* = 0$. The Hessian matrix of f at (x^*, y^*) is given by

$$\mathbf{J} = \begin{bmatrix} 2 & 0 \\ 0 & -2 \end{bmatrix}$$

Since this matrix is neither positive definite nor negative definite, the point $(x^* = 0, y^* = 0)$ is a saddle point. The function is shown graphically in Fig. 2.5. It can be seen that $f(x, y^*) = f(x, 0)$ has a relative minimum and $f(x^*, y) = f(0, y)$ has a relative maximum at the saddle point (x^*, y^*). Saddle points may exist for functions of more than two variables also. The characteristic of the saddle point stated above still holds provided that x and y are interpreted as vectors in multidimensional cases.

Example 2.5 Find the extreme points of the function

$$f(x_1, x_2) = x_1^3 + x_2^3 + 2x_1^2 + 4x_2^2 + 6$$

SOLUTION   The necessary conditions for the existence of an extreme point are

$$\frac{\partial f}{\partial x_1} = 3x_1^2 + 4x_1 = x_1(3x_1 + 4) = 0$$

$$\frac{\partial f}{\partial x_2} = 3x_2^2 + 8x_2 = x_2(3x_2 + 8) = 0$$

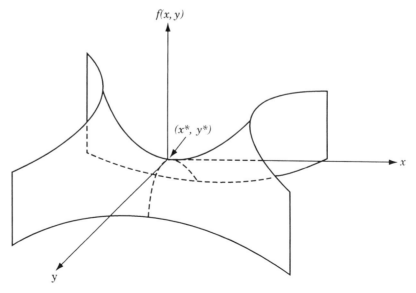

Figure 2.5   Saddle point of the function $f(x, y) = x^2 - y^2$.

These equations are satisfied at the points

$$(0, 0), \quad (0, -\tfrac{8}{3}), \quad (-\tfrac{4}{3}, 0), \quad \text{and} \quad (-\tfrac{4}{3}, -\tfrac{8}{3})$$

To find the nature of these extreme points, we have to use the sufficiency conditions. The second-order partial derivatives of f are given by

$$\frac{\partial^2 f}{\partial x_1^2} = 6x_1 + 4$$

$$\frac{\partial^2 f}{\partial x_2^2} = 6x_2 + 8$$

$$\frac{\partial^2 f}{\partial x_1 \partial x_2} = 0$$

The Hessian matrix of f is given by

$$\mathbf{J} = \begin{bmatrix} 6x_1 + 4 & 0 \\ 0 & 6x_2 + 8 \end{bmatrix}$$

If $J_1 = |6x_1 + 4|$ and $J_2 = \begin{vmatrix} 6x_1 + 4 & 0 \\ 0 & 6x_2 + 8 \end{vmatrix}$, the values of J_1 and J_2 and the nature of the extreme point are as given below:

Point $\mathbf{X}$	Value of J_1	Value of J_2	Nature of $\mathbf{J}$	Nature of $\mathbf{X}$	$f(\mathbf{X})$
$(0, 0)$	+4	+32	Positive definite	Relative minimum	6
$(0, -\tfrac{8}{3})$	+4	−32	Indefinite	Saddle point	418/27
$(-\tfrac{4}{3}, 0)$	−4	−32	Indefinite	Saddle point	194/27
$(-\tfrac{4}{3}, -\tfrac{8}{3})$	−4	+32	Negative definite	Relative maximum	50/3

2.4   MULTIVARIABLE OPTIMIZATION WITH EQUALITY CONSTRAINTS

In this section we consider the optimization of continuous functions subjected to equality constraints:

$$\text{Minimize} f = f(\mathbf{X})$$

$$\text{subject to} \tag{2.16}$$

$$g_j(\mathbf{X}) = 0, \quad j = 1, 2, \ldots, m$$

where

$$\mathbf{X} = \begin{Bmatrix} x_1 \\ x_2 \\ \vdots \\ x_n \end{Bmatrix}$$

Here m is less than or equal to n; otherwise (if $m > n$), the problem becomes overdefined and, in general, there will be no solution. There are several methods available for the solution of this problem. The methods of direct substitution, constrained variation, and Lagrange multipliers are discussed in the following sections.

2.4.1 Solution by Direct Substitution

For a problem with n variables and m equality constraints, it is theoretically possible to solve simultaneously the m equality constraints and express any set of m variables in terms of the remaining $n - m$ variables. When these expressions are substituted into the original objective function, there results a new objective function involving only $n - m$ variables. The new objective function is not subjected to any constraint, and hence its optimum can be found by using the unconstrained optimization techniques discussed in Section 2.3.

This method of direct substitution, although it appears to be simple in theory, is not convenient from a practical point of view. The reason for this is that the constraint equations will be nonlinear for most of practical problems, and often it becomes impossible to solve them and express any m variables in terms of the remaining $n - m$ variables. However, the method of direct substitution might prove to be very simple and direct for solving simpler problems, as shown by the following example.

Example 2.6 Find the dimensions of a box of largest volume that can be inscribed in a sphere of unit radius.

SOLUTION Let the origin of the Cartesian coordinate system x_1, x_2, x_3 be at the center of the sphere and the sides of the box be $2x_1$, $2x_2$, and $2x_3$. The volume of the box is given by

$$f(x_1, x_2, x_3) = 8x_1x_2x_3 \tag{E_1}$$

Since the corners of the box lie on the surface of the sphere of unit radius, x_1, x_2, and x_3 have to satisfy the constraint

$$x_1^2 + x_2^2 + x_3^2 = 1 \tag{E_2}$$

This problem has three design variables and one equality constraint. Hence the equality constraint can be used to eliminate any one of the design variables from the objective function. If we choose to eliminate x_3, Eq. (E_2) gives

$$x_3 = (1 - x_1^2 - x_2^2)^{1/2} \tag{E_3}$$

Thus the objective function becomes

$$f(x_1, x_2) = 8x_1x_2(1 - x_1^2 - x_2^2)^{1/2} \tag{E_4}$$

which can be maximized as an unconstrained function in two variables.

The necessary conditions for the maximum of f give

$$\frac{\partial f}{\partial x_1} = 8x_2 \left[(1 - x_1^2 - x_2^2)^{1/2} - \frac{x_1^2}{(1 - x_1^2 - x_2^2)^{1/2}} \right] = 0 \qquad (E_5)$$

$$\frac{\partial f}{\partial x_2} = 8x_1 \left[(1 - x_1^2 - x_2^2)^{1/2} - \frac{x_2^2}{(1 - x_1^2 - x_2^2)^{1/2}} \right] = 0 \qquad (E_6)$$

Equations (E_5) and (E_6) can be simplified to obtain

$$1 - 2x_1^2 - x_2^2 = 0$$

$$1 - x_1^2 - 2x_2^2 = 0$$

from which it follows that $x_1^* = x_2^* = 1/\sqrt{3}$ and hence $x_3^* = 1/\sqrt{3}$. This solution gives the maximum volume of the box as

$$f_{\text{max}} = \frac{8}{3\sqrt{3}}$$

To find whether the solution found corresponds to a maximum or a minimum, we apply the sufficiency conditions to $f(x_1, x_2)$ of Eq. (E_4). The second-order partial derivatives of f at (x_1^*, x_2^*) are given by

$$\frac{\partial^2 f}{\partial x_1^2} = -\frac{32}{\sqrt{3}} \text{ at } (x_1^*, x_2^*)$$

$$\frac{\partial^2 f}{\partial x_2^2} = -\frac{32}{\sqrt{3}} \text{ at } (x_1^*, x_2^*)$$

$$\frac{\partial^2 f}{\partial x_1 \partial x_2} = -\frac{16}{\sqrt{3}} \text{ at } (x_1^*, x_2^*)$$

Since

$$\frac{\partial^2 f}{\partial x_1^2} < 0 \quad \text{and} \quad \frac{\partial^2 f}{\partial x_1^2} \frac{\partial^2 f}{\partial x_2^2} - \left(\frac{\partial^2 f}{\partial x_1 \partial x_2} \right)^2 > 0$$

the Hessian matrix of f is negative definite at (x_1^*, x_2^*). Hence the point (x_1^*, x_2^*) corresponds to the maximum of f.

2.4.2   Solution by the Method of Constrained Variation

The basic idea used in the method of constrained variation is to find a closed-form expression for the first-order differential of f (df) at all points at which the constraints $g_j(\mathbf{X}) = 0$, $j = 1, 2, \ldots, m$, are satisfied. The desired optimum points are then obtained by setting the differential df equal to zero. Before presenting the general method, we indicate its salient features through the following simple problem with $n = 2$ and $m = 1$:

$$\text{Minimize } f(x_1, x_2) \tag{2.17}$$

subject to

$$g(x_1, x_2) = 0 \tag{2.18}$$

A necessary condition for f to have a minimum at some point (x_1^*, x_2^*) is that the total derivative of $f(x_1, x_2)$ with respect to x_1 must be zero at (x_1^*, x_2^*). By setting the total differential of $f(x_1, x_2)$ equal to zero, we obtain

$$df = \frac{\partial f}{\partial x_1} dx_1 + \frac{\partial f}{\partial x_2} dx_2 = 0 \tag{2.19}$$

Since $g(x_1^*, x_2^*) = 0$ at the minimum point, any variations dx_1 and dx_2 taken about the point (x_1^*, x_2^*) are called *admissible variations* provided that the new point lies on the constraint:

$$g(x_1^* + dx_1, x_2^* + dx_2) = 0 \tag{2.20}$$

The Taylor's series expansion of the function in Eq. (2.20) about the point (x_1^*, x_2^*) gives

$$g(x_1^* + dx_1, x_2^* + dx_2)$$
$$\simeq g(x_1^*, x_2^*) + \frac{\partial g}{\partial x_1}(x_1^*, x_2^*) dx_1 + \frac{\partial g}{\partial x_2}(x_1^*, x_2^*) dx_2 = 0 \tag{2.21}$$

where dx_1 and dx_2 are assumed to be small. Since $g(x_1^*, x_2^*) = 0$, Eq. (2.21) reduces to

$$dg = \frac{\partial g}{\partial x_1} dx_1 + \frac{\partial g}{\partial x_2} dx_2 = 0 \quad \text{at} \quad (x_1^*, x_2^*) \tag{2.22}$$

Thus Eq. (2.22) has to be satisfied by all admissible variations. This is illustrated in Fig. 2.6, where PQ indicates the curve at each point of which Eq. (2.18) is satisfied. If A is taken as the base point (x_1^*, x_2^*), the variations in x_1 and x_2 leading to points B and C are called *admissible variations*. On the other hand, the variations in x_1 and x_2 representing point D are not admissible since point D does not

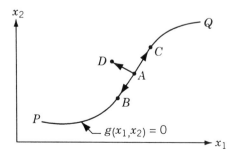

Figure 2.6   Variations about A.

lie on the constraint curve, $g(x_1, x_2) = 0$. Thus any set of variations (dx_1, dx_2) that does not satisfy Eq. (2.22) leads to points such as D, which do not satisfy constraint Eq. (2.18).

Assuming that $\partial g / \partial x_2 \neq 0$, Eq. (2.22) can be rewritten as

$$dx_2 = -\frac{\partial g / \partial x_1}{\partial g / \partial x_2}(x_1^*, x_2^*)dx_1 \tag{2.23}$$

This relation indicates that once the variation in $x_1(dx_1)$ is chosen arbitrarily, the variation in x_2 (dx_2) is decided automatically in order to have dx_1 and dx_2 as a set of admissible variations. By substituting Eq. (2.23) in Eq. (2.19), we obtain

$$df = \left(\frac{\partial f}{\partial x_1} - \frac{\partial g / \partial x_1}{\partial g / \partial x_2} \frac{\partial f}{\partial x_2} \right)\bigg|_{(x_1^*, x_2^*)} dx_1 = 0 \tag{2.24}$$

The expression on the left-hand side is called the *constrained variation* of f. Note that Eq. (2.24) has to be satisfied for all values of dx_1. Since dx_1 can be chosen arbitrarily, Eq. (2.24) leads to

$$\left(\frac{\partial f}{\partial x_1} \frac{\partial g}{\partial x_2} - \frac{\partial f}{\partial x_2} \frac{\partial g}{\partial x_1} \right)\bigg|_{(x_1^*, x_2^*)} = 0 \tag{2.25}$$

Equation (2.25) represents a necessary condition in order to have (x_1^*, x_2^*) as an extreme point (minimum or maximum).

Example 2.7   A beam of uniform rectangular cross section is to be cut from a log having a circular cross section of diameter $2a$. The beam has to be used as a cantilever beam (the length is fixed) to carry a concentrated load at the free end. Find the dimensions of the beam that correspond to the maximum tensile (bending) stress carrying capacity.

SOLUTION   From elementary strength of materials, we know that the tensile stress induced in a rectangular beam (σ) at any fiber located a distance y from the neutral axis is given by

$$\frac{\sigma}{y} = \frac{M}{I}$$

where M is the bending moment acting and I is the moment of inertia of the cross section about the x axis. If the width and depth of the rectangular beam shown in Fig. 2.7 are $2x$ and $2y$, respectively, the maximum tensile stress induced is given by

$$\sigma_{\max} = \frac{M}{I}y = \frac{My}{\frac{1}{12}(2x)(2y)^3} = \frac{3}{4}\frac{M}{xy^2}$$

Thus for any specified bending moment, the beam is said to have maximum tensile stress carrying capacity if the maximum induced stress $(\sigma_{\max})$ is a minimum. Hence we need to minimize k/xy^2 or maximize Kxy^2, where $k = 3M/4$ and $K = 1/k$, subject to the constraint

$$x^2 + y^2 = a^2$$

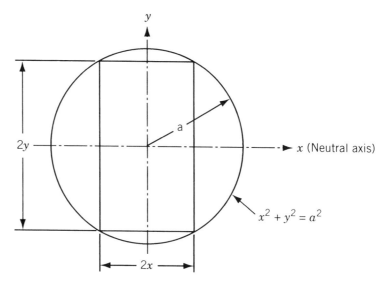

Figure 2.7   Cross section of the log.

This problem has two variables and one constraint; hence Eq. (2.25) can be applied for finding the optimum solution. Since

$$f = kx^{-1}y^{-2} \tag{E_1}$$

$$g = x^2 + y^2 - a^2 \tag{E_2}$$

we have

$$\frac{\partial f}{\partial x} = -kx^{-2}y^{-2}$$

$$\frac{\partial f}{\partial y} = -2kx^{-1}y^{-3}$$

$$\frac{\partial g}{\partial x} = 2x$$

$$\frac{\partial g}{\partial y} = 2y$$

Equation (2.25) gives

$$-kx^{-2}y^{-2}(2y) + 2kx^{-1}y^{-3}(2x) = 0 \quad \text{at} \quad (x^*, y^*)$$

that is,

$$y^* = \sqrt{2}x^* \tag{E_3}$$

Thus the beam of maximum tensile stress carrying capacity has a depth of $\sqrt{2}$ times its breadth. The optimum values of x and y can be obtained from Eqs. (E_3) and (E_2) as

$$x^* = \frac{a}{\sqrt{3}} \quad \text{and} \quad y^* = \sqrt{2}\frac{a}{\sqrt{3}}$$

Necessary Conditions for a General Problem. The procedure indicated above can be generalized to the case of a problem in n variables with m constraints. In this case, each constraint equation $g_j(\mathbf{X}) = 0$, $j = 1, 2, \ldots, m$, gives rise to a linear equation in the variations dx_i, $i = 1, 2, \ldots, n$. Thus there will be in all m linear equations in n variations. Hence any m variations can be expressed in terms of the remaining $n - m$ variations. These expressions can be used to express the differential of the objective function, df, in terms of the $n - m$ independent variations. By letting the coefficients of the independent variations vanish in the equation $df = 0$, one obtains the necessary conditions for the constrained optimum of the given function. These conditions can be expressed as [2.6]

$$J \left(\frac{f, g_1, g_2, \ldots, g_m}{x_k, x_1, x_2, x_3, \ldots, x_m} \right) = \begin{vmatrix} \dfrac{\partial f}{\partial x_k} & \dfrac{\partial f}{\partial x_1} & \dfrac{\partial f}{\partial x_2} & \cdots & \dfrac{\partial f}{\partial x_m} \\[2mm] \dfrac{\partial g_1}{\partial x_k} & \dfrac{\partial g_1}{\partial x_1} & \dfrac{\partial g_1}{\partial x_2} & \cdots & \dfrac{\partial g_1}{\partial x_m} \\[2mm] \dfrac{\partial g_2}{\partial x_k} & \dfrac{\partial g_2}{\partial x_1} & \dfrac{\partial g_2}{\partial x_2} & \cdots & \dfrac{\partial g_2}{\partial x_m} \\[2mm] \vdots & & & & \\[2mm] \dfrac{\partial g_m}{\partial x_k} & \dfrac{\partial g_m}{\partial x_1} & \dfrac{\partial g_m}{\partial x_2} & \cdots & \dfrac{\partial g_m}{\partial x_m} \end{vmatrix} = 0 \qquad (2.26)$$

where $k = m + 1, m + 2, \ldots, n$. It is to be noted that the variations of the first m variables $(dx_1, dx_2, \ldots, dx_m)$ have been expressed in terms of the variations of the remaining $n - m$ variables $(dx_{m+1}, dx_{m+2}, \ldots, dx_n)$ in deriving Eqs. (2.26). This implies that the following relation is satisfied:

$$J \left(\frac{g_1, g_2, \ldots, g_m}{x_1, x_2, \ldots, x_m} \right) \neq 0 \qquad (2.27)$$

The $n - m$ equations given by Eqs. (2.26) represent the necessary conditions for the extremum of $f(\mathbf{X})$ under the m equality constraints, $g_j(\mathbf{X}) = 0$, $j = 1, 2, \ldots, m$.

Example 2.8

$$\text{Minimize } f(\mathbf{Y}) = \tfrac{1}{2}(y_1^2 + y_2^2 + y_3^2 + y_4^2) \qquad (E_1)$$

subject to

$$g_1(\mathbf{Y}) = y_1 + 2y_2 + 3y_3 + 5y_4 - 10 = 0 \qquad (E_2)$$

$$g_2(\mathbf{Y}) = y_1 + 2y_2 + 5y_3 + 6y_4 - 15 = 0 \qquad (E_3)$$

SOLUTION   This problem can be solved by applying the necessary conditions given by Eqs. (2.26). Since $n = 4$ and $m = 2$, we have to select two variables as independent variables. First we show that any arbitrary set of variables cannot be chosen as independent variables since the remaining (dependent) variables have to satisfy the condition of Eq. (2.27).

In terms of the notation of our equations, let us take the independent variables as

$$x_3 = y_3 \quad \text{and} \quad x_4 = y_4 \qquad \text{so that} \quad x_1 = y_1 \quad \text{and} \quad x_2 = y_2$$

Then the Jacobian of Eq. (2.27) becomes

$$J\left(\frac{g_1, g_2}{x_1, x_2}\right) = \begin{vmatrix} \dfrac{\partial g_1}{\partial y_1} & \dfrac{\partial g_1}{\partial y_2} \\[2mm] \dfrac{\partial g_2}{\partial y_1} & \dfrac{\partial g_2}{\partial y_2} \end{vmatrix} = \begin{vmatrix} 1 & 2 \\ 1 & 2 \end{vmatrix} = 0$$

and hence the necessary conditions of Eqs. (2.26) cannot be applied.

Next, let us take the independent variables as $x_3 = y_2$ and $x_4 = y_4$ so that $x_1 = y_1$ and $x_2 = y_3$. Then the Jacobian of Eq. (2.27) becomes

$$J\left(\frac{g_1, g_2}{x_1, x_2}\right) = \begin{vmatrix} \dfrac{\partial g_1}{\partial y_1} & \dfrac{\partial g_1}{\partial y_3} \\[2mm] \dfrac{\partial g_2}{\partial y_1} & \dfrac{\partial g_2}{\partial y_3} \end{vmatrix} = \begin{vmatrix} 1 & 3 \\ 1 & 5 \end{vmatrix} = 2 \neq 0$$

and hence the necessary conditions of Eqs. (2.26) can be applied. Equations (2.26) give for $k = m + 1 = 3$

$$\begin{vmatrix} \dfrac{\partial f}{\partial x_3} & \dfrac{\partial f}{\partial x_1} & \dfrac{\partial f}{\partial x_2} \\[2mm] \dfrac{\partial g_1}{\partial x_3} & \dfrac{\partial g_1}{\partial x_1} & \dfrac{\partial g_1}{\partial x_2} \\[2mm] \dfrac{\partial g_2}{\partial x_3} & \dfrac{\partial g_2}{\partial x_1} & \dfrac{\partial g_2}{\partial x_2} \end{vmatrix} = \begin{vmatrix} \dfrac{\partial f}{\partial y_2} & \dfrac{\partial f}{\partial y_1} & \dfrac{\partial f}{\partial y_3} \\[2mm] \dfrac{\partial g_1}{\partial y_2} & \dfrac{\partial g_1}{\partial y_1} & \dfrac{\partial g_1}{\partial y_3} \\[2mm] \dfrac{\partial g_2}{\partial y_2} & \dfrac{\partial g_2}{\partial y_1} & \dfrac{\partial g_2}{\partial y_3} \end{vmatrix}$$

$$= \begin{vmatrix} y_2 & y_1 & y_3 \\ 2 & 1 & 3 \\ 2 & 1 & 5 \end{vmatrix}$$

$$= y_2(5 - 3) - y_1(10 - 6) + y_3(2 - 2)$$

$$= 2y_2 - 4y_1 = 0 \qquad\qquad (E_4)$$

and for $k = m + 2 = n = 4$,

$$\begin{vmatrix} \dfrac{\partial f}{\partial x_4} & \dfrac{\partial f}{\partial x_1} & \dfrac{\partial f}{\partial x_2} \\[2mm] \dfrac{\partial g_1}{\partial x_4} & \dfrac{\partial g_1}{\partial x_1} & \dfrac{\partial g_1}{\partial x_2} \\[2mm] \dfrac{\partial g_2}{\partial x_4} & \dfrac{\partial g_2}{\partial x_1} & \dfrac{\partial g_2}{\partial x_2} \end{vmatrix} = \begin{vmatrix} \dfrac{\partial f}{\partial y_4} & \dfrac{\partial f}{\partial y_1} & \dfrac{\partial f}{\partial y_3} \\[2mm] \dfrac{\partial g_1}{\partial y_4} & \dfrac{\partial g_1}{\partial y_1} & \dfrac{\partial g_1}{\partial y_3} \\[2mm] \dfrac{\partial g_2}{\partial y_4} & \dfrac{\partial g_2}{\partial y_1} & \dfrac{\partial g_2}{\partial y_3} \end{vmatrix}$$

$$= \begin{vmatrix} y_4 & y_1 & y_3 \\ 5 & 1 & 3 \\ 6 & 1 & 5 \end{vmatrix}$$

$$= y_4(5-3) - y_1(25-18) + y_3(5-6)$$

$$= 2y_4 - 7y_1 - y_3 = 0 \qquad\qquad (E_5)$$

Equations (E$_4$) and (E$_5$) give the necessary conditions for the minimum or the maximum of f as

$$y_1 = \tfrac{1}{2}y_2$$
$$y_3 = 2y_4 - 7y_1 = 2y_4 - \tfrac{7}{2}y_2 \qquad\qquad (E_6)$$

When Eqs. (E$_6$) are substituted, Eqs. (E$_2$) and (E_3) take the form

$$-8y_2 + 11y_4 = 10$$
$$-15y_2 + 16y_4 = 15$$

from which the desired optimum solution can be obtained as

$$y_1^* = -\tfrac{5}{74}$$
$$y_2^* = -\tfrac{5}{37}$$
$$y_3^* = \tfrac{155}{74}$$
$$y_4^* = \tfrac{30}{37}$$

Sufficiency Conditions for a General Problem.   By eliminating the first m variables, using the m equality constraints (this is possible, at least in theory), the objective function f can be made to depend only on the remaining variables, $x_{m+1}, x_{m+2}, \ldots, x_n$. Then the Taylor's series expansion of f, in terms of these variables, about the extreme point $\mathbf{X}^*$ gives

$$f(\mathbf{X}^* + d\mathbf{X}) \simeq f(\mathbf{X}^*) + \sum_{i=m+1}^{n} \left(\frac{\partial f}{\partial x_i}\right)_g dx_i$$

$$+ \frac{1}{2!} \sum_{i=m+1}^{n} \sum_{j=m+1}^{n} \left(\frac{\partial^2 f}{\partial x_i \partial x_j}\right)_g dx_i \, dx_j \qquad (2.28)$$

where $(\partial f/\partial x_i)_g$ is used to denote the partial derivative of f with respect to x_i (holding all the other variables $x_{m+1}, x_{m+2}, \ldots, x_{i-1}, x_{i+1}, x_{i+2}, \ldots, x_n$ constant) when $x_1, x_2, \ldots, x_m$ are allowed to change so that the constraints $g_j(\mathbf{X}^* + d\mathbf{X}) = 0$, $j = 1, 2, \ldots, m$, are satisfied; the second derivative, $(\partial^2 f/\partial x_i \partial x_j)_g$, is used to denote a similar meaning.

As an example, consider the problem of minimizing

$$f(\mathbf{X}) = f(x_1, x_2, x_3)$$

subject to the only constraint

$$g_1(\mathbf{X}) = x_1^2 + x_2^2 + x_3^2 - 8 = 0$$

Since $n = 3$ and $m = 1$ in this problem, one can think of any of the m variables, say x_1, to be dependent and the remaining $n - m$ variables, namely x_2 and x_3, to be independent. Here the constrained partial derivative $(\partial f / \partial x_2)_g$, for example, means the rate of change of f with respect to x_2 (holding the other independent variable x_3 constant) and at the same time allowing x_1 to change about $\mathbf{X}^*$ so as to satisfy the constraint $g_1(\mathbf{X}) = 0$. In the present case, this means that dx_1 has to be chosen to satisfy the relation

$$g_1(\mathbf{X}^* + d\mathbf{X}) \simeq g_1(\mathbf{X}^*) + \frac{\partial g_1}{\partial x_1}(\mathbf{X}^*)dx_1 + \frac{\partial g_1}{\partial x_2}(\mathbf{X}^*)dx_2 + \frac{\partial g_1}{\partial x_3}(\mathbf{X}^*)dx_3 = 0$$

that is,

$$2x_1^* \, dx_1 + 2x_2^* \, dx_2 = 0$$

since $g_1(\mathbf{X}^*) = 0$ at the optimum point and $dx_3 = 0$ (x_3 is held constant).

Notice that $(\partial f / \partial x_i)_g$ has to be zero for $i = m + 1, m + 2, \ldots, n$ since the dx_i appearing in Eq. (2.28) are all independent. Thus the necessary conditions for the existence of constrained optimum at $\mathbf{X}^*$ can also be expressed as

$$\left(\frac{\partial f}{\partial x_i}\right)_g = 0, \qquad i = m + 1, m + 2, \ldots, n \tag{2.29}$$

Of course, with little manipulation, one can show that Eqs. (2.29) are nothing but Eqs. (2.26). Further, as in the case of optimization of a multivariable function with no constraints, one can see that a sufficient condition for $\mathbf{X}^*$ to be a constrained relative minimum (maximum) is that the quadratic form Q defined by

$$Q = \sum_{i=m+1}^{n} \sum_{j=m+1}^{n} \left(\frac{\partial^2 f}{\partial x_i \, \partial x_j}\right)_g dx_i \, dx_j \tag{2.30}$$

is positive (negative) for all nonvanishing variations dx_i. As in Theorem 2.4, the matrix

$$\begin{bmatrix} \left(\dfrac{\partial^2 f}{\partial x_{m+1}^2}\right)_g & \left(\dfrac{\partial^2 f}{\partial x_{m+1} \, \partial x_{m+2}}\right)_g & \cdots & \left(\dfrac{\partial^2 f}{\partial x_{m+1} \, \partial x_n}\right)_g \\ \vdots & & & \\ \left(\dfrac{\partial^2 f}{\partial x_n \, \partial x_{m+1}}\right)_g & \left(\dfrac{\partial^2 f}{\partial x_n \, \partial x_{m+2}}\right)_g & \cdots & \left(\dfrac{\partial^2 f}{\partial x_n^2}\right)_g \end{bmatrix}$$

has to be positive (negative) definite to have Q positive (negative) for all choices of dx_i. It is evident that computation of the constrained derivatives $(\partial^2 f / \partial x_i \, \partial x_j)_g$ is a difficult task and may be prohibitive for problems with more than three constraints. Thus the method of constrained variation, although it appears to be simple in theory, is very difficult to apply since the necessary conditions themselves involve evaluation of determinants of order $m + 1$. This is the reason that the method of Lagrange multipliers, discussed in the following section, is more commonly used to solve a multivariable optimization problem with equality constraints.

2.4.3 Solution by the Method of Lagrange Multipliers

The basic features of the Lagrange multiplier method is given initially for a simple problem of two variables with one constraint. The extension of the method to a general problem of n variables with m constraints is given later.

Problem with Two Variables and One Constraint. Consider the problem

$$\text{Minimize } f(x_1, x_2) \tag{2.31}$$

subject to

$$g(x_1, x_2) = 0$$

For this problem, the necessary condition for the existence of an extreme point at $\mathbf{X} = \mathbf{X}^*$ was found in Section 2.4.2 to be

$$\left(\frac{\partial f}{\partial x_1} - \frac{\partial f / \partial x_2}{\partial g / \partial x_2} \frac{\partial g}{\partial x_1} \right) \Bigg|_{(x_1^*, x_2^*)} = 0 \tag{2.32}$$

By defining a quantity λ, called the *Lagrange multiplier*, as

$$\lambda = - \left(\frac{\partial f / \partial x_2}{\partial g / \partial x_2} \right) \Bigg|_{(x_1^*, x_2^*)} \tag{2.33}$$

Equation (2.32) can be expressed as

$$\left(\frac{\partial f}{\partial x_1} + \lambda \frac{\partial g}{\partial x_1} \right) \Bigg|_{(x_1^*, x_2^*)} = 0 \tag{2.34}$$

and Eq. (2.33) can be written as

$$\left(\frac{\partial f}{\partial x_2} + \lambda \frac{\partial g}{\partial x_2} \right) \Bigg|_{(x_1^*, x_2^*)} = 0 \tag{2.35}$$

In addition, the constraint equation has to be satisfied at the extreme point, that is,

$$g(x_1, x_2) \big|_{(x_1^*, x_2^*)} = 0 \tag{2.36}$$

Thus Eqs. (2.34) to (2.36) represent the necessary conditions for the point (x_1^*, x_2^*) to be an extreme point.

Notice that the partial derivative $(\partial g / \partial x_2) \big|_{(x_1^*, x_2^*)}$ has to be nonzero to be able to define λ by Eq. (2.33). This is because the variation dx_2 was expressed in terms of dx_1 in the derivation of Eq. (2.32) [see Eq. (2.23)]. On the other hand, if we choose to express dx_1 in terms of dx_2, we would have obtained the requirement that $(\partial g/\partial x_1)|_{(x_1^*, x_2^*)}$ be nonzero to define λ. Thus the derivation of the necessary conditions by the method of Lagrange multipliers requires that at least one of the partial derivatives of $g(x_1, x_2)$ be nonzero at an extreme point.

The necessary conditions given by Eqs. (2.34) to (2.36) are more commonly generated by constructing a function L, known as the Lagrange function, as

$$L(x_1, x_2, \lambda) = f(x_1, x_2) + \lambda g(x_1, x_2) \tag{2.37}$$

By treating L as a function of the three variables x_1, x_2, and λ, the necessary conditions for its extremum are given by

$$\frac{\partial L}{\partial x_1}(x_1, x_2, \lambda) = \frac{\partial f}{\partial x_1}(x_1, x_2) + \lambda \frac{\partial g}{\partial x_1}(x_1, x_2) = 0$$

$$\frac{\partial L}{\partial x_2}(x_1, x_2, \lambda) = \frac{\partial f}{\partial x_2}(x_1, x_2) + \lambda \frac{\partial g}{\partial x_2}(x_1, x_2) = 0 \tag{2.38}$$

$$\frac{\partial L}{\partial \lambda}(x_1, x_2, \lambda) = g(x_1, x_2) = 0$$

Equations (2.38) can be seen to be same as Eqs. (2.34) to (2.36). The sufficiency conditions are given later.

Example 2.9 Find the solution of Example 2.7 using the Lagrange multiplier method:

$$\text{Minimize } f(x, y) = kx^{-1}y^{-2}$$

subject to

$$g(x, y) = x^2 + y^2 - a^2 = 0$$

SOLUTION The Lagrange function is

$$L(x, y, \lambda) = f(x, y) + \lambda g(x, y) = kx^{-1}y^{-2} + \lambda(x^2 + y^2 - a^2)$$

The necessary conditions for the minimum of $f(x, y)$ [Eqs. (2.38)] give

$$\frac{\partial L}{\partial x} = -kx^{-2}y^{-2} + 2x\lambda = 0 \tag{E$_1$}$$

$$\frac{\partial L}{\partial y} = -2kx^{-1}y^{-3} + 2y\lambda = 0 \tag{E$_2$}$$

$$\frac{\partial L}{\partial \lambda} = x^2 + y^2 - a^2 = 0 \tag{E$_3$}$$

Equations (E$_1$) and (E$_2$) yield

$$2\lambda = \frac{k}{x^3 y^2} = \frac{2k}{xy^4}$$

from which the relation $x^* = (1/\sqrt{2})y^*$ can be obtained. This relation, along with Eq. (E$_3$), gives the optimum solution as

$$x^* = \frac{a}{\sqrt{3}} \quad \text{and} \quad y^* = \sqrt{2}\frac{a}{\sqrt{3}}$$

Necessary Conditions for a General Problem.   The equations derived above can be extended to the case of a general problem with n variables and m equality constraints:

$$\text{Minimize } f(\mathbf{X}) \tag{2.39}$$

subject to

$$g_j(\mathbf{X}) = 0, \quad j = 1, 2, \ldots, m$$

The Lagrange function, L, in this case is defined by introducing one Lagrange multiplier λ_j for each constraint $g_j(\mathbf{X})$ as

$$\begin{aligned} L(x_1, x_2, \ldots, x_n, \lambda_1, \lambda_2, \ldots, \lambda_m) \\ = f(\mathbf{X}) + \lambda_1 g_1(\mathbf{X}) + \lambda_2 g_2(\mathbf{X}) + \cdots + \lambda_m g_m(\mathbf{X}) \end{aligned} \tag{2.40}$$

By treating L as a function of the $n + m$ unknowns, $x_1, x_2, \ldots, x_n, \lambda_1, \lambda_2, \ldots, \lambda_m$, the necessary conditions for the extremum of L, which also correspond to the solution of the original problem stated in Eq. (2.39), are given by

$$\frac{\partial L}{\partial x_i} = \frac{\partial f}{\partial x_i} + \sum_{j=1}^{m} \lambda_j \frac{\partial g_j}{\partial x_i} = 0, \quad i = 1, 2, \ldots, n \tag{2.41}$$

$$\frac{\partial L}{\partial \lambda_j} = g_j(\mathbf{X}) = 0, \quad j = 1, 2, \ldots, m \tag{2.42}$$

Equations (2.41) and (2.42) represent $n + m$ equations in terms of the $n + m$ unknowns, x_i and λ_j. The solution of Eqs. (2.41) and (2.42) gives

$$\mathbf{X}^* = \begin{Bmatrix} x_1^* \\ x_2^* \\ \vdots \\ x_n^* \end{Bmatrix} \quad \text{and} \quad \lambda^* = \begin{Bmatrix} \lambda_1^* \\ \lambda_2^* \\ \vdots \\ \lambda_m^* \end{Bmatrix}$$

The vector $\mathbf{X}^*$ corresponds to the relative constrained minimum of $f(\mathbf{X})$ (sufficient conditions are to be verified) while the vector λ^* provides the sensitivity information, as discussed in the next subsection.

Sufficiency Conditions for a General Problem.   A sufficient condition for $f(\mathbf{X})$ to have a constrained relative minimum at $\mathbf{X}^*$ is given by the following theorem.

Theorem 2.6 Sufficient Condition   A sufficient condition for $f(\mathbf{X})$ to have a relative minimum at $\mathbf{X}^*$ is that the quadratic, Q, defined by

$$Q = \sum_{i=1}^{n} \sum_{j=1}^{n} \frac{\partial^2 L}{\partial x_i \, \partial x_j} dx_i \, dx_j \tag{2.43}$$

evaluated at $\mathbf{X} = \mathbf{X}^*$ must be positive definite for all values of $d\mathbf{X}$ for which the constraints are satisfied.

Proof: The proof is similar to that of Theorem 2.4.

Notes:

1. If

$$Q = \sum_{i=1}^{n} \sum_{j=1}^{n} \frac{\partial^2 L}{\partial x_i \, \partial x_j}(\mathbf{X}^*, \boldsymbol{\lambda}^*) dx_i \, dx_j$$

is negative for all choices of the admissible variations dx_i, $\mathbf{X}^*$ will be a con-strained maximum of $f(\mathbf{X})$.

2. It has been shown by Hancock [2.1] that a necessary condition for the quadratic form Q, defined by Eq. (2.43), to be positive (negative) definite for all admissi-ble variations $d\mathbf{X}$ is that each root of the polynomial z_i, defined by the following determinantal equation, be positive (negative):

$$\begin{vmatrix} L_{11} - z & L_{12} & L_{13} & \cdots & L_{1n} & g_{11} & g_{21} & \cdots & g_{m1} \\ L_{21} & L_{22} - z & L_{23} & \cdots & L_{2n} & g_{12} & g_{22} & \cdots & g_{m2} \\ \vdots & & & & & & & & \\ L_{n1} & L_{n2} & L_{n3} & \cdots & L_{nn} - z & g_{1n} & g_{2n} & \cdots & g_{mn} \\ g_{11} & g_{12} & g_{13} & \cdots & g_{1n} & 0 & 0 & \cdots & 0 \\ g_{21} & g_{22} & g_{23} & \cdots & g_{2n} & 0 & 0 & \cdots & 0 \\ \vdots & & & & & & & & \\ g_{m1} & g_{m2} & g_{m3} & \cdots & g_{mn} & 0 & 0 & \cdots & 0 \end{vmatrix} = 0 \qquad (2.44)$$

where

$$L_{ij} = \frac{\partial^2 L}{\partial x_i \, \partial x_j}(\mathbf{X}^*, \boldsymbol{\lambda}^*) \qquad (2.45)$$

$$g_{ij} = \frac{\partial g_i}{\partial x_j}(\mathbf{X}^*) \qquad (2.46)$$

3. Equation (2.44), on expansion, leads to an $(n - m)$th-order polynomial in z. If some of the roots of this polynomial are positive while the others are negative, the point $\mathbf{X}^*$ is not an extreme point.

The application of the necessary and sufficient conditions in the Lagrange multiplier method is illustrated with the help of the following example.

Example 2.10   Find the dimensions of a cylindrical tin (with top and bottom) made up of sheet metal to maximize its volume such that the total surface area is equal to $A_0 = 24\pi$.

SOLUTION   If x_1 and x_2 denote the radius of the base and length of the tin, respec-tively, the problem can be stated as

$$\text{Maximize } f(x_1, x_2) = \pi x_1^2 x_2$$

subject to

$$2\pi x_1^2 + 2\pi x_1 x_2 = A_0 = 24\pi$$

The Lagrange function is

$$L(x_1, x_2, \lambda) = \pi x_1^2 x_2 + \lambda(2\pi x_1^2 + 2\pi x_1 x_2 - A_0)$$

and the necessary conditions for the maximum of f give

$$\frac{\partial L}{\partial x_1} = 2\pi x_1 x_2 + 4\pi \lambda x_1 + 2\pi \lambda x_2 = 0 \qquad (E_1)$$

$$\frac{\partial L}{\partial x_2} = \pi x_1^2 + 2\pi \lambda x_1 = 0 \qquad (E_2)$$

$$\frac{\partial L}{\partial \lambda} = 2\pi x_1^2 + 2\pi x_1 x_2 - A_0 = 0 \qquad (E_3)$$

Equations (E_1) and (E_2) lead to

$$\lambda = -\frac{x_1 x_2}{2x_1 + x_2} = -\frac{1}{2}x_1$$

that is,

$$x_1 = \tfrac{1}{2}x_2 \qquad (E_4)$$

and Eqs. (E_3) and (E_4) give the desired solution as

$$x_1^* = \left(\frac{A_0}{6\pi}\right)^{1/2}, \quad x_2^* = \left(\frac{2A_0}{3\pi}\right)^{1/2}, \quad \text{and } \lambda^* = -\left(\frac{A_0}{24\pi}\right)^{1/2}$$

This gives the maximum value of f as

$$f^* = \left(\frac{A_0^3}{54\pi}\right)^{1/2}$$

If $A_0 = 24\pi$, the optimum solution becomes

$$x_1^* = 2, \quad x_2^* = 4, \quad \lambda^* = -1, \quad \text{and} \quad f^* = 16\pi$$

To see that this solution really corresponds to the maximum of f, we apply the sufficiency condition of Eq. (2.44). In this case

$$L_{11} = \left.\frac{\partial^2 L}{\partial x_1^2}\right|_{(\mathbf{X}^*, \lambda^*)} = 2\pi x_2^* + 4\pi \lambda^* = 4\pi$$

$$L_{12} = \left.\frac{\partial^2 L}{\partial x_1 \partial x_2}\right|_{(\mathbf{X}^*, \lambda^*)} = L_{21} = 2\pi x_1^* + 2\pi \lambda^* = 2\pi$$

$$L_{22} = \left.\frac{\partial^2 L}{\partial x_2^2}\right|_{(\mathbf{X}^*,\lambda^*)} = 0$$

$$g_{11} = \left.\frac{\partial g_1}{\partial x_1}\right|_{(\mathbf{X}^*,\lambda^*)} = 4\pi x_1^* + 2\pi x_2^* = 16\pi$$

$$g_{12} = \left.\frac{\partial g_1}{\partial x_2}\right|_{(\mathbf{X}^*,\lambda^*)} = 2\pi x_1^* = 4\pi$$

Thus Eq. (2.44) becomes

$$\begin{vmatrix} 4\pi - z & 2\pi & 16\pi \\ 2\pi & 0 - z & 4\pi \\ 16\pi & 4\pi & 0 \end{vmatrix} = 0$$

that is,

$$272\pi^2 z + 192\pi^3 = 0$$

This gives

$$z = -\tfrac{12}{17}\pi$$

Since the value of z is negative, the point (x_1^*, x_2^*) corresponds to the maximum of f.

Interpretation of the Lagrange Multipliers.   To find the physical meaning of the Lagrange multipliers, consider the following optimization problem involving only a single equality constraint:

$$\text{Minimize } f(\mathbf{X}) \tag{2.47}$$

subject to

$$g(\mathbf{X}) = b \quad \text{or} \quad g(\mathbf{X}) = b - g(\mathbf{X}) = 0 \tag{2.48}$$

where b is a constant. The necessary conditions to be satisfied for the solution of the problem are

$$\frac{\partial f}{\partial x_i} + \lambda \frac{\partial g}{\partial x_i} = 0, \qquad i = 1, 2, \ldots, n \tag{2.49}$$

$$g = 0 \tag{2.50}$$

Let the solution of Eqs. (2.49) and (2.50) be given by $\mathbf{X}^*$, λ^*, and $f^* = f(\mathbf{X}^*)$. Suppose that we want to find the effect of a small relaxation or tightening of the constraint on the optimum value of the objective function (i.e., we want to find the effect of a small change in b on f^*). For this we differentiate Eq. (2.48) to obtain

$$db - dg = 0$$

or

$$db = dg = \sum_{i=1}^{n} \frac{\partial g}{\partial x_i} dx_i \qquad (2.51)$$

Equation (2.49) can be rewritten as

$$\frac{\partial f}{\partial x_i} + \lambda \frac{\partial g}{\partial x_i} = \frac{\partial f}{\partial x_i} - \lambda \frac{\partial g}{\partial x_i} = 0 \qquad (2.52)$$

or

$$\frac{\partial g}{\partial x_i} = \frac{\partial f / \partial x_i}{\lambda}, \qquad i = 1, 2, \dots, n \qquad (2.53)$$

Substituting Eq. (2.53) into Eq. (2.51), we obtain

$$db = \sum_{i=1}^{n} \frac{1}{\lambda} \frac{\partial f}{\partial x_i} dx_i = \frac{df}{\lambda} \qquad (2.54)$$

since

$$df = \sum_{i=1}^{n} \frac{\partial f}{\partial x_i} dx_i \qquad (2.55)$$

Equation (2.54) gives

$$\lambda = \frac{df}{db} \quad \text{or} \quad \lambda^* = \frac{df^*}{db} \qquad (2.56)$$

or

$$df^* = \lambda^* db \qquad (2.57)$$

Thus λ^* denotes the sensitivity (or rate of change) of f with respect to b or the marginal or incremental change in f^* with respect to b at x^*. In other words, λ^* indicates how tightly the constraint is binding at the optimum point. Depending on the value of λ^* (positive, negative, or zero), the following physical meaning can be attributed to λ^*:

1. $\lambda^* > 0$. In this case, a unit decrease in b is positively valued since one gets a smaller minimum value of the objective function f. In fact, the decrease in f^* will be exactly equal to λ^* since $df = \lambda^*(-1) = -\lambda^* < 0$. Hence λ^* may be interpreted as the marginal gain (further reduction) in f^* due to the tightening of the constraint. On the other hand, if b is increased by 1 unit, f will also increase to a new optimum level, with the amount of increase in f^* being determined by the magnitude of λ^* since $df = \lambda^*(+1) > 0$. In this case, λ^* may be thought of as the marginal cost (increase) in f^* due to the relaxation of the constraint.

2. $\lambda^* < 0$. Here a unit increase in b is positively valued. This means that it decreases the optimum value of f. In this case the marginal gain (reduction) in f^* due to a relaxation of the constraint by 1 unit is determined by the value of λ^* as $df^* = \lambda^*(+1) < 0$. If b is decreased by 1 unit, the marginal cost (increase) in f^* by the tightening of the constraint is $df^* = \lambda^*(-1) > 0$ since, in this case, the minimum value of the objective function increases.

3. $\lambda^* = 0$. In this case, any incremental change in b has absolutely no effect on the optimum value of f and hence the constraint will not be binding. This means that the optimization of f subject to $g = 0$ leads to the same optimum point $\mathbf{X}^*$ as with the unconstrained optimization of f.

In economics and operations research, Lagrange multipliers are known as *shadow prices* of the constraints since they indicate the changes in optimal value of the objective function per unit change in the right-hand side of the equality constraints.

Example 2.11   Find the maximum of the function $f(\mathbf{X}) = 2x_1 + x_2 + 10$ subject to $g(\mathbf{X}) = x_1 + 2x_2^2 = 3$ using the Lagrange multiplier method. Also find the effect of changing the right-hand side of the constraint on the optimum value of f.

SOLUTION   The Lagrange function is given by

$$L(X, \lambda) = 2x_1 + x_2 + 10 + \lambda(3 - x_1 - 2x_2^2) \tag{E_1}$$

The necessary conditions for the solution of the problem are

$$\frac{\partial L}{\partial x_1} = 2 - \lambda = 0$$

$$\frac{\partial L}{\partial x_2} = 1 - 4\lambda x_2 = 0 \tag{E_2}$$

$$\frac{\partial L}{\partial \lambda} = 3 - x_1 - 2x_2^2 = 0$$

The solution of Eqs. (E_2) is

$$\mathbf{X}^* = \begin{Bmatrix} x_1^* \\ x_2^* \end{Bmatrix} = \begin{Bmatrix} 2.97 \\ 0.13 \end{Bmatrix} \tag{E_3}$$

$$\lambda^* = 2.0$$

The application of the sufficiency condition of Eq. (2.44) yields

$$\begin{vmatrix} L_{11} - z & L_{12} & g_{11} \\ L_{21} & L_{22} - z & g_{12} \\ g_{11} & g_{12} & 0 \end{vmatrix} = 0$$

$$\begin{vmatrix} -z & 0 & -1 \\ 0 & -4\lambda - z & -4x_2 \\ -1 & -4x_2 & 0 \end{vmatrix} = \begin{vmatrix} -z & 0 & -1 \\ 0 & -8 - z & -0.52 \\ -1 & -0.52 & 0 \end{vmatrix} = 0$$

$$0.2704z + 8 + z = 0$$

$$z = -6.2972$$

Hence $\mathbf{X}^*$ will be a maximum of f with $f^* = f(\mathbf{X}^*) = 16.07$.

One procedure for finding the effect on f^* of changes in the value of b (right-hand side of the constraint) would be to solve the problem all over with the new value of b. Another procedure would involve the use of the value of λ^*. When the original constraint is tightened by 1 unit (i.e., $db = -1$), Eq. (2.57) gives

$$df^* = \lambda^* db = 2(-1) = -2$$

Thus the new value of f^* is $f^* + df^* = 14.07$. On the other hand, if we relax the original constraint by 2 units (i.e., $db = 2$), we obtain

$$df^* = \lambda^* db = 2(+2) = 4$$

and hence the new value of f^* is $f^* + df^* = 20.07$.

2.5 MULTIVARIABLE OPTIMIZATION WITH INEQUALITY CONSTRAINTS

This section is concerned with the solution of the following problem:

Minimize $f(\mathbf{X})$

subject to

$$g_j(\mathbf{X}) \leq 0, \quad j = 1, 2, \ldots, m \tag{2.58}$$

The inequality constraints in Eq. (2.58) can be transformed to equality constraints by adding nonnegative slack variables, y_j^2, as

$$g_j(\mathbf{X}) + y_j^2 = 0, \quad j = 1, 2, \ldots, m \tag{2.59}$$

where the values of the slack variables are yet unknown. The problem now becomes

Minimize $f(\mathbf{X})$

subject to

$$G_j(\mathbf{X}, \mathbf{Y}) = g_j(\mathbf{X}) + y_j^2 = 0, \quad j = 1, 2, \ldots, m \tag{2.60}$$

where $\mathbf{Y} = \{y_1, y_2, \ldots, y_m\}^{\mathrm{T}}$ is the vector of slack variables.

This problem can be solved conveniently by the method of Lagrange multipliers. For this, we construct the Lagrange function L as

$$L(\mathbf{X}, \mathbf{Y}, \boldsymbol{\lambda}) = f(\mathbf{X}) + \sum_{j=1}^{m} \lambda_j G_j(\mathbf{X}, \mathbf{Y}) \tag{2.61}$$

where $\boldsymbol{\lambda} = \{\lambda_1, \lambda_2, \ldots, \lambda_m\}^{\mathrm{T}}$ is the vector of Lagrange multipliers. The stationary points of the Lagrange function can be found by solving the following equations (necessary conditions):

$$\frac{\partial L}{\partial x_i}(\mathbf{X}, \mathbf{Y}, \boldsymbol{\lambda}) = \frac{\partial f}{\partial x_i}(\mathbf{X}) + \sum_{j=1}^{m} \lambda_j \frac{\partial g_j}{\partial x_i}(\mathbf{X}) = 0, \qquad i = 1, 2, \ldots, n \tag{2.62}$$

$$\frac{\partial L}{\partial \lambda_j}(\mathbf{X}, \mathbf{Y}, \boldsymbol{\lambda}) = G_j(\mathbf{X}, \mathbf{Y}) = g_j(\mathbf{X}) + y_j^2 = 0, \qquad j = 1, 2, \ldots, m \tag{2.63}$$

$$\frac{\partial L}{\partial y_j}(\mathbf{X}, \mathbf{Y}, \boldsymbol{\lambda}) = 2\lambda_j y_j = 0, \qquad j = 1, 2, \ldots, m \tag{2.64}$$

It can be seen that Eqs. (2.62) to (2.64) represent $(n + 2m)$ equations in the $(n + 2m)$ unknowns, $\mathbf{X}$, $\boldsymbol{\lambda}$, and $\mathbf{Y}$. The solution of Eqs. (2.62) to (2.64) thus gives the optimum solution vector, $\mathbf{X}^*$; the Lagrange multiplier vector, $\boldsymbol{\lambda}^*$; and the slack variable vector, $\mathbf{Y}^*$.

Equations (2.63) ensure that the constraints $g_j(\mathbf{X}) \leq 0$, $j = 1, 2, \ldots, m$, are satisfied, while Eqs. (2.64) imply that either $\lambda_j = 0$ or $y_j = 0$. If $\lambda_j = 0$, it means that the jth constraint is inactive[†] and hence can be ignored. On the other hand, if $y_j = 0$, it means that the constraint is active ($g_j = 0$) at the optimum point. Consider the division of the constraints into two subsets, J_1 and J_2, where $J_1 + J_2$ represent the total set of constraints. Let the set J_1 indicate the indices of those constraints that are active at the optimum point and J_2 include the indices of all the inactive constraints.

Thus for $j \in J_1$,[‡] $y_j = 0$ (constraints are active), for $j \in J_2$, $\lambda_j = 0$ (constraints are inactive), and Eqs. (2.62) can be simplified as

$$\frac{\partial f}{\partial x_i} + \sum_{j \in J_1} \lambda_j \frac{\partial g_j}{\partial x_i} = 0, \qquad i = 1, 2, \ldots, n \tag{2.65}$$

Similarly, Eqs. (2.63) can be written as

$$g_j(\mathbf{X}) = 0, \qquad j \in J_1 \tag{2.66}$$

$$g_j(\mathbf{X}) + y_j^2 = 0, \qquad j \in J_2 \tag{2.67}$$

Equations (2.65) to (2.67) represent $n + p + (m - p) = n + m$ equations in the $n + m$ unknowns $x_i (i = 1, 2, \ldots, n)$, $\lambda_j (j \in J_1)$, and $y_j (j \in J_2)$, where p denotes the number of active constraints.

Assuming that the first p constraints are active, Eqs. (2.65) can be expressed as

$$-\frac{\partial f}{\partial x_i} = \lambda_1 \frac{\partial g_1}{\partial x_i} + \lambda_2 \frac{\partial g_2}{\partial x_i} + \ldots + \lambda_p \frac{\partial g_p}{\partial x_i}, \qquad i = 1, 2, \ldots, n \tag{2.68}$$

These equations can be written collectively as

$$-\nabla f = \lambda_1 \nabla g_1 + \lambda_2 \nabla g_2 + \cdots + \lambda_p \nabla g_p \tag{2.69}$$

[†]Those constraints that are satisfied with an equality sign, $g_j = 0$, at the optimum point are called the *active constraints*, while those that are satisfied with a strict inequality sign, $g_j < 0$, are termed *inactive constraints*.

[‡]The symbol $\in$ is used to denote the meaning "belongs to" or "element of".

where ∇f and ∇g_j are the gradients of the objective function and the jth constraint, respectively:

$$\nabla f = \begin{Bmatrix} \partial f/\partial x_1 \\ \partial f/\partial x_2 \\ \vdots \\ \partial f/\partial x_n \end{Bmatrix} \quad \text{and} \quad \nabla g_j = \begin{Bmatrix} \partial g_j/\partial x_1 \\ \partial g_j/\partial x_2 \\ \vdots \\ \partial g_j/\partial x_n \end{Bmatrix}$$

Equation (2.69) indicates that the negative of the gradient of the objective function can be expressed as a linear combination of the gradients of the active constraints at the optimum point.

Further, we can show that in the case of a minimization problem, the λ_j values ($j \in J_1$) have to be positive. For simplicity of illustration, suppose that only two constraints are active ($p = 2$) at the optimum point. Then Eq. (2.69) reduces to

$$-\nabla f = \lambda_1 \nabla g_1 + \lambda_2 \nabla g_2 \tag{2.70}$$

Let $\mathbf{S}$ be a feasible direction[†] at the optimum point. By premultiplying both sides of Eq. (2.70) by $\mathbf{S}^T$, we obtain

$$-\mathbf{S}^T \nabla f = \lambda_1 \mathbf{S}^T \nabla g_1 + \quad \lambda_2 \mathbf{S}^T \nabla g_2 \tag{2.71}$$

where the superscript T denotes the transpose. Since $\mathbf{S}$ is a feasible direction, it should satisfy the relations

$$\mathbf{S}^T \nabla g_1 < 0$$

$$\mathbf{S}^T \nabla g_2 < 0 \tag{2.72}$$

Thus if $\lambda_1 > 0$ and $\lambda_2 > 0$, the quantity $\mathbf{S}^T \nabla f$ can be seen always to be positive. As ∇f indicates the gradient direction, along which the value of the function increases at the maximum rate,[‡] $\mathbf{S}^T \nabla f$ represents the component of the increment of f along the direction $\mathbf{S}$. If $\mathbf{S}^T \nabla f > 0$, the function value increases as we move along the direction $\mathbf{S}$. Hence if λ_1 and λ_2 are positive, we will not be able to find any direction in the feasible domain along which the function value can be decreased further. Since the point at which Eq. (2.72) is valid is assumed to be optimum, λ_1 and λ_2 have to be positive. This reasoning can be extended to cases where there are more than two constraints active. By proceeding in a similar manner, one can show that the λ_j values have to be negative for a maximization problem.

[†]A vector $\mathbf{S}$ is called a *feasible direction* from a point $\mathbf{X}$ if at least a small step can be taken along $\mathbf{S}$ that does not immediately leave the feasible region. Thus for problems with sufficiently smooth constraint surfaces, vector $\mathbf{S}$ satisfying the relation

$$\mathbf{S}^T \nabla g_j < 0$$

can be called a feasible direction. On the other hand, if the constraint is either linear or concave, as shown in Fig. 2.8b and c, any vector satisfying the relation

$$\mathbf{S}^T \nabla g_j \leq 0$$

can be called a feasible direction. The geometric interpretation of a feasible direction is that the vector $\mathbf{S}$ makes an obtuse angle with all the constraint normals, except that for the linear or outward-curving (concave) constraints, the angle may go to as low as $90°$.

[‡]See Section 6.10.2 for a proof of this statement.

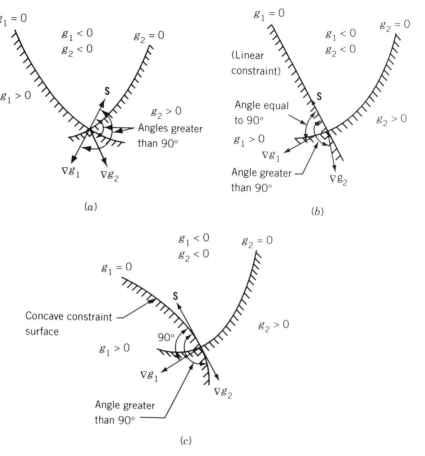

Figure 2.8 Feasible direction **S**.

Example 2.12 Consider the following optimization problem:

$$\text{Minimize } f(x_1, x_2) = x_1^2 + x_2^2$$

subject to

$$x_1 + 2x_2 \leq 15$$

$$1 \leq x_i \leq 10; \quad i = 1, 2$$

Derive the conditions to be satisfied at the point $\mathbf{X}_1 = \{1, 7\}^{\mathrm{T}}$ by the search direction $\mathbf{S} = \{s_1, s_2\}^{\mathrm{T}}$ if it is a (a) usable direction, and (b) feasible direction.

SOLUTION The objective function and the constraints can be stated as

$$f(x_1, x_2) = x_1^2 + x_2^2$$

$$g_1(\mathbf{X}) = x_1 + 2x_2 \leq 15$$

$$g_2(\mathbf{X}) = 1 - x_1 \le 0$$

$$g_3(\mathbf{X}) = 1 - x_2 \le 0$$

$$g_4(\mathbf{X}) = x_1 - 10 \le 0$$

$$g_5(\mathbf{X}) = x_2 - 10 \le 0$$

At the given point $\mathbf{X}_1 = \{1, 7\}^T$, all the constraints can be seen to be satisfied with g_1 and g_2 being active. The gradients of the objective and active constraint functions at point $\mathbf{X}_1 = \{1, 7\}^T$ are given by

$$\nabla f = \left\{ \begin{array}{c} \dfrac{\partial f}{\partial x_1} \\[2mm] \dfrac{\partial f}{\partial x_2} \end{array} \right\}_{\mathbf{X}_1} = \left\{ \begin{array}{c} 2x_1 \\ 2x_2 \end{array} \right\}_{\mathbf{X}_1} = \left\{ \begin{array}{c} 2 \\ 14 \end{array} \right\}$$

$$\nabla g_1 = \left\{ \begin{array}{c} \dfrac{\partial g_1}{\partial x_1} \\[2mm] \dfrac{\partial g_1}{\partial x_2} \end{array} \right\}_{\mathbf{X}_1} = \left\{ \begin{array}{c} 1 \\ 2 \end{array} \right\}$$

$$\nabla g_2 = \left\{ \begin{array}{c} \dfrac{\partial g_2}{\partial x_1} \\[2mm] \dfrac{\partial g_2}{\partial x_2} \end{array} \right\}_{\mathbf{X}_1} = \left\{ \begin{array}{c} -1 \\ 0 \end{array} \right\}$$

For the search direction $\mathbf{S} = \{s_1, s_2\}^T$, the usability and feasibility conditions can be expressed as

(a) Usability condition:

$$\mathbf{S}^T \nabla f \le 0 \quad \text{or} \quad (s_1 \quad s_2) \left\{ \begin{array}{c} 2 \\ 14 \end{array} \right\} \le 0 \quad \text{or} \quad 2s_1 + 14s_2 \le 0 \qquad (E_1)$$

(b) Feasibility conditions:

$$\mathbf{S}^T \nabla g_1 \le 0 \quad \text{or} \quad (s_1 \quad s_2) \left\{ \begin{array}{c} 1 \\ 2 \end{array} \right\} \le 0 \quad \text{or} \quad s_1 + 2s_2 \le 0 \qquad (E_2)$$

$$\mathbf{S}^T \nabla g_2 \le 0 \quad \text{or} \quad (s_1 \quad s_2) \left\{ \begin{array}{c} -1 \\ 0 \end{array} \right\} \le 0 \quad \text{or} \quad -s_1 \le 0 \qquad (E_3)$$

Note: Any two numbers for s_1 and s_2 that satisfy the inequality (E_1) will constitute a usable direction $\mathbf{S}$. For example, $s_1 = 1$ and $s_2 = -1$ gives the usable direction $\mathbf{S} = \{1, -1\}^T$. This direction can also be seen to be a feasible direction because it satisfies the inequalities (E_2) and (E_3).

2.5.1 Kuhn–Tucker Conditions

As shown above, the conditions to be satisfied at a constrained minimum point, $\mathbf{X}^*$, of the problem stated in Eq. (2.58) can be expressed as

$$\frac{\partial f}{\partial x_i} + \sum_{j \in J_1} \lambda_j \frac{\partial g_j}{\partial x_i} = 0, \qquad i = 1, 2, \ldots, n \tag{2.73}$$

$$\lambda_j > 0, \qquad j \in J_1 \tag{2.74}$$

These are called *Kuhn–Tucker conditions* after the mathematicians who derived them as the necessary conditions to be satisfied at a relative minimum of $f(\mathbf{X})$ [2.8]. These conditions are, in general, not sufficient to ensure a relative minimum. However, there is a class of problems, called *convex programming problems*,[†] for which the Kuhn–Tucker conditions are necessary and sufficient for a global minimum.

If the set of active constraints is not known, the Kuhn–Tucker conditions can be stated as follows:

$$\frac{\partial f}{\partial x_i} + \sum_{j=1}^{m} \lambda_j \frac{\partial g_j}{\partial x_i} = 0, \qquad i = 1, 2, \ldots, n$$

$$\lambda_j g_j = 0,^{[‡]} \qquad j = 1, 2, \ldots, m \tag{2.75}$$

$$g_j \leq 0, \qquad j = 1, 2, \ldots, m$$

$$\lambda_j \geq 0, \qquad j = 1, 2, \ldots, m$$

Note that if the problem is one of maximization or if the constraints are of the type $g_j \geq 0$, the λ_j have to be nonpositive in Eqs. (2.75). On the other hand, if the problem is one of maximization with constraints in the form $g_j \geq 0$, the λ_j have to be nonnegative in Eqs. (2.75).

2.5.2 Constraint Qualification

When the optimization problem is stated as

$$\text{Minimize } f(\mathbf{X})$$

subject to

$$g_j(\mathbf{X}) \leq 0, \qquad j = 1, 2, \ldots, m$$
$$h_k(\mathbf{X}) = 0 \qquad k = 1, 2, \ldots, p \tag{2.76}$$

the Kuhn–Tucker conditions become

$$\nabla f + \sum_{j=1}^{m} \lambda_j \nabla g_j - \sum_{k=1}^{p} \beta_k \nabla h_k = \mathbf{0}$$

$$\lambda_j g_j = 0, \qquad j = 1, 2, \ldots, m$$

[†]See Sections 2.6 and 7.14 for a detailed discussion of convex programming problems.
[‡]This condition is the same as Eq. (2.64).

$$g_j \leq 0, \qquad j = 1, 2, \ldots, m$$
$$h_k = 0, \qquad k = 1, 2, \ldots, p \qquad\qquad (2.77)$$
$$\lambda_j \geq 0, \qquad j = 1, 2, \ldots, m$$

where λ_j and β_k denote the Lagrange multipliers associated with the constraints $g_j \leq 0$ and $h_k = 0$, respectively. Although we found qualitatively that the Kuhn–Tucker conditions represent the necessary conditions of optimality, the following theorem gives the precise conditions of optimality.

Theorem 2.7 Let $\mathbf{X}^*$ be a feasible solution to the problem of Eqs. (2.76). If $\nabla g_j(\mathbf{X}^*)$, $j \in J_1$ and $\nabla h_k(\mathbf{X}^*)$, $k = 1, 2, \ldots, p$, are linearly independent, there exist $\boldsymbol{\lambda}^*$ and $\boldsymbol{\beta}^*$ such that $(\mathbf{X}^*, \boldsymbol{\lambda}^*, \boldsymbol{\beta}^*)$ satisfy Eqs. (2.77).

Proof: See Ref. [2.11].

The requirement that $\nabla g_j(\mathbf{X}^*)$, $j \in J_1$ and $\nabla h_k(\mathbf{X}^*)$, $k = 1, 2, \ldots, p$, be linearly independent is called the *constraint qualification*. If the constraint qualification is violated at the optimum point, Eqs. (2.77) may or may not have a solution. It is difficult to verify the constraint qualification without knowing $\mathbf{X}^*$ beforehand. However, the constraint qualification is always satisfied for problems having any of the following characteristics:

1. All the inequality and equality constraint functions are linear.
2. All the inequality constraint functions are convex, all the equality constraint functions are linear, and at least one feasible vector $\tilde{\mathbf{X}}$ exists that lies strictly inside the feasible region, so that

$$g_j(\tilde{\mathbf{X}}) < 0, \quad j = 1, 2, \ldots, m \quad \text{and} \quad h_k(\tilde{\mathbf{X}}) = 0, \ k = 1, 2, \ldots, p$$

Example 2.13   Consider the following problem:

$$\text{Minixize } f(x_1, \ x_2) = (x_1 - 1)^2 + x_2^2 \qquad\qquad (E_1)$$

subject to

$$g_1(x_1, x_2) = x_1^3 - 2x_2 \leq 0 \qquad\qquad (E_2)$$

$$g_2(x_1, x_2) = x_1^3 + 2x_2 \leq 0 \qquad\qquad (E_3)$$

Determine whether the constraint qualification and the Kuhn–Tucker conditions are satisfied at the optimum point.

SOLUTION   The feasible region and the contours of the objective function are shown in Fig. 2.9. It can be seen that the optimum solution is $(0, 0)$. Since g_1 and g_2 are both active at the optimum point $(0, 0)$, their gradients can be computed as

$$\nabla g_1(\mathbf{X}^*) = \begin{Bmatrix} 3x_1^2 \\ -2 \end{Bmatrix}_{(0, 0)} = \begin{Bmatrix} 0 \\ -2 \end{Bmatrix} \quad \text{and} \quad \nabla g_2(\mathbf{X}^*) = \begin{Bmatrix} 3x_1^2 \\ 2 \end{Bmatrix}_{(0, 0)} = \begin{Bmatrix} 0 \\ 2 \end{Bmatrix}$$

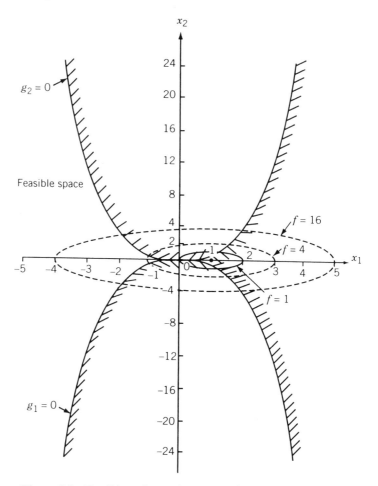

Figure 2.9   Feasible region and contours of the objective function.

It is clear that $\nabla g_1(\mathbf{X}^*)$ and $\nabla g_2(\mathbf{X}^*)$ are not linearly independent. Hence the constraint qualification is not satisfied at the optimum point. Noting that

$$\nabla f(\mathbf{X}^*) = \begin{Bmatrix} 2(x_1 - 1) \\ 2x_2 \end{Bmatrix}_{(0,\,0)} = \begin{Bmatrix} -2 \\ 0 \end{Bmatrix}$$

the Kuhn–Tucker conditions can be written, using Eqs. (2.73) and (2.74), as

$$-2 + \lambda_1(0) + \lambda_2(0) = 0 \tag{E_4}$$

$$0 + \lambda_1(-2) + \lambda_2(2) = 0 \tag{E_5}$$

$$\lambda_1 > 0 \tag{E_6}$$

$$\lambda_2 > 0 \tag{E_7}$$

Since Eq. (E$_4$) is not satisfied and Eq. (E$_5$) can be satisfied for negative values of $\lambda_1 = \lambda_2$ also, the Kuhn–Tucker conditions are not satisfied at the optimum point.

Example 2.14   A manufacturing firm producing small refrigerators has entered into a contract to supply 50 refrigerators at the end of the first month, 50 at the end of the second month, and 50 at the end of the third. The cost of producing x refrigerators in any month is given by $\$(x^2 + 1000)$. The firm can produce more refrigerators in any month and carry them to a subsequent month. However, it costs $\$20$ per unit for any refrigerator carried over from one month to the next. Assuming that there is no initial inventory, determine the number of refrigerators to be produced in each month to minimize the total cost.

SOLUTION   Let x_1, x_2, and x_3 represent the number of refrigerators produced in the first, second, and third month, respectively. The total cost to be minimized is given by

$$\text{total cost} = \text{production cost} + \text{holding cost}$$

or

$$f(x_1, x_2, x_3) = (x_1^2 + 1000) + (x_2^2 + 1000) + (x_3^2 + 1000) + 20(x_1 - 50)$$
$$+ 20(x_1 + x_2 - 100)$$
$$= x_1^2 + x_2^2 + x_3^2 + 40x_1 + 20x_2$$

The constraints can be stated as

$$g_1(x_1, x_2, x_3) = x_1 - 50 \geq 0$$
$$g_2(x_1, x_2, x_3) = x_1 + x_2 - 100 \geq 0$$
$$g_3(x_1, x_2, x_3) = x_1 + x_2 + x_3 - 150 \geq 0$$

The Kuhn–Tucker conditions are given by

$$\frac{\partial f}{\partial x_i} + \lambda_1 \frac{\partial g_1}{\partial x_i} + \lambda_2 \frac{\partial g_2}{\partial x_i} + \lambda_3 \frac{\partial g_3}{\partial x_i} = 0, \qquad i = 1, 2, 3$$

that is,

$$2x_1 + 40 + \lambda_1 + \lambda_2 + \lambda_3 = 0 \tag{E_1}$$
$$2x_2 + 20 + \lambda_2 + \lambda_3 = 0 \tag{E_2}$$
$$2x_3 + \lambda_3 = 0 \tag{E_3}$$
$$\lambda_j g_j = 0, \qquad j = 1, 2, 3$$

that is,

$$\lambda_1(x_1 \quad - 50) = 0 \tag{E_4}$$
$$\lambda_2(x_1 + x_2 - 100) = 0 \tag{E_5}$$
$$\lambda_3(x_1 + x_2 + x_3 - 150) = 0 \tag{E_6}$$
$$g_j \geq 0, \qquad j = 1, 2, 3$$

that is,

$$x_1 - 50 \geq 0 \tag{E$_7$}$$

$$x_1 + x_2 - 100 \geq 0 \tag{E$_8$}$$

$$x_1 + x_2 + x_3 - 150 \geq 0 \tag{E$_9$}$$

$$\lambda_j \leq 0, \quad j = 1, 2, 3$$

that is,

$$\lambda_1 \leq 0 \tag{E$_{10}$}$$

$$\lambda_2 \leq 0 \tag{E$_{11}$}$$

$$\lambda_3 \leq 0 \tag{E$_{12}$}$$

The solution of Eqs. (E$_1$) to (E$_{12}$) can be found in several ways. We proceed to solve these equations by first nothing that either $\lambda_1 = 0$ or $x_1 = 50$ according to Eq. (E$_4$). Using this information, we investigate the following cases to identify the optimum solution of the problem.

Case 1: $\lambda_1 = 0$.

Equations (E$_1$) to (E$_3$) give

$$x_3 = -\frac{\lambda_3}{2}$$

$$x_2 = -10 - \frac{\lambda_2}{2} - \frac{\lambda_3}{2} \tag{E$_{13}$}$$

$$x_1 = -20 - \frac{\lambda_2}{2} - \frac{\lambda_3}{2}$$

Substituting Eqs. (E$_{13}$) in Eqs. (E$_5$) and (E$_6$), we obtain

$$\lambda_2(-130 - \lambda_2 - \lambda_3) = 0$$

$$\lambda_3(-180 - \lambda_2 - \tfrac{3}{2}\lambda_3) = 0 \tag{E$_{14}$}$$

The four possible solutions of Eqs. (E$_{14}$) are

1. $\lambda_2 = 0, -180 - \lambda_2 - \dfrac{3}{2}\lambda_3 = 0$. These equations, along with Eqs. (E$_{13}$), yield the solution

$$\lambda_2 = 0, \quad \lambda_3 = -120, \quad x_1 = 40, \quad x_2 = 50, \quad x_3 = 60$$

This solution satisfies Eqs. (E$_{10}$) to (E$_{12}$) but violates Eqs. (E$_7$) and (E$_8$) and hence cannot be optimum.

2. $\lambda_3 = 0, -130 - \lambda_2 - \lambda_3 = 0$. The solution of these equations leads to

$$\lambda_2 = -130, \quad \lambda_3 = 0, \quad x_1 = 45, \quad x_2 = 55, x_3 = 0$$

This solution can be seen to satisfy Eqs. (E$_{10}$) to (E$_{12}$) but violate Eqs. (E$_7$) and (E$_9$).

3. $\lambda_2 = 0$, $\lambda_3 = 0$. Equations (E$_{13}$) give

$$x_1 = -20, \quad x_2 = -10, \quad x_3 = 0$$

This solution satisfies Eqs. (E$_{10}$) to (E$_{12}$) but violates the constraints, Eqs. (E$_7$) to (E$_9$).

4. $-130 - \lambda_2 - \lambda_3 = 0, -180 - \lambda_2 - \frac{3}{2}\lambda_3 = 0$. The solution of these equations and Eqs. (E$_{13}$) yields

$$\lambda_2 = -30, \quad \lambda_3 = -100, \quad x_1 = 45, \quad x_2 = 55, \quad x_3 = 50$$

This solution satisfies Eqs. (E$_{10}$) to (E$_{12}$) but violates the constraint, Eq. (E$_7$).

Case 2: $x_1 = 50$.

In this case, Eqs. (E$_1$) to (E$_3$) give

$$\lambda_3 = -2x_3$$
$$\lambda_2 = -20 - 2x_2 - \lambda_3 = -20 - 2x_2 + 2x_3 \qquad (E_{15})$$
$$\lambda_1 = -40 - 2x_1 - \lambda_2 - \lambda_3 = -120 + 2x_2$$

Substitution of Eqs. (E$_{15}$) in Eqs. (E$_5$) and (E$_6$) leads to

$$(-20 - 2x_2 + 2x_3)(x_1 + x_2 - 100) = 0$$
$$(-2x_3)(x_1 + x_2 + x_3 - 150) = 0 \qquad (E_{16})$$

Once again, it can be seen that there are four possible solutions to Eqs. (E$_{16}$), as indicated below:

1. $-20 - 2x_2 + 2x_3 = 0$, $x_1 + x_2 + x_3 - 150 = 0$: The solution of these equations yields
$$x_1 = 50, \quad x_2 = 45, \quad x_3 = 55$$

This solution can be seen to violate Eq. (E$_8$).

2. $-20 - 2x_2 + 2x_3 = 0$, $-2x_3 = 0$: These equations lead to the solution
$$x_1 = 50, \quad x_2 = -10, \quad x_3 = 0$$

This solution can be seen to violate Eqs. (E$_8$) and (E$_9$).

3. $x_1 + x_2 - 100 = 0$, $-2x_3 = 0$: These equations give
$$x_1 = 50, \quad x_2 = 50, \quad x_3 = 0$$

This solution violates the constraint Eq. (E$_9$).

4. $x_1 + x_2 - 100 = 0$, $x_1 + x_2 + x_3 - 150 = 0$: The solution of these equations yields

$$x_1 = 50, \quad x_2 = 50, \quad x_3 = 50$$

This solution can be seen to satisfy all the constraint Eqs. (E$_7$) to (E$_9$). The values of λ_1, λ_2, and λ_3 corresponding to this solution can be obtained from Eqs. (E$_{15}$) as

$$\lambda_1 = -20, \quad \lambda_2 = -20, \quad \lambda_3 = -100$$

Since these values of λ_i satisfy the requirements [Eqs. (E$_{10}$) to (E$_{12}$)], this solution can be identified as the optimum solution. Thus

$$x_1^* = 50, \quad x_2^* = 50, \quad x_3^* = 50$$

2.6   CONVEX PROGRAMMING PROBLEM

The optimization problem stated in Eq. (2.58) is called a *convex programming problem* if the objective function $f(\mathbf{X})$ and the constraint functions $g_j(\mathbf{X})$ are convex. The definition and properties of a convex function are given in Appendix A. Suppose that $f(\mathbf{X})$ and $g_j(\mathbf{X})$, $j = 1, 2, \ldots, m$, are convex functions. The Lagrange function of Eq. (2.61) can be written as

$$L(\mathbf{X}, \mathbf{Y}, \boldsymbol{\lambda}) = f(\mathbf{X}) + \sum_{j=1}^{m} \lambda_j [g_j(\mathbf{X}) + y_j^2] \tag{2.78}$$

If $\lambda_j \geq 0$, then $\lambda_j g_j(\mathbf{X})$ is convex, and since $\lambda_j y_j = 0$ from Eq. (2.64), $L(\mathbf{X}, \mathbf{Y}, \boldsymbol{\lambda})$ will be a convex function. As shown earlier, a necessary condition for $f(\mathbf{X})$ to be a relative minimum at $\mathbf{X}^*$ is that $L(\mathbf{X}, \mathbf{Y}, \boldsymbol{\lambda})$ have a stationary point at $\mathbf{X}^*$. However, if $L(\mathbf{X}, \mathbf{Y}, \boldsymbol{\lambda})$ is a convex function, its derivative vanishes only at one point, which must be an absolute minimum of the function $f(\mathbf{X})$. Thus the Kuhn–Tucker conditions are both necessary and sufficient for an absolute minimum of $f(\mathbf{X})$ at $\mathbf{X}^*$.

Notes:

1. If the given optimization problem is known to be a convex programming problem, there will be no relative minima or saddle points, and hence the extreme point found by applying the Kuhn–Tucker conditions is guaranteed to be an absolute minimum of $f(\mathbf{X})$. However, it is often very difficult to ascertain whether the objective and constraint functions involved in a practical engineering problem are convex.

2. The derivation of the Kuhn–Tucker conditions was based on the development given for equality constraints in Section 2.4. One of the requirements for these conditions was that at least one of the Jacobians composed of the m constraints and m of the $n + m$ variables $(x_1, x_2, \ldots, x_n; y_1, y_2, \ldots, y_m)$ be nonzero. This requirement is implied in the derivation of the Kuhn–Tucker conditions.

REFERENCES AND BIBLIOGRAPHY

2.1 H. Hancock, *Theory of Maxima and Minima*, Dover, New York, 1960.

2.2 M. E. Levenson, *Maxima and Minima*, Macmillan, New York, 1967.

2.3 G. B. Thomas, Jr., *Calculus and Analytic Geometry*, Addison-Wesley, Reading, MA, 1967.

2.4 A. E. Richmond, *Calculus for Electronics*, McGraw-Hill, New York, 1972.

2.5 B. Kolman and W. F. Trench, *Elementary Multivariable Calculus*, Academic Press, New York, 1971.

2.6 G. S. G. Beveridge and R. S. Schechter, *Optimization: Theory and Practice*, McGraw-Hill, New York, 1970.

2.7 R. Gue and M. E. Thomas, *Mathematical Methods of Operations Research*, Macmillan, New York, 1968.

2.8 H. W. Kuhn and A. Tucker, Nonlinear Programming, in *Proceedings of the 2nd Berkeley Symposium on Mathematical Statistics and Probability*, University of California Press, Berkeley, 1951.

2.9 F. Ayres, Jr., *Theory and Problems of Matrices*, Schaum's Outline Series, Schaum, New York, 1962.

2.10 M. J. Panik, *Classical Optimization: Foundations and Extensions*, North-Holland, Amsterdam, 1976.

2.11 M. S. Bazaraa and C. M. Shetty, *Nonlinear Programming: Theory and Algorithms*, Wiley, New York, 1979.

2.12 D. M. Simmons, *Nonlinear Programming for Operations Research*, Prentice Hall, Englewood Cliffs, NJ, 1975.

2.13 J. R. Howell and R. O. Buckius, *Fundamentals of Engineering Thermodynamics*, 2nd ed., McGraw-Hill, New York, 1992.

REVIEW QUESTIONS

2.1 State the necessary and sufficient conditions for the minimum of a function $f(x)$.

2.2 Under what circumstances can the condition $df(x)/dx = 0$ not be used to find the minimum of the function $f(x)$?

2.3 Define the rth differential, $d^r f(\mathbf{X})$, of a multivariable function $f(\mathbf{X})$.

2.4 Write the Taylor's series expansion of a function $f(\mathbf{X})$.

2.5 State the necessary and sufficient conditions for the maximum of a multivariable function $f(\mathbf{X})$.

2.6 What is a quadratic form?

2.7 How do you test the positive, negative, or indefiniteness of a square matrix $[A]$?

2.8 Define a saddle point and indicate its significance.

2.9 State the various methods available for solving a multivariable optimization problem with equality constraints.

2.10 State the principle behind the method of constrained variation.

2.11 What is the Lagrange multiplier method?

2.12 What is the significance of Lagrange multipliers?

2.13 Convert an inequality constrained problem into an equivalent unconstrained problem.

2.14 State the Kuhn–Tucker conditions.

2.15 What is an active constraint?

2.16 Define a usable feasible direction.

2.17 What is a convex programming problem? What is its significance?

2.18 Answer whether each of the following quadratic forms is positive definite, negative definite, or neither:

(a) $f = x_1^2 - x_2^2$
(b) $f = 4x_1x_2$
(c) $f = x_1^2 + 2x_2^2$
(d) $f = -x_1^2 + 4x_1x_2 + 4x_2^2$
(e) $f = -x_1^2 + 4x_1x_2 - 9x_2^2 + 2x_1x_3 + 8x_2x_3 - 4x_3^2$

2.19 State whether each of the following functions is convex, concave, or neither:

(a) $f = -2x^2 + 8x + 4$
(b) $f = x^2 + 10x + 1$
(c) $f = x_1^2 - x_2^2$
(d) $f = -x_1^2 + 4x_1x_2$
(e) $f = e^{-x}, \ x > 0$
(f) $f = \sqrt{x}, \ x > 0$
(g) $f = x_1x_2$
(h) $f = (x_1 - 1)^2 + 10(x_2 - 2)^2$

2.20 Match the following equations and their characteristics:

(a) $f = 4x_1 - 3x_2 + 2$ Relative maximum at (1, 2)
(b) $f = (2x_1 - 2)^2 + (x_2 - 2)^2$ Saddle point at origin
(c) $f = -(x_1 - 1)^2 - (x_2 - 2)^2$ No minimum
(d) $f = x_1x_2$ Inflection point at origin
(e) $f = x^3$ Relative minimum at (1, 2)

PROBLEMS

2.1 A dc generator has an internal resistance R ohms and develops an open-circuit voltage of V volts (Fig. 2.10). Find the value of the load resistance r for which the power delivered by the generator will be a maximum.

2.2 Find the maxima and minima, if any, of the function

$$f(x) = \frac{x^4}{(x-1)(x-3)^3}$$

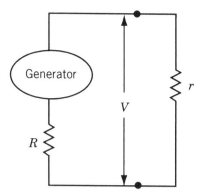

Figure 2.10   Electric generator with load.

2.3 Find the maxima and minima, if any, of the function

$$f(x) = 4x^3 - 18x^2 + 27x - 7$$

2.4 The efficiency of a screw jack is given by

$$\eta = \frac{\tan \alpha}{\tan(\alpha + \phi)}$$

where α is the lead angle and ϕ is a constant. Prove that the efficiency of the screw jack will be maximum when $\alpha = 45^\circ - \phi/2$ with $\eta_{max} = (1 - \sin\phi)/(1 + \sin\phi)$.

2.5 Find the minimum of the function

$$f(x) = 10x^6 - 48x^5 + 15x^4 + 200x^3 - 120x^2 - 480x + 100$$

2.6 Find the angular orientation of a cannon to maximize the range of the projectile.

2.7 In a submarine telegraph cable the speed of signaling varies as $x^2 \log(1/x)$, where x is the ratio of the radius of the core to that of the covering. Show that the greatest speed is attained when this ratio is $1 : \sqrt{e}$.

2.8 The horsepower generated by a Pelton wheel is proportional to $u(V - u)$, where u is the velocity of the wheel, which is variable, and V is the velocity of the jet, which is fixed. Show that the efficiency of the Pelton wheel will be maximum when $u = V/2$.

2.9 A pipe of length l and diameter D has at one end a nozzle of diameter d through which water is discharged from a reservoir. The level of water in the reservoir is maintained at a constant value h above the center of nozzle. Find the diameter of the nozzle so that the kinetic energy of the jet is a maximum. The kinetic energy of the jet can be expressed as

$$\frac{1}{4}\pi\rho d^2 \left(\frac{2g D^5 h}{D^5 + 4fld^4} \right)^{3/2}$$

where ρ is the density of water, f the friction coefficient and g the gravitational constant.

2.10 An electric light is placed directly over the center of a circular plot of lawn 100 m in diameter. Assuming that the intensity of light varies directly as the sine of the angle at which it strikes an illuminated surface, and inversely as the square of its distance from the surface, how high should the light be hung in order that the intensity may be as great as possible at the circumference of the plot?

2.11 If a crank is at an angle θ from dead center with $\theta = \omega t$, where ω is the angular velocity and t is time, the distance of the piston from the end of its stroke (x) is given by

$$x = r(1 - \cos\theta) + \frac{r^2}{4l}(1 - \cos 2\theta)$$

where r is the length of the crank and l is the length of the connecting rod. For $r = 1$ and $l = 5$, find (**a**) the angular position of the crank at which the piston moves with maximum velocity, and (**b**) the distance of the piston from the end of its stroke at that instant.

Determine whether each of the matrices in Problems 2.12–2.14 is positive definite, negative definite, or indefinite by finding its eigenvalues.

2.12 $[A] = \begin{bmatrix} 3 & 1 & -1 \\ 1 & 3 & -1 \\ -1 & -1 & 5 \end{bmatrix}$

2.13 $[B] = \begin{bmatrix} 4 & 2 & -4 \\ 2 & 4 & -2 \\ -4 & -2 & 4 \end{bmatrix}$

2.14 $[C] = \begin{bmatrix} -1 & -1 & -1 \\ -1 & -2 & -2 \\ -1 & -2 & -3 \end{bmatrix}$

Determine whether each of the matrices in Problems 2.15–2.17 is positive definite, negative definite, or indefinite by evaluating the signs of its submatrices.

2.15 $[A] = \begin{bmatrix} 3 & 1 & -1 \\ 1 & 3 & -1 \\ -1 & -1 & 5 \end{bmatrix}$

2.16 $[B] = \begin{bmatrix} 4 & 2 & -4 \\ 2 & 4 & -2 \\ -4 & -2 & 4 \end{bmatrix}$

2.17 $[C] = \begin{bmatrix} -1 & -1 & -1 \\ -1 & -2 & -2 \\ -1 & -2 & -3 \end{bmatrix}$

2.18 Express the function

$$f(x_1, x_2, x_3) = -x_1^2 - x_2^2 + 2x_1 x_2 - x_3^2 + 6x_1 x_3 + 4x_1 - 5x_3 + 2$$

in matrix form as

$$f(\mathbf{X}) = \tfrac{1}{2}\mathbf{X}^\mathrm{T}[A]\,\mathbf{X} + \mathbf{B}^\mathrm{T}\,\mathbf{X} + C$$

and determine whether the matrix $[A]$ is positive definite, negative definite, or indefinite.

2.19 Determine whether the following matrix is positive or negative definite:

$$[A] = \begin{bmatrix} 4 & -3 & 0 \\ -3 & 0 & 4 \\ 0 & 4 & 2 \end{bmatrix}$$

2.20 Determine whether the following matrix is positive definite:

$$[A] = \begin{bmatrix} -14 & 3 & 0 \\ 3 & -1 & 4 \\ 0 & 4 & 2 \end{bmatrix}$$

2.21 The potential energy of the two-bar truss shown in Fig. 2.11 is given by

$$f(x_1, x_2) = \frac{EA}{s}\left(\frac{1}{2s}\right)^2 x_1^2 + \frac{EA}{s}\left(\frac{h}{s}\right)^2 x_2^2 - Px_1\cos\theta - Px_2\sin\theta$$

where E is Young's modulus, A the cross-sectional area of each member, l the span of the truss, s the length of each member, h the height of the truss, P the applied load, θ the angle at which the load is applied, and x_1 and x_2 are, respectively, the horizontal and vertical displacements of the free node. Find the values of x_1 and x_2 that minimize the potential energy when $E = 207 \times 10^9$ Pa, $A = 10^{-5}$ m^2, $l = 1.5$ m, $h = 4.0$ m, $P = 10^4$ N, and $\theta = 30°$.

2.22 The profit per acre of a farm is given by

$$20x_1 + 26x_2 + 4x_1x_2 - 4x_1^2 - 3x_2^2$$

where x_1 and x_2 denote, respectively, the labor cost and the fertilizer cost. Find the values of x_1 and x_2 to maximize the profit.

2.23 The temperatures measured at various points inside a heated wall are as follows:

Distance from the heated surface as a percentage of wall thickness, d	0	25	50	75	100
Temperature, $t(°C)$	380	200	100	20	0

It is decided to approximate this table by a linear equation (graph) of the form $t = a + bd$, where a and b are constants. Find the values of the constants a and b that minimize the sum of the squares of all differences between the graph values and the tabulated values.

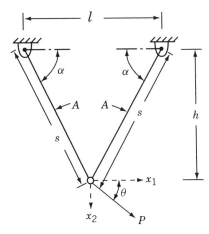

Figure 2.11   Two-bar truss.

2.24 Find the second-order Taylor's series approximation of the function

$$f(x_1, x_2) = (x_1 - 1)^2 e^{x_2} + x_1$$

at the points (**a**) (0,0) and (**b**) (1,1).

2.25 Find the third-order Taylor's series approximation of the function

$$f(x_1, x_2, x_3) = x_2^2 x_3 + x_1 e^{x_3}$$

at point (1, 0, −2).

2.26 The volume of sales (f) of a product is found to be a function of the number of newspaper advertisements (x) and the number of minutes of television time (y) as

$$f = 12xy - x^2 - 3y^2$$

Each newspaper advertisement or each minute on television costs $1000. How should the firm allocate $48,000 between the two advertising media for maximizing its sales?

2.27 Find the value of x^* at which the following function attains its maximum:

$$f(x) = \frac{1}{10\sqrt{2\pi}} e^{-(1/2)[(x-100)/10]^2}$$

2.28 It is possible to establish the nature of stationary points of an objective function based on its quadratic approximation. For this, consider the quadratic approximation of a two-variable function as

$$f(\mathbf{X}) \approx a + \mathbf{b}^T \mathbf{X} + \tfrac{1}{2} \mathbf{X}^T [c] \mathbf{X}$$

where

$$\mathbf{X} = \begin{Bmatrix} x_1 \\ x_2 \end{Bmatrix}, \quad \mathbf{b} = \begin{Bmatrix} b_1 \\ b_2 \end{Bmatrix}, \quad \text{and} \quad [c] = \begin{bmatrix} c_{11} & c_{12} \\ c_{12} & c_{22} \end{bmatrix}$$

If the eigenvalues of the Hessian matrix, [c], are denoted as β_1 and β_2, identify the nature of the contours of the objective function and the type of stationary point in each of the following situations.

(**a**) $\beta_1 = \beta_2$; both positive
(**b**) $\beta_1 > \beta_2$; both positive
(**c**) $|\beta_1| = |\beta_2|$; β_1 and β_2 have opposite signs
(**d**) $\beta_1 > 0, \beta_2 = 0$

Plot the contours of each of the following functions and identify the nature of its stationary point.

2.29 $f = 2 - x^2 - y^2 + 4xy$

2.30 $f = 2 + x^2 - y^2$

2.31 $f = xy$

2.32 $f = x^3 - 3xy^2$

2.33 Find the admissible and constrained variations at the point $\mathbf{X} = \{0, 4\}^T$ for the following problem:

$$\text{Minimize } f = x_1^2 + (x_2 - 1)^2$$

subject to

$$-2x_1^2 + x_2 = 4$$

2.34 Find the diameter of an open cylindrical can that will have the maximum volume for a given surface area, S.

2.35 A rectangular beam is to be cut from a circular log of radius r. Find the cross-sectional dimensions of the beam to (**a**) maximize the cross-sectional area of the beam, and (**b**) maximize the perimeter of the beam section.

2.36 Find the dimensions of a straight beam of circular cross section that can be cut from a conical log of height h and base radius r to maximize the volume of the beam.

2.37 The deflection of a rectangular beam is inversely proportional to the width and the cube of depth. Find the cross-sectional dimensions of a beam, which corresponds to minimum deflection, that can be cut from a cylindrical log of radius r.

2.38 A rectangular box of height a and width b is placed adjacent to a wall (Fig. 2.12). Find the length of the shortest ladder that can be made to lean against the wall.

2.39 Show that the right circular cylinder of given surface (including the ends) and maximum volume is such that its height is equal to the diameter of the base.

2.40 Find the dimensions of a closed cylindrical soft drink can that can hold soft drink of volume V for which the surface area (including the top and bottom) is a minimum.

2.41 An open rectangular box is to be manufactured from a given amount of sheet metal (area S). Find the dimensions of the box to maximize the volume.

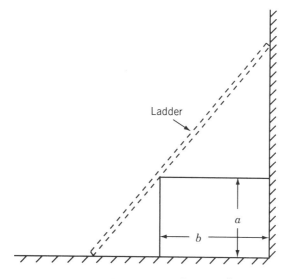

Figure 2.12 Ladder against a wall.

2.42   Find the dimensions of an open rectangular box of volume V for which the amount of material required for manufacture (surface area) is a minimum.

2.43   A rectangular sheet of metal with sides a and b has four equal square portions (of side d) removed at the corners, and the sides are then turned up so as to form an open rectangular box. Find the depth of the box that maximizes the volume.

2.44   Show that the cone of the greatest volume that can be inscribed in a given sphere has an altitude equal to two-thirds of the diameter of the sphere. Also prove that the curved surface of the cone is a maximum for the same value of the altitude.

2.45   Prove Theorem 2.6.

2.46   A log of length l is in the form of a frustum of a cone whose ends have radii a and $b(a > b)$. It is required to cut from it a beam of uniform square section. Prove that the beam of greatest volume that can be cut has a length of $al/[3(a-b)]$.

2.47   It has been decided to leave a margin of 30 mm at the top and 20 mm each at the left side, right side, and the bottom on the printed page of a book. If the area of the page is specified as 5×10^4 mm^2, determine the dimensions of a page that provide the largest printed area.

2.48 $$\text{Minimize } f = 9 - 8x_1 - 6x_2 - 4x_3 + 2x_1^2$$
$$+ 2x_2^2 + x_3^2 + 2x_1x_2 + 2x_1x_3$$

subject to
$$x_1 + x_2 + 2x_3 = 3$$

by (**a**) direct substitution, (**b**) constrained variation, and (**c**) Lagrange multiplier method.

2.49 $$\text{Minimize } f(\mathbf{X}) = \tfrac{1}{2}(x_1^2 + x_2^2 + x_3^2)$$

subject to
$$g_1(\mathbf{X}) = x_1 - x_2 = 0$$
$$g_2(\mathbf{X}) = x_1 + x_2 + x_3 - 1 = 0$$

by (**a**) direct substitution, (**b**) constrained variation, and (**c**) Lagrange multiplier method.

2.50   Find the values of x, y, and z that maximize the function

$$f(x, y, z) = \frac{6xyz}{x + 2y + 2z}$$

when x, y, and z are restricted by the relation $xyz = 16$.

2.51   A tent on a square base of side $2a$ consists of four vertical sides of height b surmounted by a regular pyramid of height h. If the volume enclosed by the tent is V, show that the area of canvas in the tent can be expressed as

$$\frac{2V}{a} - \frac{8ah}{3} + 4a\sqrt{h^2 + a^2}$$

Also show that the least area of the canvas corresponding to a given volume V, if a and h can both vary, is given by

$$a = \frac{\sqrt{5}h}{2} \quad \text{and} \quad h = 2b$$

2.52 A department store plans to construct a one-story building with a rectangular planform. The building is required to have a floor area of $22,500\,\text{ft}^2$ and a height of 18 ft. It is proposed to use brick walls on three sides and a glass wall on the fourth side. Find the dimensions of the building to minimize the cost of construction of the walls and the roof assuming that the glass wall costs twice as much as that of the brick wall and the roof costs three times as much as that of the brick wall per unit area.

2.53 Find the dimensions of the rectangular building described in Problem 2.52 to minimize the heat loss, assuming that the relative heat losses per unit surface area for the roof, brick wall, glass wall, and floor are in the proportion 4:2:5:1.

2.54 A funnel, in the form of a right circular cone, is to be constructed from a sheet metal. Find the dimensions of the funnel for minimum lateral surface area when the volume of the funnel is specified as $200\,\text{in}^3$.

2.55 Find the effect on f^* when the value of A_0 is changed to (a) 25π and (b) 22π in Example 2.10 using the property of the Lagrange multiplier.

2.56 (a) Find the dimensions of a rectangular box of volume $V = 1000\,\text{in}^3$ for which the total length of the 12 edges is a minimum using the Lagrange multiplier method.

  (b) Find the change in the dimensions of the box when the volume is changed to $1200\,\text{in}^3$ by using the value of λ^* found in part (a).

  (c) Compare the solution found in part (b) with the exact solution.

2.57 Find the effect on f^* of changing the constraint to (a) $x + x_2 + 2x_3 = 4$ and (b) $x + x_2 + 2x_3 = 2$ in Problem 2.48. Use the physical meaning of Lagrange multiplier in finding the solution.

2.58 A real estate company wants to construct a multistory apartment building on a 500×500-ft lot. It has been decided to have a total floor space of $8 \times 10^5\,\text{ft}^2$. The height of each story is required to be 12 ft, the maximum height of the building is to be restricted to 75 ft, and the parking area is required to be at least 10 % of the total floor area according to the city zoning rules. If the cost of the building is estimated at $\$(500,000h + 2000F + 500P)$, where h is the height in feet, F is the floor area in square feet, and P is the parking area in square feet. Find the minimum cost design of the building.

2.59 The Brinell hardness test is used to measure the indentation hardness of materials. It involves penetration of an indenter, in the form of a ball of diameter D (mm), under a load P (kg_f), as shown in Fig. 2.13a. The Brinell hardness number (BHN) is defined as

$$\text{BHN} = \frac{P}{A} \equiv \frac{2P}{\pi D(D - \sqrt{D^2 - d^2})} \tag{1}$$

where A (in mm^2) is the spherical surface area and d (in mm) is the diameter of the crater or indentation formed. The diameter d and the depth h of indentation are related by (Fig. 2.13b)

$$d = 2\sqrt{h(D - h)} \tag{2}$$

It is desired to find the size of indentation, in terms of the values of d and h, when a tungsten carbide ball indenter of diameter 10 mm is used under a load of $P = 3000\,\text{kg}_\text{f}$ on a stainless steel test specimen of BHN 1250. Find the values of d and h by formulating and solving the problem as an unconstrained minimization problem.

Hint: Consider the objective function as the sum of squares of the equations implied by Eqs. (1) and (2).

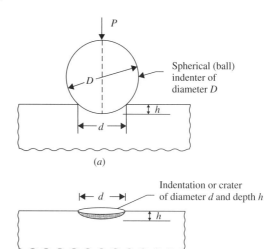

(a)

(b)

Figure 2.13   Brinell hardness test.

2.60   A manufacturer produces small refrigerators at a cost of $60 per unit and sells them to a retailer in a lot consisting of a minimum of 100 units. The selling price is set at $80 per unit if the retailer buys 100 units at a time. If the retailer buys more than 100 units at a time, the manufacturer agrees to reduce the price of all refrigerators by 10 cents for each unit bought over 100 units. Determine the number of units to be sold to the retailer to maximize the profit of the manufacturer.

2.61   Consider the following problem:

$$\text{Minimize} f = (x_1 - 2)^2 + (x_2 - 1)^2$$

subject to

$$2 \geq x_1 + x_2$$

$$x_2 \geq x_1^2$$

Using Kuhn–Tucker conditions, find which of the following vectors are local minima:

$$\mathbf{X}_1 = \begin{Bmatrix} 1.5 \\ 0.5 \end{Bmatrix}, \quad \mathbf{X}_2 = \begin{Bmatrix} 1 \\ 1 \end{Bmatrix}, \quad \mathbf{X}_3 = \begin{Bmatrix} 2 \\ 0 \end{Bmatrix}$$

2.62   Using Kuhn–Tucker conditions, find the value(s) of β for which the point $x_1^* = 1, x_2^* = 2$ will be optimal to the problem:

$$\text{Maximize } f(x_1, x_2) = 2x_1 + \beta x_2$$

subject to

$$g_1(x_1, x_2) = x_1^2 + x_2^2 - 5 \leq 0$$

$$g_2(x_1, x_2) = x_1 - x_2 - 2 \leq 0$$

Verify your result using a graphical procedure.

2.63 Consider the following optimization problem:

$$\text{Maximize } f = -x_1 - x_2$$

subject to

$$x_1^2 + x_2 \geq 2$$
$$4 \leq x_1 + 3x_2$$
$$x_1 + x_2^4 \leq 30$$

(a) Find whether the design vector $\mathbf{X} = \{1, 1\}^T$ satisfies the Kuhn–Tucker conditions for a constrained optimum.

(b) What are the values of the Lagrange multipliers at the given design vector?

2.64 Consider the following problem:

$$\text{Maximize } f(\mathbf{X}) = x_1^2 + x_2^2 + x_3^2$$

subject to

$$x_1 + x_2 + x_3 \geq 5$$
$$2 - x_2 x_3 \leq 0$$
$$x_1 \geq 0, \quad x_2 \geq 0, \quad x_3 \geq 2$$

Determine whether the Kuhn–Tucker conditions are satisfied at the following points:

$$\mathbf{X}_1 = \begin{Bmatrix} \frac{3}{2} \\ \frac{3}{2} \\ 2 \end{Bmatrix}, \quad X_2 = \begin{Bmatrix} \frac{4}{3} \\ \frac{2}{3} \\ 3 \end{Bmatrix}, \quad X_3 = \begin{Bmatrix} 2 \\ 1 \\ 2 \end{Bmatrix}$$

2.65 Find a usable and feasible direction $\mathbf{S}$ at (a) $\mathbf{X}_1 = \{-1, 5\}^T$ and (b) $\mathbf{X}_2 = \{2, 3\}$ for the following problem:

$$\text{Minimize } f(\mathbf{X}) = (x_1 - 1)^2 + (x_2 - 5)^2$$

subject to

$$g_1(X) = -x_1^2 + x_2 - 4 \leq 0$$
$$g_2(X) = -(x_1 - 2)^2 + x_2 - 3 \leq 0$$

2.66 Consider the following problem:

$$\text{Maximize } f = x_1^2 - x_2$$

subject to

$$26 \geq x_1^2 + x_2^2$$
$$x_1 + x_2 \geq 6$$
$$x_1 \geq 0$$

Determine whether the following search direction is usable, feasible, or both at the design vector $X = \left\{ \begin{smallmatrix} 5 \\ 1 \end{smallmatrix} \right\}$:

$$S = \left\{ \begin{matrix} 0 \\ 1 \end{matrix} \right\}, \quad S = \left\{ \begin{matrix} -1 \\ 1 \end{matrix} \right\}, \quad S = \left\{ \begin{matrix} 1 \\ 0 \end{matrix} \right\}, \quad S = \left\{ \begin{matrix} -1 \\ 2 \end{matrix} \right\}$$

2.67 Consider the following problem:

$$\text{Minimize } f = x_1^3 - 6x_1^2 + 11x_1 + x_3$$

subject to

$$x_1^2 + x_2^2 - x_3^2 \le 0$$
$$4 - x_1^2 - x_2^2 - x_3^2 \le 0$$
$$x_i \ge 0, \quad i = 1, 2, 3, \quad x_3 \le 5$$

Determine whether the following vector represents an optimum solution:

$$X = \left\{ \begin{matrix} 0 \\ \sqrt{2} \\ \sqrt{2} \end{matrix} \right\}$$

2.68 $$\text{Minimize } f = x_1^2 + 2x_2^2 + 3x_3^2$$

subject to the constraints

$$g_1 = x_1 - x_2 - 2x_3 \le 12$$
$$g_2 = x_1 + 2x_2 - 3x_3 \le 8$$

using Kuhn–Tucker conditions.

2.69 $$\text{Minimize } f(x_1, \ x_2) = (x_1 - 1)^2 + (x_2 - 5)^2$$

subject to

$$-x_1^2 + x_2 \le 4$$
$$-(x_1 - 2)^2 + x_2 \le 3$$

by **(a)** the graphical method and **(b)** Kuhn–Tucker conditions.

2.70 $$\text{Maximize } f = 8x_1 + 4x_2 + x_1 x_2 - x_1^2 - x_2^2$$

subject to

$$2x_1 + 3x_2 \le 24$$
$$-5x_1 + 12x_2 \le 24$$
$$x_2 \le 5$$

by applying Kuhn–Tucker conditions.

2.71 Consider the following problem:

$$\text{Maximize } f(x) = (x - 1)^2$$

subject to

$$-2 \leq x \leq 4$$

Determine whether the constraint qualification and Kuhn–Tucker conditions are satisfied at the optimum point.

2.72 Consider the following problem:

$$\text{Minimize } f = (x_1 - 1)^2 + (x_2 - 1)^2$$

subject to

$$2x_2 - (1 - x_1)^3 \leq 0$$

$$x_1 \geq 0$$

$$x_2 \geq 0$$

Determine whether the constraint qualification and the Kuhn–Tucker conditions are satisfied at the optimum point.

2.73 Verify whether the following problem is convex:

$$\text{Minimize } f(\mathbf{X}) = -4x_1 + x_1^2 - 2x_1x_2 + 2x_2^2$$

subject to

$$2x_1 + x_2 \leq 6$$

$$x_1 - 4x_2 \leq 0$$

$$x_1 \geq 0, \quad x_2 \geq 0$$

2.74 Check the convexity of the following problems.

(a)
$$\text{Minimize } f(\mathbf{X}) = 2x_1 + 3x_2 - x_1^3 - 2x_2^2$$

subject to

$$x_1 + 3x_2 \leq 6$$

$$5x_1 + 2x_2 \leq 10$$

$$x_1 \geq 0, \quad x_2 \geq 0$$

(b)
$$\text{Minimize } f(\mathbf{X}) = 9x_1^2 - 18x_1x_2 + 13x_1 - 4$$

subject to

$$x_1^2 + x_2^2 + 2x_1 \geq 16$$

2.75 Identify the optimum point among the given design vectors, $\mathbf{X}_1$, $\mathbf{X}_2$, and $\mathbf{X}_3$, by applying the Kuhn–Tlucker conditions to the following problem:

$$\text{Minimize } f(\mathbf{X}) = 100(x_2 - x_1^2)^2 + (1 - x_1)^2$$

subject to

$$x_2^2 - x_1 \geq 0$$

$$x_1^2 - x_2 \geq 0$$

$$-\tfrac{1}{2} \leq x_1 \leq \tfrac{1}{2}, \quad x_2 \leq 1$$

$$\mathbf{X}_1 = \begin{Bmatrix} 0 \\ 0 \end{Bmatrix}, \quad \mathbf{X}_2 = \begin{Bmatrix} 0 \\ -1 \end{Bmatrix}, \quad \mathbf{X}_3 = \begin{Bmatrix} -\tfrac{1}{2} \\ \tfrac{1}{4} \end{Bmatrix}$$

2.76 Consider the following optimization problem:

$$\text{Minimize } f = -x_1^2 - x_2^2 + x_1 x_2 + 7x_1 + 4x_2$$

subject to

$$2x_1 + 3x_2 \leq 24$$

$$-5x_1 + 12x_2 \leq 24$$

$$x_1 \geq 0, \quad x_2 \geq 0, \quad x_2 \leq 4$$

Find a usable feasible direction at each of the following design vectors:

$$\mathbf{X}_1 = \begin{Bmatrix} 1 \\ 1 \end{Bmatrix}, \quad \mathbf{X}_2 = \begin{Bmatrix} 6 \\ 4 \end{Bmatrix}$$

3

Linear Programming I: Simplex Method

3.1 INTRODUCTION

Linear programming is an optimization method applicable for the solution of problems in which the objective function and the constraints appear as linear functions of the decision variables. The constraint equations in a linear programming problem may be in the form of equalities or inequalities. The linear programming type of optimization problem was first recognized in the 1930s by economists while developing methods for the optimal allocation of resources. During World War II the U.S. Air Force sought more effective procedures of allocating resources and turned to linear programming. George B. Dantzig, who was a member of the Air Force group, formulated the general linear programming problem and devised the simplex method of solution in 1947. This has become a significant step in bringing linear programming into wider use. Afterward, much progress was made in the theoretical development and in the practical applications of linear programming. Among all the works, the theoretical contributions made by Kuhn and Tucker had a major impact in the development of the duality theory in LP. The works of Charnes and Cooper were responsible for industrial applications of LP.

Linear programming is considered a revolutionary development that permits us to make optimal decisions in complex situations. At least four Nobel Prizes were awarded for contributions related to linear programming. For example, when the Nobel Prize in Economics was awarded in 1975 jointly to L. V. Kantorovich of the former Soviet Union and T. C. Koopmans of the United States, the citation for the prize mentioned their contributions on the application of LP to the economic problem of allocating resources [3.14]. George Dantzig, the inventor of LP, was awarded the National Medal of Science by President Gerald Ford in 1976.

Although several other methods have been developed over the years for solving LP problems, the simplex method continues to be the most efficient and popular method for solving general LP problems. Among other methods, Karmarkar's method, developed in 1984, has been shown to be up to 50 times as fast as the simplex algorithm of Dantzig. In this chapter we present the theory, development, and applications of the simplex method for solving LP problems. Additional topics, such as the revised simplex method, duality

119

theory, decomposition method, postoptimality analysis, and Karmarkar's method, are considered in Chapter 4.

3.2 APPLICATIONS OF LINEAR PROGRAMMING

The number of applications of linear programming has been so large that it is not possible to describe all of them here. Only the early applications are mentioned here and the exercises at the end of this chapter give additional example applications of linear programming. One of the early industrial applications of linear programming was made in the petroleum refineries. In general, an oil refinery has a choice of buying crude oil from several different sources with differing compositions and at differing prices. It can manufacture different products, such as aviation fuel, diesel fuel, and gasoline, in varying quantities. The constraints may be due to the restrictions on the quantity of the crude oil available from a particular source, the capacity of the refinery to produce a particular product, and so on. A mix of the purchased crude oil and the manufactured products is sought that gives the maximum profit.

The optimal production plan in a manufacturing firm can also be decided using linear programming. Since the sales of a firm fluctuate, the company can have various options. It can build up an inventory of the manufactured products to carry it through the period of peak sales, but this involves an inventory holding cost. It can also pay overtime rates to achieve higher production during periods of higher demand. Finally, the firm need not meet the extra sales demand during the peak sales period, thus losing a potential profit. Linear programming can take into account the various cost and loss factors and arrive at the most profitable production plan.

In the food-processing industry, linear programming has been used to determine the optimal shipping plan for the distribution of a particular product from different manufacturing plants to various warehouses. In the iron and steel industry, linear programming is used to decide the types of products to be made in their rolling mills to maximize the profit. Metalworking industries use linear programming for shop loading and for determining the choice between producing and buying a part. Paper mills use it to decrease the amount of trim losses. The optimal routing of messages in a communication network and the routing of aircraft and ships can also be decided using linear programming.

Linear programming has also been applied to formulate and solve several types of engineering design problems, such as the plastic design of frame structures, as illustrated in the following example.

Example 3.1   In the limit design of steel frames, it is assumed that plastic hinges will be developed at points with peak moments. When a sufficient number of hinges develop, the structure becomes an unstable system referred to as a *collapse mechanism*. Thus a design will be safe if the energy-absorbing capacity of the frame (U) is greater than the energy imparted by the externally applied loads (E) in each of the deformed shapes as indicated by the various collapse mechanisms [3.9].

For the rigid frame shown in Fig. 3.1, plastic moments may develop at the points of peak moments (numbered 1 through 7 in Fig. 3.1). Four possible collapse mechanisms are shown in Fig. 3.2 for this frame. Assuming that the weight is a linear function

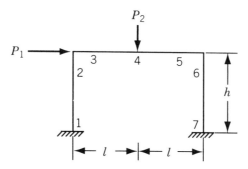

Figure 3.1 Rigid frame.

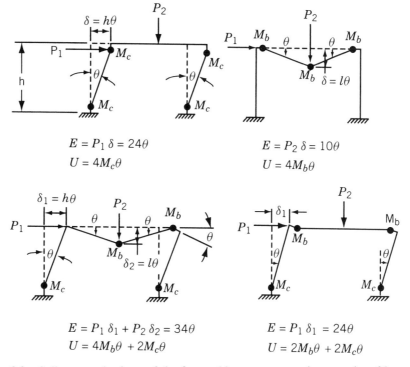

$$E = P_1 \delta = 24\theta$$
$$U = 4M_c\theta$$

$$E = P_2 \delta = 10\theta$$
$$U = 4M_b\theta$$

$$E = P_1 \delta_1 + P_2 \delta_2 = 34\theta$$
$$U = 4M_b\theta + 2M_c\theta$$

$$E = P_1 \delta_1 = 24\theta$$
$$U = 2M_b\theta + 2M_c\theta$$

Figure 3.2 Collapse mechanisms of the frame. M_b, moment carrying capacity of beam; M_c, moment carrying capacity of column [3.9].

of the plastic moment capacities, find the values of the ultimate moment capacities M_b and M_c for minimum weight. Assume that the two columns are identical and that $P_1 = 3$, $P_2 = 1$, $h = 8$, and $l = 10$.

SOLUTION   The objective function can be expressed as

$$f(M_b, M_c) = \text{weight of beam} + \text{weight of columns}$$
$$= \alpha(2lM_b + 2hM_c)$$

where α is a constant indicating the weight per unit length of the member with a unit plastic moment capacity. Since a constant multiplication factor does not affect the result, f can be taken as

$$f = 2lM_b + 2hM_c = 20M_b + 16M_c \tag{E_1}$$

The constraints $(U \geq E)$ from the four collapse mechanisms can be expressed as

$$M_c \geq 6$$
$$M_b \geq 2.5$$
$$2M_b + M_c \geq 17$$
$$M_b + M_c \geq 12 \tag{E_2}$$

3.3   STANDARD FORM OF A LINEAR PROGRAMMING PROBLEM

The general linear programming problem can be stated in the following standard forms:

Scalar Form

$$\text{Minimize } f(x_1, x_2, \ldots, x_n) = c_1 x_1 + c_2 x_2 + \cdots + c_n x_n \tag{3.1a}$$

subject to the constraints

$$\begin{aligned}
a_{11}x_1 + a_{12}x_2 + \cdots + a_{1n}x_n &= b_1 \\
a_{21}x_1 + a_{22}x_2 + \cdots + a_{2n}x_n &= b_2 \\
&\vdots \\
a_{m1}x_1 + a_{m2}x_2 + \cdots + a_{mn}x_n &= b_m
\end{aligned} \tag{3.2a}$$

$$\begin{aligned}
x_1 &\geq 0 \\
x_2 &\geq 0 \\
&\vdots \\
x_n &\geq 0
\end{aligned} \tag{3.3a}$$

where c_j, b_j, and $a_{ij} (i = 1, 2, \ldots, m; j = 1, 2, \ldots, n)$ are known constants, and x_j are the decision variables.

Matrix Form

$$\text{Minimize } f(\mathbf{X}) = \mathbf{c}^T \mathbf{X} \tag{3.1b}$$

subject to the constraints

$$\mathbf{aX} = \mathbf{b} \tag{3.2b}$$
$$\mathbf{X} \geq \mathbf{0} \tag{3.3b}$$

where

$$\mathbf{X} = \begin{Bmatrix} x_1 \\ x_2 \\ \vdots \\ x_n \end{Bmatrix}, \quad \mathbf{b} = \begin{Bmatrix} b_1 \\ b_2 \\ \vdots \\ b_m \end{Bmatrix}, \quad \mathbf{c} = \begin{Bmatrix} c_1 \\ c_2 \\ \vdots \\ c_n \end{Bmatrix},$$

$$\mathbf{a} = \begin{bmatrix} a_{11} & a_{12} & \cdots & a_{1n} \\ a_{21} & a_{22} & \cdots & a_{2n} \\ \vdots & & & \\ a_{m1} & a_{m2} & \cdots & a_{mn} \end{bmatrix}$$

The characteristics of a linear programming problem, stated in standard form, are

1. The objective function is of the minimization type.
2. All the constraints are of the equality type.
3. All the decision variables are nonnegative.

It is now shown that any linear programming problem can be expressed in standard form by using the following transformations.

1. The maximization of a function $f(x_1, x_2, \ldots, x_n)$ is equivalent to the minimization of the negative of the same function. For example, the objective function

$$\text{minimize } f = c_1 x_1 + c_2 x_2 + \cdots + c_n x_n$$

is equivalent to

$$\text{maximize } f' = -f = -c_1 x_1 - c_2 x_2 - \cdots - c_n x_n$$

Consequently, the objective function can be stated in the minimization form in any linear programming problem.

2. In most engineering optimization problems, the decision variables represent some physical dimensions, and hence the variables x_j will be nonnegative. However, a variable may be unrestricted in sign in some problems. In such cases, an unrestricted variable (which can take a positive, negative, or zero value) can be written as the difference of two nonnegative variables. Thus if x_j is unrestricted in sign, it can be written as $x_j = x'_j - x''_j$, where

$$x'_j \geq 0 \quad \text{and} \quad x''_j \geq 0$$

It can be seen that x_j will be negative, zero, or positive, depending on whether x''_j is greater than, equal to, or less than x'_j.

3. If a constraint appears in the form of a "less than or equal to" type of inequality as

$$a_{k1} x_1 + a_{k2} x_2 + \cdots + a_{kn} x_n \leq b_k$$

it can be converted into the equality form by adding a nonnegative slack variable x_{n+1} as follows:

$$a_{k1} x_1 + a_{k2} x_2 + \cdots + a_{kn} x_n + x_{n+1} = b_k$$

Similarly, if the constraint is in the form of a "greater than or equal to" type of inequality as

$$a_{k1}x_1 + a_{k2}x_2 + \cdots + a_{kn}x_n \geq b_k$$

it can be converted into the equality form by subtracting a variable as

$$a_{k1}x_1 + a_{k2}x_2 + \cdots + a_{kn}x_n - x_{n+1} = b_k$$

where x_{n+1} is a nonnegative variable known as a *surplus variable*.

It can be seen that there are m equations in n decision variables in a linear programming problem. We can assume that $m < n$; for if $m > n$, there would be $m - n$ redundant equations that could be eliminated. The case $n = m$ is of no interest, for then there is either a unique solution $\mathbf{X}$ that satisfies Eqs. (3.2) and (3.3) (in which case there can be no optimization) or no solution, in which case the constraints are inconsistent. The case $m < n$ corresponds to an underdetermined set of linear equations, which, if they have one solution, have an infinite number of solutions. The problem of linear programming is to find one of these solutions that satisfies Eqs. (3.2) and (3.3) and yields the minimum of f.

3.4 GEOMETRY OF LINEAR PROGRAMMING PROBLEMS

A linear programming problem with only two variables presents a simple case for which the solution can be obtained by using a rather elementary graphical method. Apart from the solution, the graphical method gives a physical picture of certain geometrical characteristics of linear programming problems. The following example is considered to illustrate the graphical method of solution.

Example 3.2 A manufacturing firm produces two machine parts using lathes, milling machines, and grinding machines. The different machining times required for each part, the machining times available on different machines, and the profit on each machine part are given in the following table.

Type of machine	Machining time required (min)		Maximum time available per week (min)
	Machine part I	Machine part II	
Lathes	10	5	2500
Milling machines	4	10	2000
Grinding machines	1	1.5	450
Profit per unit	$50	$100	

Determine the number of parts I and II to be manufactured per week to maximize the profit.

SOLUTION Let the number of machine parts I and II manufactured per week be denoted by x and y, respectively. The constraints due to the maximum time limitations on the various machines are given by

$$10x + 5y \le 2500 \tag{E_1}$$

$$4x + 10y \le 2000 \tag{E_2}$$

$$x + 1.5y \le 450 \tag{E_3}$$

Since the variables x and y cannot take negative values, we have

$$x \ge 0$$
$$y \ge 0 \tag{E_4}$$

The total profit is given by

$$f(x, y) = 50x + 100y \tag{E_5}$$

Thus the problem is to determine the nonnegative values of x and y that satisfy the constraints stated in Eqs. (E_1) to (E_3) and maximize the objective function given by Eq. (E_5). The inequalities (E_1) to (E_4) can be plotted in the xy plane and the feasible region identified as shown in Fig. 3.3 Our objective is to find at least one point out of the infinite points in the shaded region of Fig. 3.3 that maximizes the profit function (E_5).

The contours of the objective function, f, are defined by the linear equation

$$50x + 100y = k = \text{constant}$$

As k is varied, the objective function line is moved parallel to itself. The maximum value of f is the largest k whose objective function line has at least one point in common with the feasible region. Such a point can be identified as point G in Fig. 3.4. The optimum solution corresponds to a value of $x^* = 187.5$, $y^* = 125.0$ and a profit of $21,875.00.

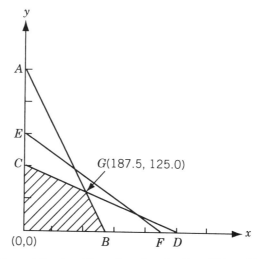

Figure 3.3   Feasible region given by Eqs. (E_1) to (E_4).

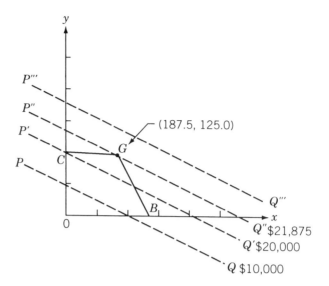

Figure 3.4   Contours of objective function.

In some cases, the optimum solution may not be unique. For example, if the profit rates for the machine parts I and II are $40 and $100 instead of $50 and $100, respectively, the contours of the profit function will be parallel to side CG of the feasible region as shown in Fig. 3.5. In this case, line $P''Q''$, which coincides with the boundary line CG, will correspond to the maximum (feasible) profit. Thus there is no unique optimal solution to the problem and any point between C and G on line $P''Q''$

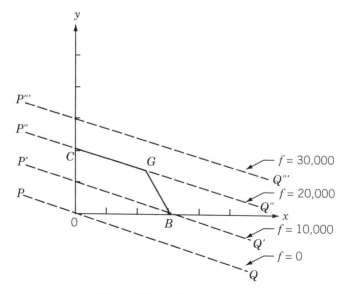

Figure 3.5   Infinite solutions.

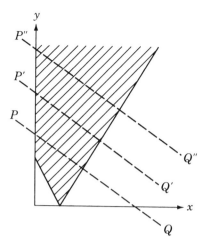

Figure 3.6   Unbounded solution.

can be taken as an optimum solution with a profit value of $20,000. There are three other possibilities. In some problems, the feasible region may not be a closed convex polygon. In such a case, it may happen that the profit level can be increased to an infinitely large value without leaving the feasible region, as shown in Fig. 3.6. In this case the solution of the linear programming problem is said to be unbounded. On the other extreme, the constraint set may be empty in some problems. This could be due to the inconsistency of the constraints; or, sometimes, even though the constraints may be consistent, no point satisfying the constraints may also satisfy the nonnegativity restrictions. The last possible case is when the feasible region consists of a single point. This can occur only if the number of constraints is at least equal to the number of variables. A problem of this kind is of no interest to us since there is only one feasible point and there is nothing to be optimized.

Thus a linear programming problem may have (1) a unique and finite optimum solution, (2) an infinite number of optimal solutions, (3) an unbounded solution, (4) no solution, or (5) a unique feasible point. Assuming that the linear programming problem is properly formulated, the following general geometrical characteristics can be noted from the graphical solution:

1. The feasible region is a convex polygon.[†]
2. The optimum value occurs at an extreme point or vertex of the feasible region.

3.5   DEFINITIONS AND THEOREMS

The geometrical characteristics of a linear programming problem stated in Section 3.4 can be proved mathematically. Some of the more powerful methods of solving linear programming problems take advantage of these characteristics. The terminology used in linear programming and some of the important theorems are presented in this section.

[†]A convex polygon consists of a set of points having the property that the line segment joining any two points in the set is entirely in the convex set. In problems having more than two decision variables, the feasible region is called a *convex polyhedron*, which is defined in the next section.

Definitions

1. *Point in n-dimensional space.* A point $\mathbf{X}$ in an n-dimensional space is characterized by an ordered set of n values or coordinates $(x_1, x_2, \ldots, x_n)$. The coordinates of $\mathbf{X}$ are also called the *components* of $\mathbf{X}$.

2. *Line segment in n dimensions (L).* If the coordinates of two points A and B are given by $x_j^{(1)}$ and $x_j^{(2)}(j = 1, 2, \ldots, n)$, the line segment (L) joining these points is the collection of points $\mathbf{X}(\lambda)$ whose coordinates are given by $x_j = \lambda x_j^{(1)} + (1 - \lambda)x_j^{(2)}$, $j = 1, 2, \ldots, n$, with $0 \leq \lambda \leq 1$. Thus

$$L = \{\mathbf{X} \mid \mathbf{X} = \lambda \mathbf{X}^{(1)} + (1 - \lambda)\mathbf{X}^{(2)}\} \tag{3.4}$$

In one dimension, for example, it is easy to see that the definition is in accordance with out experience (Fig. 3.7):

$$x^{(2)} - x(\lambda) = \lambda[x^{(2)} - x^{(1)}], \quad 0 \leq \lambda \leq 1 \tag{3.5}$$

whence

$$x(\lambda) = \lambda x^{(1)} + (1 - \lambda)x^{(2)}, \quad 0 \leq \lambda \leq 1 \tag{3.6}$$

3. *Hyperplane.* In n-dimensional space, the set of points whose coordinates satisfy a linear equation

$$a_1 x_1 + \cdots + a_n x_n = \mathbf{a}^T \mathbf{X} = b \tag{3.7}$$

is called a hyperplane. A hyperplane, H, is represented as

$$H(\mathbf{a}, b) = \{\mathbf{X} \mid \mathbf{a}^T \mathbf{X} = b\} \tag{3.8}$$

A hyperplane has $n - 1$ dimensions in an n-dimensional space. For example, in three-dimensional space it is a plane, and in two-dimensional space it is a line. The set of points whose coordinates satisfy a linear inequality like $a_1 x_1 + \cdots + a_n x_n \leq b$ is called a *closed half-space*, closed due to the inclusion of an equality sign in the inequality above. A hyperplane partitions the n-dimensional space (E^n) into two closed half-spaces, so that

$$H^+ = \{\mathbf{X} \mid \mathbf{a}^T \mathbf{X} \geq b\} \tag{3.9}$$

$$H^- = \{\mathbf{X} \mid \mathbf{a}^T \mathbf{X} \leq b\} \tag{3.10}$$

This is illustrated in Fig. 3.8 in the case of a two-dimensional space (E^2).

Figure 3.7   Line segment.

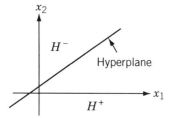

Figure 3.8 Hyperplane in two dimensions.

4. *Convex set.* A convex set is a collection of points such that if $\mathbf{X}^{(1)}$ and $\mathbf{X}^{(2)}$ are any two points in the collection, the line segment joining them is also in the collection. A convex set, S, can be defined mathematically as follows:

$$\text{If } \mathbf{X}^{(1)}, \mathbf{X}^{(2)} \in S, \quad \text{then} \quad \mathbf{X} \in S$$

where

$$\mathbf{X} = \lambda \mathbf{X}^{(1)} + (1 - \lambda)\mathbf{X}^{(2)}, \quad 0 \le \lambda \le 1$$

A set containing only one point is always considered to be convex. Some examples of convex sets in two dimensions are shown shaded in Fig. 3.9. On the other hand, the sets depicted by the shaded region in Fig. 3.10 are not convex. The L-shaped region, for example, is not a convex set because it is possible to find two points a and b in the set such that not all points on the line joining them belong to the set.

5. *Convex polyhedron and convex polytope.* A convex polyhedron is a set of points common to one or more half-spaces. A convex polyhedron that is bounded is called a convex polytope.

   Figure 3.11a and b represents convex polytopes in two and three dimensions, and Fig. 3.11c and d denotes convex polyhedra in two and three dimensions. It

Figure 3.9 Convex sets.

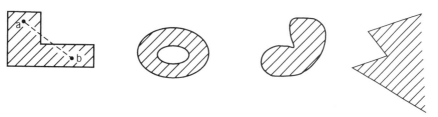

Figure 3.10 Nonconvex sets.

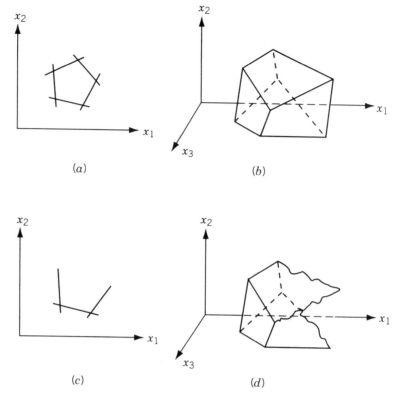

Figure 3.11   Convex polytopes in two and three dimensions (a, b) and convex polyhedra in two and three dimensions (c, d).

can be seen that a convex polygon, shown in Fig. 3.11a and c, can be considered as the intersection of one or more half-planes.

6. *Vertex or extreme point.*   This is a point in the convex set that does not lie on a line segment joining two other points of the set. For example, every point on the circumference of a circle and each corner point of a polygon can be called a vertex or extreme point.

7. *Feasible solution.*   In a linear programming problem, any solution that satisfies the constraints

$$\mathbf{aX} = \mathbf{b} \qquad (3.2)$$

$$\mathbf{X} \geq \mathbf{0} \qquad (3.3)$$

is called a feasible solution.

8. *Basic solution.*   A basic solution is one in which $n - m$ variables are set equal to zero. A basic solution can be obtained by setting $n - m$ variables to zero and solving the constraint Eqs. (3.2) simultaneously.

9. *Basis.*   The collection of variables not set equal to zero to obtain the basic solution is called the basis.

10. *Basic feasible solution.* This is a basic solution that satisfies the nonnegativity conditions of Eq. (3.3).
11. *Nondegenerate basic feasible solution.* This is a basic feasible solution that has got exactly m positive x_i.
12. *Optimal solution.* A feasible solution that optimizes the objective function is called an optimal solution.
13. *Optimal basic solution.* This is a basic feasible solution for which the objective function is optimal.

Theorems. The basic theorems of linear programming can now be stated and proved [†].

Theorem 3.1 The intersection of any number of convex sets is also convex.

Proof: Let the given convex sets be represented as $R_i (i = 1, 2, \ldots, K)$ and their intersection as R, so that[‡]

$$R = \bigcap_{i=1}^{K} R_i$$

If the points $\mathbf{X}^{(1)}, \mathbf{X}^{(2)} \in R$, then from the definition of intersection,

$$\mathbf{X} = \lambda \mathbf{X}^{(1)} + (1 - \lambda)\mathbf{X}^{(2)} \in R_i \quad (i = 1, 2, \ldots, K)$$
$$0 \leq \lambda \leq 1$$

Thus

$$\mathbf{X} \in R = \bigcap_{i=1}^{K} R_i$$

and the theorem is proved. Physically, the theorem states that if there are a number of convex sets represented by $R_1, R_2, \ldots$, the set of points R common to all these sets will also be convex. Figure 3.12 illustrates the meaning of this theorem for the case of two convex sets.

Theorem 3.2 The feasible region of a linear programming problem is convex.

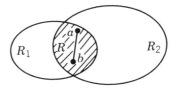

Figure 3.12 Intersection of two convex sets.

[†]The proofs of the theorems are not needed for an understanding of the material presented in subsequent sections.
[‡]The symbol ∩ represents the intersection of sets.

Proof: The feasible region S of a standard linear programming problem is defined as

$$S = \{X \mid aX = b, X \geq 0\} \tag{3.11}$$

Let the points X_1 and X_2 belong to the feasible set S so that

$$aX_1 = b, \qquad X_1 \geq 0 \tag{3.12}$$

$$aX_2 = b, \qquad X_2 \geq 0 \tag{3.13}$$

Multiply Eq. (3.12) by λ and Eq. (3.13) by $(1 - \lambda)$ and add them to obtain

$$a[\lambda X_1 + (1 - \lambda)X_2] = \lambda b + (1 - \lambda)b = b$$

that is,

$$aX_\lambda = b$$

where

$$X_\lambda = \lambda X_1 + (1 - \lambda)X_2$$

Thus the point X_λ satisfies the constraints and if

$$0 \leq \lambda \leq 1, \quad X_\lambda \geq 0$$

Hence the theorem is proved.

Theorem 3.3 Any local minimum solution is global for a linear programming problem.

Proof: In the case of a function of one variable, the minimum (maximum) of a function $f(x)$ is obtained at a value x at which the derivative is zero. This may be a point like $A(x = x_1)$ in Fig. 3.13, where $f(x)$ is only a relative (local) minimum, or a point like $B(x = x_2)$, where $f(x)$ is a global minimum. Any solution that is a local minimum solution is also a global minimum solution for the linear programming problem. To see this, let A be the local minimum solution and assume that it is not a global minimum solution so that there is another point B at which $f_B < f_A$. Let the coordinates of A and B be given by $\{x_1, x_2, \ldots, x_n\}^T$ and $\{y_1, y_2, \ldots, y_n\}^T$, respectively. Then any point $C = \{z_1, z_2, \ldots, z_n\}^T$ that lies on the line segment joining the two points A and B is

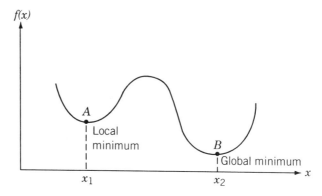

Figure 3.13 Local and global minima.

a feasible solution and $f_C = \lambda f_A + (1 - \lambda) f_B$. In this case, the value of f decreases uniformly from f_A to f_B, and thus all points on the line segment between A and B (including those in the neighborhood of A) have f values less than f_A and correspond to feasible solutions. Hence it is not possible to have a local minimum at A and at the same time another point B such that $f_A > f_B$. This means that for all B, $f_A \leq f_B$, so that f_A is the global minimum value.

The generalized version of this theorem is proved in Appendix A so that it can be applied to nonlinear programming problems also.

Theorem 3.4 Every basic feasible solution is an extreme point of the convex set of feasible solutions.

Theorem 3.5 Let S be a closed, bounded convex polyhedron with $\mathbf{X}_i^e$, $i = 1$ to p, as the set of its extreme points. Then any vector $\mathbf{X} \in S$ can be written as

$$\mathbf{X} = \sum_{i=1}^{p} \lambda_i \mathbf{X}_i^e$$

$$\lambda_i \geq 0$$

$$\sum_{i=1}^{p} \lambda_i = 1$$

Theorem 3.6 Let S be a closed convex polyhedron. Then the minimum of a linear function over S is attained at an extreme point of S.

The proofs of Theorems 3.4 to 3.6 can be found in Ref. [3.1].

3.6 SOLUTION OF A SYSTEM OF LINEAR SIMULTANEOUS EQUATIONS

Before studying the most general method of solving a linear programming problem, it will be useful to review the methods of solving a system of linear equations. Hence in the present section we review some of the elementary concepts of linear equations. Consider the following system of n equations in n unknowns:

$$\begin{aligned}
a_{11}x_1 + a_{12}x_2 + \cdots + a_{1n}x_n &= b_1 \quad (E_1) \\
a_{21}x_1 + a_{22}x_2 + \cdots + a_{2n}x_n &= b_2 \quad (E_2) \\
a_{31}x_1 + a_{32}x_2 + \cdots + a_{3n}x_n &= b_3 \quad (E_3) \\
&\vdots \qquad \vdots \\
a_{n1}x_1 + a_{n2}x_2 + \cdots + a_{nn}x_n &= b_n \quad (E_n)
\end{aligned} \qquad (3.14)$$

Assuming that this set of equations possesses a unique solution, a method of solving the system consists of reducing the equations to a form known as *canonical form*.

It is well known from elementary algebra that the solution of Eqs. (3.14) will not be altered under the following elementary operations: (1) any equation E_r is replaced by the equation kE_r, where k is a nonzero constant, and (2) any equation E_r is replaced by the equation $E_r + kE_s$, where E_s is any other equation of the system. By making use of these elementary operations, the system of Eqs. (3.14) can be reduced to a convenient equivalent form as follows. Let us select some variable x_i and try to eliminate it from all the equations except the jth one (for which a_{ji} is nonzero). This can be accomplished by dividing the jth equation by a_{ji} and subtracting a_{ki} times the result from each of the other equations, $k = 1, 2, \ldots, j-1, j+1, \ldots, n$. The resulting system of equations can be written as

$$a'_{11}x_1 + a'_{12}x_2 + \cdots + a'_{1,i-1}x_{i-1} + 0x_i + a'_{1,i+1}x_{i+1} + \cdots$$
$$+ a'_{1n}x_n = b'_1$$
$$a'_{21}x_1 + a'_{22}x_2 + \cdots + a'_{2,i-1}x_{i-1} + 0x_i + a'_{2,i+1}x_{i+1} + \cdots$$
$$+ a'_{2n}x_n = b'_2$$
$$\vdots$$
$$a'_{j-1,1}x_1 + a'_{j-1,2}x_2 + \cdots + a'_{j-1,i-1} + 0x_i + a'_{j-1,i+1}x_{i+1}$$
$$+ \cdots + a'_{j-1,n}x_n = b'_{j-1}$$
$$a'_{j1}x_1 + a'_{j2}x_2 + \cdots + a'_{j,i-1}x_{i-1} + 1x_i + a'_{j,i+1}x_{i+1}$$
$$+ \cdots + a'_{jn}x_n = b'_j$$
$$a'_{j+1,1}x_1 + a'_{j+1,2}x_2 + \cdots + a'_{j+1,i-1}x_{i-1} + 0x_i + a'_{j+1,i+1}x_{i+1}$$
$$+ \cdots + a'_{j+1,n}x_n = b'_{j+1}$$
$$\vdots$$
$$a'_{n1}x_1 + a'_{n2}x_2 + \cdots + a'_{n,i-1}x_{i-1} + 0x_i + a'_{n,i+1}x_{i+1} + \cdots$$
$$+ a'_{nn}x_n = b'_n \tag{3.15}$$

where the primes indicate that the a'_{ij} and b'_j are changed from the original system. This procedure of eliminating a particular variable from all but one equations is called a *pivot operation*. The system of Eqs. (3.15) produced by the pivot operation have exactly the same solution as the original set of Eqs. (3.14). That is, the vector $\mathbf{X}$ that satisfies Eqs. (3.14) satisfies Eqs. (3.15), and vice versa.

Next time, if we take the system of Eqs. (3.15) and perform a new pivot operation by eliminating x_s, $s \neq i$, in all the equations except the tth equation, $t \neq j$, the zeros or the 1 in the ith column will not be disturbed. The pivotal operations can be repeated by using a different variable and equation each time until the system of Eqs. (3.14) is reduced to the form

$$1x_1 + 0x_2 + 0x_3 + \cdots + 0x_n = b''_1$$
$$0x_1 + 1x_2 + 0x_3 + \cdots + 0x_n = b''_2$$

$$0x_1 + 0x_2 + 1x_3 + \cdots + 0x_n = b_3'' \tag{3.16}$$

$$\vdots$$

$$0x_1 + 0x_2 + 0x_3 + \cdots + 1x_n = b_n''$$

This system of Eqs. (3.16) is said to be in canonical form and has been obtained after carrying out n pivot operations. From the canonical form, the solution vector can be directly obtained as

$$x_i = b_i'', \quad i = 1, 2, \ldots, n \tag{3.17}$$

Since the set of Eqs. (3.16) has been obtained from Eqs. (3.14) only through elementary operations, the system of Eqs. (3.16) is equivalent to the system of Eqs. (3.14). Thus the solution given by Eqs. (3.17) is the desired solution of Eqs. (3.14).

3.7   PIVOTAL REDUCTION OF A GENERAL SYSTEM OF EQUATIONS

Instead of a square system, let us consider a system of m equations in n variables with $n \geq m$. This system of equations is assumed to be consistent so that it will have at least one solution:

$$\begin{aligned}
a_{11}x_1 + a_{12}x_2 + \cdots + a_{1n}x_n &= b_1 \\
a_{21}x_1 + a_{22}x_2 + \cdots + a_{2n}x_n &= b_2 \\
&\vdots \\
a_{m1}x_1 + a_{m2}x_2 + \cdots + a_{mn}x_n &= b_m
\end{aligned} \tag{3.18}$$

The solution vector(s) $\mathbf{X}$ that satisfy Eqs. (3.18) are not evident from the equations. However, it is possible to reduce this system to an equivalent canonical system from which at least one solution can readily be deduced. If pivotal operations with respect to any set of m variables, say, $x_1, x_2, \ldots, x_m$, are carried, the resulting set of equations can be written as follows:

Canonical system with pivotal variables $x_1, x_2, \ldots, x_m$

$$1x_1 + 0x_2 + \cdots + 0x_m + a_{1,m+1}''x_{m+1} + \cdots + a_{1n}''x_n = b_1''$$

$$0x_1 + 1x_2 + \cdots + 0x_m + a_{2,m+1}''x_{m+1} + \cdots + a_{2n}''x_n = b_2'' \tag{3.19}$$

$$\vdots$$

$$0x_1 + 0x_2 + \cdots + 1x_m + a_{m,m+1}''x_{m+1} + \cdots + a_{mn}''x_n = b_m''$$

Pivotal variables	Nonpivotal or independent variables	Constants

---

One special solution that can always be deduced from the system of Eqs. (3.19) is

$$x_i = \begin{cases} b_i'', & i = 1, 2, \ldots, m \\ 0, & i = m+1, \, m+2, \ldots, \, n \end{cases} \qquad (3.20)$$

This solution is called a *basic solution* since the solution vector contains no more than m nonzero terms. The pivotal variables x_i, $i = 1, 2, \ldots, m$, are called the *basic variables* and the other variables x_i, $i = m+1, m+2, \ldots, n$, are called the *nonbasic variables*. Of course, this is not the only solution, but it is the one most readily deduced from Eqs. (3.19). If all b_i'', $i = 1, 2, \ldots, m$, in the solution given by Eqs. (3.20) are nonnegative, it satisfies Eqs. (3.3) in addition to Eqs. (3.2), and hence it can be called a *basic feasible solution*.

It is possible to obtain the other basic solutions from the canonical system of Eqs. (3.19). We can perform an additional pivotal operation on the system after it is in canonical form, by choosing a_{pq}'' (which is nonzero) as the pivot term, $q > m$, and using any row p (among $1, 2, \ldots, m$). The new system will still be in canonical form but with x_q as the pivotal variable in place of x_p. The variable x_p, which was a basic variable in the original canonical form, will no longer be a basic variable in the new canonical form. This new canonical system yields a new basic solution (which may or may not be feasible) similar to that of Eqs. (3.20). It is to be noted that the values of all the basic variables change, in general, as we go from one basic solution to another, but only one zero variable (which is nonbasic in the original canonical form) becomes nonzero (which is basic in the new canonical system), and vice versa.

Example 3.3   Find all the basic solutions corresponding to the system of equations

$$2x_1 + 3x_2 - 2x_3 - 7x_4 = 1 \qquad (I_0)$$

$$x_1 + x_2 + x_3 + 3x_4 = 6 \qquad (II_0)$$

$$x_1 - x_2 + x_3 + 5x_4 = 4 \qquad (III_0)$$

SOLUTION   First we reduce the system of equations into a canonical form with x_1, x_2, and x_3 as basic variables. For this, first we pivot on the element $a_{11} = 2$ to obtain

$$x_1 + \tfrac{3}{2}x_2 - x_3 - \tfrac{7}{2}x_4 = \tfrac{1}{2} \qquad I_1 = \tfrac{1}{2}I_0$$

$$0 - \tfrac{1}{2}x_2 + 2x_3 + \tfrac{13}{2}x_4 = \tfrac{11}{2} \qquad II_1 = II_0 - I_1$$

$$0 - \tfrac{5}{2}x_2 + 2x_3 + \tfrac{17}{2}x_4 = \tfrac{7}{2} \qquad III_1 = III_0 - I_1$$

Then we pivot on $a_{22}' = -\tfrac{1}{2}$, to obtain

$$x_1 + 0 + 5x_3 + 16x_4 = 17 \qquad I_2 = I_1 - \tfrac{3}{2}II_2$$

$$0 + x_2 - 4x_3 - 13x_4 = -11 \qquad II_2 = -2II_1$$

$$0 + 0 - 8x_3 - 24x_4 = -24 \qquad III_2 = III_1 + \tfrac{5}{2}II_2$$

Finally we pivot on a'_{33} to obtain the required canonical form as

$$
\begin{aligned}
x_1 \quad\quad + x_4 &= 2 \quad\quad \text{I}_3 = \text{I}_2 - 5\text{III}_3 \\
x_2 \quad - x_4 &= 1 \quad\quad \text{II}_3 = \text{II}_2 + 4\text{III}_3 \\
x_3 + 3x_4 &= 3 \quad\quad \text{III}_3 = -\tfrac{1}{8}\,\text{III}_2
\end{aligned}
$$

From this canonical form, we can readily write the solution of x_1, x_2, and x_3 in terms of the other variable x_4 as

$$
x_1 = 2 - x_4
$$
$$
x_2 = 1 + x_4
$$
$$
x_3 = 3 - 3x_4
$$

If Eqs. (I_0), (II_0), and (III_0) are the constraints of a linear programming problem, the solution obtained by setting the independent variable equal to zero is called a basic solution. In the present case, the basic solution is given by

$$
x_1 = 2, \quad x_2 = 1, \quad x_3 = 3 \quad \text{(basic variables)}
$$

and $x_4 = 0$ (nonbasic or independent variable). Since this basic solution has all $x_j \geq 0$ ($j = 1, 2, 3, 4$), it is a basic feasible solution.

If we want to move to a neighboring basic solution, we can proceed from the canonical form given by Eqs. (I_3), (II_3), and (III_3). Thus if a canonical form in terms of the variables x_1, x_2, and x_4 is required, we have to bring x_4 into the basis in place of the original basic variable x_3. Hence we pivot on a''_{34} in Eq. (III_3). This gives the desired canonical form as

$$
\begin{aligned}
x_1 \quad\quad - \tfrac{1}{3}x_3 &= 1 \quad\quad \text{I}_4 = \text{I}_3 - \text{III}_4 \\
x_2 \quad + \tfrac{1}{3}x_3 &= 2 \quad\quad \text{II}_4 = \text{II}_3 + \text{III}_4 \\
x_4 + \tfrac{1}{3}x_3 &= 1 \quad\quad \text{III}_4 = \tfrac{1}{3}\text{III}_3
\end{aligned}
$$

This canonical system gives the solution of x_1, x_2, and x_4 in terms of x_3 as

$$
x_1 = 1 + \tfrac{1}{3}x_3
$$
$$
x_2 = 2 - \tfrac{1}{3}x_3
$$
$$
x_4 = 1 - \tfrac{1}{3}x_3
$$

and the corresponding basic solution is given by

$$
x_1 = 1, \quad x_2 = 2, \quad x_4 = 1 \quad \text{(basic variables)}
$$
$$
x_3 = 0 \quad \text{(nonbasic variable)}
$$

This basic solution can also be seen to be a basic feasible solution. If we want to move to the next basic solution with x_1, x_3, and x_4 as basic variables, we have to bring x_3

into the current basis in place of x_2. Thus we have to pivot a''_{23} in Eq. (II$_4$). This leads to the following canonical system:

$$x_1 \quad + x_2 \quad = 3 \qquad I_5 = I_4 + \tfrac{1}{3}II_5$$
$$x_3 \quad + 3x_2 = 6 \qquad II_5 = 3II_4$$
$$x_4 - x_2 \quad = -1 \quad III_5 = III_4 - \tfrac{1}{3}II_5$$

The solution for x_1, x_3, and x_4 is given by

$$x_1 = 3 - x_2$$
$$x_3 = 6 - 3x_2$$
$$x_4 = -1 + x_2$$

from which the basic solution can be obtained as

$$x_1 = 3, \quad x_3 = 6, \quad x_4 = -1 \quad \text{(basic variables)}$$
$$x_2 = 0 \quad \text{(nonbasic variable)}$$

Since all the x_j are not nonnegative, this basic solution is not feasible.

Finally, to obtain the canonical form in terms of the basic variables x_2, x_3, and x_4, we pivot on a''_{12} in Eq. (I$_5$), thereby bringing x_2 into the current basis in place of x_1. This gives

$$x_2 \quad + x_1 \quad = 3 \qquad I_6 = I_5$$
$$x_3 \quad - 3x_1 = -3 \quad II_6 = II_5 - 3I_6$$
$$x_4 + x_1 \quad = 2 \qquad III_6 = III_5 + I_6$$

This canonical form gives the solution for x_2, x_3, and x_4 in terms of x_1 as

$$x_2 = 3 - x_1$$
$$x_3 = -3 + 3x_1$$
$$x_4 = 2 - x_1$$

and the corresponding basic solution is

$$x_2 = 3, \quad x_3 = -3, \quad x_4 = 2 \quad \text{(basic variables)}$$
$$x_1 = 0 \quad \text{(nonbasic variable)}$$

This basic solution can also be seen to be infeasible due to the negative value for x_3.

3.8 MOTIVATION OF THE SIMPLEX METHOD

Given a system in canonical form corresponding to a basic solution, we have seen how to move to a neighboring basic solution by a pivot operation. Thus one way to find the optimal solution of the given linear programming problem is to generate all the basic solutions and pick the one that is feasible and corresponds to the optimal value of the objective function. This can be done because the optimal solution, if one exists, always occurs at an extreme point or vertex of the feasible domain. If there are m equality constraints in n variables with $n \geq m$, a basic solution can be obtained by setting any of the $n - m$ variables equal to zero. The number of basic solutions to be inspected is thus equal to the number of ways in which m variables can be selected from a set of n variables, that is,

$$\binom{n}{m} = \frac{n!}{(n - m)! \, m!}$$

For example, if $n = 10$ and $m = 5$, we have 252 basic solutions, and if $n = 20$ and $m = 10$, we have 184,756 basic solutions. Usually, we do not have to inspect all these basic solutions since many of them will be infeasible. However, for large values of n and m, this is still a very large number to inspect one by one. Hence what we really need is a computational scheme that examines a sequence of basic feasible solutions, each of which corresponds to a lower value of f until a minimum is reached. The simplex method of Dantzig is a powerful scheme for obtaining a basic feasible solution; if the solution is not optimal, the method provides for finding a neighboring basic feasible solution that has a lower or equal value of f. The process is repeated until, in a finite number of steps, an optimum is found.

The first step involved in the simplex method is to construct an auxiliary problem by introducing certain variables known as *artificial variables* into the standard form of the linear programming problem. The primary aim of adding the artificial variables is to bring the resulting auxiliary problem into a canonical form from which its basic feasible solution can be obtained immediately. Starting from this canonical form, the optimal solution of the original linear programming problem is sought in two phases. The first phase is intended to find a basic feasible solution to the original linear programming problem. It consists of a sequence of pivot operations that produces a succession of different canonical forms from which the optimal solution of the auxiliary problem can be found. This also enables us to find a basic feasible solution, if one exists, of the original linear programming problem. The second phase is intended to find the optimal solution of the original linear programming problem. It consists of a second sequence of pivot operations that enables us to move from one basic feasible solution to the next of the original linear programming problem. In this process, the optimal solution of the problem, if one exists, will be identified. The sequence of different canonical forms that is necessary in both the phases of the simplex method is generated according to the simplex algorithm described in the next section. That is, the simplex algorithm forms the main subroutine of the simplex method.

3.9   SIMPLEX ALGORITHM

The starting point of the simplex algorithm is always a set of equations, which includes the objective function along with the equality constraints of the problem in canonical form. Thus the objective of the simplex algorithm is to find the vector $\mathbf{X} \geq 0$ that minimizes the function $f(\mathbf{X})$ and satisfies the equations:

$$
\begin{aligned}
1x_1 + 0x_2 + \cdots + 0x_m + a''_{1,m+1}x_{m+1} + \cdots + a''_{1n}x_n &= b''_1 \\
0x_1 + 1x_2 + \cdots + 0x_m + a''_{2,m+1}x_{m+1} + \cdots + a''_{2n}x_n &= b''_2 \\
&\ \ \vdots \\
0x_1 + 0x_2 + \cdots + 1x_m + a''_{m,m+1}x_{m+1} + \cdots + a''_{mn}x_n &= b''_m \\
0x_1 + 0x_2 + \cdots + 0x_m - f \qquad\qquad\qquad \\
+\, c''_{m+1}x_{m+1} + \cdots + c''_{mn}x_n &= -f''_0
\end{aligned}
\tag{3.21}
$$

where a''_{ij}, c''_j, b''_i, and f''_0 are constants. Notice that $(-f)$ is treated as a basic variable in the canonical form of Eqs. (3.21). The basic solution that can readily be deduced from Eqs. (3.21) is

$$
\begin{aligned}
x_i &= b''_i, \quad i = 1, 2, \ldots, m \\
f &= f^n_0 \\
x_i &= 0, \quad i = m+1, m+2, \ldots, n
\end{aligned}
\tag{3.22}
$$

If the basic solution is also feasible, the values of x_i, $i = 1, 2, \ldots, n$, are nonnegative and hence

$$
b''_i \geq 0, \qquad i = 1, 2, \ldots, m
\tag{3.23}
$$

In phase I of the simplex method, the basic solution corresponding to the canonical form obtained after the introduction of the artificial variables will be feasible for the auxiliary problem. As stated earlier, phase II of the simplex method starts with a basic feasible solution of the original linear programming problem. Hence the initial canonical form at the start of the simplex algorithm will always be a basic feasible solution.

We know from Theorem 3.6 that the optimal solution of a linear programming problem lies at one of the basic feasible solutions. Since the simplex algorithm is intended to move from one basic feasible solution to the other through pivotal operations, before moving to the next basic feasible solution, we have to make sure that the present basic feasible solution is not the optimal solution. By merely glancing at the numbers c''_j, $j = 1, 2, \ldots, n$, we can tell whether or not the present basic feasible solution is optimal. Theorem 3.7 provides a means of identifying the optimal point.

3.9.1 Identifying an Optimal Point

Theorem 3.7 A basic feasible solution is an optimal solution with a minimum objective function value of f''_0 if all the cost coefficients c''_j, $j = m+1, m+2, \ldots, n$, in Eqs. (3.21) are nonnegative.

Proof: From the last row of Eqs. (3.21), we can write that

$$
f''_0 + \sum_{i=m+1}^{n} c''_i x_i = f
\tag{3.24}
$$

Since the variables $x_{m+1}, x_{m+2}, \ldots, x_n$ are presently zero and are constrained to be nonnegative, the only way any one of them can change is to become positive. But if $c_i'' > 0$ for $i = m + 1, m + 2, \ldots, n$, then increasing any x_i cannot decrease the value of the objective function f. Since no change in the nonbasic variables can cause f to decrease, the present solution must be optimal with the optimal value of f equal to f_0''.

A glance over c_i'' can also tell us if there are multiple optima. Let all $c_i'' > 0$, $i = m + 1, m + 2, \ldots, k - 1, k + 1, \ldots, n$, and let $c_k'' = 0$ for some nonbasic variable x_k. Then if the constraints allow that variable to be made positive (from its present value of zero), no change in f results, and there are multiple optima. It is possible, however, that the variable may not be allowed by the constraints to become positive; this may occur in the case of degenerate solutions. Thus as a corollary to the discussion above, we can state that a basic feasible solution is the unique optimal feasible solution if $c_j'' > 0$ for all nonbasic variables x_j, $j = m + 1, m + 2, \ldots, n$. If, after testing for optimality, the current basic feasible solution is found to be nonoptimal, an improved basic solution is obtained from the present canonical form as follows.

3.9.2 Improving a Nonoptimal Basic Feasible Solution

From the last row of Eqs. (3.21), we can write the objective function as

$$f = f_0'' + \sum_{i=1}^{m} c_i'' x_i + \sum_{j=m+1}^{n} c_j'' x_j \tag{3.25}$$

$$= f_0'' \quad \text{for the solution given by Eqs. (3.22)}$$

If at least one c_j'' is negative, the value of f can be reduced by making the corresponding $x_j > 0$. In other words, the nonbasic variable x_j, for which the cost coefficient c_j'' is negative, is to be made a basic variable in order to reduce the value of the objective function. At the same time, due to the pivotal operation, one of the current basic variables will become nonbasic and hence the values of the new basic variables are to be adjusted in order to bring the value of f less than f_0''. If there are more than one $c_j'' < 0$, the index s of the nonbasic variable x_s which is to be made basic is chosen such that

$$c_s'' = \text{minimum } c_j'' < 0 \tag{3.26}$$

Although this may not lead to the greatest possible decrease in f (since it may not be possible to increase x_s very far), this is intuitively at least a good rule for choosing the variable to become basic. It is the one generally used in practice because it is simple and it usually leads to fewer iterations than just choosing any $c_j'' < 0$. If there is a tie-in applying Eq. (3.26), (i.e., if more than one c_j'' has the same minimum value), we select one of them arbitrarily as c_s''.

Having decided on the variable x_s to become basic, we increase it from zero, holding all other nonbasic variables zero, and observe the effect on the current basic variables. From Eqs. (3.21), we can obtain

$$x_1 = b_1'' - a_{1s}'' x_s, \quad b_1'' \geq 0$$
$$x_2 = b_2'' - a_{2s}'' x_s, \quad b_2'' \geq 0 \tag{3.27}$$

$$\vdots$$

$$x_m = b_m'' - a_{ms}'' x_s, \qquad b_m'' \geq 0$$

$$f = f_0'' + c_s'' x_s, \qquad c_s'' < 0 \tag{3.28}$$

Since $c_s'' < 0$, Eq. (3.28) suggests that the value of x_s should be made as large as possible in order to reduce the value of f as much as possible. However, in the process of increasing the value of x_s, some of the variables $x_i (i = 1, 2, \ldots, m)$ in Eqs. (3.27) may become negative. It can be seen that if all the coefficients $a_{is}'' \leq 0$, $i = 1, 2, \ldots, m$, then x_s can be made infinitely large without making any $x_i < 0$, $i = 1, 2, \ldots, m$. In such a case, the minimum value of f is minus infinity and the linear programming problem is said to have an *unbounded solution*.

On the other hand, if at least one a_{is}'' is positive, the maximum value that x_s can take without making x_i negative is b_i''/a_{is}''. If there are more than one $a_{is}'' > 0$, the largest value x_s^* that x_s can take is given by the minimum of the ratios b_i''/a_{is}'' for which $a_{is}'' > 0$. Thus

$$x_s^* = \frac{b_r''}{a_{rs}''} = \underset{a_{is}'' > 0}{\text{minimum}} \left(\frac{b_i''}{a_{is}''} \right) \tag{3.29}$$

The choice of r in the case of a tie, assuming that all $b_i'' > 0$, is arbitrary. If any b_i'' for which $a_{is}'' > 0$ is zero in Eqs. (3.27), x_s cannot be increased by any amount. Such a solution is called a *degenerate solution*.

In the case of a nondegenerate basic feasible solution, a new basic feasible solution can be constructed with a lower value of the objective function as follows. By substituting the value of x_s^* given by Eq. (3.29) into Eqs. (3.27) and (3.28), we obtain

$$x_s = x_s^*$$

$$x_i = b_i'' - a_{is}'' x_s^* \geq 0, \quad i = 1, 2, \ldots, m \quad \text{and} \quad i \neq r \tag{3.30}$$

$$x_r = 0$$

$$x_j = 0, \quad j = m + 1, m + 2, \ldots, n \quad \text{and} \quad j \neq s$$

$$f = f_0'' + c_s'' x_s^* \leq f_0'' \tag{3.31}$$

which can readily be seen to be a feasible solution different from the previous one. Since $a_{rs}'' > 0$ in Eq. (3.29), a single pivot operation on the element a_{rs}'' in the system of Eqs. (3.21) will lead to a new canonical form from which the basic feasible solution of Eqs. (3.30) can easily be deduced. Also, Eq. (3.31) shows that this basic feasible solution corresponds to a lower objective function value compared to that of Eqs. (3.22). This basic feasible solution can again be tested for optimality by seeing whether all $c_i'' > 0$ in the new canonical form. If the solution is not optimal, the entire procedure of moving to another basic feasible solution from the present one has to be repeated. In the simplex algorithm, this procedure is repeated in an iterative manner until the algorithm finds either (1) a class of feasible solutions for which $f \to -\infty$ or (2) an optimal basic feasible solution with all $c_i'' \geq 0$, $i = 1, 2, \ldots, n$. Since there are only a finite number of ways to choose a set of m basic variables out of n variables, the iterative process of the simplex algorithm will terminate in a finite number of cycles. The iterative process of the simplex algorithm is shown as a flowchart in Fig. 3.14.

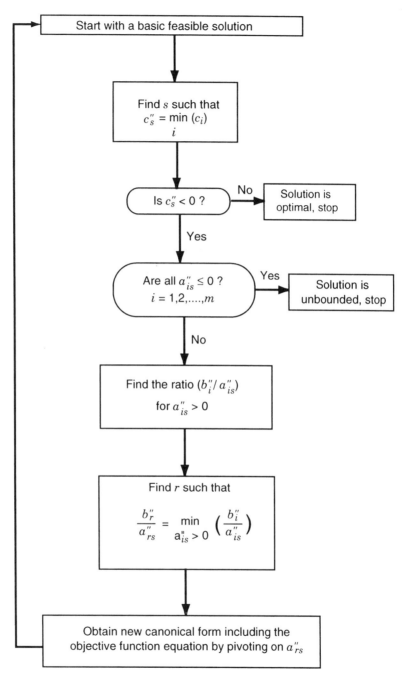

Figure 3.14   Flowchart for finding the optimal solution by the simplex algorithm.

Example 3.4

$$\text{Maximize } F = x_1 + 2x_2 + x_3$$

subject to

$$2x_1 + x_2 - x_3 \leq 2$$
$$-2x_1 + x_2 - 5x_3 \geq -6$$
$$4x_1 + x_2 + x_3 \leq 6$$
$$x_i \geq 0, \quad i = 1, 2, 3$$

SOLUTION   We first change the sign of the objective function to convert it to a minimization problem and the signs of the inequalities (where necessary) so as to obtain nonnegative values of b_i (to see whether an initial basic feasible solution can be obtained readily). The resulting problem can be stated as

$$\text{Minimize } f = -x_1 - 2x_2 - x_3$$

subject to

$$2x_1 + x_2 - x_3 \leq 2$$
$$2x_1 - x_2 + 5x_3 \leq 6$$
$$4x_1 + x_2 + x_3 \leq 6$$
$$x_i \geq 0, \quad i = 1 \text{ to } 3$$

By introducing the slack variables $x_4 \geq 0$, $x_5 \geq 0$, and $x_6 \geq 0$, the system of equations can be stated in canonical form as

$$
\begin{aligned}
2x_1 + x_2 - x_3 + x_4 &= 2 \\
2x_1 - x_2 + 5x_3 + x_5 &= 6 \\
4x_1 + x_2 + x_3 + x_6 &= 6 \\
-x_1 - 2x_2 - x_3 \qquad\quad -f &= 0
\end{aligned}
\qquad (\text{E}_1)
$$

where x_4, x_5, x_6, and $-f$ can be treated as basic variables. The basic solution corresponding to Eqs. (E$_1$) is given by

$$x_4 = 2, \quad x_5 = 6, \quad x_6 = 6 \quad \text{(basic variables)}$$

$$x_1 = x_2 = x_3 = 0 \quad \text{(nonbasic variables)} \qquad (\text{E}_2)$$

$$f = 0$$

which can be seen to be feasible.

Since the cost coefficients corresponding to nonbasic variables in Eqs. (E$_1$) are negative ($c_1'' = -1$, $c_2'' = -2$, $c_3'' = -1$), the present solution given by Eqs. (E$_2$) is not optimum. To improve the present basic feasible solution, we first decide the variable (x_s) to be brought into the basis as

$$c_s'' = \min(c_j'' < 0) = c_2'' = -2$$

Thus x_2 enters the next basic set. To obtain the new canonical form, we select the pivot element a_{rs}'' such that

$$\frac{b_r''}{a_{rs}''} = \min_{a_{is}'' > 0} \left(\frac{b_i''}{a_{is}''} \right)$$

In the present case, $s = 2$ and a_{12}'' and a_{32}'' are ≥ 0. Since $b_1''/a_{12}'' = 2/1$ and $b_3''/a_{32}'' = 6/1$, $x_r = x_1$. By pivoting an a_{12}'', the new system of equations can be obtained as

$$
\begin{aligned}
2x_1 + 1x_2 - x_3 + x_4 &= 2 \\
4x_1 + 0x_2 + 4x_3 + x_4 + x_5 &= 8 \\
2x_1 + 0x_2 + 2x_3 - x_4 + x_6 &= 4 \\
3x_1 + 0x_2 - 3x_3 + 2x_4 \quad - f &= 4
\end{aligned}
\qquad \text{(E}_3\text{)}
$$

The basic feasible solution corresponding to this canonical form is

$$
\begin{aligned}
x_2 = 2, \quad x_5 = 8, \quad x_6 &= 4 \quad \text{(basic variables)} \\
x_1 = x_3 = x_4 &= 0 \quad \text{(nonbasic variables)} \\
f &= -4
\end{aligned}
\qquad \text{(E}_4\text{)}
$$

Since $c_3'' = -3$, the present solution is not optimum. As $c_s'' = \min(c_i'' < 0) = c_3''$, $x_s = x_3$ enters the next basis.

To find the pivot element a_{rs}'', we find the ratios b_i''/a_{is}'' for $a_{is}'' > 0$. In Eqs. (E$_3$), only a_{23}'' and a_{33}'' are > 0, and hence

$$\frac{b_2''}{a_{23}''} = \frac{8}{4} \quad \text{and} \quad \frac{b_3''}{a_{33}''} = \frac{4}{2}$$

Since both these ratios are same, we arbitrarily select a_{23}'' as the pivot element. Pivoting on a_{23}'' gives the following canonical system of equations:

$$
\begin{aligned}
3x_1 + 1x_2 + 0x_3 + \tfrac{5}{4}x_4 + \tfrac{1}{4}x_5 &= 4 \\
1x_1 + 0x_2 + 1x_3 + \tfrac{1}{4}x_4 + \tfrac{1}{4}x_5 &= 2 \\
0x_1 + 0x_2 + 0x_3 - \tfrac{3}{2}x_4 - \tfrac{1}{2}x_5 + x_6 &= 0 \\
6x_1 + 0x_2 + 0x_3 + \tfrac{11}{4}x_4 + \tfrac{3}{4}x_5 - f &= 10
\end{aligned}
\qquad \text{(E}_5\text{)}
$$

The basic feasible solution corresponding to this canonical system is given by

$$
\begin{aligned}
x_2 = 4, \quad x_3 = 2, \quad x_6 &= 0 \quad \text{(basic variables)} \\
x_1 = x_4 = x_5 &= 0 \quad \text{(nonbasic variables)} \\
f &= -10
\end{aligned}
\qquad \text{(E}_6\text{)}
$$

Since all c_i'' are ≥ 0 in the present canonical form, the solution given in (E$_6$) will be optimum. Usually, starting with Eqs. (E$_1$), all the computations are done in a tableau form as shown below:

Basic variables	x_1	x_2	x_3	x_4	x_5	x_6	$-f$	b_i''	b_i''/a_{is}'' for $a_{is}'' > 0$
x_4	2	1 (Pivot element)	-1	1	0	0	0	2	2 ← Smaller one (x_4 drops from next basis)
x_5	2	-1	5	0	1	0	0	6	
x_6	4	1	1	0	0	1	0	6	6
$-f$	-1	-2 ↑	-1	0	0	0	1	0	

Most negative c_i'' (x_2 enters next basis)

Result of pivoting:

x_2	2	1	-1	1	0	0	0	2	
x_5	4	0	4 (Pivot element)	1	1	0	0	8	2 (Select this arbitrarily. x_5 drops from next basis)
x_6	2	0	2	-1	0	1	0	4	2
$-f$	3	0	-3	2	0	0	1	4	

↑
Most negative c_i'' (x_3 enters the next basis)

Result of pivoting:

x_2	3	1	0	$\frac{5}{4}$	$\frac{1}{4}$	0	0	4	
x_3	1	0	1	$\frac{1}{4}$	$\frac{1}{4}$	0	0	2	
x_6	0	0	0	$-\frac{3}{2}$	$-\frac{1}{2}$	1	0	0	
$-f$	6	0	0	$\frac{11}{4}$	$\frac{3}{4}$	0	1	10	

All c_i'' are ≥ 0 and hence the present solution is optimum.

Example 3.5 Unbounded Solution

$$\text{Minimize } f = -3x_1 - 2x_2$$

subject to

$$x_1 - x_2 \leq 1$$
$$3x_1 - 2x_2 \leq 6$$
$$x_1 \geq 0, \quad x_2 \geq 0$$

SOLUTION   Introducing the slack variables $x_3 \geq 0$ and $x_4 \geq 0$, the given system of equations can be written in canonical form as

$$
\begin{aligned}
x_1 - x_2 + x_3 &= 1 \\
3x_1 - 2x_2 + x_4 &= 6 \\
-3x_1 - 2x_2 - f &= 0
\end{aligned}
\qquad (\text{E}_1)
$$

The basic feasible solution corresponding to this canonical form is given by

$$
x_3 = 1, \quad x_4 = 6 \quad \text{(basic variables)}
$$
$$
x_1 = x_2 = 0 \quad \text{(nonbasic variables)} \qquad (\text{E}_2)
$$
$$
f = 0
$$

Since the cost coefficients corresponding to the nonbasic variables are negative, the solution given by Eq. (E$_2$) is not optimum. Hence the simplex procedure is applied to the canonical system of Eqs. (E$_1$) starting from the solution, Eqs. (E$_2$). The computations are done in tableau form as shown below:

Basic variables	Variables				$-f$	b_i''	b_i''/a_{is}'' for $a_{is}'' > 0$
	x_1	x_2	x_3	x_4			
x_3	$\boxed{1}$ Pivot element	-1	1	0	0	1	$1 \leftarrow$ Smaller value (x_3 leaves the basis)
x_4	3	-2	0	1	0	6	2
$-f$	-3	-2	0	0	1	0	

$$\uparrow$$
Most negative c_i'' (x_1 enters the next basis)

Result of pivoting:

x_1	1	-1	1	0	0	1	
x_4	0	$\boxed{1}$ Pivot element	-3	1	0	3	3 (x_4 leaves the basis)
$-f$	0	-5	3	0	1	3	

$$\uparrow$$
Most negative c_i'' (x_2 enters the next basis)

Result of pivoting:

x_1	1	0	-2	1	0	4	Both a_{is}'' are negative (i.e., no variable leaves the basis)
x_2	0	1	-3	1	0	3	
$-f$	0	0	-12	5	1	18	

$$\uparrow$$
Most negative c_i'' (x_3 enters the basis)

At this stage we notice that x_3 has the most negative cost coefficient and hence it should be brought into the next basis. However, since all the coefficients a''_{i3} are negative, the value of f can be decreased indefinitely without violating any of the constraints if we bring x_3 into the basis. Hence the problem has no bounded solution. In general, if all the coefficients of the entering variable $x_s(a''_{is})$ have negative or zero values at any iteration, we can conclude that the problem has an unbounded solution.

Example 3.6 Infinite Number of Solutions   To demonstrate how a problem having infinite number of solutions can be solved, Example 3.2 is again considered with a modified objective function:

$$\text{Minimize } f = -40x_1 - 100x_2$$

subject to

$$10x_1 + 5x_2 \leq 2500$$
$$4x_1 + 10x_2 \leq 2000$$
$$2x_1 + 3x_2 \leq 900$$
$$x_1 \geq 0, \quad x_2 \geq 0$$

SOLUTION   By adding the slack variables $x_3 \geq 0$, $x_4 \geq 0$ and $x_5 \geq 0$, the equations can be written in canonical form as follows:

$$10x_1 + 5x_2 + x_3 \qquad\qquad = 2500$$
$$4x_1 + 10x_2 \qquad + x_4 \qquad = 2000$$
$$2x_1 + 3x_2 \qquad\qquad + x_5 \quad = 900$$
$$-40x_1 - 100x_2 \qquad\qquad\quad - f = 0$$

The computations can be done in tableau form as shown below:

Basic variables	Variables x_1	x_2	x_3	x_4	x_5	$-f$	b''_i	b''_i/a''_{is} for $a''_{is} > 0$
x_3	10	5	1	0	0	0	2,500	500
x_4	4	[10] Pivot element	0	1	0	0	2,000	200 ← Smaller value (x_4 leaves the basis)
x_5	2	3	0	0	1	0	900	300
$-f$	-40	-100	0	0	0	1	0	

↑
Most negative c''_i (x_2 enters the basis)

Result of pivoting:

x_3	8	0	1	$-\frac{1}{2}$	0	0	1,500
x_2	$\frac{4}{10}$	1	0	$\frac{1}{10}$	0	0	200
x_5	$\frac{8}{10}$	0	0	$-\frac{3}{10}$	1	0	300
$-f$	0	0	0	10	0	1	20,000

Since all $c_i'' \geq 0$, the present solution is optimum. The optimum values are given by

$$x_2 = 200, \quad x_3 = 1500, \quad x_5 = 300 \quad \text{(basic variables)}$$

$$x_1 = x_4 = 0 \quad \text{(nonbasic variables)}$$

$$f_{\min} = -20,000$$

Important note: It can be observed from the last row of the preceding tableau that the cost coefficient corresponding to the nonbasic variable $x_1 (c_1'')$ is zero. This is an indication that an alternative solution exists. Here x_1 can be brought into the basis and the resulting new solution will also be an optimal basic feasible solution. For example, introducing x_1 into the basis in place of x_3 (i.e., by pivoting on a_{13}''), we obtain the new canonical system of equations as shown in the following tableau:

Basic variables	Variables					$-f$	b_i''	b_i''/a_{is}'' for $a_{is}'' > 0$
	x_1	x_2	x_3	x_4	x_5			
x_1	1	0	$\frac{1}{8}$	$-\frac{1}{16}$	0	0	$\frac{1500}{8}$	
x_2	0	1	$-\frac{1}{20}$	$\frac{1}{8}$	0	0	125	
x_5	0	0	$-\frac{1}{10}$	$-\frac{1}{4}$	1	0	150	
$-f$	0	0	0	10	0	1	20,000	

The solution corresponding to this canonical form is given by

$$x_1 = \frac{1500}{8}, \quad x_2 = 125, \quad x_5 = 150 \quad \text{(basic variables)}$$

$$x_3 = x_4 = 0 \quad \text{(nonbasic variables)}$$

$$f_{\min} = -20,000$$

Thus the value of f has not changed compared to the preceding value since x_1 has a zero cost coefficient in the last row of the preceding tableau. Once two basic (optimal) feasible solutions, namely,

$$\mathbf{X}_1 = \begin{Bmatrix} 0 \\ 200 \\ 1500 \\ 0 \\ 300 \end{Bmatrix} \quad \text{and} \quad \mathbf{X}_2 = \begin{Bmatrix} \frac{1500}{8} \\ 125 \\ 0 \\ 0 \\ 150 \end{Bmatrix}$$

are known, an infinite number of nonbasic (optimal) feasible solutions can be obtained by taking any weighted average of the two solutions as

$$\mathbf{X}^* = \lambda \mathbf{X}_1 + (1 - \lambda)\mathbf{X}_2$$

$$\mathbf{X}^* = \begin{Bmatrix} x_1^* \\ x_2^* \\ x_3^* \\ x_4^* \\ x_5^* \end{Bmatrix} = \begin{Bmatrix} (1-\lambda)\frac{1500}{8} \\ 200\lambda + (1-\lambda)125 \\ 1500\lambda \\ 0 \\ 300\lambda + (1-\lambda)150 \end{Bmatrix} = \begin{Bmatrix} (1-\lambda)\frac{1500}{8} \\ 125 + 75\lambda \\ 1500\lambda \\ 0 \\ 150 + 150\lambda \end{Bmatrix}$$

$$0 \le \lambda \le 1$$

It can be verified that the solution $\mathbf{X}^*$ will always give the same value of $-20,000$ for f for all $0 \le \lambda \le 1$.

3.10 TWO PHASES OF THE SIMPLEX METHOD

The problem is to find nonnegative values for the variables $x_1, x_2, \ldots, x_n$ that satisfy the equations

$$\begin{aligned} a_{11}x_1 + a_{12}x_2 + \cdots + a_{1n}x_n &= b_1 \\ a_{21}x_1 + a_{22}x_2 + \cdots + a_{2n}x_n &= b_2 \\ &\vdots \\ a_{m1}x_1 + a_{m2}x_2 + \cdots + a_{mn}x_n &= b_m \end{aligned} \tag{3.32}$$

and minimize the objective function given by

$$c_1x_1 + c_2x_2 + \cdots + c_nx_n = f \tag{3.33}$$

The general problems encountered in solving this problem are

1. An initial feasible canonical form may not be readily available. This is the case when the linear programming problem does not have slack variables for some of the equations or when the slack variables have negative coefficients.
2. The problem may have redundancies and/or inconsistencies, and may not be solvable in nonnegative numbers.

The two-phase simplex method can be used to solve the problem.

Phase I of the simplex method uses the simplex algorithm itself to find whether the linear programming problem has a feasible solution. If a feasible solution exists, it provides a basic feasible solution in canonical form ready to initiate phase II of the method. Phase II, in turn, uses the simplex algorithm to find whether the problem has a bounded optimum. If a bounded optimum exists, it finds the basic feasible solution that is optimal. The simplex method is described in the following steps.

1. Arrange the original system of Eqs. (3.32) so that all constant terms b_i are positive or zero by changing, where necessary, the signs on both sides of any of the equations.

2. Introduce to this system a set of artificial variables $y_1, y_2, \ldots, y_m$ (which serve as basic variables in phase I), where each $y_i \geq 0$, so that it becomes

$$
\begin{aligned}
a_{11}x_1 + a_{12}x_2 + \cdots + a_{1n}x_n + y_1 \qquad\qquad &= b_1 \\
a_{21}x_1 + a_{22}x_2 + \cdots + a_{2n}x_n \qquad + y_2 \qquad &= b_2 \\
&\vdots \\
a_{m1}x_1 + a_{m2}x_2 + \cdots + a_{mn}x_n \qquad\quad + y_m &= b_m \\
b_i &\geq 0
\end{aligned}
\tag{3.34}
$$

Note that in Eqs. (3.34), for a particular i, the a_{ij}'s and the b_i may be the negative of what they were in Eq. (3.32) because of step 1.

The objective function of Eq. (3.33) can be written as

$$
c_1x_1 + c_2x_2 + \cdots + c_nx_n + (-f) = 0
\tag{3.35}
$$

3. *Phase I of the method*. Define a quantity w as the sum of the artificial variables

$$
w = y_1 + y_2 + \cdots + y_m
\tag{3.36}
$$

and use the simplex algorithm to find $x_i \geq 0$ ($i = 1, 2, \ldots, n$) and $y_i \geq 0$ ($i = 1, 2, \ldots, m$) which minimize w and satisfy Eqs. (3.34) and (3.35). Consequently, consider the array

$$
\begin{aligned}
a_{11}x_1 + a_{12}x_2 + \cdots + a_{1n}x_n + y_1 \qquad\qquad\qquad &= b_1 \\
a_{21}x_1 + a_{22}x_2 + \cdots + a_{2n}x_n \qquad + y_2 \qquad\qquad &= b_2 \\
\vdots \qquad\qquad\qquad\qquad &\;\;\vdots \\
a_{m1}x_1 + a_{m2}x_2 + \cdots + a_{mn}x_n \qquad\quad + y_m \qquad &= b_m \\
c_1x_1 + c_2x_2 + \cdots + c_nx_n \qquad\qquad\quad + (-f) &= 0 \\
y_1 + y_2 + \cdots + y_m \qquad\quad + (-w) &= 0
\end{aligned}
\tag{3.37}
$$

This array is not in canonical form; however, it can be rewritten as a canonical system with basic variables $y_1, y_2, \ldots, y_m, -f$, and $-w$ by subtracting the sum of the first m equations from the last to obtain the new system

$$
\begin{aligned}
a_{11}x_1 + a_{12}x_2 + \cdots + a_{1n}x_n + y_1 \qquad\qquad\qquad &= b_1 \\
a_{21}x_1 + a_{22}x_2 + \cdots + a_{2n}x_n \qquad + y_2 \qquad\qquad &= b_2 \\
\vdots \qquad\qquad\qquad\qquad &\;\;\vdots \\
a_{m1}x_1 + a_{m2}x_2 + \cdots + a_{mn}x_n \qquad\quad + y_m \qquad &= b_m \\
c_1x_1 + c_2x_2 + \cdots + c_nx_n \qquad\qquad\quad + (-f) &= 0 \\
d_1x_1 + d_2x_2 + \cdots + d_nx_n \qquad\qquad\quad + (-w) &= -w_0
\end{aligned}
\tag{3.38}
$$

where

$$d_i = -(a_{1i} + a_{2i} + \cdots + a_{mi}), \quad i = 1, 2, \ldots, n \tag{3.39}$$

$$-w_0 = -(b_1 + b_2 + \cdots + b_m) \tag{3.40}$$

Equations (3.38) provide the initial basic feasible solution that is necessary for starting phase I.

4. In Eq. (3.37), the expression of w, in terms of the artificial variables $y_1, y_2, \ldots, y_m$ is known as the *infeasibility form*. w has the property that if as a result of phase I, with a minimum of $w > 0$, no feasible solution exists for the original linear programming problem stated in Eqs. (3.32) and (3.33), and thus the procedure is terminated. On the other hand, if the minimum of $w = 0$, the resulting array will be in canonical form and hence initiate phase II by eliminating the w equation as well as the columns corresponding to each of the artificial variables $y_1, y_2, \ldots, y_m$ from the array.

5. *Phase II of the method*. Apply the simplex algorithm to the adjusted canonical system at the end of phase I to obtain a solution, if a finite one exists, which optimizes the value of f.

The flowchart for the two-phase simplex method is given in Fig. 3.15.

Example 3.7

$$\text{Minimize } f = 2x_1 + 3x_2 + 2x_3 - x_4 + x_5$$

subject to the constraints

$$3x_1 - 3x_2 + 4x_3 + 2x_4 - x_5 = 0$$
$$x_1 + x_2 + x_3 + 3x_4 + x_5 = 2$$
$$x_i \geq 0, \quad i = 1 \text{ to } 5$$

SOLUTION

Step 1 As the constants on the right-hand side of the constraints are already nonnegative, the application of step 1 is unnecessary.

Step 2 Introducing the artificial variables $y_1 \geq 0$ and $y_2 \geq 0$, the equations can be written as follows:

$$\begin{aligned}
3x_1 - 3x_2 + 4x_3 + 2x_4 - x_5 + y_1 &= 0 \\
x_1 + x_2 + x_3 + 3x_4 + x_5 \quad\quad + y_2 &= 2 \\
2x_1 + 3x_2 + 2x_3 - x_4 + x_5 \quad\quad\quad\quad - f &= 0
\end{aligned} \tag{E_1}$$

Step 3 By defining the infeasibility form w as

$$w = y_1 + y_2$$

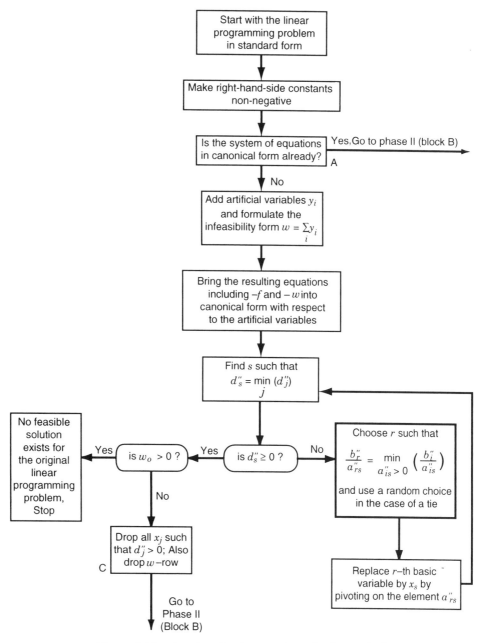

Figure 3.15   Flowchart for the two-phase simplex method.

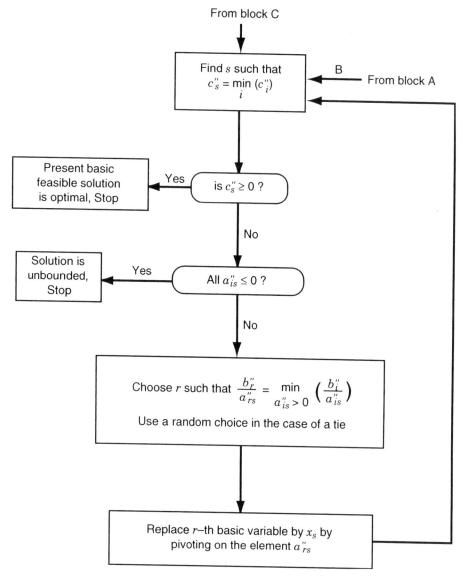

Figure 3.15   (*continued*)

the complete array of equations can be written as

$$3x_1 - 3x_2 + 4x_3 + 2x_4 - x_5 + y_1 \qquad\qquad = 0$$
$$x_1 + x_2 + x_3 + 3x_4 + x_5 \qquad\quad + y_2 \quad\; = 2$$
$$2x_1 + 3x_2 + 2x_3 - x_4 + x_5 \qquad\qquad - f \;\; = 0 \qquad (E_2)$$
$$y_1 + y_2 - w = 0$$

This array can be rewritten as a canonical system with basic variables as y_1, y_2, $-f$, and $-w$ by subtracting the sum of the first two equations of (E$_2$) from the last equation of (E$_2$). Thus the last equation of (E$_2$) becomes

$$-4x_1 + 2x_2 - 5x_3 - 5x_4 + 0x_5 - w = -2 \qquad \text{(E}_3\text{)}$$

Since this canonical system [first three equations of (E$_2$), and (E$_3$)] provides an initial basic feasible solution, phase I of the simplex method can be started. The phase I computations are shown below in tableau form.

Basic variables	Admissible variables					Artificial variables		b_i''	Value of b_i''/a_{is}'' for $a_{is}'' > 0$	
	x_1	x_2	x_3	x_4	x_5	y_1	y_2			
y_1	3	-3	4	2	-1	1	0	0	0	$\leftarrow$ Smaller value
				Pivot element						(y_1 drops from next basis)
y_2	1	1	1	3	1	0	1	2	$\frac{2}{3}$	
$-f$	2	3	2	-1	1	0	0	0		
$-w$	-4	2	-5	-5	0	0	0	-2		
			$\uparrow$	$\uparrow$						
			Most negative							

Since there is a tie between d_3'' and d_4'', d_4'' is selected arbitrarily as the most negative d_i'' for pivoting (x_4 enters the next basis).
Result of pivoting:

	x_1	x_2	x_3	x_4	x_5	y_1	y_2			
x_4	$\frac{3}{2}$	$-\frac{3}{2}$	2	1	$-\frac{1}{2}$	$\frac{1}{2}$	0	0		
y_2	$-\frac{7}{2}$	$\boxed{\frac{11}{2}}$	-5	0	$\frac{5}{2}$	$-\frac{3}{2}$	1	2	$\frac{1}{11}$	$\leftarrow$ y_2 drops from next basis
		Pivot element								
$-f$	$\frac{7}{2}$	$\frac{3}{2}$	4	0	$\frac{1}{2}$	$\frac{1}{2}$	0	0		
$-w$	$\frac{7}{2}$	$-\frac{11}{2}$	5	0	$-\frac{5}{2}$	$\frac{5}{2}$	0	-2		
		$\uparrow$								

Most negative d_i'' (x_2 enters next basis)

Result of pivoting (since y_1 and y_2 are dropped from basis, the columns corresponding to them need not be filled):

	x_1	x_2	x_3	x_4	x_5	y_1	y_2			
x_4	$\frac{6}{11}$	0	$\frac{7}{11}$	1	$\frac{2}{11}$	Dropped		$\frac{6}{11}$	$\frac{6}{2}$	
x_2	$-\frac{7}{11}$	1	$-\frac{10}{11}$	0	$\frac{5}{11}$			$\frac{4}{11}$	$\frac{4}{5}$	
$-f$	$\frac{98}{22}$	0	$\frac{118}{22}$	0	$-\frac{4}{22}$			$-\frac{6}{11}$		
$-w$	0	0	0	0	0			0		

Step 4 At this stage we notice that the present basic feasible solution does not contain any of the artificial variables y_1 and y_2, and also the value of w is reduced to 0. This indicates that phase I is completed.

Step 5 Now we start phase II computations by dropping the w row from further consideration. The results of phase II are again shown in tableau form:

Basic variables	Original variables					Constant b_i''	Value of b_i''/a_{is}'' for $a_{is}'' > 0$
	x_1	x_2	x_3	x_4	x_5		
x_4	$\frac{6}{11}$	0	$\frac{7}{11}$	1	$\frac{2}{11}$	$\frac{6}{11}$	$\frac{6}{2}$
x_2	$-\frac{7}{11}$	1	$-\frac{10}{11}$	0	$\boxed{\frac{5}{11}}$ Pivot element	$\frac{4}{11}$	$\frac{4}{5}$ ← Smaller value (x_2 drops from next basis)
$-f$	$\frac{98}{22}$	0	$\frac{118}{22}$	0	$-\frac{4}{22}$	$-\frac{6}{11}$	

$$\uparrow$$
Most negative c_i'' (x_5 enters next basis)

Result of pivoting:

x_4	$\frac{4}{5}$	$-\frac{2}{5}$	1	1	0	$\frac{2}{5}$
x_5	$-\frac{7}{5}$	$\frac{11}{5}$	-2	0	1	$\frac{4}{5}$
$-f$	$\frac{21}{5}$	$\frac{2}{5}$	5	0	0	$-\frac{2}{5}$

Now, since all c_i'' are nonnegative, phase II is completed. The (unique) optimal solution is given by

$$x_1 = x_2 = x_3 = 0 \quad \text{(nonbasic variables)}$$
$$x_4 = \tfrac{2}{5}, \quad x_5 = \tfrac{4}{5} \quad \text{(basic variables)}$$
$$f_{\min} = \tfrac{2}{5}$$

3.11 MATLAB SOLUTION OF LP PROBLEMS

The solution of linear programming problems, using simplex method, can be found as illustrated by the following example.

Example 3.8 Find the solution of the following linear programming problem using MATLAB (simplex method):

$$\text{Minimize } f = -x_1 - 2x_2 - x_3$$

subject to

$$2x_1 + x_2 - x_3 \leq 2$$
$$2x_1 - x_2 + 5x_3 \leq 6$$

$$4x_1 + x_2 + x_3 \leq 6$$

$$x_i \geq 0; \quad i = 1, 2, 3$$

SOLUTION

Step 1 Express the objective function in the form $f(x) = f^T x$ and identify the vectors x and f as

$$x = \begin{Bmatrix} x_1 \\ x_2 \\ x_3 \end{Bmatrix} \quad \text{and} \quad f = \begin{Bmatrix} -1 \\ -2 \\ -1 \end{Bmatrix}$$

Express the constraints in the form $Ax \leq b$ and identify the matrix A and the vector b as

$$A = \begin{bmatrix} 2 & 1 & -1 \\ 2 & -1 & 5 \\ 4 & 1 & 1 \end{bmatrix} \quad \text{and} \quad b = \begin{Bmatrix} 2 \\ 6 \\ 6 \end{Bmatrix}$$

Step 2 Use the command for executing linear programming program using simplex method as indicated below:

```
clc
clear all
f=[-1;-2;-1];
A=[2 1 - 1;
   2 -1 5;
   4 1 1];
b=[2;6;6];
lb=zeros(3,1);
Aeq=[];
beq=[];
options = optimset('LargeScale', 'off', 'Simplex', 'on');
[x,fval,exitflag,output] = linprog(f,A,b,Aeq,beq,lb,[],[],
   optimset('Display','iter'))
```

This produces the solution or output as follows:

```
Optimization terminated.
x=
   0
   4
   2
fval =
  -10
exitflag =
   1
output =
   iterations:3
   algorithm: 'medium scale: simplex'
```

```
cgiterations: []
message: 'Optimization terminated.'
```

REFERENCES AND BIBLIOGRAPHY

3.1   G. B. Dantzig, *Linear Programming and Extensions*, Princeton University Press, Princeton, NJ, 1963.

3.2   W. J. Adams, A. Gewirtz, and L. V. Quintas, *Elements of Linear Programming*, Van Nostrand Reinhold, New York, 1969.

3.3   W.W. Garvin, *Introduction to Linear Programming*, McGraw-Hill, New York, 1960.

3.4   S. I. Gass, *Linear Programming: Methods and Applications*, 5th ed., McGraw-Hill, New York, 1985.

3.5   G. Hadley, *Linear Programming*, Addison-Wesley, Reading, MA, 1962.

3.6   S. Vajda, *An Introduction to Linear Programming and the Theory of Games*, Wiley, New York, 1960.

3.7   W. Orchard-Hays, *Advanced Linear Programming Computing Techniques*, McGraw-Hill, New York, 1968.

3.8   S. I. Gass, *An Illustrated Guide to Linear Programming*, McGraw-Hill, New York, 1970.

3.9   M. F. Rubinstein and J. Karagozian, Building design using linear programming, *Journal of the Structural Division, Proceedings of ASCE*, Vol. 92, No. ST6, pp. 223–245, Dec. 1966.

3.10   T. Au, *Introduction to Systems Engineering: Deterministic Models*, Addison-Wesley, Reading, MA, 1969.

3.11   H. A. Taha, *Operations Research: An Introduction*, 5th ed., Macmillan, New York, 1992.

3.12   W. F. Stoecker, *Design of Thermal Systems*, 3rd ed., McGraw-Hill, New York, 1989.

3.13   K. G. Murty, *Linear Programming*, Wiley, New York, 1983.

3.14   W. L. Winston, *Operations Research: Applications and Algorithms*, 2nd ed., PWS-Kent, Boston, 1991.

3.15   R. M. Stark and R. L. Nicholls, *Mathematical Foundations for Design: Civil Engineering Systems*, McGraw-Hill, New York, 1972.

3.16   N. Karmarkar, A new polynomial-time algorithm for linear programming, *Combinatorica*, Vol. 4, No. 4, pp. 373–395, 1984.

3.17   A. Maass   et al., *Design of Water Resources Systems*, Harvard University Press, Cambridge, MA, 1962.

REVIEW QUESTIONS

3.1   Define a line segment in n-dimensional space.

3.2   What happens when $m = n$ in a (standard) LP problem?

3.3   How many basic solutions can an LP problem have?

3.4   State an LP problem in standard form.

3.5   State four applications of linear programming.

3.6   Why is linear programming important in several types of industries?

3.7   Define the following terms: point, hyperplane, convex set, extreme point.

3.8   What is a basis?

3.9   What is a pivot operation?

3.10 What is the difference between a convex polyhedron and a convex polytope?

3.11 What is a basic degenerate solution?

3.12 What is the difference between the simplex algorithm and the simplex method?

3.13 How do you identify the optimum solution in the simplex method?

3.14 Define the infeasibility form.

3.15 What is the difference between a slack and a surplus variable?

3.16 Can a slack variable be part of the basis at the optimum solution of an LP problem?

3.17 Can an artificial variable be in the basis at the optimum point of an LP problem?

3.18 How do you detect an unbounded solution in the simplex procedure?

3.19 How do you identify the presence of multiple optima in the simplex method?

3.20 What is a canonical form?

3.21 Answer true or false:

   (a) The feasible region of an LP problem is always bounded.

   (b) An LP problem will have infinite solutions whenever a constraint is redundant.

   (c) The optimum solution of an LP problem always lies at a vertex.

   (d) A linear function is always convex.

   (e) The feasible space of some LP problems can be nonconvex.

   (f) The variables must be nonnegative in a standard LP problem.

   (g) The optimal solution of an LP problem can be called the optimal basic solution.

   (h) Every basic solution represents an extreme point of the convex set of feasible solutions.

   (i) We can generate all the basic solutions of an LP problem using pivot operations.

   (j) The simplex algorithm permits us to move from one basic solution to another basic solution.

   (k) The slack and surplus variables can be unrestricted in sign.

   (l) An LP problem will have an infinite number of feasible solutions.

   (m) An LP problem will have an infinite number of basic feasible solutions.

   (n) The right-hand-side constants can assume negative values during the simplex procedure.

   (o) All the right-hand-side constants can be zero in an LP problem.

   (p) The cost coefficient corresponding to a nonbasic variable can be positive in a basic feasible solution.

   (q) If all elements in the pivot column are negative, the LP problem will not have a feasible solution.

   (r) A basic degenerate solution can have negative values for some of the variables.

   (s) If a greater-than or equal-to type of constraint is active at the optimum point, the corresponding surplus variable must have a positive value.

   (t) A pivot operation brings a nonbasic variable into the basis.

(**u**) The optimum solution of an LP problem cannot contain slack variables in the basis.

(**v**) If the infeasibility form has a nonzero value at the end of phase I, it indicates an unbounded solution to the LP problem.

(**w**) The solution of an LP problem can be a local optimum.

(**x**) In a standard LP problem, all the cost coefficients will be positive.

(**y**) In a standard LP problem, all the right-hand-side constants will be positive.

(**z**) In a LP problem, the number of inequality constraints cannot exceed the number of variables.

(**aa**) A basic feasible solution cannot have zero value for any of the variables.

PROBLEMS

3.1 State the following LP problem in standard form:

$$\text{Maximize } f = -2x_1 - x_2 + 5x_3$$

subject to

$$x_1 - 2x_2 + x_3 \le 8$$
$$3x_1 - 2x_2 \ge -18$$
$$2x_1 + x_2 - 2x_3 \le -4$$

3.2 State the following LP problem in standard form:

$$\text{Maximize } f = x_1 - 8x_2$$

subject to

$$3x_1 + 2x_2 \ge 6$$
$$9x_1 + 7x_2 \le 108$$
$$2x_1 - 5x_2 \ge -35$$
$$x_1, \ x_2 \text{ unrestricted in sign}$$

3.3 Solve the following system of equations using pivot operations:

$$6x_1 - 2x_2 + 3x_3 = 11$$
$$4x_1 + 7x_2 + x_3 = 21$$
$$5x_1 + 8x_2 + 9x_3 = 48$$

3.4 It is proposed to build a reservoir of capacity x_1 to better control the supply of water to an irrigation district [3.15, 3.17]. The inflow to the reservoir is expected to be 4.5×10^6 acre-ft during the wet (rainy) season and 1.1×10^6 acre-ft during the dry (summer) season. Between the reservoir and the irrigation district, one stream (A) adds water to and another stream (B) carries water away from the main stream, as shown in Fig. 3.16. Stream A adds 1.2×10^6 and 0.3×10^6 acre-ft of water during the wet and dry seasons, respectively. Stream B takes away 0.5×10^6 and 0.2×10^6 acre-ft of water during the

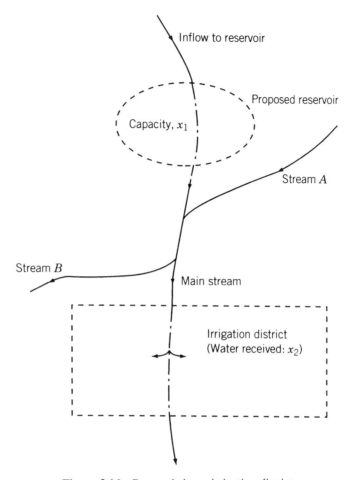

Figure 3.16   Reservoir in an irrigation district.

wet and dry seasons, respectively. Of the total amount of water released to the irrigation district per year (x_2), 30% is to be released during the wet season and 70% during the dry season. The yearly cost of diverting the required amount of water from the main stream to the irrigation district is given by $18(0.3x_2) + 12(0.7x_2)$. The cost of building and maintaining the reservoir, reduced to an yearly basis, is given by $25x_1$. Determine the values of x_1 and x_2 to minimize the total yearly cost.

3.5   Solve the following system of equations using pivot operations:

$$4x_1 - 7x_2 + 2x_3 = -8$$
$$3x_1 + 4x_2 - 5x_3 = -8$$
$$5x_1 + x_2 - 8x_3 = -34$$

3.6   What elementary operations can be used to transform

$$2x_1 + x_2 + x_3 = 9$$
$$x_1 + x_2 + x_3 = 6$$
$$2x_1 + 3x_2 + x_3 = 13$$

into

$$x_1 = 3$$
$$x_2 = 2$$
$$x_1 + 3x_2 + x_3 = 10$$

Find the solution of this system by reducing into canonical form.

3.7   Find the solution of the following LP problem graphically:

$$\text{Maximize } f = 2x_1 + 6x_2$$

subject to

$$-x_1 + x_2 \leq 1$$
$$2x_1 + x_2 \leq 2$$
$$x_1 \geq 0, \quad x_2 \geq 0$$

3.8   Find the solution of the following LP problem graphically:

$$\text{Minimize } f = -3x_1 + 2x_2$$

subject to

$$0 \leq x_1 \leq 4$$
$$1 \leq x_2 \leq 6$$
$$x_1 + x_2 \leq 5$$

3.9   Find the solution of the following LP problem graphically:

$$\text{Minimize } f = 3x_1 + 2x_2$$

subject to

$$8x_1 + x_2 \geq 8$$
$$2x_1 + x_2 \geq 6$$
$$x_1 + 3x_2 \geq 6$$
$$x_1 + 6x_2 \geq 8$$
$$x_1 \geq 0, \quad x_2 \geq 0$$

3.10 Find the solution of the following problem by the graphical method:

$$\text{Minimize } f = x_1^2 x_2^2$$

subject to

$$x_1^3 x_2^2 \geq e^3$$

$$x_1 x_2^4 \geq e^4$$

$$x_1^2 x_2^3 \leq e$$

$$x_1 \geq 0, \quad x_2 \geq 0$$

where e is the base of natural logarithms.

3.11 Prove Theorem 3.6.

For Problems 3.12 to 3.42, use a graphical procedure to identify (**a**) the feasible region, (**b**) the region where the slack (or surplus) variables are zero, and (**c**) the optimum solution.

3.12

$$\text{Maximize } f = 6x + 7y$$

subject to

$$7x + 6y \leq 42$$

$$5x + 9y \leq 45$$

$$x - y \leq 4$$

$$x \geq 0, \quad y \geq 0$$

3.13 Rework Problem 3.12 when x and y are unrestricted in sign.

3.14

$$\text{Maximize } f = 19x + 7y$$

subject to

$$7x + 6y \leq 42$$

$$5x + 9y \leq 45$$

$$x - y \leq 4$$

$$x \geq 0, \quad y \geq 0$$

3.15 Rework Problem 3.14 when x and y are unrestricted in sign.

3.16

$$\text{Maximize } f = x + 2y$$

subject to

$$x - y \geq -8$$

$$5x - y \geq 0$$

$$x + y \geq 8$$

$$-x + 6y \geq 12$$
$$5x + 2y \leq 68$$
$$x \leq 10$$
$$x \geq 0, \quad y \geq 0$$

3.17   Rework Problem 3.16 by changing the objective to Minimize $f = x - y$.

3.18   Maximize $f = x + 2y$

subject to

$$x - y \geq -8$$
$$5x - y \geq 0$$
$$x + y \geq 8$$
$$-x + 6y \geq 12$$
$$5x + 2y \geq 68$$
$$x \leq 10$$
$$x \geq 0, \quad y \geq 0$$

3.19   Rework Problem 3.18 by changing the objective to Minimize $f = x - y$.

3.20   Maximize $f = x + 3y$

subject to

$$-4x + 3y \leq 12$$
$$x + y \leq 7$$
$$x - 4y \leq 2$$
$$x \geq 0, \quad y \geq 0$$

3.21   Minimize $f = x + 3y$

subject to

$$-4x + 3y \leq 12$$
$$x + y \leq 7$$
$$x - 4y \leq 2$$

x and y are unrestricted in sign

3.22   Rework Problem 3.20 by changing the objective to Maximize $f = x + y$.

3.23   Maximize $f = x + 3y$

subject to

$$-4x + 3y \leq 12$$
$$x + y \leq 7$$
$$x - 4y \geq 2$$
$$x \geq 0, \quad y \geq 0$$

3.24   Minimize $f = x - 8y$

subject to

$$3x + 2y \geq 6$$
$$x - y \leq 6$$
$$9x + 7y \leq 108$$
$$3x + 7y \leq 70$$
$$2x - 5y \geq -35$$
$$x \geq 0, \quad y \geq 0$$

3.25 Rework Problem 3.24 by changing the objective to Maximize $f = x - 8y$.

3.26   Maximize $f = x - 8y$

subject to

$$3x + 2y \geq 6$$
$$x - y \leq 6$$
$$9x + 7y \leq 108$$
$$3x + 7y \leq 70$$
$$2x - 5y \geq -35$$
$$x \geq 0, \quad y \text{ is unrestricted in sign}$$

3.27   Maximize $f = 5x - 2y$

subject to

$$3x + 2y \geq 6$$
$$x - y \leq 6$$

$$9x + 7y \leq 108$$
$$3x + 7y \leq 70$$
$$2x - 5y \geq -35$$
$$x \geq 0, \quad y \geq 0$$

3.28   Minimize $f = x - 4y$

subject to

$$x - y \geq -4$$
$$4x + 5y \leq 45$$
$$5x - 2y \leq 20$$
$$5x + 2y \leq 10$$
$$x \geq 0, \quad y \geq 0$$

3.29   Maximize $f = x - 4y$

subject to

$$x - y \geq -4$$
$$4x + 5y \leq 45$$
$$5x - 2y \leq 20$$
$$5x + 2y \geq 10$$
$$x \geq 0, \ y \text{ is unrestricted in sign}$$

3.30   Minimize $f = x - 4y$

subject to

$$x - y \geq -4$$
$$4x + 5y \leq 45$$
$$5x - 2y \leq 20$$
$$5x + 2y \geq 10$$
$$x \geq 0, \quad y \geq 0$$

3.31 Rework Problem 3.30 by changing the objective to Maximize $f = x - 4y$.

3.32   Minimize $f = 4x + 5y$

subject to

$$10x + y \geq 10$$
$$5x + 4y \geq 20$$
$$3x + 7y \geq 21$$
$$x + 12y \geq 12$$
$$x \geq 0, \quad y \geq 0$$

3.33 Rework Problem 3.32 by changing the objective to Maximize $f = 4x + 5y$.

3.34 Rework Problem 3.32 by changing the objective to Minimize $f = 6x + 2y$.

3.35   Minimize $f = 6x + 2y$

subject to

$$10x + y \geq 10$$
$$5x + 4y \geq 20$$
$$3x + 7y \geq 21$$
$$x + 12y \geq 12$$

x and y are unrestricted in sign

3.36   Minimize $f = 5x + 2y$

subject to

$$3x + 4y \leq 24$$
$$x - y \leq 3$$
$$x + 4y \geq 4$$
$$3x + y \geq 3$$
$$x \geq 0, \quad y \geq 0$$

3.37 Rework Problem 3.36 by changing the objective to Maximize $f = 5x + 2y$.

3.38 Rework Problem 3.36 when x is unrestricted in sign and $y \geq 0$.

3.39   Maximize $f = 5x + 2y$

subject to

$$3x + 4y \le 24$$
$$x - y \le 3$$
$$x + 4y \le 4$$
$$3x + y \ge 3$$
$$x \ge 0, \quad y \ge 0$$

3.40   Maximize $f = 3x + 2y$

subject to

$$9x + 10y \le 330$$
$$21x - 4y \ge -36$$
$$x + 2y \ge 6$$
$$6x - y \le 72$$
$$3x + y \le 54$$
$$x \ge 0, \quad y \ge 0$$

3.41   Rework Problem 3.40 by changing the constraint $x + 2y \ge 6$ to $x + 2y \le 6$.

3.42   Maximize $f = 3x + 2y$

subject to

$$9x + 10y \le 330$$
$$21x - 4y \ge -36$$
$$x + 2y \le 6$$
$$6x - y \le 72$$
$$3x + y \ge 54$$
$$x \ge 0, \quad y \ge 0$$

3.43   Maximize $f = 3x + 2y$

subject to

$$21x - 4y \ge -36$$
$$x + 2y \ge 6$$
$$6x - y \le 72$$
$$x \ge 0, \quad y \ge 0$$

3.44 Reduce the system of equations

$$2x_1 + 3x_2 - 2x_3 - 7x_4 = 2$$

$$x_1 + x_2 - x_3 + 3x_4 = 12$$

$$x_1 - x_2 + x_3 + 5x_4 = 8$$

into a canonical system with x_1, x_2, and x_3 as basic variables. From this derive all other canonical forms.

3.45   Maximize $f = 240x_1 + 104x_2 + 60x_3 + 19x_4$

subject to

$$20x_1 + 9x_2 + 6x_3 + x_4 \leq 20$$

$$10x_1 + 4x_2 + 2x_3 + x_4 \leq 10$$

$$x_i \geq 0, \quad i = 1 \text{ to } 4$$

Find all the basic feasible solutions of the problem and identify the optimal solution.

3.46 A progressive university has decided to keep its library open round the clock and gathered that the following number of attendants are required to reshelve the books:

Time of day (hours)	Minimum number of attendants required
0–4	4
4–8	7
8–12	8
12–16	9
16–20	14
20–24	3

If each attendant works eight consecutive hours per day, formulate the problem of finding the minimum number of attendants necessary to satisfy the requirements above as a LP problem.

3.47 A paper mill received an order for the supply of paper rolls of widths and lengths as indicated below:

Number of rolls ordered	Width of roll (m)	Length (m)
1	6	100
1	8	300
1	9	200

The mill produces rolls only in two standard widths, 10 and 20 m. The mill cuts the standard rolls to size to meet the specifications of the orders. Assuming that there is no limit on the lengths of the standard rolls, find the cutting pattern that minimizes the trim losses while satisfying the order above.

3.48 Solve the LP problem stated in Example 1.6 for the following data: $l = 2$ m, $W_1 = 3000$ N, $W_2 = 2000$ N, $W_3 = 1000$ N, and $w_1 = w_2 = w_3 = 200$ N.

3.49 Find the solution of Problem 1.1 using the simplex method.

3.50 Find the solution of Problem 1.15 using the simplex method.

3.51 Find the solution of Example 3.1 using (**a**) the graphical method and (**b**) the simplex method.

3.52 In the scaffolding system shown in Fig. 3.17, loads x_1 and x_2 are applied on beams 2 and 3, respectively. Ropes A and B can carry a load of $W_1 = 300$ lb each; the middle ropes, C and D, can withstand a load of $W_2 = 200$ lb each, and ropes E and F are capable of supporting a load $W_3 = 100$ lb each. Formulate the problem of finding the loads x_1 and x_2 and their location parameters x_3 and x_4 to maximize the total load carried by the system, $x_1 + x_2$, by assuming that the beams and ropes are weightless.

3.53 A manufacturer produces three machine parts, A, B, and C. The raw material costs of parts A, B, and C are $5, $10, and $15 per unit, and the corresponding prices of the finished parts are $50, $75, and $100 per unit. Part A requires turning and drilling operations, while part B needs milling and drilling operations. Part C requires turning and milling operations. The number of parts that can be produced on various machines per day and the daily costs of running the machines are given below:

| Machine part | Number of parts that can be produced on | | |
	Turning lathes	Drilling machines	Milling machines
A	15	15	
B		20	30
C	25		10
Cost of running the machines per day	$250	$200	$300

Formulate the problem of maximizing the profit.

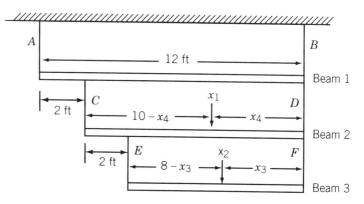

Figure 3.17   Scaffolding system with three beams.

Solve Problems 3.54–3.90 by the simplex method.

3.54 Problem 1.22

3.55 Problem 1.23

3.56 Problem 1.24

3.57 Problem 1.25

3.58 Problem 3.7

3.59 Problem 3.12

3.60 Problem 3.13

3.61 Problem 3.14

3.62 Problem 3.15

3.63 Problem 3.16

3.64 Problem 3.17

3.65 Problem 3.18

3.66 Problem 3.19

3.67 Problem 3.20

3.68 Problem 3.21

3.69 Problem 3.22

3.70 Problem 3.23

3.71 Problem 3.24

3.72 Problem 3.25

3.73 Problem 3.26

3.74 Problem 3.27

3.75 Problem 3.28

3.76 Problem 3.29

3.77 Problem 3.30

3.78 Problem 3.31

3.79 Problem 3.32

3.80 Problem 3.33

3.81 Problem 3.34

3.82 Problem 3.35

3.83 Problem 3.36

3.84 Problem 3.37

3.85 Problem 3.38

3.86 Problem 3.39

3.87 Problem 3.40

3.88 Problem 3.41

3.89 Problem 3.42

3.90 Problem 3.43

3.91 The temperatures measured at various points inside a heated wall are given below:

Distance from the heated surface as a percentage of wall thickness, x_i	0	20	40	60	80	100
Temperature, t_i (°C)	400	350	250	175	100	50

It is decided to use a linear model to approximate the measured values as

$$t = a + bx \tag{1}$$

where t is the temperature, x the percentage of wall thickness, and a and b the coefficients that are to be estimated. Obtain the best estimates of a and b using linear programming with the following objectives.

(a) Minimize the sum of absolute deviations between the measured values and those given by Eq. (1): $\Sigma_i |a + bx_i - t_i|$.

(b) Minimize the maximum absolute deviation between the measured values and those given by Eq. (1):

$$\underset{i}{\text{Max}} |a + bx_i - t_i|$$

3.92 A snack food manufacturer markets two kinds of mixed nuts, labeled A and B. Mixed nuts A contain 20% almonds, 10% cashew nuts, 15% walnuts, and 55% peanuts. Mixed nuts B contain 10% almonds, 20% cashew nuts, 25% walnuts, and 45% peanuts. A customer wants to use mixed nuts A and B to prepare a new mix that contains at least 4 lb of almonds, 5 lb of cashew nuts, and 6 lb of walnuts, for a party. If mixed nuts A and B cost $2.50 and $3.00 per pound, respectively, determine the amounts of mixed nuts A and B to be used to prepare the new mix at a minimum cost.

3.93 A company produces three types of bearings, B_1, B_2, and B_3, on two machines, A_1 and A_2. The processing times of the bearings on the two machines are indicated in the following table:

Machine	Processing time (min) for bearing:		
	B_1	B_2	B_3
A_1	10	6	12
A_2	8	4	4

The times available on machines A_1 and A_2 per day are 1200 and 1000 minutes, respectively. The profits per unit of B_1, B_2, and B_3 are \$4, \$2, and \$3, respectively. The maximum number of units the company can sell are 500, 400, and 600 for B_1, B_2, and B_3, respectively. Formulate and solve the problem for maximizing the profit.

3.94 Two types of printed circuit boards A and B are produced in a computer manufacturing company. The component placement time, soldering time, and inspection time required in producing each unit of A and B are given below:

Circuit board	Time required per unit (min) for:		
	Component placement	Soldering	Inspection
A	16	10	4
B	10	12	8

If the amounts of time available per day for component placement, soldering, and inspection are 1500, 1000, and 500 person-minutes, respectively, determine the number of units of A and B to be produced for maximizing the production. If each unit of A and B contributes a profit of \$10 and \$15, respectively, determine the number of units of A and B to be produced for maximizing the profit.

3.95 A paper mill produces paper rolls in two standard widths; one with width 20 in. and the other with width 50 in. It is desired to produce new rolls with different widths as indicated below:

Width (in.)	Number of rolls required
40	150
30	200
15	50
6	100

The new rolls are to be produced by cutting the rolls of standard widths to minimize the trim loss. Formulate the problem as an LP problem.

3.96 A manufacturer produces two types of machine parts, P_1 and P_2, using lathes and milling machines. The machining times required by each part on the lathe and the milling machine and the profit per unit of each part are given below:

Machine part	Machine time (hr) required by each unit on:		Cost per unit
	Lathe	Milling machine	
P_1	5	2	\$200
P_2	4	4	\$300

If the total machining times available in a week are 500 hours on lathes and 400 hours on milling machines, determine the number of units of P_1 and P_2 to be produced per week to maximize the profit.

3.97   A bank offers four different types of certificates of deposits (CDs) as indicated below:

CD type	Duration (yr)	Total interest at maturity (%)
1	0.5	5
2	1.0	7
3	2.0	10
4	4.0	15

If a customer wants to invest $50,000 in various types of CDs, determine the plan that yields the maximum return at the end of the fourth year.

3.98   The production of two machine parts A and B requires operations on a lathe (L), a shaper (S), a drilling machine (D), a milling machine (M), and a grinding machine (G). The machining times required by A and B on various machines are given below.

Machine part	Machine time required (hours per unit) on:				
	L	S	D	M	G
A	0.6	0.4	0.1	0.5	0.2
B	0.9	0.1	0.2	0.3	0.3

The number of machines of different types available is given by $L : 10, S : 3, D : 4, M :$ 6, and G: 5. Each machine can be used for 8 hours a day for 30 days in a month.

(a) Determine the production plan for maximizing the output in a month

(b) If the number of units of A is to be equal to the number of units of B, find the optimum production plan.

3.99   A salesman sells two types of vacuum cleaners, A and B. He receives a commission of 20% on all sales, provided that at least 10 units each of A and B are sold per month. The salesman needs to make telephone calls to make appointments with customers and demonstrate the products in order to sell the products. The selling price of the products, the average money to be spent on telephone calls, the time to be spent on demonstrations, and the probability of a potential customer buying the product are given below:

Vacuum cleaner	Selling price per unit	Money to be spent on telephone calls to find a potential customer	Time to be spent in demonstrations to a potential customer (hr)	Probability of a potential customer buying the product
A	$250	$3	3	0.4
B	$100	$1	1	0.8

In a particular month, the salesman expects to sell at most 25 units of A and 45 units of B. If he plans to spend a maximum of 200 hours in the month, formulate the problem of determining the number of units of A and B to be sold to maximize his income.

3.100   An electric utility company operates two thermal power plants, A and B, using three different grades of coal, C_1, C_2, and C_3. The minimum power to be generated at plants A and B is 30 and 80 MWh, respectively. The quantities of various grades of coal required to generate 1 MWh of power at each power plant, the pollution caused by the various grades of coal at each power plant, and the costs of coal are given in the following table:

Coal type	Quantity of coal required to generate 1 MWh at the power plant (tons)		Pollution caused at power plant		Cost of coal at power plant	
	A	B	A	B	A	B
C_1	2.5	1.5	1.0	1.5	20	18
C_2	1.0	2.0	1.5	2.0	25	28
C_3	3.0	2.5	2.0	2.5	18	12

Formulate the problem of determining the amounts of different grades of coal to be used at each power plant to minimize (**a**) the total pollution level, and (**b**) the total cost of operation.

3.101 A grocery store wants to buy five different types of vegetables from four farms in a month. The prices of the vegetables at different farms, the capacities of the farms, and the minimum requirements of the grocery store are indicated in the following table:

Farm	Price ($/ton) of vegetable type					Maximum (of all types combined) they can supply
	1 (Potato)	2 (Tomato)	3 (Okra)	4 (Eggplant)	5 (Spinach)	
1	200	600	1600	800	1200	180
2	300	550	1400	850	1100	200
3	250	650	1500	700	1000	100
4	150	500	1700	900	1300	120
Minimum amount required (tons)	100	60	20	80	40	

Formulate the problem of determining the buying scheme that corresponds to a minimum cost.

3.102 A steel plant produces steel using four different types of processes. The iron ore, coal, and labor required, the amounts of steel and side products produced, the cost information, and the physical limitations on the system are given below:

Process type	Iron ore required (tons/day)	Coal required (tons/day)	Labor required (person-days)	Steel Produced (tons/day)	Side products Produced (tons/day)
1	5	3	6	4	1
2	8	5	12	6	2
3	3	2	5	2	1
4	10	7	12	6	4
Cost	$50/ton	$10/ton	$150/person-day	$350/ton	$100/ton
Limitations	600 tons available per month	250 tons available per month	No limitations on availability of labor	All steel produced can be sold	Only 200 tons can be sold per month

Assuming that a particular process can be employed for any number of days in a 30-day month, determine the operating schedule of the plant for maximizing the profit.

3.103 Solve Example 3.7 using MATLAB (simplex method).

3.104 Solve Problem 3.12 using MATLAB (simplex method).

3.105 Solve Problem 3.24 using MATLAB (simplex method).

3.106 Find the optimal solution of the LP problem stated in Problem 3.45 using MATLAB (simplex method).

3.107 Find the optimal solution of the LP problem described in Problem 3.101 using MATLAB.

4

Linear Programming II: Additional Topics and Extensions

4.1 INTRODUCTION

If a LP problem involving several variables and constraints is to be solved by using the simplex method described in Chapter 3, it requires a large amount of computer storage and time. Some techniques, which require less computational time and storage space compared to the original simplex method, have been developed. Among these techniques, the revised simplex method is very popular. The principal difference between the original simplex method and the revised one is that in the former we transform all the elements of the simplex tableau, while in the latter we need to transform only the elements of an inverse matrix. Associated with every LP problem, another LP problem, called the *dual*, can be formulated. The solution of a given LP problem, in many cases, can be obtained by solving its dual in a much simpler manner.

As stated above, one of the difficulties in certain practical LP problems is that the number of variables and/or the number of constraints is so large that it exceeds the storage capacity of the available computer. If the LP problem has a special structure, a principle known as the *decomposition principle* can be used to solve the problem more efficiently. In many practical problems, one will be interested not only in finding the optimum solution to a LP problem, but also in finding how the optimum solution changes when some parameters of the problem, such as cost coefficients change. Hence the sensitivity or postoptimality analysis becomes very important.

An important special class of LP problems, known as *transportation problems*, occurs often in practice. These problems can be solved by algorithms that are more efficient (for this class of problems) than the simplex method. Karmarkar's method is an interior method and has been shown to be superior to the simplex method of Dantzig for large problems. The quadratic programming problem is the best-behaved nonlinear programming problem. It has a quadratic objective function and linear constraints and is convex (for minimization problems). Hence the quadratic programming problem can be solved by suitably modifying the linear programming techniques. All these topics are discussed in this chapter.

4.2 REVISED SIMPLEX METHOD

We notice that the simplex method requires the computing and recording of an entirely new tableau at each iteration. But much of the information contained in the tableau is not used; only the following items are needed.

1. The relative cost coefficients $\bar{c}_j$ to compute[†]

$$\bar{c}_s = \min(\bar{c}_j) \tag{4.1}$$

$\bar{c}_s$ determines the variable x_s that has to be brought into the basis in the next iteration.

2. By assuming that $\bar{c}_s < 0$, the elements of the updated column

$$\overline{\mathbf{A}}_s = \begin{Bmatrix} \bar{a}_{1s} \\ \bar{a}_{2s} \\ \vdots \\ \bar{a}_{ms} \end{Bmatrix}$$

and the values of the basic variables

$$\overline{\mathbf{X}}_B = \begin{Bmatrix} \bar{b}_1 \\ \bar{b}_2 \\ \vdots \\ \bar{b}_m \end{Bmatrix}$$

have to be calculated. With this information, the variable x_r that has to be removed from the basis is found by computing the quantity

$$\frac{\bar{b}_r}{\bar{a}_{rs}} = \min_{\bar{a}_{is} > 0} \left\{ \frac{\bar{b}_i}{\bar{a}_{is}} \right\} \tag{4.2}$$

and a pivot operation is performed on $\bar{a}_{rs}$. Thus only one nonbasic column $\overline{\mathbf{A}}_s$ of the current tableau is useful in finding x_r. Since most of the linear programming problems involve many more variables (columns) than constraints (rows), considerable effort and storage is wasted in dealing with the $\overline{\mathbf{A}}_j$ for $j \neq s$. Hence it would be more efficient if we can generate the modified cost coefficients $\bar{c}_j$ and the column $\overline{\mathbf{A}}_s$, from the original problem data itself. The revised simplex method is used for this purpose; it makes use of the inverse of the current basis matrix in generating the required quantities.

Theoretical Development.   Although the revised simplex method is applicable for both phase I and phase II computations, the method is initially developed by considering linear programming in phase II for simplicity. Later, a step-by-step procedure is given to solve the general linear programming problem involving both phases I and II.

Let the given linear programming problem (phase II) be written in column form as

Minimize

$$f(\mathbf{X}) = c_1 x_1 + c_2 x_2 + \cdots + c_n x_n \tag{4.3}$$

[†]The modified values of b_i, a_{ij}, and c_j are denoted by overbars in this chapter (they were denoted by primes in Chapter 3).

subject to

$$\mathbf{AX} = \mathbf{A}_1 x_1 + \mathbf{A}_2 x_2 + \cdots + \mathbf{A}_n x_n = \mathbf{b} \qquad (4.4)$$

$$\underset{n \times 1}{\mathbf{X}} \geq \underset{n \times 1}{\mathbf{0}} \qquad (4.5)$$

where the jth column of the coefficient matrix $\mathbf{A}$ is given by

$$\underset{m \times 1}{\mathbf{A}_j} = \begin{Bmatrix} a_{1j} \\ a_{2j} \\ \vdots \\ a_{mj} \end{Bmatrix}$$

Assuming that the linear programming problem has a solution, let

$$\mathbf{B} = [\mathbf{A}_{j1} \ \mathbf{A}_{j2} \ \cdots \ \mathbf{A}_{jm}]$$

be a basis matrix with

$$\underset{m \times 1}{\mathbf{X}_B} = \begin{Bmatrix} x_{j1} \\ x_{j2} \\ \vdots \\ x_{jm} \end{Bmatrix} \quad \text{and} \quad \underset{m \times 1}{\mathbf{c}_B} = \begin{Bmatrix} c_{j1} \\ c_{j2} \\ \vdots \\ c_{jm} \end{Bmatrix}$$

representing the corresponding vectors of basic variables and cost coefficients, respectively. If $\mathbf{X}_B$ is feasible, we have

$$\mathbf{X}_B = \mathbf{B}^{-1}\mathbf{b} = \overline{\mathbf{b}} \geq \mathbf{0}$$

As in the regular simplex method, the objective function is included as the $(m+1)$th equation and $-f$ is treated as a permanent basic variable. The augmented system can be written as

$$\sum_{j=1}^{n} \mathbf{P}_j x_j + \mathbf{P}_{n+1}(-f) = \mathbf{q} \qquad (4.6)$$

where

$$\mathbf{P}_j = \begin{Bmatrix} a_{1j} \\ a_{2j} \\ \vdots \\ a_{mj} \\ c_j \end{Bmatrix}, \ j = 1 \text{ to } n, \quad \mathbf{P}_{n+1} = \begin{Bmatrix} 0 \\ 0 \\ \vdots \\ 0 \\ 1 \end{Bmatrix} \quad \text{and} \quad \mathbf{q} = \begin{Bmatrix} b_1 \\ b_2 \\ \vdots \\ b_m \\ 0 \end{Bmatrix}$$

Since $\mathbf{B}$ is a feasible basis for the system of Eqs. (4.4), the matrix $\mathbf{D}$ defined by

$$\underset{m+1 \times m+1}{\mathbf{D}} = [\mathbf{P}_{j1} \ \mathbf{P}_{j2} \ \cdots \ \mathbf{P}_{jm} \ \mathbf{P}_{n+1}] = \begin{bmatrix} \mathbf{B} & \mathbf{0} \\ \mathbf{c}_B^{\mathrm{T}} & 1 \end{bmatrix}$$

will be a feasible basis for the augmented system of Eqs. (4.6). The inverse of $\mathbf{D}$ can be found to be

$$\mathbf{D}^{-1} = \begin{bmatrix} \mathbf{B}^{-1} & \mathbf{0} \\ -\mathbf{c}_B^{\mathrm{T}}\mathbf{B}^{-1} & 1 \end{bmatrix}$$

Definition.   The row vector

$$\mathbf{c}_B^T \mathbf{B}^{-1} = \boldsymbol{\pi}^T = \begin{Bmatrix} \pi_1 \\ \pi_2 \\ \vdots \\ \pi_m \end{Bmatrix}^T \tag{4.7}$$

is called the vector of simplex multipliers relative to the f equation. If the computations correspond to phase I, two vectors of simplex multipliers, one relative to the f equation, and the other relative to the w equation are to be defined as

$$\boldsymbol{\pi}^T = \mathbf{c}_B^T \mathbf{B}^{-1} = \begin{Bmatrix} \pi_1 \\ \pi_2 \\ \vdots \\ \pi_m \end{Bmatrix}^T$$

$$\boldsymbol{\sigma}^T = \mathbf{d}_B^T \mathbf{B}^{-1} = \begin{Bmatrix} \sigma_1 \\ \sigma_2 \\ \vdots \\ \sigma_m \end{Bmatrix}^T$$

By premultiplying each column of Eq. (4.6) by $\mathbf{D}^{-1}$, we obtain the following canonical system of equations[†]:

$$\begin{matrix} x_{j1} \\ x_{j2} \\ \vdots \quad + \sum_{j\,\text{nonbasic}} \overline{\mathbf{A}}_j x_j = \quad \vdots \\ x_{jm} \\ -f + \sum_{j\,\text{nonbasic}} \overline{c}_j x_j = -f_0 \end{matrix} \quad \begin{matrix} \overline{b}_1 \\ \overline{b}_2 \\ \\ b_m \end{matrix}$$

where

$$\begin{Bmatrix} \overline{\mathbf{A}}_j \\ \overline{c}_j \end{Bmatrix} = \mathbf{D}^{-1} \mathbf{P}_j = \begin{bmatrix} \mathbf{B}^{-1} & \mathbf{0} \\ -\boldsymbol{\pi}^T & 1 \end{bmatrix} \begin{Bmatrix} \mathbf{A}_j \\ c_j \end{Bmatrix} \tag{4.8}$$

From Eq. (4.8), the updated column $\overline{\mathbf{A}}_j$ can be identified as

$$\overline{\mathbf{A}}_j = \mathbf{B}^{-1} \mathbf{A}_j \tag{4.9}$$

[†]Premultiplication of $\mathbf{P}_j x_j$ by $\mathbf{D}^{-1}$ gives

$$\mathbf{D}^{-1} \mathbf{P}_j x_j = \begin{bmatrix} \mathbf{B}^{-1} & \mathbf{0} \\ -\boldsymbol{\pi}^T & 1 \end{bmatrix} \begin{Bmatrix} \mathbf{A}_j \\ c_j \end{Bmatrix} x_j$$

$$= \begin{Bmatrix} \mathbf{B}^{-1}\mathbf{A}_j \\ -\boldsymbol{\pi}^T\mathbf{A}_j + c_j \end{Bmatrix} x_j = \begin{cases} x_j & \text{if } x_j \text{ is a basic variable} \\ \mathbf{D}^{-1}\mathbf{P}_j x_j & \text{if } x_j \text{ is not a basic variable} \end{cases}$$

and the modified cost coefficient $\bar{c}_j$ as

$$\bar{c}_j = c_j - \boldsymbol{\pi}^{\mathrm{T}} \mathbf{A}_j \tag{4.10}$$

Equations (4.9) and (4.10) can be used to perform a simplex iteration by generating $\overline{\mathbf{A}}_j$ and $\bar{c}_j$ from the original problem data, $\mathbf{A}_j$ and c_j.

Once $\overline{\mathbf{A}}_j$ and $\bar{c}_j$ are computed, the pivot element $\bar{a}_{rs}$ can be identified by using Eqs. (4.1) and (4.2). In the next step, $\mathbf{P}_s$ is introduced into the basis and $\mathbf{P}_{jr}$ is removed. This amounts to generating the inverse of the new basis matrix. The computational procedure can be seen by considering the matrix:

$$
\begin{bmatrix}
\underbrace{\mathbf{P}_{j1}\ \mathbf{P}_{j2}\ \cdots\ \mathbf{P}_{jm}\ \mathbf{P}_{n+1}}_{\substack{\mathbf{D} \\ m+1 \times m+1}} & \underbrace{\mathbf{e}_1\ \mathbf{e}_2\ \cdots\ \mathbf{e}_{m+1}}_{\substack{\mathbf{I} \\ m+1 \times m+1}} & \overset{\overbrace{\mathbf{P}_s}}{\begin{matrix} a_{1s} \\ a_{2s} \\ \vdots \\ a_{ms} \\ c_s \end{matrix}}
\end{bmatrix} \tag{4.11}
$$

where $\mathbf{e}_i$ is a $(m+1)$-dimensional unit vector with a one in the ith row. Premultiplication of the above matrix by $\mathbf{D}^{-1}$ yields

$$
\begin{bmatrix}
\underbrace{\mathbf{e}_1\ \mathbf{e}_2\ \cdots\ \mathbf{e}_r\ \cdots\ \mathbf{e}_{m+1}}_{\substack{\mathbf{I} \\ m+1 \times m+1}} & \substack{\mathbf{D}^{-1} \\ m+1 \times m+1} & \begin{matrix} \bar{a}_{1s} \\ \bar{a}_{2s} \\ \vdots \\ \bar{a}_{rs} \\ \text{Pivot} \\ \text{element} \\ \vdots \\ \bar{a}_{ms} \\ \bar{c}_s \\ m+1 \times 1 \end{matrix}
\end{bmatrix} \tag{4.12}
$$

By carrying out a pivot operation on $\bar{a}_{rs}$, this matrix transforms to

$$[[\mathbf{e}_1\ \mathbf{e}_2 \cdots \mathbf{e}_{r-1}\ \boldsymbol{\beta}\ \mathbf{e}_{r+1}\ \cdots\ \mathbf{e}_{m+1}]\quad \mathbf{D}_{\text{new}}^{-1}\quad \mathbf{e}_r] \tag{4.13}$$

where all the elements of the vector $\boldsymbol{\beta}$ are, in general, nonzero and the second partition gives the desired matrix $\mathbf{D}_{\text{new}}^{-1}.^{\dagger}$ It can be seen that the first partition (matrix $\mathbf{I}$) is included

---

†This can be verified by comparing the matrix of Eq. (4.13) with the one given in Eq. (4.11). The columns corresponding to the new basis matrix are given by

$$\mathbf{D}_{\text{new}} = [\mathbf{P}_{j1}\ \mathbf{P}_{j2} \cdots \mathbf{P}_{j_{r-1}}\ \underset{\substack{\text{brought in} \\ \text{place of } \mathbf{P}_r}}{\mathbf{P}_s}\ \mathbf{P}_{j_{r+1}} \cdots \mathbf{P}_{jm}\ \mathbf{P}_{n+1}]$$

These columns are modified and can be seen to form a unit matrix in Eq. (4.13). The sequence of pivot operations that did this must be equivalent to multiplying the original matrix, Eq. (4.11), by $\mathbf{D}_{\text{new}}^{-1}$. Thus the second partition of the matrix in Eq. (4.13) gives the desired $\mathbf{D}_{\text{new}}^{-1}$.

only to illustrate the transformation, and it can be dropped in actual computations. Thus in practice, we write the $m + 1 \times m + 2$ matrix

$$
\begin{bmatrix}
 & \bar{a}_{1s} \\
 & \bar{a}_{2s} \\
 & \vdots \\
\mathbf{D}^{-1} & \boxed{\bar{a}_{rs}} \\
 & \vdots \\
 & \bar{a}_{ms} \\
 & \bar{c}_s
\end{bmatrix}
$$

and carry out a pivot operation on $\bar{a}_{rs}$. The first $m + 1$ columns of the resulting matrix will give us the desired matrix $\mathbf{D}_{\text{new}}^{-1}$.

Procedure.   The detailed iterative procedure of the revised simplex method to solve a general linear programming problem is given by the following steps.

1. Write the given system of equations in canonical form, by adding the artificial variables $x_{n+1}, x_{n+2}, \ldots, x_{n+m}$, and the infeasibility form for phase I as shown below:

$$
\begin{aligned}
a_{11}x_1 + a_{12}x_2 + \cdots + a_{1n}x_n + x_{n+1} \qquad\qquad\qquad\qquad &= b_1 \\
a_{21}x_1 + a_{22}x_2 + \cdots + a_{2n}x_n \qquad +x_{n+2} \qquad\qquad\qquad &= b_2 \\
\vdots \qquad\qquad\qquad\qquad\qquad\qquad& \\
a_{m1}x_1 + a_{m2}x_2 + \cdots + a_{mn}x_n \qquad\qquad\qquad +x_{n+m} \qquad\quad &= b_m \\
c_1x_1 + c_2x_2 + \cdots + c_nx_n \qquad\qquad\qquad\qquad\quad -f \qquad &= 0 \\
d_1x_1 + d_2x_2 + \cdots + d_nx_n \qquad\qquad\qquad\qquad\qquad -w &= -w_0
\end{aligned}
$$

$$(4.14)$$

Here the constants b_i, $i = 1$ to m, are made nonnegative by changing, if necessary, the signs of all terms in the original equations before the addition of the artificial variables x_{n+i}, $i = 1$ to m. Since the original infeasibility form is given by

$$w = x_{n+1} + x_{n+2} + \cdots + x_{n+m} \qquad\qquad (4.15)$$

the artificial variables can be eliminated from Eq. (4.15) by adding the first m equations of Eqs. (4.14) and subtracting the result from Eq. (4.15). The resulting equation is shown as the last equation in Eqs. (4.14) with

$$d_j = -\sum_{i=1}^{m} a_{ij} \quad \text{and} \quad w_0 = \sum_{i=1}^{m} b_i \qquad\qquad (4.16)$$

Equations (4.14) are written in tableau form as shown in Table 4.1.

2. The iterative procedure (cycle 0) is started with $x_{n+1}, x_{n+2}, \ldots, x_{n+m}, -f$, and $-w$ as the basic variables. A tableau is opened by entering the coefficients of the basic variables and the constant terms as shown in Table 4.2. The starting basis matrix is, from Table 4.1, $\mathbf{B} = \mathbf{I}$, and its inverse $\mathbf{B}^{-1} = [\beta_{ij}]$ can also be

Table 4.1 Original System of Equations

Admissible (original) variable					Artificial variable				Objective variable		Constant
x_1	x_2	$\cdots x_j$		$\cdots x_n$	x_{n+1}	x_{n+2}	$\cdots$	x_{n+m}	$-f$	$-w$	
					$\longleftarrow$ Initial basis $\longrightarrow$						
a_{11}	a_{12}	a_{1j}		a_{1n}	1						b_1
a_{21}	a_{22}	a_{2j}		a_{2n}		1					b_2
$\cdots$	$\vdots$	$\vdots$		$\vdots$							
a_{m1}	a_{m2}	a_{mj}		a_{mn}				1			b_m
c_1	c_2	c_j		c_n	0	0		0	1	0	0
d_1	d_2	d_j		d_n	0	0		0	0	1	$-w_0$

$\mathbf{A}_1$ $\mathbf{A}_2$ $\mathbf{A}_j$ $\mathbf{A}_n$

Table 4.2   Tableau at the Beginning of Cycle 0

Basic variables	Columns of the canonical form									Value of the basic variable	$x_s{}^a$
	x_{n+1}	x_{n+2}	$\cdots$	x_{n+r}	$\cdots$	x_{n+m}	$-f$	$-w$			
x_{n+1}	1									b_1	
x_{n+2}		1								b_2	
$\vdots$										$\vdots$	
x_{n+r}				1						b_r	
$\vdots$										$\vdots$	
x_{n+m}						1				b_m	
			$\longleftarrow$ Inverse of the basis $\longleftarrow$								
$-f$	0	0	$\cdots$	0	$\cdots$	0	1			0	
$-w$	0	0	$\cdots$	0	$\cdots$	0		1		$-w_0 = -\sum\limits_{i=1}^{m} b_i$	

^{a}This column is blank at the beginning of cycle 0 and filled up only at the end of cycle 0.

seen to be an identity matrix in Table 4.2. The rows corresponding to $-f$ and $-w$ in Table 4.2 give the negative of simplex multipliers π_i and σ_i ($i = 1$ to m), respectively. These are also zero since $\mathbf{c}_B = \mathbf{d}_B = \mathbf{0}$ and hence

$$\boldsymbol{\pi}^\mathrm{T} = \mathbf{c}_B^\mathrm{T}\mathbf{B}^{-1} = \mathbf{0}$$

$$\boldsymbol{\sigma}^\mathrm{T} = \mathbf{d}_B^\mathrm{T}\mathbf{B}^{-1} = \mathbf{0}$$

In general, at the start of some cycle k ($k = 0$ to start with) we open a tableau similar to Table 4.2, as shown in Table 4.4. This can also be interpreted as composed of the inverse of the current basis, $\mathbf{B}^{-1} = [\beta_{ij}]$, two rows for the simplex multipliers π_i and σ_i, a column for the values of the basic variables in the basic solution, and a column for the variable x_s. At the start of any cycle, all entries in the tableau, except the last column, are known.

3. The values of the relative cost factors $\overline{d}_j$ (for phase I) or $\overline{c}_j$ (for phase II) are computed as

$$\overline{d}_j = d_j - \boldsymbol{\sigma}^\mathrm{T}\mathbf{A}_j$$

$$\overline{c}_j = c_j - \boldsymbol{\pi}^\mathrm{T}\mathbf{A}_j$$

and entered in a tableau form as shown in Table 4.3. For cycle 0, $\boldsymbol{\sigma}^T = \mathbf{0}$ and hence $\overline{d}_j \equiv d_j$.

4. If the current cycle corresponds to phase I, find whether all $\overline{d}_j \geq 0$. If all $\overline{d}_j \geq 0$ and $\overline{w}_0 > 0$, there is no feasible solution to the linear programming problem, so the process is terminated. If all $\overline{d}_j \geq 0$ and $\overline{w}_0 = 0$, the current basic solution is a basic feasible solution to the linear programming problem and hence phase II is started by (a) dropping all variables x_j with $\overline{d}_j > 0$, (b) dropping the w row of the tableau, and (c) restarting the cycle (step 3) using phase II rules.

Table 4.3 Relative Cost Factor $\bar{d}_j$ or $\bar{c}_j$

Cycle number	x_1	x_2	$\cdots$	x_n	x_{n+1}	x_{n+2}	$\cdots$	x_{n+m}
				Variable x_j				
0	d_1	d_2	$\cdots$	d_n	0	0	$\cdots$	0

Phase I $\begin{cases} 0 \\ 1 \\ \vdots \\ l \end{cases}$

Use the values of σ_i (if phase I) or π_i (if phase II) of the current cycle and compute
$$\bar{d}_j = d_j - (\sigma_1 a_{1j} + \sigma_2 a_{2j} + \cdots + \sigma_m a_{mj})$$
or
$$\bar{c}_j = c_j - (\pi_1 a_{1j} + \pi_2 a_{2j} + \cdots + \pi_m a_{mj})$$

Phase II $\begin{cases} l+1 \\ l+2 \\ \vdots \end{cases}$

Enter $\bar{d}_j$ or $\bar{c}_j$ in the row corresponding to the current cycle and choose the pivot column s such that $\bar{d}_s = \min \bar{d}_j$ (if phase I) or $\bar{c}_s = \min \bar{c}_j$ (if phase II)

Table 4.4 Tableau at the Beginning of Cycle k

Basic variable	$x_{n+1} \cdots x_{n+m}$	$-f$	$-w$	Value of the basic variable	$x_s{}^a$
	Columns of the original canonical form				
	$[\beta_{ij}] = [\bar{a}_{i,n+j}]$ $\leftarrow$ Inverse of the basis $\rightarrow$				
x_{j1}	$\beta_{11} \cdots \beta_{1m}$			$\bar{b}_1$	$\bar{a}_{1s} = \sum_{i=1}^{m} \beta_{1i} a_{is}$
$\vdots$	$\vdots \quad \vdots$			$\vdots$	
x_{jr}	$\beta_{r1} \cdots \beta_{rm}$			$\bar{b}_r$	$\bar{a}_{rs} = \sum_{i=1}^{m} \beta_{ri} a_{is}$
$\vdots$	$\vdots \quad \vdots$			$\vdots$	
x_{jm}	$\beta_{m1} \cdots \beta_{mm}$			$\bar{b}_m$	$\bar{a}_{ms} = \sum_{i=1}^{m} \beta_{mi} a_{is}$
$-f$	$-\pi_1 \cdots -\pi_m$ $(-\pi_j = +\bar{c}_{n+j})$	1		$-\bar{f}_0$	$\bar{c}_s = c_s - \sum_{i=1}^{m} \pi_i a_{is}$
$-w$	$-\sigma_1 \cdots -\sigma_m$ $(-\sigma_j = +\bar{d}_{n+j})$		1	$-\bar{w}_0$	$\bar{d}_s = d_s - \sum_{i=1}^{m} \sigma_i a_{is}$

^{a}This column is blank at the start of cycle k and is filled up only at the end of cycle k.

If some $\bar{d}_j < 0$, choose x_s as the variable to enter the basis in the next cycle in place of the present rth basic variable (r will be determined later) such that

$$\bar{d}_s = \min(\bar{d}_j < 0)$$

On the other hand, if the current cycle corresponds to phase II, find whether all $\bar{c}_j \geq 0$. If all $\bar{c}_j \geq 0$, the current basic feasible solution is also an optimal solution and hence terminate the process. If some $\bar{c}_j < 0$, choose x_s to enter the basic set in the next cycle in place of the rth basic variable (r to be found later), such that

$$\bar{c}_s = \min(\bar{c}_j < 0)$$

5. Compute the elements of the x_s column from Eq. (4.9) as

$$\bar{\mathbf{A}}_s = \mathbf{B}^{-1}\mathbf{A}_s = \bar{\beta}_{ij}\mathbf{A}_s$$

that is,

$$\bar{a}_{1s} = \beta_{11}a_{1s} + \beta_{12}a_{2s} + \cdots + \beta_{1m}a_{ms}$$
$$\bar{a}_{2s} = \beta_{21}a_{1s} + \beta_{22}a_{2s} + \cdots + \beta_{2m}a_{ms}$$
$$\vdots$$
$$\bar{a}_{ms} = \beta_{m1}a_{1s} + \beta_{m2}a_{2s} + \cdots + \beta_{mm}a_{ms}$$

and enter in the last column of Table 4.2 (if cycle 0) or Table 4.4 (if cycle k).

6. Inspect the signs of all entries $\bar{a}_{is}$, $i = 1$ to m. If all $\bar{a}_{is} \leq 0$, the class of solutions

$$x_s \geq 0 \text{ arbitrary}$$

$x_{ji} = \bar{b}_i - \bar{a}_{is} \cdot x_s$ if x_{ji} is a basic variable, and $x_j = 0$ if x_j is a nonbasic variable ($j \neq s$), satisfies the original system and has the property

$$f = \bar{f}_0 + \bar{c}_s x_s \to -\infty \qquad \text{as} \quad x_s \to +\infty$$

Hence terminate the process. On the other hand, if some $\bar{a}_{is} > 0$, select the variable x_r that can be dropped in the next cycle as

$$\frac{\bar{b}_r}{\bar{a}_{rs}} = \min_{\bar{a}_{is} > 0} (\bar{b}_i / \bar{a}_{is})$$

In the case of a tie, choose r at random.

7. To bring x_s into the basis in place of x_r, carry out a pivot operation on the element $\bar{a}_{rs}$ in Table 4.4 and enter the result as shown in Table 4.5. As usual, the last column of Table 4.5 will be left blank at the beginning of the current cycle $k + 1$. Also, retain the list of basic variables in the first column of Table 4.5 the same as in Table 4.4, except that j_r is changed to the value of s determined in step 4.

8. Go to step 3 to initiate the next cycle, $k + 1$.

Example 4.1

$$\text{Maximize } F = x_1 + 2x_2 + x_3$$

subject to

$$2x_1 + x_2 - x_3 \leq 2$$
$$-2x_1 + x_2 - 5x_3 \geq -6$$

Table 4.5 Tableau at the Beginning of Cycle $k + 1$

Basic variables	Columns of the canonical form					Value of the basic variable	$x_s{}^a$
	x_{n+1}	$\cdots$	x_{n+m}	$-f$	$-w$		
x_{j1}	$\beta_{11} - a_{1s}\beta_{r1}^*$	$\cdots$	$\beta_{1m} - \overline{a}_{1s}\beta_{rm}^*$			$\overline{b}_1 - \overline{a}_{1s}\overline{b}_r{}^*$	
$\vdots$							
x_s	β_{r1}^*	$\cdots$	β_{rm}^*			$\overline{b}_r{}^*$	
$\vdots$							
x_{jm}	$\beta_{m1} - \overline{a}_{ms}\beta_{r1}^*$	$\cdots$	$\beta_{mm} - \overline{a}_{ms}\beta_{rm}^*$			$\overline{b}_m - \overline{a}_{ms}\overline{b}_r^*$	
$-f$	$-\pi_1 - \overline{c}_s\beta_{r1}^*$	$\cdots$	$-\pi_m - \overline{c}_s\beta_{rm}^*$	1		$-\overline{f}_0 - \overline{c}_s\overline{b}_r^*$	
$-w$	$-\sigma_1 - \overline{d}_s\beta_{r1}^*$	$\cdots$	$-\sigma_m - \overline{d}_s\beta_{rm}^*$		1	$-\overline{w}_0 - \overline{d}_s\overline{b}_r^*$	

$$\beta_{ri}^* = \frac{\beta_{ri}}{\overline{a}_{rs}}(i = 1 \text{ to } m) \quad \text{and} \quad \overline{b}_r^* = \frac{\overline{b}_r}{\overline{a}_{rs}}$$

^{a}This column is blank at the start of the cycle.

$$4x_1 + x_2 + x_3 \le 6$$

$$x_1 \ge 0, \ x_2 \ge 0, \ x_3 \ge 0$$

SOLUTION This problem can be stated in standard form as (making all the constants b_i positive and then adding the slack variables):

Minimize

$$f = -x_1 - 2x_2 - x_3 \tag{E$_1$}$$

subject to

$$\begin{aligned}
2x_1 + x_2 - x_3 + x_4 \qquad\qquad &= 2 \\
2x_1 - x_2 + 5x_3 \qquad + x_5 \qquad &= 6 \\
4x_1 + x_2 + x_3 \qquad\qquad + x_6 &= 6 \\
x_i \ge 0, \quad i = 1 \text{ to } 6
\end{aligned} \tag{E$_2$}$$

where x_4, x_5, and x_6 are slack variables. Since the set of equations (E$_2$) are in canonical form with respect to x_4, x_5, and x_6, $x_i = 0$ $(i = 1, 2, 3)$ and $x_4 = 2$, $x_5 = 6$, and $x_6 = 6$ can be taken as an initial basic feasible solution and hence there is no need for phase I.

Step 1 All the equations (including the objective function) can be written in canonical form as

$$\begin{aligned}
2x_1 + x_2 - x_3 + x_4 \qquad\qquad &= 2 \\
2x_1 - x_2 + 5x_3 \qquad + x_5 \qquad &= 6 \\
4x_1 + x_2 + x_3 \qquad\qquad + x_6 \qquad &= 6 \\
-x_1 - 2x_2 - x_3 \qquad\qquad\qquad -f &= 0
\end{aligned} \tag{E$_3$}$$

These equations are written in tableau form in Table 4.6.

Table 4.6 Detached Coefficients of the Original System

	Admissible variables							
x_1	x_2	x_3	x_4	x_5	x_6	$-f$	Constants
2	1	-1	1	0	0		2
2	-1	5	0	1	0		6
4	1	1	0	0	1		6
-1	-2	-1	0	0	0	1	0

Table 4.7 Tableau at the Beginning of Cycle 0

Basic variables	Columns of the canonical form				Value of the basic variable (constant)	$x_2{}^a$
	x_4	x_5	x_6	$-f$		
x_4	1	0	0	0	2	$\boxed{\bar{a}_{42} = 1}$
						Pivot element
x_5	0	1	0	0	6	$\bar{a}_{52} = -1$
x_6	0	0	1	0	6	$\bar{a}_{62} = 1$
	Inverse of the basis $= [\beta_{ij}]$					
$-f$	0	0	0	1	0	$\bar{c}_2 = -2$

^{a}This column is entered at the end of step 5.

Step 2 The iterative procedure (cycle 0) starts with x_4, x_5, x_6, and $-f$ as basic variables. A tableau is opened by entering the coefficients of the basic variables and the constant terms as shown in Table 4.7. Since the basis matrix is $\mathbf{B} = \mathbf{I}$, its inverse $\mathbf{B}^{-1} = [\beta_{ij}] = \mathbf{I}$. The row corresponding to $-f$ in Table 4.7 gives the negative of simplex multipliers π_i, $i = 1, 2, 3$. These are all zero in cycle 0. The entries of the last column of the table are, of course, not yet known.

Step 3 The relative cost factors $\bar{c}_j$ are computed as

$$\bar{c}_j = c_j - \pi^T \mathbf{A}_j = c_j, \qquad j = 1 \text{ to } 6$$

since all π_i are zero. Thus

$$\bar{c}_1 = c_1 = -1$$
$$\bar{c}_2 = c_2 = -2$$
$$\bar{c}_3 = c_3 = -1$$
$$\bar{c}_4 = c_4 = 0$$
$$\bar{c}_5 = c_5 = 0$$
$$\bar{c}_6 = c_6 = 0$$

These cost coefficients are entered as the first row of a tableau (Table 4.8).

Table 4.8 Relative Cost Factors $\bar{c}_j$

Cycle number	Variable x_j					
	x_1	x_2	x_3	x_4	x_5	x_6
Phase II						
Cycle 0	-1	$\boxed{-2}$	-1	0	0	0
Cycle 1	3	0	$\boxed{-3}$	2	0	0
Cycle 2	6	0	0	$\frac{11}{4}$	$\frac{3}{4}$	0

Step 4 Find whether all $\bar{c}_j \geq 0$ for optimality. The present basic feasible solution is not optimal since some $\bar{c}_j$ are negative. Hence select a variable x_s to enter the basic set in the next cycle such that $\bar{c}_s = \min(\bar{c}_j < 0) = \bar{c}_2$ in this case. Therefore, x_2 enters the basic set.

Step 5 Compute the elements of the x_s column as

$$\overline{\mathbf{A}}_s = [\beta_{ij}]\mathbf{A}_s$$

where $[\beta_{ij}]$ is available in Table 4.7 and $\mathbf{A}_s$ in Table 4.6.

$$\overline{\mathbf{A}}_2 = \mathbf{I}\mathbf{A}_2 = \begin{Bmatrix} 1 \\ -1 \\ 1 \end{Bmatrix}$$

These elements, along with the value of $\bar{c}_2$, are entered in the last column of Table 4.7.

Step 6 Select a variable (x_r) to be dropped from the current basic set as

$$\frac{\bar{b}_r}{\bar{a}_{rs}} = \min_{\bar{a}_{is} > 0} \left(\frac{\bar{b}_i}{\bar{a}_{is}} \right)$$

In this case,

$$\frac{\bar{b}_4}{\bar{a}_{42}} = \frac{2}{1} = 2$$

$$\frac{\bar{b}_6}{\bar{a}_{62}} = \frac{6}{1} = 6$$

Therefore, $x_r = x_4$.

Step 7 To bring x_2 into the basic set in place of x_4, pivot on $a_{rs} = a_{42}$ in Table 4.7. Enter the result as shown in Table 4.9, keeping its last column blank. Since a new cycle has to be started, we go to step 3.

Step 3 The relative cost factors are calculated as

$$\bar{c}_j = c_j - (\pi_1 a_{1j} + \pi_2 a_{2j} + \pi_3 a_{3j})$$

Table 4.9   Tableau at the Beginning of Cycle 1

Basic variables	Columns of the original canonical form				Value of the basic variable	$x_3{}^a$
	x_4	x_5	x_6	$-f$		
x_2	1	0	0	0	2	$\bar{a}_{23} = -1$
x_5	1	1	0	0	8	$\boxed{\bar{a}_{53} = 4}$
						Pivot element
x_6	-1	0	1	1	4	$\bar{a}_{63} = 2$
	$\leftarrow$Inverse of the basis $= [\beta_{ij}] \rightarrow$					
$-f$	$2 = -\pi_1$	$0 = -\pi_2$	$0 = -\pi_3$	1	4	$\bar{c}_3 = -3$

^{a}This column is entered at the end of step 5.

where the negative values of π_1, π_2, and π_3 are given by the row of $-f$ in Table 4.9, and a_{ij} and c_i are given in Table 4.6. Here $\pi_1 = -2, \pi_2 = 0$, and $\pi_3 = 0$.

$$\bar{c}_1 = c_1 - \pi_1 a_{11} = -1 - (-2)(2) = 3$$
$$\bar{c}_2 = c_2 - \pi_1 a_{12} = -2 - (-2)(1) = 0$$
$$\bar{c}_3 = c_3 - \pi_1 a_{13} = -1 - (-2)(-1) = -3$$
$$\bar{c}_4 = c_4 - \pi_1 a_{14} = 0 - (-2)(1) = 2$$
$$\bar{c}_5 = c_5 - \pi_1 a_{15} = 0 - (-2)(0) = 0$$
$$\bar{c}_6 = c_6 - \pi_1 a_{16} = 0 - (-2)(0) = 0$$

Enter these values in the second row of Table 4.8.

Step 4 Since all $\bar{c}_j$ are not ≥ 0, the current solution is not optimum. Hence select a variable (x_s) to enter the basic set in the next cycle such that $\bar{c}_s = \min(\bar{c}_j < 0) = \bar{c}_3$ in this case. Therefore, $x_s = x_3$.

Step 5 Compute the elements of the x_s column as

$$\overline{\mathbf{A}}_s = [\beta_{ij}]\mathbf{A}_s$$

where $[\beta_{ij}]$ is available in Table 4.9 and $\mathbf{A}_s$ in Table 4.6:

$$\overline{\mathbf{A}}_3 = \begin{Bmatrix} \bar{a}_{23} \\ \bar{a}_{53} \\ \bar{a}_{63} \end{Bmatrix} = \begin{bmatrix} 1 & 0 & 0 \\ 1 & 1 & 0 \\ -1 & 0 & 1 \end{bmatrix} \begin{Bmatrix} -1 \\ 5 \\ 1 \end{Bmatrix} = \begin{Bmatrix} -1 \\ 4 \\ 2 \end{Bmatrix}$$

Enter these elements and the value of $\bar{c}_s = \bar{c}_3 = -3$ in the last column of Table 4.9.

Step 6 Find the variable (x_r) to be dropped from the basic set in the next cycle as

$$\frac{\bar{b}_r}{\bar{a}_{rs}} = \min_{\bar{a}_{is} > 0} \left(\frac{\bar{b}_i}{\bar{a}_{is}} \right)$$

Table 4.10 Tableau at the Beginning of Cycle 2

Basic variables	Columns of the original canonical form				Value of the basic variable	x_s [a]
	x_4	x_5	x_6	$-f$		
x_2	$\frac{5}{4}$	$\frac{1}{4}$	0	0	4	
x_3	$\frac{1}{4}$	$\frac{1}{4}$	0	0	2	
x_6	$-\frac{6}{4}$	$-\frac{2}{4}$	1	1	0	
$-f$	$\frac{11}{4}$	$\frac{3}{4}$	0	1	10	

[a]This column is blank at the beginning of cycle 2.

Here

$$\frac{\bar{b}_5}{\bar{a}_{53}} = \frac{8}{4} = 2$$

$$\frac{\bar{b}_6}{\bar{a}_{63}} = \frac{4}{2} = 2$$

Since there is a tie between x_5 and x_6, we select $x_r = x_5$ arbitrarily.

Step 7 To bring x_3 into the basic set in place of x_5, pivot on $\bar{a}_{rs} = \bar{a}_{53}$ in Table 4.9. Enter the result as shown in Table 4.10, keeping its last column blank. Since a new cycle has to be started, we go to step 3.

Step 3 The simplex multipliers are given by the negative values of the numbers appearing in the row of $-f$ in Table 4.10. Therefore, $\pi_1 = -\frac{11}{4}, \pi_2 = -\frac{3}{4}$, and $\pi_3 = 0$. The relative cost factors are given by

$$\bar{c}_j = c_j = -\pi^T A_j$$

Then

$$\bar{c}_1 = c_1 - \pi_1 a_{11} - \pi_2 a_{21} = -1 - (-\tfrac{11}{4})(2) - (-\tfrac{3}{4})(2) = 6$$

$$\bar{c}_2 = c_2 - \pi_1 a_{12} - \pi_2 a_{22} = -2 - (-\tfrac{11}{4})(1) - (-\tfrac{3}{4})(-1) = 0$$

$$\bar{c}_3 = c_3 - \pi_1 a_{13} - \pi_2 a_{23} = -1 - (-\tfrac{11}{4})(-1) - (-\tfrac{3}{4})(5) = 0$$

$$\bar{c}_4 = c_4 - \pi_1 a_{14} - \pi_2 a_{24} = 0 - (-\tfrac{11}{4})(1) - (-\tfrac{3}{4})(0) = \tfrac{11}{4}$$

$$\bar{c}_5 = c_5 - \pi_1 a_{15} - \pi_2 a_{25} = 0 - (-\tfrac{11}{4})(0) - (-\tfrac{3}{4})(1) = \tfrac{3}{4}$$

$$\bar{c}_6 = c_6 - \pi_1 a_{16} - \pi_2 a_{26} = 0 - (-\tfrac{11}{4})(0) - (-\tfrac{3}{4})(0) = 0$$

These values are entered as third row in Table 4.8.

Step 4 Since all $\bar{c}_j$ are ≥ 0, the present solution will be optimum. Hence the optimum solution is given by

$$x_2 = 4, \quad x_3 = 2, \quad x_6 = 0 \text{ (basic variables)}$$

$$x_1 = x_4 = x_5 = 0 \text{ (nonbasic variables)}$$

$$f_{\min} = -10$$

4.3 DUALITY IN LINEAR PROGRAMMING

Associated with every linear programming problem, called the *primal*, there is another linear programming problem called its *dual*. These two problems possess very interesting and closely related properties. If the optimal solution to any one is known, the optimal solution to the other can readily be obtained. In fact, it is immaterial which problem is designated the primal since the dual of a dual is the primal. Because of these properties, the solution of a linear programming problem can be obtained by solving either the primal or the dual, whichever is easier. This section deals with the primal–dual relations and their application in solving a given linear programming problem.

4.3.1 Symmetric Primal–Dual Relations

A nearly symmetric relation between a primal problem and its dual problem can be seen by considering the following system of linear inequalities (rather than equations).

Primal Problem.

$$a_{11}x_1 + a_{12}x_2 + \cdots + a_{1n}x_n \geq b_1$$

$$a_{21}x_1 + a_{22}x_2 + \cdots + a_{2n}x_n \geq b_2$$

$$\vdots \tag{4.17}$$

$$a_{m1}x_1 + a_{m2}x_2 + \cdots + a_{mn}x_n \geq b_m$$

$$c_1x_1 + c_2x_2 + \cdots + c_nx_n = f$$

$$(x_i \geq 0, i = 1 \text{ to } n, \text{ and } f \text{ is to be minimized})$$

Dual Problem. As a definition, the dual problem can be formulated by transposing the rows and columns of Eq. (4.17) including the right-hand side and the objective function, reversing the inequalities and maximizing instead of minimizing. Thus by denoting the dual variables as $y_1, y_2, \ldots, y_m$, the dual problem becomes

$$a_{11}y_1 + a_{21}y_2 + \cdots + a_{m1}y_m \leq c_1$$

$$a_{12}y_1 + a_{22}y_2 + \cdots + a_{m2}x_m \leq c_2$$

$$\vdots \tag{4.18}$$

$$a_{1n}y_1 + a_{2n}y_2 + \cdots + a_{mn}y_m \leq c_n$$

$$b_1y_1 + b_2y_2 + \cdots + b_my_m = v$$

$$(y_i \geq 0, i = 1 \text{ to } m, \text{ and } v \text{ is to be maximized})$$

Equations (4.17) and (4.18) are called *symmetric primal–dual pairs* and it is easy to see from these relations that the dual of the dual is the primal.

4.3.2   General Primal–Dual Relations

Although the primal–dual relations of Section 4.3.1 are derived by considering a system of inequalities in nonnegative variables, it is always possible to obtain the primal–dual relations for a general system consisting of a mixture of equations, less than or greater than type of inequalities, nonnegative variables or variables unrestricted in sign by reducing the system to an equivalent inequality system of Eqs. (4.17). The correspondence rules that are to be applied in deriving the general primal–dual relations are given in Table 4.11 and the primal–dual relations are shown in Table 4.12.

4.3.3   Primal–Dual Relations When the Primal Is in Standard Form

If $m^* = m$ and $n^* = n$, primal problem shown in Table 4.12 reduces to the standard form and the general primal–dual relations take the special form shown in Table 4.13. It is to be noted that the symmetric primal–dual relations, discussed in Section 4.3.1, can also be obtained as a special case of the general relations by setting $m^* = 0$ and $n^* = n$ in the relations of Table 4.12.

Table 4.11   Correspondence Rules for Primal–Dual Relations

Primal quantity	Corresponding dual quantity
Objective function: Minimize $\mathbf{c}^T\mathbf{X}$	Maximize $\mathbf{Y}^T\mathbf{b}$
Variable $x_i \geq 0$	ith constraint $\mathbf{Y}^T\mathbf{A}_i \leq c_i$ (inequality)
Variable x_i unrestricted in sign	ith constraint $\mathbf{Y}^T\mathbf{A}_i = c_i$ (equality)
jth constraint, $\mathbf{A}_j\mathbf{X} = b_j$ (equality)	jth variable y_j unrestricted in sign
jth constraint, $\mathbf{A}_j\mathbf{X} \geq b_j$ (inequality)	jth variable $y_j \geq 0$
Coefficient matrix $\mathbf{A} \equiv [\mathbf{A}_1 \ldots \mathbf{A}_m]$	Coefficient matrix $\mathbf{A}^T \equiv [\mathbf{A}_1, \ldots, \mathbf{A}_m]^T$
Right-hand-side vector $\mathbf{b}$	Right-hand-side vector $\mathbf{c}$
Cost coefficients $\mathbf{c}$	Cost coefficients $\mathbf{b}$

Table 4.12   Primal–Dual Relations

Primal problem	Corresponding dual problem
Minimize $f = \sum\limits_{i=1}^{n} c_i x_i$ subject to	Maximize $v = \sum\limits_{i=1}^{m} y_i b_i$ subject to
$\sum\limits_{j=1}^{n} a_{ij}x_j = b_i, \ i = 1, 2, \ldots, m^*$	$\sum\limits_{i=1}^{m} y_i a_{ij} = c_j, \ j = n^* + 1, n^* + 2,$
$\sum\limits_{j=1}^{n} a_{ij}x_j \geq b_i, \ i = m^* + 1, m^* + 2,$	$\ldots, n$
$\ldots, m$	$\sum\limits_{i=1}^{m} y_i a_{ij} \leq c_j, j = 1, 2, \ldots, n^*$
where	where
$x_i \geq 0, i = 1, 2, \ldots, n^*$;	$y_i \geq 0, i = m^* + 1, m^* + 2, \ldots, m$;
and	and
x_i unrestricted in sign, $i = n^* + 1,$ $n^* + 2, \ldots, n$	y_i unrestricted in sign, $i = 1, 2, \ldots, m^*$

Table 4.13 Primal–Dual Relations Where $m^* = m$ and $n^* = n$

Primal problem	Corresponding dual problem
Minimize $f = \sum\limits_{i=1}^{n} c_i x_i$	Maximize $v = \sum\limits_{i=1}^{m} b_i y_i$
subject to	subject to
$\sum\limits_{j=1}^{n} a_{ij} x_j = b_i$, $i = 1, 2, \ldots, m$	$\sum\limits_{i=1}^{m} y_i a_{ij} \leq c_j$, $j = 1, 2, \ldots, n$
where	where
$\quad x_i \geq 0$, $i = 1, 2, \ldots, n$	$\quad y_i$ is unrestricted in sign, $i = 1, 2, \cdots, m$
In matrix form	*In matrix form*
$\quad$ Minimize $f = \mathbf{c}^T\mathbf{X}$	$\quad$ Maximize $v = \mathbf{Y}^T\mathbf{b}$
subject to	subject to
$\quad \mathbf{AX} = \mathbf{b}$	$\quad \mathbf{A}^T\mathbf{Y} \leq \mathbf{c}$
where	where
$\quad \mathbf{X} \geq \mathbf{0}$	$\quad \mathbf{Y}$ is unrestricted in sign

Example 4.2 Write the dual of the following linear programming problem:

$$\text{Maximize } f = 50x_1 + 100x_2$$

subject to

$$\left. \begin{array}{r} 2x_1 + x_2 \leq 1250 \\ 2x_1 + 5x_2 \leq 1000 \\ 2x_1 + 3x_2 \leq 900 \\ x_2 \leq 150 \end{array} \right\} n = 2,\ m = 4$$

where

$$x_1 \geq 0 \quad \text{and} \quad x_2 \geq 0$$

SOLUTION Let y_1, y_2, y_3, and y_4 be the dual variables. Then the dual problem can be stated as

$$\text{Minimize } v = 1250y_1 + 1000y_2 + 900y_3 + 150y_4$$

subject to

$$2y_1 + 2y_2 + 2y_3 \geq 50$$

$$y_1 + 5y_2 + 3y_3 + y_4 \geq 100$$

$$\text{where } y_1 \geq 0,\ y_2 \geq 0,\ y_3 \geq 0,\ \text{and } y_4 \geq 0.$$

Notice that the dual problem has a lesser number of constraints compared to the primal problem in this case. Since, in general, an additional constraint requires more computational effort than an additional variable in a linear programming problem, it is evident that it is computationally more efficient to solve the dual problem in the present case. This is one of the advantages of the dual problem.

4.3.4   Duality Theorems

The following theorems are useful in developing a method for solving LP problems using dual relationships. The proofs of these theorems can be found in Ref. [4.10].

Theorem 4.1 The dual of the dual is the primal.

Theorem 4.2 Any feasible solution of the primal gives an f value greater than or at least equal to the v value obtained by any feasible solution of the dual.

Theorem 4.3 If both primal and dual problems have feasible solutions, both have optimal solutions and minimum f = maximum v.

Theorem 4.4 If either the primal or the dual problem has an unbounded solution, the other problem is infeasible.

4.3.5   Dual Simplex Method

There exist a number of situations in which it is required to find the solution of a linear programming problem for a number of different right-hand-side vectors $\mathbf{b}^{(i)}$. Similarly, in some cases, we may be interested in adding some more constraints to a linear programming problem for which the optimal solution is already known. When the problem has to be solved for different vectors $\mathbf{b}^{(i)}$, one can always find the desired solution by applying the two phases of the simplex method separately for each vector $\mathbf{b}^{(i)}$. However, this procedure will be inefficient since the vectors $\mathbf{b}^{(i)}$ often do not differ greatly from one another. Hence the solution for one vector, say, $\mathbf{b}^{(1)}$ may be close to the solution for some other vector, say, $\mathbf{b}^{(2)}$. Thus a better strategy is to solve the linear programming problem for $\mathbf{b}^{(1)}$ and obtain an optimal basis matrix $\mathbf{B}$. If this basis happens to be feasible for all the right-hand-side vectors, that is, if

$$\mathbf{B}^{-1}\mathbf{b}^{(i)} \geq \mathbf{0} \quad \text{for} \quad \text{all } i \tag{4.19}$$

then it will be optimal for all cases. On the other hand, if the basis $\mathbf{B}$ is not feasible for some of the right-hand-side vectors, that is, if

$$\mathbf{B}^{-1}\mathbf{b}^{(r)} < 0 \quad \text{for} \quad \text{some } r \tag{4.20}$$

then the vector of simplex multipliers

$$\boldsymbol{\pi}^{\text{T}} = \mathbf{c}_B^{\text{T}}\mathbf{B}^{-1} \tag{4.21}$$

will form a dual feasible solution since the quantities

$$\bar{c}_j = c_j - \boldsymbol{\pi}^{\text{T}}\mathbf{A}_j \geq 0$$

are independent of the right-hand-side vector $\mathbf{b}^{(r)}$. A similar situation exists when the problem has to be solved with additional constraints.

   In both the situations discussed above, we have an infeasible basic (primal) solution whose associated dual solution is feasible. Several methods have been proposed, as variants of the regular simplex method, to solve a linear programming problem by starting from an infeasible solution to the primal. All these methods work in an iterative manner such that they force the solution to become feasible as well as optimal simultaneously at some stage. Among all the methods, the dual simplex method developed by Lemke [4.2] and the primal–dual method developed by Dantzig, Ford, and Fulkerson [4.3] have been most widely used. Both these methods have the following important characteristics:

1. They do not require the phase I computations of the simplex method. This is a desirable feature since the starting point found by phase I may be nowhere near optimal, since the objective of phase I ignores the optimality of the problem completely.
2. Since they work toward feasibility and optimality simultaneously, we can expect to obtain the solution in a smaller total number of iterations.

We shall consider only the dual simplex algorithm in this section.

Algorithm.   As stated earlier, the dual simplex method requires the availability of a dual feasible solution that is not primal feasible to start with. It is the same as the simplex method applied to the dual problem but is developed such that it can make use of the same tableau as the primal method. Computationally, the dual simplex algorithm also involves a sequence of pivot operations, but with different rules (compared to the regular simplex method) for choosing the pivot element.

Let the problem to be solved be initially in canonical form with some of the $\bar{b}_i < 0$, the relative cost coefficients corresponding to the basic variables $\bar{c}_j = 0$, and all other $\bar{c}_j \geq 0$. Since some of the $\bar{b}_i$ are negative, the primal solution will be infeasible, and since all $\bar{c}_j \geq 0$, the corresponding dual solution will be feasible. Then the simplex method works according to the following iterative steps.

1. Select row r as the pivot row such that

$$\bar{b}_r = \min \bar{b}_i < 0 \tag{4.22}$$

2. Select column s as the pivot column such that

$$\frac{\bar{c}_s}{-\bar{a}_{rs}} = \min_{\bar{a}_{rj} < 0} \left(\frac{\bar{c}_j}{-\bar{a}_{rj}} \right) \tag{4.23}$$

   If all $\bar{a}_{rj} \geq 0$, the primal will not have any feasible (optimal) solution.
3. Carry out a pivot operation on $\bar{a}_{rs}$
4. Test for optimality: If all $\bar{b}_i \geq 0$, the current solution is optimal and hence stop the iterative procedure. Otherwise, go to step 1.

Remarks:

1. Since we are applying the simplex method to the dual, the dual solution will always be maintained feasible, and hence all the relative cost factors of the primal ($\bar{c}_j$) will be nonnegative. Thus the optimality test in step 4 is valid because it guarantees that all $\bar{b}_i$ are also nonnegative, thereby ensuring a feasible solution to the primal.

2. We can see that the primal will not have a feasible solution when all $\bar{a}_{rj}$ are nonnegative from the following reasoning. Let $(x_1, x_2, \ldots, x_m)$ be the set of basic variables. Then the rth basic variable, x_r, can be expressed as

$$x_r = \bar{b}_r - \sum_{j=m+1}^{n} \bar{a}_{rj} x_j$$

It can be seen that if $\bar{b}_r < 0$ and $\bar{a}_{rj} \geq 0$ for all j, x_r cannot be made non-negative for any nonnegative value of x_j. Thus the primal problem contains an equation (the rth one) that cannot be satisfied by any set of nonnegative variables and hence will not have any feasible solution.

The following example is considered to illustrate the dual simplex method.

Example 4.3

$$\text{Minimize } f = 20x_1 + 16x_2$$

subject to

$$x_1 \geq 2.5$$
$$x_2 \geq 6$$
$$2x_1 + x_2 \geq 17$$
$$x_1 + x_2 \geq 12$$
$$x_1 \geq 0, \ x_2 \geq 0$$

SOLUTION   By introducing the surplus variables x_3, x_4, x_5, and x_6, the problem can be stated in canonical form as
Minimize f
with

$$
\begin{aligned}
-x_1 \qquad\qquad + x_3 \qquad\qquad\qquad\qquad\qquad &= -2.5 \\
- x_2 \qquad\quad + x_4 \qquad\qquad\qquad\qquad &= -6 \\
-2x_1 - x_2 \qquad\qquad\quad + x_5 \qquad\qquad &= -17 \\
-x_1 - x_2 \qquad\qquad\qquad\quad + x_6 \qquad &= -12 \\
20x_1 + 16x_2 \qquad\qquad\qquad\qquad\quad - f &= 0 \\
x_i \geq 0, \quad i = 1 \text{ to } 6 &
\end{aligned}
\tag{E_1}
$$

The basic solution corresponding to (E_1) is infeasible since $x_3 = -2.5$, $x_4 = -6$, $x_5 = -17$, and $x_6 = -12$. However, the objective equation shows optimality since the cost coefficients corresponding to the nonbasic variables are nonnegative $(\bar{c}_1 = 20, \bar{c}_2 = 16)$. This shows that the solution is infeasible to the primal but feasible to the dual. Hence the dual simplex method can be applied to solve this problem as follows.

Step 1 Write the system of equations (E_1) in tableau form:

Basic variables	Variables x_1	x_2	x_3	x_4	x_5	x_6	$-f$	$\bar{b}_i$	
x_3	-1	0	1	0	0	0	0	-2.5	
x_4	0	-1	0	1	0	0	0	-6	
x_5	$\boxed{-2}$	-1	0	0	1	0	0	-17	← Minimum, pivot row
	Pivot element								
x_6	-1	-1	0	0	0	1	0	-12	
$-f$	20	16	0	0	0	0	1	0	

Select the pivotal row r such that

$$\bar{b}_r = \min(\bar{b}_i < 0) = \bar{b}_3 = -17$$

in this case. Hence $r = 3$.

Step 2 Select the pivotal column s as

$$\frac{\bar{c}_s}{-\bar{a}_{rs}} = \min_{\bar{a}_{rj} < 0} \left(\frac{\bar{c}_j}{-\bar{a}_{rj}} \right)$$

Since

$$\frac{\bar{c}_1}{-\bar{a}_{31}} = \frac{20}{2} = 10, \quad \frac{\bar{c}_2}{-\bar{a}_{32}} = \frac{16}{1} = 16, \quad \text{and} \quad s = 1$$

Step 3 The pivot operation is carried on $\bar{a}_{31}$ in the preceding table, and the result is as follows:

Basic variables	Variables x_1	x_2	x_3	x_4	x_5	x_6	$-f$	$\bar{b}_i$	
x_3	0	$\frac{1}{2}$	1	0	$-\frac{1}{2}$	0	0	6	
x_4	0	$\boxed{-1}$	0	1	0	0	0	-6	← Minimum, pivot row
	Pivot element								
x_1	1	$\frac{1}{2}$	0	0	$-\frac{1}{2}$	0	0	$\frac{17}{2}$	
x_6	0	$-\frac{1}{2}$	0	0	$-\frac{1}{2}$	1	0	$-\frac{7}{2}$	
$-f$	0	6	0	0	10	0	1	-170	

Step 4 Since some of the $\bar{b}_i$ are < 0, the present solution is not optimum. Hence we proceed to the next iteration.

Step 1 The pivot row corresponding to minimum ($\bar{b}_i < 0$) can be seen to be 2 in the preceding table.

Step 2 Since $\bar{a}_{22}$ is the only negative coefficient, it is taken as the pivot element.
Step 3 The result of pivot operation on $\bar{a}_{22}$ in the preceding table is as follows:

Basic variables	x_1	x_2	x_3	x_4	x_5	x_6	$-f$	$\bar{b}_i$	
x_3	0	0	1	$\frac{1}{2}$	$-\frac{1}{2}$	0	0	3	
x_2	0	1	0	-1	0	0	0	6	
x_1	1	0	0	$\frac{1}{2}$	$-\frac{1}{2}$	0	0	$\frac{11}{2}$	
x_6	0	0	0	$\boxed{-\frac{1}{2}}$	$-\frac{1}{2}$	1	0	$-\frac{1}{2}$	← Minimum, pivot row
				Pivot element					
$-f$	0	0	0	6	10	0	1	-206	

Step 4 Since all $\bar{b}_i$ are not ≥ 0, the present solution is not optimum. Hence we go to the next iteration.
Step 1 The pivot row (corresponding to minimum $\bar{b}_i \leq 0$) can be seen to be the fourth row.
Step 2 Since

$$\frac{\bar{c}_4}{-\bar{a}_{44}} = 12 \quad \text{and} \quad \frac{\bar{c}_5}{-\bar{a}_{45}} = 20$$

the pivot column is selected as $s = 4$.
Step 3 The pivot operation is carried on $\bar{a}_{44}$ in the preceding table, and the result is as follows:

Basic variables	x_1	x_2	x_3	x_4	x_5	x_6	$-f$	$\bar{b}_i$
x_3	0	0	1	0	-1	1	0	$\frac{5}{2}$
x_2	0	1	0	0	1	-2	0	7
x_1	1	0	0	0	-1	1	0	5
x_4	0	0	0	1	1	-2	0	1
$-f$	0	0	0	0	4	12	1	-212

Step 4 Since all $\bar{b}_i$ are ≥ 0, the present solution is dual optimal and primal feasible. The solution is

$$x_1 = 5, \quad x_2 = 7, \quad x_3 = \tfrac{5}{2}, \quad x_4 = 1 \quad \text{(dual basic variables)}$$

$$x_5 = x_6 = 0 \quad \text{(dual nonbasic variables)}$$

$$f_{\min} = 212$$

4.4 DECOMPOSITION PRINCIPLE

Some of the linear programming problems encountered in practice may be very large in terms of the number of variables and/or constraints. If the problem has some special structure, it is possible to obtain the solution by applying the decomposition principle developed by Dantzing and Wolfe [4.4]. In the decomposition method, the original problem is decomposed into small subproblems and then these subproblems are solved almost independently. The procedure, when applicable, has the advantage of making it possible to solve large-scale problems that may otherwise be computationally very difficult or infeasible. As an example of a problem for which the decomposition principle can be applied, consider a company having two factories, producing three and two products, respectively. Each factory has its own internal resources for production, namely, workers and machines. The two factories are coupled by the fact that there is a shared resource that both use, for example, a raw material whose availability is limited. Let b_2 and b_3 be the maximum available internal resources for factory 1, and let b_4 and b_5 be the similar availabilities for factory 2. If the limitation on the common resource is b_1, the problem can be stated as follows:

$$\text{Minimize } f(x_1, x_2, x_3, y_1, y_2) = c_1 x_1 + c_2 x_2 + c_3 x_3 + c_4 y_1 + c_5 y_2$$

subject to

$$\boxed{a_{11} x_1 + a_{12} x_2 + a_{13} x_3 + a_{14} y_1 + a_{15} y_2} \leq b_1$$

$$\boxed{\begin{array}{l} a_{21} x_1 + a_{22} x_2 + a_{23} x_3 \\ a_{31} x_1 + a_{32} x_2 + a_{33} x_2 \end{array}} \quad \begin{array}{l} \leq b_2 \\ \leq b_3 \end{array} \qquad (4.24)$$

$$\boxed{\begin{array}{l} a_{41} y_1 + a_{42} y_2 \\ a_{51} y_1 + a_{52} y_2 \end{array}} \quad \begin{array}{l} \leq b_4 \\ \leq b_5 \end{array}$$

where x_i and y_j are the quantities of the various products produced by the two factories (design variables) and the a_{ij} are the quantities of resource i required to produce 1 unit of product j.

$$\underset{(i=1,2,3)}{x_i \geq 0,} \qquad \underset{(j=1,2)}{y_j \geq 0}$$

An important characteristic of the problem stated in Eqs. (4.24) is that its constraints consist of two independent sets of inequalities. The first set consists of a coupling constraint involving all the design variables, and the second set consists of two groups of constraints, each group containing the design variables of that group only. This problem can be generalized as follows:

$$\text{Minimize } f(\mathbf{X}) = \mathbf{c}_1^T \mathbf{X}_1 + \mathbf{c}_2^T \mathbf{X}_2 + \cdots + \mathbf{c}_p^T \mathbf{X}_p \qquad (4.25a)$$

subject to

$$A_1X_1 + A_2X_2 + \cdots + A_pX_p = b_0 \qquad (4.25b)$$

$$\left.\begin{array}{rl} B_1X_1 & = b_1 \\ B_2X_2 & = b_2 \\ & \vdots \\ B_pX_p & = b_p \end{array}\right\} \qquad (4.25c)$$

$$X_1 \geq 0, \quad X_2 \geq 0, \cdots, X_p \geq 0$$

where

$$X_1 = \left\{\begin{array}{c} x_1 \\ x_2 \\ \vdots \\ x_{m1} \end{array}\right\}, \quad X_2 = \left\{\begin{array}{c} x_{m1+1} \\ x_{m1+2} \\ \vdots \\ x_{m1+m2} \end{array}\right\}, \cdots,$$

$$X_p = \left\{\begin{array}{c} x_{m1+m2+\cdots+m_{p-1}+1} \\ x_{m1+m2+\cdots+m_{p-1}+2} \\ x_{m1+m2+\cdots+m_{p-1}+m_p} \end{array}\right\}$$

$$X = \left\{\begin{array}{c} X_1 \\ X_2 \\ \vdots \\ X_p \end{array}\right\}$$

It can be noted that if the size of the matrix A_k is $(r_0 \times m_k)$ and that of B_k is $(r_k \times m_k)$, the problem has $\sum_{k=0}^{p} r_k$ constraints and $\sum_{k=1}^{p} m_k$ variables.

Since there are a large number of constraints in the problem stated in Eqs. (4.25), it may not be computationally efficient to solve it by using the regular simplex method. However, the decomposition principle can be used to solve it in an efficient manner. The basic solution procedure using the decomposition principle is given by the following steps.

1. Define p subsidiary constraint sets using Eqs. (4.25) as

$$B_1X_1 = b_1$$

$$B_2X_2 = b_2$$

$$\vdots$$

$$B_kX_k = b_k \qquad (4.26)$$

$$\vdots$$

$$B_pX_p = b_p$$

The subsidiary constraint set

$$B_kX_k = b_k, \quad k = 1, 2, \ldots, p \qquad (4.27)$$

represents r_k equality constraints. These constraints along with the requirement $X_k \geq 0$ define the set of feasible solutions of Eqs. (4.27). Assuming that this set of feasible solutions is a bounded convex set, let s_k be the number of vertices of this set. By using the definition of convex combination of a set of points,[†] any point $\mathbf{X}_k$ satisfying Eqs. (4.27) can be represented as

$$\mathbf{X}_k = \mu_{k,1}\mathbf{X}_1^{(k)} + \mu_{k,2}\mathbf{X}_2^{(k)} + \cdots + \mu_{k,s_k}\mathbf{X}_{s_k}^{(k)} \tag{4.28}$$

$$\mu_{k,1} + \mu_{k,2} + \cdots + \mu_{k,s_k} = 1 \tag{4.29}$$

$$0 \le \mu_{k,i} \le 1, \qquad i = 1, 2, \ldots, s_k, \quad k = 1, 2, \ldots, p \tag{4.30}$$

where $\mathbf{X}_1^{(k)}, \mathbf{X}_2^{(k)}, \ldots, \mathbf{X}_{s_k}^{(k)}$ are the extreme points of the feasible set defined by Eqs. (4.27). These extreme points $\mathbf{X}_1^{(k)}, \mathbf{X}_2^{(k)}, \ldots, \mathbf{X}_{sk}^{(k)}; k = 1, 2, \ldots, p$, can be found by solving the Eqs. (4.27).

2. These new Eqs. (4.28) imply the complete solution space enclosed by the constraints

$$\mathbf{B}_k\mathbf{X}_k = \mathbf{b}_k$$
$$\mathbf{X}_k \ge \mathbf{0}, \qquad k = 1, 2, \ldots, p \tag{4.31}$$

By substituting Eqs. (4.28) into Eqs. (4.25), it is possible to eliminate the subsidiary constraint sets from the original problem and obtain the following equivalent form:

$$\text{Minimize } f(\mathbf{X}) = \mathbf{c}_1^{\mathrm{T}}\left(\sum_{i=1}^{s1}\mu_{1,i}\mathbf{X}_i^{(1)}\right) + \mathbf{c}_2^{\mathrm{T}}\left(\sum_{i=1}^{s2}\mu_{2,i}\mathbf{X}_i^{(2)}\right)$$
$$+ \cdots + \mathbf{c}_p^{\mathrm{T}}\left(\sum_{i=1}^{sp}\mu_{p,i}\mathbf{X}_i^{(p)}\right)$$

subject to

$$\mathbf{A}_1\left(\sum_{i=1}^{s_1}\mu_{1,i}\mathbf{X}_i^{(1)}\right) + \mathbf{A}_2\left(\sum_{i=1}^{s_2}\mu_{2,i}\mathbf{X}_i^{(2)}\right) + \cdots + \mathbf{A}_p\left(\sum_{i=1}^{s_p}\mu_{p,i}\mathbf{X}_i^{(p)}\right) = \mathbf{b}_0$$

$$\sum_{i=1}^{s_1}\mu_{1,i} = 1$$

$$\sum_{i=1}^{s_2}\mu_{2,i} = 1$$

$$\sum_{i=1}^{s_p}\mu_{p,i} = 1$$

[†]If $\mathbf{X}^{(1)}$ and $\mathbf{X}^{(2)}$ are any two points in an n-dimensional space, any point lying on the line segment joining $\mathbf{X}^{(1)}$ and $\mathbf{X}^{(2)}$ is given by a convex combination of $\mathbf{X}^{(1)}$ and $\mathbf{X}^{(2)}$ as

$$\mathbf{X}(\mu) = \mu\,\mathbf{X}^{(1)} + (1 - \mu)\,\mathbf{X}^{(2)}, \quad 0 \le \mu \le 1$$

This idea can be generalized to define the convex combination of r points $\mathbf{X}^{(1)}, \mathbf{X}^{(2)}, \ldots, \mathbf{X}^{(r)}$ as

$$\mathbf{X}(\mu_1, \mu_2, \cdots, \mu_r) = \mu_1\mathbf{X}^{(1)} + \mu_2\mathbf{X}^{(2)} + \cdots + \mu_r\mathbf{X}^{(r)}$$

where $\mu_1 + \mu_2 + \cdots + \mu_r = 1$ and $0 \le \mu_i \le 1$, $i = 1, 2, \ldots, r$.

$$\mu_{j,i} \geq 0, \quad i = 1, 2, \ldots, s_j, \quad j = 1, 2, \ldots, p \qquad (4.32)$$

Since the extreme points $\mathbf{X}_1^{(k)}, \mathbf{X}_2^{(k)}, \ldots, \mathbf{X}_{s_k}^{(k)}$ are known from the solution of the set $\mathbf{B}_k \mathbf{X}_k = \mathbf{b}_k, \mathbf{X}_k \geq \mathbf{0}, k = 1, 2, \ldots, p$, and since $\mathbf{c}_k$ and $\mathbf{A}_k$, $k = 1, 2, \ldots, p$, are known as problem data, the unknowns in Eqs. (4.32) are $\mu_{j,i}$, $i = 1, 2, \ldots, s_j$; $j = 1, 2, \ldots, p$. Hence $\mu_{j,i}$ will be the new decision variables of the modified problem stated in Eqs. (4.32).

3. Solve the linear programming problem stated in Eqs. (4.32) by any of the known techniques and find the optimal values of $\mu_{j,i}$. Once the optimal values $\mu_{j,i}^*$ are determined, the optimal solution of the original problem can be obtained as

$$\mathbf{X}^* = \left\{ \begin{array}{c} \mathbf{X}_1^* \\ \mathbf{X}_2^* \\ \vdots \\ \mathbf{X}_p^* \end{array} \right\}$$

where

$$\mathbf{X}_k^* = \sum_{i=1}^{s_k} \mu_{k,i}^* \mathbf{X}_i^{(k)}, \quad k = 1, 2, \ldots, p$$

Remarks:

1. It is to be noted that the new problem in Eqs. (4.32) has $(r_0 + p)$ equality constraints only as against $r_0 + \sum_{k=1}^{p} r_k$ in the original problem of Eq. (4.25). Thus there is a substantial reduction in the number of constraints due to the application of the decomposition principle. At the same time, the number of variables might increase from $\sum_{k=1}^{p} m_k$ to $\sum_{k=1}^{p} s_k$, depending on the number of extreme points of the different subsidiary problems defined by Eqs. (4.31). The modified problem, however, is computationally more attractive since the computational effort required for solving any linear programming problem depends primarily on the number of constraints rather than on the number of variables.

2. The procedure outlined above requires the determination of all the extreme points of every subsidiary constraint set defined by Eqs. (4.31) before the optimal values $\mu_{j,i}^*$ are found. However, this is not necessary when the revised simplex method is used to implement the decomposition algorithm [4.5].

3. If the size of the problem is small, it will be convenient to enumerate all the extreme points of the subproblems and use the simplex method to solve the problem. This procedure is illustrated in the following example.

Example 4.4 A fertilizer mixing plant produces two fertilizers, A and B, by mixing two chemicals, C_1 and C_2, in different proportions. The contents and costs of the chemicals C_1 and C_2 are as follows:

Chemical	Contents		Cost ($/lb)
	Ammonia	Phosphates	
C_1	0.70	0.30	5
C_2	0.40	0.60	4

Fertilizer A should not contain more than 60% of ammonia and B should contain at least 50% of ammonia. On the average, the plant can sell up to 1000 lb/hr and due to limitations on the production facilities, not more than 600 lb of fertilizer A can be produced per hour. The availability of chemical C_1 is restricted to 500 lb/hr. Assuming that the production costs are same for both A and B, determine the quantities of A and B to be produced per hour for maximum return if the plant sells A and B at the rates of \$6 and \$7 per pound, respectively.

SOLUTION   Let x_1 and x_2 indicate the amounts of chemicals C_1 and C_2 used in fertilizer A, and y_1 and y_2 in fertilizer B per hour. Thus the total amounts of A and B produced per hour are given by $x_1 + x_2$ and $y_1 + y_2$, respectively. The objective function to be maximized is given by

$$f = \text{selling price} \; - \; \text{cost of chemical } C_1 \text{ and } C_2$$

$$= 6(x_1 + x_2) + 7(y_1 + y_2) - 5(x_1 + y_1) - 4(x_2 + y_2)$$

The constraints are given by

$$(x_1 + x_2) + (y_1 + y_2) \leq 1000 \qquad \text{(amount that can be sold)}$$

$$x_1 \qquad + y_1 \qquad\quad \leq 500 \qquad \text{(availability of } C_1)$$

$$x_1 + x_2 \qquad\qquad\quad \leq 600 \qquad \text{(production limitations on A)}$$

$$\tfrac{7}{10}x_1 + \tfrac{4}{10}x_2 \qquad\qquad \leq \tfrac{6}{10}(x_1 + x_2) \;\; (A \text{ should not contain more} \atop \text{than 60\% of ammonia)}$$

$$\tfrac{7}{10}y_1 + \tfrac{4}{10}y_2 \qquad\qquad \geq \tfrac{5}{10}(y_1 + y_2) \;\; (B \text{ should contain at least} \atop \text{50\% of ammonia)}$$

Thus the problem can be restated as

$$\text{Maximize } f = x_1 + 2x_2 + 2y_1 + 3y_2 \qquad\qquad\qquad (E_1)$$

subject to

$$\boxed{\begin{aligned} x_1 + x_2 + y_1 + y_2 \\ x_1 \qquad + y_1 \end{aligned}} \qquad \begin{aligned} \leq 1000 \\ \leq 500 \end{aligned} \qquad (E_2)$$

$$\boxed{\begin{aligned} x_1 + x_2 \\ x_1 - 2x_2 \end{aligned}} \qquad\qquad \begin{aligned} \leq 600 \\ \leq 0 \end{aligned} \qquad (E_3)$$

$$\boxed{-2y_1 + y_2} \quad \leq 0 \qquad\qquad\qquad (E_4)$$

$$x_i \geq 0, \quad y_i \geq 0, \quad i = 1, 2$$

This problem can also be stated in matrix notation as follows:

$$\text{Maximize } f(\mathbf{X}) = \mathbf{c}_1^T \mathbf{X}_1 + \mathbf{c}_2^T \mathbf{X}_2$$

subject to

$$
\begin{aligned}
\mathbf{A}_1\mathbf{X}_1 + \mathbf{A}_2\mathbf{X}_2 &\le \mathbf{b}_0 \\
\mathbf{B}_1\mathbf{X}_1 &\le \mathbf{b}_1 \\
\mathbf{B}_2\mathbf{X}_2 &\le \mathbf{b}_2 \\
\mathbf{X}_1 \ge \mathbf{0}, \quad \mathbf{X}_2 &\ge \mathbf{0}
\end{aligned}
\qquad (\mathrm{E}_5)
$$

where

$$
\mathbf{X}_1 = \begin{Bmatrix} x_1 \\ x_2 \end{Bmatrix}, \quad
\mathbf{X}_2 = \begin{Bmatrix} y_1 \\ y_2 \end{Bmatrix}, \quad
\mathbf{c}_1 = \begin{Bmatrix} 1 \\ 2 \end{Bmatrix}, \quad
\mathbf{c}_2 = \begin{Bmatrix} 2 \\ 3 \end{Bmatrix},
$$

$$
\mathbf{A}_1 = \begin{bmatrix} 1 & 1 \\ 1 & 0 \end{bmatrix}, \quad
[\mathbf{A}_2] = \begin{bmatrix} 1 & 1 \\ 1 & 0 \end{bmatrix}, \quad
\mathbf{b}_0 = \begin{Bmatrix} 1000 \\ 500 \end{Bmatrix},
$$

$$
\mathbf{B}_1 = \begin{bmatrix} 1 & 1 \\ 1 & -2 \end{bmatrix}, \quad
\mathbf{b}_1 = \begin{Bmatrix} 600 \\ 0 \end{Bmatrix}, \quad
\mathbf{B}_2 = \{-2 \ \ 1\}, \mathbf{b}_2 = \{0\},
$$

$$
\mathbf{X} = \begin{Bmatrix} \mathbf{X}_1 \\ \mathbf{X}_2 \end{Bmatrix}
$$

Step 1 We first consider the subsidiary constraint sets

$$
\mathbf{B}_1\mathbf{X}_1 \le \mathbf{b}_1, \quad \mathbf{X}_1 \ge \mathbf{0} \qquad (\mathrm{E}_6)
$$

$$
\mathbf{B}_2\mathbf{X}_2 \le \mathbf{b}_2, \quad \mathbf{X}_2 \ge \mathbf{0} \qquad (\mathrm{E}_7)
$$

The convex feasible regions represented by (E_6) and (E_7) are shown in Fig. 4.1a and b, respectively. The vertices of the two feasible regions are given by

$$
\mathbf{X}_1^{(1)} = \text{point } P = \begin{Bmatrix} 0 \\ 0 \end{Bmatrix}
$$

$$
\mathbf{X}_2^{(1)} = \text{point } Q = \begin{Bmatrix} 0 \\ 600 \end{Bmatrix}
$$

$$
\mathbf{X}_3^{(1)} = \text{point } R = \begin{Bmatrix} 400 \\ 200 \end{Bmatrix}
$$

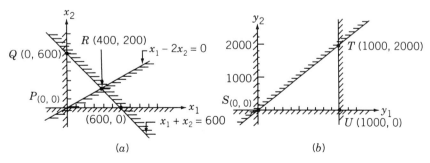

Figure 4.1   Vertices of feasible regions. To make the feasible region bounded, the constraint $y_1 \le 1000$ is added in view of Eq. (E_2).

$$\mathbf{X}_1^{(2)} = \text{point } S = \begin{Bmatrix} 0 \\ 0 \end{Bmatrix}$$

$$\mathbf{X}_2^{(2)} = \text{point } T = \begin{Bmatrix} 1000 \\ 2000 \end{Bmatrix}$$

$$\mathbf{X}_3^{(2)} = \text{point } U = \begin{Bmatrix} 1000 \\ 0 \end{Bmatrix}$$

Thus any point in the convex feasible sets defined by Eqs. (E_6) and (E_7) can be represented, respectively, as

$$\left. \begin{aligned} \mathbf{X}_1 = \mu_{11} \begin{Bmatrix} 0 \\ 0 \end{Bmatrix} + \mu_{12} \begin{Bmatrix} 0 \\ 600 \end{Bmatrix} + \mu_{13} \begin{Bmatrix} 400 \\ 200 \end{Bmatrix} = \begin{Bmatrix} 400\mu_{13} \\ 600\mu_{12} + 200\mu_{13} \end{Bmatrix} \\ \text{with} \\ \mu_{11} + \mu_{12} + \mu_{13} = 1, \qquad 0 \le \mu_{1i} \le 1, \quad i = 1, 2, 3 \end{aligned} \right\} \quad (E_8)$$

and

$$\left. \begin{aligned} \mathbf{X}_2 = \mu_{21} \begin{Bmatrix} 0 \\ 0 \end{Bmatrix} + \mu_{22} \begin{Bmatrix} 1000 \\ 2000 \end{Bmatrix} + \mu_{23} \begin{Bmatrix} 1000 \\ 0 \end{Bmatrix} \\ = \begin{Bmatrix} 1000\mu_{22} + 1000\mu_{23} \\ 2000\mu_{22} \end{Bmatrix} \\ \text{with} \\ \mu_{21} + \mu_{22} + \mu_{23} = 1; \quad 0 \le \mu_{2i} \le 1, \quad i = 1, 2, 3 \end{aligned} \right\} \quad (E_9)$$

Step 2 By substituting the relations of (E_8) and (E_9), the problem stated in Eqs. (E_5) can be rewritten as

$$\text{Maximize } f(\mu_{11}, \mu_{12}, \dots, \mu_{23}) = (1 \quad 2) \begin{Bmatrix} 400\mu_{13} \\ 600\mu_{12} + 200\mu_{13} \end{Bmatrix}$$

$$+ (2 \quad 3) \begin{Bmatrix} 1000\mu_{22} + 1000\mu_{23} \\ 2000\mu_{22} \end{Bmatrix}$$

$$= 800\mu_{13} + 1200\mu_{12} + 8000\mu_{22} + 2000\mu_{23}$$

subject to

$$\begin{bmatrix} 1 & 1 \\ 1 & 0 \end{bmatrix} \begin{Bmatrix} 400\mu_{13} \\ 600\mu_{12} + 200\mu_{13} \end{Bmatrix}$$

$$+ \begin{bmatrix} 1 & 1 \\ 1 & 0 \end{bmatrix} \begin{Bmatrix} 1000\mu_{22} + 1000\mu_{23} \\ 2000\mu_{22} \end{Bmatrix} \le \begin{Bmatrix} 1000 \\ 500 \end{Bmatrix}$$

that is,

$$600\mu_{12} + 600\mu_{13} + 3000\mu_{22} + 1000\mu_{23} \le 1000$$

$$400\mu_{13} + 1000\mu_{22} + 1000\mu_{23} \le 500$$

$$\mu_{11} + \mu_{12} + \mu_{13} = 1$$

$$\mu_{21} + \mu_{22} + \mu_{23} = 1$$

with

$$\mu_{11} \geq 0, \ \mu_{12} \geq 0, \ \mu_{13} \geq 0, \ \mu_{21} \geq 0, \ \mu_{22} \geq 0, \ \mu_{23} \geq 0$$

The optimization problem can be stated in standard form (after adding the slack variables α and β) as

$$\text{Minimize } f = -1200\mu_{12} - 800\mu_{13} - 8000\mu_{22} - 2000\mu_{23}$$

subject to

$$600\mu_{12} + 600\mu_{13} + 3000\mu_{22} + 1000\mu_{23} + \alpha = 1000$$

$$400\mu_{13} + 1000\mu_{22} + 1000\mu_{23} + \beta = 500$$

$$\mu_{11} + \mu_{12} + \mu_{13} = 1 \qquad\qquad (\text{E}_{10})$$

$$\mu_{21} + \mu_{22} + \mu_{23} = 1$$

$$\mu_{ij} \geq 0 \ (i = 1, 2; \ j = 1, 2, 3), \quad \alpha \geq 0, \quad \beta \geq 0$$

Step 3 The problem (E_{10}) can now be solved by using the simplex method.

4.5   SENSITIVITY OR POSTOPTIMALITY ANALYSIS

In most practical problems, we are interested not only in optimal solution of the LP problem, but also in how the solution changes when the parameters of the problem change. The change in the parameters may be discrete or continuous. The study of the effect of discrete parameter changes on the optimal solution is called *sensitivity analysis* and that of the continuous changes is termed *parametric programming*. One way to determine the effects of changes in the parameters is to solve a series of new problems once for each of the changes made. This is, however, very inefficient from a computational point of view. Some techniques that take advantage of the properties of the simplex solution are developed to make a sensitivity analysis. We study some of these techniques in this section. There are five basic types of parameter changes that affect the optimal solution:

1. Changes in the right-hand-side constants b_i
2. Changes in the cost coefficients c_j
3. Changes in the coefficients of the constraints a_{ij}
4. Addition of new variables
5. Addition of new constraints

In general, when a parameter is changed, it results in one of three cases:

1. The optimal solution remains unchanged; that is, the basic variables and their values remain unchanged.
2. The basic variables remain the same but their values are changed.
3. The basic variables as well as their values are changed.

4.5.1 Changes in the Right-Hand-Side Constants b_i

Suppose that we have found the optimal solution to a LP problem. Let us now change the b_i to $b_i + \Delta b_i$ so that the new problem differs from the original only on the right-hand side. Our interest is to investigate the effect of changing b_i to $b_i + \Delta b_i$ on the original optimum. We know that a basis is optimal if the relative cost coefficients corresponding to the nonbasic variables $\bar{c}_j$ are nonnegative. By considering the procedure according to which $\bar{c}_j$ are obtained, we can see that the values of $\bar{c}_j$ are not related to the b_i. The values of $\bar{c}_j$ depend only on the basis, on the coefficients of the constraint matrix, and the original coefficients of the objective function. The relation is given in Eq. (4.10):

$$\bar{c}_j = c_j - \boldsymbol{\pi}^{\mathrm{T}}\mathbf{A}_j = c_j - \mathbf{c}_B^{\mathrm{T}}\mathbf{B}^{-1}\mathbf{A}_j \qquad (4.33)$$

Thus changes in b_i will affect the values of basic variables in the optimal solution and the optimality of the basis will not be affected provided that the changes made in b_i do not make the basic solution infeasible. Thus if the new basic solution remains feasible for the new right-hand side, that is, if

$$\mathbf{X}'_B = \mathbf{B}^{-1}(\mathbf{b} + \Delta\mathbf{b}) \geq \mathbf{0} \qquad (4.34)$$

then the original optimal basis, $\mathbf{B}$, also remains optimal for the new problem. Since the original solution, say[†]

$$\mathbf{X}_B = \begin{Bmatrix} x_1 \\ x_2 \\ \vdots \\ x_m \end{Bmatrix}$$

is given by

$$\mathbf{X}_B = \mathbf{B}^{-1}\mathbf{b} \qquad (4.35)$$

Equation (4.34) can also be expressed as

$$x'_i = x_i + \sum_{j=1}^{m} \beta_{ij}\,\Delta b_j \geq 0, \qquad i = 1, 2, \ldots, m \qquad (4.36)$$

where

$$\mathbf{B}^{-1} = [\beta_{ij}] \qquad (4.37)$$

Hence the original optimal basis $\mathbf{B}$ remains optimal provided that the changes made in b_i, Δb_i, satisfy the inequalities (4.36). The change in the value of the ith optimal basic variable, Δx_i, due to the change in b_i is given by

$$\mathbf{X}'_B - \mathbf{X}_B = \Delta\mathbf{X}_B = \mathbf{B}^{-1}\Delta\mathbf{b}$$

[†]It is assumed that the variables are renumbered such that the first m variables represent the basic variables and the remaining $n - m$ the nonbasic variables.

that is,

$$\Delta x_i = \sum_{j=1}^{m} \beta_{ij} \Delta b_j, \quad i = 1, 2, \ldots, m \tag{4.38}$$

Finally, the change in the optimal value of the objective function (Δf) due to the change Δb_i can be obtained as

$$\Delta f = \mathbf{c}_B^T \Delta \mathbf{X}_B = \mathbf{c}_B^T \mathbf{B}^{-1} \Delta \mathbf{b} = \boldsymbol{\pi}^T \Delta \mathbf{b} = \sum_{j=1}^{m} \pi_j \Delta \mathbf{b}_j \tag{4.39}$$

Suppose that the changes made in $b_i (\Delta b_i)$ are such that the inequality (4.34) is violated for some variables so that these variables become infeasible for the new right-hand-side vector. Our interest in this case will be to determine the new optimal solution. This can be done without reworking the problem from the beginning by proceeding according to the following steps:

1. Replace the $\overline{b}_i$ of the original optimal tableau by the new values, $\overline{\mathbf{b}}' = \mathbf{B}^{-1}(\mathbf{b} + \Delta \mathbf{b})$ and change the signs of all the numbers that are lying in the rows in which the infeasible variables appear, that is, in rows for which $\overline{b}_i' < 0$.
2. Add artificial variables to these rows, thereby replacing the infeasible variables in the basis by the artificial variables.
3. Go through the phase I calculations to find a basic feasible solution for the problem with the new right-hand side.
4. If the solution found at the end of phase I is not optimal, we go through the phase II calculations to find the new optimal solution.

The procedure outlined above saves considerable time and effort compared to the reworking of the problem from the beginning if only a few variables become infeasible with the new right-hand side. However, if the number of variables that become infeasible are not few, the procedure above might also require as much effort as the one involved in reworking of the problem from the beginning.

Example 4.5   A manufacturer produces four products, A, B, C, and D, by using two types of machines (lathes and milling machines). The times required on the two machines to manufacture 1 unit of each of the four products, the profit per unit of the product, and the total time available on the two types of machines per day are given below:

Machine	Time required per unit (min) for product:				Total time available per day (min)
	A	B	C	D	
Lathe machine	7	10	4	9	1200
Milling machine	3	40	1	1	800
Profit per unit ($)	45	100	30	50	

Find the number of units to be manufactured of each product per day for maximizing the profit.

   Note: This is an ordinary LP problem and is given to serve as a reference problem for illustrating the sensitivity analysis.

SOLUTION Let x_1, x_2, x_3, and x_4 denote the number of units of products A, B, C, and D produced per day. Then the problem can be stated in standard form as follows:

$$\text{Minimize } f = -45x_1 - 100x_2 - 30x_3 - 50x_4$$

subject to

$$7x_1 + 10x_2 + 4x_3 + 9x_4 \leq 1200$$

$$3x_1 + 40x_2 + x_3 + x_4 \leq 800$$

$$x_i \geq 0, \quad i = 1 \text{ to } 4$$

By introducing the slack variables $x_5 \geq 0$ and $x_6 \geq 0$, the problem can be stated in canonical form and the simplex method can be applied. The computations are shown in tableau form below:

Basic variables	x_1	x_2	x_3	x_4	x_5	x_6	$-f$	$\bar{b}_i$	Ratio $\bar{b}_i/\bar{a}_{is}$ for $\bar{a}_{is} > 0$
x_5	7	10	4	9	1	0	0	1200	120
x_6	3	40	1	1	0	1	0	800	20 ← Smaller one, x_6 leaves the basis
		Pivot element							
$-f$	-45	-100	-30	-50	0	0	1	0	

Minimum $\bar{c}_j < 0$; x_2 enters the next basis

Result of pivot operation:

	x_1	x_2	x_3	x_4	x_5	x_6	$-f$	$\bar{b}_i$	
x_5	$\frac{25}{4}$	0	$\frac{15}{4}$	$\frac{35}{4}$	1	$-\frac{1}{4}$	0	1000	$\frac{4000}{35}$ ←Smaller one, x_5 leaves the basis
				Pivot element					
x_2	$\frac{3}{40}$	1	$\frac{1}{40}$	$\frac{1}{40}$	0	$\frac{1}{40}$	0	20	800
$-f$	$-\frac{75}{2}$	0	$-\frac{55}{2}$	$-\frac{95}{2}$	0	$-\frac{5}{2}$	1	2000	

Minimum $\bar{c}_j < 0$, x_4 enters the basis

Result of pivot operation:

	x_1	x_2	x_3	x_4	x_5	x_6	$-f$	$\bar{b}_i$	
x_4	$\frac{5}{7}$	0	$\frac{3}{7}$	1	$\frac{4}{35}$	$-\frac{1}{35}$	0	$\frac{4,000}{35}$	$\frac{800}{3}$ ←Smaller one, x_4 leaves the basis
			Pivot element						
x_2	$\frac{2}{35}$	1	$\frac{1}{70}$	0	$-\frac{1}{350}$	$\frac{9}{350}$	0	$\frac{120}{7}$	1200
$-f$	$-\frac{25}{7}$	0	$-\frac{50}{7}$	0	$\frac{38}{7}$	$\frac{8}{7}$	1	$\frac{52,000}{7}$	

Minimum $c_j < 0$, x_3 enters the basis

Result of pivot operation:

x_3	$\frac{5}{3}$	0	1	$\frac{7}{3}$	$\frac{4}{15}$	$-\frac{1}{15}$	0	$\frac{800}{3}$
x_2	$\frac{1}{30}$	1	0	$-\frac{1}{30}$	$-\frac{1}{150}$	$\frac{2}{75}$	0	$\frac{40}{3}$
$-f$	$\frac{25}{3}$	0	0	$\frac{50}{3}$	$\frac{22}{3}$	$\frac{2}{3}$	1	$\frac{28,000}{3}$

The optimum solution is given by

$$x_2 = \frac{40}{3}, \quad x_3 = \frac{800}{3} \text{ (basic variables)}$$

$$x_1 = x_4 = x_5 = x_6 = 0 \text{ (nonbasic variables)}$$

$$f_{\min} = \frac{-28,000}{3} \quad \text{or} \quad \text{maximum profit} = \frac{\$28,000}{3}$$

From the final tableau, one can find that

$$\mathbf{X}_B = \begin{Bmatrix} x_3 \\ x_2 \end{Bmatrix} = \begin{Bmatrix} \frac{800}{3} \\ \frac{40}{3} \end{Bmatrix} = \begin{matrix} \text{vector of basic variables in} \\ \text{the optimum solution} \end{matrix} \tag{E_1}$$

$$\mathbf{c}_B = \begin{Bmatrix} c_3 \\ c_2 \end{Bmatrix} = \begin{Bmatrix} -30 \\ -100 \end{Bmatrix} = \begin{matrix} \text{vector of original cost} \\ \text{coefficients corresponding} \\ \text{to the basic variables} \end{matrix} \tag{E_2}$$

$$\mathbf{B} = \begin{bmatrix} 4 & 10 \\ 1 & 40 \end{bmatrix} = \begin{matrix} \text{matrix of original coefficients} \\ \text{corresponding to the basic variables} \end{matrix} \tag{E_3}$$

$$\mathbf{B}^{-1} = \begin{bmatrix} \beta_{33} & \beta_{32} \\ \beta_{23} & \beta_{22} \end{bmatrix} = \begin{bmatrix} \frac{4}{15} & -\frac{1}{15} \\ -\frac{1}{150} & \frac{2}{75} \end{bmatrix} = \begin{matrix} \text{inverse of the coefficient} \\ \text{matrix } \mathbf{B}, \text{ which appears} \\ \text{in the final tableau also} \end{matrix} \tag{E_4}$$

$$\boldsymbol{\pi} = \mathbf{c}_B^T \mathbf{B}^{-1} = (-30 \; -100) \begin{bmatrix} \frac{4}{15} & -\frac{1}{15} \\ -\frac{1}{150} & \frac{2}{75} \end{bmatrix}$$

$$= \begin{Bmatrix} -\frac{22}{3} \\ -\frac{2}{3} \end{Bmatrix} = \begin{matrix} \text{simplex multipliers, the} \\ \text{negatives of which appear} \\ \text{in the final tableau also} \end{matrix} \tag{E_5}$$

Example 4.6 Find the effect of changing the total time available per day on the two machines from 1200 and 800 min to 1500 and 1000 min in Example 4.5.

SOLUTION   Equation (4.36) gives

$$x_i + \sum_{j=1}^{m} \beta_{ij} \Delta b_j \geq 0, \quad i = 1, 2, \ldots, m \tag{4.36}$$

where x_i is the optimum value of the ith basic variable. (This equation assumes that the variables are renumbered such that x_1 to x_m represent the basic variables.)

If the variables are not renumbered, Eq. (4.36) will be applicable for $i = 3$ and 2 in the present problem with $\Delta b_3 = 300$ and $\Delta b_2 = 200$. From Eqs. (E$_1$) to (E$_5$) of Example 4.5, the left-hand sides of Eq. (4.36) become

$$x_3 + \beta_{33}\Delta b_3 + \beta_{32}\Delta b_2 = \frac{800}{3} + \frac{4}{15}(300) - \frac{1}{15}(200) = \frac{5000}{15}$$

$$x_2 + \beta_{23}\Delta b_3 + \beta_{22}\Delta b_2 = \frac{40}{3} - \frac{1}{150}(300) + \frac{2}{75}(200) = \frac{2500}{150}$$

Since both these values are ≥ 0, the original optimal basis **B** remains optimal even with the new values of b_i. The new values of the (optimal) basic variables are given by Eq. (4.38) as

$$\mathbf{X}'_B = \begin{Bmatrix} x'_3 \\ x'_2 \end{Bmatrix} = \mathbf{X}_B + \Delta\mathbf{X}_B = \mathbf{X}_B + \mathbf{B}^{-1}\Delta\mathbf{b}$$

$$= \begin{Bmatrix} \frac{800}{3} \\ \frac{40}{3} \end{Bmatrix} + \begin{bmatrix} \frac{4}{15} & -\frac{1}{15} \\ -\frac{1}{150} & \frac{2}{75} \end{bmatrix} \begin{Bmatrix} 300 \\ 200 \end{Bmatrix} = \begin{Bmatrix} \frac{1000}{3} \\ \frac{50}{3} \end{Bmatrix}$$

and the optimum value of the objective function by Eq. (4.39) as

$$f'_{\min} = f_{\min} + \Delta f = f_{\min} + \mathbf{c}_B^{\mathrm{T}}\Delta\mathbf{X}_B = -\frac{28,000}{3} + (-30 - 100)\begin{Bmatrix} \frac{200}{3} \\ \frac{10}{3} \end{Bmatrix}$$

$$= -\frac{35,000}{3}$$

Thus the new profit will be $35,000/3.

4.5.2 Changes in the Cost Coefficients c_j

The problem here is to find the effect of changing the cost coefficients from c_j to $c_j + \Delta c_j$ on the optimal solution obtained with c_j. The relative cost coefficients corresponding to the nonbasic variables, $x_{m+1}, x_{m+2}, \ldots, x_n$ are given by Eq. (4.10):

$$\bar{c}_j = c_j - \boldsymbol{\pi}^{\mathrm{T}}\mathbf{A}_j = c_j - \sum_{i=1}^{m} \pi_i a_{ij}, \quad j = m+1, m+2, \ldots, n \qquad (4.40)$$

where the simplex multipliers π_i are related to the cost coefficients of the basic variables by the relation

$$\boldsymbol{\pi}^{\mathrm{T}} = \mathbf{c}_B^{\mathrm{T}}\mathbf{B}^{-1}$$

that is,

$$\pi_i = \sum_{k=1}^{m} c_k \beta_{ki}, \quad i = 1, 2, \cdots, m \qquad (4.41)$$

From Eqs. (4.40) and (4.41), we obtain

$$\bar{c}_j = c_j - \sum_{i=1}^{m} a_{ij}\left(\sum_{k=1}^{m} c_k \beta_{ki}\right) = c_j - \sum_{k=1}^{m} c_k\left(\sum_{i=1}^{m} a_{ij}\beta_{ki}\right),$$

$$i = m+1, m+2, \ldots, n \qquad (4.42)$$

If the c_j are changed to $c_j + \Delta c_j$, the original optimal solution remains optimal, provided that the new values of $\overline{c}_j$, $\overline{c}'_j$ satisfy the relation

$$\overline{c}'_j = c_j + \Delta c_j - \sum_{k=1}^{m} (c_k + \Delta c_k) \left(\sum_{i=1}^{m} a_{ij} \beta_{ki} \right) \geq 0$$

$$= \overline{c}_j + \Delta c_j - \sum_{k=1}^{m} \Delta c_k \left(\sum_{i=1}^{m} a_{ij} \beta_{ki} \right) \geq 0,$$

$$j = m + 1, m + 2, \cdots, n \tag{4.43}$$

where $\overline{c}_j$ indicate the values of the relative cost coefficients corresponding to the original optimal solution.

In particular, if changes are made only in the cost coefficients of the nonbasic variables, Eq. (4.43) reduces to

$$\overline{c}_j + \Delta c_j \geq 0, \quad j = m + 1, m + 2, \ldots, n \tag{4.44}$$

If Eq. (4.43) is satisfied, the changes made in c_j, Δc_j, will not affect the optimal basis and the values of the basic variables. The only change that occurs is in the optimal value of the objective function according to

$$\Delta f = \sum_{j=1}^{m} x_j \Delta c_j \tag{4.45}$$

and this change will be zero if only the c_j of nonbasic variables are changed.

Suppose that Eq. (4.43) is violated for some of the nonbasic variables. Then it is possible to improve the value of the objective function by bringing any nonbasic variable that violates Eq. (4.43) into the basis provided that it can be assigned a nonzero value. This can be done easily with the help of the previous optimal tableau. Since some of the $\overline{c}'_j$ are negative, we start the optimization procedure again by using the old optimum as an initial feasible solution. We continue the iterative process until the new optimum is found. As in the case of changing the right-hand-side b_i, the effectiveness of this procedure depends on the number of violations made in Eq. (4.43) by the new values $c_j + \Delta c_j$.

In some of the practical problems, it may become necessary to solve the optimization problem with a series of objective functions. This can be accomplished without reworking the entire problem for each new objective function. Assume that the optimum solution for the first objective function is found by the regular procedure. Then consider the second objective function as obtained by changing the first one and evaluate Eq. (4.43). If the resulting $\overline{c}'_j \geq 0$, the old optimum still remains as optimum and one can proceed to the next objective function in the same manner. On the other hand, if one or more of the resulting $\overline{c}'_j < 0$, we can adopt the procedure outlined above and continue the iterative process using the old optimum as the starting feasible solution. After the optimum is found, we switch to the next objective function.

Example 4.7   Find the effect of changing c_3 from -30 to -24 in Example 4.5.

SOLUTION   Here $\Delta c_3 = 6$ and Eq. (4.43) gives that

$$\overline{c}_1' = \overline{c}_1 + \Delta c_1 - \Delta c_3[a_{21}\beta_{32} + a_{31}\beta_{33}] = \tfrac{25}{3} + 0 - 6[3(-\tfrac{1}{15}) + 7(\tfrac{4}{15})] = -\tfrac{5}{3}$$

$$\overline{c}_4' = \overline{c}_4 + \Delta c_4 - \Delta c_3[a_{24}\beta_{32} + a_{34}\beta_{33}] = \tfrac{50}{3} + 0 - 6[1(-\tfrac{1}{15}) + 9(\tfrac{4}{15})] = \tfrac{8}{3}$$

$$\overline{c}_5' = \overline{c}_5 + \Delta c_5 - \Delta c_3[a_{25}\beta_{32} + a_{35}\beta_{33}] = \tfrac{22}{3} + 0 - 6[0(-\tfrac{1}{15}) + 1(\tfrac{4}{15})] = \tfrac{86}{15}$$

$$\overline{c}_6' = \overline{c}_6 + \Delta c_6 - \Delta c_3[a_{26}\beta_{32} + a_{36}\beta_{33}] = \tfrac{2}{3} + 0 - 6[1(-\tfrac{1}{15}) + 0(\tfrac{4}{15})] = \tfrac{16}{15}$$

The change in the value of the objective function is given by Eq. (4.45) as

$$\Delta f = \Delta c_3 x_3 = \frac{4800}{3} \quad \text{so that} \quad f = -\frac{28{,}000}{3} + \frac{4800}{3} = -\frac{23{,}200}{3}$$

Since $\overline{c}_1'$ is negative, we can bring x_1 into the basis. Thus we start with the optimal tableau of the original problem with the new values of relative cost coefficients and improve the solution according to the regular procedure.

Basic variables	Variables								Ratio $\overline{b}_i/\overline{a}_{ij}$
	x_1	x_2	x_3	x_4	x_5	x_6	$-f$	$\overline{b}_i$	for $\overline{a}_{ij} > 0$
x_3	$\boxed{\tfrac{5}{3}}$	0	1	$\tfrac{7}{3}$	$\tfrac{4}{15}$	$-\tfrac{1}{15}$	0	$\tfrac{800}{3}$	$160 \leftarrow$
	Pivot element								
x_2	$\tfrac{1}{30}$	1	0	$-\tfrac{1}{30}$	$-\tfrac{1}{150}$	$\tfrac{2}{75}$	0	$\tfrac{40}{3}$	400
$-f$	$-\tfrac{5}{3}$	0	0	$\tfrac{8}{3}$	$\tfrac{86}{15}$	$\tfrac{16}{15}$	1	$\tfrac{23{,}200}{3}$	
	$\uparrow$								
x_1	1	0	$\tfrac{3}{5}$	$\tfrac{7}{5}$	$\tfrac{4}{25}$	$-\tfrac{1}{25}$	0	160	
x_2	0	1	$-\tfrac{1}{50}$	$-\tfrac{2}{25}$	$-\tfrac{3}{250}$	$\tfrac{7}{250}$	0	8	
$-f$	0	0	1	5	6	1	1	8000	

Since all the relative cost coefficients are nonnegative, the present solution is optimum with

$$x_1 = 160, \quad x_2 = 8 \text{ (basic variables)}$$

$$x_3 = x_4 = x_5 = x_6 = 0 \text{ (nonbasic variables)}$$

$$f_{\min} = -8000 \quad \text{and maximum profit} = \$8000$$

4.5.3   Addition of New Variables

Suppose that the optimum solution of a LP problem with n variables $x_1, x_2, \ldots, x_n$ has been found and we want to examine the effect of adding some more variables x_{n+k}, $k = 1, 2, \ldots$, on the optimum solution. Let the constraint coefficients and the cost coefficients corresponding to the new variables x_{n+k} be denoted by $a_{i,n+k}, i = 1$ to m and c_{n+k}, respectively. If the new variables are treated as additional nonbasic variables in the old optimum solution, the corresponding relative cost coefficients are given by

$$\bar{c}_{n+k} = c_{n+k} - \sum_{i=1}^{m} \pi_i a_{i,n+k} \qquad (4.46)$$

where $\pi_1, \pi_2, \ldots, \pi_m$ are the simplex multipliers corresponding to the original optimum solution. The original optimum remains optimum for the new problem also provided that $\bar{c}_{n+k} \geq 0$ for all k. However, if one or more $\bar{c}_{n+k} < 0$, it pays to bring some of the new variables into the basis provided that they can be assigned a nonzero value. For bringing a new variable into the basis, we first have to transform the coefficients $a_{i,n+k}$ into $\bar{a}_{i,n+k}$ so that the columns of the new variables correspond to the canonical form of the old optimal basis. This can be done by using Eq. (4.9) as

$$\underset{m \times 1}{\bar{\mathbf{A}}_{n+k}} = \underset{m \times m}{\mathbf{B}^{-1}} \underset{m \times 1}{\mathbf{A}_{n+k}}$$

that is,

$$\bar{a}_{i,n+k} = \sum_{j=1}^{m} \beta_{ij} a_{j,n+k}, \qquad i = 1 \text{ to } m \qquad (4.47)$$

where $\mathbf{B}^{-1} = [\beta_{ij}]$ is the inverse of the old optimal basis. The rules for bringing a new variable into the basis, finding a new basic feasible solution, testing this solution for optimality, and the subsequent procedure is same as the one outlined in the regular simplex method.

Example 4.8   In Example 4.5, if a new product, E, which requires 15 min of work on the lathe and 10 min on the milling machine per unit, is available, will it be worthwhile to manufacture it if the profit per unit is \$40?

SOLUTION   Let x_k be the number of units of product E manufactured per day. Then $c_k = -40$, $a_{1k} = 15$, and $a_{2k} = 10$; therefore,

$$\bar{c}_k = c_k - \pi_1 a_{1k} - \pi_2 a_{2k} = -40 + (\tfrac{22}{3})(15) + (\tfrac{2}{3})(10) = \tfrac{230}{3} \geq 0$$

Since the relative cost coefficient $\bar{c}_k$ is nonnegative, the original optimum solution remains optimum for the new problem also and the variable x_k will remain as a nonbasic variable. This means that it is not worth manufacturing product E.

4.5.4   Changes in the Constraint Coefficients a_{ij}

Here the problem is to investigate the effect of changing the coefficient a_{ij} to $a_{ij} + \Delta a_{ij}$ after finding the optimum solution with a_{ij}. There are two possibilities in this case. The first possibility occurs when all the coefficients a_{ij}, in which changes are made, belong to the columns of those variables that are nonbasic in the old optimal solution. In this case, the effect of changing a_{ij} on the optimal solution can be investigated by adopting the procedure outlined in the preceding section. The second possibility occurs when the coefficients changed a_{ij} correspond to a basic variable, say, x_{j0} of the old optimal solution. The following procedure can be adopted to examine the effect of changing $a_{i,j0}$ to $a_{i,j0} + \Delta a_{i,j0}$.

1. Introduce a new variable x_{n+1} to the original system with constraint coefficients

$$a_{i,n+1} = a_{i,j0} + \Delta a_{i,j0} \tag{4.48}$$

and cost coefficient

$$c_{n+1} = c_{j0} \text{ (original value itself)} \tag{4.49}$$

2. Transform the coefficients $a_{i,n+1}$ to $\overline{a}_{i,n+1}$ by using the inverse of the old optimal basis, $\mathbf{B}^{-1} = [\beta_{ij}]$, as

$$\overline{a}_{i,n+1} = \sum_{j=1}^{m} \beta_{ij} a_{j,n+1}, \qquad i = 1 \text{ to } m \tag{4.50}$$

3. Replace the original cost coefficient (c_{j0}) of x_{j0} by a large positive number N, but keep c_{n+1} equal to the old value c_{j0}.
4. Compute the modified cost coefficients using Eq. (4.43):

$$\overline{c}_j' = \overline{c}_j + \Delta c_j - \sum_{k=1}^{m} \Delta c_k \left(\sum_{i=1}^{m} a_{ij} \beta_{ki} \right),$$

$$j = m+1, m+2, \cdots, n, n+1 \tag{4.51}$$

where $\Delta c_k = 0$ for $k = 1, 2, \ldots, j_0 - 1, j_0 + 1, \ldots, m$ and $\Delta c_{j0} = N - c_{j0}$.
5. Carry the regular iterative procedure of simplex method with the new objective function and the augmented matrix found in Eqs. (4.50) and (4.51) until the new optimum is found.

Remarks:

1. The number N has to be taken sufficiently large to ensure that x_{j0} cannot be contained in the new optimal basis that is ultimately going to be found.
2. The procedure above can easily be extended to cases where changes in coefficients a_{ij} of more than one column are made.
3. The present procedure will be computationally efficient (compared to reworking of the problem from the beginning) only for cases where there are not too many number of basic columns in which the a_{ij} are changed.

Example 4.9 Find the effect of changing $\mathbf{A}_1$ from $\{^7_3\}$ to $\{^6_{10}\}$ in Example 4.5 (i.e., changes are made in the coefficients a_{ij} of nonbasic variables only).

SOLUTION   The relative cost coefficients of the nonbasic variables (of the original optimum solution) corresponding to the new a_{ij} are given by

$$\overline{c}_j = c_j - \boldsymbol{\pi}^{\mathrm{T}} \mathbf{A}_j, \qquad j = \text{nonbasic } (1, 4, 5, 6)$$

Since $\mathbf{A}_1$ is changed, we have

$$\bar{c}_1 = c_1 - \boldsymbol{\pi}^T \mathbf{A}_1 = -45 - (-\tfrac{22}{3} - \tfrac{2}{3}) \begin{Bmatrix} 6 \\ 10 \end{Bmatrix} = \tfrac{17}{3}$$

As $\bar{c}_1$ is positive, the original optimum solution remains optimum for the new problem also.

Example 4.10   Find the effect of changing $\mathbf{A}_1$ from $\begin{Bmatrix} 7 \\ 3 \end{Bmatrix}$ to $\begin{Bmatrix} 5 \\ 6 \end{Bmatrix}$ in Example 4.5.

SOLUTION   The relative cost coefficient of the nonbasic variable x_1 for the new $\mathbf{A}_1$ is given by

$$\bar{c}_1 = c_1 - \boldsymbol{\pi}^T \mathbf{A}_1 = -45 - (-\tfrac{22}{3} - \tfrac{2}{3}) \begin{Bmatrix} 5 \\ 6 \end{Bmatrix} = -\tfrac{13}{3}$$

Since $\bar{c}_1$ is negative, x_1 can be brought into the basis to reduce the objective function further. For this we start with the original optimum tableau with the new values of $\overline{\mathbf{A}}_1$ given by

$$\overline{\mathbf{A}}_1 = \mathbf{B}^{-1} \mathbf{A}_1 = \begin{bmatrix} \tfrac{4}{15} & -\tfrac{1}{15} \\ -\tfrac{1}{150} & \tfrac{2}{75} \end{bmatrix} \begin{Bmatrix} 5 \\ 6 \end{Bmatrix} = \begin{bmatrix} \tfrac{20}{15} & -\tfrac{6}{15} \\ -\tfrac{1}{30} & +\tfrac{4}{25} \end{bmatrix} = \begin{Bmatrix} \tfrac{14}{15} \\ \tfrac{19}{150} \end{Bmatrix}$$

Basic variables	Variables						$-f$	$\bar{b}_i$	$(\bar{b}_i/\bar{a}_{is})$
	x_1	x_2	x_3	x_4	x_5	x_6			
x_3	$\tfrac{14}{15}$	0	1	$\tfrac{7}{3}$	$\tfrac{4}{15}$	$-\tfrac{1}{15}$	0	$\tfrac{800}{3}$	$\tfrac{4000}{14}$
x_2	$\boxed{\tfrac{19}{150}}$	1	0	$-\tfrac{1}{30}$	$-\tfrac{1}{150}$	$\tfrac{2}{75}$	0	$\tfrac{40}{3}$	$\tfrac{2000}{19}$ ←
	Pivot element								
$-f$	$-\tfrac{13}{3}$	0	0	$\tfrac{50}{3}$	$\tfrac{22}{3}$	$\tfrac{2}{3}$	1	$\tfrac{28,000}{3}$	
	↑								
x_3	0	$-\tfrac{140}{19}$	1	$\tfrac{49}{19}$	$\tfrac{6}{19}$	$-\tfrac{5}{19}$	0	$\tfrac{3,200}{19}$	
x_1	1	$\tfrac{150}{19}$	0	$-\tfrac{5}{19}$	$-\tfrac{1}{19}$	$\tfrac{4}{19}$	0	$\tfrac{2,000}{19}$	
$-f$	0	$\tfrac{650}{19}$	0	$\tfrac{295}{19}$	$\tfrac{135}{19}$	$\tfrac{30}{19}$	1	$\tfrac{186,000}{19}$	

Since all $\bar{c}_j$ are nonnegative, the present tableau gives the new optimum solution as

$$x_1 = 2000/19, \quad x_3 = 3200/19 \quad \text{(basic variables)}$$

$$x_2 = x_4 = x_5 = x_6 = 0 \quad \text{(nonbasic variables)}$$

$$f_{\min} = -\frac{186,000}{19} \quad \text{and} \quad \text{maximum profit} = \frac{\$186,000}{19}$$

4.5.5   Addition of Constraints

Suppose that we have solved a LP problem with m constraints and obtained the optimal solution. We want to examine the effect of adding some more inequality constraints on the original optimum solution. For this we evaluate the new constraints by substituting the old optimal solution and see whether they are satisfied. If they are satisfied, it means that the inclusion of the new constraints in the old problem would not have affected the old optimum solution, and hence the old optimal solution remains optimal for the new problem also. On the other hand, if one or more of the new constraints are not satisfied by the old optimal solution, we can solve the problem without reworking the entire problem by proceeding as follows.

1. The simplex tableau corresponding to the old optimum solution expresses all the basic variables in terms of the nonbasic ones. With this information, eliminate the basic variables from the new constraints.
2. Transform the constraints thus obtained by multiplying throughout by -1.
3. Add the resulting constraints to the old optimal tableau and introduce one artificial variable for each new constraint added. Thus the enlarged system of equations will be in canonical form since the old basic variables were eliminated from the new constraints in step 1. Hence a new basis, consisting of the old optimal basis plus the artificial variables in the new constraint equations, will be readily available from this canonical form.
4. Go through phase I computations to eliminate the artificial variables.
5. Go through phase II computations to find the new optimal solution.

Example 4.11   If each of the products A, B, C, and D require, respectively, 2, 5, 3, and 4 min of time per unit on grinding machine in addition to the operations specified in Example 4.5, find the new optimum solution. Assume that the total time available on grinding machine per day is 600 min and all this time has to be utilized fully.

SOLUTION   The present data correspond to the addition of a constraint that can be stated as

$$2x_1 + 5x_2 + 3x_3 + 4x_4 = 600 \tag{E_1}$$

By substituting the original optimum solution,

$$x_2 = \tfrac{40}{3}, \quad x_3 = \tfrac{800}{3}, \quad x_1 = x_4 = x_5 = x_6 = 0$$

the left-hand side of Eq. (E_1) gives

$$2(0) + 5(\tfrac{40}{3}) + 3(\tfrac{800}{3}) + 4(0) = \tfrac{2600}{3} \neq 600$$

Thus the new constraint is not satisfied by the original optimum solution. Hence we proceed as follows.

Step 1   From the original optimum tableau, we can express the basic variables as

$$x_3 = \tfrac{800}{3} - \tfrac{5}{3}x_1 - \tfrac{7}{3}x_4 - \tfrac{4}{15}x_5 + \tfrac{1}{15}x_6$$

$$x_2 = \tfrac{40}{3} - \tfrac{1}{30}x_1 + \tfrac{1}{30}x_4 + \tfrac{1}{150}x_5 - \tfrac{1}{75}x_6$$

Thus Eq. (E₁) can be expressed as

$$2x_1 + 5(\tfrac{40}{3} - \tfrac{1}{30}x_1 + \tfrac{1}{30}x_4 + \tfrac{1}{150}x_5 - \tfrac{2}{75}x_6)$$

$$+ 3(\tfrac{800}{3} - \tfrac{5}{3}x_1 - \tfrac{7}{3}x_4 - \tfrac{4}{15}x_5 + \tfrac{1}{15}x_6) + 4x_4 = 600$$

that is,

$$-\tfrac{19}{6}x_1 - \tfrac{17}{6}x_4 - \tfrac{23}{30}x_5 + \tfrac{1}{15}x_6 = -\tfrac{800}{3} \tag{E₂}$$

Step 2 Transform this constraint such that the right-hand side becomes positive, that is,

$$\tfrac{19}{6}x_1 + \tfrac{17}{6}x_4 + \tfrac{23}{30}x_5 - \tfrac{1}{15}x_6 = \tfrac{800}{3} \tag{E₃}$$

Step 3 Add an artifical variable, say, x_k, the new constraint given by Eq. (E₃) and the infeasibility form $w = x_k$ into the original optimum tableau to obtain the new canonical system as follows:

Basic variables	Variables							$-f$	$-w$	$\bar{b}_i$	$(\bar{b}_i/\bar{a}_{is})$
	x_1	x_2	x_3	x_4	x_5	x_6	x_k				
x_3	$\frac{5}{3}$	0	1	$\frac{7}{3}$	$\frac{4}{5}$	$-\frac{1}{15}$	0	0	0	$\frac{800}{3}$	160
x_2	$\frac{1}{30}$	1	0	$-\frac{1}{30}$	$-\frac{1}{150}$	$\frac{2}{75}$	0	0	0	$\frac{40}{3}$	400
x_k	$\boxed{\frac{19}{6}}$	0	0	$\frac{17}{6}$	$\frac{23}{30}$	$-\frac{1}{15}$	1	0	0	$\frac{800}{3}$	$\frac{1600}{19}$
	Pivot element										
$-f$	$\frac{25}{3}$	0	0	$\frac{50}{3}$	$\frac{22}{3}$	$\frac{2}{3}$	0	1	0	$\frac{28,000}{3}$	
$-w$	$-\frac{19}{6}$	0	0	$-\frac{17}{6}$	$-\frac{23}{30}$	$\frac{1}{15}$	0	0	1	$-\frac{800}{3}$	

$\uparrow$

Step 4 Eliminate the artificial variable by applying the phase I procedure:

Basic variables	Variables							$-f$	$-w$	$\bar{b}_i$
	x_1	x_2	x_3	x_4	x_5	x_6	x_k			
x_3	0	0	1	$\frac{16}{19}$	$\frac{113}{285}$	$-\frac{3}{95}$	$-\frac{10}{19}$	0	0	$\frac{2,400}{19}$
x_2	0	1	0	$-\frac{6}{95}$	$-\frac{7}{475}$	$\frac{13}{475}$	$-\frac{1}{95}$	0	0	$\frac{200}{19}$
x_1	1	0	0	$\frac{17}{19}$	$\frac{23}{95}$	$-\frac{2}{95}$	$\frac{6}{19}$	0	0	$\frac{1,600}{19}$
$-f$	0	0	0	$\frac{175}{19}$	$\frac{101}{19}$	$\frac{16}{19}$	$-\frac{50}{19}$	1	0	$\frac{164,000}{19}$
$-w$	0	0	0	0	0	0	0	0	1	0

Thus the new optimum solution is given by

$$x_1 = \tfrac{1600}{19}, \quad x_2 = \tfrac{200}{19}, \quad x_3 = \tfrac{2400}{19} \quad \text{(basic variables)}$$

$$x_4 = x_5 = x_6 = 0 \quad \text{(nonbasic variables)}$$

$$f_{\min} = -\frac{164,000}{19} \quad \text{and} \quad \text{maximum profit} = \frac{\$164,000}{19}$$

4.6 TRANSPORTATION PROBLEM

This section deals with an important class of LP problems called the transportation problem. As the name indicates, a *transportation problem* is one in which the objective for minimization is the cost of transporting a certain commodity from a number of origins to a number of destinations. Although the transportation problem can be solved using the regular simplex method, its special structure offers a more convenient procedure for solving this type of problems. This procedure is based on the same theory of the simplex method, but it makes use of some shortcuts that yield a simpler computational scheme.

Suppose that there are m origins $R_1, R_2, \cdots, R_m$ (e.g., warehouses) and n destinations, $D_1, D_2, \cdots, D_n$ (e.g., factories). Let a_i be the amount of a commodity available at origin i $(i = 1, 2, \ldots, m)$ and b_j be the amount required at destination j $(j = 1, 2, \ldots, n)$. Let c_{ij} be the cost per unit of transporting the commodity from origin i to destination j. The objective is to determine the amount of commodity (x_{ij}) transported from origin i to destination j such that the total transportation costs are minimized. This problem can be formulated mathematically as

$$\text{Minimize } f = \sum_{i=1}^{m} \sum_{j=1}^{n} c_{ij} \tag{4.52}$$

subject to

$$\sum_{j=1}^{n} x_{ij} = a_i, \qquad i = 1, 2, \ldots, m \tag{4.53}$$

$$\sum_{i=1}^{m} x_{ij} = b_j, \qquad j = 1, 2, \ldots, n \tag{4.54}$$

$$x_{ij} \geq 0, \quad i = 1, 2, \ldots, m, \quad j = 1, 2, \ldots, n \tag{4.55}$$

Clearly, this is a LP problem in mn variables and $m + n$ equality constraints.

Equations (4.53) state that the total amount of the commodity transported from the origin i to the various destinations must be equal to the amount available at origin i $(i = 1, 2, \ldots, m)$, while Eqs. (4.54) state that the total amount of the commodity received by destination j from all the sources must be equal to the amount required at the destination j $(j = 1, 2, \ldots, n)$. The nonnegativity conditions Eqs. (4.55) are added since negative values for any x_{ij} have no physical meaning. It is assumed that the total demand equals the total supply, that is,

$$\sum_{i=1}^{m} a_i = \sum_{j=1}^{n} b_j \tag{4.56}$$

Equation (4.56), called the *consistency condition*, must be satisfied if a solution is to exist. This can be seen easily since

$$\sum_{i=1}^{m} a_i = \sum_{i=1}^{m} \left(\sum_{j=1}^{n} x_{ij} \right) = \sum_{j=1}^{n} \left(\sum_{i=1}^{m} x_{ij} \right) = \sum_{j=1}^{n} b_j \tag{4.57}$$

The problem stated in Eqs. (4.52) to (4.56) was originally formulated and solved by Hitchcock in 1941 [4.6]. This was also considered independently by Koopmans in 1947 [4.7]. Because of these early investigations the problem is sometimes called the *Hitchcock-Koopmans transportation problem*. The special structure of the transportation matrix can be seen by writing the equations in standard form:

$$
\begin{aligned}
x_{11} + x_{12} + \cdots + x_{1n} &= a_1 \\
x_{21} + x_{22} + \cdots + x_{2n} &= a_2 \\
&\vdots \\
x_{m1} + x_{m2} + \cdots + x_{mn} &= a_m
\end{aligned}
\qquad (4.58a)
$$

$$
\begin{aligned}
x_{11} \quad + x_{21} \quad + x_{m1} \quad &= b_1 \\
x_{12} \quad + x_{22} \quad + x_{m2} \quad &= b_2 \\
\vdots \qquad \vdots \qquad \vdots \qquad & \\
x_{1n} \quad + x_{2n} \quad + x_{mn} \quad &= b_n
\end{aligned}
\qquad (4.58b)
$$

$$
\begin{aligned}
c_{11}x_{11} + c_{12}x_{12} + \cdots + c_{1n}x_{1n} + c_{21}x_{21} + \cdots + c_{2n}x_{2n} + \cdots \\
+ c_{m1}x_{m1} + \cdots + c_{mn}x_{mn} = f
\end{aligned}
\qquad (4.58c)
$$

We notice the following properties from Eqs. (4.58):

1. All the nonzero coefficients of the constraints are equal to 1.
2. The constraint coefficients appear in a triangular form.
3. Any variable appears only once in the first m equations and once in the next n equations.

These are the special properties of the transportation problem that allow development of the *transportation technique*. To facilitate the identification of a starting solution, the system of equations (4.58) is represented in the form of an array, called the *transportation array*, as shown in Fig. 4.2. In all the techniques developed for solving the transportation problem, the calculations are made directly on the transportation array.

Computational Procedure. The solution of a LP problem, in general, requires a calculator or, if the problem is large, a high-speed digital computer. On the other hand, the solution of a transportation problem can often be obtained with the use of a pencil and paper since additions and subtractions are the only calculations required. The basic steps involved in the solution of a transportation problem are

1. Determine a starting basic feasible solution.
2. Test the current basic feasible solution for optimality. If the current solution is optimal, stop the iterative process; otherwise, go to step 3.
3. Select a variable to enter the basis from among the current nonbasic variables.

To From		Destination j					Amount available a_i
		1	2	3	$\cdots$	n	
Origin i	1	x_{11} c_{11}	x_{12} c_{12}	x_{13} c_{13}	$\cdots$	x_{1n} c_{1n}	a_1
	2	x_{21} c_{21}	x_{22} c_{22}	x_{23} c_{23}	$\cdots$	x_{2n} c_{2n}	a_2
	3	x_{31} c_{31}	x_{32} c_{32}	x_{33} c_{33}	$\cdots$	x_{3n} c_{3n}	a_3
	$\vdots$	$\vdots$	$\vdots$	$\vdots$	$\vdots$	$\vdots$	$\vdots$
	m	x_{m1} c_{m1}	x_{m2} c_{m2}	x_{m3} c_{m3}	$\cdots$	x_{mn} c_{mn}	a_m
Amount required b_j		b_1	b_2	b_3	$\cdots$	b_n	

Figure 4.2   Transportation array.

4. Select a variable to leave from the basis from among the current basic variables (using the feasibility condition).
5. Find a new basic feasible solution and return to step 2.

The details of these steps are given in Ref. [4.10].

4.7 KARMARKAR'S INTERIOR METHOD

Karmarkar proposed a new method in 1984 for solving large-scale linear programming problems very efficiently. The method is known as an *interior method* since it finds improved search directions strictly in the interior of the feasible space. This is in contrast with the simplex method, which searches along the boundary of the feasible space by moving from one feasible vertex to an adjacent one until the optimum point is found. For large LP problems, the number of vertices will be quite large and hence the simplex method would become very expensive in terms of computer time. Along with many other applications, Karmarkar's method has been applied to aircraft route scheduling problems. It was reported [4.19] that Karmarkar's method solved problems involving 150,000 design variables and 12,000 constraints in 1 hour while the simplex method required 4 hours for solving a smaller problem involving only 36,000 design variables and 10,000 constraints. In fact, it was found that Karmarkar's method is as much as 50 times faster than the simplex method for large problems.

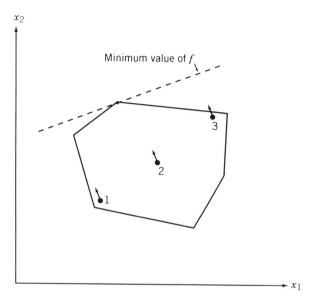

Figure 4.3   Improvement of objective function from different points of a polytope.

Karmarkar's method is based on the following two observations:

1. If the current solution is near the center of the polytope, we can move along the steepest descent direction to reduce the value of f by a maximum amount. From Fig. 4.3, we can see that the current solution can be improved substantially by moving along the steepest descent direction if it is near the center (point 2) but not near the boundary point (points 1 and 3).

2. The solution space can always be transformed without changing the nature of the problem so that the current solution lies near the center of the polytope.

It is well known that in many numerical problems, by changing the units of data or rescaling (e.g., using feet instead of inches), we may be able to reduce the numerical instability. In a similar manner, Karmarkar observed that the variables can be transformed (in a more general manner than ordinary rescaling) so that straight lines remain straight lines while angles and distances change for the feasible space.

4.7.1   Statement of the Problem

Karmarkar's method requires the LP problem in the following form:

$$\text{Minimize } f = \mathbf{c}^{\mathrm{T}}\mathbf{X}$$

subject to

$$[a]\mathbf{X} = \mathbf{0}$$

$$x_1 + x_2 + \cdots + x_n = 1 \qquad (4.59)$$

$$\mathbf{X} \geq \mathbf{0}$$

where $\mathbf{X} = \{x_1, x_2, \ldots, x_n\}^{\mathrm{T}}$, $\mathbf{c} = \{c_1, c_2, \ldots, c_n\}^{\mathrm{T}}$, and $[a]$ is an $m \times n$ matrix. In addition, an interior feasible starting solution to Eqs. (4.59) must be known. Usually,

$$\mathbf{X} = \left\{\frac{1}{n}, \frac{1}{n}, \cdots \frac{1}{n}\right\}^{\mathrm{T}}$$

is chosen as the starting point. In addition, the optimum value of f must be zero for the problem. Thus

$$\mathbf{X}^{(1)} = \left\{\frac{1}{n} \quad \frac{1}{n} \cdots \frac{1}{n}\right\}^{\mathrm{T}} = \text{interior feasible}$$

$$f_{\min} = 0 \tag{4.60}$$

Although most LP problems may not be available in the form of Eq. (4.59) while satisfying the conditions of Eq. (4.60), it is possible to put any LP problem in a form that satisfies Eqs. (4.59) and (4.60) as indicated below.

4.7.2   Conversion of an LP Problem into the Required Form

Let the given LP problem be of the form

$$\text{Minimize } \mathbf{d}^{\mathrm{T}}\mathbf{X}$$

subject to

$$[\alpha]\mathbf{X} = \mathbf{b}$$

$$\mathbf{X} \geq \mathbf{0} \tag{4.61}$$

To convert this problem into the form of Eq. (4.59), we use the procedure suggested in Ref. [4.20] and define integers m and n such that $\mathbf{X}$ will be an $(n - 3)$-component vector and $[\alpha]$ will be a matrix of order $m - 1 \times n - 3$. We now define the vector $\bar{\mathbf{z}} = \{z_1, z_2, \cdots, z_{n-3}\}^{\mathrm{T}}$ as

$$\bar{\mathbf{z}} = \frac{\mathbf{X}}{\beta} \tag{4.62}$$

where β is a constant chosen to have a sufficiently large value such that

$$\beta > \sum_{i=1}^{n-3} x_i \tag{4.63}$$

for any feasible solution $\mathbf{X}$ (assuming that the solution is bounded). By using Eq. (4.62), the problem of Eq. (4.61) can be stated as follows:

$$\text{Minimize } \beta \mathbf{d}^{\mathrm{T}}\bar{\mathbf{z}}$$

subject to

$$[\alpha]\bar{\mathbf{z}} = \frac{1}{\beta}\mathbf{b}$$

$$\bar{\mathbf{z}} \geq \mathbf{0} \tag{4.64}$$

We now define a new vector $\mathbf{z}$ as

$$\mathbf{z} = \left\{ \begin{array}{c} \bar{\mathbf{z}} \\ z_{n-2} \\ z_{n-1} \\ z_n \end{array} \right\}$$

and solve the following related problem instead of the problem in Eqs. (4.64):

$$\text{Minimize } \{\beta \mathbf{d}^T \quad 0 \quad 0 \quad M\} \mathbf{z}$$

subject to

$$\left[\begin{array}{cccc} [\alpha] & \mathbf{0} & -\dfrac{n}{\beta}\mathbf{b} & \left(\dfrac{n}{\beta}\mathbf{b} - [\alpha]\mathbf{e}\right) \\ 0 & 0 & n & 0 \end{array} \right] \mathbf{z} = \left\{ \begin{array}{c} \mathbf{0} \\ 1 \end{array} \right\}$$

$$\mathbf{e}^T \bar{\mathbf{z}} + z_{n-2} + z_{n-1} + z_n = 1$$

$$\mathbf{z} \geq \mathbf{0}$$

(4.65)

where $\mathbf{e}$ is an $(m-1)$-component vector whose elements are all equal to 1, z_{n-2} is a slack variable that absorbs the difference between 1 and the sum of other variables, z_{n-1} is constrained to have a value of $1/n$, and M is given a large value (corresponding to the artificial variable z_n) to force z_n to zero when the problem stated in Eqs. (4.61) has a feasible solution. Equations (4.65) are developed such that if $\mathbf{z}$ is a solution to these equations, $\mathbf{X} = \beta \bar{\mathbf{z}}$ will be a solution to Eqs. (4.61) if Eqs. (4.61) have a feasible solution. Also, it can be verified that the interior point $\mathbf{z} = (1/n)\mathbf{e}$ is a feasible solution to Eqs. (4.65). Equations (4.65) can be seen to be the desired form of Eqs. (4.61) except for a 1 on the right-hand side. This can be eliminated by subtracting the last constraint from the next-to-last constraint, to obtain the required form:

$$\text{Minimize } \{\beta \mathbf{d}^T \quad 0 \quad \text{-}0 \quad M\} \mathbf{z}$$

subject to

$$\left[\begin{array}{cccc} [\alpha] & \mathbf{0} & -\dfrac{n}{\beta}\mathbf{b} & \left(\dfrac{n}{\beta}\mathbf{b} - [\alpha]\mathbf{e}\right) \\ -\mathbf{e}^T & -1 & (n-1) & -1 \end{array} \right] \mathbf{z} = \left\{ \begin{array}{c} \mathbf{0} \\ 0 \end{array} \right\}$$

$$\mathbf{e}^T \bar{\mathbf{z}} + z_{n-2} + z_{n-1} + z_n = 1$$

$$\mathbf{z} \geq \mathbf{0}$$

(4.66)

Note: When Eqs. (4.66) are solved, if the value of the artificial variable $z_n > 0$, the original problem in Eqs. (4.61) is infeasible. On the other hand, if the value of the slack variable $z_{n-2} = 0$, the solution of the problem given by Eqs. (4.61) is unbounded.

Example 4.12 Transform the following LP problem into a form required by Karmarkar's method:

$$\text{Minimize } 2x_1 + 3x_2$$

subject to

$$3x_1 + x_2 - 2x_3 = 3$$
$$5x_1 - 2x_2 = 2$$
$$x_i \geq 0, \quad i = 1, 2, 3$$

SOLUTION   It can be seen that

$$\mathbf{d} = \{2 \quad 3 \quad 0\}^T, \ [\alpha] = \begin{bmatrix} 3 & 1 & -2 \\ 5 & -2 & 0 \end{bmatrix}, \ \mathbf{b} = \begin{Bmatrix} 3 \\ 2 \end{Bmatrix}, \text{ and } \mathbf{X} = \begin{Bmatrix} x_1 \\ x_2 \\ x_3 \end{Bmatrix}$$

We define the integers m and n as $n = 6$ and $m = 3$ and choose $\beta = 10$ so that

$$\bar{\mathbf{z}} = \frac{1}{10} \begin{Bmatrix} z_1 \\ z_2 \\ z_3 \end{Bmatrix}$$

Noting that $\mathbf{e} = \{1, 1, 1\}^T$, Eqs. (4.66) can be expressed as

$$\text{Minimize } \{20 \quad 30 \quad 0 \quad 0 \quad 0 \quad M\} \mathbf{z}$$

subject to

$$\left[\begin{bmatrix} 3 & 1 & -2 \\ 5 & -2 & 0 \end{bmatrix} \begin{Bmatrix} 0 \\ 0 \end{Bmatrix} - \frac{6}{10} \begin{Bmatrix} 3 \\ 2 \end{Bmatrix} \right.$$
$$\left. \times \left(\frac{6}{10} \begin{Bmatrix} 3 \\ 2 \end{Bmatrix} - \begin{bmatrix} 3 & 1 & -2 \\ 5 & -2 & 0 \end{bmatrix} \begin{Bmatrix} 1 \\ 1 \\ 1 \end{Bmatrix} \right) \right] \mathbf{z} = \mathbf{0}$$
$$\{-\{1 \quad 1 \quad 1\} \quad -1 \quad 5 \quad -1\} \mathbf{z} = 0$$
$$z_1 + z_2 + z_3 + z_4 + z_5 + z_6 \quad = 1$$
$$\mathbf{z} = \{z_1 \quad z_2 \quad z_3 \quad z_4 \quad z_5 \quad z_6\}^T \geq \mathbf{0}$$

where M is a very large number. These equations can be seen to be in the desired form.

4.7.3 Algorithm

Starting from an interior feasible point $\mathbf{X}^{(1)}$, Karmarkar's method finds a sequence of points $\mathbf{X}^{(2)}, \mathbf{X}^{(3)}, \cdots$ using the following iterative procedure:

1. Initialize the iterative process. Begin with the center point of the simplex as the initial feasible point

$$\mathbf{X}^{(1)} = \left\{ \frac{1}{n} \ \frac{1}{n} \cdots \frac{1}{n} \right\}^T.$$

Set the iteration number as $k = 1$.

2. Test for optimality. Since $f = 0$ at the optimum point, we stop the procedure if the following convergence criterion is satisfied:

$$||\mathbf{c}^T \mathbf{X}^{(k)}|| \leq \varepsilon \tag{4.67}$$

where ε is a small number. If Eq. (4.67) is not satisfied, go to step 3.

3. Compute the next point, $\mathbf{X}^{(k+1)}$. For this, we first find a point $\mathbf{Y}^{(k+1)}$ in the transformed unit simplex as

$$
\begin{aligned}
\mathbf{Y}^{(k+1)} = &\left\{ \frac{1}{n} \quad \frac{1}{n} \cdots \frac{1}{n} \right\}^T \\
&- \frac{\alpha([I] - [P]^T([P][P]^T)^{-1}[P])[D(\mathbf{X}^{(k)})]\mathbf{c}}{||\mathbf{c}|| \sqrt{n(n-1)}}
\end{aligned}
\tag{4.68}
$$

where $||\mathbf{c}||$ is the length of the vector $\mathbf{c}$, $[I]$ the identity matrix of order n, $[D(\mathbf{X}^{(k)})]$ an $n \times n$ matrix with all off-diagonal entries equal to 0, and diagonal entries equal to the components of the vector $\mathbf{X}^{(k)}$ as

$$[D(\mathbf{X}^{(k)})]_{ii} = x_i^{(k)}, \qquad i = 1, 2, \ldots, n \tag{4.69}$$

$[P]$ is an $(m+1) \times n$ matrix whose first m rows are given by $[a] [D(\mathbf{X}^{(k)})]$ and the last row is composed of 1's:

$$[P] = \begin{bmatrix} [a][D(\mathbf{X}^{(k)})] \\ 1 \qquad 1 \quad \cdots \quad 1 \end{bmatrix} \tag{4.70}$$

and the value of the parameter α is usually chosen as $\alpha = \frac{1}{4}$ to ensure convergence. Once $\mathbf{Y}^{(k+1)}$ is found, the components of the new point $\mathbf{X}^{(k+1)}$ are determined as

$$x_i^{(k+1)} = \frac{x_i^{(k)} y_i^{(k+1)}}{\sum_{r=1}^n x_r^{(k)} y_r^{(k+1)}}, \qquad i = 1, 2, \ldots, n \tag{4.71}$$

Set the new iteration number as $k = k + 1$ and go to step 2.

Example 4.13 Find the solution of the following problem using Karmarkar's method:

$$\text{Minimize } f = 2x_1 + x_2 - x_3$$

subject to

$$x_2 - x_3 = 0$$

$$x_1 + x_2 + x_3 = 1$$

$$x_i \geq 0, \quad i = 1, 2, 3 \tag{E.1}$$

Use the value of $\varepsilon = 0.05$ for testing the convergence of the procedure.

SOLUTION   The problem is already in the required form of Eq. (4.59), and hence the following iterative procedure can be used to find the solution of the problem.

Step 1 We choose the initial feasible point as

$$\mathbf{X}^{(1)} = \begin{Bmatrix} \frac{1}{3} \\ \frac{1}{3} \\ \frac{1}{3} \end{Bmatrix}$$

and set $k = 1$.

Step 2 Since $|f(\mathbf{X}^{(1)})| = |\frac{2}{3}| > 0.05$, we go to step 3.

Step 3 Since $[a] = \{0,\ 1,\ -1\}, \mathbf{c} = \{2, 1, -1\}^T, \|\mathbf{c}\| = \sqrt{(2)^2 + (1)^2 + (-1)^2} = \sqrt{6}$, we find that

$$[D(\mathbf{X}^{(1)})] = \begin{bmatrix} \frac{1}{3} & 0 & 0 \\ 0 & \frac{1}{3} & 0 \\ 0 & 0 & \frac{1}{3} \end{bmatrix}$$

$$[a][D(\mathbf{X}^{(1)})] = \{0 \quad \frac{1}{3} \quad -\frac{1}{3}\}$$

$$[P] = \begin{bmatrix} [a][D(\mathbf{X}^{(1)})] \\ 1 \quad 1 \quad 1 \end{bmatrix} = \begin{bmatrix} 0 & \frac{1}{3} & -\frac{1}{3} \\ 1 & 1 & 1 \end{bmatrix}$$

$$([P][P]^T)^{-1} = \begin{bmatrix} \frac{2}{9} & 0 \\ 0 & 3 \end{bmatrix}^{-1} = \begin{bmatrix} \frac{9}{2} & 0 \\ 0 & \frac{1}{3} \end{bmatrix}$$

$$[D(\mathbf{X}^{(1)})]\mathbf{c} = \begin{bmatrix} \frac{1}{3} & 0 & 0 \\ 0 & \frac{1}{3} & 0 \\ 0 & 0 & \frac{1}{3} \end{bmatrix} \begin{Bmatrix} 2 \\ 1 \\ -1 \end{Bmatrix} = \begin{Bmatrix} \frac{2}{3} \\ \frac{1}{3} \\ -\frac{1}{3} \end{Bmatrix}$$

$$([I] - [P]^T([P][P]^T)^{-1}[P])[D(\mathbf{X}^{(1)})]\mathbf{c}$$

$$= \left(\begin{bmatrix} 1 & 0 & 0 \\ 0 & 1 & 0 \\ 0 & 0 & 1 \end{bmatrix} - \begin{bmatrix} 0 & 1 \\ \frac{1}{3} & 1 \\ -\frac{1}{3} & 1 \end{bmatrix} \begin{bmatrix} \frac{9}{2} & 0 \\ 0 & \frac{1}{3} \end{bmatrix} \begin{bmatrix} 0 & \frac{1}{3} & -\frac{1}{3} \\ 1 & 1 & 1 \end{bmatrix} \right) \begin{Bmatrix} \frac{2}{3} \\ \frac{1}{3} \\ -\frac{1}{3} \end{Bmatrix}$$

$$= \begin{bmatrix} \frac{2}{3} & -\frac{1}{3} & -\frac{1}{3} \\ -\frac{1}{3} & \frac{1}{6} & \frac{1}{6} \\ -\frac{1}{3} & \frac{1}{6} & \frac{1}{6} \end{bmatrix} \begin{Bmatrix} \frac{2}{3} \\ \frac{1}{3} \\ -\frac{1}{3} \end{Bmatrix} = \begin{Bmatrix} \frac{4}{9} \\ -\frac{2}{9} \\ -\frac{2}{9} \end{Bmatrix}$$

Using $\alpha = \frac{1}{4}$, Eq. (4.68) gives

$$\mathbf{Y}^{(2)} = \begin{Bmatrix} \frac{1}{3} \\ \frac{1}{3} \\ \frac{1}{3} \end{Bmatrix} - \frac{1}{4} \begin{Bmatrix} \frac{4}{9} \\ -\frac{2}{9} \\ -\frac{2}{9} \end{Bmatrix} \frac{1}{\sqrt{3(2)}\sqrt{6}} = \begin{Bmatrix} \frac{34}{108} \\ \frac{37}{108} \\ \frac{37}{108} \end{Bmatrix}$$

Noting that

$$\sum_{r=1}^{n} x_r^{(1)} y_r^{(2)} = \tfrac{1}{3}\left(\tfrac{34}{108}\right) + \tfrac{1}{3}\left(\tfrac{37}{108}\right) + \tfrac{1}{3}\left(\tfrac{37}{108}\right) = \tfrac{1}{3}$$

Eq. (4.71) can be used to find

$$\{x_i^{(2)}\} = \left\{ \frac{x_i^{(1)} y_i^{(2)}}{\sum\limits_{r=1}^{3} x_r^{(1)} y_r^{(2)}} \right\} = 3 \left\{ \begin{matrix} \tfrac{34}{324} \\ \tfrac{37}{324} \\ \tfrac{37}{324} \end{matrix} \right\} = \left\{ \begin{matrix} \tfrac{34}{108} \\ \tfrac{37}{108} \\ \tfrac{37}{108} \end{matrix} \right\}$$

Set the new iteration number as $k = k + 1 = 2$ and go to step 2. The procedure is to be continued until convergence is achieved.

Notes:

1. Although $\mathbf{X}^{(2)} = \mathbf{Y}^{(2)}$ in this example, they need not be, in general, equal to one another.
2. The value of f at $\mathbf{X}^{(2)}$ is

$$f(\mathbf{X}^{(2)}) = 2\left(\tfrac{34}{108}\right) + \tfrac{37}{108} - \tfrac{37}{108} = \tfrac{17}{27} < f(\mathbf{X}^{(1)}) = \tfrac{18}{27}$$

4.8   QUADRATIC PROGRAMMING

A quadratic programming problem can be stated as

$$\text{Minimize } f(\mathbf{X}) = \mathbf{C}^T \mathbf{X} + \tfrac{1}{2} \mathbf{X}^T \mathbf{D} \mathbf{X} \tag{4.72}$$

subject to

$$\mathbf{A}\mathbf{X} \leq \mathbf{B} \tag{4.73}$$

$$\mathbf{X} \geq \mathbf{0} \tag{4.74}$$

where

$$\mathbf{X} = \left\{ \begin{matrix} x_1 \\ x_2 \\ \vdots \\ x_n \end{matrix} \right\}, \quad \mathbf{C} = \left\{ \begin{matrix} c_1 \\ c_2 \\ \vdots \\ c_n \end{matrix} \right\}, \quad \mathbf{B} = \left\{ \begin{matrix} b_1 \\ b_2 \\ \vdots \\ b_m \end{matrix} \right\},$$

$$\mathbf{D} = \begin{bmatrix} d_{11} & d_{12} & \cdots & d_{1n} \\ d_{21} & d_{22} & \cdots & d_{2n} \\ \vdots & & & \\ d_{n1} & d_{n2} & \cdots & d_{nn} \end{bmatrix}, \quad \text{and} \quad \mathbf{A} = \begin{bmatrix} a_{11} & a_{12} & \cdots & a_{1n} \\ a_{21} & a_{22} & \cdots & a_{2n} \\ \vdots & & & \\ a_{m1} & a_{m2} & \cdots & a_{mn} \end{bmatrix}$$

In Eq. (4.72) the term $\mathbf{X}^T \mathbf{D} \mathbf{X}/2$ represents the quadratic part of the objective function with $\mathbf{D}$ being a symmetric positive-definite matrix. If $\mathbf{D} = \mathbf{0}$, the problem reduces to a LP problem. The solution of the quadratic programming problem stated in Eqs. (4.72) to (4.74) can be obtained by using the Lagrange multiplier technique. By introducing the slack variables s_i^2, $i = 1, 2, \ldots, m$, in Eqs. (4.73) and the surplus variables t_j^2, $j = 1, 2, \ldots, n$, in Eqs. (4.74), the quadratic programming problem can be written as

$$\text{Minimize } f(\mathbf{X}) = \mathbf{C}^\mathrm{T}\mathbf{X} + \tfrac{1}{2}\mathbf{X}^\mathrm{T}\mathbf{D}\mathbf{X} \tag{4.72}$$

subject to the equality constraints

$$\mathbf{A}_i^\mathrm{T}\mathbf{X} + s_i^2 = b_i, \quad i = 1, 2, \ldots, m \tag{4.75}$$

$$-x_j + t_j^2 = 0, \quad j = 1, 2, \ldots, n \tag{4.76}$$

where

$$\mathbf{A}_i = \begin{Bmatrix} a_{i1} \\ a_{i2} \\ \vdots \\ a_{in} \end{Bmatrix}$$

The Lagrange function can be written as

$$L(\mathbf{X}, \mathbf{S}, \mathbf{T}, \boldsymbol{\lambda}, \boldsymbol{\theta}) = \mathbf{C}^\mathrm{T}\mathbf{X} + \tfrac{1}{2}\mathbf{X}^\mathrm{T}\mathbf{D}\mathbf{X} + \sum_{i=1}^{m} \lambda_i(\mathbf{A}_i^\mathrm{T}\mathbf{X} + s_i^2 - b_i)$$

$$+ \sum_{j=1}^{n} \theta_j(-x_j + t_j^2) \tag{4.77}$$

The necessary conditions for the stationariness of L give

$$\frac{\partial L}{\partial x_j} = c_j + \sum_{i=1}^{n} d_{ij}x_i + \sum_{i=1}^{m} \lambda_i a_{ij} - \theta_j = 0, \qquad j = 1, 2, \ldots, n \tag{4.78}$$

$$\frac{\partial L}{\partial s_i} = 2\lambda_i s_i = 0, \qquad i = 1, 2, \ldots, m \tag{4.79}$$

$$\frac{\partial L}{\partial t_j} = 2\theta_j t_j = 0, \qquad j = 1, 2, \ldots, n \tag{4.80}$$

$$\frac{\partial L}{\partial \lambda_i} = \mathbf{A}_i^\mathrm{T}\mathbf{X} + s_i^2 - b_i = 0, \qquad i = 1, 2, \ldots, m \tag{4.81}$$

$$\frac{\partial L}{\partial \theta_j} = -x_j + t_j^2 = 0, \qquad j = 1, 2, \ldots, n \tag{4.82}$$

By defining a set of new variables Y_i as

$$Y_i = s_i^2 \geq 0, \qquad i = 1, 2, \ldots, m \tag{4.83}$$

Equations (4.81) can be written as

$$\mathbf{A}_i^\mathrm{T}\mathbf{X} - b_i = -s_i^2 = -Y_i, \qquad i = 1, 2, \ldots, m \tag{4.84}$$

Multiplying Eq. (4.79) by s_i and Eq. (4.80) by t_j, we obtain

$$\lambda_i s_i^2 = \lambda_i Y_i = 0, \qquad i = 1, 2, \ldots, m \tag{4.85}$$

$$\theta_j t_j^2 = 0, \qquad j = 1, 2, \ldots, n \tag{4.86}$$

Combining Eqs. (4.84) and (4.85), and Eqs. (4.82) and (4.86), we obtain

$$\lambda_i (\mathbf{A}_i^T \mathbf{X} - b_i) = 0, \qquad i = 1, 2, \ldots, m \tag{4.87}$$

$$\theta_j x_j = 0, \qquad j = 1, 2, \ldots, n \tag{4.88}$$

Thus the necessary conditions can be summarized as follows:

$$c_j - \theta_j + \sum_{i=1}^{n} x_i d_{ij} + \sum_{i=1}^{m} \lambda_i a_{ij} = 0, \qquad j = 1, 2, \ldots, n \tag{4.89}$$

$$\mathbf{A}_i^T \mathbf{X} - b_i = -Y_i, \qquad i = 1, 2, \ldots, m \tag{4.90}$$

$$x_j \geq 0, \qquad j = 1, 2, \ldots, n \tag{4.91}$$

$$Y_i \geq 0, \qquad i = 1, 2, \ldots, m \tag{4.92}$$

$$\lambda_i \geq 0, \qquad i = 1, 2, \ldots, m \tag{4.93}$$

$$\theta_j \geq 0, \qquad j = 1, 2, \ldots, n \tag{4.94}$$

$$\lambda_i Y_i = 0, \qquad i = 1, 2, \ldots, m \tag{4.95}$$

$$\theta_j x_j = 0, \qquad j = 1, 2, \ldots, n \tag{4.96}$$

We can notice one important thing in Eqs. (4.89) to (4.96). With the exception of Eqs. (4.95) and (4.96), the necessary conditions are linear functions of the variables x_j, Y_i, λ_i, and θ_j. Thus the solution of the original quadratic programming problem can be obtained by finding a nonnegative solution to the set of $m + n$ linear equations given by Eqs. (4.89) and (4.90), which also satisfies the $m + n$ equations stated in Eqs. (4.95) and (4.96).

Since $\mathbf{D}$ is a positive-definite matrix, $f(\mathbf{X})$ will be a strictly convex function,[†] and the feasible space is convex (because of linear equations), any local minimum of the problem will be the global minimum. Further, it can be seen that there are $2(n + m)$ variables and $2(n + m)$ equations in the necessary conditions stated in Eqs. (4.89) to (4.96). Hence the solution of the Eqs. (4.89), (4.90), (4.95), and (4.96) must be unique. Thus the feasible solution satisfying all the Eqs. (4.89) to (4.96), if it exists, must give the optimum solution of the quadratic programming problem directly. The solution of the system of equations above can be obtained by using phase I of the simplex method. The only restriction here is that the satisfaction of the nonlinear relations, Eqs. (4.95) and (4.96), has to be maintained all the time. Since our objective is just to find a feasible solution to the set of Eqs. (4.89) to (4.96), there is no necessity of phase II computations. We shall follow the procedure developed by Wolfe [4.21] to apply

---

[†]See Appendix A for the definition and properties of a convex function.

phase I. This procedure involves the introduction of n nonnegative artificial variables z_i into the Eqs. (4.89) so that

$$c_j - \theta_j + \sum_{i=1}^{n} x_i d_{ij} + \sum_{i=1}^{m} \lambda_i a_{ij} + z_j = 0, \qquad j = 1, 2, \ldots, n \qquad (4.97)$$

Then we minimize

$$F = \sum_{j=1}^{n} z_j \qquad (4.98)$$

subject to the constraints

$$c_j - \theta_j + \sum_{i=1}^{n} x_i d_{ij} + \sum_{i=1}^{m} \lambda_i a_{ij} + z_j = 0, \qquad j = 1, 2, \ldots, n$$

$$\mathbf{A}_i^T \mathbf{X} + Y_i = b_i, \qquad i = 1, 2, \ldots, m$$

$$\mathbf{X} \geq \mathbf{0}, \quad \mathbf{Y} \geq \mathbf{0}, \quad \boldsymbol{\lambda} \geq \mathbf{0}, \quad \boldsymbol{\theta} \geq \mathbf{0}$$

While solving this problem, we have to take care of the additional conditions

$$\begin{aligned} \lambda_i Y_i &= 0, & j &= 1, 2, \ldots, m \\ \theta_j x_j &= 0, & j &= 1, 2, \ldots, n \end{aligned} \qquad (4.99)$$

Thus when deciding whether to introduce Y_i into the basic solution, we first have to ensure that either λ_i is not in the solution or λ_i will be removed when Y_i enters the basis. Similar care has to be taken regarding the variables θ_j and x_j. These additional checks are not very difficult to make during the solution procedure.

Example 4.14

$$\text{Minimize } f = -4x_1 + x_1^2 - 2x_1 x_2 + 2x_2^2$$

subject to

$$2x_1 + x_2 \leq 6$$

$$x_1 - 4x_2 \leq 0$$

$$x_1 \geq 0, \quad x_2 \geq 0$$

SOLUTION   By introducing the slack variables $Y_1 = s_1^2$ and $Y_2 = s_2^2$ and the surplus variables $\theta_1 = t_1^2$ and $\theta_2 = t_2^2$, the problem can be stated as follows:

$$\text{Minimize } f = (-4 \ \ 0) \begin{Bmatrix} x_1 \\ x_2 \end{Bmatrix} + \frac{1}{2} (x_1 \ \ x_2) \begin{bmatrix} 2 & -2 \\ -2 & 4 \end{bmatrix} \begin{Bmatrix} x_1 \\ x_2 \end{Bmatrix}$$

subject to

$$\begin{bmatrix} 2 & 1 \\ 1 & -4 \end{bmatrix} \begin{Bmatrix} x_1 \\ x_2 \end{Bmatrix} + \begin{Bmatrix} Y_1 \\ Y_2 \end{Bmatrix} = \begin{Bmatrix} 6 \\ 0 \end{Bmatrix}$$

$$-x_1 + \theta_1 = 0 \qquad (E_1)$$

$$-x_2 + \theta_2 = 0$$

By comparing this problem with the one stated in Eqs. (4.72) to (4.74), we find that

$$c_1 = -4, \quad c_2 = 0, \quad \mathbf{D} = \begin{bmatrix} 2 & -2 \\ -2 & 4 \end{bmatrix}, \quad \mathbf{A} = \begin{bmatrix} 2 & 1 \\ 1 & -4 \end{bmatrix},$$

$$\mathbf{A}_1 = \begin{Bmatrix} 2 \\ 1 \end{Bmatrix}, \quad \mathbf{A}_2 = \begin{Bmatrix} 1 \\ -4 \end{Bmatrix}, \quad \text{and} \quad \mathbf{B} = \begin{Bmatrix} 6 \\ 0 \end{Bmatrix}$$

The necessary conditions for the solution of the problem stated in Eqs. (E_1) can be obtained, using Eqs. (4.89) to (4.96), as

$$-4 - \theta_1 + 2x_1 - 2x_2 + 2\lambda_1 + \lambda_2 = 0$$
$$0 - \theta_2 - 2x_1 + 4x_2 + \lambda_1 - 4\lambda_2 = 0$$
$$2x_1 + x_2 - 6 = -Y_1 \tag{E_2}$$
$$x_1 - 4x_2 - 0 = -Y_2$$

$$x_1 \geq 0, \ x_2 \geq 0, \ Y_1 \geq 0, \ Y_2 \geq 0, \ \lambda_1 \geq 0,$$
$$\lambda_2 \geq 0, \ \theta_1 \geq 0, \ \theta_2 \geq 0 \tag{E_3}$$

$$\lambda_1 Y_1 = 0, \quad \theta_1 x_1 = 0$$
$$\lambda_2 Y_2 = 0, \quad \theta_2 x_2 = 0 \tag{E_4}$$

(If Y_i is in the basis, λ_i cannot be in the basis, and if x_j is in the basis, θ_j cannot be in the basis to satisfy these equations.) Equations (E_2) can be rewritten as

$$2x_1 - 2x_2 + 2\lambda_1 + \lambda_2 - \theta_1 + z_1 = 4$$
$$-2x_1 + 4x_2 + \lambda_1 - 4\lambda_2 - \theta_2 + z_2 = 0$$
$$2x_1 + x_2 + Y_1 = 6 \tag{E_5}$$
$$x_1 - 4x_2 + Y_2 = 0$$

where z_1 and z_2 are artificial variables. To find a feasible solution to Eqs. (E_2) to (E_4) by using phase I of simplex method, we minimize $w = z_1 + z_2$ with constraints stated in Eqs. (E_5), (E_3), and (E_4). The initial simplex tableau is shown below:

Basic variables	x_1	x_2	λ_1	λ_2	θ_1	θ_2	Y_1	Y_2	z_1	z_2	w	$\bar{b}_i$	$\bar{b}_i/\bar{a}_{is}$ for $\bar{a}_{is} > 0$
Y_1	2	1	0	0	0	0	1	0	0	0	0	6	6
Y_2	1	−4	0	0	0	0	0	1	0	0	0	0	
z_1	2	−2	2	1	−1	0	0	0	1	0	0	4	
z_2	−2	4	1	−4	0	−1	0	0	0	1	0	0	0 ← Smaller one
$-w$	0	−2	−3	3	1	1	0	0	0	0	1	−4	

x_2 selected for entering next basis   Most negative

According to the regular procedure of simplex method, λ_1 enters the next basis since the cost coefficient of λ_1 is most negative and z_2 leaves the basis since the ratio $\bar{b}_i/\bar{a}_{is}$ is smaller for z_2. However, λ_1 cannot enter the basis, as Y_1 is already in the basis [to satisfy Eqs. (E$_4$)]. Hence we select x_2 for entering the next basis. According to this choice, z_2 leaves the basis. By carrying out the required pivot operation, we obtain the following tableau:

Basic variables	\multicolumn{11}{c}{Variables}			$\bar{b}_i/\bar{a}_{is}$
	x_1	x_2	λ_1	λ_2	θ_1	θ_2	Y_1	Y_2	z_1	z_2	w	$\bar{b}_i$	for $\bar{a}_{is} > 0$
Y_1	$\boxed{\tfrac{5}{2}}$	0	$-\tfrac{1}{4}$	1	0	$\tfrac{1}{4}$	1	0	0	$-\tfrac{1}{4}$	0	6	$\tfrac{12}{5}$ ←Smaller one
Y_2	-1	0	1	-4	0	-1	0	1	0	1	0	0	
z_1	1	0	$\tfrac{5}{2}$	-1	-1	$-\tfrac{1}{2}$	0	0	1	$\tfrac{1}{2}$	0	4	4
x_2	$-\tfrac{1}{2}$	1	$\tfrac{1}{4}$	-1	0	$-\tfrac{1}{4}$	0	0	0	$\tfrac{1}{4}$	0	0	
$-w$	-1	0	$-\tfrac{5}{2}$	1	1	$\tfrac{1}{2}$	0	0	0	$\tfrac{1}{2}$	1	-4	

↑ (under x_1)   ↑ (under λ_1)

x_1 selected to enter the basis   Most negative

This tableau shows that λ_1 has to enter the basis and Y_2 or x_2 has to leave the basis. However, λ_1 cannot enter the basis since Y_1 is already in the basis [to satisfy the requirement of Eqs. (E$_4$)]. Hence x_1 is selected to enter the basis and this gives Y_1 as the variable that leaves the basis. The pivot operation on the element $\tfrac{5}{2}$ results in the following tableau:

Basic variables	\multicolumn{11}{c}{Variables}			$\bar{b}_i/\bar{a}_{is}$
	x_1	x_2	λ_1	λ_2	θ_1	θ_2	Y_1	Y_2	z_1	z_2	w	$\bar{b}_i$	for $\bar{a}_{is} > 0$
x_1	1	0	$-\tfrac{1}{10}$	$\tfrac{2}{5}$	0	$\tfrac{1}{10}$	$\tfrac{2}{5}$	0	0	$-\tfrac{1}{10}$	0	$\tfrac{12}{5}$	
Y_2	0	0	$\tfrac{9}{10}$	$-\tfrac{18}{5}$	0	$-\tfrac{9}{10}$	$\tfrac{2}{5}$	1	0	$\tfrac{9}{10}$	0	$\tfrac{12}{5}$	$\tfrac{8}{3}$
z_1	0	0	$\boxed{\tfrac{13}{5}}$	$-\tfrac{7}{5}$	-1	$-\tfrac{3}{5}$	$-\tfrac{2}{5}$	0	1	$\tfrac{3}{5}$	0	$\tfrac{8}{5}$	$\tfrac{8}{13}$ ←Smaller one
x_2	0	1	$\tfrac{1}{5}$	$-\tfrac{4}{5}$	0	$-\tfrac{1}{5}$	$\tfrac{1}{5}$	0	0	$\tfrac{1}{5}$	0	$\tfrac{6}{5}$	6
$-w$	0	0	$-\tfrac{13}{5}$	$\tfrac{7}{5}$	1	$\tfrac{3}{5}$	$\tfrac{2}{5}$	0	0	$\tfrac{2}{5}$	1	$-\tfrac{8}{5}$	

↑ (under λ_1)

Most negative

From this tableau we find that λ_1 enters the basis (this can be permitted this time since Y_1 is not in the basis) and z_1 leaves the basis. The necessary pivot operation gives the following tableau:

Basic variables	Variables											$\bar{b}_i$	$\bar{b}_i/\bar{a}_{is}$ for $\bar{a}_{is}>0$
	x_1	x_2	λ_1	λ_2	θ_1	θ_2	Y_1	Y_2	z_1	z_2	w	$\bar{b}_i$	
x_1	1	0	0	$\frac{9}{26}$	$-\frac{1}{26}$	$\frac{1}{13}$	$\frac{5}{13}$	0	$\frac{1}{26}$	$-\frac{1}{13}$	0	$\frac{32}{13}$	
Y_2	0	0	0	$-\frac{81}{26}$	$\frac{9}{26}$	$-\frac{9}{13}$	$\frac{7}{13}$	1	$-\frac{9}{26}$	$\frac{9}{13}$	0	$\frac{24}{13}$	
λ_1	0	0	1	$-\frac{7}{13}$	$-\frac{5}{13}$	$-\frac{3}{13}$	$-\frac{2}{13}$	0	$\frac{5}{13}$	$\frac{3}{13}$	0	$\frac{8}{13}$	
x_2	0	1	0	$-\frac{9}{13}$	$\frac{1}{13}$	$-\frac{2}{13}$	$\frac{3}{13}$	0	$-\frac{1}{13}$	$\frac{2}{13}$	0	$\frac{14}{13}$	
$-w$	0	0	0	0	0	0	0	0	1	1	1	0	

Since both the artificial variables z_1 and z_2 are driven out of the basis, the present tableau gives the desired solution as $x_1 = \frac{32}{13}$, $x_2 = \frac{14}{13}$, $Y_2 = \frac{24}{13}$, $\lambda_1 = \frac{8}{13}$ (basic variables), $\lambda_2 = 0$, $Y_1 = 0$, $\theta_1 = 0$, $\theta_2 = 0$ (nonbasic variables). Thus the solution of the original quadratic programming problem is given by

$$x_1^* = \tfrac{32}{13}, \quad x_2^* = \tfrac{14}{13}, \quad \text{and} \quad f_{\min} = f(x_1^*, x_2^*) = -\tfrac{88}{13}$$

4.9 MATLAB SOLUTIONS

The solutions of linear programming problems, based on interior point method, and quadratic programming problems using MATLAB are illustrated by the following examples.

Example 4.15 Find the solution of the following linear programming problem using MATLAB (interior point method):

$$\text{Minimize } f = -x_1 - 2x_2 - x_3$$

subject to

$$2x_1 + x_2 - x_3 \le 2$$
$$2x_1 - x_2 + 5x_3 \le 6$$
$$4x_1 + x_2 + x_3 \le 6$$
$$x_i \ge 0 \; ; \; i = 1, 2, 3$$

SOLUTION

Step 1 Express the objective function in the form $f(x) = f^T x$ and identify the vectors x and f as

$$x = \begin{Bmatrix} x_1 \\ x_2 \\ x_3 \end{Bmatrix} \quad \text{and} \quad f = \begin{Bmatrix} -1 \\ -2 \\ -1 \end{Bmatrix}$$

Express the constraints in the form $A\,x \le b$ and identify the matrix A and the vector b as

$$A = \begin{bmatrix} 2 & 1 & -1 \\ 2 & -1 & 5 \\ 4 & 1 & 1 \end{bmatrix} \quad \text{and} \quad b = \begin{Bmatrix} 2 \\ 6 \\ 6 \end{Bmatrix}$$

Step 2 Use the command for executing linear programming program using interior point method as indicated below:

```
clc
clear all
f=[-1;-2;-1];
A=[2 1-1;
  2-1 5;
  4 1 1];
b=[2;6;6];
lb=zeros(3,1);
Aeq=[];
beq=[];
options = optimset('Display', 'iter');
[x,fval,exitflag,output] = linprog(f,A,b,Aeq,beq,lb,[],[],
options)
```

This produces the solution or ouput as follows:

```
Iter 0: 1.03e+003 7.97e+000 1.50e+003 4.00e+002
Iter 1: 4.11e+002 2.22e-016 2.78e+002 4.72e+001
Iter 2: 1.16e-013 1.90e-015 2.85e+000 2.33e-001
Iter 3: 1.78e-015 1.80e-015 3.96e-002 3.96e-003
Iter 4: 7.48e-014 1.02e-015 1.99e-006 1.99e-007
Iter 5: 2.51e-015 4.62e-015 1.99e-012 1.98e-013
Optimization terminated.
x =
 0.0000
 4.0000
 2.0000
fval = -10.0000
exitflag =   1
output =
   iterations: 5
   algorithm: 'large-scale: interior point'
   cgiterations: 0
   message: 'Optimization terminated.'
```

Example 4.16 Find the solution of the following quadratic programming problem using MATLAB:

$$\text{Minimize } f = -4x_1 + x_1^2 - 2x_1x_2 + 2x_2^2$$

$$\text{subject to} \quad 2x_1 + x_2 \leq 6, \ x_1 - 4x_2 \leq 0, \ x_1 \geq 0, \ x_2 \geq 0$$

SOLUTION

Step 1 Express the objective function in the form $f(x) = \frac{1}{2}x^T H x + f^T x$ and identify the matrix H and vectors f and x:

$$H = \begin{pmatrix} 2 & -2 \\ -2 & 4 \end{pmatrix} \qquad f = \begin{pmatrix} -4 \\ 0 \end{pmatrix} \qquad x = \begin{pmatrix} x_1 \\ x_2 \end{pmatrix}$$

Step 2 State the constraints in the form: $Ax \leq b$ and identify the matrix A and vector b:

$$A = \begin{pmatrix} 2 & 1 \\ 1 & -4 \end{pmatrix} \qquad b = \begin{pmatrix} 6 \\ 0 \end{pmatrix}$$

Step 3 Use the command for executing quadratic programming as

```
[x,fval] = quadprog(H,f,A,b)
```

which returns the solution vector x that minimizes

$$f = \frac{1}{2}x^T H x + f^T x \quad \text{subject to} \quad Ax \leq b$$

The MATLAB solution is given below:

```
clear;clc;
H = [2−2;−2 4];
f = [−4 0];
A = [2 1;1−4];
b = [6; 0];
[x,fval] = quadprog(H,f,A,b)
Warning: Large-scale method does not currently solve this
problem formulation, switching to medium-scale method.
x =
   2.4615
   1.0769
fval =
   -6.7692
```

REFERENCES AND BIBLIOGRAPHY

4.1 S. Gass, *Linear Programming*, McGraw-Hill, New York, 1964.

4.2 C. E. Lemke, The dual method of solving the linear programming problem, *Naval Research and Logistics Quarterly*, Vol. 1, pp. 36–47, 1954.

4.3   G. B. Dantzig, L. R. Ford, and D. R. Fulkerson, A primal–dual algorithm for linear programs, pp. 171–181 in *Linear Inequalities and Related Systems*, H. W. Kuhn and A. W. Tucker, Eds., Annals of Mathematics Study No. 38, Princeton University Press, Princeton, NJ, 1956.

4.4   G. B. Dantzig and P. Wolfe, Decomposition principle for linear programming, *Operations Research*, Vol. 8, pp. 101–111, 1960.

4.5   L. S. Lasdon, *Optimization Theory for Large Systems*, Macmillan, New York, 1970.

4.6   F. L. Hitchcock, The distribution of a product from several sources to numerous localities, *Journal of Mathematical Physics*, Vol. 20, pp. 224–230, 1941.

4.7   T. C. Koopmans, Optimum utilization of the transportation system, *Proceedings of the International Statistical Conference*, Washington, DC, 1947.

4.8   S. Zukhovitskiy and L. Avdeyeva, *Linear and Convex Programming*, W. B. Saunders, Philadelphia, pp. 147–155, 1966.

4.9   W.W. Garvin, *Introduction to Linear Programming*, McGraw-Hill, New York, 1960.

4.10 G. B. Dantzig, *Linear Programming and Extensions*, Princeton University Press, Princeton, NJ, 1963.

4.11 C. E. Lemke, On complementary pivot theory, in *Mathematics of the Decision Sciences*, G. B. Dantzig and A. F. Veinott, Eds., Part 1, pp. 95–136, American Mathematical Society, Providence, RI, 1968.

4.12 K. Murty, *Linear and Combinatorial Programming*, Wiley, New York, 1976.

4.13 G. R. Bitran and A. G. Novaes, Linear programming with a fractional objective function, *Operations Research*, Vol. 21, pp. 22–29, 1973.

4.14 E. U. Choo and D. R. Atkins, Bicriteria linear fractional programming, *Journal of Optimization Theory and Applications*, Vol. 36, pp. 203–220, 1982.

4.15 C. Singh, Optimality conditions in fractional programming, *Journal of Optimization Theory and Applications*, Vol. 33, pp. 287–294, 1981.

4.16 J. B. Lasserre, A property of certain multistage linear programs and some applications, *Journal of Optimization Theory and Applications*, Vol. 34, pp. 197–205, 1981.

4.17 G. Cohen, Optimization by decomposition and coordination: a unified approach, *IEEE Transactions on Automatic Control*, Vol. AC-23, pp. 222–232, 1978.

4.18 N. Karmarkar, A new polynomial-time algorithm for linear programming, *Combinatorica*, Vol. 4, No. 4, pp. 373–395, 1984.

4.19 W. L. Winston, *Operations Research Applications and Algorithms*, 2nd ed. , PWS-Kent, Boston, 1991.

4.20 J. N. Hooker, Karmarkar's linear programming algorithm, *Interfaces*, Vol. 16, No. 4, pp. 75–90, 1986.

4.21 P. Wolfe, The simplex method for quadratic programming, *Econometrica*, Vol. 27, pp. 382–398, 1959.

4.22 J.C.G. Boot, *Quadratic Programming*, North-Holland, Amsterdam, 1964.

4.23 C. Van de Panne, *Methods for Linear and Quadratic Programming*, North-Holland, Amsterdam, 1974.

4.24 C. Van de Panne and A. Whinston, The symmetric formulation of the simplex method for quadratic programming, *Econometrica*, Vol. 37, pp. 507–527, 1969.

REVIEW QUESTIONS

4.1 Is the decomposition method efficient for all LP problems?

4.2 What is the scope of postoptimality analysis?

4.3 Why is Karmarkar's method called an interior method?

4.4 What is the major difference between the simplex and Karmarkar methods?

4.5 State the form of LP problem required by Karmarkar's method.

4.6 What are the advantages of the revised simplex method?

4.7 Match the following terms and descriptions:

(a) Karmarkar's method	Moves from one vertex to another
(b) Simplex method	Interior point algorithm
(c) Quadratic programming	Phase I computations not required
(d) Dual simplex method	Dantzig and Wolfe method
(e) Decomposition method	Wolfe's method

4.8 Answer true or false:

(a) The quadratic programming problem is a convex programming problem.

(b) It is immaterial whether a given LP problem is designated the primal or dual.

(c) If the primal problem involves minimization of f subject to greater-than constraints, its dual deals with the minimization of f subject to less-than constraints.

(d) If the primal problem has an unbounded solution, its dual will also have an unbounded solution.

(e) The transportation problem can be solved by simplex method.

4.9 Match the following in the context of duality theory:

(a) x_i is nonnegative	ith constraint is of less-than or equal-to type
(b) x_i is unrestricted	Maximization type
(c) ith constraint is of equality type	ith variable is unrestricted
(d) ith constraint is of greater-than or equal-to type	ith variable is nonnegative
(e) Minimization type	ith constraint is of equality type

PROBLEMS

Solve LP problems 4.1 to 4.3 by the revised simplex method.

4.1   Minimize $f = -5x_1 + 2x_2 + 5x_3 - 3x_4$

subject to

$$2x_1 + x_2 - x_3 = 6$$

$$3x_1 + 8x_3 + x_4 = 7$$

$$x_i \geq 0, \quad i = 1 \text{ to } 4$$

4.2   Maximize $f = 15x_1 + 6x_2 + 9x_3 + 2x_4$

subject to

$$10x_1 + 5x_2 + 25x_3 + 3x_4 \leq 50$$
$$12x_1 + 4x_2 + 12x_3 + x_4 \leq 48$$
$$7x_1 + x_4 \qquad\qquad \leq 35$$
$$x_i \geq 0, \quad i = 1 \text{ to } 4$$

4.3   Minimize $f = 2x_1 + 3x_2 + 2x_3 - x_4 + x_5$

subject to

$$3x_1 - 3x_2 + 4x_3 + 2x_4 - x_5 = 0$$
$$x_1 + x_2 + x_3 + 3x_4 + x_5 = 2$$
$$x_i \geq 0, \quad i = 1, 2, \ldots, 5$$

4.4 Discuss the relationships between the regular simplex method and the revised simplex method.

4.5 Solve the following LP problem graphically and by the revised simplex method:

$$\text{Maximize } f = x_2$$

subject to

$$-x_1 + x_2 \leq 0$$
$$-2x_1 - 3x_2 \leq 6$$
$$x_1, x_2 \text{ unrestricted in sign}$$

4.6 Consider the following LP problem:

$$\text{Minimize } f = 3x_1 + x_3 + 2x_5$$

subject to

$$x_1 + x_3 - x_4 + x_5 = -1$$
$$x_2 - 2x_3 + 3x_4 + 2x_5 = -2$$
$$x_i \geq 0, \quad i = 1 \text{ to } 5$$

Solve this problem using the dual simplex method.

4.7   Maximize $f = 4x_1 + 2x_2$

subject to

$$x_1 - 2x_2 \geq 2$$
$$x_1 + 2x_2 = 8$$

$$x_1 - x_2 \leq 11$$

$$x_1 \geq 0, \quad x_2 \text{ unrestricted in sign}$$

(a) Write the dual of this problem.

(b) Find the optimum solution of the dual.

(c) Verify the solution obtained in part (b) by solving the primal problem graphically.

4.8 A water resource system consisting of two reservoirs is shown in Fig. 4.4. The flows and storages are expressed in a consistent set of units. The following data are available:

Quantity	Stream 1 ($i = 1$)	Stream 2 ($i = 2$)
Capacity of reservoir i	9	7
Available release from reservoir i	9	6
Capacity of channel below reservoir i	4	4
Actual release from reservoir i	x_1	x_2

The capacity of the main channel below the confluence of the two streams is 5 units. If the benefit is equivalent to \$2 $\times$ 10^6 and \$3 $\times$ 10^6 per unit of water released from reservoirs 1 and 2, respectively, determine the releases x_1 and x_2 from the reserovirs to maximize the benefit. Solve this problem using duality theory.

4.9 Solve the following LP problem by the dual simplex method:

$$\text{Minimize } f = 2x_1 + 9x_2 + 24x_3 + 8x_4 + 5x_5$$

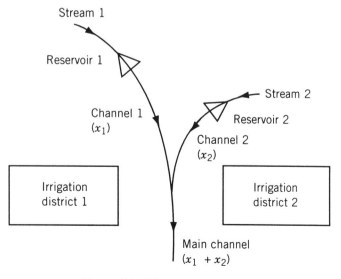

Figure 4.4 Water resource system.

subject to

$$x_1 + x_2 + 2x_3 - x_5 - x_6 = 1$$

$$-2x_1 + x_3 + x_4 + x_5 - x_7 = 2$$

$$x_i \geq 0, \quad i = 1 \text{ to } 7$$

4.10   Solve Problem 3.1 by solving its dual.

4.11   Show that neither the primal nor the dual of the problem

$$\text{Maximize } f = -x_1 + 2x_2$$

subject to

$$-x_1 + x_2 \leq -2$$

$$x_1 - x_2 \leq 1$$

$$x_1 \geq 0, \quad x_2 \geq 0$$

has a feasible solution. Verify your result graphically.

4.12   Solve the following LP problem by decomposition principle, and verify your result by solving it by the revised simplex method:

$$\text{Maximize } f = 8x_1 + 3x_2 + 8x_3 + 6x_4$$

subject to

$$4x_1 + 3x_2 + x_3 + 3x_4 \leq 16$$

$$4x_1 - x_2 + x_3 \leq 12$$

$$x_1 + 2x_2 \leq 8$$

$$3x_1 + x_2 \leq 10$$

$$2x_3 + 3x_4 \leq 9$$

$$4x_3 + x_4 \leq 12$$

$$x_i \geq 0, \quad i = 1 \text{ to } 4$$

4.13   Apply the decomposition principle to the dual of the following problem and solve it:

$$\text{Minimize } f = 10x_1 + 2x_2 + 4x_3 + 8x_4 + x_5$$

subject to

$$x_1 + 4x_2 - x_3 \geq 16$$

$$2x_1 + x_2 + x_3 \geq 4$$

$$3x_1 + x_4 + x_5 \geq 8$$

$$x_1 + 2x_4 - x_5 \geq 20$$

$$x_i \geq 0, \quad i = 1 \text{ to } 5$$

4.14 Express the dual of the following LP problem:

$$\text{Maximize } f = 2x_1 + x_2$$

subject to

$$x_1 - 2x_2 \geq 2$$

$$x_1 + 2x_2 = 8$$

$$x_1 - x_2 \leq 11$$

$$x_1 \geq 0, \quad x_2 \text{ is unrestricted in sign}$$

4.15 Find the effect of changing $\mathbf{b} = \left\{ {1200 \atop 800} \right\}$ to $\left\{ {1180 \atop 120} \right\}$ in Example 4.5 using sensitivity analysis.

4.16 Find the effect of changing the cost coefficients c_1 and c_4 from -45 and -50 to -40 and -60, respectively, in Example 4.5 using sensitivity analysis.

4.17 Find the effect of changing c_1 from -45 to -40 and c_2 from -100 to -90 in Example 4.5 using sensitivity analysis.

4.18 If a new product, E, which requires 10 min of work on lathe and 10 min of work on milling machine per unit, with a profit of \$120 per unit is available in Example 4.5, determine whether it is worth manufacturing E.

4.19 A metallurgical company produces four products, A, B, C, and D, by using copper and zinc as basic materials. The material requirements and the profit per unit of each of the four products, and the maximum quantities of copper and zinc available are given below:

	Product				Maximum quantity
	A	B	C	D	available
Copper (lb)	4	9	7	10	6000
Zinc (lb)	2	1	3	20	4000
Profit per unit (\$)	15	25	20	60	

Find the number of units of the various products to be produced for maximizing the profit.

Solve Problems 4.20–4.28 using the data of Problem 4.19.

4.20 Find the effect of changing the profit per unit of product D to \$30.

4.21 Find the effect of changing the profit per unit of product A to \$10, and of product B to \$20.

4.22 Find the effect of changing the profit per unit of product B to \$30 and of product C to \$25.

4.23 Find the effect of changing the available quantities of copper and zinc to 4000 and 6000 lb, respectively.

4.24 What is the effect of introducing a new product, E, which requires 6 lb of copper and 3 lb of zinc per unit if it brings a profit of \$30 per unit?

4.25 Assume that products A, B, C, and D require, in addition to the stated amounts of copper and zinc, 4, 3, 2 and 5 lb of nickel per unit, respectively. If the total quantity of nickel available is 2000 lb, in what way the original optimum solution is affected?

4.26 If product A requires 5 lb of copper and 3 lb of zinc (instead of 4 lb of copper and 2 lb of zinc) per unit, find the change in the optimum solution.

4.27 If product C requires 5 lb of copper and 4 lb of zinc (instead of 7 lb of copper and 3 lb of zinc) per unit, find the change in the optimum solution.

4.28 If the available quantities of copper and zinc are changed to 8000 lb and 5000 lb, respectively, find the change in the optimum solution.

4.29 Solve the following LP problem:

$$\text{Minimize } f = 8x_1 - 2x_2$$

subject to

$$-4x_1 + 2x_2 \leq 1$$
$$5x_1 - 4x_2 \leq 3$$
$$x_1 \geq 0, \quad x_2 \geq 0$$

Investigate the change in the optimum solution of Problem 4.29 when the following changes are made **(a)** by using sensitivity analysis and **(b)** by solving the new problem graphically:

4.30	$b_1 = 2$		**4.33**	$c_2 = -4$
4.31	$b_2 = 4$		**4.34**	$a_{11} = -5$
4.32	$c_1 = 10$		**4.35**	$a_{22} = -2$

4.36 Perform one iteration of Karmarkar's method for the LP problem:

$$\text{Minimize } f = 2x_1 - 2x_2 + 5x_3$$

subject to

$$x_1 - x_2 = 0$$
$$x_1 + x_2 + x_3 = 1$$
$$x_i \geq 0, \quad i = 1, 2, 3$$

4.37 Perform one iteration of Karmarkar's method for the following LP problem:

$$\text{Minimize } f = 3x_1 + 5x_2 - 3x_3$$

subject to

$$x_1 - x_3 = 0$$
$$x_1 + x_2 + x_3 = 1$$
$$x_i \geq 0, \quad i = 1, 2, 3$$

4.38 Transform the following LP problem into the form required by Karmarkar's method:

$$\text{Minimize } f = \quad x_1 + x_2 + x_3$$

subject to

$$x_1 + x_2 - x_3 = 4$$

$$3x_1 - x_2 = 0$$

$$x_i \geq 0, \quad i = 1, 2, 3$$

4.39 A contractor has three sets of heavy construction equipment available at both New York and Los Angeles. He has construction jobs in Seattle, Houston, and Detroit that require two, three, and one set of equipment, respectively. The shipping costs per set between cities i and j (c_{ij}) are shown in Fig. 4.5. Formulate the problem of finding the shipping pattern that minimizes the cost.

4.40
$$\text{Minimize } f(\mathbf{X}) = 3x_1^2 + 2x_2^2 + 5x_3^3 - 4x_1x_2 - 2x_1x_3 - 2x_2x_3$$

subject to

$$3x_1 + 5x_2 + 2x_3 \geq 10$$

$$3x_1 + 5x_3 \qquad \leq 15$$

$$x_i \geq 0, \quad i = 1, 2, 3$$

by quadratic programming.

4.41 Find the solution of the quadratic programming problem stated in Example 1.5.

4.42 According to elastic–plastic theory, a frame structure fails (collapses) due to the formation of a plastic hinge mechanism. The various possible mechanisms in which a portal frame (Fig. 4.6) can fail are shown in Fig. 4.7. The reserve strengths of the frame in various failure mechanisms (Z_i) can be expressed in terms of the plastic moment capacities of the hinges as indicated in Fig. 4.7. Assuming that the cost of the frame is proportional to 200 times each of the moment capacities M_1, M_2, M_6, and M_7, and 100 times each of the moment capacities M_3, M_4, and M_5, formulate the problem of minimizing the total cost

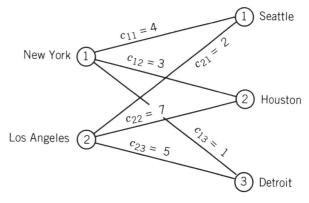

Figure 4.5 Shipping costs between cities.

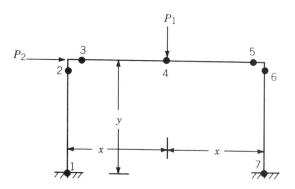

Figure 4.6 Plastic hinges in a frame.

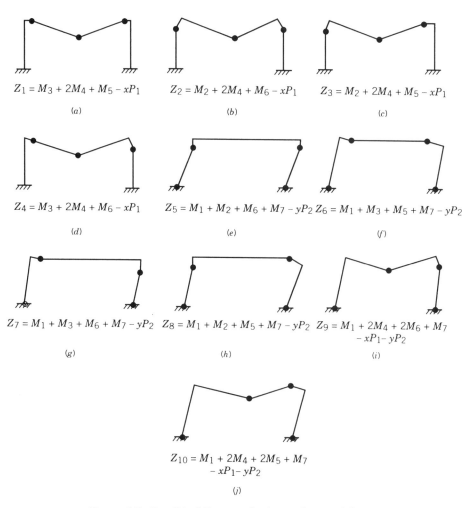

$Z_1 = M_3 + 2M_4 + M_5 - xP_1$

(a)

$Z_2 = M_2 + 2M_4 + M_6 - xP_1$

(b)

$Z_3 = M_2 + 2M_4 + M_5 - xP_1$

(c)

$Z_4 = M_3 + 2M_4 + M_6 - xP_1$

(d)

$Z_5 = M_1 + M_2 + M_6 + M_7 - yP_2$

(e)

$Z_6 = M_1 + M_3 + M_5 + M_7 - yP_2$

(f)

$Z_7 = M_1 + M_3 + M_6 + M_7 - yP_2$

(g)

$Z_8 = M_1 + M_2 + M_5 + M_7 - yP_2$

(h)

$Z_9 = M_1 + 2M_4 + 2M_6 + M_7 - xP_1 - yP_2$

(i)

$Z_{10} = M_1 + 2M_4 + 2M_5 + M_7 - xP_1 - yP_2$

(j)

Figure 4.7 Possible failure mechanisms of a portal frame.

to ensure nonzero reserve strength in each failure mechanism. Also, suggest a suitable technique for solving the problem. Assume that the moment capacities are restricted as $0 \le M_i \le 2 \times 10^5$ lb-in., $i = 1, 2, \ldots, 7$. Data: $x = 100$ in., $y = 150$ in., $P_1 = 1000$ lb, and $P_2 = 500$ lb.

4.43   Solve the LP problem stated in Problem 4.9 using MATLAB (interior method).

4.44   Solve the LP problem stated in Problem 4.12 using MATLAB (interior method).

4.45   Solve the LP problem stated in Problem 4.13 using MATLAB (interior method).

4.46   Solve the LP problem stated in Problem 4.36 using MATLAB (interior method).

4.47   Solve the LP problem stated in Problem 4.37 using MATLAB (interior method).

4.48   Solve the following quadratic programming problem using MATLAB:

Maximize $f = 2x_1 + x_2 - x_1^2$

subject to $2x_1 + 3x_2 \le 6$, $2x_1 + x_2 \le 4$, $x_1 \ge 0$, $x_2 \ge 0$

4.49   Solve the following quadratic programming problem using MATLAB:

Maximize $f = 4x_1 + 6x_2 - x_1^2 - x_2^2$

subject to $x_1 + x_2 \le 2$, $x_1 \ge 0$, $x_2 \ge 0$

4.50   Solve the following quadratic programming problem using MATLAB:

Minimize $f = (x_1 - 1)^2 + x_2 - 2$

subject to $-x_1 + x_2 - 1 = 0$, $x_1 + x_2 - 2 \le 0$, $x_1 \ge 0$, $x_2 \ge 0$

4.51   Solve the following quadratic programming problem using MATLAB:

Minimize $f = x_1^2 + x_2^2 - 3x_1x_2 - 6x_1 + 5x_2$

subject to $x_1 + x_2 \le 4$, $3x_1 + 6x_2 \le 20$, $x_1 \ge 0$, $x_2 \ge 0$

5

Nonlinear Programming I: One-Dimensional Minimization Methods

5.1 INTRODUCTION

In Chapter 2 we saw that if the expressions for the objective function and the constraints are fairly simple in terms of the design variables, the classical methods of optimization can be used to solve the problem. On the other hand, if the optimization problem involves the objective function and/or constraints that are not stated as explicit functions of the design variables or which are too complicated to manipulate, we cannot solve it by using the classical analytical methods. The following example is given to illustrate a case where the constraints cannot be stated as explicit functions of the design variables. Example 5.2 illustrates a case where the objective function is a complicated one for which the classical methods of optimization are difficult to apply.

Example 5.1 Formulate the problem of designing the planar truss shown in Fig. 5.1 for minimum weight subject to the constraint that the displacement of any node, in either the vertical or the horizontal direction, should not exceed a value δ.

SOLUTION Let the density ρ and Young's modulus E of the material, the length of the members l, and the external loads Q, R, and S be known as design data. Let the member areas $A_1, A_2, \ldots, A_{11}$ be taken as the design variables $x_1, x_2, \ldots, x_{11}$, respectively. The equations of equilibrium can be derived in terms of the unknown nodal displacements $u_1, u_2, \ldots, u_{10}$ as[†] (the displacements $u_{11}, u_{12}, u_{13},$ and u_{14} are

---

[†]According to the matrix methods of structural analysis, the equilibrium equations for the jth member are given by [5.1]

$$\underset{4\times4}{[\mathbf{k}_j]} \ \underset{4\times1}{\mathbf{u}_j} = \underset{4\times1}{\mathbf{P}_j}$$

where the stiffness matrix can be expressed as

$$[\mathbf{k}_j] = \frac{A_j E_j}{l_j} \begin{bmatrix} \cos^2\theta_j & \cos\theta_j\sin\theta_j & -\cos^2\theta_j & -\cos\theta_j\sin\theta_j \\ \cos\theta_j\sin\theta_j & \sin^2\theta_j & -\cos\theta_j\sin\theta_j & -\sin^2\theta_j \\ -\cos^2\theta_j & -\cos\theta_j\sin\theta_j & \cos^2\theta_j & \cos\theta_j\sin\theta_j \\ -\cos\theta_j\sin\theta_j & -\sin^2\theta_j & \cos\theta_j\sin\theta_j & \sin^2\theta_j \end{bmatrix}$$

where θ_j is the inclination of the jth member with respect to the x-axis, A_j the cross-sectional area of the jth member, l_j the length of the jth member, $\mathbf{u}_j$ the vector of displacements for the jth member, and $\mathbf{P}_j$

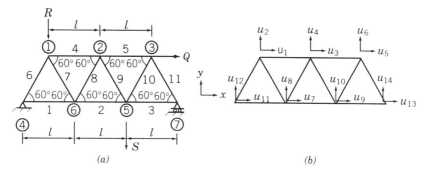

Figure 5.1 Planar truss: (*a*) nodal and member numbers; (*b*) nodal degrees of freedom.

zero, as they correspond to the fixed nodes)

$$(4x_4 + x_6 + x_7)u_1 + \sqrt{3}(x_6 - x_7)u_2 - 4x_4u_3 - x_7u_7 + \sqrt{3}x_7u_8 = 0 \qquad (E_1)$$

$$\sqrt{3}(x_6 - x_7)u_1 + 3(x_6 + x_7)u_2 + \sqrt{3}x_7u_7 - 3x_7u_8 = -\frac{4Rl}{E} \qquad (E_2)$$

$$-4x_4u_1 + (4x_4 + 4x_5 + x_8 + x_9)u_3 + \sqrt{3}(x_8 - x_9)u_4 - 4x_5u_5$$
$$- x_8u_7 - \sqrt{3}x_8u_8 - x_9u_9 + \sqrt{3}x_9u_{10} = 0 \qquad (E_3)$$

$$\sqrt{3}(x_8 - x_9)u_3 + 3(x_8 + x_9)u_4 - \sqrt{3}x_8u_7$$
$$- 3x_8u_8 + \sqrt{3}x_9u_9 - 3x_9u_{10} = 0 \qquad (E_4)$$

$$-4x_5u_3 + (4x_5 + x_{10} + x_{11})u_5 + \sqrt{3}(x_{10} - x_{11})u_6$$
$$- x_{10}u_9 - \sqrt{3}x_{10}u_{10} = \frac{4Ql}{E} \qquad (E_5)$$

$$\sqrt{3}(x_{10} - x_{11})u_5 + 3(x_{10} + x_{11})u_6 - \sqrt{3}x_{10}u_9 - 3x_{10}u_{10} = 0 \qquad (E_6)$$

$$- x_7u_1 + \sqrt{3}x_7u_2 - x_8u_3 - \sqrt{3}x_8u_4 + (4x_1 + 4x_2$$
$$+ x_7 + x_8)u_7 - \sqrt{3}(x_7 - x_8)u_8 - 4x_2u_9 = 0 \qquad (E_7)$$

$$\sqrt{3}x_7u_1 - 3x_7u_2 - \sqrt{3}x_8u_3 - 3x_8u_4 - \sqrt{3}(x_7 - x_8)u_7$$
$$+ 3(x_7 + x_8)u_8 = 0 \qquad (E_8)$$

$$- x_9u_3 + \sqrt{3}x_9u_4 - x_{10}u_5 - \sqrt{3}x_{10}u_6 - 4x_2u_7$$
$$+ (4x_2 + 4x_3 + x_9 + x_{10})u_9 - \sqrt{3}(x_9 - x_{10})u_{10} = 0 \qquad (E_9)$$

$$\sqrt{3}x_9u_3 - 3x_9u_4 - \sqrt{3}x_{10}u_5 - 3x_{10}u_6 - \sqrt{3}(x_9 - x_{10})u_9$$
$$+ 3(x_9 + x_{10})u_{10} = -\frac{4Sl}{E} \qquad (E_{10})$$

the vector of loads for the jth member. The formulation of the equilibrium equations for the complete truss follows fairly standard procedure [5.1].

It is important to note that an explicit closed-form solution cannot be obtained for the displacements as the number of equations becomes large. However, given any vector $\mathbf{X}$, the system of Eqs. (E_1) to (E_{10}) can be solved numerically to find the nodal displacement $u_1, u_2, \ldots, u_{10}$.

The optimization problem can be stated as follows:

$$\text{Minimize } f(\mathbf{X}) = \sum_{i=1}^{11} \rho x_i l_i \tag{E_{11}}$$

subject to the constraints

$$g_j(\mathbf{X}) = |u_j(\mathbf{X})| - \delta \leq 0, \qquad j = 1, 2, \ldots, 10 \tag{E_{12}}$$

$$x_i \geq 0, \qquad i = 1, 2, \ldots, 11 \tag{E_{13}}$$

The objective function of this problem is a straightforward function of the design variables as given in Eq. (E_{11}). The constraints, although written by the abstract expressions $g_j(\mathbf{X})$, cannot easily be written as explicit functions of the components of $\mathbf{X}$. However, given any vector $\mathbf{X}$ we can calculate $g_j(\mathbf{X})$ numerically. Many engineering design problems possess this characteristic (i.e., the objective and/or the constraints cannot be written explicitly in terms of the design variables). In such cases we need to use the numerical methods of optimization for solution.

Example 5.2 The shear stress induced along the z-axis when two spheres are in contact with each other is given by

$$\frac{\tau_{zx}}{p_{max}} = \frac{1}{2} \left[\frac{3}{2\left\{ 1 + \left(\frac{z}{a}\right)^2 \right\}} - (1 + v) \left\{ 1 - \frac{z}{a} \tan^{-1}\left(\frac{1}{\frac{z}{a}}\right) \right\} \right] \tag{E_1}$$

where a is the radius of the contact area and p_{max} is the maximum pressure developed at the center of the contact area (Fig. 5.2):

$$a = \left\{ \frac{3F}{8} \frac{\dfrac{1 - v_1^2}{E_1} + \dfrac{1 - v_2^2}{E_2}}{\dfrac{1}{d_1} + \dfrac{1}{d_2}} \right\}^{1/3} \tag{E_2}$$

$$p_{max} = \frac{3F}{2\pi a^2} \tag{E_3}$$

where F is the contact force, E_1 and E_2 are Young's moduli of the two spheres, v_1 and v_2 are Poisson's ratios of the two spheres, and d_1 and d_2 the diameters of the two spheres. In many practical applications, such as ball bearings, when the contact load (F) is large, a crack originates at the point of maximum shear stress and propagates to the surface, leading to a fatigue failure. To locate the origin of a crack, it is necessary to find the point at which the shear stress attains its maximum value. Formulate the problem of finding the location of maximum shear stress for $v = v_1 = v_2 = 0.3$.

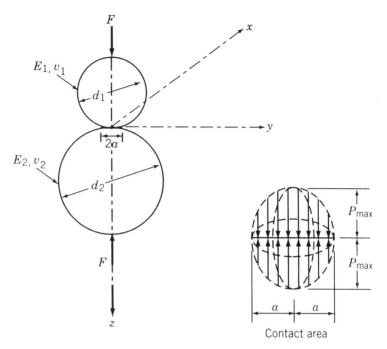

Figure 5.2 Contact stress between two spheres.

SOLUTION   For $v_1 = v_2 = 0.3$, Eq. (E$_1$) reduces to

$$f(\lambda) = \frac{0.75}{1 + \lambda^2} + 0.65\lambda \tan^{-1}\frac{1}{\lambda} - 0.65 \qquad \text{(E}_4\text{)}$$

where $f = \tau_{zx}/p_{\max}$ and $\lambda = z/a$. Since Eq. (E$_4$) is a nonlinear function of the distance, λ, the application of the necessary condition for the maximum of f, $df/d\lambda = 0$, gives rise to a nonlinear equation from which a closed-form solution for λ^* cannot easily be obtained. In such cases, numerical methods of optimization can be conveniently used to find the value of λ^*.

The basic philosophy of most of the numerical methods of optimization is to produce a sequence of improved approximations to the optimum according to the following scheme:

1. Start with an initial trial point $\mathbf{X}_1$.
2. Find a suitable direction $\mathbf{S}_i$ ($i = 1$ to start with) that points in the general direction of the optimum.
3. Find an appropriate step length λ_i^* for movement along the direction $\mathbf{S}_i$.
4. Obtain the new approximation $\mathbf{X}_{i+1}$ as

$$\mathbf{X}_{i+1} = \mathbf{X}_i + \lambda_i^* \mathbf{S}_i \qquad (5.1)$$

5. Test whether $\mathbf{X}_{i+1}$ is optimum. If $\mathbf{X}_{i+1}$ is optimum, stop the procedure. Otherwise, set a new $i = i + 1$ and repeat step (2) onward.

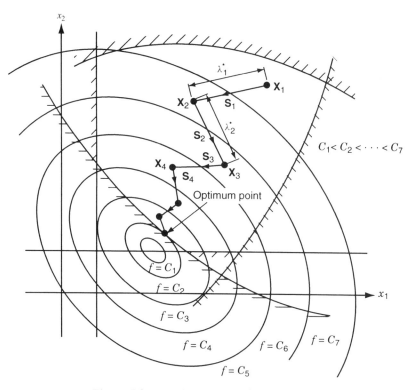

Figure 5.3   Iterative process of optimization.

The iterative procedure indicated by Eq. (5.1) is valid for unconstrained as well as constrained optimization problems. The procedure is represented graphically for a hypothetical two-variable problem in Fig. 5.3. Equation (5.1) indicates that the efficiency of an optimization method depends on the efficiency with which the quantities λ_i^* and $\mathbf{S}_i$ are determined. The methods of finding the step length λ_i^* are considered in this chapter and the methods of finding $\mathbf{S}_i$ are considered in Chapters 6 and 7.

If $f(\mathbf{X})$ is the objective function to be minimized, the problem of determining λ_i^* reduces to finding the value $\lambda_i = \lambda_i^*$ that minimizes $f(\mathbf{X}_{i+1}) = f(\mathbf{X}_i + \lambda_i \mathbf{S}_i) = f(\lambda_i)$ for fixed values of $\mathbf{X}_i$ and $\mathbf{S}_i$. Since f becomes a function of one variable λ_i only, the methods of finding λ_i^* in Eq. (5.1) are called *one-dimensional minimization methods*. Several methods are available for solving a one-dimensional minimization problem. These can be classified as shown in Table 5.1.

We saw in Chapter 2 that the differential calculus method of optimization is an analytical approach and is applicable to continuous, twice-differentiable functions. In this method, calculation of the numerical value of the objective function is virtually the last step of the process. The optimal value of the objective function is calculated after determining the optimal values of the decision variables. In the numerical methods of optimization, an opposite procedure is followed in that the values of the objective function are first found at various combinations of the decision variables and conclusions are then drawn regarding the optimal solution. The elimination methods can be used for the minimization of even discontinuous functions. The quadratic and cubic

Table 5.1 One-dimensional Minimization Methods

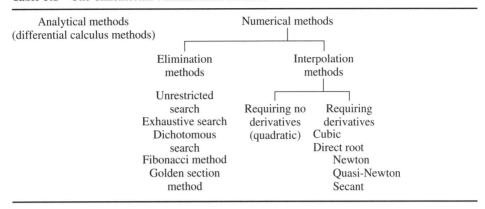

interpolation methods involve polynomial approximations to the given function. The direct root methods are root finding methods that can be considered to be equivalent to quadratic interpolation.

5.2 UNIMODAL FUNCTION

A *unimodal function* is one that has only one peak (maximum) or valley (minimum) in a given interval. Thus a function of one variable is said to be *unimodal* if, given that two values of the variable are on the same side of the optimum, the one nearer the optimum gives the better functional value (i.e., the smaller value in the case of a minimization problem). This can be stated mathematically as follows:

A function $f(x)$ is unimodal if (*i*) $x_1 < x_2 < x^*$ implies that $f(x_2) < f(x_1)$, and (ii) $x_2 > x_1 > x^*$ implies that $f(x_1) < f(x_2)$, where x^* is the minimum point.

Some examples of unimodal functions are shown in Fig. 5.4. Thus a unimodal function can be a nondifferentiable or even a discontinuous function. If a function is known to be unimodal in a given range, the interval in which the minimum lies can be narrowed down provided that the function values are known at two different points in the range.

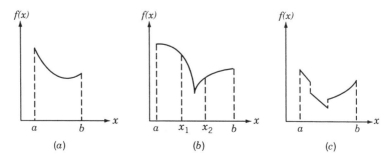

Figure 5.4 Unimodal function.

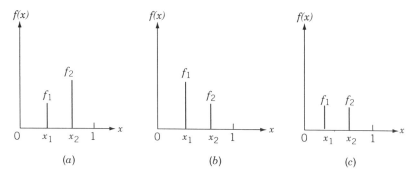

Figure 5.5 Outcome of first two experiments: (a) $f_1 < f_2$; (b) $f_1 > f_2$; (c) $f_1 = f_2$.

For example, consider the normalized interval [0, 1] and two function evaluations within the interval as shown in Fig. 5.5. There are three possible outcomes, namely, $f_1 < f_2$, $f_1 > f_2$, or $f_1 = f_2$. If the outcome is that $f_1 < f_2$, the minimizing x cannot lie to the right of x_2. Thus that part of the interval $[x_2, 1]$ can be discarded and a new smaller interval of uncertainty, $[0, x_2]$, results as shown in Fig. 5.5a. If $f(x_1) > f(x_2)$, the interval $[0, x_1]$ can be discarded to obtain a new smaller interval of uncertainty, $[x_1, 1]$ (Fig. 5.5b), while if $f(x_1) = f(x_2)$, intervals $[0, x_1]$ and $[x_2, 1]$ can both be discarded to obtain the new interval of uncertainty as $[x_1, x_2]$ (Fig. 5.5c). Further, if one of the original experiments[†] remains within the new interval, as will be the situation in Fig. 5.5a and b, only one other experiment need be placed within the new interval in order that the process be repeated. In situations such as Fig. 5.5c, two more experiments are to be placed in the new interval in order to find a reduced interval of uncertainty.

The assumption of unimodality is made in all the elimination techniques. If a function is known to be *multimodal* (i.e., having several valleys or peaks), the range of the function can be subdivided into several parts and the function treated as a unimodal function in each part.

Elimination Methods

5.3 UNRESTRICTED SEARCH

In most practical problems, the optimum solution is known to lie within restricted ranges of the design variables. In some cases this range is not known, and hence the search has to be made with no restrictions on the values of the variables.

5.3.1 Search with Fixed Step Size

The most elementary approach for such a problem is to use a fixed step size and move from an initial guess point in a favorable direction (positive or negative). The step size

---

[†]Each function evaluation is termed as an experiment or a trial in the elimination methods.

used must be small in relation to the final accuracy desired. Although this method is very simple to implement, it is not efficient in many cases. This method is described in the following steps:

1. Start with an initial guess point, say, x_1.
2. Find $f_1 = f(x_1)$.
3. Assuming a step size s, find $x_2 = x_1 + s$.
4. Find $f_2 = f(x_2)$.
5. If $f_2 < f_1$, and if the problem is one of minimization, the assumption of unimodality indicates that the desired minimum cannot lie at $x < x_1$. Hence the search can be continued further along points $x_3, x_4, \ldots$ using the unimodality assumption while testing each pair of experiments. This procedure is continued until a point, $x_i = x_1 + (i-1)s$, shows an increase in the function value.
6. The search is terminated at x_i, and either x_{i-1} or x_i can be taken as the optimum point.
7. Originally, if $f_2 > f_1$, the search should be carried in the reverse direction at points $x_{-2}, x_{-3}, \ldots$, where $x_{-j} = x_1 - (j-1)s$.
8. If $f_2 = f_1$, the desired minimum lies in between x_1 and x_2, and the minimum point can be taken as either x_1 or x_2.
9. If it happens that both f_2 and f_{-2} are greater than f_1, it implies that the desired minimum will lie in the double interval $x_{-2} < x < x_2$.

5.3.2   Search with Accelerated Step Size

Although the search with a fixed step size appears to be very simple, its major limitation comes because of the unrestricted nature of the region in which the minimum can lie. For example, if the minimum point for a particular function happens to be $x_{\text{opt}} = 50,000$ and, in the absence of knowledge about the location of the minimum, if x_1 and s are chosen as 0.0 and 0.1, respectively, we have to evaluate the function 5,000,001 times to find the minimum point. This involves a large amount of computational work. An obvious improvement can be achieved by increasing the step size gradually until the minimum point is bracketed. A simple method consists of doubling the step size as long as the move results in an improvement of the objective function. Several other improvements of this method can be developed. One possibility is to reduce the step length after bracketing the optimum in (x_{i-1}, x_i). By starting either from x_{i-1} or x_i, the basic procedure can be applied with a reduced step size. This procedure can be repeated until the bracketed interval becomes sufficiently small. The following example illustrates the search method with accelerated step size.

Example 5.3   Find the minimum of $f = x(x - 1.5)$ by starting from 0.0 with an initial step size of 0.05.

SOLUTION   The function value at x_1 is $f_1 = 0.0$. If we try to start moving in the negative x direction, we find that $x_{-2} = -0.05$ and $f_{-2} = 0.0775$. Since $f_{-2} > f_1$, the assumption of unimodality indicates that the minimum cannot lie toward the left of x_{-2}. Thus we start moving in the positive x direction and obtain the following results:

i	Value of s	$x_i = x_1 + s$	$f_i = f(x_i)$	Is $f_i > f_{i-1}$?
1	—	0.0	0.0	—
2	0.05	0.05	−0.0725	No
3	0.10	0.10	−0.140	No
4	0.20	0.20	−0.260	No
5	0.40	0.40	−0.440	No
6	0.80	0.80	−0.560	No
7	1.60	1.60	+0.160	Yes

From these results, the optimum point can be seen to be $x_{\text{opt}} \approx x_6 = 0.8$. In this case, the points x_6 and x_7 do not really bracket the minimum point but provide information about it. If a better approximation to the minimum is desired, the procedure can be restarted from x_5 with a smaller step size.

5.4   EXHAUSTIVE SEARCH

The exhaustive search method can be used to solve problems where the interval in which the optimum is known to lie is finite. Let x_s and x_f denote, respectively, the starting and final points of the interval of uncertainty.[†] The *exhaustive search method* consists of evaluating the objective function at a predetermined number of equally spaced points in the interval (x_s, x_f), and reducing the interval of uncertainty using the assumption of unimodality. Suppose that a function is defined on the interval (x_s, x_f) and let it be evaluated at eight equally spaced interior points x_1 to x_8. Assuming that the function values appear as shown in Fig. 5.6, the minimum point must lie, according to the assumption of unimodality, between points x_5 and x_7. Thus the interval (x_5, x_7) can be considered as the final interval of uncertainty.

In general, if the function is evaluated at n equally spaced points in the original interval of uncertainty of length $L_0 = x_f - x_s$, and if the optimum value of the function (among the n function values) turns out to be at point x_j, the final interval of uncertainty

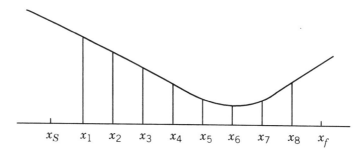

Figure 5.6   Exhaustive search.

[†]Since the interval (x_s, x_f), but not the exact location of the optimum in this interval, is known to us, the interval (x_s, x_f) is called the *interval of uncertainty*.

is given by

$$L_n = x_{j+1} - x_{j-1} = \frac{2}{n+1} L_0 \qquad (5.2)$$

The final interval of uncertainty obtainable for different number of trials in the exhaustive search method is given below:

Number of trials	2	3	4	5	6	· · ·	n
L_n/L_0	2/3	2/4	2/5	2/6	2/7	· · ·	$2/(n+1)$

Since the function is evaluated at all n points simultaneously, this method can be called a *simultaneous search method*. This method is relatively inefficient compared to the sequential search methods discussed next, where the information gained from the initial trials is used in placing the subsequent experiments.

Example 5.4   Find the minimum of $f = x(x - 1.5)$ in the interval (0.0, 1.00) to within 10% of the exact value.

SOLUTION   If the middle point of the final interval of uncertainty is taken as the approximate optimum point, the maximum deviation could be $1/(n + 1)$ times the initial interval of uncertainty. Thus to find the optimum within 10% of the exact value, we should have

$$\frac{1}{n+1} \leq \frac{1}{10} \quad \text{or} \quad n \geq 9$$

By taking $n = 9$, the following function values can be calculated:

i	1	2	3	4	5	6	7	8	9
x_i	0.1	0.2	0.3	0.4	0.5	0.6	0.7	0.8	0.9
$f_i = f(x_i)$	−0.14	−0.26	−0.36	−0.44	−0.50	−0.54	−0.56	−0.56	−0.54

Since $x_7 = x_8$, the assumption of unimodality gives the final interval of uncertainty as $L_9 = (0.7, 0.8)$. By taking the middle point of L_9 (i.e., 0.75) as an approximation to the optimum point, we find that it is, in fact, the true optimum point.

5.5   DICHOTOMOUS SEARCH

The exhaustive search method is a simultaneous search method in which all the experiments are conducted before any judgment is made regarding the location of the optimum point. The *dichotomous search method*, as well as the Fibonacci and the golden section methods discussed in subsequent sections, are sequential search methods in which the result of any experiment influences the location of the subsequent experiment.

In the dichotomous search, two experiments are placed as close as possible at the center of the interval of uncertainty. Based on the relative values of the objective

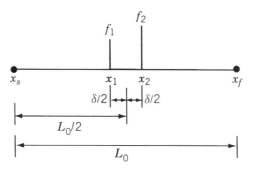

Figure 5.7 Dichotomous search.

function at the two points, almost half of the interval of uncertainty is eliminated. Let the positions of the two experiments be given by (Fig. 5.7)

$$x_1 = \frac{L_0}{2} - \frac{\delta}{2}$$

$$x_2 = \frac{L_0}{2} + \frac{\delta}{2}$$

where δ is a small positive number chosen so that the two experiments give significantly different results. Then the new interval of uncertainty is given by $(L_0/2 + \delta/2)$. The building block of dichotomous search consists of conducting a pair of experiments at the center of the current interval of uncertainty. The next pair of experiments is, therefore, conducted at the center of the remaining interval of uncertainty. This results in the reduction of the interval of uncertainty by nearly a factor of 2. The intervals of uncertainty at the end of different pairs of experiments are given in the following table:

Number of experiments	2	4	6
Final interval of uncertainty	$\frac{1}{2}(L_0 + \delta)$	$\frac{1}{2}\left(\dfrac{L_0 + \delta}{2}\right) + \dfrac{\delta}{2}$	$\frac{1}{2}\left(\dfrac{L_0 + \delta}{4}\right) + \dfrac{\delta}{2}\right) + \dfrac{\delta}{2}$

In general, the final interval of uncertainty after conducting n experiments (n even) is given by

$$L_n = \frac{L_0}{2^{n/2}} + \delta\left(1 - \frac{1}{2^{n/2}}\right) \tag{5.3}$$

The following example is given to illustrate the method of search.

Example 5.5 Find the minimum of $f = x(x - 1.5)$ in the interval $(0.0, 1.00)$ to within 10% of the exact value.

SOLUTION The ratio of final to initial intervals of uncertainty is given by [from Eq. (5.3)]

$$\frac{L_n}{L_0} = \frac{1}{2^{n/2}} + \frac{\delta}{L_0}\left(1 - \frac{1}{2^{n/2}}\right)$$

where δ is a small quantity, say 0.001, and n is the number of experiments. If the middle point of the final interval is taken as the optimum point, the requirement can be stated as

$$\frac{1}{2}\frac{L_n}{L_0} \leq \frac{1}{10}$$

i.e.,

$$\frac{1}{2^{n/2}} + \frac{\delta}{L_0}\left(1 - \frac{1}{2^{n/2}}\right) \leq \frac{1}{5}$$

Since $\delta = 0.001$ and $L_0 = 1.0$, we have

$$\frac{1}{2^{n/2}} + \frac{1}{1000}\left(1 - \frac{1}{2^{n/2}}\right) \leq \frac{1}{5}$$

i.e.,

$$\frac{999}{1000}\frac{1}{2^{n/2}} \leq \frac{995}{5000} \quad \text{or} \quad 2^{n/2} \geq \frac{999}{199} \simeq 5.0$$

Since n has to be even, this inequality gives the minimum admissible value of n as 6. The search is made as follows. The first two experiments are made at

$$x_1 = \frac{L_0}{2} - \frac{\delta}{2} = 0.5 - 0.0005 = 0.4995$$

$$x_2 = \frac{L_0}{2} + \frac{\delta}{2} = 0.5 + 0.0005 = 0.5005$$

with the function values given by

$$f_1 = f(x_1) = 0.4995(-1.0005) \simeq -0.49975$$
$$f_2 = f(x_2) = 0.5005(-0.9995) \simeq -0.50025$$

Since $f_2 < f_1$, the new interval of uncertainty will be (0.4995, 1.0). The second pair of experiments is conducted at

$$x_3 = \left(0.4995 + \frac{1.0 - 0.4995}{2}\right) - 0.0005 = 0.74925$$

$$x_4 = \left(0.4995 + \frac{1.0 - 0.4995}{2}\right) + 0.0005 = 0.75025$$

which give the function values as

$$f_3 = f(x_3) = 0.74925(-0.75075) = -0.5624994375$$
$$f_4 = f(x_4) = 0.75025(-0.74975) = -0.5624999375$$

Since $f_3 > f_4$, we delete (0.4995, x_3) and obtain the new interval of uncertainty as

$$(x_3, 1.0) = (0.74925, 1.0)$$

The final set of experiments will be conducted at

$$x_5 = \left(0.74925 + \frac{1.0 - 0.74925}{2}\right) - 0.0005 = 0.874125$$

$$x_6 = \left(0.74925 + \frac{1.0 - 0.74925}{2}\right) + 0.0005 = 0.875125$$

The corresponding function values are

$$f_5 = f(x_5) = 0.874125(-0.625875) = -0.5470929844$$

$$f_6 = f(x_6) = 0.875125(-0.624875) = -0.5468437342$$

Since $f_5 < f_6$, the new interval of uncertainty is given by $(x_3, x_6) = (0.74925, 0.875125)$. The middle point of this interval can be taken as optimum, and hence

$$x_{\text{opt}} \simeq 0.8121875 \quad \text{and} \quad f_{\text{opt}} \simeq -0.5586327148$$

5.6 INTERVAL HALVING METHOD

In the *interval halving method*, exactly one-half of the current interval of uncertainty is deleted in every stage. It requires three experiments in the first stage and two experiments in each subsequent stage. The procedure can be described by the following steps:

1. Divide the initial interval of uncertainty $L_0 = [a, b]$ into four equal parts and label the middle point x_0 and the quarter-interval points x_1 and x_2.
2. Evaluate the function $f(x)$ at the three interior points to obtain $f_1 = f(x_1)$, $f_0 = f(x_0)$, and $f_2 = f(x_2)$.
3. (a) If $f_2 > f_0 > f_1$ as shown in Fig. 5.8a, delete the interval (x_0, b), label x_1 and x_0 as the new x_0 and b, respectively, and go to step 4.
   (b) If $f_2 < f_0 < f_1$ as shown in Fig. 5.8b, delete the interval (a, x_0), label x_2 and x_0 as the new x_0 and a, respectively, and go to step 4.
   (c) If $f_1 > f_0$ and $f_2 > f_0$ as shown in Fig. 5.8c, delete both the intervals (a, x_1) and (x_2, b), label x_1 and x_2 as the new a and b, respectively, and go to step 4.
4. Test whether the new interval of uncertainty, $L = b - a$, satisfies the convergence criterion $L \leq \varepsilon$, where ε is a small quantity. If the convergence criterion is satisfied, stop the procedure. Otherwise, set the new $L_0 = L$ and go to step 1.

Remarks:

1. In this method, the function value at the middle point of the interval of uncertainty, f_0, will be available in all the stages except the first stage.
2. The interval of uncertainty remaining at the end of n experiments ($n \geq 3$ and odd) is given by

$$L_n = \left(\frac{1}{2}\right)^{(n-1)/2} L_0 \tag{5.4}$$

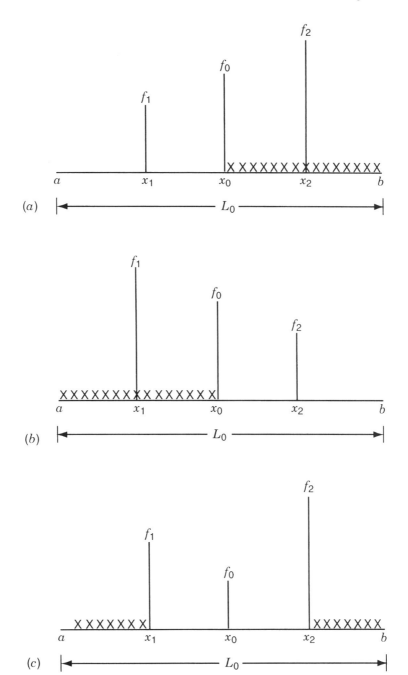

Figure 5.8   Possibilities in the interval halving method: (a) $f_2 > f_0 > f_1$; (b) $f_1 > f_0 > f_2$; (c) $f_1 > f_0$ and $f_2 > f_0$.

Example 5.6   Find the minimum of $f = x(x - 1.5)$ in the interval $(0.0, 1.0)$ to within 10% of the exact value.

SOLUTION   If the middle point of the final interval of uncertainty is taken as the optimum point, the specified accuracy can be achieved if

$$\frac{1}{2}L_n \leq \frac{L_0}{10} \quad \text{or} \quad \left(\frac{1}{2}\right)^{(n-1)/2} L_0 \leq \frac{L_0}{5} \tag{E_1}$$

Since $L_0 = 1$, Eq. (E_1) gives

$$\frac{1}{2^{(n-1)/2}} \leq \frac{1}{5} \quad \text{or} \quad 2^{(n-1)/2} \geq 5 \tag{E_2}$$

Since n has to be odd, inequality (E_2) gives the minimum permissible value of n as 7. With this value of $n = 7$, the search is conducted as follows. The first three experiments are placed at one-fourth points of the interval $L_0 = [a = 0, b = 1]$ as

$$x_1 = 0.25, \qquad f_1 = 0.25(-1.25) = -0.3125$$
$$x_0 = 0.50, \qquad f_0 = 0.50(-1.00) = -0.5000$$
$$x_2 = 0.75, \qquad f_2 = 0.75(-0.75) = -0.5625$$

Since $f_1 > f_0 > f_2$, we delete the interval $(a, x_0) = (0.0, 0.5)$, label x_2 and x_0 as the new x_0 and a so that $a = 0.5$, $x_0 = 0.75$, and $b = 1.0$. By dividing the new interval of uncertainty, $L_3 = (0.5, 1.0)$ into four equal parts, we obtain

$$x_1 = 0.625, \qquad f_1 = 0.625(-0.875) = -0.546875$$
$$x_0 = 0.750, \qquad f_0 = 0.750(-0.750) = -0.562500$$
$$x_2 = 0.875, \qquad f_2 = 0.875(-0.625) = -0.546875$$

Since $f_1 > f_0$ and $f_2 > f_0$, we delete both the intervals (a, x_1) and (x_2, b), and label x_1, x_0, and x_2 as the new a, x_0, and b, respectively. Thus the new interval of uncertainty will be $L_5 = (0.625, 0.875)$. Next, this interval is divided into four equal parts to obtain

$$x_1 = 0.6875, \qquad f_1 = 0.6875(-0.8125) = -0.558594$$
$$x_0 = 0.75, \qquad f_0 = 0.75(-0.75) = -0.5625$$
$$x_2 = 0.8125, \qquad f_2 = 0.8125(-0.6875) = -0.558594$$

Again we note that $f_1 > f_0$ and $f_2 > f_0$ and hence we delete both the intervals (a, x_1) and (x_2, b) to obtain the new interval of uncertainty as $L_7 = (0.6875, 0.8125)$. By taking the middle point of this interval (L_7) as optimum, we obtain

$$x_{\text{opt}} \approx 0.75 \quad \text{and} \quad f_{\text{opt}} \approx -0.5625$$

(This solution happens to be the exact solution in this case.)

5.7 FIBONACCI METHOD

As stated earlier, the *Fibonacci method* can be used to find the minimum of a function of one variable even if the function is not continuous. This method, like many other elimination methods, has the following limitations:

1. The initial interval of uncertainty, in which the optimum lies, has to be known.
2. The function being optimized has to be unimodal in the initial interval of uncertainty.
3. The exact optimum cannot be located in this method. Only an interval known as the *final interval of uncertainty* will be known. The final interval of uncertainty can be made as small as desired by using more computations.
4. The number of function evaluations to be used in the search or the resolution required has to be specified beforehand.

This method makes use of the sequence of Fibonacci numbers, $\{F_n\}$, for placing the experiments. These numbers are defined as

$$F_0 = F_1 = 1$$

$$F_n = F_{n-1} + F_{n-2}, \qquad n = 2, 3, 4, \ldots$$

which yield the sequence 1, 1, 2, 3, 5, 8, 13, 21, 34, 55, 89,....

Procedure.   Let L_0 be the initial interval of uncertainty defined by $a \le x \le b$ and n be the total number of experiments to be conducted. Define

$$L_2^* = \frac{F_{n-2}}{F_n} L_0 \tag{5.5}$$

and place the first two experiments at points x_1 and x_2, which are located at a distance of L_2^* from each end of L_0.[†] This gives[‡]

$$x_1 = a + L_2^* = a + \frac{F_{n-2}}{F_n} L_0$$

$$x_2 = b - L_2^* = b - \frac{F_{n-2}}{F_n} L_0 = a + \frac{F_{n-1}}{F_n} L_0 \tag{5.6}$$

Discard part of the interval by using the unimodality assumption. Then there remains a smaller interval of uncertainty L_2 given by[§]

$$L_2 = L_0 - L_2^* = L_0 \left(1 - \frac{F_{n-2}}{F_n}\right) = \frac{F_{n-1}}{F_n} L_0 \tag{5.7}$$

[†]If an experiment is located at a distance of $(F_{n-2}/F_n)L_0$ from one end, it will be at a distance of $(F_{n-1}/F_n)L_0$ from the other end. Thus $L_2^* = (F_{n-1}/F_n)L_0$ will yield the same result as with $L_2^* = (F_{n-2}/F_n)L_0$.

[‡]It can be seen that

$$L_2^* = \frac{F_{n-2}}{F_n} L_0 \le \frac{1}{2} L_0 \quad \text{for} \quad n \ge 2$$

[§]The symbol L_j is used to denote the interval of uncertainty remaining after conducting j experiments, while the symbol L_j^* is used to define the position of the jth experiment.

and with one experiment left in it. This experiment will be at a distance of

$$L_2^* = \frac{F_{n-2}}{F_n}L_0 = \frac{F_{n-2}}{F_{n-1}}L_2 \tag{5.8}$$

from one end and

$$L_2 - L_2^* = \frac{F_{n-3}}{F_n}L_0 = \frac{F_{n-3}}{F_{n-1}}L_2 \tag{5.9}$$

from the other end. Now place the third experiment in the interval L_2 so that the current two experiments are located at a distance of

$$L_3^* = \frac{F_{n-3}}{F_n}L_0 = \frac{F_{n-3}}{F_{n-1}}L_2 \tag{5.10}$$

from each end of the interval L_2. Again the unimodality property will allow us to reduce the interval of uncertainty to L_3 given by

$$L_3 = L_2 - L_3^* = L_2 - \frac{F_{n-3}}{F_{n-1}}L_2 = \frac{F_{n-2}}{F_{n-1}}L_2 = \frac{F_{n-2}}{F_n}L_0 \tag{5.11}$$

This process of discarding a certain interval and placing a new experiment in the remaining interval can be continued, so that the location of the jth experiment and the interval of uncertainty at the end of j experiments are, respectively, given by

$$L_j^* = \frac{F_{n-j}}{F_{n-(j-2)}}L_{j-1} \tag{5.12}$$

$$L_j = \frac{F_{n-(j-1)}}{F_n}L_0 \tag{5.13}$$

The ratio of the interval of uncertainty remaining after conducting j of the n predetermined experiments to the initial interval of uncertainty becomes

$$\frac{L_j}{L_0} = \frac{F_{n-(j-1)}}{F_n} \tag{5.14}$$

and for $j = n$, we obtain

$$\frac{L_n}{L_0} = \frac{F_1}{F_n} = \frac{1}{F_n} \tag{5.15}$$

The ratio L_n/L_0 will permit us to determine n, the required number of experiments, to achieve any desired accuracy in locating the optimum point. Table 5.2 gives the reduction ratio in the interval of uncertainty obtainable for different number of experiments.

Table 5.2 Reduction Ratios

Value of n	Fibonacci number, F_n	Reduction ratio, L_n/L_0
0	1	1.0
1	1	1.0
2	2	0.5
3	3	0.3333
4	5	0.2
5	8	0.1250
6	13	0.07692
7	21	0.04762
8	34	0.02941
9	55	0.01818
10	89	0.01124
11	144	0.006944
12	233	0.004292
13	377	0.002653
14	610	0.001639
15	987	0.001013
16	1,597	0.0006406
17	2,584	0.0003870
18	4,181	0.0002392
19	6,765	0.0001479
20	10,946	0.00009135

Position of the Final Experiment. In this method the last experiment has to be placed with some care. Equation (5.12) gives

$$\frac{L_n^*}{L_{n-1}} = \frac{F_0}{F_2} = \frac{1}{2} \quad \text{for} \quad \text{all } n \tag{5.16}$$

Thus after conducting $n - 1$ experiments and discarding the appropriate interval in each step, the remaining interval will contain one experiment precisely at its middle point. However, the final experiment, namely, the nth experiment, is also to be placed at the center of the present interval of uncertainty. That is, the position of the nth experiment will be same as that of $(n - 1)$th one, and this is true for whatever value we choose for n. Since no new information can be gained by placing the nth experiment exactly at the same location as that of the $(n - 1)$th experiment, we place the nth experiment very close to the remaining valid experiment, as in the case of the dichotomous search method. This enables us to obtain the final interval of uncertainty to within $\frac{1}{2}L_{n-1}$. A flowchart for implementing the Fibonacci method of minimization is given in Fig. 5.9.

Example 5.7 Minimize $f(x) = 0.65 - [0.75/(1 + x^2)] - 0.65x \tan^{-1}(1/x)$ in the interval [0,3] by the Fibonacci method using $n = 6$. (Note that this objective is equivalent to the one stated in Example 5.2.)

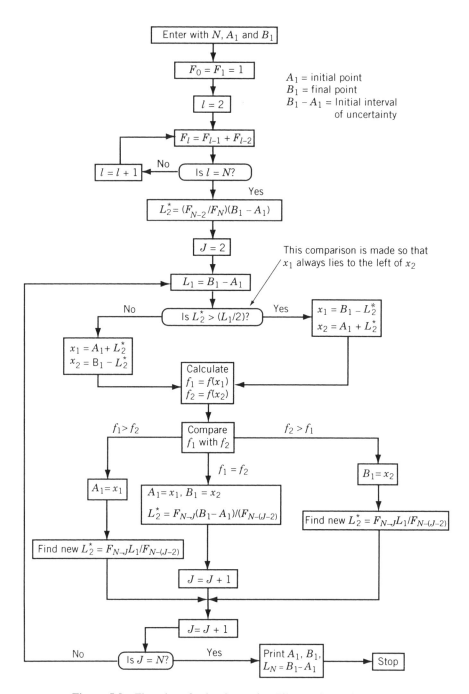

Figure 5.9 Flowchart for implementing Fibonacci search method.

SOLUTION   Here $n = 6$ and $L_0 = 3.0$, which yield

$$L_2^* = \frac{F_{n-2}}{F_n} L_0 = \frac{5}{13}(3.0) = 1.153846$$

Thus the positions of the first two experiments are given by $x_1 = 1.153846$ and $x_2 = 3.0 - 1.153846 = 1.846154$ with $f_1 = f(x_1) = -0.207270$ and $f_2 = f(x_2) = -0.115843$. Since f_1 is less than f_2, we can delete the interval $[x_2, 3.0]$ by using the unimodality assumption (Fig. 5.10a). The third experiment is placed at $x_3 = 0 + (x_2 - x_1) = 1.846154 - 1.153846 = 0.692308$, with the corresponding function value of $f_3 = -0.291364$.

Since $f_1 > f_3$, we delete the interval $[x_1, x_2]$ (Fig. 5.10b). The next experiment is located at $x_4 = 0 + (x_1 - x_3) = 1.153846 - 0.692308 = 0.461538$ with $f_4 = -0.309811$. Nothing that $f_4 < f_3$, we delete the interval $[x_3, x_1]$ (Fig. 5.10c). The location of the next experiment can be obtained as $x_5 = 0 + (x_3 - x_4) = 0.692308 - 0.461538 = 0.230770$ with the corresponding objective function value of $f_5 = -0.263678$. Since $f_5 > f_4$, we delete the interval $[0, x_5]$ (Fig. 5.10d). The final experiment is positioned at $x_6 = x_5 + (x_3 - x_4) = 0.230770 + (0.692308 - 0.461538) = 0.461540$ with $f_6 = -0.309810$. (Note that, theoretically, the value of x_6 should be same as that of x_4; however, it is slightly different from x_4, due to round-off error).

Since $f_6 > f_4$, we delete the interval $[x_6, x_3]$ and obtain the final interval of uncertainty as $L_6 = [x_5, x_6] = [0.230770, 0.461540]$ (Fig. 5.10e). The ratio of the final to the initial interval of uncertainty is

$$\frac{L_6}{L_0} = \frac{0.461540 - 0.230770}{3.0} = 0.076923$$

This value can be compared with Eq. (5.15), which states that if n experiments ($n = 6$) are planned, a resolution no finer than $1/F_n = 1/F_6 = \frac{1}{13} = 0.076923$ can be expected from the method.

5.8 GOLDEN SECTION METHOD

The *golden section method* is same as the Fibonacci method except that in the Fibonacci method the total number of experiments to be conducted has to be specified before beginning the calculation, whereas this is not required in the golden section method. In the Fibonacci method, the location of the first two experiments is determined by the total number of experiments, N. In the golden section method we start with the assumption that we are going to conduct a large number of experiments. Of course, the total number of experiments can be decided during the computation.

The intervals of uncertainty remaining at the end of different number of experiments can be computed as follows:

$$L_2 = \lim_{N \to \infty} \frac{F_{N-1}}{F_N} L_0 \tag{5.17}$$

$$L_3 = \lim_{N \to \infty} \frac{F_{N-2}}{F_N} L_0 = \lim_{N \to \infty} \frac{F_{N-2}}{F_{N-1}} \frac{F_{N-1}}{F_N} L_0$$

$$\simeq \lim_{N \to \infty} \left(\frac{F_{N-1}}{F_N} \right)^2 L_0 \tag{5.18}$$

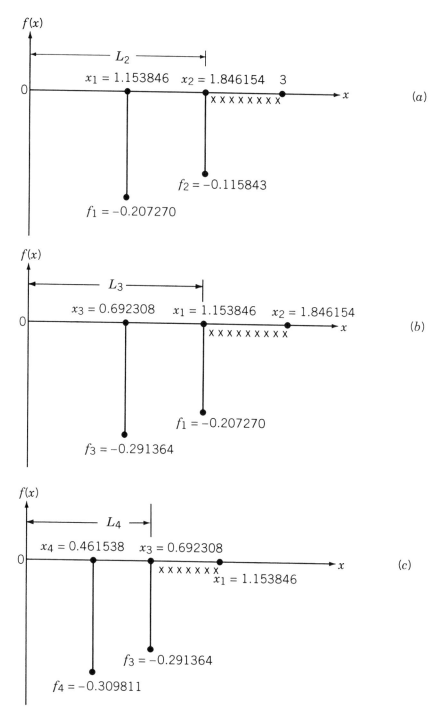

Figure 5.10   Graphical representation of the solution of Example 5.7.

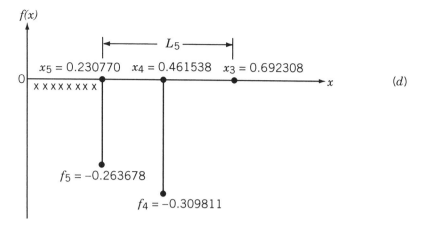

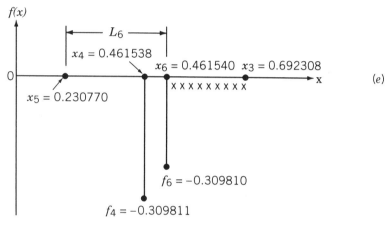

Figure 5.10   (*continued*)

This result can be generalized to obtain

$$L_k = \lim_{N \to \infty} \left(\frac{F_{N-1}}{F_N} \right)^{k-1} L_0 \tag{5.19}$$

Using the relation

$$F_N = F_{N-1} + F_{N-2} \tag{5.20}$$

we obtain, after dividing both sides by F_{N-1},

$$\frac{F_N}{F_{N-1}} = 1 + \frac{F_{N-2}}{F_{N-1}} \tag{5.21}$$

By defining a ratio γ as

$$\gamma = \lim_{N \to \infty} \frac{F_N}{F_{N-1}} \tag{5.22}$$

Eq. (5.21) can be expressed as

$$\gamma \simeq \frac{1}{\gamma} + 1$$

that is,

$$\gamma^2 - \gamma - 1 = 0 \qquad (5.23)$$

This gives the root $\gamma = 1.618$, and hence Eq. (5.19) yields

$$L_k = \left(\frac{1}{\gamma}\right)^{k-1} L_0 = (0.618)^{k-1} L_0 \qquad (5.24)$$

In Eq. (5.18) the ratios F_{N-2}/F_{N-1} and F_{N-1}/F_N have been taken to be same for large values of N. The validity of this assumption can be seen from the following table:

Value of N	2	3	4	5	6	7	8	9	10	∞
Ratio $\dfrac{F_{N-1}}{F_N}$	0.5	0.667	0.6	0.625	0.6156	0.619	0.6177	0.6181	0.6184	0.618

The ratio γ has a historical background. Ancient Greek architects believed that a building having the sides d and b satisfying the relation

$$\frac{d+b}{d} = \frac{d}{b} = \gamma \qquad (5.25)$$

would have the most pleasing properties (Fig. 5.11). The origin of the name, *golden section method*, can also be traced to the Euclid's geometry. In Euclid's geometry, when a line segment is divided into two unequal parts so that the ratio of the whole to the larger part is equal to the ratio of the larger to the smaller, the division is called the golden section and the ratio is called the golden mean.

Procedure. The procedure is same as the Fibonacci method except that the location of the first two experiments is defined by

$$L_2^* = \frac{F_{N-2}}{F_N} L_0 = \frac{F_{N-2}}{F_{N-1}} \frac{F_{N-1}}{F_N} L_0 = \frac{L_0}{\gamma^2} = 0.382 L_0 \qquad (5.26)$$

The desired accuracy can be specified to stop the procedure.

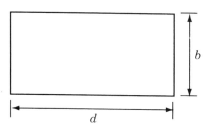

Figure 5.11   Rectangular building of sides b and d.

Example 5.8   Minimize the function

$$f(x) = 0.65 - [0.75/(1 + x^2)] - 0.65x \tan^{-1}(1/x)$$

using the golden section method with $n = 6$.

SOLUTION   The locations of the first two experiments are defined by $L_2^* = 0.382L_0 = (0.382)(3.0) = 1.1460$. Thus $x_1 = 1.1460$ and $x_2 = 3.0 - 1.1460 = 1.8540$ with $f_1 = f(x_1) = -0.208654$ and $f_2 = f(x_2) = -0.115124$. Since $f_1 < f_2$, we delete the interval $[x_2, 3.0]$ based on the assumption of unimodality and obtain the new interval of uncertainty as $L_2 = [0, x_2] = [0.0, 1.8540]$. The third experiment is placed at $x_3 = 0 + (x_2 - x_1) = 1.8540 - 1.1460 = 0.7080$. Since $f_3 = -0.288943$ is smaller than $f_1 = -0.208654$, we delete the interval $[x_1, x_2]$ and obtain the new interval of uncertainty as $[0.0, x_1] = [0.0, 1.1460]$. The position of the next experiment is given by $x_4 = 0 + (x_1 - x_3) = 1.1460 - 0.7080 = 0.4380$ with $f_4 = -0.308951$.

Since $f_4 < f_3$, we delete $[x_3, x_1]$ and obtain the new interval of uncertainty as $[0, x_3] = [0.0, 0.7080]$. The next experiment is placed at $x_5 = 0 + (x_3 - x_4) = 0.7080 - 0.4380 = 0.2700$. Since $f_5 = -0.278434$ is larger than $f_4 = -0.308951$, we delete the interval $[0, x_5]$ and obtain the new interval of uncertainty as $[x_5, x_3] = [0.2700, 0.7080]$. The final experiment is placed at $x_6 = x_5 + (x_3 - x_4) = 0.2700 + (0.7080 - 0.4380) = 0.5400$ with $f_6 = -0.308234$. Since $f_6 > f_4$, we delete the interval $[x_6, x_3]$ and obtain the final interval of uncertainty as $[x_5, x_6] = [0.2700, 0.5400]$. Note that this final interval of uncertainty is slightly larger than the one found in the Fibonacci method, $[0.461540, 0.230770]$. The ratio of the final to the initial interval of uncertainty in the present case is

$$\frac{L_6}{L_0} = \frac{0.5400 - 0.2700}{3.0} = \frac{0.27}{3.0} = 0.09$$

5.9   COMPARISON OF ELIMINATION METHODS

The efficiency of an elimination method can be measured in terms of the ratio of the final and the initial intervals of uncertainty, L_n/L_0. The values of this ratio achieved in various methods for a specified number of experiments ($n = 5$ and $n = 10$) are compared in Table 5.3. It can be seen that the Fibonacci method is the most efficient method, followed by the golden section method, in reducing the interval of uncertainty.

A similar observation can be made by considering the number of experiments (or function evaluations) needed to achieve a specified accuracy in various methods. The results are compared in Table 5.4 for maximum permissible errors of 0.1 and 0.01. It can be seen that to achieve any specified accuracy, the Fibonacci method requires the least number of experiments, followed by the golden section method.

Interpolation Methods

The interpolation methods were originally developed as one-dimensional searches within multivariable optimization techniques, and are generally more efficient than Fibonacci-type approaches. The aim of all the one-dimensional minimization methods

Table 5.3   Final Intervals of Uncertainty

Method	Formula	$n = 5$	$n = 10$
Exhaustive search	$L_n = \dfrac{2}{n+1} L_0$	$0.33333 L_0$	$0.18182 L_0$
Dichotomous search ($\delta = 0.01$ and $n =$ even)	$L_n = \dfrac{L_0}{2^{n/2}} + \delta \left(1 - \dfrac{1}{2^{n/2}}\right)$	$\frac{1}{4}L_0 + 0.0075$ with $n = 4$, $\frac{1}{8}L_0 + 0.00875$ with $n = 6$	$0.03125 L_0 + 0.0096875$
Interval halving ($n \geq 3$ and odd)	$L_n = (\frac{1}{2})^{(n-1)/2} L_0$	$0.25 L_0$	$0.0625 L_0$ with $n = 9$, $0.03125 L_0$ with $n = 11$
Fibonacci	$L_n = \dfrac{1}{F_n} L_0$	$0.125 L_0$	$0.01124 L_0$
Golden section	$L_n = (0.618)^{n-1} L_0$	$0.1459 L_0$	$0.01315 L_0$

Table 5.4   Number of Experiments for a Specified Accuracy

Method	Error: $\dfrac{1}{2}\dfrac{L_n}{L_0} \leq 0.1$	Error: $\dfrac{1}{2}\dfrac{L_n}{L_0} \leq 0.01$
Exhaustive search	$n \geq 9$	$n \geq 99$
Dichotomous search ($\delta = 0.01$, $L_0 = 1$)	$n \geq 6$	$n \geq 14$
Interval halving ($n \geq 3$ and odd)	$n \geq 7$	$n \geq 13$
Fibonacci	$n \geq 4$	$n \geq 9$
Golden section	$n \geq 5$	$n \geq 10$

is to find λ^*, the smallest nonnegative value of λ, for which the function

$$f(\lambda) = f(\mathbf{X} + \lambda \mathbf{S}) \tag{5.27}$$

attains a local minimum. Hence if the original function $f(\mathbf{X})$ is expressible as an explicit function of x_i ($i = 1, 2, \ldots, n$), we can readily write the expression for $f(\lambda) = f(\mathbf{X} + \lambda \mathbf{S})$ for any specified vector $\mathbf{S}$, set

$$\frac{df}{d\lambda}(\lambda) = 0 \tag{5.28}$$

and solve Eq. (5.28) to find λ^* in terms of $\mathbf{X}$ and $\mathbf{S}$. However, in many practical problems, the function $f(\lambda)$ cannot be expressed explicitly in terms of λ (as shown in Example 5.1). In such cases the interpolation methods can be used to find the value of λ^*.

Example 5.9   Derive the one-dimensional minimization problem for the following case:

$$\text{Minimize } f(\mathbf{X}) = (x_1^2 - x_2)^2 + (1 - x_1)^2 \tag{E\(_1\)}$$

from the starting point $\mathbf{X}_1 = \left\{ \begin{matrix} -2 \\ -2 \end{matrix} \right\}$ along the search direction $\mathbf{S} = \left\{ \begin{matrix} 1.00 \\ 0.25 \end{matrix} \right\}$.

SOLUTION   The new design point $\mathbf{X}$ can be expressed as

$$\mathbf{X} = \begin{Bmatrix} x_1 \\ x_2 \end{Bmatrix} = \mathbf{X}_1 + \lambda \mathbf{S} = \begin{Bmatrix} -2 + \lambda \\ -2 + 0.25\lambda \end{Bmatrix}$$

By substituting $x_1 = -2 + \lambda$ and $x_2 = -2 + 0.25\lambda$ in Eq. (E$_1$), we obtain f as a function of λ as

$$f(\lambda) = f\left(\begin{matrix} -2 + \lambda \\ -2 + 0.25\lambda \end{matrix} \right) = [(-2 + \lambda)^2 - (-2 + 0.25\lambda)]^2$$

$$+ [1 - (-2 + \lambda)]^2 = \lambda^4 - 8.5\lambda^3 + 31.0625\lambda^2 - 57.0\lambda + 45.0$$

The value of λ at which $f(\lambda)$ attains a minimum gives λ^*.

In the following sections, we discuss three different interpolation methods with reference to one-dimensional minimization problems that arise during multivariable optimization problems.

5.10   QUADRATIC INTERPOLATION METHOD

The quadratic interpolation method uses the function values only; hence it is useful to find the minimizing step (λ^*) of functions $f(\mathbf{X})$ for which the partial derivatives with respect to the variables x_i are not available or difficult to compute [5.2, 5.5]. This method finds the minimizing step length λ^* in three stages. In the first stage the $\mathbf{S}$-vector is normalized so that a step length of $\lambda = 1$ is acceptable. In the second stage the function $f(\lambda)$ is approximated by a quadratic function $h(\lambda)$ and the minimum, $\tilde{\lambda}^*$, of $h(\lambda)$ is found. If $\tilde{\lambda}^*$ is not sufficiently close to the true minimum λ^*, a third stage is used. In this stage a new quadratic function (refit) $h'(\lambda) = a' + b'\lambda + c'\lambda^2$ is used to approximate $f(\lambda)$, and a new value of $\tilde{\lambda}^*$ is found. This procedure is continued until a $\tilde{\lambda}^*$ that is sufficiently close to λ^* is found.

Stage 1.   In this stage,[†] the $\mathbf{S}$ vector is normalized as follows: Find $\Delta = \max_{i}|s_i|$, where s_i is the ith component of $\mathbf{S}$ and divide each component of $\mathbf{S}$ by Δ. Another method of normalization is to find $\Delta = (s_1^2 + s_2^2 + \cdots + s_n^2)^{1/2}$ and divide each component of $\mathbf{S}$ by Δ.

Stage 2.   Let

$$h(\lambda) = a + b\lambda + c\lambda^2 \tag{5.29}$$

be the quadratic function used for approximating the function $f(\lambda)$. It is worth noting at this point that a quadratic is the lowest-order polynomial for which a finite minimum can exist. The necessary condition for the minimum of $h(\lambda)$ is that

$$\frac{dh}{d\lambda} = b + 2c\lambda = 0$$

[†]This stage is not required if the one-dimensional minimization problem has not arisen within a multivariable minimization problem.

that is,

$$\tilde{\lambda}^* = -\frac{b}{2c} \tag{5.30}$$

The sufficiency condition for the minimum of $h(\lambda)$ is that

$$\left.\frac{d^2 h}{d\lambda^2}\right|_{\tilde{\lambda}^*} > 0$$

that is,

$$c > 0 \tag{5.31}$$

To evaluate the constants a, b, and c in Eq. (5.29), we need to evaluate the function $f(\lambda)$ at three points. Let $\lambda = A$, $\lambda = B$, and $\lambda = C$ be the points at which the function $f(\lambda)$ is evaluated and let f_A, f_B, and f_C be the corresponding function values, that is,

$$f_A = a + bA + cA^2$$

$$f_B = a + bB + cB^2$$

$$f_C = a + bC + cC^2 \tag{5.32}$$

The solution of Eqs. (5.32) gives

$$a = \frac{f_A BC(C - B) + f_B CA(A - C) + f_C AB(B - A)}{(A - B)(B - C)(C - A)} \tag{5.33}$$

$$b = \frac{f_A(B^2 - C^2) + f_B(C^2 - A^2) + f_C(A^2 - B^2)}{(A - B)(B - C)(C - A)} \tag{5.34}$$

$$c = -\frac{f_A(B - C) + f_B(C - A) + f_C(A - B)}{(A - B)(B - C)(C - A)} \tag{5.35}$$

From Eqs. (5.30), (5.34), and (5.35), the minimum of $h(\lambda)$ can be obtained as

$$\tilde{\lambda}^* = \frac{-b}{2c} = \frac{f_A(B^2 - C^2) + f_B(C^2 - A^2) + f_C(A^2 - B^2)}{2[f_A(B - C) + f_B(C - A) + f_C(A - B)]} \tag{5.36}$$

provided that c, as given by Eq. (5.35), is positive.

To start with, for simplicity, the points A, B, and C can be chosen as 0, t, and $2t$, respectively, where t is a preselected trial step length. By this procedure, we can save one function evaluation since $f_A = f(\lambda = 0)$ is generally known from the previous iteration (of a multivariable search). For this case, Eqs. (5.33) to (5.36) reduce to

$$a = f_A \tag{5.37}$$

$$b = \frac{4f_B - 3f_A - f_C}{2t} \tag{5.38}$$

$$c = \frac{f_C + f_A - 2f_B}{2t^2} \tag{5.39}$$

$$\tilde{\lambda}^* = \frac{4f_B - 3f_A - f_C}{4f_B - 2f_C - 2f_A} t \tag{5.40}$$

provided that

$$c = \frac{f_C + f_A - 2f_B}{2t^2} > 0 \tag{5.41}$$

The inequality (5.41) can be satisfied if

$$\frac{f_A + f_C}{2} > f_B \tag{5.42}$$

(i.e., the function value f_B should be smaller than the average value of f_A and f_C). This can be satisfied if f_B lies below the line joining f_A and f_C as shown in Fig. 5.12.

The following procedure can be used not only to satisfy the inequality (5.42) but also to ensure that the minimum $\tilde{\lambda}^*$ lies in the interval $0 < \tilde{\lambda}^* < 2t$.

1. Assuming that $f_A = f(\lambda = 0)$ and the initial step size t_0 are known, evaluate the function f at $\lambda = t_0$ and obtain $f_1 = f(\lambda = t_0)$. The possible outcomes are shown in Fig. 5.13.
2. If $f_1 > f_A$ is realized (Fig. 5.13c), set $f_C = f_1$ and evaluate the function f at $\lambda = t_0/2$ and $\tilde{\lambda}^*$ using Eq. (5.40) with $t = t_0/2$.
3. If $f_1 \leq f_A$ is realized (Fig. 5.13a or b), set $f_B = f_1$, and evaluate the function f at $\lambda = 2t_0$ to find $f_2 = f(\lambda = 2t_0)$. This may result in any one of the situations shown in Fig. 5.14.
4. If f_2 turns out to be greater than f_1 (Fig. 5.14b or c), set $f_C = f_2$ and compute $\tilde{\lambda}^*$ according to Eq. (5.40) with $t = t_0$.
5. If f_2 turns out to be smaller than f_1, set new $f_1 = f_2$ and $t_0 = 2t_0$, and repeat steps 2 to 4 until we are able to find $\tilde{\lambda}^*$.

Stage 3. The $\tilde{\lambda}^*$ found in stage 2 is the minimum of the approximating quadratic $h(\lambda)$ and we have to make sure that this $\tilde{\lambda}^*$ is sufficiently close to the true minimum λ^* of $f(\lambda)$ before taking $\lambda^* \simeq \tilde{\lambda}^*$. Several tests are possible to ascertain this. One possible test is to compare $f(\tilde{\lambda}^*)$ with $h(\tilde{\lambda}^*)$ and consider $\tilde{\lambda}^*$ a sufficiently good approximation

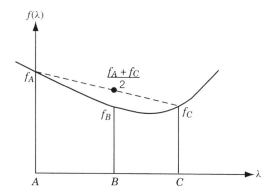

Figure 5.12 f_B smaller than $(f_A + f_C)/2$.

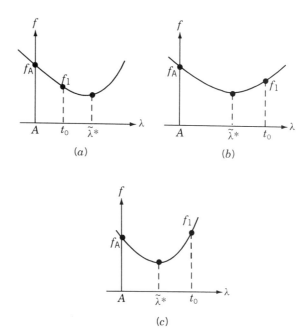

Figure 5.13   Possible outcomes when the function is evaluated at $\lambda = t_0$: (a) $f_1 < f_A$ and $t_0 < \tilde{\lambda}^*$; (b) $f_1 < f_A$ and $t_0 > \tilde{\lambda}^*$; (c) $f_1 > f_A$ and $t_0 > \tilde{\lambda}^*$.

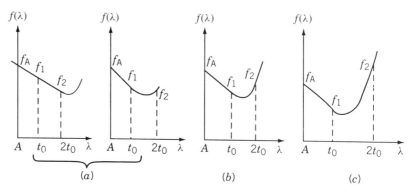

Figure 5.14   Possible outcomes when function is evaluated at $\lambda = t_0$ and $2t_0$: (a) $f_2 < f_1$ and $f_2 < f_A$; (b) $f_2 < f_A$ and $f_2 > f_1$; (c) $f_2 > f_A$ and $f_2 > f_1$.

if they differ not more than by a small amount. This criterion can be stated as

$$\left| \frac{h(\tilde{\lambda}^*) - f(\tilde{\lambda}^*)}{f(\tilde{\lambda}^*)} \right| \leq \varepsilon_1 \tag{5.43}$$

Another possible test is to examine whether $df/d\lambda$ is close to zero at $\tilde{\lambda}^*$. Since the derivatives of f are not used in this method, we can use a finite-difference formula for

$df/d\lambda$ and use the criterion

$$\left| \frac{f(\tilde{\lambda}^* + \Delta\tilde{\lambda}^*) - f(\tilde{\lambda}^* - \Delta\tilde{\lambda}^*)}{2\Delta\tilde{\lambda}^*} \right| \leq \varepsilon_2 \tag{5.44}$$

to stop the procedure. In Eqs. (5.43) and (5.44), ε_1 and ε_2 are small numbers to be specified depending on the accuracy desired.

If the convergence criteria stated in Eqs. (5.43) and (5.44) are not satisfied, a new quadratic function

$$h'(\lambda) = a' + b'\lambda + c'\lambda^2$$

is used to approximate the function $f(\lambda)$. To evaluate the constants a', b', and c', the three best function values of the current $f_A = f(\lambda = 0)$, $f_B = f(\lambda = t_0)$, $f_C = f(\lambda = 2t_0)$, and $\tilde{f} = f(\lambda = \tilde{\lambda}^*)$ are to be used. This process of trying to fit another polynomial to obtain a better approximation to $\tilde{\lambda}^*$ is known as *refitting* the polynomial.

For refitting the quadratic, we consider all possible situations and select the best three points of the present A, B, C, and $\tilde{\lambda}^*$. There are four possibilities, as shown in Fig. 5.15. The best three points to be used in refitting in each case are given in Table 5.5. A new value of $\tilde{\lambda}^*$ is computed by using the general formula, Eq. (5.36). If this $\tilde{\lambda}^*$ also does not satisfy the convergence criteria stated in Eqs. (5.43) and (5.44), a new quadratic has to be refitted according to the scheme outlined in Table 5.5.

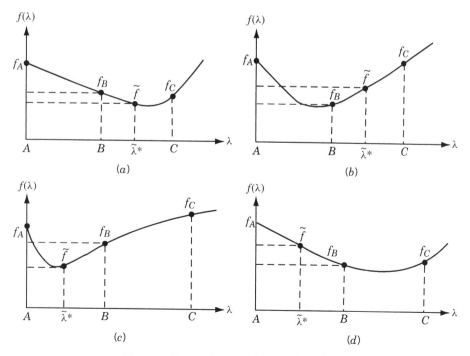

Figure 5.15   Various possibilities for refitting.

Table 5.5 Refitting Scheme

Case	Characteristics	New points for refitting New	Old
1	$\tilde{\lambda}^* > B$	A	B
	$\tilde{f} < f_B$	B	$\tilde{\lambda}^*$
		C	C
		Neglect old A	
2	$\tilde{\lambda}^* > B$	A	A
	$\tilde{f} > f_B$	B	B
		C	$\tilde{\lambda}^*$
		Neglect old C	
3	$\tilde{\lambda}^* < B$	A	A
	$\tilde{f} < f_B$	B	$\tilde{\lambda}^*$
		C	B
		Neglect old C	
4	$\tilde{\lambda}^* < B$	A	$\tilde{\lambda}^*$
	$\tilde{f} > f_B$	B	B
		C	C
		Neglect old A	

Example 5.10   Find the minimum of $f = \lambda^5 - 5\lambda^3 - 20\lambda + 5$.

SOLUTION   Since this is not a multivariable optimization problem, we can proceed directly to stage 2. Let the initial step size be taken as $t_0 = 0.5$ and $A = 0$.

Iteration 1

$$f_A = f(\lambda = 0) = 5$$

$$f_1 = f(\lambda = t_0) = 0.03125 - 5(0.125) - 20(0.5) + 5 = -5.59375$$

Since $f_1 < f_A$, we set $f_B = f_1 = -5.59375$, and find that

$$f_2 = f(\lambda = 2t_0 = 1.0) = -19.0$$

As $f_2 < f_1$, we set new $t_0 = 1$ and $f_1 = -19.0$. Again we find that $f_1 < f_A$ and hence set $f_B = f_1 = -19.0$, and find that $f_2 = f(\lambda = 2t_0 = 2) = -43$. Since $f_2 < f_1$, we again set $t_0 = 2$ and $f_1 = -43$. As this $f_1 < f_A$, set $f_B = f_1 = -43$ and evaluate $f_2 = f(\lambda = 2t_0 = 4) = 629$. This time $f_2 > f_1$ and hence we set $f_C = f_2 = 629$ and compute $\tilde{\lambda}^*$ from Eq. (5.40) as

$$\tilde{\lambda}^* = \frac{4(-43) - 3(5) - 629}{4(-43) - 2(629) - 2(5)}(2) = \frac{1632}{1440} = 1.135$$

Convergence test: Since $A = 0$, $f_A = 5$, $B = 2$, $f_B = -43$, $C = 4$, and $f_C = 629$, the values of a, b, and c can be found to be

$$a = 5, \quad b = -204, \quad c = 90$$

and

$$h(\tilde{\lambda}^*) = h(1.135) = 5 - 204(1.135) + 90(1.135)^2 = -110.9$$

Since

$$\tilde{f} = f(\tilde{\lambda}^*) = (1.135)^5 - 5(1.135)^3 - 20(1.135) + 5.0 = -23.127$$

we have

$$\left| \frac{h(\tilde{\lambda}^*) - f(\tilde{\lambda}^*)}{f(\tilde{\lambda}^*)} \right| = \left| \frac{-116.5 + 23.127}{-23.127} \right| = 3.8$$

As this quantity is very large, convergence is not achieved and hence we have to use *refitting*.

Iteration 2

Since $\tilde{\lambda}^* < B$ and $\tilde{f} > f_B$, we take the new values of A, B, and C as

$$A = 1.135, \qquad f_A = -23.127$$
$$B = 2.0, \qquad f_B = -43.0$$
$$C = 4.0, \qquad f_C = 629.0$$

and compute new $\tilde{\lambda}^*$, using Eq. (5.36), as

$$\tilde{\lambda}^* = \frac{(-23.127)(4.0 - 16.0) + (-43.0)(16.0 - 1.29)}{2[(-23.127)(2.0 - 4.0) + (-43.0)(4.0 - 1.135)}{+ (629.0)(1.135 - 2.0)]} = 1.661$$

Convergence test: To test the convergence, we compute the coefficients of the quadratic as

$$a = 288.0, \quad b = -417.0, \quad c = 125.3$$

As

$$h(\tilde{\lambda}^*) = h(1.661) = 288.0 - 417.0(1.661) + 125.3(1.661)^2 = -59.7$$
$$\tilde{f} = f(\tilde{\lambda}^*) = 12.8 - 5(4.59) - 20(1.661) + 5.0 = -38.37$$

we obtain

$$\left| \frac{h(\tilde{\lambda}^*) - f(\tilde{\lambda}^*)}{f(\tilde{\lambda}^*)} \right| = \left| \frac{-59.70 + 38.37}{-38.37} \right| = 0.556$$

Since this quantity is not sufficiently small, we need to proceed to the next refit.

5.11 CUBIC INTERPOLATION METHOD

The cubic interpolation method finds the minimizing step length λ^* in four stages [5.5, 5.11]. It makes use of the derivative of the function f:

$$f'(\lambda) = \frac{df}{d\lambda} = \frac{d}{d\lambda} f(\mathbf{X} + \lambda \mathbf{S}) = \mathbf{S}^T \nabla f(\mathbf{X} + \lambda \mathbf{S})$$

The first stage normalizes the $\mathbf{S}$ vector so that a step size $\lambda = 1$ is acceptable. The second stage establishes bounds on λ^*, and the third stage finds the value of $\tilde{\lambda}^*$ by approximating $f(\lambda)$ by a cubic polynomial $h(\lambda)$. If the $\tilde{\lambda}^*$ found in stage 3 does not satisfy the prescribed convergence criteria, the cubic polynomial is refitted in the fourth stage.

Stage 1. Calculate $\Delta = \max_i |s_i|$, where $|s_i|$ is the absolute value of the ith component of $\mathbf{S}$, and divide each component of $\mathbf{S}$ by Δ. An alternative method of normalization is to find

$$\Delta = (s_1^2 + s_2^2 + \cdots + s_n^2)^{1/2}$$

and divide each component of $\mathbf{S}$ by Δ.

Stage 2. To establish lower and upper bounds on the optimal step size λ^*, we need to find two points A and B at which the slope $df/d\lambda$ has different signs. We know that at $\lambda = 0$,

$$\left. \frac{df}{d\lambda} \right|_{\lambda=0} = \mathbf{S}^T \nabla f(\mathbf{X}) < 0$$

since $\mathbf{S}$ is presumed to be a direction of descent.[†]

Hence to start with we can take $A = 0$ and try to find a point $\lambda = B$ at which the slope $df/d\lambda$ is positive. Point B can be taken as the first value out of $t_0, 2t_0, 4t_0, 8t_0, \ldots$ at which f' is nonnegative, where t_0 is a preassigned initial step size. It then follows that λ^* is bounded in the interval $A < \lambda^* \leq B$ (Fig. 5.16).

Stage 3. If the cubic equation

$$h(\lambda) = a + b\lambda + c\lambda^2 + \lambda^3 \tag{5.45}$$

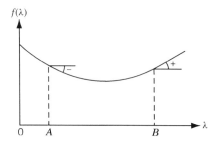

Figure 5.16 Minimum of $f(\lambda)$ lies between A and B.

[†]In this case the angle between the direction of steepest descent and $\mathbf{S}$ will be less than $90°$.

is used to approximate the function $f(\lambda)$ between points A and B, we need to find the values $f_A = f(\lambda = A)$, $f'_A = df/d\lambda(\lambda = A)$, $f_B = f(\lambda = B)$, and $f'_B = df/d\lambda(\lambda = B)$ in order to evaluate the constants, a, b, c, and d in Eq. (5.45). By assuming that $A \neq 0$, we can derive a general formula for $\tilde{\lambda}^*$. From Eq. (5.45) we have

$$f_A = a + bA + cA^2 + dA^3$$
$$f_B = a + bB + cB^2 + dB^3$$
$$f'_A = b + 2cA + 3dA^2$$
$$f'_B = b + 2cB + 3dB^2 \tag{5.46}$$

Equations (5.46) can be solved to find the constants as

$$a = f_A - bA - cA^2 - dA^3 \tag{5.47}$$

with

$$b = \frac{1}{(A-B)^2}(B^2 f'_A + A^2 f'_B + 2ABZ) \tag{5.48}$$

$$c = -\frac{1}{(A-B)^2}[(A+B)Z + Bf'_A + Af'_B] \tag{5.49}$$

and

$$d = \frac{1}{3(A-B)^2}(2Z + f'_A + f'_B) \tag{5.50}$$

where

$$Z = \frac{3(f_A - f_B)}{B - A} + f'_A + f'_B \tag{5.51}$$

The necessary condition for the minimum of $h(\lambda)$ given by Eq. (5.45) is that

$$\frac{dh}{d\lambda} = b + 2c\lambda + 3d\lambda^2 = 0$$

that is,

$$\tilde{\lambda}^* = \frac{-c \pm (c^2 - 3bd)^{1/2}}{3d} \tag{5.52}$$

The application of the sufficiency condition for the minimum of $h(\lambda)$ leads to the relation

$$\left.\frac{d^2 h}{d\lambda^2}\right|_{\tilde{\lambda}^*} = 2c + 6d\tilde{\lambda}^* > 0 \tag{5.53}$$

By substituting the expressions for b, c, and d given by Eqs. (5.48) to (5.50) into Eqs. (5.52) and (5.53), we obtain

$$\tilde{\lambda}^* = A + \frac{f'_A + Z \pm Q}{f'_A + f'_B + 2Z}(B - A) \tag{5.54}$$

where

$$Q = (Z^2 - f'_A f'_B)^{1/2} \tag{5.55}$$

$$2(B - A)(2Z + f'_A + f'_B)(f'_A + Z \pm Q)$$

$$-2(B - A)(f'^2_A + Zf'_B + 3Zf'_A + 2Z^2)$$

$$-2(B + A)f'_A f'_B > 0 \tag{5.56}$$

By specializing Eqs. (5.47) to (5.56) for the case where $A = 0$, we obtain

$$a = f_A$$

$$b = f'_A$$

$$c = -\frac{1}{B}(Z + f'_A)$$

$$d = \frac{1}{3B^2}(2Z + f'_A + f'_B)$$

$$\tilde{\lambda}^* = B\frac{f'_A + Z \pm Q}{f'_A + f'_B + 2Z} \tag{5.57}$$

$$Q = (Z^2 - f'_A f'_B)^{1/2} > 0 \tag{5.58}$$

where

$$Z = \frac{3(f_A - f_B)}{B} + f'_A + f'_B \tag{5.59}$$

The two values of $\tilde{\lambda}^*$ in Eqs. (5.54) and (5.57) correspond to the two possibilities for the vanishing of $h'(\lambda)$ [i.e., at a maximum of $h(\lambda)$ and at a minimum]. To avoid imaginary values of Q, we should ensure the satisfaction of the condition

$$Z^2 - f'_A f'_B \geq 0$$

in Eq. (5.55). This inequality is satisfied automatically since A and B are selected such that $f'_A < 0$ and $f'_B \geq 0$. Furthermore, the sufficiency condition (when $A = 0$) requires that $Q > 0$, which is already satisfied. Now we compute $\tilde{\lambda}^*$ using Eq. (5.57) and proceed to the next stage.

Stage 4.   The value of $\tilde{\lambda}^*$ found in stage 3 is the true minimum of $h(\lambda)$ and may not be close to the minimum of $f(\lambda)$. Hence the following convergence criteria can be used before choosing $\lambda^* \approx \tilde{\lambda}^*$:

$$\left| \frac{h(\tilde{\lambda}^*) - f(\tilde{\lambda}^*)}{f(\tilde{\lambda}^*)} \right| \leq \varepsilon_1 \tag{5.60}$$

$$\left| \frac{df}{d\lambda} \right|_{\tilde{\lambda}^*} = |\mathbf{S}^T \nabla f|_{\tilde{\lambda}^*}| \leq \varepsilon_2 \tag{5.61}$$

where ε_1 and ε_2 are small numbers whose values depend on the accuracy desired. The criterion of Eq. (5.61) can be stated in nondimensional form as

$$\left| \frac{\mathbf{S}^T \nabla f}{|\mathbf{S}||\nabla f|} \right|_{\tilde{\lambda}^*} \leq \varepsilon_2 \tag{5.62}$$

If the criteria stated in Eqs. (5.60) and (5.62) are not satisfied, a new cubic equation

$$h'(\lambda) = a' + b'\lambda + c'\lambda^2 + d'\lambda^3$$

can be used to approximate $f(\lambda)$. The constants a', b', c', and d' can be evaluated by using the function and derivative values at the best two points out of the three points currently available: A, B, and $\tilde{\lambda}^*$. Now the general formula given by Eq. (5.54) is to be used for finding the optimal step size $\tilde{\lambda}^*$. If $f'(\tilde{\lambda}^*) < 0$, the new points A and B are taken as $\tilde{\lambda}^*$ and B, respectively; otherwise [if $f'(\tilde{\lambda}^*) > 0$], the new points A and B are taken as A and $\tilde{\lambda}^*$, and Eq. (5.54) is applied to find the new value of $\tilde{\lambda}^*$. Equations (5.60) and (5.62) are again used to test for the convergence of $\tilde{\lambda}^*$. If convergence is achieved, $\tilde{\lambda}^*$ is taken as λ^* and the procedure is stopped. Otherwise, the entire procedure is repeated until the desired convergence is achieved.

The flowchart for implementing the cubic interpolation method is given in Fig. 5.17.

Example 5.11 Find the minimum of $f = \lambda^5 - 5\lambda^3 - 20\lambda + 5$ by the cubic interpolation method.

SOLUTION   Since this problem has not arisen during a multivariable optimization process, we can skip stage 1. We take $A = 0$ and find that

$$\frac{df}{d\lambda}(\lambda = A = 0) = 5\lambda^4 - 15\lambda^2 - 20 \Big|_{\lambda=0} = -20 < 0$$

To find B at which $df/d\lambda$ is nonnegative, we start with $t_0 = 0.4$ and evaluate the derivative at $t_0, 2t_0, 4t_0, \ldots$. This gives

$$f'(t_0 = 0.4) = 5(0.4)^4 - 15(0.4)^2 - 20.0 = -22.272$$

$$f'(2t_0 = 0.8) = 5(0.8)^4 - 15(0.8)^2 - 20.0 = -27.552$$

$$f'(4t_0 = 1.6) = 5(1.6)^4 - 15(1.6)^2 - 20.0 = -25.632$$

$$f'(8t_0 = 3.2) = 5(3.2)^4 - 15(3.2)^2 - 20.0 = 350.688$$

Thus we find that[†]

$$A = 0.0, \qquad f_A = 5.0, \qquad f'_A = -20.0$$
$$B = 3.2, \qquad f_B = 113.0, \qquad f'_B = 350.688$$
$$A < \lambda^* < B$$

[†]As f' has been found to be negative at $\lambda = 1.6$ also, we can take $A = 1.6$ for faster convergence.

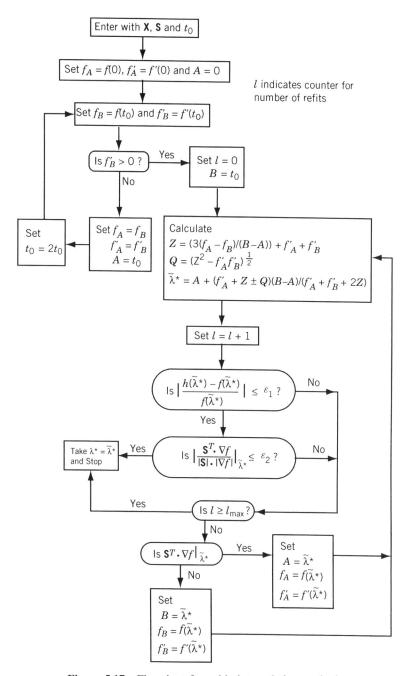

Figure 5.17   Flowchart for cubic interpolation method.

Iteration 1

To find the value of $\tilde{\lambda}^*$ and to test the convergence criteria, we first compute Z and Q as

$$Z = \frac{3(5.0 - 113.0)}{3.2} - 20.0 + 350.688 = 229.588$$

$$Q = [229.588^2 + (20.0)(350.688)]^{1/2} = 244.0$$

Hence

$$\tilde{\lambda}^* = 3.2 \left(\frac{-20.0 + 229.588 \pm 244.0}{-20.0 + 350.688 + 459.176} \right) = 1.84 \quad \text{or} \quad -0.1396$$

By discarding the negative value, we have

$$\tilde{\lambda}^* = 1.84$$

Convergence criterion: If $\tilde{\lambda}^*$ is close to the true minimum, λ^*, then $f'(\tilde{\lambda}^*) = df(\tilde{\lambda}^*)/d\lambda$ should be approximately zero. Since $f' = 5\lambda^4 - 15\lambda^2 - 20$,

$$f'(\tilde{\lambda}^*) = 5(1.84)^4 - 15(1.84)^2 - 20 = -13.0$$

Since this is not small, we go to the next iteration or refitting. As $f'(\tilde{\lambda}^*) < 0$, we take $A = \tilde{\lambda}^*$ and

$$f_A = f(\tilde{\lambda}^*) = (1.84)^5 - 5(1.84)^3 - 20(1.84) + 5 = -41.70$$

Thus

$$A = 1.84, \qquad f_A = -41.70, \qquad f'_A = -13.0$$

$$B = 3.2, \qquad f_B = 113.0, \qquad f'_B = 350.688$$

$$A < \tilde{\lambda}^* < B$$

Iteration 2

$$Z = \frac{3(-41.7 - 113.0)}{3.20 - 1.84} - 13.0 + 350.688 = -3.312$$

$$Q = [(-3.312)^2 + (13.0)(350.688)]^{1/2} = 67.5$$

Hence

$$\tilde{\lambda}^* = 1.84 + \frac{-13.0 - 3.312 \pm 67.5}{-13.0 + 350.688 - 6.624}(3.2 - 1.84) = 2.05$$

Convergence criterion:

$$f'(\tilde{\lambda}^*) = 5.0(2.05)^4 - 15.0(2.05)^2 - 20.0 = 5.35$$

Since this value is large, we go the next iteration with $B = \tilde{\lambda}^* = 2.05$ [as $f'(\tilde{\lambda}^*) > 0$] and

$$f_B = (2.05)^5 - 5.0(2.05)^3 - 20.0(2.05) + 5.0 = -42.90$$

Thus

$$A = 1.84, \qquad f_A = -41.70, \qquad f_A' = -13.00$$

$$B = 2.05, \qquad f_B = -42.90, \qquad f_B' = 5.35$$

$$A < \lambda^* < B$$

Iteration 3

$$Z = \frac{3.0(-41.70 + 42.90)}{(2.05 - 1.84)} - 13.00 + 5.35 = 9.49$$

$$Q = [(9.49)^2 + (13.0)(5.35)]^{1/2} = 12.61$$

Therefore,

$$\tilde{\lambda}^* = 1.84 + \frac{-13.00 + 9.49 \pm 12.61}{-13.00 + 5.35 + 18.98}(2.05 - 1.84) = 2.0086$$

Convergence criterion:

$$f'(\tilde{\lambda}^*) = 5.0(2.0086)^4 - 15.0(2.0086)^2 - 20.0 = 0.855$$

Assuming that this value is close to zero, we can stop the iterative process and take

$$\lambda^* \simeq \tilde{\lambda}^* = 2.0086$$

5.12 DIRECT ROOT METHODS

The necessary condition for $f(\lambda)$ to have a minimum of λ^* is that $f'(\lambda^*) = 0$. The direct root methods seek to find the root (or solution) of the equation, $f'(\lambda) = 0$. Three root-finding methods—the Newton, the quasi-Newton, and the secant methods—are discussed in this section.

5.12.1 Newton Method

Consider the quadratic approximation of the function $f(\lambda)$ at $\lambda = \lambda_i$ using the Taylor's series expansion:

$$f(\lambda) = f(\lambda_i) + f'(\lambda_i)(\lambda - \lambda_i) + \tfrac{1}{2}f''(\lambda_i)(\lambda - \lambda_i)^2 \tag{5.63}$$

By setting the derivative of Eq. (5.63) equal to zero for the minimum of $f(\lambda)$, we obtain

$$f'(\lambda) = f'(\lambda_i) + f''(\lambda_i)(\lambda - \lambda_i) = 0 \tag{5.64}$$

If λ_i denotes an approximation to the minimum of $f(\lambda)$, Eq. (5.64) can be rearranged to obtain an improved approximation as

$$\lambda_{i+1} = \lambda_i - \frac{f'(\lambda_i)}{f''(\lambda_i)} \tag{5.65}$$

Thus the *Newton method*, Eq. (5.65), is equivalent to using a quadratic approximation for the function $f(\lambda)$ and applying the necessary conditions. The iterative process given by Eq. (5.65) can be assumed to have converged when the derivative, $f'(\lambda_{i+1})$, is close to zero:

$$|f'(\lambda_{i+1})| \leq \varepsilon \tag{5.66}$$

where ε is a small quantity. The convergence process of the method is shown graphically in Fig. 5.18*a*.

Remarks:

1. The Newton method was originally developed by Newton for solving nonlinear equations and later refined by Raphson, and hence the method is also known as *Newton–Raphson method* in the literature of numerical analysis.
2. The method requires both the first- and second-order derivatives of $f(\lambda)$.
3. If $f''(\lambda_i) \neq 0$ [in Eq. (5.65)], the Newton iterative method has a powerful (fastest) convergence property, known as *quadratic convergence*.[†]
4. If the starting point for the iterative process is not close to the true solution λ^*, the Newton iterative process might diverge as illustrated in Fig. 5.18*b*.

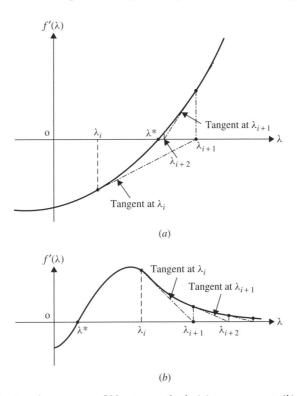

(*a*)

(*b*)

Figure 5.18   Iterative process of Newton method: (*a*) convergence; (*b*) divergence.

[†]The definition of quadratic convergence is given in Section 6.7.

Example 5.12   Find the minimum of the function

$$f(\lambda) = 0.65 - \frac{0.75}{1 + \lambda^2} - 0.65\lambda \tan^{-1} \frac{1}{\lambda}$$

using the Newton–Raphson method with the starting point $\lambda_1 = 0.1$. Use $\varepsilon = 0.01$ in Eq. (5.66) for checking the convergence.

SOLUTION   The first and second derivatives of the function $f(\lambda)$ are given by

$$f'(\lambda) = \frac{1.5\lambda}{(1 + \lambda^2)^2} + \frac{0.65\lambda}{1 + \lambda^2} - 0.65 \tan^{-1} \frac{1}{\lambda}$$

$$f''(\lambda) = \frac{1.5(1 - 3\lambda^2)}{(1 + \lambda^2)^3} + \frac{0.65(1 - \lambda^2)}{(1 + \lambda^2)^2} + \frac{0.65}{1 + \lambda^2} = \frac{2.8 - 3.2\lambda^2}{(1 + \lambda^2)^3}$$

Iteration 1

$$\lambda_1 = 0.1, \quad f(\lambda_1) = -0.188197, \quad f'(\lambda_1) = -0.744832, \quad f''(\lambda_1) = 2.68659$$

$$\lambda_2 = \lambda_1 - \frac{f'(\lambda_1)}{f''(\lambda_1)} = 0.377241$$

Convergence check: $|f'(\lambda_2)| = |-0.138230| > \varepsilon$.

Iteration 2

$$f(\lambda_2) = -0.303279, \quad f'(\lambda_2) = -0.138230, \quad f''(\lambda_2) = 1.57296$$

$$\lambda_3 = \lambda_2 - \frac{f'(\lambda_2)}{f''(\lambda_2)} = 0.465119$$

Convergence check: $|f'(\lambda_3)| = |-0.0179078| > \varepsilon$.

Iteration 3

$$f(\lambda_3) = -0.309881, \quad f'(\lambda_3) = -0.0179078, \quad f''(\lambda_3) = 1.17126$$

$$\lambda_4 = \lambda_3 - \frac{f'(\lambda_3)}{f''(\lambda_3)} = 0.480409$$

Convergence check: $|f'(\lambda_4)| = |-0.0005033| < \varepsilon$.
Since the process has converged, the optimum solution is taken as $\lambda^* \approx \lambda_4 = 0.480409$.

5.12.2   Quasi-Newton Method

If the function being minimized $f(\lambda)$ is not available in closed form or is difficult to differentiate, the derivatives $f'(\lambda)$ and $f''(\lambda)$ in Eq. (5.65) can be approximated by the finite difference formulas as

$$f'(\lambda_i) = \frac{f(\lambda_i + \Delta\lambda) - f(\lambda_i - \Delta\lambda)}{2\Delta\lambda} \tag{5.67}$$

$$f''(\lambda_i) = \frac{f(\lambda_i + \Delta\lambda) - 2f(\lambda_i) + f(\lambda_i - \Delta\lambda)}{\Delta\lambda^2} \tag{5.68}$$

where $\Delta\lambda$ is a small step size. Substitution of Eqs. (5.67) and (5.68) into Eq. (5.65) leads to

$$\lambda_{i+1} = \lambda_i - \frac{\Delta\lambda[f(\lambda_i + \Delta\lambda) - f(\lambda_i - \Delta\lambda)]}{2[f(\lambda_i + \Delta\lambda) - 2f(\lambda_i) + f(\lambda_i - \Delta\lambda)]} \tag{5.69}$$

The iterative process indicated by Eq. (5.69) is known as the *quasi-Newton method*. To test the convergence of the iterative process, the following criterion can be used:

$$|f'(\lambda_{i+1})| = \left| \frac{f(\lambda_{i+1} + \Delta\lambda) - f(\lambda_{i+1} - \Delta\lambda)}{2\Delta\lambda} \right| \le \varepsilon \tag{5.70}$$

where a central difference formula has been used for evaluating the derivative of f and ε is a small quantity.

Remarks:

1. The central difference formulas have been used in Eqs. (5.69) and (5.70). However, the forward or backward difference formulas can also be used for this purpose.
2. Equation (5.69) requires the evaluation of the function at the points $\lambda_i + \Delta\lambda$ and $\lambda_i - \Delta\lambda$ in addition to λ_i in each iteration.

Example 5.13   Find the minimum of the function

$$f(\lambda) = 0.65 - \frac{0.75}{1 + \lambda^2} - 0.65\lambda \tan^{-1}\frac{1}{\lambda}$$

using quasi-Newton method with the starting point $\lambda_1 = 0.1$ and the step size $\Delta\lambda = 0.01$ in central difference formulas. Use $\varepsilon = 0.01$ in Eq. (5.70) for checking the convergence.

SOLUTION

Iteration 1

$$\lambda_1 = 0.1, \quad \Delta\lambda = 0.01, \quad \varepsilon = 0.01, \quad f_1 = f(\lambda_1) = -0.188197,$$

$$f_1^+ = f(\lambda_1 + \Delta\lambda) = -0.195512, \quad f_1^- = f(\lambda_1 - \Delta\lambda) = -0.180615$$

$$\lambda_2 = \lambda_1 - \frac{\Delta\lambda(f_1^+ - f_1^-)}{2(f_1^+ - 2f_1 + f_1^-)} = 0.377882$$

Convergence check:

$$|f'(\lambda_2)| = \left| \frac{f_2^+ - f_2^-}{2\Delta\lambda} \right| = 0.137300 > \varepsilon$$

Iteration 2

$$f_2 = f(\lambda_2) = -0.303368, \quad f_2^+ = f(\lambda_2 + \Delta\lambda) = -0.304662,$$

$$f_2^- = f(\lambda_2 - \Delta\lambda) = -0.301916$$

$$\lambda_3 = \lambda_2 - \frac{\Delta\lambda(f_2^+ - f_2^-)}{2(f_2^+ - 2f_2 + f_2^-)} = 0.465390$$

Convergence check:

$$|f'(\lambda_3)| = \left| \frac{f_3^+ - f_3^-}{2\Delta\lambda} \right| = 0.017700 > \varepsilon$$

Iteration 3

$$f_3 = f(\lambda_3) = -0.309885, \quad f_3^+ = f(\lambda_3 + \Delta\lambda) = -0.310004,$$

$$f_3^- = f(\lambda_3 - \Delta\lambda) = -0.309650$$

$$\lambda_4 = \lambda_3 - \frac{\Delta\lambda(f_3^+ - f_3^-)}{2(f_3^+ - 2f_3 + f_3^-)} = 0.480600$$

Convergence check:

$$|f'(\lambda_4)| = \left| \frac{f_4^+ - f_4^-}{2\Delta\lambda} \right| = 0.000350 < \varepsilon$$

Since the process has converged, we take the optimum solution as $\lambda^* \approx \lambda_4 = 0.480600$.

5.12.3 Secant Method

The secant method uses an equation similar to Eq. (5.64) as

$$f'(\lambda) = f'(\lambda_i) + s(\lambda - \lambda_i) = 0 \tag{5.71}$$

where s is the slope of the line connecting the two points $(A, f'(A))$ and $(B, f'(B))$, where A and B denote two different approximations to the correct solution, λ^*. The slope s can be expressed as (Fig. 5.19)

$$s = \frac{f'(B) - f'(A)}{B - A} \tag{5.72}$$

Equation (5.71) approximates the function $f'(\lambda)$ between A and B as a linear equation (secant), and hence the solution of Eq. (5.71) gives the new approximation to the root of $f'(\lambda)$ as

$$\lambda_{i+1} = \lambda_i - \frac{f'(\lambda_i)}{s} = A - \frac{f'(A)(B - A)}{f'(B) - f'(A)} \tag{5.73}$$

The iterative process given by Eq. (5.73) is known as the *secant method* (Fig. 5.19). Since the secant approaches the second derivative of $f(\lambda)$ at A as B approaches A,

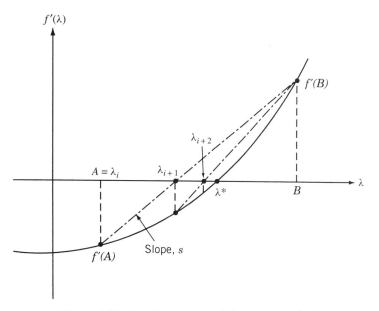

Figure 5.19 Iterative process of the secant method.

the secant method can also be considered as a quasi-Newton method. It can also be considered as a form of elimination technique since part of the interval, (A, λ_{i+1}) in Fig. 5.19, is eliminated in every iteration. The iterative process can be implemented by using the following step-by-step procedure.

1. Set $\lambda_1 = A = 0$ and evaluate $f'(A)$. The value of $f'(A)$ will be negative. Assume an initial trial step length t_0. Set $i = 1$.
2. Evaluate $f'(t_0)$.
3. If $f'(t_0) < 0$, set $A = \lambda_i = t_0$, $f'(A) = f'(t_0)$, new $t_0 = 2t_0$, and go to step 2.
4. If $f'(t_0) \geq 0$, set $B = t_0$, $f'(B) = f'(t_0)$, and go to step 5.
5. Find the new approximate solution of the problem as

$$\lambda_{i+1} = A - \frac{f'(A)(B - A)}{f'(B) - f'(A)} \tag{5.74}$$

6. Test for convergence:

$$|f'(\lambda_i + 1)| \leq \varepsilon \tag{5.75}$$

where ε is a small quantity. If Eq. (5.75) is satisfied, take $\lambda^* \approx \lambda_{i+1}$ and stop the procedure. Otherwise, go to step 7.
7. If $f'(\lambda_{i+1}) \geq 0$, set new $B = \lambda_{i+1}$, $f'(B) = f'(\lambda_{i+1})$, $i = i + 1$, and go to step 5.
8. If $f'(\lambda_{i+1}) < 0$, set new $A = \lambda_{i+1}$, $f'(A) = f'(\lambda_{i+1})$, $i = i + 1$, and go to step 5.

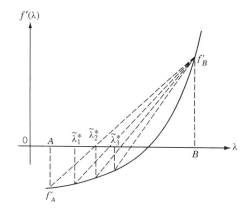

Figure 5.20   Situation when f'_A varies very slowly.

Remarks:

1. The secant method is identical to assuming a linear equation for $f'(\lambda)$. This implies that the original function, $f(\lambda)$, is approximated by a quadratic equation.

2. In some cases we may encounter a situation where the function $f'(\lambda)$ varies very slowly with λ, as shown in Fig. 5.20. This situation can be identified by noticing that the point B remains unaltered for several consecutive refits. Once such a situation is suspected, the convergence process can be improved by taking the next value of λ_{i+1} as $(A + B)/2$ instead of finding its value from Eq. (5.74).

Example 5.14   Find the minimum of the function

$$f(\lambda) = 0.65 - \frac{0.75}{1 + \lambda^2} - 0.65\lambda \tan^{-1} \frac{1}{\lambda}$$

using the secant method with an initial step size of $t_0 = 0.1$, $\lambda_1 = 0.0$, and $\varepsilon = 0.01$.

SOLUTION $\lambda_1 = A = 0.0$, $t_0 = 0.1$, $f'(A) = -1.02102$, $B = A + t_0 = 0.1$, $f'(B) = -0.744832$. Since $f'(B) < 0$, we set new $A = 0.1$, $f'(A) = -0.744832$, $t_0 = 2(0.1) = 0.2$, $B = \lambda_1 + t_0 = 0.2$, and compute $f'(B) = -0.490343$. Since $f'(B) < 0$, we set new $A = 0.2$, $f'(A) = -0.490343$, $t_0 = 2(0.2) = 0.4$, $B = \lambda_1 + t_0 = 0.4$, and compute $f'(B) = -0.103652$. Since $f'(B) < 0$, we set new $A = 0.4$, $f'(A) = -0.103652$, $t_0 = 2(0.4) = 0.8$, $B = \lambda_1 + t_0 = 0.8$, and compute $f'(B) = +0.180800$. Since $f'(B) > 0$, we proceed to find λ_2.

Iteration 1

Since $A = \lambda_1 = 0.4$, $f'(A) = -0.103652$, $B = 0.8$, $f'(B) = +0.180800$, we compute

$$\lambda_2 = A - \frac{f'(A)(B - A)}{f'(B) - f'(A)} = 0.545757$$

Convergence check: $|f'(\lambda_2)| = |+0.0105789| > \varepsilon$.

Iteration 2

Since $f'(\lambda_2) = +0.0105789 > 0$, we set new $A = 0.4$, $f'(A) = -0.103652$, $B = \lambda_2 = 0.545757$, $f'(B) = f'(\lambda_2) = +0.0105789$, and compute

$$\lambda_3 = A - \frac{f'(A)(B - A)}{f'(B) - f'(A)} = 0.490632$$

Convergence check: $|f'(\lambda_3)| = |+0.00151235| < \varepsilon$.

Since the process has converged, the optimum solution is given by $\lambda^* \approx \lambda_3 = 0.490632$.

5.13 PRACTICAL CONSIDERATIONS

5.13.1 How to Make the Methods Efficient and More Reliable

In some cases, some of the interpolation methods discussed in Sections 5.10 to 5.12 may be very slow to converge, may diverge, or may predict the minimum of the function, $f(\lambda)$, outside the initial interval of uncertainty, especially when the interpolating polynomial is not representative of the variation of the function being minimized. In such cases we can use the Fibonacci or golden section method to find the minimum. In some problems it might prove to be more efficient to combine several techniques. For example, the unrestricted search with an accelerated step size can be used to bracket the minimum and then the Fibonacci or the golden section method can be used to find the optimum point. In some cases the Fibonacci or golden section method can be used in conjunction with an interpolation method.

5.13.2 Implementation in Multivariable Optimization Problems

As stated earlier, the one-dimensional minimization methods are useful in multivariable optimization problems to find an improved design vector $\mathbf{X}_{i+1}$ from the current design vector $\mathbf{X}_i$ using the formula

$$\mathbf{X}_{i+1} = \mathbf{X}_i + \lambda_i^* \mathbf{S}_i \tag{5.76}$$

where $\mathbf{S}_i$ is the known search direction and λ_i^* is the optimal step length found by solving the one-dimensional minimization problem as

$$\lambda_i^* = \min_{\lambda_i} \left[f(\mathbf{X}_i + \lambda_i \mathbf{S}_i) \right] \tag{5.77}$$

Here the objective function f is to be evaluated at any trial step length t_0 as

$$f(t_0) = f(\mathbf{X}_i + t_0 \mathbf{S}_i) \tag{5.78}$$

Similarly, the derivative of the function f with respect to λ corresponding to the trial step length t_0 is to be found as

$$\left. \frac{df}{d\lambda} \right|_{\lambda = t_0} = \mathbf{S}_i^{\mathrm{T}} \Delta f|_{\lambda = t_0} \tag{5.79}$$

Separate function programs or subroutines can be written conveniently to implement Eqs. (5.78) and (5.79).

5.13.3   Comparison of Methods

It has been shown in Section 5.9 that the Fibonacci method is the most efficient elimination technique in finding the minimum of a function if the initial interval of uncertainty is known. In the absence of the initial interval of uncertainty, the quadratic interpolation method or the quasi-Newton method is expected to be more efficient when the derivatives of the function are not available. When the first derivatives of the function being minimized are available, the cubic interpolation method or the secant method are expected to be very efficient. On the other hand, if both the first and second derivatives of the function are available, the Newton method will be the most efficient one in finding the optimal step length, λ^*.

In general, the efficiency and reliability of the various methods are problem dependent and any efficient computer program must include many heuristic additions not indicated explicitly by the method. The heuristic considerations are needed to handle multimodal functions (functions with multiple extreme points), sharp variations in the slopes (first derivatives) and curvatures (second derivatives) of the function, and the effects of round-off errors resulting from the precision used in the arithmetic operations. A comparative study of the efficiencies of the various search methods is given in Ref. [5.10].

5.14   MATLAB SOLUTION OF ONE-DIMENSIONAL MINIMIZATION PROBLEMS

The solution of one-dimensional minimization problems, using the MATLAB program `optimset`, is illustrated by the following example.

Example 5.15   Find the minimum of the following function:

$$f(x) = 0.65 - \frac{0.75}{1+x^2} - 0.65x \tan^{-1}\left(\frac{1}{x}\right)$$

SOLUTION

Step 1: Write an M-file `objfun.m` for the objective function.

```
function f= objfun(x)
f= 0.65-(0.75/(1+x^2))-0.65*x*atan(1/x);
```

Step 2: Invoke unconstrained optimization program (write this in new MATLAB file).

```
clc
clear all
warning off
options = optimset('LargeScale','off');
[x,fval] = fminbnd(@objfun,0,0.5,options)
```

This produces the solution or ouput as follows:

```
x=
 0.4809
fval =
 -0.3100
```

REFERENCES AND BIBLIOGRAPHY

5.1   J. S. Przemieniecki, *Theory of Matrix Structural Analysis*, McGraw-Hill, New York, 1968.

5.2   M. J. D. Powell, An efficient method for finding the minimum of a function of several variables without calculating derivatives, *Computer Journal*, Vol. 7, pp. 155–162, 1964.

5.3   R. Fletcher and C. M. Reeves, Function minimization by conjugate gradients, *Computer Journal*, Vol. 7, pp. 149–154, 1964.

5.4   B. Carnahan, H. A. Luther, and J. O. Wilkes, *Applied Numerical Methods*, Wiley, New York, 1969.

5.5   R. L. Fox, *Optimization Methods for Engineering Design*, Addison-Wesley, Reading, MA, 1971.

5.6   D. J. Wilde, *Optimum Seeking Methods*, Prentice Hall, Englewood Cliffs, NJ, 1964.

5.7   A. I. Cohen, Stepsize analysis for descent methods, *Journal of Optimization Theory and Applications*, Vol. 33, pp. 187–205, 1981.

5.8   P. E. Gill, W. Murray, and M. H. Wright, *Practical Optimization*, Academic Press, New York, 1981.

5.9   J. E. Dennis and R. B. Schnabel, *Numerical Methods for Unconstrained Optimization and Nonlinear Equations*, Prentice Hall, Englewood Cliffs, NJ, 1983.

5.10 R. P. Brent, *Algorithms for Minimization Without Derivatives*, Prentice Hall, Englewood Cliffs, NJ, 1973.

5.11 W. C. Davidon, Variable metric method for minimization, Argonne National Laboratory, ANL-5990 (rev), 1959.

REVIEW QUESTIONS

5.1   What is a one-dimensional minimization problem?

5.2   What are the limitations of classical methods in solving a one-dimensional minimization problem?

5.3   What is the difference between elimination and interpolation methods?

5.4   Define Fibonacci numbers.

5.5   What is the difference between Fibonacci and golden section methods?

5.6   What is a unimodal function?

5.7   What is an interval of uncertainty?

5.8   Suggest a method of finding the minimum of a multimodal function.

5.9   What is an exhaustive search method?

5.10   What is a dichotomous search method?

5.11   Define the golden mean.

5.12   What is the difference between quadratic and cubic interpolation methods?

5.13   Why is refitting necessary in interpolation methods?

5.14   What is a direct root method?

5.15   What is the basis of the interval halving method?

5.16   What is the difference between Newton and quasi-Newton methods?

5.17   What is the secant method?

5.18   Answer true or false:

(a) A unimodal function cannot be discontinuous.
(b) All elimination methods assume the function to be unimodal.
(c) The golden section method is more accurate than the Fibonacci method.
(d) Nearly 50% of the interval of uncertainty is eliminated with each pair of experiments in the dichotomous search method.
(e) The number of experiments to be conducted is to be specified beforehand in both the Fibonacci and golden section methods.

PROBLEMS

5.1   Find the minimum of the function

$$f(x) = 0.65 - \frac{0.75}{1+x^2} - 0.65x \tan^{-1}\frac{1}{x}$$

using the following methods:

(a) Unrestricted search with a fixed step size of 0.1 from the starting point 0.0
(b) Unrestricted search with an accelerated step size using an initial step size of 0.1 and starting point of 0.0
(c) Exhaustive search method in the interval $(0, 3)$ to achieve an accuracy of within 5% of the exact value
(d) Dichotomous search method in the interval $(0, 3)$ to achieve an accuracy of within 5% of the exact value using a value of $\delta = 0.0001$
(e) Interval halving method in the interval $(0, 3)$ to achieve an accuracy of within 5% of the exact value

5.2   Find the minimum of the function given in Problem 5.1 using the quadratic interpolation method with an initial step size of 0.1.

5.3   Find the minimum of the function given in Problem 5.1 using the cubic interpolation method with an initial step size of $t_0 = 0.1$.

5.4   Plot the graph of the function $f(x)$ given in Problem 5.1 in the range $(0, 3)$ and identify its minimum.

5.5 The shear stress induced along the z-axis when two cylinders are in contact with each
other is given by

$$\frac{\tau_{zy}}{p_{\max}} = -\frac{1}{2}\left[-\frac{1}{\sqrt{1+\left(\frac{z}{b}\right)^2}} + \left\{ 2 - \frac{1}{1+\left(\frac{z}{b}\right)^2} \right\} \right.$$

$$\left. \times \sqrt{1+\left(\frac{z}{b}\right)^2} - 2\left(\frac{z}{b}\right) \right] \tag{1}$$

where $2b$ is the width of the contact area and $p_{\max}$ is the maximum pressure developed
at the center of the contact area (Fig. 5.21):

$$b = \left(\frac{2F}{\pi l} \frac{\dfrac{1-v_1^2}{E_1} + \dfrac{1-v_2^2}{E_2}}{\dfrac{1}{d_1} + \dfrac{1}{d_2}} \right)^{1/2} \tag{2}$$

$$p_{\max} = \frac{2F}{\pi b l} \tag{3}$$

F is the contact force; E_1 and E_2 are Young's moduli of the two cylinders; v_1 and v_2 are
Poisson's ratios of the two cylinders; d_1 and d_2 the diameters of the two cylinders, and l
the axial length of contact (length of the shorter cylinder). In many practical applications,

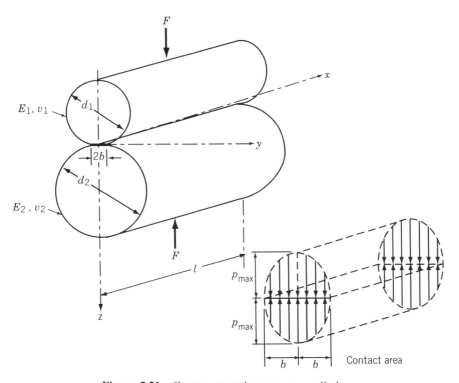

Figure 5.21 Contact stress between two cylinders.

such as roller bearings, when the contact load (F) is large, a crack originates at the point of maximum shear stress and propagates to the surface leading to a fatigue failure. To locate the origin of a crack, it is necessary to find the point at which the shear stress attains its maximum value. Show that the problem of finding the location of the maximum shear stress for $v_1 = v_2 = 0.3$ reduces to maximizing the function

$$f(\lambda) = \frac{0.5}{\sqrt{1 + \lambda^2}} - \sqrt{1 + \lambda^2} \left(1 - \frac{0.5}{1 + \lambda^2} \right) + \lambda \qquad (4)$$

where $f = \tau_{zy}/p_{max}$ and $\lambda = z/b$.

5.6 Plot the graph of the function $f(\lambda)$ given by Eq. (4) in Problem 5.5 in the range $(0, 3)$ and identify its maximum.

5.7 Find the maximum of the function given by Eq. (4) in Problem 5.5 using the following methods:

(a) Unrestricted search with a fixed step size of 0.1 from the starting point 0.0

(b) Unrestricted search with an accelerated step size using an initial step length of 0.1 and a starting point of 0.0

(c) Exhaustive search method in the interval $(0, 3)$ to achieve an accuracy of within 5% of the exact value

(d) Dichotomous search method in the interval $(0, 3)$ to achieve an accuracy of within 5% of the exact value using a value of $\delta = 0.0001$

(e) Interval halving method in the interval $(0, 3)$ to achieve an accuracy of within 5% of the exact value

5.8 Find the maximum of the function given by Eq. (4) in Problem 5.5 using the following methods:

(a) Fibonacci method with $n = 8$

(b) Golden section method with $n = 8$

5.9 Find the maximum of the function given by Eq. (4) in Problem 5.5 using the quadratic interpolation method with an initial step length of 0.1.

5.10 Find the maximum of the function given by Eq. (4) in Problem 5.5 using the cubic interpolation method with an initial step length of $t_0 = 0.1$.

5.11 Find the maximum of the function $f(\lambda)$ given by Eq. (4) in Problem 5.5 using the following methods:

(a) Newton method with the starting point 0.6

(b) Quasi-Newton method with the starting point 0.6 and a finite difference step size of 0.001

(c) Secant method with the starting point $\lambda_1 = 0.0$ and $t_0 = 0.1$

5.12 Prove that a convex function is unimodal.

5.13 Compare the ratios of intervals of uncertainty (L_n/L_0) obtainable in the following methods for $n = 2, 3, \ldots, 10$:

(a) Exhaustive search

(b) Dichotomous search with $\delta = 10^{-4}$

(c) Interval halving method

(d) Fibonacci method

(e) Golden section method

5.14 Find the number of experiments to be conducted in the following methods to obtain a value of $L_n/L_0 = 0.001$:

(a) Exhaustive search

(b) Dichotomous search with $\delta = 10^{-4}$

(c) Interval halving method

(d) Fibonacci method

(e) Golden section method

5.15 Find the value of x in the interval $(0, 1)$ which minimizes the function $f = x(x - 1.5)$ to within ± 0.05 by **(a)** the golden section method and **(b)** the Fibonacci method.

5.16 Find the minimum of the function $f = \lambda^5 - 5\lambda^3 - 20\lambda + 5$ by the following methods:

(a) Unrestricted search with a fixed step size of 0.1 starting from $\lambda = 0.0$

(b) Unrestricted search with accelerated step size from the initial point 0.0 with a starting step length of 0.1

(c) Exhaustive search in the interval $(0, 5)$

(d) Dichotomous search in the interval $(0, 5)$ with $\delta = 0.0001$

(e) Interval halving method in the interval $(0, 5)$

(f) Fibonacci search in the interval $(0, 5)$

(g) Golden section method in the interval $(0, 5)$

5.17 Find the minimum of the function $f = (\lambda/\log \lambda)$ by the following methods (take the initial trial step length as 0.1):

(a) Quadratic interpolation method

(b) Cubic interpolation method

5.18 Find the minimum of the function $f = \lambda/\log \lambda$ using the following methods:

(a) Newton method

(b) Quasi-Newton method

(c) Secant method

5.19 Consider the function

$$f = \frac{2x_1^2 + 2x_2^2 + 3x_3^2 - 2x_1x_2 - 2x_2x_3}{x_1^2 + x_2^2 + 2x_3^2}$$

Substitute $\mathbf{X} = \mathbf{X}_1 + \lambda\mathbf{S}$ into this function and derive an exact formula for the minimizing step length λ^*.

5.20 Minimize the function $f = x_1 - x_2 + 2x_1^2 + 2x_1x_2 + x_2^2$ starting from the point $\mathbf{X}_1 = \begin{Bmatrix} 0 \\ 0 \end{Bmatrix}$ along the direction $\mathbf{S} = \begin{Bmatrix} -1 \\ 0 \end{Bmatrix}$ using the quadratic interpolation method with an initial step length of 0.1.

5.21 Consider the problem

$$\text{Minimize } f(\mathbf{X}) = 100(x_2 - x_1^2)^2 + (1 - x_1)^2$$

and the starting point, $\mathbf{X}_1 = \left\{ {-1 \atop 1} \right\}$. Find the minimum of $f(\mathbf{X})$ along the direction, $\mathbf{S}_1 = \left\{ {4 \atop 0} \right\}$ using quadratic interpolation method. Use a maximum of two refits.

5.22 Solve Problem 5.21 using the cubic interpolation method. Use a maximum of two refits.

5.23 Solve Problem 5.21 using the direct root method. Use a maximum of two refits.

5.24 Solve Problem 5.21 using the Newton method. Use a maximum of two refits.

5.25 Solve Problem 5.21 using the Fibonacci method with $L_0 = (0, 0.1)$.

5.26 Write a computer program, in the form of a subroutine, to implement the Fibonacci method.

5.27 Write a computer program, in the form of a subroutine, to implement the golden section method.

5.28 Write a computer program, in the form of a subroutine, to implement the quadratic interpolation method.

5.29 Write a computer program, in the form of a subroutine, to implement the cubic interpolation method.

5.30 Write a computer program, in the form of a subroutine, to implement the secant method.

5.31 Find the maximum of the function given by Eq. (4) in Problem 5.5 using MATLAB. Assume the bounds on λ as 0 and 3.

5.32 Find the minimum of the function f(λ) given in Problem 5.16, in the range 0 and 5, using MATLAB.

5.33 Find the minimum of $f(x) = x(x - 1.5)$ in the interval $(0, 1)$ using MATLAB.

5.34 Find the minimum of the function $f(x) = \frac{x^3}{16} - \frac{27x}{4}$ in the range $(0, 10)$ using MATLAB.

5.35 Find the minimum of the function $f(x) = x^3 + x^2 - x - 2$ in the interval -4 and 4 using MATLAB.

5.36 Find the minimum of the function $f(x) = -\frac{1.5}{x} + \frac{6(10^{-6})}{x^9}$ in the interval -4 and 4 using MATLAB.

6

Nonlinear Programming II: Unconstrained Optimization Techniques

6.1 INTRODUCTION

This chapter deals with the various methods of solving the unconstrained minimization problem:

$$\text{Find } \mathbf{X} = \begin{Bmatrix} x_1 \\ x_2 \\ \vdots \\ x_n \end{Bmatrix} \text{ which minimizes } f(\mathbf{X}) \qquad (6.1)$$

It is true that rarely a practical design problem would be unconstrained; still, a study of this class of problems is important for the following reasons:

1. The constraints do not have significant influence in certain design problems.
2. Some of the powerful and robust methods of solving constrained minimization problems require the use of unconstrained minimization techniques.
3. The study of unconstrained minimization techniques provide the basic understanding necessary for the study of constrained minimization methods.
4. The unconstrained minimization methods can be used to solve certain complex engineering analysis problems. For example, the displacement response (linear or nonlinear) of any structure under any specified load condition can be found by minimizing its potential energy. Similarly, the eigenvalues and eigenvectors of any discrete system can be found by minimizing the Rayleigh quotient.

As discussed in Chapter 2, a point $\mathbf{X}^*$ will be a relative minimum of $f(\mathbf{X})$ if the necessary conditions

$$\frac{\partial f}{\partial x_i}(\mathbf{X} = \mathbf{X}^*) = 0, \quad i = 1, 2, \ldots, n \qquad (6.2)$$

are satisfied. The point $\mathbf{X}^*$ is guaranteed to be a relative minimum if the Hessian matrix is positive definite, that is,

$$\mathbf{J}_{\mathbf{X}^*} = [J]_{\mathbf{X}^*} = \left[\frac{\partial^2 f}{\partial x_i\, \partial x_j}(\mathbf{X}^*) \right] = \text{positive definite} \tag{6.3}$$

Equations (6.2) and (6.3) can be used to identify the optimum point during numerical computations. However, if the function is not differentiable, Eqs. (6.2) and (6.3) cannot be applied to identify the optimum point. For example, consider the function

$$f(x) = \begin{cases} ax & \text{for} \quad x \geq 0 \\ -bx & \text{for} \quad x \leq 0 \end{cases}$$

where $a > 0$ and $b > 0$. The graph of this function is shown in Fig. 6.1. It can be seen that this function is not differentiable at the minimum point, $x^* = 0$, and hence Eqs. (6.2) and (6.3) are not applicable in identifying x^*. In all such cases, the commonly understood notion of a minimum, namely, $f(\mathbf{X}^*) < f(\mathbf{X})$ for all $\mathbf{X}$, can be used only to identify a minimum point. The following example illustrates the formulation of a typical analysis problem as an unconstrained minimization problem.

Example 6.1   A cantilever beam is subjected to an end force P_0 and an end moment M_0 as shown in Fig. 6.2a. By using a one-finite-element model indicated in Fig. 6.2b, the transverse displacement, $w(x)$, can be expressed as [6.1]

$$w(x) = \{N_1(x) \quad N_2(x) \quad N_3(x) \quad N_4(x)\} \begin{Bmatrix} u_1 \\ u_2 \\ u_3 \\ u_4 \end{Bmatrix} \tag{E_1}$$

where $N_i(x)$ are called *shape functions* and are given by

$$N_1(x) = 2\alpha^3 - 3\alpha^2 + 1 \tag{E_2}$$

$$N_2(x) = (\alpha^3 - 2\alpha^2 + \alpha)l \tag{E_3}$$

$$N_3(x) = -2\alpha^3 + 3\alpha^2 \tag{E_4}$$

$$N_4(x) = (\alpha^3 - \alpha^2)l \tag{E_5}$$

$\alpha = x/l$, and u_1, u_2, u_3, and u_4 are the end displacements (or slopes) of the beam. The deflection of the beam at point A can be found by minimizing the potential energy

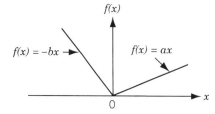

Figure 6.1   Function is not differentiable at minimum point.

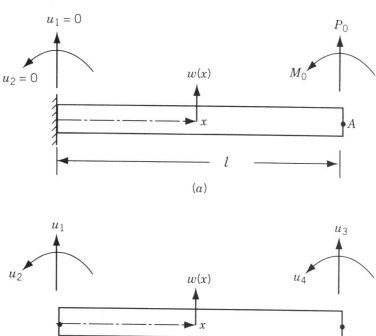

Figure 6.2   Finite-element model of a cantilever beam.

of the beam (F), which can be expressed as [6.1]

$$F = \frac{1}{2} \int_0^l EI \left(\frac{d^2w}{dx^2}\right)^2 dx - P_0 u_3 - M_0 u_4 \tag{E$_6$}$$

where E is Young's modulus and I is the area moment of inertia of the beam. Formulate the optimization problem in terms of the variables $x_1 = u_3$ and $x_2 = u_4 l$ for the case $P_0 l^3 / EI = 1$ and $M_0 l^2 / EI = 2$.

SOLUTION   Since the boundary conditions are given by $u_1 = u_2 = 0$, $w(x)$ can be expressed as

$$w(x) = (-2\alpha^3 + 3\alpha^2)u_3 + (\alpha^3 - \alpha^2)l u_4 \tag{E$_7$}$$

so that

$$\frac{d^2w}{dx^2} = \frac{6u_3}{l^2}(-2\alpha + 1) + \frac{2u_4}{l}(3\alpha - 1) \tag{E$_8$}$$

Equation (E_6) can be rewritten as

$$F = \frac{1}{2} \int_0^1 EI \left(\frac{d^2 w}{dx^2} \right)^2 l \, d\alpha - P_0 u_3 - M_0 u_4$$

$$= \frac{EIl}{2} \int_0^1 \left[\frac{6u_3}{l^2} (-2\alpha + 1) + \frac{2u_4}{l} (3\alpha - 1) \right]^2 d\alpha - P_0 u_3 - M_0 u_4$$

$$= \frac{EI}{l^3} (6u_3^2 + 2u_4^2 l^2 - 6u_3 u_4 l) - P_0 u_3 - M_0 u_4 \tag{E_9}$$

By using the relations $u_3 = x_1$, $u_4 l = x_2$, $P_0 l^3 / EI = 1$, and $M_0 l^2 / EI = 2$, and introducing the notation $f = Fl^3 / EI$, Eq. (E_9) can be expressed as

$$f = 6x_1^2 - 6x_1 x_2 + 2x_2^2 - x_1 - 2x_2 \tag{E_{10}}$$

Thus the optimization problem is to determine x_1 and x_2, which minimize the function f given by Eq. (E_{10}).

6.1.1 Classification of Unconstrained Minimization Methods

Several methods are available for solving an unconstrained minimization problem. These methods can be classified into two broad categories as direct search methods and descent methods as indicated in Table 6.1. The direct search methods require only the objective function values but not the partial derivatives of the function in finding the minimum and hence are often called the *nongradient methods*. The direct search methods are also known as *zeroth-order methods* since they use zeroth-order derivatives of the function. These methods are most suitable for simple problems involving a relatively small number of variables. These methods are, in general, less efficient than the descent methods. The descent techniques require, in addition to the function values, the first and in some cases the second derivatives of the objective function. Since more information about the function being minimized is used (through the use of derivatives), descent methods are generally more efficient than direct search techniques. The descent methods are known as *gradient methods*. Among the gradient methods,

Table 6.1   Unconstrained Minimization Methods

Direct search methods[a]	Descent methods[b]
Random search method	Steepest descent (Cauchy) method
Grid search method	Fletcher–Reeves method
Univariate method	Newton's method
Pattern search methods	Marquardt method
Powell's method	Quasi-Newton methods
	Davidon–Fletcher–Powell method
	Broyden–Fletcher–Goldfarb–Shanno method
Simplex method	

[a]Do not require the derivatives of the function.
[b]Require the derivatives of the function.

those requiring only first derivatives of the function are called *first-order methods*; those requiring both first and second derivatives of the function are termed *second-order methods*.

6.1.2 General Approach

All the unconstrained minimization methods are iterative in nature and hence they start from an initial trial solution and proceed toward the minimum point in a sequential manner as shown in Fig. 5.3. The iterative process is given by

$$\mathbf{X}_{i+1} = \mathbf{X}_i + \lambda_i^* \mathbf{S}_i \tag{6.4}$$

where $\mathbf{X}_i$ is the starting point, $\mathbf{S}_i$ is the search direction, λ_i^* is the optimal step length, and $\mathbf{X}_{i+1}$ is the final point in iteration i. It is important to note that all the unconstrained minimization methods (1) require an initial point $\mathbf{X}_1$ to start the iterative procedure, and (2) differ from one another only in the method of generating the new point $\mathbf{X}_{i+1}$ (from $\mathbf{X}_i$) and in testing the point $\mathbf{X}_{i+1}$ for optimality.

6.1.3 Rate of Convergence

Different iterative optimization methods have different rates of convergence. In general, an optimization method is said to have convergence of order p if [6.2]

$$\frac{\|\mathbf{X}_{i+1} - \mathbf{X}^*\|}{\|\mathbf{X}_i - \mathbf{X}^*\|^p} \le k, \quad k \ge 0, \ p \ge 1 \tag{6.5}$$

where $\mathbf{X}_i$ and $\mathbf{X}_{i+1}$ denote the points obtained at the end of iterations i and $i+1$, respectively, $\mathbf{X}^*$ represents the optimum point, and $\|\mathbf{X}\|$ denotes the length or norm of the vector $\mathbf{X}$:

$$\|\mathbf{X}\| = \sqrt{x_1^2 + x_2^2 + \cdots + x_n^2}$$

If $p = 1$ and $0 \le k \le 1$, the method is said to be *linearly convergent* (corresponds to slow convergence). If $p = 2$, the method is said to be *quadratically convergent* (corresponds to fast convergence). An optimization method is said to have *superlinear convergence* (corresponds to fast convergence) if

$$\lim_{i \to \infty} \frac{\|\mathbf{X}_{i+1} - \mathbf{X}^*\|}{\|\mathbf{X}_i - \mathbf{X}^*\|} \to 0 \tag{6.6}$$

The definitions of rates of convergence given in Eqs. (6.5) and (6.6) are applicable to single-variable as well as multivariable optimization problems. In the case of single-variable problems, the vector, $\mathbf{X}_i$, for example, degenerates to a scalar, x_i.

6.1.4 Scaling of Design Variables

The rate of convergence of most unconstrained minimization methods can be improved by scaling the design variables. For a quadratic objective function, the scaling of the design variables changes the condition number[†] of the Hessian matrix. When the condition number of the Hessian matrix is 1, the steepest descent method, for example, finds the minimum of a quadratic objective function in one iteration.

If $f = \frac{1}{2}\mathbf{X}^T[A]\mathbf{X}$ denotes a quadratic term, a transformation of the form

$$\mathbf{X} = [R]\mathbf{Y} \quad \text{or} \quad \begin{Bmatrix} x_1 \\ x_2 \end{Bmatrix} = \begin{bmatrix} r_{11} & r_{12} \\ r_{21} & r_{22} \end{bmatrix} \begin{Bmatrix} y_1 \\ y_2 \end{Bmatrix} \tag{6.7}$$

can be used to obtain a new quadratic term as

$$\frac{1}{2}\mathbf{Y}^T[\tilde{A}]\mathbf{Y} = \frac{1}{2}\mathbf{Y}^T[R]^T[A][R]\mathbf{Y} \tag{6.8}$$

The matrix $[R]$ can be selected to make $[\tilde{A}] = [R]^T[A][R]$ diagonal (i.e., to eliminate the mixed quadratic terms). For this, the columns of the matrix $[R]$ are to be chosen as the eigenvectors of the matrix $[A]$. Next the diagonal elements of the matrix $[\tilde{A}]$ can be reduced to 1 (so that the condition number of the resulting matrix will be 1) by using the transformation

$$\mathbf{Y} = [S]\mathbf{Z} \quad \text{or} \quad \begin{Bmatrix} y_1 \\ y_2 \end{Bmatrix} = \begin{bmatrix} s_{11} & 0 \\ 0 & s_{22} \end{bmatrix} \begin{Bmatrix} z_1 \\ z_2 \end{Bmatrix} \tag{6.9}$$

where the matrix $[S]$ is given by

$$[S] = \begin{bmatrix} s_{11} = \dfrac{1}{\sqrt{\tilde{a}_{11}}} & 0 \\ 0 & s_{22} = \dfrac{1}{\sqrt{\tilde{a}_{22}}} \end{bmatrix} \tag{6.10}$$

Thus the complete transformation that reduces the Hessian matrix of f to an identity matrix is given by

$$\mathbf{X} = [R][S]\mathbf{Z} \equiv [T]\mathbf{Z} \tag{6.11}$$

so that the quadratic term $\frac{1}{2}\mathbf{X}^T[A]\mathbf{X}$ reduces to $\frac{1}{2}\mathbf{Z}^T[I]\mathbf{Z}$.

If the objective function is not a quadratic, the Hessian matrix and hence the transformations vary with the design vector from iteration to iteration. For example,

---

[†]The condition number of an $n \times n$ matrix, $[A]$, is defined as

$$\text{cond}([A]) = \|[A]\| \; \|[A]^{-1}\| \geq 1$$

where $\|[A]\|$ denotes a norm of the matrix $[A]$. For example, the infinite norm of $[A]$ is defined as the maximum row sum given by

$$\|[A]\|_\infty = \max_{1 \leq i \leq n} \sum_{j=1}^{n} |a_{ij}|$$

If the condition number is close to 1, the round-off errors are expected to be small in dealing with the matrix $[A]$. For example, if cond$[A]$ is large, the solution vector $\mathbf{X}$ of the system of equations $[A]\mathbf{X} = \mathbf{B}$ is expected to be very sensitive to small variations in $[A]$ and $\mathbf{B}$. If cond$[A]$ is close to 1, the matrix $[A]$ is said to be *well behaved* or *well conditioned*. On the other hand, if cond$[A]$ is significantly greater than 1, the matrix $[A]$ is said to be *not well behaved* or *ill conditioned*.

the second-order Taylor's series approximation of a general nonlinear function at the design vector $\mathbf{X}_i$ can be expressed as

$$f(\mathbf{X}) = c + \mathbf{B}^\mathrm{T}\mathbf{X} + \tfrac{1}{2}\mathbf{X}^\mathrm{T}[A]\mathbf{X} \tag{6.12}$$

where

$$c = f(\mathbf{X}_i) \tag{6.13}$$

$$B = \left\{ \begin{array}{c} \dfrac{\partial f}{\partial x_1}\bigg|_{\mathbf{X}_i} \\ \vdots \\ \dfrac{\partial f}{\partial x_n}\bigg|_{\mathbf{X}_i} \end{array} \right\} \tag{6.14}$$

$$[A] = \begin{bmatrix} \dfrac{\partial^2 f}{\partial x_1^2}\bigg|_{\mathbf{X}_i} & \cdots & \dfrac{\partial^2 f}{\partial x_1 \partial x_n}\bigg|_{\mathbf{X}_i} \\ \vdots & & \vdots \\ \dfrac{\partial^2 f}{\partial x_n \partial x_1}\bigg|_{\mathbf{X}_i} & \cdots & \dfrac{\partial^2 f}{\partial x_n^2}\bigg|_{\mathbf{X}_i} \end{bmatrix} \tag{6.15}$$

The transformations indicated by Eqs. (6.7) and (6.9) can be applied to the matrix $[A]$ given by Eq. (6.15). The procedure of scaling the design variables is illustrated with the following example.

Example 6.2   Find a suitable scaling (or transformation) of variables to reduce the condition number of the Hessian matrix of the following function to 1:

$$f(x_1, x_2) = 6x_1^2 - 6x_1 x_2 + 2x_2^2 - x_1 - 2x_2 \tag{E_1}$$

SOLUTION   The quadratic function can be expressed as

$$f(\mathbf{X}) = \mathbf{B}^\mathrm{T}\mathbf{X} + \tfrac{1}{2}\mathbf{X}^\mathrm{T}[A]\mathbf{X} \tag{E_2}$$

where

$$\mathbf{X} = \begin{Bmatrix} x_1 \\ x_2 \end{Bmatrix}, \quad \mathbf{B} = \begin{Bmatrix} -1 \\ -2 \end{Bmatrix}, \quad \text{and} \quad [A] = \begin{bmatrix} 12 & -6 \\ -6 & 4 \end{bmatrix}$$

As indicated above, the desired scaling of variables can be accomplished in two stages.

Stage 1: Reducing [A] to a Diagonal Form, [Ã]

The eigenvectors of the matrix $[A]$ can be found by solving the eigenvalue problem

$$[[A] - \lambda_i[I]]\, \mathbf{u}_i = \mathbf{0} \tag{E_3}$$

where λ_i is the ith eigenvalue and $\mathbf{u}_i$ is the corresponding eigenvector. In the present case, the eigenvalues, λ_i, are given by

$$\begin{vmatrix} 12 - \lambda_i & -6 \\ -6 & 4 - \lambda_i \end{vmatrix} = \lambda_i^2 - 16\lambda_i + 12 = 0 \tag{E_4}$$

which yield $\lambda_1 = 8 + \sqrt{52} = 15.2111$ and $\lambda_2 = 8 - \sqrt{52} = 0.7889$. The eigenvector $\mathbf{u}_i$ corresponding to λ_i can be found by solving Eq. (E_3):

$$\begin{bmatrix} 12 - \lambda_1 & -6 \\ -6 & 4 - \lambda_1 \end{bmatrix} \begin{Bmatrix} u_{11} \\ u_{21} \end{Bmatrix} = \begin{Bmatrix} 0 \\ 0 \end{Bmatrix} \quad \text{or} \quad (12 - \lambda_1)u_{11} - 6u_{21} = 0$$

$$\text{or} \quad u_{21} = -0.5332u_{11}$$

that is,

$$\mathbf{u}_1 = \begin{Bmatrix} u_{11} \\ u_{21} \end{Bmatrix} = \begin{Bmatrix} 1.0 \\ -0.5332 \end{Bmatrix}$$

and

$$\begin{bmatrix} 12 - \lambda_2 & -6 \\ -6 & 4 - \lambda_2 \end{bmatrix} \begin{Bmatrix} u_{12} \\ u_{22} \end{Bmatrix} = \begin{Bmatrix} 0 \\ 0 \end{Bmatrix} \quad \text{or} \quad (12 - \lambda_2)u_{12} - 6u_{22} = 0$$

$$\text{or} \quad u_{22} = 1.8685u_{12}$$

that is,

$$\mathbf{u}_2 = \begin{Bmatrix} u_{12} \\ u_{22} \end{Bmatrix} = \begin{Bmatrix} 1.0 \\ 1.8685 \end{Bmatrix}$$

Thus the transformation that reduces $[A]$ to a diagonal form is given by

$$\mathbf{X} = [R]\mathbf{Y} = [\mathbf{u}_1 \quad \mathbf{u}_2]\mathbf{Y} = \begin{bmatrix} 1 & 1 \\ -0.5352 & 1.8685 \end{bmatrix} \begin{Bmatrix} y_1 \\ y_2 \end{Bmatrix} \tag{E_5}$$

that is,

$$x_1 = y_1 + y_2$$
$$x_2 = -0.5352y_1 + 1.8685y_2$$

This yields the new quadratic term as $\frac{1}{2}\mathbf{Y}^T[\tilde{A}]\mathbf{Y}$, where

$$[\tilde{A}] = [R]^T[A][R] = \begin{bmatrix} 19.5682 & 0.0 \\ 0.0 & 3.5432 \end{bmatrix}$$

and hence the quadratic function becomes

$$f(y_1, y_2) = \mathbf{B}^T[R]\mathbf{Y} + \frac{1}{2}\mathbf{Y}^T[\tilde{A}]\mathbf{Y}$$

$$= 0.0704y_1 - 4.7370y_2 + \frac{1}{2}(19.8682)y_1^2 + \frac{1}{2}(3.5432)y_2^2 \tag{E_6}$$

Stage 2: Reducing [Ã] to a Unit Matrix

The transformation is given by $\mathbf{Y} = [S]\mathbf{Z}$, where

$$[S] = \begin{bmatrix} \dfrac{1}{\sqrt{19.5682}} & 0 \\ 0 & \dfrac{1}{\sqrt{3.5432}} \end{bmatrix} = \begin{bmatrix} 0.2262 & 0.0 \\ 0.0 & 0.5313 \end{bmatrix}$$

Stage 3: Complete Transformation

The total transformation is given by

$$\mathbf{X} = [R]\mathbf{Y} = [R][S]\mathbf{Z} = [T]\mathbf{Z} \tag{E_7}$$

where

$$\begin{aligned} [T] = [R][S] &= \begin{bmatrix} 1 & 1 \\ -0.5352 & 1.8685 \end{bmatrix} \begin{bmatrix} 0.2262 & 0 \\ 0 & 0.5313 \end{bmatrix} \\ &= \begin{bmatrix} 0.2262 & 0.5313 \\ -0.1211 & 0.9927 \end{bmatrix} \end{aligned} \tag{E_8}$$

or

$$x_1 = 0.2262z_1 + 0.5313z_2$$

$$x_2 = -0.1211z_1 + 0.9927z_2$$

With this transformation, the quadratic function of Eq. (E_1) becomes

$$\begin{aligned} f(z_1, z_2) &= \mathbf{B}^{\mathrm{T}}[T]\mathbf{Z} + \tfrac{1}{2}\mathbf{Z}^{\mathrm{T}}[T]^{\mathrm{T}}[A][T]\mathbf{Z} \\ &= 0.0160z_1 - 2.5167z_2 + \tfrac{1}{2}z_1^2 + \tfrac{1}{2}z_2^2 \end{aligned} \tag{E_9}$$

The contours of the quadratic functions given by Eqs. (E_1), (E_6), and (E_9) are shown in Fig. 6.3*a*, *b*, and *c*, respectively.

Direct Search Methods

6.2   RANDOM SEARCH METHODS

Random search methods are based on the use of random numbers in finding the minimum point. Since most of the computer libraries have random number generators, these methods can be used quite conveniently. Some of the best known random search methods are presented in this section.

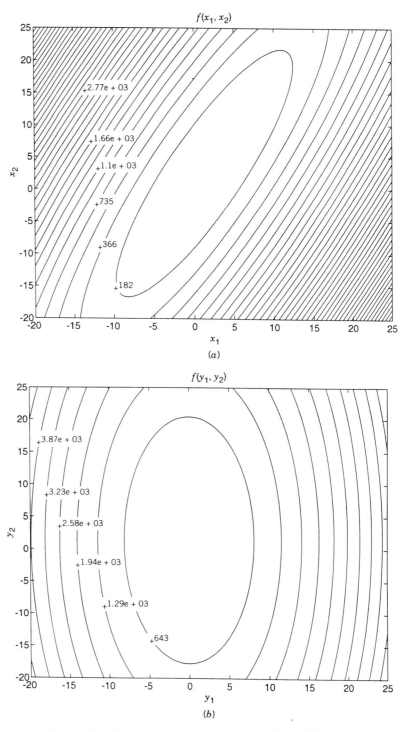

Figure 6.3   Contours of the original and transformed functions.

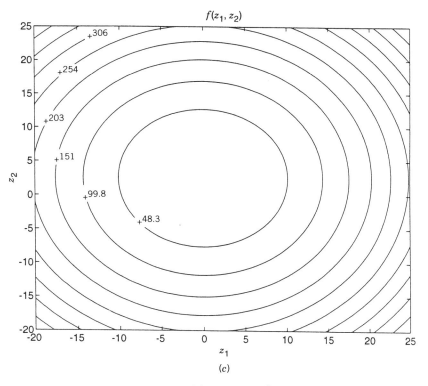

$f(z_1, z_2)$

Figure 6.3   (*continued*).

6.2.1   Random Jumping Method

Although the problem is an unconstrained one, we establish the bounds l_i and u_i for each design variable $x_i, i = 1, 2, \ldots, n$, for generating the random values of x_i:

$$l_i \leq x_i \leq u_i, \quad i = 1, 2, \ldots, n \tag{6.16}$$

In the random jumping method, we generate sets of n random numbers, $(r_1, r_2, \ldots, r_n)$, that are uniformly distributed between 0 and 1. Each set of these numbers is used to find a point, $\mathbf{X}$, inside the hypercube defined by Eqs. (6.16) as

$$\mathbf{X} = \begin{Bmatrix} x_1 \\ x_2 \\ \vdots \\ x_n \end{Bmatrix} = \begin{Bmatrix} l_1 + r_1(u_1 - l_1) \\ l_2 + r_2(u_2 - l_2) \\ \vdots \\ l_n + r_n(u_n - l_n) \end{Bmatrix} \tag{6.17}$$

and the value of the function is evaluated at this point $\mathbf{X}$. By generating a large number of random points $\mathbf{X}$ and evaluating the value of the objective function at each of these points, we can take the smallest value of $f(\mathbf{X})$ as the desired minimum point.

6.2.2   Random Walk Method

The *random walk method* is based on generating a sequence of improved approximations to the minimum, each derived from the preceding approximation. Thus if $\mathbf{X}_i$ is the approximation to the minimum obtained in the $(i-1)$th stage (or step or iteration), the new or improved approximation in the ith stage is found from the relation

$$\mathbf{X}_{i+1} = \mathbf{X}_i + \lambda \mathbf{u}_i \qquad (6.18)$$

where λ is a prescribed scalar step length and $\mathbf{u}_i$ is a unit random vector generated in the ith stage. The detailed procedure of this method is given by the following steps [6.3]:

1. Start with an initial point $\mathbf{X}_1$, a sufficiently large initial step length λ, a minimum allowable step length ε, and a maximum permissible number of iterations N.
2. Find the function value $f_1 = f(\mathbf{X}_1)$.
3. Set the iteration number as $i = 1$.
4. Generate a set of n random numbers $r_1, r_2, \ldots, r_n$ each lying in the interval $[-1, 1]$ and formulate the unit vector $\mathbf{u}$ as

$$\mathbf{u} = \frac{1}{(r_1^2 + r_2^2 + \cdots + r_n^2)^{1/2}} \begin{Bmatrix} r_1 \\ r_2 \\ \vdots \\ r_n \end{Bmatrix} \qquad (6.19)$$

   The directions generated using Eq. (6.19) are expected to have a bias toward the diagonals of the unit hypercube [6.3]. To avoid such a bias, the length of the vector, R, is computed as

$$R = (r_1^2 + r_2^2 + \cdots + r_n^2)^{1/2}$$

   and the random numbers generated $(r_1, r_2, \ldots, r_n)$ are accepted only if $R \leq 1$ but are discarded if $R > 1$. If the random numbers are accepted, the unbiased random vector $\mathbf{u}_i$ is given by Eq. (6.19).
5. Compute the new vector and the corresponding function value as $\mathbf{X} = \mathbf{X}_1 + \lambda \mathbf{u}$ and $f = f(\mathbf{X})$.
6. Compare the values of f and f_1. If $f < f_1$, set the new values as $\mathbf{X}_1 = \mathbf{X}$ and $f_1 = f$ and go to step 3. If $f \geq f_1$, go to step 7.
7. If $i \leq N$, set the new iteration number as $i = i + 1$ and go to step 4. On the other hand, if $i > N$, go to step 8.
8. Compute the new, reduced, step length as $\lambda = \lambda/2$. If the new step length is smaller than or equal to ε, go to step 9. Otherwise (i.e., if the new step length is greater than ε), go to step 4.
9. Stop the procedure by taking $\mathbf{X}_{\text{opt}} \approx \mathbf{X}_1$ and $f_{\text{opt}} \approx f_1$.

This method is illustrated with the following example.

Example 6.3   Minimize $f(x_1, x_2) = x_1 - x_2 + 2x_1^2 + 2x_1 x_2 + x_2^2$ using random walk method from the point $\mathbf{X}_1 = \begin{Bmatrix} 0.0 \\ 0.0 \end{Bmatrix}$ with a starting step length of $\lambda = 1.0$. Take $\varepsilon = 0.05$ and $N = 100$.

Table 6.2 Minimization of f by Random Walk Method

Step length, λ	Number of trials required[a]	Components of $\mathbf{X}_1 + \lambda \mathbf{u}$ 1	Components of $\mathbf{X}_1 + \lambda \mathbf{u}$ 2	Current objective function value, $f_1 = f(\mathbf{X}_1 + \lambda \mathbf{u})$
1.0	1	−0.93696	0.34943	−0.06329
1.0	2	−1.15271	1.32588	−1.11986
		Next 100 trials did not reduce the function value.		
0.5	1	−1.34361	1.78800	−1.12884
0.5	3	−1.07318	1.36744	−1.20232
		Next 100 trials did not reduce the function value.		
0.25	4	−0.86419	1.23025	−1.21362
0.25	2	−0.86955	1.48019	−1.22074
0.25	8	−1.10661	1.55958	−1.23642
0.25	30	−0.94278	1.37074	−1.24154
0.25	6	−1.08729	1.57474	−1.24222
0.25	50	−0.92606	1.38368	−1.24274
0.25	23	−1.07912	1.58135	−1.24374
		Next 100 trials did not reduce the function value.		
0.125	1	−0.97986	1.50538	−1.24894
		Next 100 trials did not reduce the function value.		
0.0625	100 trials did not reduce the function value.			
0.03125	As this step length is smaller than ϵ, the program is terminated.			

[a]Out of the directions generated that satisfy $R \le 1$, number of trials required to find a direction that also reduces the value of f.

SOLUTION   The results are summarized in Table 6.2, where only the trials that produced an improvement are shown.

6.2.3 Random Walk Method with Direction Exploitation

In the random walk method described in Section 6.2.2, we proceed to generate a new unit random vector $\mathbf{u}_{i+1}$ as soon as we find that $\mathbf{u}_i$ is successful in reducing the function value for a fixed step length λ. However, we can expect to achieve a further decrease in the function value by taking a longer step length along the direction $\mathbf{u}_i$. Thus the random walk method can be improved if the maximum possible step is taken along each successful direction. This can be achieved by using any of the one-dimensional minimization methods discussed in Chapter 5. According to this procedure, the new vector $\mathbf{X}_{i+1}$ is found as

$$\mathbf{X}_{i+1} = \mathbf{X}_i + \lambda_i^* \mathbf{u}_i \tag{6.20}$$

where λ_i^* is the optimal step length found along the direction $\mathbf{u}_i$ so that

$$f_{i+1} = f(\mathbf{X}_i + \lambda_i^* \mathbf{u}_i) = \min_{\lambda_i} f(\mathbf{X}_i + \lambda_i \mathbf{u}_i) \tag{6.21}$$

The search method incorporating this feature is called the *random walk method with direction exploitation*.

6.2.4 Advantages of Random Search Methods

1. These methods can work even if the objective function is discontinuous and nondifferentiable at some of the points.

2. The random methods can be used to find the global minimum when the objective function possesses several relative minima.

3. These methods are applicable when other methods fail due to local difficulties such as sharply varying functions and shallow regions.

4. Although the random methods are not very efficient by themselves, they can be used in the early stages of optimization to detect the region where the global minimum is likely to be found. Once this region is found, some of the more efficient techniques can be used to find the precise location of the global minimum point.

6.3 GRID SEARCH METHOD

This method involves setting up a suitable grid in the design space, evaluating the objective function at all the gird points, and finding the grid point corresponding to the lowest function value. For example, if the lower and upper bounds on the ith design variable are known to be l_i and u_i, respectively, we can divide the range (l_i, u_i) into $p_i - 1$ equal parts so that $x_i^{(1)}, x_i^{(2)}, \ldots, x_i^{(pi)}$ denote the grid points along the x_i axis $(i = 1, 2, \ldots, n)$. This leads to a total of $p_1 p_2 \cdots p_n$ grid points in the design space. A grid with $p_i = 4$ is shown in a two-dimensional design space in Fig. 6.4. The grid points can also be chosen based on methods of experimental design [6.4, 6.5]. It can be seen that the grid method requires prohibitively large number of function evaluations in most practical problems. For example, for a problem with 10 design

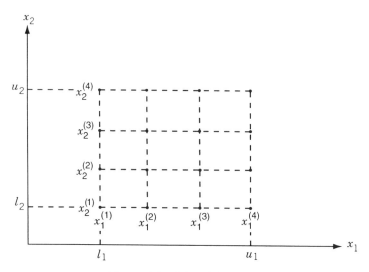

Figure 6.4 Grid with $p_i = 4$.

variables ($n = 10$), the number of grid points will be $3^{10} = 59,049$ with $p_i = 3$ and $4^{10} = 1,048,576$ with $p_i = 4$. However, for problems with a small number of design variables, the grid method can be used conveniently to find an approximate minimum. Also, the grid method can be used to find a good starting point for one of the more efficient methods.

6.4   UNIVARIATE METHOD

In this method we change only one variable at a time and seek to produce a sequence of improved approximations to the minimum point. By starting at a base point $\mathbf{X}_i$ in the ith iteration, we fix the values of $n - 1$ variables and vary the remaining variable. Since only one variable is changed, the problem becomes a one-dimensional minimization problem and any of the methods discussed in Chapter 5 can be used to produce a new base point $\mathbf{X}_{i+1}$. The search is now continued in a new direction. This new direction is obtained by changing any one of the $n - 1$ variables that were fixed in the previous iteration. In fact, the search procedure is continued by taking each coordinate direction in turn. After all the n directions are searched sequentially, the first cycle is complete and hence we repeat the entire process of sequential minimization. The procedure is continued until no further improvement is possible in the objective function in any of the n directions of a cycle. The univariate method can be summarized as follows:

1. Choose an arbitrary staring point $\mathbf{X}_1$ and set $i = 1$.
2. Find the search direction $\mathbf{S}_i$ as

$$\mathbf{S}_i^{\mathrm{T}} = \begin{cases} (1, 0, 0, \ldots, 0) & \text{for} \quad i = 1, n+1, 2n+1, \ldots \\ (1, 0, 0, \ldots, 0) & \text{for} \quad i = 2, n+2, 2n+2, \ldots \\ (0, 0, 1, \ldots, 0) & \text{for} \quad i = 3, n+3, 2n+3, \ldots \\ \quad \vdots & \\ (0, 0, 0, \ldots, 1) & \text{for} \quad i = n, 2n, 3n, \ldots \end{cases} \tag{6.22}$$

3. Determine whether λ_i should be positive or negative. For the current direction $\mathbf{S}_i$, this means find whether the function value decreases in the positive or negative direction. For this we take a small probe length (ε) and evaluate $f_i = f(\mathbf{X}_i)$, $f^+ = f(\mathbf{X}_i + \varepsilon \mathbf{S}_i)$, and $f^- = f(\mathbf{X}_i - \varepsilon \mathbf{S}_i)$. If $f^+ < f_i$, $\mathbf{S}_i$ will be the correct direction for decreasing the value of f and if $f^- < f_i$, $-\mathbf{S}_i$ will be the correct one. If both f^+ and f^- are greater than f_i, we take $\mathbf{X}_i$ as the minimum along the direction $\mathbf{S}_i$.
4. Find the optimal step length λ_i^* such that

$$f(\mathbf{X}_i \pm \lambda_i^* \mathbf{S}_i) = \min_{\lambda_i}(\mathbf{X}_i \pm \lambda_i \mathbf{S}_i) \tag{6.23}$$

where $+$ or $-$ sign has to be used depending upon whether $\mathbf{S}_i$ or $-\mathbf{S}_i$ is the direction for decreasing the function value.
5. Set $\mathbf{X}_{i+1} = \mathbf{X}_i \pm \lambda_i^* \mathbf{S}_i$ depending on the direction for decreasing the function value, and $f_{i+1} = f(\mathbf{X}_{i+1})$.
6. Set the new value of $i = i + 1$ and go to step 2. Continue this procedure until no significant change is achieved in the value of the objective function.

The univariate method is very simple and can be implemented easily. However, it will not converge rapidly to the optimum solution, as it has a tendency to oscillate with steadily decreasing progress toward the optimum. Hence it will be better to stop the computations at some point near to the optimum point rather than trying to find the precise optimum point. In theory, the univariate method can be applied to find the minimum of any function that possesses continuous derivatives. However, if the function has a steep valley, the method may not even converge. For example, consider the contours of a function of two variables with a valley as shown in Fig. 6.5. If the univariate search starts at point P, the function value cannot be decreased either in the direction $\pm\mathbf{S}_1$ or in the direction $\pm\mathbf{S}_2$. Thus the search comes to a halt and one may be misled to take the point P, which is certainly not the optimum point, as the optimum point. This situation arises whenever the value of the probe length ε needed for detecting the proper direction ($\pm\mathbf{S}_1$ or $\pm\mathbf{S}_2$) happens to be less than the number of significant figures used in the computations.

Example 6.4   Minimize $f(x_1, x_2) = x_1 - x_2 + 2x_1^2 + 2x_1x_2 + x_2^2$ with the starting point $(0, 0)$.

SOLUTION   We will take the probe length (ε) as 0.01 to find the correct direction for decreasing the function value in step 3. Further, we will use the differential calculus method to find the optimum step length λ_i^* along the direction $\pm\mathbf{S}_i$ in step 4.

Iteration i = 1

Step 2: Choose the search direction $\mathbf{S}_1$ as $\mathbf{S}_1 = \left\{ \begin{smallmatrix} 1 \\ 0 \end{smallmatrix} \right\}$.

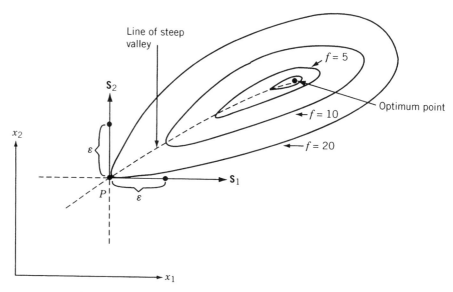

Figure 6.5   Failure of the univariate method on a steep valley.

Step 3: To find whether the value of f decreases along $\mathbf{S}_1$ or $-\mathbf{S}_1$, we use the probe
length ε. Since

$$f_1 = f(\mathbf{X}_1) = f(0, 0) = 0,$$

$$f^+ = f(\mathbf{X}_1 + \varepsilon\mathbf{S}_1) = f(\varepsilon, 0) = 0.01 - 0 + 2(0.0001)$$

$$+ 0 + 0 = 0.0102 > f_1$$

$$f^- = f(\mathbf{X}_1 - \varepsilon\mathbf{S}_1) = f(-\varepsilon, 0) = -0.01 - 0 + 2(0.0001)$$

$$+ 0 + 0 = -0.0098 < f_1,$$

$-\mathbf{S}_1$ is the correct direction for minimizing f from $\mathbf{X}_1$.

Step 4: To find the optimum step length λ_1^*, we minimize

$$f(\mathbf{X}_1 - \lambda_1\mathbf{S}_1) = f(-\lambda_1, 0)$$

$$= (-\lambda_1) - 0 + 2(-\lambda_1)^2 + 0 + 0 = 2\lambda_1^2 - \lambda_1$$

As $df/d\lambda_1 = 0$ at $\lambda_1 = \frac{1}{4}$, we have $\lambda_1^* = \frac{1}{4}$.

Step 5: Set

$$\mathbf{X}_2 = \mathbf{X}_1 - \lambda_1^*\mathbf{S}_1 = \begin{Bmatrix} 0 \\ 0 \end{Bmatrix} - \frac{1}{4}\begin{Bmatrix} 1 \\ 0 \end{Bmatrix} = \begin{Bmatrix} -\frac{1}{4} \\ 0 \end{Bmatrix}$$

$$f_2 = f(\mathbf{X}_2) = f(-\tfrac{1}{4}, 0) = -\tfrac{1}{8}.$$

Iteration $i = 2$

Step 2: Choose the search direction $\mathbf{S}_2$ as $\mathbf{S}_2 = \begin{Bmatrix} 0 \\ 1 \end{Bmatrix}$.

Step 3: Since $f_2 = f(\mathbf{X}_2) = -0.125$,

$$f^+ = f(\mathbf{X}_2 + \varepsilon\mathbf{S}_2) = f(-0.25, 0.01) = -0.1399 < f_2$$

$$f^- = f(\mathbf{X}_2 + \varepsilon\mathbf{S}_2) = f(-0.25, -0.01) = -0.1099 > f_2$$

$\mathbf{S}_2$ is the correct direction for decreasing the value of f from $\mathbf{X}_2$.

Step 4: We minimize $f(\mathbf{X}_2 + \lambda_2\mathbf{S}_2)$ to find λ_2^*.
Here

$$f(\mathbf{X}_2 + \lambda_2\mathbf{S}_2) = f(-0.25, \lambda_2)$$

$$= -0.25 - \lambda_2 + 2(0.25)^2 - 2(0.25)(\lambda_2) + \lambda_2^2$$

$$= \lambda_2^2 - 1.5\lambda_2 - 0.125$$

$$\frac{df}{d\lambda_2} = 2\lambda_2 - 1.5 = 0 \quad \text{at} \quad \lambda_2^* = 0.75$$

Step 5: Set

$$\mathbf{X}_3 = \mathbf{X}_2 + \lambda_2^*\mathbf{S}_2 = \begin{Bmatrix} -0.25 \\ 0 \end{Bmatrix} + 0.75\begin{Bmatrix} 0 \\ 1 \end{Bmatrix} = \begin{Bmatrix} -0.25 \\ 0.75 \end{Bmatrix}$$

$$f_3 = f(\mathbf{X}_3) = -0.6875$$

Next we set the iteration number as $i = 3$, and continue the procedure until the optimum solution $\mathbf{X}^* = \begin{Bmatrix} -1.0 \\ 1.5 \end{Bmatrix}$ with $f(\mathbf{X}^*) = -1.25$ is found.

Note: If the method is to be computerized, a suitable convergence criterion has to be used to test the point $\mathbf{X}_{i+1} (i = 1, 2, \ldots)$ for optimality.

6.5 PATTERN DIRECTIONS

In the univariate method, we search for the minimum along directions parallel to the coordinate axes. We noticed that this method may not converge in some cases, and that even if it converges, its convergence will be very slow as we approach the optimum point. These problems can be avoided by changing the directions of search in a favorable manner instead of retaining them always parallel to the coordinate axes. To understand this idea, consider the contours of the function shown in Fig. 6.6. Let the points $1, 2, 3, \ldots$ indicate the successive points found by the univariate method. It can be noticed that the lines joining the alternate points of the search (e.g., 1, 3; 2, 4; 3, 5; 4, 6; ...) lie in the general direction of the minimum and are known as *pattern directions*. It can be proved that if the objective function is a quadratic in two variables, all such lines pass through the minimum. Unfortunately, this property will not be valid for multivariable functions even when they are quadratics. However, this idea can still be used to achieve rapid convergence while finding the minimum of an n-variable function. Methods that use pattern directions as search directions are known as *pattern search methods*.

One of the best-known pattern search methods, the Powell's method, is discussed in Section 6.6. In general, a pattern search method takes n univariate steps, where n

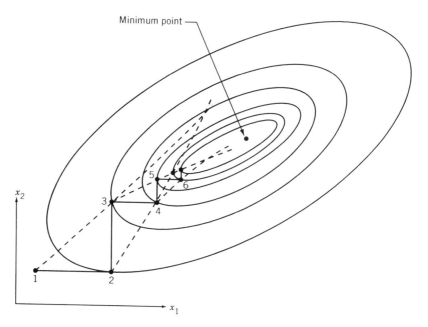

Figure 6.6   Lines defined by the alternate points lie in the general direction of the minimum.

denotes the number of design variables and then searches for the minimum along the pattern direction $\mathbf{S}_i$, defined by

$$\mathbf{S}_i = \mathbf{X}_i - \mathbf{X}_{i-n} \tag{6.24}$$

where $\mathbf{X}_i$ is the point obtained at the end of n univariate steps and $\mathbf{X}_{i-n}$ is the starting point before taking the n univariate steps. In general, the directions used prior to taking a move along a pattern direction need not be univariate directions.

6.6   POWELL'S METHOD

Powell's method is an extension of the basic pattern search method. It is the most widely used direct search method and can be proved to be a method of conjugate directions [6.7]. A conjugate directions method will minimize a quadratic function in a finite number of steps. Since a general nonlinear function can be approximated reasonably well by a quadratic function near its minimum, a conjugate directions method is expected to speed up the convergence of even general nonlinear objective functions. The definition, a method of generation of conjugate directions, and the property of quadratic convergence are presented in this section.

6.6.1   Conjugate Directions

Definition: Conjugate Directions.   Let $\mathbf{A} = [A]$ be an $n \times n$ symmetric matrix. A set of n vectors (or directions) $\{\mathbf{S}_i\}$ is said to be conjugate (more accurately $\mathbf{A}$-conjugate) if

$$\mathbf{S}_i^T \mathbf{A} \mathbf{S}_j = 0 \quad \text{for} \quad \text{all } i \neq j, \quad i = 1, 2, \ldots, n, \quad j = 1, 2, \ldots, n \tag{6.25}$$

It can be seen that *orthogonal directions* are a special case of conjugate directions (obtained with $[A] = [I]$ in Eq. (6.25)).

Definition: Quadratically Convergent Method.   If a minimization method, using exact arithmetic, can find the minimum point in n steps while minimizing a quadratic function in n variables, the method is called a *quadratically convergent method*.

Theorem 6.1 Given a quadratic function of n variables and two parallel hyperplanes 1 and 2 of dimension $k < n$. Let the constrained stationary points of the quadratic function in the hyperplanes be $\mathbf{X}_1$ and $\mathbf{X}_2$, respectively. Then the line joining $\mathbf{X}_1$ and $\mathbf{X}_2$ is conjugate to any line parallel to the hyperplanes.

Proof: Let the quadratic function be expressed as

$$Q(\mathbf{X}) = \tfrac{1}{2}\mathbf{X}^T \mathbf{A} \mathbf{X} + \mathbf{B}^T \mathbf{X} + C \tag{6.26}$$

The gradient of Q is given by

$$\nabla Q(\mathbf{X}) = \mathbf{A}\mathbf{X} + \mathbf{B}$$

and hence

$$\nabla Q(\mathbf{X}_1) - \nabla Q(\mathbf{X}_2) = \mathbf{A}(\mathbf{X}_1 - \mathbf{X}_2) \tag{6.27}$$

If $\mathbf{S}$ is any vector parallel to the hyperplanes, it must be orthogonal to the gradients $\nabla Q(\mathbf{X}_1)$ and $\nabla Q(\mathbf{X}_2)$. Thus

$$\mathbf{S}^\mathrm{T}\nabla Q(\mathbf{X}_1) = \mathbf{S}^\mathrm{T}\mathbf{A}\mathbf{X}_1 + \mathbf{S}^\mathrm{T}\mathbf{B} = 0 \tag{6.28}$$

$$\mathbf{S}^\mathrm{T}\nabla Q(\mathbf{X}_2) = \mathbf{S}^\mathrm{T}\mathbf{A}\mathbf{X}_2 + \mathbf{S}^\mathrm{T}\mathbf{B} = 0 \tag{6.29}$$

By subtracting Eq. (6.29) from Eq. (6.28), we obtain

$$\mathbf{S}^T\mathbf{A}(\mathbf{X}_1 - \mathbf{X}_2) = \mathbf{0} \tag{6.30}$$

Hence $\mathbf{S}$ and $(\mathbf{X}_1 - \mathbf{X}_2)$ are $\mathbf{A}$-conjugate.

   The meaning of this theorem is illustrated in a two-dimensional space in Fig. 6.7. If $\mathbf{X}_1$ and $\mathbf{X}_2$ are the minima of Q obtained by searching along the direction $\mathbf{S}$ from two

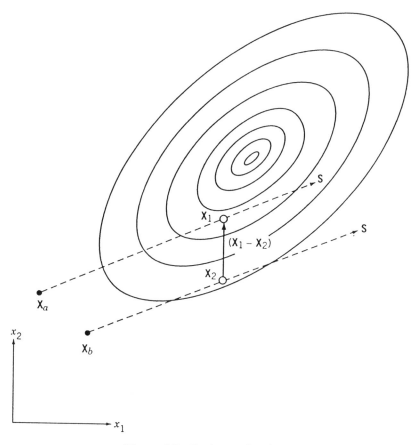

Figure 6.7   Conjugate directions.

different starting points $\mathbf{X}_a$ and $\mathbf{X}_b$, respectively, the line $(\mathbf{X}_1 - \mathbf{X}_2)$ will be conjugate to the search direction $\mathbf{S}$.

Theorem 6.2 If a quadratic function

$$Q(\mathbf{X}) = \tfrac{1}{2}\mathbf{X}^{\mathrm{T}}\mathbf{A}\mathbf{X} + \mathbf{B}^{\mathrm{T}}\mathbf{X} + C \tag{6.31}$$

is minimized sequentially, once along each direction of a set of n mutually conjugate directions, the minimum of the function Q will be found at or before the nth step irrespective of the starting point.

Proof: Let $\mathbf{X}^*$ minimize the quadratic function $Q(\mathbf{X})$. Then

$$\nabla Q(\mathbf{X}^*) = \mathbf{B} + \mathbf{A}\mathbf{X}^* = \mathbf{0} \tag{6.32}$$

Given a point $\mathbf{X}_1$ and a set of linearly independent directions $\mathbf{S}_1, \mathbf{S}_2, \ldots, \mathbf{S}_n$, constants β_i can always be found such that

$$\mathbf{X}^* = \mathbf{X}_1 + \sum_{i=1}^{n} \beta_i \mathbf{S}_i \tag{6.33}$$

where the vectors $\mathbf{S}_1, \mathbf{S}_2, \ldots, \mathbf{S}_n$ have been used as basis vectors. If the directions $\mathbf{S}_i$ are $\mathbf{A}$-conjugate and none of them is zero, the $\mathbf{S}_i$ can easily be shown to be linearly independent and the β_i can be determined as follows.

Equations (6.32) and (6.33) lead to

$$\mathbf{B} + \mathbf{A}\mathbf{X}_1 + \mathbf{A}\left(\sum_{i=1}^{n} \beta_i \mathbf{S}_i\right) = \mathbf{0} \tag{6.34}$$

Multiplying this equation throughout by $\mathbf{S}_j^T$, we obtain

$$\mathbf{S}_j^T(\mathbf{B} + \mathbf{A}\mathbf{X}_1) + \mathbf{S}_j^{\mathrm{T}}\mathbf{A}\left(\sum_{i=1}^{n} \beta_i \mathbf{S}_i\right) = 0 \tag{6.35}$$

Equation (6.35) can be rewritten as

$$(\mathbf{B} + \mathbf{A}\mathbf{X}_1)^{\mathrm{T}}\mathbf{S}_j + \beta_j \mathbf{S}_j^T\mathbf{A}\mathbf{S}_j = \mathbf{0} \tag{6.36}$$

that is,

$$\beta_j = -\frac{(\mathbf{B} + \mathbf{A}\mathbf{X}_1)^{\mathrm{T}}\mathbf{S}_j}{\mathbf{S}_j^T\mathbf{A}\mathbf{S}_j} \tag{6.37}$$

Now consider an iterative minimization procedure starting at point $\mathbf{X}_1$, and successively minimizing the quadratic $Q(\mathbf{X})$ in the directions $\mathbf{S}_1, \mathbf{S}_2, \ldots, \mathbf{S}_n$, where these directions satisfy Eq. (6.25). The successive points are determined by the relation

$$\mathbf{X}_{i+1} = \mathbf{X}_i + \lambda_i^*\mathbf{S}_i, \quad i = 1 \text{ to } n \tag{6.38}$$

where λ_i^* is found by minimizing $Q(\mathbf{X}_i + \lambda_i \mathbf{S}_i)$ so that[†]

$$\mathbf{S}_i^T \nabla Q(\mathbf{X}_{i+1}) = 0 \tag{6.39}$$

Since the gradient of Q at the point $\mathbf{X}_{i+1}$ is given by

$$\nabla Q(\mathbf{X}_{i+1}) = \mathbf{B} + \mathbf{A}\mathbf{X}_{i+1} \tag{6.40}$$

Eq. (6.39) can be written as

$$\mathbf{S}_i^T \{\mathbf{B} + \mathbf{A}(\mathbf{X}_i + \lambda_i^* \mathbf{S}_i)\} = 0 \tag{6.41}$$

This equation gives

$$\lambda_i^* = -\frac{(\mathbf{B} + \mathbf{A}\mathbf{X}_i)^T \mathbf{S}_i}{\mathbf{S}_i^T \mathbf{A}\mathbf{S}_i} \tag{6.42}$$

From Eq. (6.38), we can express $\mathbf{X}_i$ as

$$\mathbf{X}_i = \mathbf{X}_1 + \sum_{j=1}^{i-1} \lambda_j^* \mathbf{S}_j \tag{6.43}$$

so that

$$\mathbf{X}_i^T \mathbf{A}\mathbf{S}_i = \mathbf{X}_1^T \mathbf{A}\mathbf{S}_i + \sum_{j=1}^{i-1} \lambda_j^* \mathbf{S}_j^T \mathbf{A}\mathbf{S}_i$$

$$= \mathbf{X}_1^T \mathbf{A}\mathbf{S}_i \tag{6.44}$$

using the relation (6.25). Thus Eq. (6.42) becomes

$$\lambda_i^* = -(\mathbf{B} + \mathbf{A}\mathbf{X}_1)^T \frac{\mathbf{S}_i}{\mathbf{S}_i^T \mathbf{A}\mathbf{S}_i} \tag{6.45}$$

which can be seen to be identical to Eq. (6.37). Hence the minimizing step lengths are given by β_i or λ_i^*. Since the optimal point $\mathbf{X}^*$ is originally expressed as a sum of n quantities $\beta_1, \beta_2, \ldots, \beta_n$, which have been shown to be equivalent to the minimizing step lengths, the minimization process leads to the minimum point in n steps or less. Since we have not made any assumption regarding $\mathbf{X}_1$ and the order of $\mathbf{S}_1, \mathbf{S}_2, \ldots, \mathbf{S}_n$, the process converges in n steps or less, independent of the starting point as well as the order in which the minimization directions are used.

[†]$\mathbf{S}_i^T \nabla Q(\mathbf{X}_{i+1}) = 0$ is equivalent to $dQ/d\lambda_i = 0$ at $\mathbf{Y} = \mathbf{X}_{i+1}$:

$$\frac{dQ}{d\lambda_i} = \sum_{j=1}^{n} \frac{\partial Q}{\partial y_j} \frac{\partial y_j}{\partial \lambda_i}$$

where y_j are the components of $\mathbf{Y} = \mathbf{X}_{i+1}$.

Example 6.5 Consider the minimization of the function

$$f(x_1, x_2) = 6x_1^2 + 2x_2^2 - 6x_1x_2 - x_1 - 2x_2$$

If $S_1 = \{^1_2\}$ denotes a search direction, find a direction S_2 that is conjugate to the direction S_1.

SOLUTION   The objective function can be expressed in matrix form as

$$f(\mathbf{X}) = \mathbf{B}^T\mathbf{X} + \frac{1}{2}\mathbf{X}^T[A]\mathbf{X}$$

$$= \{-1 \;\; -2\}\begin{Bmatrix}x_1\\x_2\end{Bmatrix} + \frac{1}{2}\{x_1 \;\; x_2\}\begin{bmatrix}12 & -6\\-6 & 4\end{bmatrix}\begin{Bmatrix}x_1\\x_2\end{Bmatrix}$$

and the Hessian matrix $[A]$ can be identified as

$$[A] = \begin{bmatrix}12 & -6\\-6 & 4\end{bmatrix}$$

The direction $S_2 = \{^{s_1}_{s_2}\}$ will be conjugate to $S_1 = \{^1_2\}$ if

$$S_1^T[A]S_2 = (1 \;\; 2)\begin{bmatrix}12 & -6\\-6 & 4\end{bmatrix}\begin{Bmatrix}s_1\\s_2\end{Bmatrix} = 0$$

which upon expansion gives $2s_2 = 0$ or $s_1 =$ arbitrary and $s_2 = 0$. Since s_1 can have any value, we select $s_1 = 1$ and the desired conjugate direction can be expressed as $S_2 = \{^1_0\}$.

6.6.2 Algorithm

The basic idea of Powell's method is illustrated graphically for a two-variable function in Fig. 6.8. In this figure the function is first minimized once along each of the coordinate directions starting with the second coordinate direction and then in the corresponding pattern direction. This leads to point 5. For the next cycle of minimization, we discard one of the coordinate directions (the x_1 direction in the present case) in favor of the pattern direction. Thus we minimize along $\mathbf{u}_2$ and S_1 and obtain point 7. Then we generate a new pattern direction S_2 as shown in the figure. For the next cycle of minimization, we discard one of the previously used coordinate directions (the x_2 direction in this case) in favor of the newly generated pattern direction. Then, by starting from point 8, we minimize along directions S_1 and S_2, thereby obtaining points 9 and 10, respectively. For the next cycle of minimization, since there is no coordinate direction to discard, we restart the whole procedure by minimizing along the x_2 direction. This procedure is continued until the desired minimum point is found.

The flow diagram for the version of Powell's method described above is given in Fig. 6.9. Note that the search will be made sequentially in the directions S_n; $S_1, S_2, S_3, \ldots, S_{n-1}, S_n$; $S_p^{(1)}$; $S_2, S_3, \ldots, S_{n-1}, S_n, S_p^{(1)}$; $S_p^{(2)}$; $S_3, S_4, \ldots, S_{n-1}, S_n$, $S_p^{(1)}, S_p^{(2)}; S_p^{(3)}, \ldots$ until the minimum point is found. Here S_i indicates the coordinate direction $\mathbf{u}_i$ and $S_p^{(j)}$ the jth pattern direction. In Fig. 6.9, the previous base point

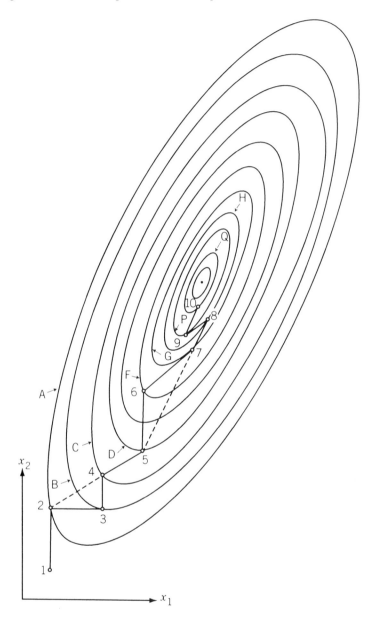

Figure 6.8 Progress of Powell's method.

is stored as the vector **Z** in block A, and the pattern direction is constructed by subtracting the previous base point from the current one in block B. The pattern direction
is then used as a minimization direction in blocks C and D. For the next cycle, the
first direction used in the previous cycle is discarded in favor of the current pattern
direction. This is achieved by updating the numbers of the search directions as shown
in block E. Thus both points **Z** and **X** used in block B for the construction of pattern

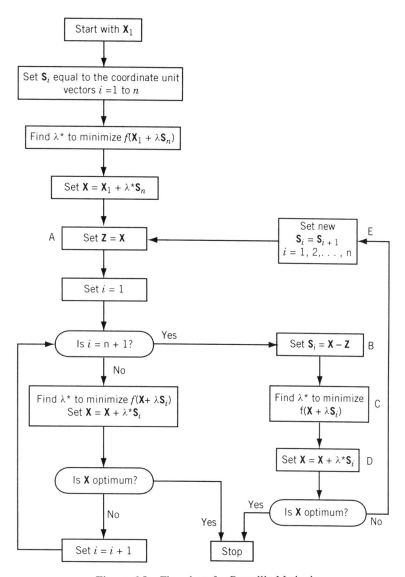

Figure 6.9 Flowchart for Powell's Method.

direction are points that are minima along $\mathbf{S}_n$ in the first cycle, the first pattern direction $\mathbf{S}_p^{(1)}$ in the second cycle, the second pattern direction $\mathbf{S}_p^{(2)}$ in the third cycle, and so on.

Quadratic Convergence. It can be seen from Fig. 6.9 that the pattern directions $\mathbf{S}_p^{(1)}, \mathbf{S}_p^{(2)}, \mathbf{S}_p^{(3)}, \ldots$ are nothing but the lines joining the minima found along the directions $\mathbf{S}_n, \mathbf{S}_p^{(1)}, \mathbf{S}_p^{(2)}, \ldots$, respectively. Hence by Theorem 6.1, the pairs of directions $(\mathbf{S}_n, \mathbf{S}_p^{(1)})$, $(\mathbf{S}_p^{(1)}, \mathbf{S}_p^{(2)})$, and so on, are $\mathbf{A}$-conjugate. Thus all the directions

$\mathbf{S}_n, \mathbf{S}_p^{(1)}, \mathbf{S}_p^{(2)}, \ldots$ are **A**-conjugate. Since, by Theorem 6.2, any search method involving minimization along a set of conjugate directions is quadratically convergent, Powell's method is quadratically convergent. From the method used for constructing the conjugate directions $\mathbf{S}_p^{(1)}, \mathbf{S}_p^{(2)}, \ldots$, we find that n minimization cycles are required to complete the construction of n conjugate directions. In the ith cycle, the minimization is done along the already constructed i conjugate directions and the $n - i$ nonconjugate (coordinate) directions. Thus after n cycles, all the n search directions are mutually conjugate and a quadratic will theoretically be minimized in n^2 one-dimensional minimizations. This proves the *quadratic convergence* of Powell's method.

It is to be noted that as with most of the numerical techniques, the convergence in many practical problems may not be as good as the theory seems to indicate. Powell's method may require a lot more iterations to minimize a function than the theoretically estimated number. There are several reasons for this:

1. Since the number of cycles n is valid only for quadratic functions, it will take generally greater than n cycles for nonquadratic functions.

2. The proof of *quadratic convergence* has been established with the assumption that the exact minimum is found in each of the one-dimensional minimizations. However, the actual minimizing step lengths λ_i^* will be only approximate, and hence the subsequent directions will not be conjugate. Thus the method requires more number of iterations for achieving the overall convergence.

3. Powell's method, described above, can break down before the minimum point is found. This is because the search directions $\mathbf{S}_i$ might become dependent or almost dependent during numerical computation.

Convergence Criterion.   The convergence criterion one would generally adopt in a method such as Powell's method is to stop the procedure whenever a minimization cycle produces a change in all variables less than one-tenth of the required accuracy. However, a more elaborate convergence criterion, which is more likely to prevent premature termination of the process, was given by Powell [6.7].

Example 6.6   Minimize $f(x_1, x_2) = x_1 - x_2 + 2x_1^2 + 2x_1 x_2 + x_2^2$ from the starting point $\mathbf{X}_1 = \begin{Bmatrix} 0 \\ 0 \end{Bmatrix}$ using Powell's method.

SOLUTION
Cycle 1: Univariate Search

We minimize f along $\mathbf{S}_2 = \mathbf{S}_n = \begin{Bmatrix} 0 \\ 1 \end{Bmatrix}$ from $\mathbf{X}_1$. To find the correct direction ($+\mathbf{S}_2$ or $-\mathbf{S}_2$) for decreasing the value of f, we take the probe length as $\varepsilon = 0.01$. As $f_1 = f(\mathbf{X}_1) = 0.0$, and

$$f^+ = f(\mathbf{X}_1 + \varepsilon \mathbf{S}_2) = f(0.0, 0.01) = -0.0099 < f_1$$

f decreases along the direction $+\mathbf{S}_2$. To find the minimizing step length λ^* along $\mathbf{S}_2$, we minimize

$$f(\mathbf{X}_1 + \lambda \mathbf{S}_2) = f(0.0, \lambda) = \lambda^2 - \lambda$$

As $df/d\lambda = 0$ at $\lambda^* = \frac{1}{2}$, we have $\mathbf{X}_2 = \mathbf{X}_1 + \lambda^* \mathbf{S}_2 = \left\{ \begin{matrix} 0 \\ 0.5 \end{matrix} \right\}$.

Next we minimize f along $\mathbf{S}_1 = \left\{ \begin{matrix} 1 \\ 0 \end{matrix} \right\}$ from $\mathbf{X}_2 = \left\{ \begin{matrix} 0.5 \\ 0.0 \end{matrix} \right\}$. Since

$$f_2 = f(\mathbf{X}_2) = f(0.0, 0.5) = -0.25$$

$$f^+ = f(\mathbf{X}_2 + \varepsilon \mathbf{S}_1) = f(0.01, 0.50) = -0.2298 > f_2$$

$$f^- = f(\mathbf{X}_2 - \varepsilon \mathbf{S}_1) = f(-0.01, 0.50) = -0.2698$$

f decreases along $-\mathbf{S}_1$. As $f(\mathbf{X}_2 - \lambda \mathbf{S}_1) = f(-\lambda, 0.50) = 2\lambda^2 - 2\lambda - 0.25$, $df/d\lambda = 0$ at $\lambda^* = \frac{1}{2}$. Hence $\mathbf{X}_3 = \mathbf{X}_2 - \lambda^* \mathbf{S}_1 = \left\{ \begin{matrix} -0.5 \\ 0.5 \end{matrix} \right\}$.

Now we minimize f along $\mathbf{S}_2 = \left\{ \begin{matrix} 0 \\ 1 \end{matrix} \right\}$ from $\mathbf{X}_3 = \left\{ \begin{matrix} -0.5 \\ 0.5 \end{matrix} \right\}$. As $f_3 = f(\mathbf{X}_3) = -0.75$, $f^+ = f(\mathbf{X}_3 + \varepsilon \mathbf{S}_2) = f(-0.5, 0.51) = -0.7599 < f_3$, f decreases along $+\mathbf{S}_2$ direction. Since

$$f(\mathbf{X}_3 + \lambda \mathbf{S}_2) = f(-0.5, 0.5 + \lambda) = \lambda^2 - \lambda - 0.75, \qquad \frac{df}{d\lambda} = 0 \ \text{at} \ \lambda^* = \frac{1}{2}$$

This gives

$$\mathbf{X}_4 = \mathbf{X}_3 + \lambda^* \mathbf{S}_2 = \left\{ \begin{matrix} -0.5 \\ 1.0 \end{matrix} \right\}$$

Cycle 2: Pattern Search

Now we generate the first pattern direction as

$$\mathbf{S}_p^{(1)} = \mathbf{X}_4 - \mathbf{X}_2 = \left\{ \begin{matrix} -\frac{1}{2} \\ 1 \end{matrix} \right\} - \left\{ \begin{matrix} 0 \\ \frac{1}{2} \end{matrix} \right\} = \left\{ \begin{matrix} -0.5 \\ 0.5 \end{matrix} \right\}$$

and minimize f along $\mathbf{S}_p^{(1)}$ from $\mathbf{X}_4$. Since

$$f_4 = f(\mathbf{X}_4) = -1.0$$

$$f^+ = f(\mathbf{X}_4 + \varepsilon \mathbf{S}_p^{(1)}) = f(-0.5 - 0.005, 1 + 0.005)$$

$$= f(-0.505, 1.005) = -1.004975$$

f decreases in the positive direction of $\mathbf{S}_p^{(1)}$. As

$$f(\mathbf{X}_4 + \lambda \mathbf{S}_p^{(1)}) = f(-0.5 - 0.5\lambda, 1.0 + 0.5\lambda)$$

$$= 0.25\lambda^2 - 0.50\lambda - 1.00,$$

$\dfrac{df}{d\lambda} = 0$ at $\lambda^* = 1.0$ and hence

$$\mathbf{X}_5 = \mathbf{X}_4 + \lambda^* \mathbf{S}_p^{(1)} = \left\{ \begin{matrix} -\frac{1}{2} \\ 1 \end{matrix} \right\} + 1.0 \left\{ \begin{matrix} -\frac{1}{2} \\ \frac{1}{2} \end{matrix} \right\} = \left\{ \begin{matrix} -1.0 \\ 1.5 \end{matrix} \right\}$$

The point $\mathbf{X}_5$ can be identified to be the optimum point.

If we do not recognize $\mathbf{X}_5$ as the optimum point at this stage, we proceed to minimize f along the direction $S_2 = \left\{ {0 \atop 1} \right\}$ from $\mathbf{X}_5$. Then we would obtain

$$f_5 = f(\mathbf{X}_5) = -1.25, \quad f^+ = f(\mathbf{X}_5 + \varepsilon S_2) > f_5,$$
$$\text{and} \quad f^- = f(\mathbf{X}_5 - \varepsilon S_2) > f_5$$

This shows that f cannot be minimized along S_2, and hence $\mathbf{X}_5$ will be the optimum point. In this example the convergence has been achieved in the second cycle itself. This is to be expected in this case, as f is a quadratic function, and the method is a quadratically convergent method.

6.7 SIMPLEX METHOD

Definition: Simplex.   The geometric figure formed by a set of $n + 1$ points in an n-dimensional space is called a *simplex*. When the points are equidistant, the simplex is said to be *regular*. Thus in two dimensions, the simplex is a triangle, and in three dimensions, it is a tetrahedron.

The basic idea in the simplex method[†] is to compare the values of the objective function at the $n + 1$ vertices of a general simplex and move the simplex gradually toward the optimum point during the iterative process. The following equations can be used to generate the vertices of a regular simplex (equilateral triangle in two-dimensional space) of size a in the n-dimensional space [6.10]:

$$\mathbf{X}_i = \mathbf{X}_0 + p\mathbf{u}_i + \sum_{j=1, j \neq i}^{n} q\mathbf{u}_j, \quad i = 1, 2, \ldots, n \qquad (6.46)$$

where

$$p = \frac{a}{n\sqrt{2}}(\sqrt{n+1} + n - 1) \quad \text{and} \quad q = \frac{a}{n\sqrt{2}}(\sqrt{n+1} - 1) \qquad (6.47)$$

where $\mathbf{X}_0$ is the initial base point and $\mathbf{u}_j$ is the unit vector along the jth coordinate axis. This method was originally given by Spendley, Hext, and Himsworth [6.10] and was developed later by Nelder and Mead [6.11]. The movement of the simplex is achieved by using three operations, known as reflection, contraction, and expansion.

6.7.1 Reflection

If $\mathbf{X}_h$ is the vertex corresponding to the highest value of the objective function among the vertices of a simplex, we can expect the point $\mathbf{X}_r$ obtained by reflecting the point $\mathbf{X}_h$ in the opposite face to have the smallest value. If this is the case, we can construct a new simplex by rejecting the point $\mathbf{X}_h$ from the simplex and including the new point $\mathbf{X}_r$. This process is illustrated in Fig. 6.10. In Fig. 6.10a, the points $\mathbf{X}_1$, $\mathbf{X}_2$, and $\mathbf{X}_3$ form the original simplex, and the points $\mathbf{X}_1$, $\mathbf{X}_2$, and $\mathbf{X}_r$ form the new one. Similarly, in Fig. 6.10b, the original simplex is given by points $\mathbf{X}_1$, $\mathbf{X}_2$, $\mathbf{X}_3$, and $\mathbf{X}_4$, and the new one by $\mathbf{X}_1$, $\mathbf{X}_2$, $\mathbf{X}_3$, and $\mathbf{X}_r$. Again we can construct a new simplex from the present one

---

[†]This simplex method should not be confused with the simplex method of linear programming.

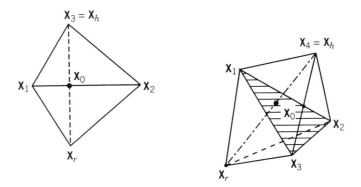

Figure 6.10   Reflection.

by rejecting the vertex corresponding to the highest function value. Since the direction of movement of the simplex is always away from the worst result, we will be moving in a favorable direction. If the objective function does not have steep valleys, repetitive application of the reflection process leads to a zigzag path in the general direction of the minimum as shown in Fig. 6.11. Mathematically, the reflected point $\mathbf{X}_r$ is given by

$$\mathbf{X}_r = (1 + \alpha)\mathbf{X}_0 - \alpha\mathbf{X}_h \qquad (6.48)$$

where $\mathbf{X}_h$ is the vertex corresponding to the maximum function value:

$$f(\mathbf{X}_h) = \max_{i=1 \text{ to } n+1} f(\mathbf{X}_i), \qquad (6.49)$$

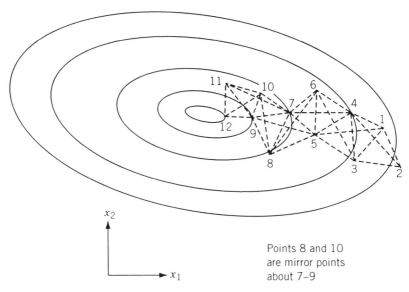

Points 8 and 10 are mirror points about 7–9

Figure 6.11   Progress of the reflection process.

$\mathbf{X}_0$ is the centroid of all the points $\mathbf{X}_i$ except $i = h$:

$$\mathbf{X}_0 = \frac{1}{n} \sum_{\substack{i = 1 \\ i \neq h}}^{n+1} \mathbf{X}_i \tag{6.50}$$

and $\alpha > 0$ is the reflection coefficient defined as

$$\alpha = \frac{\text{distance between } \mathbf{X}_r \text{ and } \mathbf{X}_0}{\text{distance between } \mathbf{X}_h \text{ and } \mathbf{X}_0} \tag{6.51}$$

Thus $\mathbf{X}_r$ will lie on the line joining $\mathbf{X}_h$ and $\mathbf{X}_0$, on the far side of $\mathbf{X}_0$ from $\mathbf{X}_h$ with $|\mathbf{X}_r - \mathbf{X}_0| = \alpha |\mathbf{X}_h - \mathbf{X}_0|$. If $f(\mathbf{X}_r)$ lies between $f(\mathbf{X}_h)$ and $f(\mathbf{X}_l)$, where $\mathbf{X}_l$ is the vertex corresponding to the minimum function value,

$$f(\mathbf{X}_l) = \min_{i=1 \text{ to } n+1} f(\mathbf{X}_i) \tag{6.52}$$

$\mathbf{X}_h$ is replaced by $\mathbf{X}_r$ and a new simplex is started.

If we use only the reflection process for finding the minimum, we may encounter certain difficulties in some cases. For example, if one of the simplexes (triangles in two dimensions) straddles a valley as shown in Fig. 6.12 and if the reflected point $\mathbf{X}_r$ happens to have an objective function value equal to that of the point $\mathbf{X}_h$, we will enter into a closed cycle of operations. Thus if $\mathbf{X}_2$ is the worst point in the simplex defined by the vertices $\mathbf{X}_1$, $\mathbf{X}_2$, and $\mathbf{X}_3$, the reflection process gives the new simplex with vertices $\mathbf{X}_1$, $\mathbf{X}_3$, and $\mathbf{X}_r$. Again, since $\mathbf{X}_r$ has the highest function value out of the vertices $\mathbf{X}_1$, $\mathbf{X}_3$, and $\mathbf{X}_r$, we obtain the old simplex itself by using the reflection process. Thus the optimization process is stranded over the valley and there is no way of moving toward the optimum point. This trouble can be overcome by making a rule that no return can be made to points that have just been left.

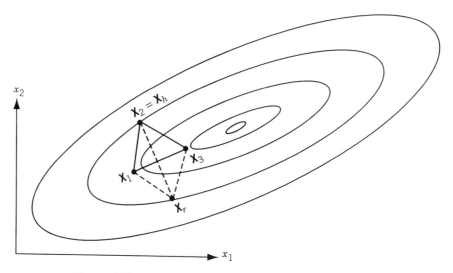

Figure 6.12 Reflection process not leading to a new simplex.

Whenever such situation is encountered, we reject the vertex corresponding to the second worst value instead of the vertex corresponding to the worst function value. This method, in general, leads the process to continue toward the region of the desired minimum. However, the final simplex may again straddle the minimum, or it may lie within a distance of the order of its own size from the minimum. In such cases it may not be possible to obtain a new simplex with vertices closer to the minimum compared to those of the previous simplex, and the pattern may lead to a cyclic process, as shown in Fig. 6.13. In this example the successive simplexes formed from the simplex 123 are 234, 245, 456, 467, 478, 348, 234, 245, …, [†] which can be seen to be forming a cyclic process. Whenever this type of cycling is observed, one can take the vertex that is occurring in every simplex (point 4 in Fig. 6.13) as the best approximation to the optimum point. If more accuracy is desired, the simplex has to be contracted or reduced in size, as indicated later.

6.7.2   Expansion

If a reflection process gives a point $\mathbf{X}_r$ for which $f(\mathbf{X}_r) < f(\mathbf{X}_l)$, (i.e., if the reflection produces a new minimum), one can generally expect to decrease the function value further by moving along the direction pointing from $\mathbf{X}_0$ to $\mathbf{X}_r$. Hence we expand $\mathbf{X}_r$

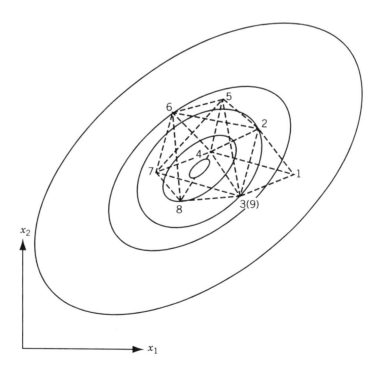

Figure 6.13   Reflection process leading to a cyclic process.

[†]Simplexes 456, 467, and 234 are formed by reflecting the second-worst point to avoid the difficulty mentioned earlier.

to $\mathbf{X}_e$ using the relation

$$\mathbf{X}_e = \gamma \mathbf{X}_r + (1 - \gamma)\mathbf{X}_0 \tag{6.53}$$

where γ is called the *expansion coefficient*, defined as

$$\gamma = \frac{\text{distance between } \mathbf{X}_e \text{ and } \mathbf{X}_0}{\text{distance between } \mathbf{X}_r \text{ and } \mathbf{X}_0} > 1$$

If $f(\mathbf{X}_e) < f(\mathbf{X}_l)$, we replace the point $\mathbf{X}_h$ by $\mathbf{X}_e$ and restart the process of reflection. On the other hand, if $f(\mathbf{X}_e) > f(\mathbf{X}_l)$, it means that the expansion process is not successful and hence we replace point $\mathbf{X}_h$ by $\mathbf{X}_r$ and start the reflection process again.

6.7.3 Contraction

If the reflection process gives a point $\mathbf{X}_r$ for which $f(\mathbf{X}_r) > f(\mathbf{X}_i)$ for all i except $i = h$, and $f(\mathbf{X}_r) < f(\mathbf{X}_h)$, we replace point $\mathbf{X}_h$ by $\mathbf{X}_r$. Thus the new $\mathbf{X}_h$ will be $\mathbf{X}_r$. In this case we contract the simplex as follows:

$$\mathbf{X}_c = \beta \mathbf{X}_h + (1 - \beta)\mathbf{X}_0 \tag{6.54}$$

where β is called the *contraction coefficient* $(0 \le \beta \le 1)$ and is defined as

$$\beta = \frac{\text{distance between } \mathbf{X}_e \text{ and } \mathbf{X}_0}{\text{distance between } \mathbf{X}_h \text{ and } \mathbf{X}_0}$$

If $f(\mathbf{X}_r) > f(\mathbf{X}_h)$, we still use Eq. (6.54) without changing the previous point $\mathbf{X}_h$. If the contraction process produces a point $\mathbf{X}_c$ for which $f(\mathbf{X}_c) < \min[f(\mathbf{X}_h), f(\mathbf{X}_r)]$, we replace the point $\mathbf{X}_h$ in $\mathbf{X}_1, \mathbf{X}_2, \dots, \mathbf{X}_{n+1}$ by $\mathbf{X}_c$ and proceed with the reflection process again. On the other hand, if $f(\mathbf{X}_c) \ge \min[f(\mathbf{X}_h), f(\mathbf{X}_r)]$, the contraction process will be a failure, and in this case we replace all $\mathbf{X}_i$ by $(\mathbf{X}_i + \mathbf{X}_l)/2$ and restart the reflection process.

The method is assumed to have converged whenever the standard deviation of the function at the $n + 1$ vertices of the current simplex is smaller than some prescribed small quantity ε, that is,

$$Q = \left\{ \sum_{i=1}^{n+1} \frac{[f(\mathbf{X}_i) - f(\mathbf{X}_0)]^2}{n + 1} \right\}^{1/2} \le \varepsilon \tag{6.55}$$

Example 6.7   Minimize $f(x_1, x_2) = x_1 - x_2 + 2x_1^2 + 2x_1 x_2 + x_2^2$. Take the points defining the initial simplex as

$$\mathbf{X}_1 = \begin{Bmatrix} 4.0 \\ 4.0 \end{Bmatrix}, \quad \mathbf{X}_2 = \begin{Bmatrix} 5.0 \\ 4.0 \end{Bmatrix}, \quad \text{and} \quad \mathbf{X}_3 = \begin{Bmatrix} 4.0 \\ 5.0 \end{Bmatrix}$$

and $\alpha = 1.0$, $\beta = 0.5$, and $\gamma = 2.0$. For convergence, take the value of ε as 0.2.

SOLUTION

Iteration 1

Step 1: The function value at each of the vertices of the current simplex is given by

$$f_1 = f(\mathbf{X}_1) = 4.0 - 4.0 + 2(16.0) + 2(16.0) + 16.0 = 80.0$$
$$f_2 = f(\mathbf{X}_2) = 5.0 - 4.0 + 2(25.0) + 2(20.0) + 16.0 = 107.0$$
$$f_3 = f(\mathbf{X}_3) = 4.0 - 5.0 + 2(16.0) + 2(20.0) + 25.0 = 96.0$$

Therefore,

$$\mathbf{X}_h = \mathbf{X}_2 = \begin{Bmatrix} 5.0 \\ 4.0 \end{Bmatrix}, \quad f(\mathbf{X}_h) = 107.0,$$

$$\mathbf{X}_l = \mathbf{X}_1 = \begin{Bmatrix} 4.0 \\ 4.0 \end{Bmatrix}, \quad \text{and} \quad f(\mathbf{X}_l) = 80.0$$

Step 2: The centroid $\mathbf{X}_0$ is obtained as

$$\mathbf{X}_0 = \frac{1}{2}(\mathbf{X}_1 + \mathbf{X}_3) = \frac{1}{2}\begin{Bmatrix} 4.0 + 4.0 \\ 4.0 + 5.0 \end{Bmatrix} = \begin{Bmatrix} 4.0 \\ 4.5 \end{Bmatrix} \quad \text{with} \quad f(\mathbf{X}_0) = 87.75$$

Step 3: The reflection point is found as

$$\mathbf{X}_r = 2\mathbf{X}_0 - \mathbf{X}_h = \begin{Bmatrix} 8.0 \\ 9.0 \end{Bmatrix} - \begin{Bmatrix} 5.0 \\ 4.0 \end{Bmatrix} = \begin{Bmatrix} 3.0 \\ 5.0 \end{Bmatrix}$$

Then

$$f(\mathbf{X}_r) = 3.0 - 5.0 + 2(9.0) + 2(15.0) + 25.0 = 71.0$$

Step 4: As $f(\mathbf{X}_r) < f(\mathbf{X}_l)$, we find $\mathbf{X}_e$ by expansion as

$$\mathbf{X}_e = 2\mathbf{X}_r - \mathbf{X}_0 = \begin{Bmatrix} 6.0 \\ 10.0 \end{Bmatrix} - \begin{Bmatrix} 4.0 \\ 4.5 \end{Bmatrix} = \begin{Bmatrix} 2.0 \\ 5.5 \end{Bmatrix}$$

Then

$$f(\mathbf{X}_e) = 2.0 - 5.5 + 2(4.0) + 2(11.0) + 30.25 = 56.75$$

Step 5: Since $f(\mathbf{X}_e) < f(\mathbf{X}_l)$, we replace $\mathbf{X}_h$ by $\mathbf{X}_e$ and obtain the vertices of the new simplex as

$$\mathbf{X}_1 = \begin{Bmatrix} 4.0 \\ 4.0 \end{Bmatrix}, \quad \mathbf{X}_2 = \begin{Bmatrix} 2.0 \\ 5.5 \end{Bmatrix}, \quad \text{and} \quad \mathbf{X}_3 = \begin{Bmatrix} 4.0 \\ 5.0 \end{Bmatrix}$$

Step 6: To test for convergence, we compute

$$Q = \left[\frac{(80.0 - 87.75)^2 + (56.75 - 87.75)^2 + (96.0 - 87.75)^2}{3} \right]^{1/2}$$

$$= 19.06$$

As this quantity is not smaller than ε, we go to the next iteration.

Iteration 2

Step 1: As $f(\mathbf{X}_1) = 80.0$, $f(\mathbf{X}_2) = 56.75$, and $f(\mathbf{X}_3) = 96.0$,

$$\mathbf{X}_h = \mathbf{X}_3 = \begin{Bmatrix} 4.0 \\ 5.0 \end{Bmatrix} \quad \text{and} \quad \mathbf{X}_l = \mathbf{X}_2 = \begin{Bmatrix} 2.0 \\ 5.5 \end{Bmatrix}$$

Step 2: The centroid is

$$\mathbf{X}_0 = \frac{1}{2}(\mathbf{X}_1 + \mathbf{X}_2) = \frac{1}{2} \begin{Bmatrix} 4.0 + 2.0 \\ 4.0 + 5.5 \end{Bmatrix} = \begin{Bmatrix} 3.0 \\ 4.75 \end{Bmatrix}$$

$$f(\mathbf{X}_0) = 67.31$$

Step 3:

$$\mathbf{X}_r = 2\mathbf{X}_0 - \mathbf{X}_h = \begin{Bmatrix} 6.0 \\ 9.5 \end{Bmatrix} - \begin{Bmatrix} 4.0 \\ 5.0 \end{Bmatrix} = \begin{Bmatrix} 2.0 \\ 4.5 \end{Bmatrix}$$

$$f(\mathbf{X}_r) = 2.0 - 4.5 + 2(4.0) + 2(9.0) + 20.25 = 43.75$$

Step 4: As $f(\mathbf{X}_r) < f(\mathbf{X}_l)$, we find $\mathbf{X}_e$ as

$$\mathbf{X}_e = 2\mathbf{X}_r - \mathbf{X}_0 = \begin{Bmatrix} 4.0 \\ 9.0 \end{Bmatrix} - \begin{Bmatrix} 3.0 \\ 4.75 \end{Bmatrix} = \begin{Bmatrix} 1.0 \\ 4.25 \end{Bmatrix}$$

$$f(\mathbf{X}_e) = 1.0 - 4.25 + 2(1.0) + 2(4.25) + 18.0625 = 25.3125$$

Step 5: As $f(\mathbf{X}_e) < f(\mathbf{X}_l)$, we replace $\mathbf{X}_h$ by $\mathbf{X}_e$ and obtain the new vertices as

$$\mathbf{X}_1 = \begin{Bmatrix} 4.0 \\ 4.0 \end{Bmatrix}, \quad \mathbf{X}_2 = \begin{Bmatrix} 2.0 \\ 5.5 \end{Bmatrix}, \quad \text{and} \quad \mathbf{X}_3 = \begin{Bmatrix} 1.0 \\ 4.25 \end{Bmatrix}$$

Step 6: For convergence, we compute Q as

$$Q = \left[\frac{(80.0 - 67.31)^2 + (56.75 - 67.31)^2 + (25.3125 - 67.31)^2}{3} \right]^{1/2}$$

$$= 26.1$$

Since $Q > \varepsilon$, we go to the next iteration.

This procedure can be continued until the specified convergence is satisfied. When the convergence is satisfied, the centroid $\mathbf{X}_0$ of the latest simplex can be taken as the optimum point.

Indirect Search (Descent) Methods

6.8   GRADIENT OF A FUNCTION

The gradient of a function is an n-component vector given by

$$\nabla f_{n \times 1} = \begin{Bmatrix} \partial f/\partial x_1 \\ \partial f/\partial x_2 \\ \vdots \\ \partial f/\partial x_n \end{Bmatrix} \tag{6.56}$$

The gradient has a very important property. If we move along the gradient direction from any point in n-dimensional space, the function value increases at the fastest rate. Hence the gradient direction is called the *direction of steepest ascent*. Unfortunately, the direction of steepest ascent is a local property and not a global one. This is illustrated in Fig. 6.14, where the gradient vectors ∇f evaluated at points 1, 2, 3, and 4 lie along the directions $11'$, $22'$, $33'$, and $44'$, respectively. Thus the function value increases at the fastest rate in the direction $11'$ at point 1, but not at point 2. Similarly, the function value increases at the fastest rate in direction $22'(33')$ at point 2 (3), but not at point 3 (4). In other words, the direction of steepest ascent generally varies from point to point, and if we make infinitely small moves along the direction of steepest ascent, the path will be a curved line like the curve $1-2-3-4$ in Fig. 6.14.

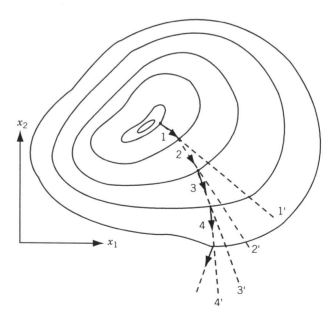

Figure 6.14   Steepest ascent directions.

Since the gradient vector represents the direction of steepest ascent, the negative of the gradient vector denotes the direction of steepest descent. Thus any method that makes use of the gradient vector can be expected to give the minimum point faster than one that does not make use of the gradient vector. All the descent methods make use of the gradient vector, either directly or indirectly, in finding the search directions. Before considering the descent methods of minimization, we prove that the gradient vector represents the direction of steepest ascent.

Theorem 6.3 The gradient vector represents the direction of steepest ascent.

Proof: Consider an arbitary point $\mathbf{X}$ in the n-dimensional space. Let f denote the value of the objective function at the point $\mathbf{X}$. Consider a neighboring point $\mathbf{X} + d\mathbf{X}$ with

$$
d\mathbf{X} = \begin{Bmatrix} dx_1 \\ dx_2 \\ \vdots \\ dx_n \end{Bmatrix} \tag{6.57}
$$

where $dx_1, dx_2, \ldots, dx_n$ represent the components of the vector $d\mathbf{X}$. The magnitude of the vector $d\mathbf{X}$, ds, is given by

$$
d\mathbf{X}^T d\mathbf{X} = (ds)^2 = \sum_{i=1}^{n} (dx_i)^2 \tag{6.58}
$$

If $f + df$ denotes the value of the objective function at $\mathbf{X} + d\mathbf{X}$, the change in f, df, associated with $d\mathbf{X}$ can be expressed as

$$
df = \sum_{i=1}^{n} \frac{\partial f}{\partial x_i} dx_i = \nabla f^T d\mathbf{X} \tag{6.59}
$$

If $\mathbf{u}$ denotes the unit vector along the direction $d\mathbf{X}$ and ds the length of $d\mathbf{X}$, we can write

$$
d\mathbf{X} = \mathbf{u} \, ds \tag{6.60}
$$

The rate of change of the function with respect to the step length ds is given by Eq. (6.59) as

$$
\frac{df}{ds} = \sum_{i=1}^{n} \frac{\partial f}{\partial x_i} \frac{dx_i}{ds} = \nabla f^T \frac{d\mathbf{X}}{ds} = \nabla f^T \mathbf{u} \tag{6.61}
$$

The value of df/ds will be different for different directions and we are interested in finding the particular step $d\mathbf{X}$ along which the value of df/ds will be maximum. This will give the direction of steepest ascent.[†] By using the definition of the dot product,

---

[†]In general, if $df/ds = \nabla f^T \mathbf{u} > 0$ along a vector $d\mathbf{X}$, it is called a direction of *ascent*, and if $df/ds < 0$, it is called a direction of *descent*.

Eq. (6.61) can be rewritten as

$$\frac{df}{ds} = \|\nabla f\| \; \|\mathbf{u}\| \cos\theta \tag{6.62}$$

where $\|\nabla f\|$ and $\|\mathbf{u}\|$ denote the lengths of the vectors ∇f and $\mathbf{u}$, respectively, and θ indicates the angle between the vectors ∇f and $\mathbf{u}$. It can be seen that df/ds will be maximum when $\theta = 0°$ and minimum when $\theta = 180°$. This indicates that the function value increases at a maximum rate in the direction of the gradient (i.e., when $\mathbf{u}$ is along ∇f).

Theorem 6.4 The maximum rate of change of f at any point $\mathbf{X}$ is equal to the magnitude of the gradient vector at the same point.

Proof: The rate of change of the function f with respect to the step length s along a direction $\mathbf{u}$ is given by Eq. (6.62). Since df/ds is maximum when $\theta = 0°$ and $\mathbf{u}$ is a unit vector, Eq. (6.62) gives

$$\left.\left(\frac{df}{ds}\right)\right|_{\max} = \|\nabla f\|$$

which proves the theorem.

6.8.1   Evaluation of the Gradient

The evaluation of the gradient requires the computation of the partial derivatives $\partial f/\partial x_i$, $i = 1, 2, \ldots, n$. There are three situations where the evaluation of the gradient poses certain problems:

1. The function is differentiable at all the points, but the calculation of the components of the gradient, $\partial f/\partial x_i$, is either impractical or impossible.
2. The expressions for the partial derivatives $\partial f/\partial x_i$ can be derived, but they require large computational time for evaluation.
3. The gradient ∇f is not defined at all the points.

In the first case, we can use the forward finite-difference formula

$$\left.\frac{\partial f}{\partial x_i}\right|_{\mathbf{X}_m} \simeq \frac{f(\mathbf{X}_m + \Delta x_i \mathbf{u}_i) - f(\mathbf{X}_m)}{\Delta x_i}, \quad i = 1, 2, \ldots, n \tag{6.63}$$

to approximate the partial derivative $\partial f/\partial x_i$ at $\mathbf{X}_m$. If the function value at the base point $\mathbf{X}_m$ is known, this formula requires one additional function evaluation to find $(\partial f/\partial x_i)|_{Xm}$. Thus it requires n additional function evaluations to evaluate the approximate gradient $\nabla f|_{\mathbf{X}m}$. For better results we can use the central finite difference formula to find the approximate partial derivative $\partial f/\partial x_i|_{\mathbf{X}m}$:

$$\left.\frac{\partial f}{\partial x_i}\right|_{\mathbf{X}_m} \simeq \frac{f(\mathbf{X}_m + \Delta x_i \mathbf{u}_i) - f(\mathbf{X}_m - \Delta x_i \mathbf{u}_i)}{2\Delta x_i}, \quad i = 1, 2, \ldots, n \tag{6.64}$$

This formula requires two additional function evaluations for each of the partial derivatives. In Eqs. (6.63) and (6.64), Δx_i is a small scalar quantity and $\mathbf{u}_i$ is a vector of order n whose ith component has a value of 1, and all other components have a value of zero. In practical computations, the value of Δx_i has to be chosen with some care. If Δx_i is too small, the difference between the values of the function evaluated at $(\mathbf{X}_m + \Delta x_i \mathbf{u}_i)$ and $(\mathbf{X}_m - \Delta x_i \mathbf{u}_i)$ may be very small and numerical round-off error may predominate. On the other hand, if Δx_i is too large, the truncation error may predominate in the calculation of the gradient.

In the second case also, the use of finite-difference formulas is preferred whenever the exact gradient evaluation requires more computational time than the one involved in using Eq. (6.63) or (6.64).

In the third case, we cannot use the finite-difference formulas since the gradient is not defined at all the points. For example, consider the function shown in Fig. 6.15. If Eq. (6.64) is used to evaluate the derivative df/ds at $\mathbf{X}_m$, we obtain a value of α_1 for a step size Δx_1 and a value of α_2 for a step size Δx_2. Since, in reality, the derivative does not exist at the point $\mathbf{X}_m$, use of finite-difference formulas might lead to a complete breakdown of the minimization process. In such cases the minimization can be done only by one of the direct search techniques discussed earlier.

6.8.2 Rate of Change of a Function along a Direction

In most optimization techniques, we are interested in finding the rate of change of a function with respect to a parameter λ along a specified direction, $\mathbf{S}_i$, away from a point $\mathbf{X}_i$. Any point in the specified direction away from the given point $\mathbf{X}_i$ can be expressed as $\mathbf{X} = \mathbf{X}_i + \lambda \mathbf{S}_i$. Our interest is to find the rate of change of the function along the direction $\mathbf{S}_i$ (characterized by the parameter λ), that is,

$$\frac{df}{d\lambda} = \sum_{j=1}^{n} \frac{\partial f}{\partial x_j} \frac{\partial x_j}{\partial \lambda} \qquad (6.65)$$

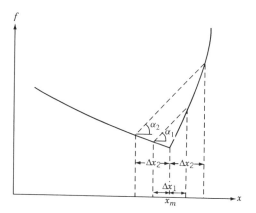

Figure 6.15 Gradient not defined at x_m.

where x_j is the jth component of $\mathbf{X}$. But

$$\frac{\partial x_j}{\partial \lambda} = \frac{\partial}{\partial \lambda}(x_{ij} + \lambda s_{ij}) = s_{ij} \tag{6.66}$$

where x_{ij} and s_{ij} are the jth components of $\mathbf{X}_i$ and $\mathbf{S}_i$, respectively. Hence

$$\frac{df}{d\lambda} = \sum_{j=1}^{n} \frac{\partial f}{\partial x_j} s_{ij} = \nabla f^{\mathrm{T}} \mathbf{S}_i \tag{6.67}$$

If λ^* minimizes f in the direction $\mathbf{S}_i$, we have

$$\frac{df}{d\lambda}\bigg|_{\lambda=\lambda^*} = \nabla f|_{\lambda^*}^{\mathrm{T}} \mathbf{S}_i = 0 \tag{6.68}$$

at the point $\mathbf{X}_i + \lambda^* \mathbf{S}_i$.

6.9 STEEPEST DESCENT (CAUCHY) METHOD

The use of the negative of the gradient vector as a direction for minimization was first made by Cauchy in 1847 [6.12]. In this method we start from an initial trial point $\mathbf{X}_1$ and iteratively move along the steepest descent directions until the optimum point is found. The steepest descent method can be summarized by the following steps:

1. Start with an arbitrary initial point $\mathbf{X}_1$. Set the iteration number as $i = 1$.
2. Find the search direction $\mathbf{S}_i$ as

$$\mathbf{S}_i = -\nabla f_i = -\nabla f(\mathbf{X}_i) \tag{6.69}$$

3. Determine the optimal step length λ_i^* in the direction $\mathbf{S}_i$ and set

$$\mathbf{X}_{i+1} = \mathbf{X}_i + \lambda_i^* \mathbf{S}_i = \mathbf{X}_i - \lambda_i^* \nabla f_i \tag{6.70}$$

4. Test the new point, $\mathbf{X}_{i+1}$, for optimality. If $\mathbf{X}_{i+1}$ is optimum, stop the process. Otherwise, go to step 5.
5. Set the new iteration number $i = i + 1$ and go to step 2.

The method of steepest descent may appear to be the *best unconstrained minimization* technique since each one-dimensional search starts in the "best" direction. However, owing to the fact that the steepest descent direction is a local property, the method is not really effective in most problems.

Example 6.8 Minimize $f(x_1, x_2) = x_1 - x_2 + 2x_1^2 + 2x_1x_2 + x_2^2$ starting from the point $\mathbf{X}_1 = \begin{Bmatrix} 0 \\ 0 \end{Bmatrix}$.

SOLUTION

Iteration 1

The gradient of f is given by

$$\nabla f = \left\{ \begin{matrix} \partial f/\partial x_1 \\ \partial f/\partial x_2 \end{matrix} \right\} = \left\{ \begin{matrix} 1 + 4x_1 + 2x_2 \\ -1 + 2x_1 + 2x_2 \end{matrix} \right\}$$

$$\nabla f_1 = \nabla f(\mathbf{X}_1) = \left\{ \begin{matrix} 1 \\ -1 \end{matrix} \right\}$$

Therefore,

$$\mathbf{S}_1 = -\nabla f_1 = \left\{ \begin{matrix} -1 \\ 1 \end{matrix} \right\}$$

To find $\mathbf{X}_2$, we need to find the optimal step length λ_1^*. For this, we minimize $f(\mathbf{X}_1 + \lambda_1 \mathbf{S}_1) = f(-\lambda_1, \lambda_1) = \lambda_1^2 - 2\lambda_1$ with respect to λ_1. Since $df/d\lambda_1 = 0$ at $\lambda_1^* = 1$, we obtain

$$\mathbf{X}_2 = \mathbf{X}_1 + \lambda_1^* \mathbf{S}_1 = \left\{ \begin{matrix} 0 \\ 0 \end{matrix} \right\} + 1 \left\{ \begin{matrix} -1 \\ 1 \end{matrix} \right\} = \left\{ \begin{matrix} -1 \\ 1 \end{matrix} \right\}$$

As $\nabla f_2 = \nabla f(\mathbf{X}_2) = \left\{ \begin{matrix} -1 \\ -1 \end{matrix} \right\} \neq \left\{ \begin{matrix} 0 \\ 0 \end{matrix} \right\}$, $\mathbf{X}_2$ is not optimum.

Iteration 2

$$\mathbf{S}_2 = -\nabla f_2 = \left\{ \begin{matrix} 1 \\ 1 \end{matrix} \right\}$$

To minimize

$$f(\mathbf{X}_2 + \lambda_2 \mathbf{S}_2) = f(-1 + \lambda_2, 1 + \lambda_2)$$

$$= 5\lambda_2^2 - 2\lambda_2 - 1$$

we set $df/d\lambda_2 = 0$. This gives $\lambda_2^* = \frac{1}{5}$, and hence

$$\mathbf{X}_3 = \mathbf{X}_2 + \lambda_2^* \mathbf{S}_2 = \left\{ \begin{matrix} -1 \\ 1 \end{matrix} \right\} + \frac{1}{5} \left\{ \begin{matrix} 1 \\ 1 \end{matrix} \right\} = \left\{ \begin{matrix} -0.8 \\ 1.2 \end{matrix} \right\}$$

Since the components of the gradient at $\mathbf{X}_3$, $\nabla f_3 = \left\{ \begin{matrix} 0.2 \\ -0.2 \end{matrix} \right\}$, are not zero, we proceed to the next iteration.

Iteration 3

$$\mathbf{S}_3 = -\nabla f_3 = \left\{ \begin{matrix} -0.2 \\ 0.2 \end{matrix} \right\}$$

As

$$f(\mathbf{X}_3 + \lambda_3 \mathbf{S}_3) = f(-0.8 - 0.2\lambda_3, 1.2 + 0.2\lambda_3)$$

$$= 0.04\lambda_3^2 - 0.08\lambda_3 - 1.20, \quad \frac{df}{d\lambda_3} = 0 \text{ at } \lambda_3^* = 1.0$$

Therefore,

$$\mathbf{X}_4 = \mathbf{X}_3 + \lambda_3^* \mathbf{S}_3 = \begin{Bmatrix} -0.8 \\ 1.2 \end{Bmatrix} + 1.0 \begin{Bmatrix} -0.2 \\ 0.2 \end{Bmatrix} = \begin{Bmatrix} -1.0 \\ 1.4 \end{Bmatrix}$$

The gradient at $\mathbf{X}_4$ is given by

$$\nabla f_4 = \begin{Bmatrix} -0.20 \\ -0.20 \end{Bmatrix}$$

Since $\nabla f_4 \neq \begin{Bmatrix} 0 \\ 0 \end{Bmatrix}$, $\mathbf{X}_4$ is not optimum and hence we have to proceed to the next iteration. This process has to be continued until the optimum point, $\mathbf{X}^* = \begin{Bmatrix} -1.0 \\ 1.5 \end{Bmatrix}$, is found.

Convergence Criteria: The following criteria can be used to terminate the iterative process.

1. When the change in function value in two consecutive iterations is small:

$$\left| \frac{f(\mathbf{X}_{i+1}) - f(\mathbf{X}_i)}{f(\mathbf{X}_i)} \right| \leq \varepsilon_1 \tag{6.71}$$

2. When the partial derivatives (components of the gradient) of f are small:

$$\left| \frac{\partial f}{\partial x_i} \right| \leq \varepsilon_2, \quad i = 1, 2, \ldots, n \tag{6.72}$$

3. When the change in the design vector in two consecutive iterations is small:

$$|\mathbf{X}_{i+1} - \mathbf{X}_i| \leq \varepsilon_3 \tag{6.73}$$

6.10   CONJUGATE GRADIENT (FLETCHER–REEVES) METHOD

The convergence characteristics of the steepest descent method can be improved greatly by modifying it into a conjugate gradient method (which can be considered as a conjugate directions method involving the use of the gradient of the function). We saw (in Section 6.6.) that any minimization method that makes use of the conjugate directions is quadratically convergent. This property of quadratic convergence is very useful because it ensures that the method will minimize a quadratic function in n steps or less. Since any general function can be approximated reasonably well by a quadratic near the optimum point, any quadratically convergent method is expected to find the optimum point in a finite number of iterations.

We have seen that Powell's conjugate direction method requires n single-variable minimizations per iteration and sets up a new conjugate direction at the end of each iteration. Thus it requires, in general, n^2 single-variable minimizations to find the minimum of a quadratic function. On the other hand, if we can evaluate the gradients of the objective function, we can set up a new conjugate direction after every one-dimensional minimization, and hence we can achieve faster convergence. The construction of conjugate directions and development of the Fletcher–Reeves method are discussed in this section.

6.10.1 Development of the Fletcher–Reeves Method

The Fletcher–Reeves method is developed by modifying the steepest descent method to make it quadratically convergent. Starting from an arbitrary point $\mathbf{X}_1$, the quadratic function

$$f(\mathbf{X}) = \tfrac{1}{2}\mathbf{X}^T[\mathbf{A}]\mathbf{X} + \mathbf{B}^T\mathbf{X} + C \tag{6.74}$$

can be minimized by searching along the search direction $\mathbf{S}_1 = -\nabla f_1$ (steepest descent direction) using the step length (see Problem 6.40):

$$\lambda_1^* = -\frac{\mathbf{S}_1^T \nabla f_1}{\mathbf{S}_1^T \mathbf{A}\mathbf{S}_1} \tag{6.75}$$

The second search direction $\mathbf{S}_2$ is found as a linear combination of $\mathbf{S}_1$ and $-\nabla f_2$:

$$\mathbf{S}_2 = -\nabla f_2 + \beta_2 \mathbf{S}_1 \tag{6.76}$$

where the constant β_2 can be determined by making $\mathbf{S}_1$ and $\mathbf{S}_2$ conjugate with respect to $[A]$. This leads to (see Problem 6.41):

$$\beta_2 = -\frac{\nabla f_2^T \nabla f_2}{\nabla f_1^T \mathbf{S}_1} = \frac{\nabla f_2^T \nabla f_2}{\nabla f_1^T \nabla f_1} \tag{6.77}$$

This process can be continued to obtain the general formula for the ith search direction as

$$\mathbf{S}_i = -\nabla f_i + \beta_i \mathbf{S}_{i-1} \tag{6.78}$$

where

$$\beta_i = \frac{\nabla f_i^T \nabla f_i}{\nabla f_{i-1}^T \nabla f_{i-1}} \tag{6.79}$$

Thus the Fletcher–Reeves algorithm can be stated as follows.

6.10.2   Fletcher–Reeves Method

The iterative procedure of Fletcher–Reeves method can be stated as follows:

1. Start with an arbitrary initial point $\mathbf{X}_1$.
2. Set the first search direction $\mathbf{S}_1 = -\nabla f(\mathbf{X}_1) = -\nabla f_1$.
3. Find the point $\mathbf{X}_2$ according to the relation

$$\mathbf{X}_2 = \mathbf{X}_1 + \lambda_1^* \mathbf{S}_1 \tag{6.80}$$

   where λ_1^* is the optimal step length in the direction $\mathbf{S}_1$. Set $i = 2$ and go to the next step.
4. Find $\nabla f_i = \nabla f(\mathbf{X}_i)$, and set

$$\mathbf{S}_i = -\nabla f_i + \frac{|\nabla f_i|^2}{|\nabla f_{i-1}|^2} \mathbf{S}_{i-1} \tag{6.81}$$

5. Compute the optimum step length λ_i^* in the direction $\mathbf{S}_i$, and find the new point

$$\mathbf{X}_{i+1} = \mathbf{X}_i + \lambda_i^* \mathbf{S}_i \tag{6.82}$$

6. Test for the optimality of the point $\mathbf{X}_{i+1}$. If $\mathbf{X}_{i+1}$ is optimum, stop the process. Otherwise, set the value of $i = i + 1$ and go to step 4.

Remarks:

1. The Fletcher–Reeves method was originally proposed by Hestenes and Stiefel [6.14] as a method for solving systems of linear equations derived from the stationary conditions of a quadratic. Since the directions $\mathbf{S}_i$ used in this method are $\mathbf{A}$-conjugate, the process should converge in n cycles or less for a quadratic function. However, for ill-conditioned quadratics (whose contours are highly eccentric and distorted), the method may require much more than n cycles for convergence. The reason for this has been found to be the cumulative effect of rounding errors. Since $\mathbf{S}_i$ is given by Eq. (6.81), any error resulting from the inaccuracies involved in the determination of λ_i^*, and from the round-off error involved in accumulating the successive $|\nabla f_i|^2 \mathbf{S}_{i-1} / |\nabla f_{i-1}|^2$ terms, is carried forward through the vector $\mathbf{S}_i$. Thus the search directions $\mathbf{S}_i$ will be progressively contaminated by these errors. Hence it is necessary, in practice, to restart the method periodically after every, say, m steps by taking the new search direction as the steepest descent direction. That is, after every m steps, $\mathbf{S}_{m+1}$ is set equal to $-\nabla f_{m+1}$ instead of the usual form. Fletcher and Reeves have recommended a value of $m = n + 1$, where n is the number of design variables.

2. Despite the limitations indicated above, the Fletcher–Reeves method is vastly superior to the steepest descent method and the pattern search methods, but it turns out to be rather less efficient than the Newton and the quasi-Newton (variable metric) methods discussed in the latter sections.

Example 6.9   Minimize $f(x_1, x_2) = x_1 - x_2 + 2x_1^2 + 2x_1 x_2 + x_2^2$ starting from the point $\mathbf{X}_1 = \left\{ {0 \atop 0} \right\}$.

SOLUTION

Iteration 1

$$\nabla f = \begin{Bmatrix} \partial f / \partial x_1 \\ \partial f / \partial x_2 \end{Bmatrix} = \begin{Bmatrix} 1 + 4x_1 + 2x_2 \\ -1 + 2x_1 + 2x_2 \end{Bmatrix}$$

$$\nabla f_1 = \nabla f(\mathbf{X}_1) = \begin{Bmatrix} 1 \\ -1 \end{Bmatrix}$$

The search direction is taken as $\mathbf{S}_1 = -\nabla f_1 = \begin{Bmatrix} -1 \\ 1 \end{Bmatrix}$. To find the optimal step length λ_1^* along $\mathbf{S}_1$, we minimize $f(\mathbf{X}_1 + \lambda_1 \mathbf{S}_1)$ with respect to λ_1. Here

$$f(\mathbf{X}_1 + \lambda_1 \mathbf{S}_1) = f(-\lambda_1, +\lambda_1) = \lambda_1^2 - 2\lambda_1$$

$$\frac{df}{d\lambda_1} = 0 \quad \text{at} \quad \lambda_1^* = 1$$

Therefore,

$$\mathbf{X}_2 = \mathbf{X}_1 + \lambda_1^* \mathbf{S}_1 = \begin{Bmatrix} 0 \\ 0 \end{Bmatrix} + 1 \begin{Bmatrix} -1 \\ 1 \end{Bmatrix} = \begin{Bmatrix} -1 \\ 1 \end{Bmatrix}$$

Iteration 2

Since $\nabla f_2 = \nabla f(\mathbf{X}_2) = \begin{Bmatrix} -1 \\ -1 \end{Bmatrix}$, Eq. (6.81) gives the next search direction as

$$\mathbf{S}_2 = -\nabla f_2 + \frac{|\nabla f_2|^2}{|\nabla f_1|^2} \mathbf{S}_1$$

where

$$|\nabla f_1|^2 = 2 \quad \text{and} \quad |\nabla f_2|^2 = 2$$

Therefore,

$$\mathbf{S}_2 = -\begin{Bmatrix} -1 \\ -1 \end{Bmatrix} + \left(\frac{2}{2}\right) \begin{Bmatrix} -1 \\ 1 \end{Bmatrix} = \begin{Bmatrix} 0 \\ +2 \end{Bmatrix}$$

To find λ_2^*, we minimize

$$f(\mathbf{X}_2 + \lambda_2 \mathbf{S}_2) = f(-1, 1 + 2\lambda_2)$$

$$= -1 - (1 + 2\lambda_2) + 2 - 2(1 + 2\lambda_2) + (1 + 2\lambda_2)^2$$

$$= 4\lambda_2^2 - 2\lambda_2 - 1$$

with respect to λ_2. As $df/d\lambda_2 = 8\lambda_2 - 2 = 0$ at $\lambda_2^* = \frac{1}{4}$, we obtain

$$\mathbf{X}_3 = \mathbf{X}_2 + \lambda_2^* \mathbf{S}_2 = \begin{Bmatrix} -1 \\ 1 \end{Bmatrix} + \frac{1}{4} \begin{Bmatrix} 0 \\ 2 \end{Bmatrix} = \begin{Bmatrix} -1 \\ 1.5 \end{Bmatrix}$$

Thus the optimum point is reached in two iterations. Even if we do not know this point to be optimum, we will not be able to move from this point in the next iteration. This can be verified as follows.

Iteration 3

Now

$$\nabla f_3 = \nabla f(\mathbf{X}_3) = \begin{Bmatrix} 0 \\ 0 \end{Bmatrix}, \quad |\nabla f_2|^2 = 2, \quad \text{and} \quad |\nabla f_3|^2 = 0.$$

Thus

$$\mathbf{S}_3 = -\nabla f_3 + (|\nabla f_3|^2/|\nabla f_2|^2)\mathbf{S}_2 = -\begin{Bmatrix} 0 \\ 0 \end{Bmatrix} + \left(\frac{0}{2}\right)\begin{Bmatrix} 0 \\ 0 \end{Bmatrix} = \begin{Bmatrix} 0 \\ 0 \end{Bmatrix}$$

This shows that there is no search direction to reduce f further, and hence $\mathbf{X}_3$ is optimum.

6.11 NEWTON'S METHOD

Newton's method presented in Section 5.12.1 can be extended for the minimization of multivariable functions. For this, consider the quadratic approximation of the function $f(\mathbf{X})$ at $\mathbf{X} = \mathbf{X}_i$ using the Taylor's series expansion

$$f(\mathbf{X}) = f(\mathbf{X}_i) + \nabla f_i^{\mathrm{T}}(\mathbf{X} - \mathbf{X}_i) + \tfrac{1}{2}(\mathbf{X} - \mathbf{X}_i)^{\mathrm{T}}[J_i](\mathbf{X} - \mathbf{X}_i) \qquad (6.83)$$

where $[J_i] = [J]|_{\mathbf{X}_i}$ is the matrix of second partial derivatives (Hessian matrix) of f evaluated at the point $\mathbf{X}_i$. By setting the partial derivatives of Eq. (6.83) equal to zero for the minimum of $f(\mathbf{X})$, we obtain

$$\frac{\partial f(\mathbf{X})}{\partial x_j} = 0, \qquad j = 1, 2, \ldots, n \qquad (6.84)$$

Equations (6.84) and (6.83) give

$$\nabla f = \nabla f_i + [J_i](\mathbf{X} - \mathbf{X}_i) = \mathbf{0} \qquad (6.85)$$

If $[J_i]$ is nonsingular, Eqs. (6.85) can be solved to obtain an improved approximation $(\mathbf{X} = \mathbf{X}_{i+1})$ as

$$\mathbf{X}_{i+1} = \mathbf{X}_i - [J_i]^{-1} \ \nabla f_i \qquad (6.86)$$

Since higher-order terms have been neglected in Eq. (6.83), Eq. (6.86) is to be used iteratively to find the optimum solution $\mathbf{X}^*$.

The sequence of points $\mathbf{X}_1, \mathbf{X}_2, \ldots, \mathbf{X}_{i+1}$ can be shown to converge to the actual solution $\mathbf{X}^*$ from any initial point $\mathbf{X}_1$ sufficiently close to the solution $\mathbf{X}^*$, provided that $[J_1]$ is nonsingular. It can be seen that Newton's method uses the second partial derivatives of the objective function (in the form of the matrix $[J_i]$) and hence is a second-order method.

Example 6.10   Show that the Newton's method finds the minimum of a quadratic function in one iteration.

SOLUTION   Let the quadratic function be given by

$$f(\mathbf{X}) = \tfrac{1}{2}\mathbf{X}^{\mathrm{T}}[A]\mathbf{X} + \mathbf{B}^{\mathrm{T}}\mathbf{X} + C$$

The minimum of $f(\mathbf{X})$ is given by

$$\nabla f = [A]\mathbf{X} + \mathbf{B} = \mathbf{0}$$

or

$$\mathbf{X}^* = -[A]^{-1}\mathbf{B}$$

The iterative step of Eq. (6.86) gives

$$\mathbf{X}_{i+1} = \mathbf{X}_i - [A]^{-1}([A]\mathbf{X}_i + \mathbf{B}) \tag{E_1}$$

where $\mathbf{X}_i$ is the starting point for the ith iteration. Thus Eq. (E_1) gives the exact solution

$$\mathbf{X}_{i+1} = \mathbf{X}^* = -[A]^{-1}\mathbf{B}$$

Figure 6.16 illustrates this process.

Example 6.11   Minimize $f(x_1, x_2) = x_1 - x_2 + 2x_1^2 + 2x_1x_2 + x_2^2$ by taking the starting point as $\mathbf{X}_1 = \{{0 \atop 0}\}$.

SOLUTION   To find $\mathbf{X}_2$ according to Eq. (6.86), we require $[J_1]^{-1}$, where

$$[J_1] = \begin{bmatrix} \dfrac{\partial^2 f}{\partial x_1^2} & \dfrac{\partial^2 f}{\partial x_1 \partial x_2} \\[3mm] \dfrac{\partial^2 f}{\partial x_2 \partial x_1} & \dfrac{\partial^2 f}{\partial x_2^2} \end{bmatrix}_{\mathbf{X}_1} = \begin{bmatrix} 4 & 2 \\ 2 & 2 \end{bmatrix}$$

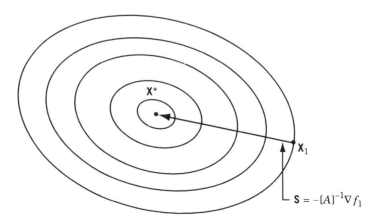

Figure 6.16   Minimization of a quadratic function in one step.

Therefore,

$$[J_1]^{-1} = \frac{1}{4}\begin{bmatrix} +2 & -2 \\ -2 & 4 \end{bmatrix} = \begin{bmatrix} \frac{1}{2} & -\frac{1}{2} \\ -\frac{1}{2} & 1 \end{bmatrix}$$

As

$$\mathbf{g}_1 = \begin{Bmatrix} \partial f/\partial x_1 \\ \partial f/\partial x_2 \end{Bmatrix}_{\mathbf{X}_1} = \begin{Bmatrix} 1 + 4x_1 + 2x_2 \\ -1 + 2x_1 + 2x_2 \end{Bmatrix}_{(0,0)} = \begin{Bmatrix} 1 \\ -1 \end{Bmatrix}$$

Equation (6.86) gives

$$\mathbf{X}_2 = \mathbf{X}_1 - [J_1]^{-1}\mathbf{g}_1 = \begin{Bmatrix} 0 \\ 0 \end{Bmatrix} - \begin{bmatrix} \frac{1}{2} & -\frac{1}{2} \\ -\frac{1}{2} & 1 \end{bmatrix}\begin{Bmatrix} 1 \\ -1 \end{Bmatrix} = \begin{Bmatrix} -1 \\ \frac{3}{2} \end{Bmatrix}$$

To see whether or not $\mathbf{X}_2$ is the optimum point, we evaluate

$$\mathbf{g}_2 = \begin{Bmatrix} \partial f/\partial x_1 \\ \partial f/\partial x_2 \end{Bmatrix}_{\mathbf{X}_2} = \begin{Bmatrix} 1 + 4x_1 + 2x_2 \\ -1 + 2x_1 + 2x_2 \end{Bmatrix}_{(-1,3/2)} = \begin{Bmatrix} 0 \\ 0 \end{Bmatrix}$$

As $\mathbf{g}_2 = \mathbf{0}$, $\mathbf{X}_2$ is the optimum point. Thus the method has converged in one iteration for this quadratic function.

If $f(\mathbf{X})$ is a nonquadratic function, Newton's method may sometimes diverge, and it may converge to saddle points and relative maxima. This problem can be avoided by modifying Eq. (6.86) as

$$\mathbf{X}_{i+1} = \mathbf{X}_i + \lambda_i^*\mathbf{S}_i = \mathbf{X}_i - \lambda_i^*[J_i]^{-1}\nabla f_i \qquad (6.87)$$

where λ_i^* is the minimizing step length in the direction $\mathbf{S}_i = -[J_i]^{-1}\nabla f_i$. The modification indicated by Eq. (6.87) has a number of advantages. First, it will find the minimum in lesser number of steps compared to the original method. Second, it finds the minimum point in all cases, whereas the original method may not converge in some cases. Third, it usually avoids convergence to a saddle point or a maximum. With all these advantages, this method appears to be the most powerful minimization method. Despite these advantages, the method is not very useful in practice, due to the following features of the method:

1. It requires the storing of the $n \times n$ matrix $[J_i]$.
2. It becomes very difficult and sometimes impossible to compute the elements of the matrix $[J_i]$.
3. It requires the inversion of the matrix $[J_i]$ at each step.
4. It requires the evaluation of the quantity $[J_i]^{-1}\nabla f_i$ at each step.

These features make the method impractical for problems involving a complicated objective function with a large number of variables.

6.12   MARQUARDT METHOD

The steepest descent method reduces the function value when the design vector $\mathbf{X}_i$ is away from the optimum point $\mathbf{X}^*$. The Newton method, on the other hand, converges fast when the design vector $\mathbf{X}_i$ is close to the optimum point $\mathbf{X}^*$. The Marquardt method [6.15] attempts to take advantage of both the steepest descent and Newton methods. This method modifies the diagonal elements of the Hessian matrix, $[J_i]$, as

$$[\tilde{J}_i] = [J_i] + \alpha_i[I] \tag{6.88}$$

where $[I]$ is an identity matrix and α_i is a positive constant that ensures the positive definiteness of $[\tilde{J}_i]$ when $[J_i]$ is not positive definite. It can be noted that when α_i is sufficiently large (on the order of 10^4), the term $\alpha_i[I]$ dominates $[J_i]$ and the inverse of the matrix $[\tilde{J}_i]$ becomes

$$[\tilde{J}_i]^{-1} = [[J_i] + \alpha_i[I]]^{-1} \approx [\alpha_i[I]]^{-1} = \frac{1}{\alpha_i}[I] \tag{6.89}$$

Thus if the search direction $\mathbf{S}_i$ is computed as

$$\mathbf{S}_i = -[\tilde{J}_i]^{-1}\nabla f_i \tag{6.90}$$

$\mathbf{S}_i$ becomes a steepest descent direction for large values of α_i. In the Marquardt method, the value of α_i is taken to be large at the beginning and then reduced to zero gradually as the iterative process progresses. Thus as the value of α_i decreases from a large value to zero, the characteristics of the search method change from those of a steepest descent method to those of the Newton method. The iterative process of a modified version of Marquardt method can be described as follows.

1. Start with an arbitrary initial point $\mathbf{X}_1$ and constants α_1 (on the order of 10^4), $c_1(0 < c_1 < 1)$, $c_2(c_2 > 1)$, and ε (on the order of 10^{-2}). Set the iteration number as $i = 1$.
2. Compute the gradient of the function, $\nabla f_i = \nabla f(\mathbf{X}_i)$.
3. Test for optimality of the point $\mathbf{X}_i$. If $\|\nabla f_i\| = \|\nabla f(\mathbf{X}_i)\| \le \varepsilon$, $\mathbf{X}_i$ is optimum and hence stop the process. Otherwise, go to step 4.
4. Find the new vector $\mathbf{X}_{i+1}$ as

$$\mathbf{X}_{i+1} = \mathbf{X}_i + \mathbf{S}_i = \mathbf{X}_i - [[J_i]] + \alpha_i[I]]^{-1} \quad \nabla f_i \tag{6.91}$$

5. Compare the values of f_{i+1} and f_i. If $f_{i+1} < f_i$, go to, step 6. If $f_{i+1} \ge f_i$, go to step 7.
6. Set $\alpha_{i+1} = c_1\alpha_i$, $i = i + 1$, and go to step 2.
7. Set $\alpha_i = c_2\alpha_i$ and go to step 4.

An advantage of this method is the absence of the step size λ_i along the search direction $\mathbf{S}_i$. In fact, the algorithm above can be modified by introducing an optimal step length in Eq. (6.91) as

$$\mathbf{X}_{i+1} = \mathbf{X}_i + \lambda_i^* \mathbf{S}_i = \mathbf{X}_i - \lambda_i^*[[J_i] + \alpha_i[I]]^{-1}\nabla f_i \tag{6.92}$$

where λ_i^* is found using any of the one-dimensional search methods described in Chapter 5.

Example 6.12   Minimize $f(x_1, x_2) = x_1 - x_2 + 2x_1^2 + 2x_1x_2 + x_2^2$ from the starting point $\mathbf{X}_1 = \left\{ \begin{smallmatrix} 0 \\ 0 \end{smallmatrix} \right\}$ using Marquardt method with $\alpha_1 = 10^4$, $c_1 = \frac{1}{4}$, $c_2 = 2$, and $\varepsilon = 10^{-2}$.

SOLUTION

Iteration 1 (i = 1)

Here $f_1 = f(\mathbf{X}_1) = 0.0$ and

$$\nabla f_1 = \left\{ \begin{array}{c} \dfrac{\partial f}{\partial x_1} \\[2mm] \dfrac{\partial f}{\partial x_2} \end{array} \right\}_{(0,0)} = \left\{ \begin{array}{c} 1 + 4x_1 + 2x_2 \\ -1 + 2x_1 + 2x_2 \end{array} \right\}_{(0,0)} = \left\{ \begin{array}{c} 1 \\ -1 \end{array} \right\}$$

Since $\|\nabla f_1\| = 1.4142 > \varepsilon$, we compute

$$[J_1] = \begin{bmatrix} \dfrac{\partial^2 f}{\partial x_1^2} & \dfrac{\partial^2 f}{\partial x_1 x_2} \\[3mm] \dfrac{\partial^2}{\partial x_1 x_2} & \dfrac{\partial^2 f}{\partial x_2^2} \end{bmatrix}_{(0,0)} = \begin{bmatrix} 4 & 2 \\ 2 & 2 \end{bmatrix}$$

$$\mathbf{X}_2 = \mathbf{X}_1 - [[J_1] + \alpha_1[I]]^{-1}\nabla f_1$$

$$= \left\{ \begin{array}{c} 0 \\ 0 \end{array} \right\} - \begin{bmatrix} 4 + 10^4 & 2 \\ 2 & 2 + 10^4 \end{bmatrix}^{-1} \left\{ \begin{array}{c} 1 \\ -1 \end{array} \right\} = \left\{ \begin{array}{c} -0.9998 \\ 1.0000 \end{array} \right\} 10^{-4}$$

As $f_2 = f(\mathbf{X}_2) = -1.9997 \times 10^{-4} < f_1$, we set $\alpha_2 = c_1\alpha_1 = 2500$, $i = 2$, and proceed to the next iteration.

Iteration 2 (i = 2)

The gradient vector corresponding to $\mathbf{X}_2$ is given by $\nabla f_2 = \left\{ \begin{smallmatrix} 0.9998 \\ -1.0000 \end{smallmatrix} \right\}$, $\|\nabla f_2\| = 1.4141 > \varepsilon$, and hence we compute

$$\mathbf{X}_3 = \mathbf{X}_2 - [[J_2] + \alpha_2[I]]^{-1}\nabla f_2$$

$$= \left\{ \begin{array}{c} -0.9998 \times 10^{-4} \\ 1.0000 \times 10^{-4} \end{array} \right\} - \begin{bmatrix} 2504 & 2 \\ 2 & 2502 \end{bmatrix}^{-1} \left\{ \begin{array}{c} 0.9998 \\ -1.0000 \end{array} \right\}$$

$$= \left\{ \begin{array}{c} -4.9958 \times 10^{-4} \\ 5.0000 \times 10^{-4} \end{array} \right\}$$

Since $f_3 = f(\mathbf{X}_3) = -0.9993 \times 10^{-3} < f_2$, we set $\alpha_3 = c_1\alpha_2 = 625$, $i = 3$, and proceed to the next iteration. The iterative process is to be continued until the convergence criterion, $\|\nabla f_i\| < \varepsilon$, is satisfied.

6.13 QUASI-NEWTON METHODS

The basic iterative process used in the Newton's method is given by Eq. (6.86):

$$\mathbf{X}_{i+1} = \mathbf{X}_i - [J_i]^{-1} \nabla f(\mathbf{X}_i) \tag{6.93}$$

where the Hessian matrix $[J_i]$ is composed of the second partial derivatives of f and varies with the design vector $\mathbf{X}_i$ for a nonquadratic (general nonlinear) objective function f. The basic idea behind the quasi-Newton or variable metric methods is to approximate either $[J_i]$ by another matrix $[A_i]$ or $[J_i]^{-1}$ by another matrix $[B_i]$, using only the first partial derivatives of f. If $[J_i]^{-1}$ is approximated by $[B_i]$, Eq. (6.93) can be expressed as

$$\mathbf{X}_{i+1} = \mathbf{X}_i - \lambda_i^* [B_i] \nabla f(\mathbf{X}_i) \tag{6.94}$$

where λ_i^* can be considered as the optimal step length along the direction

$$\mathbf{S}_i = -[B_i] \nabla f(\mathbf{X}_i) \tag{6.95}$$

It can be seen that the steepest descent direction method can be obtained as a special case of Eq. (6.95) by setting $[B_i] = [I]$.

Computation of $[\boldsymbol{B}_i]$**.**   To implement Eq. (6.94), an approximate inverse of the Hessian matrix, $[B_i] \equiv [A_i]^{-1}$, is to be computed. For this, we first expand the gradient of f about an arbitrary reference point, $\mathbf{X}_0$, using Taylor's series as

$$\nabla f(\mathbf{X}) \approx \nabla f(\mathbf{X}_0) + [J_0](\mathbf{X} - \mathbf{X}_0) \tag{6.96}$$

If we pick two points $\mathbf{X}_i$ and $\mathbf{X}_{i+1}$ and use $[A_i]$ to approximate $[J_0]$, Eq. (6.96) can be rewritten as

$$\nabla f_{i+1} = \nabla f(\mathbf{X}_0) + [A_i](\mathbf{X}_{i+1} - \mathbf{X}_0) \tag{6.97}$$

$$\nabla f_i = \nabla f(\mathbf{X}_0) + [A_i](\mathbf{X}_i - \mathbf{X}_0) \tag{6.98}$$

Subtracting Eq. (6.98) from (6.97) yields

$$[A_i]\mathbf{d}_i = \mathbf{g}_i \tag{6.99}$$

where

$$\mathbf{d}_i = \mathbf{X}_{i+1} - \mathbf{X}_i \tag{6.100}$$

$$\mathbf{g}_i = \nabla f_{i+1} - \nabla f_i \tag{6.101}$$

The solution of Eq. (6.99) for $\mathbf{d}_i$ can be written as

$$\mathbf{d}_i = [B_i]\mathbf{g}_i \tag{6.102}$$

where $[B_i] = [A_i]^{-1}$ denotes an approximation to the inverse of the Hessian matrix, $[J_0]^{-1}$. It can be seen that Eq. (6.102) represents a system of n equations in n^2 unknown elements of the matrix $[B_i]$. Thus for $n > 1$, the choice of $[B_i]$ is not unique and one would like to choose $[B_i]$ that is closest to $[J_0]^{-1}$, in some sense. Numerous techniques have been suggested in the literature for the computation of $[B_i]$ as the iterative process progresses (i.e., for the computation of $[B_{i+1}]$ once $[B_i]$ is known). A major concern is that in addition to satisfying Eq. (6.102), the symmetry and positive definiteness of the matrix $[B_i]$ is to be maintained; that is, if $[B_i]$ is symmetric and positive definite, $[B_{i+1}]$ must remain symmetric and positive definite.

6.13.1   Rank 1 Updates

The general formula for updating the matrix $[B_i]$ can be written as

$$[B_{i+1}] = [B_i] + [\Delta B_i] \tag{6.103}$$

where $[\Delta B_i]$ can be considered to be the update (or correction) matrix added to $[B_i]$. Theoretically, the matrix $[\Delta B_i]$ can have its rank as high as n. However, in practice, most updates, $[\Delta B_i]$, are only of rank 1 or 2. To derive a rank 1 update, we simply choose a scaled outer product of a vector $\mathbf{z}$ for $[\Delta B_i]$ as

$$[\Delta B_i] = c\mathbf{z}\mathbf{z}^{\mathrm{T}} \tag{6.104}$$

where the constant c and the n-component vector $\mathbf{z}$ are to be determined. Equations (6.103) and (6.104) lead to

$$[B_{i+1}] = [B_i] + c\mathbf{z}\mathbf{z}^{\mathrm{T}} \tag{6.105}$$

By forcing Eq. (6.105) to satisfy the quasi-Newton condition, Eq. (6.102),

$$\mathbf{d}_i = [B_{i+1}]\mathbf{g}_i \tag{6.106}$$

we obtain

$$\mathbf{d}_i = ([B_i] + c\mathbf{z}\mathbf{z}^T)\mathbf{g}_i = [B_i]\mathbf{g}_i + c\mathbf{z}(\mathbf{z}^T\mathbf{g}_i) \tag{6.107}$$

Since $(\mathbf{z}^T\mathbf{g}_i)$ in Eq. (6.107) is a scalar, we can rewrite Eq. (6.107) as

$$c\mathbf{z} = \frac{\mathbf{d}_i - [B_i]\mathbf{g}_i}{\mathbf{z}^T\mathbf{g}_i} \tag{6.108}$$

Thus a simple choice for $\mathbf{z}$ and c would be

$$\mathbf{z} = \mathbf{d}_i - [B_i]\mathbf{g}_i \tag{6.109}$$

$$c = \frac{1}{\mathbf{z}^T\mathbf{g}_i} \tag{6.110}$$

This leads to the unique rank 1 update formula for $[B_{i+1}]$:

$$[B_{i+1}] = [B_i] + [\Delta B_i] \equiv [B_i] + \frac{(\mathbf{d}_i - [B_i]\mathbf{g}_i)(\mathbf{d}_i - [B_i]\mathbf{g}_i)^{\mathrm{T}}}{(\mathbf{d}_i - [B_i]\mathbf{g}_i)^T\mathbf{g}_i} \tag{6.111}$$

This formula has been attributed to Broyden [6.16]. To implement Eq. (6.111), an initial symmetric positive definite matrix is selected for $[B_1]$ at the start of the algorithm, and the next point $\mathbf{X}_2$ is computed using Eq. (6.94). Then the new matrix $[B_2]$ is computed using Eq. (6.111) and the new point $\mathbf{X}_3$ is determined from Eq. (6.94). This iterative process is continued until convergence is achieved. If $[B_i]$ is symmetric, Eq. (6.111) ensures that $[B_{i+1}]$ is also symmetric. However, there is no guarantee that $[B_{i+1}]$ remains positive definite even if $[B_i]$ is positive definite. This might lead to a breakdown of the procedure, especially when used for the optimization of nonquadratic functions. It can be verified easily that the columns of the matrix $[\Delta B_i]$ given by Eq. (6.111) are multiples of each other. Thus the updating matrix has only one independent column and hence the rank of the matrix will be 1. This is the reason why Eq. (6.111) is considered to be a rank 1 updating formula. Although the Broyden formula, Eq. (6.111), is not robust, it has the property of quadratic convergence [6.17]. The rank 2 update formulas given next guarantee both symmetry and positive definiteness of the matrix $[B_{i+1}]$ and are more robust in minimizing general nonlinear functions, hence are preferred in practical applications.

6.13.2   Rank 2 Updates

In rank 2 updates we choose the update matrix $[\Delta B_i]$ as the sum of two rank 1 updates as

$$[\Delta B_i] = c_1 \mathbf{z}_1 \mathbf{z}_1^{\mathrm{T}} + c_2 \mathbf{z}_2 \mathbf{z}_2^{\mathrm{T}} \tag{6.112}$$

where the constants c_1 and c_2 and the n-component vectors $\mathbf{z}_1$ and $\mathbf{z}_2$ are to be determined. Equations (6.103) and (6.112) lead to

$$[B_{i+1}] = [B_i] + c_1 \mathbf{z}_1 \mathbf{z}_1^{\mathrm{T}} + c_2 \mathbf{z}_2 \mathbf{z}_2^{\mathrm{T}} \tag{6.113}$$

By forcing Eq. (6.113) to satisfy the quasi-Newton condition, Eq. (6.106), we obtain

$$\mathbf{d}_i = [B_i]\mathbf{g}_i + c_1 \mathbf{z}_1 (\mathbf{z}_1^{\mathrm{T}} \mathbf{g}_i) + c_2 \mathbf{z}_2 (\mathbf{z}_2^{\mathrm{T}} \mathbf{g}_i) \tag{6.114}$$

where $(\mathbf{z}_1^{\mathrm{T}} \mathbf{g}_i)$ and $(\mathbf{z}_2^{\mathrm{T}} \mathbf{g}_i)$ can be identified as scalars. Although the vectors $\mathbf{z}_1$ and $\mathbf{z}_2$ in Eq. (6.114) are not unique, the following choices can be made to satisfy Eq. (6.114):

$$\mathbf{z}_1 = \mathbf{d}_i \tag{6.115}$$

$$\mathbf{z}_2 = [B_i]\mathbf{g}_i \tag{6.116}$$

$$c_1 = \frac{1}{\mathbf{z}_1^{\mathrm{T}} \mathbf{g}_i} \tag{6.117}$$

$$c_2 = -\frac{1}{\mathbf{z}_2^{\mathrm{T}} \mathbf{g}_i} \tag{6.118}$$

Thus the rank 2 update formula can be expressed as

$$[B_{i+1}] = [B_i] + [\Delta B_i] \equiv [B_i] + \frac{\mathbf{d}_i \mathbf{d}_i^{\mathrm{T}}}{\mathbf{d}_i^{T} \mathbf{g}_i} - \frac{([B_i]\mathbf{g}_i)([B_i]\mathbf{g}_i)^{\mathrm{T}}}{([B_i]\mathbf{g}_i)^{\mathrm{T}} \mathbf{g}_i} \tag{6.119}$$

This equation is known as the Davidon–Fletcher–Powell (DFP) formula [6.20, 6.21]. Since

$$\mathbf{X}_{i+1} = \mathbf{X}_i + \lambda_i^* \mathbf{S}_i \tag{6.120}$$

where $\mathbf{S}_i$ is the search direction, $\mathbf{d}_i = \mathbf{X}_{i+1} - \mathbf{X}_i$ can be rewritten as

$$\mathbf{d}_i = \lambda_i^* \mathbf{S}_i \tag{6.121}$$

Thus Eq. (6.119) can be expressed as

$$[B_{i+1}] = [B_i] + \frac{\lambda_i^* \mathbf{S}_i \mathbf{S}_i^{\mathrm{T}}}{\mathbf{S}_i^{\mathrm{T}} \mathbf{g}_i} - \frac{[B_i] \mathbf{g}_i \mathbf{g}_i^{\mathrm{T}} [B_i]}{\mathbf{g}_i^T [B_i] \mathbf{g}_i} \tag{6.122}$$

Remarks:

1. Equations (6.111) and (6.119) are known as *inverse update formulas* since these equations approximate the inverse of the Hessian matrix of f.

2. It is possible to derive a family of direct update formulas in which approximations to the Hessian matrix itself are considered. For this we express the quasi-Newton condition as [see Eq. (6.99)]

$$\mathbf{g}_i = [A_i] \mathbf{d}_i \tag{6.123}$$

The procedure used in deriving Eqs. (6.111) and (6.119) can be followed by using $[A_i]$, $\mathbf{d}_i$, and $\mathbf{g}_i$ in place of $[B_i]$, $\mathbf{g}_i$, and $\mathbf{d}_i$, respectively. This leads to the rank 2 update formula (similar to Eq. (6.119), known as the Broydon–Fletcher–Goldfarb–Shanno (BFGS) formula [6.22–6.25]:

$$[A_{i+1}] = [A_i] + \frac{\mathbf{g}_i \mathbf{g}_i^{\mathrm{T}}}{\mathbf{g}_i^T \mathbf{d}_i} - \frac{([A_i] \mathbf{d}_i)([A_i] \mathbf{d}_i)^{\mathrm{T}}}{([A_i] \mathbf{d}_i)^T \mathbf{d}_i} \tag{6.124}$$

In practical computations, Eq. (6.124) is rewritten more conveniently in terms of $[B_i]$, as

$$[B_{i+1}] = [B_i] + \frac{\mathbf{d}_i \mathbf{d}_i^{\mathrm{T}}}{\mathbf{d}_i^T \mathbf{g}_i} \left(1 + \frac{\mathbf{g}_i^{\mathrm{T}} [B_i] \mathbf{g}_i}{\mathbf{d}_i^T \mathbf{g}_i}\right) - \frac{[B_i] \mathbf{g}_i \mathbf{d}_i^{\mathrm{T}}}{\mathbf{d}_i^T \mathbf{g}_i} - \frac{\mathbf{d}_i \mathbf{g}_i^{\mathrm{T}} [B_i]}{\mathbf{d}_i^T \mathbf{g}_i} \tag{6.125}$$

3. The DFP and the BFGS formulas belong to a family of rank 2 updates known as *Huang's family of updates* [6.18], which can be expressed for updating the inverse of the Hessian matrix as

$$[B_{i+1}] = \rho_i \left([B_i] - \frac{[B_i] \mathbf{g}_i \mathbf{g}_i^{\mathrm{T}} [B_i]}{\mathbf{g}_i^T [B_i] \mathbf{g}_i} + \theta_i \mathbf{y}_i \mathbf{y}_i^T\right) + \frac{\mathbf{d}_i \mathbf{d}_i^{\mathrm{T}}}{\mathbf{d}_i^T \mathbf{g}_i} \tag{6.126}$$

where

$$\mathbf{y}_i = (\mathbf{g}_i^{\mathrm{T}} [\mathbf{B}_i] \mathbf{g}_i)^{1/2} \left(\frac{\mathbf{d}_i}{\mathbf{d}_i^T \mathbf{g}_i} - \frac{[B_i] \mathbf{g}_i}{\mathbf{g}_i^T [B_i] \mathbf{g}_i}\right) \tag{6.127}$$

and ρ_i and θ_i are constant parameters. It has been shown [6.18] that Eq. (6.126) maintains the symmetry and positive definiteness of $[B_{i+1}]$ if $[B_i]$ is symmetric and positive definite. Different choices of ρ_i and θ_i in Eq. (6.126) lead to different algorithms. For example, when $\rho_i = 1$ and $\theta_i = 0$, Eq. (6.126) gives the DFP formula, Eq. (6.119). When $\rho_i = 1$ and $\theta_i = 1$, Eq. (6.126) yields the BFGS formula, Eq. (6.125).

4. It has been shown that the BFGS method exhibits superlinear convergence near $\mathbf{X}^*$ [6.17].

5. Numerical experience indicates that the BFGS method is the best unconstrained variable metric method and is less influenced by errors in finding λ_i^* compared to the DFP method.

6. The methods discussed in this section are also known as secant methods since Eqs. (6.99) and (6.102) can be considered as secant equations (see Section 5.12).

The DFP and BFGS iterative methods are described in detail in the following sections.

6.14 DAVIDON–FLETCHER–POWELL METHOD

The iterative procedure of the Davidon–Fletcher–Powell (DFP) method can be described as follows:

1. Start with an initial point $\mathbf{X}_1$ and a $n \times n$ positive definite symmetric matrix $[B_1]$ to approximate the inverse of the Hessian matrix of f. Usually, $[B_1]$ is taken as the identity matrix $[I]$. Set the iteration number as $i = 1$.

2. Compute the gradient of the function, ∇f_i, at point $\mathbf{X}_i$, and set

$$\mathbf{S}_i = -[B_i]\nabla f_i \tag{6.128}$$

3. Find the optimal step length λ_i^* in the direction $\mathbf{S}_i$ and set

$$\mathbf{X}_{i+1} = \mathbf{X}_i + \lambda_i^*\mathbf{S}_i \tag{6.129}$$

4. Test the new point $\mathbf{X}_{i+1}$ for optimality. If $\mathbf{X}_{i+1}$ is optimal, terminate the iterative process. Otherwise, go to step 5.

5. Update the matrix $[B_i]$ using Eq. (6.119) as

$$[B_{i+1}] = [B_i] + [M_i] + [N_i] \tag{6.130}$$

where

$$[M_i] = \lambda_i^* \frac{\mathbf{S}_i \mathbf{S}_i^{\mathrm{T}}}{\mathbf{S}_i^T \mathbf{g}_i} \tag{6.131}$$

$$[N_i] = -\frac{([B_i]\mathbf{g}_i)([B_i]\mathbf{g}_i)^{\mathrm{T}}}{\mathbf{g}_i^T [B_i]\mathbf{g}_i} \tag{6.132}$$

$$\mathbf{g}_i = \nabla f(\mathbf{X}_{i+1}) - \nabla f(\mathbf{X}_i) = \nabla f_{i+1} - \nabla f_i \tag{6.133}$$

6. Set the new iteration number as $i = i + 1$, and go to step 2.

Note: The matrix $[B_{i+1}]$, given by Eq. (6.130), remains positive definite only if λ_i^* is found accurately. Thus if λ_i^* is not found accurately in any iteration, the matrix $[B_i]$ should not be updated. There are several alternatives in such a case. One possibility is to compute a better value of λ_i^* by using more number of refits in the one-dimensional minimization procedure (until the product $\mathbf{S}_i^T \nabla f_{i+1}$ becomes sufficiently small). However, this involves more computational effort. Another possibility is to specify a maximum number of refits in the one-dimensional minimization method and to skip the updating of $[B_i]$ if λ_i^* could not be found accurately in the specified number of refits. The last possibility is to continue updating the matrix $[B_i]$ using the approximate values of λ_i^* found, but restart the whole procedure after certain number of iterations, that is, restart with $i = 1$ in step 2 of the method.

Example 6.13   Show that the DFP method is a conjugate gradient method.

SOLUTION   Consider the quadratic function

$$f(\mathbf{X}) = \tfrac{1}{2}\mathbf{X}^{\mathrm{T}}[A]\mathbf{X} + \mathbf{B}^{\mathrm{T}}\mathbf{X} + C \tag{E_1}$$

for which the gradient is given by

$$\nabla f = [A]\mathbf{X} + \mathbf{B} \tag{E_2}$$

Equations (6.133) and (E$_2$) give

$$\mathbf{g}_i = \nabla f_{i+1} - \nabla f_i = [A](\mathbf{X}_{i+1} - \mathbf{X}_i) \tag{E_3}$$

Since

$$\mathbf{X}_{i+1} = \mathbf{X}_i + \lambda_i^*\mathbf{S}_i \tag{E_4}$$

Eq. (E$_3$) becomes

$$\mathbf{g}_i = \lambda_i^*[A]\mathbf{S}_i \tag{E_5}$$

or

$$[A]\mathbf{S}_i = \frac{1}{\lambda_i^*}\mathbf{g}_i \tag{E_6}$$

Premultiplication of Eq. (E$_6$) by $[B_{i+1}]$ leads to

$$[B_{i+1}][A]\mathbf{S}_i = \frac{1}{\lambda_i^*}([\mathbf{B}_i] + [M_i] + [N_i])\mathbf{g}_i \tag{E_7}$$

Equations (6.131) and (E$_5$) yield

$$[M_i]\mathbf{g}_i = \lambda_i^*\frac{\mathbf{S}_i\mathbf{S}_i^{\mathrm{T}}\mathbf{g}_i}{\mathbf{S}_i^T\mathbf{g}_i} = \lambda_i^*\mathbf{S}_i \tag{E_8}$$

Equation (6.132) can be used to obtain

$$[N_i]\mathbf{g}_i = -\frac{([B_i]\mathbf{g}_i)(\mathbf{g}_i^{\mathrm{T}}[B_i]^{\mathrm{T}}\mathbf{g}_i)}{\mathbf{g}_i^T[B_i]\mathbf{g}_i} = -[B_i]\mathbf{g}_i \tag{E_9}$$

since $[B_i]$ is symmetric. By substituting Eqs. (E$_8$) and (E$_9$) into Eq. (E$_7$), we obtain

$$[B_{i+1}][A]\mathbf{S}_i = \frac{1}{\lambda_i^*}([B_i]\mathbf{g}_i + \lambda_i^*\mathbf{S}_i - [B_i]\mathbf{g}_i) = \mathbf{S}_i \tag{E_{10}}$$

The quantity $\mathbf{S}_{i+1}^T[A]\mathbf{S}_i$ can be written as

$$\mathbf{S}_{i+1}^{\mathrm{T}}[A]\mathbf{S}_i = -([B_{i+1}]\nabla f_{i+1})^{\mathrm{T}}[A]\mathbf{S}_i$$

$$= -\nabla f_{i+1}^T[B_{i+1}][A]\mathbf{S}_i = -\nabla f_{i+1}^T\mathbf{S}_i = 0 \qquad (E_{11})$$

since λ_i^* is the minimizing step in the direction $\mathbf{S}_i$. Equation (E_{11}) proves that the successive directions generated in the DFP method are $[A]$-conjugate and hence the method is a conjugate gradient method.

Example 6.14 Minimize $f(x_1, x_2) = 100(x_1^2 - x_2)^2 + (1 - x_1)^2$ taking $\mathbf{X}_1 = \left\{ {-2 \atop -2} \right\}$ as the starting point. Use cubic interpolation method for one-dimensional minimization.

SOLUTION   Since this method requires the gradient of f, we find that

$$\nabla f = \left\{ {\partial f/\partial x_1 \atop \partial f/\partial x_2} \right\} = \left\{ {400x_1(x_1^2 - x_2) - 2(1 - x_1) \atop -200(x_1^2 - x_2)} \right\}$$

Iteration 1

We take

$$[B_1] = \begin{bmatrix} 1 & 0 \\ 0 & 1 \end{bmatrix}$$

At $\mathbf{X}_1 = \left\{ {-2 \atop -2} \right\}$, $\nabla f_1 = \nabla f(\mathbf{X}_1) = \left\{ {-4806 \atop -1200} \right\}$ and $f_1 = 3609$. Therefore,

$$\mathbf{S}_1 = -[B_1]\nabla f_1 = \left\{ {4806 \atop 1200} \right\}$$

By normalizing, we obtain

$$\mathbf{S}_1 = \frac{1}{[(4806)^2 + (1200)^2]^{1/2}} \left\{ {4806 \atop 1200} \right\} = \left\{ {0.970 \atop 0.244} \right\}$$

To find λ_i^*, we minimize

$$f(\mathbf{X}_1 + \lambda_1\mathbf{S}_1) = f(-2 + 0.970\lambda_1, -2 + 0.244\lambda_1)$$

$$= 100(6 - 4.124\lambda_1 + 0.938\lambda_1^2)^2 + (3 - 0.97\lambda_1)^2 \qquad (E_1)$$

with respect to λ_1. Equation (E_1) gives

$$\frac{df}{d\lambda_1} = 200(6 - 4.124\lambda_1 + 0.938\lambda_1^2)(1.876\lambda_1 - 4.124) - 1.94(3 - 0.97\lambda_1)$$

Since the solution of the equation $df/d\lambda_1 = 0$ cannot be obtained in a simple manner, we use the cubic interpolation method for finding λ_i^*.

Cubic Interpolation Method (First Fitting)

Stage 1: As the search direction $\mathbf{S}_1$ is normalized already, we go to stage 2.

Stage 2: To establish lower and upper bounds on the optimal step size λ_1^*, we have to find two points A and B at which the slope $df/d\lambda_1$ has different signs. We take $A = 0$ and choose an initial step size of $t_0 = 0.25$ to find B.

At $\lambda_1 = A = 0$:

$$f_A = f(\lambda_1 = A = 0) = 3609$$

$$f_A' = \left.\frac{df}{d\lambda_1}\right|_{\lambda_1=A=0} = -4956.64$$

At $\lambda_1 = t_0 = 0.25$:

$$f = 2535.62$$

$$\frac{df}{d\lambda_1} = -3680.82$$

As $df/d\lambda_1$ is negative, we accelerate the search by taking $\lambda_1 = 4t_0 = 1.00$.

At $\lambda_1 = 1.00$:

$$f = 795.98$$

$$\frac{df}{d\lambda_1} = -1269.18$$

Since $df/d\lambda_1$ is still negative, we take $\lambda_1 = 2.00$.

At $\lambda_1 = 2.00$:

$$f = 227.32$$

$$\frac{df}{d\lambda_1} = -113.953$$

Although $df/d\lambda_1$ is still negative, it appears to have come close to zero and hence we take the next value of λ_1 as 2.50.

At $\lambda_1 = 2.50$:

$$f = 241.51$$

$$\frac{df}{d\lambda_1} = 174.684 = \text{positive}$$

Since $df/d\lambda_1$ is negative at $\lambda_1 = 2.0$ and positive at $\lambda_1 = 2.5$, we take $A = 2.0$ (instead of zero for faster convergence) and $B = 2.5$. Therefore,

$$A = 2.0, \quad f_A = 227.32, \quad f_A' = -113.95$$
$$B = 2.5, \quad f_B = 241.51, \quad f_B' = 174.68$$

Stage 3: To find the optimal step length $\tilde{\lambda}_1^*$ using Eq. (5.54), we compute

$$Z = \frac{3(227.32 - 241.51)}{2.5 - 2.0} - 113.95 + 174.68 = -24.41$$

$$Q = [(24.41)^2 + (113.95)(174.68)]^{1/2} = 143.2$$

Therefore,

$$\tilde{\lambda}_i^* = 2.0 + \frac{-113.95 - 24.41 + 143.2}{-113.95 + 174.68 - 48.82}(2.5 - 2.0)$$

$$= 2.2$$

Stage 4: To find whether $\tilde{\lambda}_1^*$ is close to λ_1^*, we test the value of $df/d\lambda_1$.

$$\left.\frac{df}{d\lambda_1}\right|_{\tilde{\lambda}_1^*} = -0.818$$

Also,

$$f(\lambda_1 = \tilde{\lambda}_1^*) = 216.1$$

Since $df/d\lambda_1$ is not close to zero at $\tilde{\lambda}_1^*$, we use a refitting technique.

Second Fitting: Now we take $A = \tilde{\lambda}_1^*$ since $df/d\lambda_1$ is negative at $\tilde{\lambda}_1^*$ and $B = 2.5$. Thus

$$A = 2.2, \quad f_A = 216.10, \quad f_A' = -0.818$$

$$B = 2.5, \quad f_B = 241.51, \quad f_B' = 174.68$$

With these values we find that

$$Z = \frac{3(216.1 - 241.51)}{2.5 - 2.2} - 2.818 + 174.68 = -80.238$$

$$Q = [(80.238)^2 + (0.818)(174.68)]^{1/2} = 81.1$$

$$\tilde{\lambda}_1^* = 2.2 + \frac{-0.818 - 80.238 + 81.1}{-0.818 + 174.68 - 160.476}(2.5 - 2.2) = 2.201$$

To test for convergence, we evaluate $df/d\lambda$ at $\tilde{\lambda}_1^*$. Since $df/d\lambda|_{\lambda_1 = \tilde{\lambda}_1^*} = -0.211$, it can be assumed to be sufficiently close to zero and hence we take $\lambda_1^* \simeq \tilde{\lambda}_1^* = 2.201$. This gives

$$\mathbf{X}_2 = \mathbf{X}_1 + \lambda_1^*\mathbf{S}_1 = \begin{Bmatrix} -2 + 0.970\lambda_1^* \\ -2 + 0.244\lambda_1^* \end{Bmatrix} = \begin{Bmatrix} 0.135 \\ -1.463 \end{Bmatrix}$$

Testing $\mathbf{X}_2$ for convergence: To test whether the D-F-P method has converged, we compute the gradient of f at $\mathbf{X}_2$:

$$\nabla f_2 = \begin{Bmatrix} \partial f/\partial x_1 \\ \partial f/\partial x_2 \end{Bmatrix}_{\mathbf{X}_2} = \begin{Bmatrix} 78.29 \\ -296.24 \end{Bmatrix}$$

As the components of this vector are not close to zero, $\mathbf{X}_2$ is not optimum and hence the procedure has to be continued until the optimum point is found.

Example 6.15 Minimize $f(x_1, x_2) = x_1 - x_2 + 2x_1^2 + 2x_1x_2 + x_2^2$ from the starting point $\mathbf{X}_1 = \begin{Bmatrix} 0 \\ 0 \end{Bmatrix}$ using the DFP method with

$$[B_1] = \begin{bmatrix} 1 & 0 \\ 0 & 1 \end{bmatrix} \quad \varepsilon = 0.01$$

SOLUTION

Iteration 1 (i = 1)

Here

$$\nabla f_1 = \nabla f(\mathbf{X}_1) = \left\{ \begin{array}{c} 1 + 4x_1 + 2x_2 \\ -1 + 2x_1 + 2x_2 \end{array} \right\} \bigg|_{(0,0)} = \left\{ \begin{array}{c} 1 \\ -1 \end{array} \right\}$$

and hence

$$\mathbf{S}_1 = -[B_1]\nabla f_1 = - \begin{bmatrix} 1 & 0 \\ 0 & 1 \end{bmatrix} \left\{ \begin{array}{c} 1 \\ -1 \end{array} \right\} = \left\{ \begin{array}{c} -1 \\ 1 \end{array} \right\}$$

To find the minimizing step length λ_1^* along $\mathbf{S}_1$, we minimize

$$f(\mathbf{X}_1 + \lambda_1 \mathbf{S}_1) = f\left(\left\{ \begin{array}{c} 0 \\ 0 \end{array} \right\} + \lambda_1 \left\{ \begin{array}{c} -1 \\ 1 \end{array} \right\} \right) = f(-\lambda_1, \lambda_1) = \lambda_1^2 - 2\lambda_1$$

with respect to λ_1. Since $df/d\lambda_1 = 0$ at $\lambda_1^* = 1$, we obtain

$$\mathbf{X}_2 = \mathbf{X}_1 + \lambda_1^* \mathbf{S}_1 = \left\{ \begin{array}{c} 0 \\ 0 \end{array} \right\} + 1 \left\{ \begin{array}{c} -1 \\ 1 \end{array} \right\} = \left\{ \begin{array}{c} -1 \\ 1 \end{array} \right\}$$

Since $\nabla f_2 = \nabla f(\mathbf{X}_2) = \left\{ \begin{array}{c} -1 \\ -1 \end{array} \right\}$ and $\|\nabla f_2\| = 1.4142 > \varepsilon$, we proceed to update the matrix $[B_i]$ by computing

$$\mathbf{g}_1 = \nabla f_2 - \nabla f_1 = \left\{ \begin{array}{c} -1 \\ -1 \end{array} \right\} - \left\{ \begin{array}{c} 1 \\ -1 \end{array} \right\} = \left\{ \begin{array}{c} -2 \\ 0 \end{array} \right\}$$

$$\mathbf{S}_1^T \mathbf{g}_1 = \{-1 \quad 1\} \left\{ \begin{array}{c} -2 \\ 0 \end{array} \right\} = 2$$

$$\mathbf{S}_1 \mathbf{S}_1^T = \left\{ \begin{array}{c} -1 \\ 1 \end{array} \right\} \{-1 \quad 1\} = \begin{bmatrix} 1 & -1 \\ -1 & 1 \end{bmatrix}$$

$$[B_1]\mathbf{g}_1 = \begin{bmatrix} 1 & 0 \\ 0 & 1 \end{bmatrix} \left\{ \begin{array}{c} -2 \\ 0 \end{array} \right\} = \left\{ \begin{array}{c} -2 \\ 0 \end{array} \right\}$$

$$([B_1]\mathbf{g}_1)^T = \left\{ \begin{array}{c} -2 \\ 0 \end{array} \right\}^T = \{-2 \quad 0\}$$

$$\mathbf{g}_1^T [B_1]\mathbf{g}_1 = \{-2 \quad 0\} \begin{bmatrix} 1 & 0 \\ 0 & 1 \end{bmatrix} \left\{ \begin{array}{c} -2 \\ 0 \end{array} \right\} = \{-2 \quad 0\} \left\{ \begin{array}{c} -2 \\ 0 \end{array} \right\} = 4$$

$$[M_1] = \lambda_1^* \frac{\mathbf{S}_1 \mathbf{S}_1^T}{\mathbf{S}_1^T \mathbf{g}_1} = 1 \left(\frac{1}{2} \right) \begin{bmatrix} 1 & -1 \\ -1 & 1 \end{bmatrix} = \begin{bmatrix} \dfrac{1}{2} & -\dfrac{1}{2} \\ -\dfrac{1}{2} & \dfrac{1}{2} \end{bmatrix}$$

$$[N_1] = -\frac{([B_1]\mathbf{g}_1)([B_1]\mathbf{g}_1)^{\mathrm{T}}}{\mathbf{g}_1^T[B_1]\mathbf{g}_1} = -\frac{\begin{Bmatrix} -2 \\ 0 \end{Bmatrix}\{-2\ 0\}}{4} = -\frac{1}{4}\begin{bmatrix} 4 & 0 \\ 0 & 0 \end{bmatrix} = -\begin{bmatrix} 1 & 0 \\ 0 & 0 \end{bmatrix}$$

$$[B_2] = [B_1] + [M_1] + [N_1] = \begin{bmatrix} 1 & 0 \\ 0 & 1 \end{bmatrix} + \begin{bmatrix} \dfrac{1}{2} & -\dfrac{1}{2} \\ -\dfrac{1}{2} & \dfrac{1}{2} \end{bmatrix} + \begin{bmatrix} -1 & 0 \\ 0 & 0 \end{bmatrix} = \begin{bmatrix} 0.5 & -0.5 \\ -0.5 & 1.5 \end{bmatrix}$$

Iteration 2 (i = 2)

The next search direction is determined as

$$\mathbf{S}_2 = -[B_2]\nabla f_2 = -\begin{bmatrix} 0.5 & -0.5 \\ -0.5 & 1.5 \end{bmatrix}\begin{Bmatrix} -1 \\ -1 \end{Bmatrix} = \begin{Bmatrix} 0 \\ 1 \end{Bmatrix}$$

To find the minimizing step length λ_2^* along $\mathbf{S}_2$, we minimize

$$f(\mathbf{X}_2 + \lambda_2\mathbf{S}_2) = f\left(\begin{Bmatrix} -1 \\ 1 \end{Bmatrix} + \lambda_2\begin{Bmatrix} 0 \\ 1 \end{Bmatrix}\right) = f\left(\begin{Bmatrix} -1 \\ 1 + \lambda_2 \end{Bmatrix}\right)$$

$$= -1 - (1 + \lambda_2) + 2(-1)^2 + 2(-1)(1 + \lambda_2) + (1 + \lambda_2)^2$$

$$= \lambda_2^2 - \lambda_2 - 1$$

with respect to λ_2. Since $df/d\lambda_2 = 0$ at $\lambda_2^* = \frac{1}{2}$, we obtain

$$\mathbf{X}_3 = \mathbf{X}_2 + \lambda_2^* = \begin{Bmatrix} -1 \\ 1 \end{Bmatrix} + \frac{1}{2}\begin{Bmatrix} 0 \\ 1 \end{Bmatrix} = \begin{Bmatrix} -1 \\ 1.5 \end{Bmatrix}$$

This point can be identified to be optimum since

$$\nabla f_3 = \begin{Bmatrix} 0 \\ 0 \end{Bmatrix} \quad \text{and} \quad \|\nabla f_3\| = 0 < \varepsilon$$

6.15 BROYDEN–FLETCHER–GOLDFARB–SHANNO METHOD

As stated earlier, a major difference between the DFP and BFGS methods is that in the BFGS method, the Hessian matrix is updated iteratively rather than the inverse of the Hessian matrix. The BFGS method can be described by the following steps.

1. Start with an initial point $\mathbf{X}_1$ and a $n \times n$ positive definite symmetric matrix $[B_1]$ as an initial estimate of the inverse of the Hessian matrix of f. In the absence of additional information, $[B_1]$ is taken as the identity matrix $[I]$. Compute the gradient vector $\nabla f_1 = \nabla f(\mathbf{X}_1)$ and set the iteration number as $i = 1$.

2. Compute the gradient of the function, ∇f_i, at point $\mathbf{X}_i$, and set

$$\mathbf{S}_i = -[B_i]\nabla f_i \tag{6.134}$$

3. Find the optimal step length λ_i^* in the direction $\mathbf{S}_i$ and set

$$\mathbf{X}_{i+1} = \mathbf{X}_i + \lambda_i^* \mathbf{S}_i \qquad (6.135)$$

4. Test the point $\mathbf{X}_{i+1}$ for optimality. If $\|\nabla f_{i+1}\| \leq \varepsilon$, where ε is a small quantity, take $\mathbf{X}^* \approx \mathbf{X}_{i+1}$ and stop the process. Otherwise, go to step 5.

5. Update the Hessian matrix as

$$[B_{i+1}] = [B_i] + \left(1 + \frac{\mathbf{g}_i^T [B_i] \mathbf{g}_i}{\mathbf{d}_i^T \mathbf{g}_i}\right) \frac{\mathbf{d}_i \mathbf{d}_i^T}{\mathbf{d}_i^T \mathbf{g}_i} - \frac{\mathbf{d}_i \mathbf{g}_i^T [B_i]}{\mathbf{d}_i^T \mathbf{g}_i} - \frac{[B_i] \mathbf{g}_i \mathbf{d}_i^T}{\mathbf{d}_i^T \mathbf{g}_i} \qquad (6.136)$$

where

$$\mathbf{d}_i = \mathbf{X}_{i+1} - \mathbf{X}_i = \lambda_i^* \mathbf{S}_i \qquad (6.137)$$

$$\mathbf{g}_i = \nabla f_{i+1} - \nabla f_i = \nabla f(\mathbf{X}_{i+1}) - \nabla f(\mathbf{X}_i) \qquad (6.138)$$

6. Set the new iteration number as $i = i + 1$ and go to step 2.

Remarks:

1. The BFGS method can be considered as a quasi-Newton, conjugate gradient, and variable metric method.

2. Since the inverse of the Hessian matrix is approximated, the BFGS method can be called an indirect update method.

3. If the step lengths λ_i^* are found accurately, the matrix, $[B_i]$, retains its positive definiteness as the value of i increases. However, in practical application, the matrix $[B_i]$ might become indefinite or even singular if λ_i^* are not found accurately. As such, periodical resetting of the matrix $[B_i]$ to the identity matrix $[I]$ is desirable. However, numerical experience indicates that the BFGS method is less influenced by errors in λ_i^* than is the DFP method.

4. It has been shown that the BFGS method exhibits superlinear convergence near $\mathbf{X}^*$ [6.19].

Example 6.16 Minimize $f(x_1, x_2) = x_1 - x_2 + 2x_1^2 + 2x_1 x_2 + x_2^2$ from the starting point $\mathbf{X}_1 = \left\{ {0 \atop 0} \right\}$ using the BFGS method with

$$[B_1] = \begin{bmatrix} 1 & 0 \\ 0 & 1 \end{bmatrix} \qquad \varepsilon = 0.01.$$

SOLUTION

Iteration 1 ($i = 1$)

Here

$$\nabla f_1 = \nabla f(\mathbf{X}_1) = \left\{ {1 + 4x_1 + 2x_2 \atop -1 + 2x_1 + 2x_2} \right\}\bigg|_{(0,0)} = \left\{ {1 \atop -1} \right\}$$

and hence

$$\mathbf{S}_1 = -[B_1] \nabla f_1 = -\begin{bmatrix} 1 & 0 \\ 0 & 1 \end{bmatrix} \left\{ {1 \atop -1} \right\} = \left\{ {-1 \atop 1} \right\}$$

To find the minimizing step length λ_1^* along $\mathbf{S}_1$, we minimize

$$f(\mathbf{X}_1 + \lambda_1 \mathbf{S}_1) = f\left(\begin{Bmatrix} 0 \\ 0 \end{Bmatrix} + \lambda_1 \begin{Bmatrix} -1 \\ 1 \end{Bmatrix}\right) = f(-\lambda_1, \lambda_1) = \lambda_1^2 - 2\lambda_1$$

with respect to λ_1. Since $df/d\lambda_1 = 0$ at $\lambda_1^* = 1$, we obtain

$$\mathbf{X}_2 = \mathbf{X}_1 + \lambda_1^* \mathbf{S}_1 = \begin{Bmatrix} 0 \\ 0 \end{Bmatrix} + 1 \begin{Bmatrix} -1 \\ 1 \end{Bmatrix} = \begin{Bmatrix} -1 \\ 1 \end{Bmatrix}$$

Since $\nabla f_2 = \nabla f(\mathbf{X}_2) = \begin{Bmatrix} -1 \\ -1 \end{Bmatrix}$ and $\|\nabla f_2\| = 1.4142 > \varepsilon$, we proceed to update the matrix $[B_i]$ by computing

$$\mathbf{g}_1 = \nabla f_2 - \nabla f_1 = \begin{Bmatrix} -1 \\ -1 \end{Bmatrix} - \begin{Bmatrix} 1 \\ -1 \end{Bmatrix} = \begin{Bmatrix} -2 \\ 0 \end{Bmatrix}$$

$$\mathbf{d}_1 = \lambda_1^* \mathbf{S}_1 = 1 \begin{Bmatrix} -1 \\ 1 \end{Bmatrix} = \begin{Bmatrix} -1 \\ 1 \end{Bmatrix}$$

$$\mathbf{d}_1 \mathbf{d}_1^T = \begin{Bmatrix} -1 \\ 1 \end{Bmatrix} \{-1 \quad 1\} = \begin{bmatrix} 1 & -1 \\ -1 & 1 \end{bmatrix}$$

$$\mathbf{d}_1^T \mathbf{g}_1 = \{-1 \quad 1\} \begin{Bmatrix} -2 \\ 0 \end{Bmatrix} = 2$$

$$\mathbf{d}_1 \mathbf{g}_1^T = \begin{Bmatrix} -1 \\ 1 \end{Bmatrix} \{-2 \quad 0\} = \begin{bmatrix} 2 & 0 \\ -2 & 0 \end{bmatrix}$$

$$\mathbf{g}_1 \mathbf{d}_1^T = \begin{Bmatrix} -2 \\ 0 \end{Bmatrix} \{-1 \quad 1\} = \begin{bmatrix} 2 & -2 \\ 0 & 0 \end{bmatrix}$$

$$\mathbf{g}_1^T [B_1] \mathbf{g}_1 = \{-2 \quad 0\} \begin{bmatrix} 1 & 0 \\ 0 & 1 \end{bmatrix} \begin{Bmatrix} -2 \\ 0 \end{Bmatrix} = \{-2 \quad 0\} \begin{Bmatrix} -2 \\ 0 \end{Bmatrix} = 4$$

$$\mathbf{d}_1 \mathbf{g}_1^T [B_1] = \begin{bmatrix} 2 & 0 \\ -2 & 0 \end{bmatrix} \begin{bmatrix} 1 & 0 \\ 0 & 1 \end{bmatrix} = \begin{bmatrix} 2 & 0 \\ -2 & 0 \end{bmatrix}$$

$$[B_1] \mathbf{g}_1 \mathbf{d}_1^T = \begin{bmatrix} 1 & 0 \\ 0 & 1 \end{bmatrix} \begin{bmatrix} 2 & -2 \\ 0 & 0 \end{bmatrix} = \begin{bmatrix} 2 & -2 \\ 0 & 0 \end{bmatrix}$$

Equation (6.136) gives

$$[B_2]| = \begin{bmatrix} 1 & 0 \\ 0 & 1 \end{bmatrix} + \left(1 + \frac{4}{2}\right) \frac{1}{2} \begin{bmatrix} 1 & -1 \\ -1 & 1 \end{bmatrix} - \frac{1}{2} \begin{bmatrix} 2 & 0 \\ -2 & 0 \end{bmatrix} - \frac{1}{2} \begin{bmatrix} 2 & -2 \\ 0 & 0 \end{bmatrix}$$

$$= \begin{bmatrix} 1 & 0 \\ 0 & 1 \end{bmatrix} + \begin{bmatrix} \frac{3}{2} & -\frac{3}{2} \\ -\frac{3}{2} & \frac{3}{2} \end{bmatrix} - \begin{bmatrix} 1 & 0 \\ -1 & 0 \end{bmatrix} - \begin{bmatrix} 1 & -1 \\ 0 & 0 \end{bmatrix} = \begin{bmatrix} \frac{1}{2} & -\frac{1}{2} \\ -\frac{1}{2} & \frac{5}{2} \end{bmatrix}$$

Iteration 2 (i = 2)

The next search direction is determined as

$$\mathbf{S}_2 = -[B_2]\nabla f_2 = -\begin{bmatrix} \frac{1}{2} & -\frac{1}{2} \\ -\frac{1}{2} & \frac{5}{2} \end{bmatrix}\begin{Bmatrix} -1 \\ -1 \end{Bmatrix} = \begin{Bmatrix} 0 \\ 2 \end{Bmatrix}$$

To find the minimizing step length λ_2^* along $\mathbf{S}_2$, we minimize

$$f(\mathbf{X}_2 + \lambda_2\mathbf{S}_2) = f\left(\begin{Bmatrix} -1 \\ 1 \end{Bmatrix} + \lambda_2\begin{Bmatrix} 0 \\ 2 \end{Bmatrix}\right) = f(-1, 1 + 2\lambda_2) = 4\lambda_2^2 - 2\lambda_2 - 1$$

with respect to λ_2. Since $df/d\lambda_2 = 0$ at $\lambda_2^* = \frac{1}{4}$, we obtain

$$\mathbf{X}_3 = \mathbf{X}_2 + \lambda_2^*\mathbf{S}_2 = \begin{Bmatrix} -1 \\ 1 \end{Bmatrix} + \frac{1}{4}\begin{Bmatrix} 0 \\ 2 \end{Bmatrix} = \begin{Bmatrix} -1 \\ \frac{3}{2} \end{Bmatrix}$$

This point can be identified to be optimum since

$$\nabla f_3 = \begin{Bmatrix} 0 \\ 0 \end{Bmatrix} \quad \text{and} \quad \|\nabla f_3\| = 0 < \varepsilon$$

6.16   TEST FUNCTIONS

The efficiency of an optimization algorithm is studied using a set of standard functions. Several functions, involving different number of variables, representing a variety of complexities have been used as test functions. Almost all the test functions presented in the literature are nonlinear least squares; that is, each function can be represented as

$$f(x_1, x_2, \ldots, x_n) = \sum_{i=1}^{m} f_i(x_1, x_2, \ldots, x_n)^2 \tag{6.139}$$

where n denotes the number of variables and m indicates the number of functions (f_i) that define the least-squares problem. The purpose of testing the functions is to show how well the algorithm works compared to other algorithms. Usually, each test function is minimized from a standard starting point. The total number of function evaluations required to find the optimum solution is usually taken as a measure of the efficiency of the algorithm. References [6.29] to [6.32] present a comparative study of the various unconstrained optimization techniques. Some of the commonly used test functions are given below.

1. Rosenbrock's parabolic valley [6.8]:

$$f(x_1, x_2) = 100(x_2 - x_1^2)^2 + (1 - x_1)^2 \tag{6.140}$$

$$\mathbf{X}_1 = \begin{Bmatrix} -1.2 \\ 1.0 \end{Bmatrix}, \quad \mathbf{X}^* = \begin{Bmatrix} 1 \\ 1 \end{Bmatrix}$$

$$f_1 = 24.0, \quad f^* = 0.0$$

2. A quadratic function:

$$f(x_1, x_2) = (x_1 + 2x_2 - 7)^2 + (2x_1 + x_2 - 5)^2 \qquad (6.141)$$

$$\mathbf{X}_1 = \begin{Bmatrix} 0 \\ 0 \end{Bmatrix}, \quad \mathbf{X}^* = \begin{Bmatrix} 1 \\ 3 \end{Bmatrix}$$

$$f_1 = 7.40, \quad f^* = 0.0$$

3. Powell's quartic function [6.7]:

$$f(x_1, x_2, x_3, x_4) = (x_1 + 10x_2)^2 + 5(x_3 - x_4)^2$$
$$+ (x_2 - 2x_3)^4 + 10(x_1 - x_4)^4 \qquad (6.142)$$

$$\mathbf{X}_1^{\mathrm{T}} = \{x_1 \ x_2 \ x_3 \ x_4\}_1 = \{3 \ -1 \ 0 \ 1\}, \quad \mathbf{X}^{*\mathrm{T}} = \{0 \ 0 \ 0 \ 0\}$$

$$f_1 = 215.0, \quad f^* = 0.0$$

4. Fletcher and Powell's helical valley [6.21]:

$$f(x_1, x_2, x_3) = 100 \left\{ [x_3 - 10\theta(x_1, x_2)]^2 + [\sqrt{x_1^2 + x_2^2} - 1]^2 \right\} + x_3^2 \qquad (6.143)$$

where

$$2\pi\theta(x_1, x_2) = \begin{cases} \arctan \dfrac{x_2}{x_1} & \text{if } x_1 > 0 \\[2mm] \pi + \arctan \dfrac{x_2}{x_1} & \text{if } x_1 < 0 \end{cases}$$

$$\mathbf{X}_1 = \begin{Bmatrix} -1 \\ 0 \\ 0 \end{Bmatrix}, \quad \mathbf{X}^* = \begin{Bmatrix} 1 \\ 0 \\ 0 \end{Bmatrix}$$

$$f_1 = 25,000.0, \quad f^* = 0.0$$

5. A nonlinear function of three variables [6.7]:

$$f(x_1, x_2, x_3) = \frac{1}{1 + (x_1 - x_2)^2} + \sin\left(\frac{1}{2}\pi x_2 x_3\right)$$
$$+ \exp\left[-\left(\frac{x_1 + x_3}{x_2} - 2 \right)^2 \right] \qquad (6.144)$$

$$\mathbf{X}_1 = \begin{Bmatrix} 0 \\ 1 \\ 2 \end{Bmatrix}, \quad \mathbf{X}^* = \begin{Bmatrix} 1 \\ 1 \\ 1 \end{Bmatrix}$$

$$f_1 = 1.5, \quad f^* = f_{\max} = 3.0$$

6. Freudenstein and Roth function [6.27]:

$$f(x_1, x_2) = \{-13 + x_1 + [(5 - x_2)x_2 - 2]x_2\}^2$$
$$+ \{-29 + x_1 + [(x_2 + 1)x_2 - 14]x_2\}^2 \qquad (6.145)$$

$$\mathbf{X}_1 = \left\{ \begin{array}{c} 0.5 \\ -2 \end{array} \right\}, \quad \mathbf{X}^* = \left\{ \begin{array}{c} 5 \\ 4 \end{array} \right\}, \quad \mathbf{X}^*_{\text{alternate}} = \left\{ \begin{array}{c} 11.41\ldots \\ -0.8968\ldots \end{array} \right\}$$

$$f_1 = 400.5, \quad f^* = 0.0, \quad f^*_{\text{alternate}} = 48.9842\ldots$$

7. Powell's badly scaled function [6.28]:

$$f(x_1, x_2) = (10{,}000x_1x_2 - 1)^2 + [\exp(-x_1) + \exp(-x_2) - 1.0001]^2 \quad (6.146)$$

$$\mathbf{X}_1 = \left\{ \begin{array}{c} 0 \\ 1 \end{array} \right\}, \quad \mathbf{X}^* = \left\{ \begin{array}{c} 1.098\ldots \times 10^{-5} \\ 9.106\ldots \end{array} \right\}$$

$$f_1 = 1.1354, \quad f^* = 0.0$$

8. Brown's badly scaled function [6.29]:

$$f(x_1, x_2) = (x_1 - 10^6)^2 + (x_2 - 2 \times 10^{-6})^2 + (x_1x_2 - 2)^2 \quad (6.147)$$

$$\mathbf{X}_1 = \left\{ \begin{array}{c} 1 \\ 1 \end{array} \right\}, \quad \mathbf{X}^* = \left\{ \begin{array}{c} 10^6 \\ 2 \times 10^{-6} \end{array} \right\}$$

$$f_1 \approx 10^{12}, \quad f^* = 0.0$$

9. Beale's function [6.29]:

$$f(x_1, x_2) = [1.5 - x_1(1 - x_2)]^2 + [2.25 - x_1(1 - x_2^2)]^2$$
$$+ [2.625 - x_1(1 - x_2^3)]^2 \quad (6.148)$$

$$\mathbf{X}_1 = \left\{ \begin{array}{c} 1 \\ 1 \end{array} \right\}, \quad \mathbf{X}^* = \left\{ \begin{array}{c} 3 \\ 0.5 \end{array} \right\}$$

$$f_1 = 14.203125, \quad f^* = 0.0$$

10. Wood's function [6.30]:

$$f(x_1, x_2, x_3, x_4) = [10(x_2 - x_1^2)]^2 + (1 - x_1)^2 + 90(x_4 - x_3^2)^2$$
$$+ (1 - x_3)^2 + 10(x_2 + x_4 - 2)^2 + 0.1(x_2 - x_4) \quad (6.149)$$

$$\mathbf{X}_1 = \left\{ \begin{array}{c} -3 \\ -1 \\ -3 \\ -1 \end{array} \right\}, \quad \mathbf{X}^* = \left\{ \begin{array}{c} 1 \\ 1 \\ 1 \\ 1 \end{array} \right\}$$

$$f_1 = 19192.0, \quad f^* = 0.0$$

6.17   MATLAB SOLUTION OF UNCONSTRAINED OPTIMIZATION PROBLEMS

The solution of multivariable unconstrained minimization problems using the MATLAB function `fminunc` is illustrated in this section.

Example 6.17   Find the minimum of the Rosenbrock's parabolic valley function, given by Eq. (6.140), starting from initial point $\mathbf{X}_1 = \{-1.2 \ 1.0\}^{\text{T}}$.

SOLUTION

Step 1: Write an M-file `objfun.m` for the objective function.

```
function f= objfun (x)
f= 100* (x(2)-x(1) *x(1))^2+(1-x(1))^2;
```

Step 2: Invoke unconstrained optimization program (write this in new MATLAB file).

```
clc
clear all
warning off
x0 = [-1.2,1.0]; % Starting guess
fprintf ('The values of function value at starting
pointn');
f=objfun(x0)
options = optimset('LargeScale', 'off');
[x, fval] = fminunc (@objfun,x0,options)
```

This produces the solution or ouput as follows:

```
The values of function value at starting point
f=
 24.2000
Optimization terminated: relative infinity-norm of gradi-
ent less than options.TolFun.
x=
 1.0000 1.0000
fval=
 2.8336e-011
```

REFERENCES AND BIBLIOGRAPHY

6.1   S. S. Rao, *The Finite Element Method in Engineering*, 4th ed., Elsevier Butterworth Heinemann, Burlington, MA, 2005.

6.2   T. F. Edgar and D. M. Himmelblau, *Optimization of Chemical Processes*, McGraw-Hill, New York, 1988.

6.3   R. L. Fox, *Optimization Methods for Engineering Design*, Addison-Wesley, Reading, MA, 1971.

6.4   W. E. Biles and J. J. Swain, *Optimization and Industrial Experimentation*, Wiley, New York, 1980.

6.5   C. R. Hicks, *Fundamental Concepts in the Design of Experiments*, Saunders College Publishing, Fort Worth, TX, 1993.

6.6   R. Hooke and T. A. Jeeves, Direct search solution of numerical and statistical problems, *Journal of the ACM*, Vol. 8, No. 2, pp. 212–229, 1961.

6.7   M.J.D. Powell, An efficient method for finding the minimum of a function of several variables without calculating derivatives, *Computer Journal*, Vol. 7, No. 4, pp. 303–307, 1964.

6.8 H. H. Rosenbrock, An automatic method for finding the greatest or least value of a function, *Computer Journal*, Vol. 3, No. 3, pp. 175–184, 1960.

6.9 S. S. Rao, *Optimization: Theory and Applications*, 2nd ed., Wiley Eastern, New Delhi, 1984.

6.10 W. Spendley, G. R. Hext, and F. R. Himsworth, Sequential application of simplex designs in optimization and evolutionary operation, *Technometrics*, Vol. 4, p. 441, 1962.

6.11 J. A. Nelder and R. Mead, A simplex method for function minimization, *Computer Journal*, Vol. 7, p. 308, 1965.

6.12 A. L. Cauchy, Méthode générale pour la résolution des systèmes d'équations simultanées, *Comptes Rendus de l'Academie des Sciences*, Paris, Vol. 25, pp. 536–538, 1847.

6.13 R. Fletcher and C. M. Reeves, Function minimization by conjugate gradients, *Computer Journal*, Vol. 7, No. 2, pp. 149–154, 1964.

6.14 M. R. Hestenes and E. Stiefel, *Methods of Conjugate Gradients for Solving Linear Systems*, Report 1659, National Bureau of Standards, Washington, DC, 1952.

6.15 D. Marquardt, An algorithm for least squares estimation of nonlinear parameters, *SIAM Journal of Applied Mathematics*, Vol. 11, No. 2, pp. 431–441, 1963.

6.16 C. G. Broyden, Quasi-Newton methods and their application to function minimization, *Mathematics of Computation*, Vol. 21, p. 368, 1967.

6.17 C. G. Broyden, J. E. Dennis, and J. J. More, On the local and superlinear convergence of quasi-Newton methods, *Journal of the Institute of Mathematics and Its Applications*, Vol. 12, p. 223, 1975.

6.18 H. Y. Huang, Unified approach to quadratically convergent algorithms for function minimization, *Journal of Optimization Theory and Applications*, Vol. 5, pp. 405–423, 1970.

6.19 J. E. Dennis, Jr., and J. J. More, Quasi-Newton methods, motivation and theory, *SIAM Review*, Vol. 19, No. 1, pp. 46–89, 1977.

6.20 W. C. Davidon, *Variable Metric Method of Minimization*, Report ANL-5990, Argonne National Laboratory, Argonne, IL, 1959.

6.21 R. Fletcher and M.J.D. Powell, A rapidly convergent descent method for minimization, *Computer Journal*, Vol. 6, No. 2, pp. 163–168, 1963.

6.22 G. G. Broyden, The convergence of a class of double-rank minimization algorithms, Parts I and II, *Journal of the Institute of Mathematics and Its Applications*, Vol. 6, pp. 76–90, 222-231, 1970.

6.23 R. Fletcher, A new approach to variable metric algorithms, *Computer Journal*, Vol. 13, pp. 317–322, 1970.

6.24 D. Goldfarb, A family of variable metric methods derived by variational means, *Mathematics of Computation*, Vol. 24, pp. 23–26, 1970.

6.25 D. F. Shanno, Conditioning of quasi-Newton methods for function minimization, *Mathematics of Computation*, Vol. 24, pp. 647–656, 1970.

6.26 M.J.D. Powell, An iterative method for finding stationary values of a function of several variables, *Computer Journal*, Vol. 5, pp. 147–151, 1962.

6.27 F. Freudenstein and B. Roth, Numerical solution of systems of nonlinear equations, *Journal of ACM*, Vol. 10, No. 4, pp. 550–556, 1963.

6.28 M.J.D. Powell, A hybrid method for nonlinear equations, pp. 87–114 in *Numerical Methods for Nonlinear Algebraic Equations*, P. Rabinowitz, Ed., Gordon & Breach, New York, 1970.

6.29 J. J. More, B. S. Garbow, and K. E. Hillstrom, Testing unconstrained optimization software, *ACM Transactions on Mathematical Software*, Vol. 7, No. 1, pp. 17–41, 1981.

6.30   A. R. Colville, *A Comparative Study of Nonlinear Programming Codes*, Report 320-2949, IBM New York Scientific Center, 1968.

6.31   E. D. Eason and R. G. Fenton, A comparison of numerical optimization methods for engineering design, *ASME Journal of Engineering Design*, Vol. 96, pp. 196–200, 1974.

6.32   R.W.H. Sargent and D. J. Sebastian, Numerical experience with algorithms for unconstrained minimization, pp. 45–113 in *Numerical Methods for Nonlinear Optimization*, F. A. Lootsma, Ed., Academic Press, London, 1972.

6.33   D. F. Shanno, Recent advances in gradient based unconstrained optimization techniques for large problems, *ASME Journal of Mechanisms, Transmissions, and Automation in Design*, Vol. 105, pp. 155–159, 1983.

6.34   S. S. Rao, *Mechanical Vibrations*, 4th ed., Pearson Prentice Hall, Upper Saddle River, NJ, 2004.

6.35   R. T. Haftka and Z. Gürdal, *Elements of Structural Optimization*, 3rd ed., Kluwer Academic, Dordrecht, The Netherlands, 1992.

6.36   J. Kowalik and M. R. Osborne, *Methods for Unconstrained Optimization Problems*, American Elsevier, New York, 1968.

REVIEW QUESTIONS

6.1   State the necessary and sufficient conditions for the unconstrained minimum of a function.

6.2   Give three reasons why the study of unconstrained minimization methods is important.

6.3   What is the major difference between zeroth-, first-, and second-order methods?

6.4   What are the characteristics of a direct search method?

6.5   What is a descent method?

6.6   Define each term:

(a) Pattern directions

(b) Conjugate directions

(c) Simplex

(d) Gradient of a function

(e) Hessian matrix of a function

6.7   State the iterative approach used in unconstrained optimization.

6.8   What is quadratic convergence?

6.9   What is the difference between linear and superlinear convergence?

6.10   Define the condition number of a square matrix.

6.11   Why is the scaling of variables important?

6.12   What is the difference between random jumping and random walk methods?

6.13   Under what conditions are the processes of reflection, expansion, and contraction used in the simplex method?

6.14   When is the grid search method preferred in minimizing an unconstrained function?

6.15   Why is a quadratically convergent method considered to be superior for the minimization of a nonlinear function?

6.16   Why is Powell's method called a pattern search method?

6.17   What are the roles of univariate and pattern moves in the Powell's method?

6.18   What is univariate method?

6.19   Indicate a situation where a central difference formula is not as accurate as a forward difference formula.

6.20   Why is a central difference formula more expensive than a forward or backward difference formula in finding the gradient of a function?

6.21   What is the role of one-dimensional minimization methods in solving an unconstrained minimization problem?

6.22   State possible convergence criteria that can be used in direct search methods.

6.23   Why is the steepest descent method not efficient in practice, although the directions used are the best directions?

6.24   What are rank 1 and rank 2 updates?

6.25   How are the search directions generated in the Fletcher–Reeves method?

6.26   Give examples of methods that require n^2, n, and 1 one-dimensional minimizations for minimizing a quadratic in n variables.

6.27   What is the reason for possible divergence of Newton's method?

6.28   Why is a conjugate directions method preferred in solving a general nonlinear problem?

6.29   What is the difference between Newton and quasi-Newton methods?

6.30   What is the basic difference between DFP and BFGS methods?

6.31   Why are the search directions reset to the steepest descent directions periodically in the DFP method?

6.32   What is a metric? Why is the DFP method considered as a variable metric method?

6.33   Answer true or false:

    (a) A conjugate gradient method can be called a conjugate directions method.
    (b) A conjugate directions method can be called a conjugate gradient method.
    (c) In the DFP method, the Hessian matrix is sequentially updated directly.
    (d) In the BFGS method, the inverse of the Hessian matrix is sequentially updated.
    (e) The Newton method requires the inversion of an $n \times n$ matrix in each iteration.
    (f) The DFP method requires the inversion of an $n \times n$ matrix in each iteration.
    (g) The steepest descent directions are the best possible directions.
    (h) The central difference formula always gives a more accurate value of the gradient than does the forward or backward difference formula.
    (i) Powell's method is a conjugate directions method.
    (j) The univariate method is a conjugate directions method.

PROBLEMS

6.1 A bar is subjected to an axial load, P_0, as shown in Fig. 6.17. By using a one-finite-element model, the axial displacement, $u(x)$, can be expressed as [6.1]

$$u(x) = \{N_1(x) \quad N_2(x)\} \begin{Bmatrix} u_1 \\ u_2 \end{Bmatrix}$$

where $N_i(x)$ are called the shape functions:

$$N_1(x) = 1 - \frac{x}{l}, \quad N_2(x) = \frac{x}{l}$$

and u_1 and u_2 are the end displacements of the bar. The deflection of the bar at point Q can be found by minimizing the potential energy of the bar (f), which can be expressed as

$$f = \frac{1}{2} \int_0^l EA \left(\frac{\partial u}{\partial x}\right)^2 dx - P_0 u_2$$

where E is Young's modulus and A is the cross-sectional area of the bar. Formulate the optimization problem in terms of the variables u_1 and u_2 for the case $P_0 l / EA = 1$.

6.2 The natural frequencies of the tapered cantilever beam (ω) shown in Fig. 6.18, based on the Rayleigh-Ritz method, can be found by minimizing the function [6.34]:

$$f(c_1, c_2) = \frac{\dfrac{Eh^3}{3l^2}\left(\dfrac{c_1^2}{4} + \dfrac{c_2^2}{10} + \dfrac{c_1 c_2}{5}\right)}{\rho h l \left(\dfrac{c_1^2}{30} + \dfrac{c_2^2}{280} + \dfrac{2 c_1 c_2}{105}\right)}$$

with respect to c_1 and c_2, where $f = \omega^2$, E is Young's modulus, and ρ is the density. Plot the graph of $3 f \rho l^3 / E h^2$ in (c_1, c_2) space and identify the values of ω_1 and ω_2.

6.3 The Rayleigh's quotient corresponding to the three-degree-of-freedom spring–mass system shown in Fig. 6.19 is given by [6.34]

$$R(\mathbf{X}) = \frac{\mathbf{X}^T[K]\mathbf{X}}{\mathbf{X}^T[M]\mathbf{X}}$$

where

$$[K] = k \begin{bmatrix} 2 & -1 & 0 \\ -1 & 2 & -1 \\ 0 & -1 & 1 \end{bmatrix}, \quad [M] = \begin{bmatrix} 1 & 0 & 0 \\ 0 & 1 & 0 \\ 0 & 0 & 1 \end{bmatrix}, \quad \mathbf{X} = \begin{Bmatrix} x_1 \\ x_2 \\ x_3 \end{Bmatrix}$$

It is known that the fundamental natural frequency of vibration of the system can be found by minimizing $R(\mathbf{X})$. Derive the expression of $R(\mathbf{X})$ in terms of $x_1, x_2,$ and x_3 and suggest a suitable method for minimizing the function $R(\mathbf{X})$.

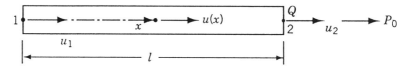

Figure 6.17   Bar subjected to an axial load.

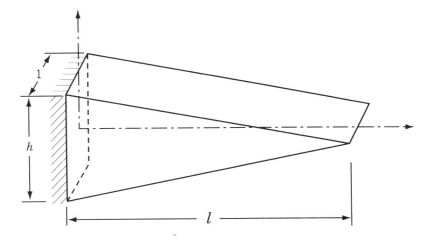

Figure 6.18 Tapered cantilever beam.

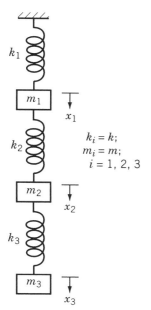

Figure 6.19 Three-degree-of-freedom spring–mass system.

6.4 The steady-state temperatures at points 1 and 2 of the one-dimensional fin (x_1 and x_2) shown in Fig. 6.20 correspond to the minimum of the function [6.1]:

$$f(x_1, x_2) = 0.6382x_1^2 + 0.3191x_2^2 - 0.2809x_1x_2$$

$$- 67.906x_1 - 14.290x_2$$

Plot the function f in the (x_1, x_2) space and identify the steady-state temperatures of points 1 and 2 of the fin.

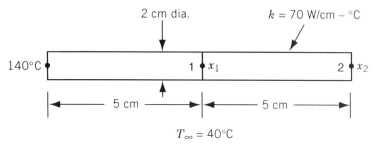

Figure 6.20   Straight fin.

6.5   Figure 6.21 shows two bodies, A and B, connected by four linear springs. The springs are at their natural positions when there is no force applied to the bodies. The displacements x_1 and x_2 of the bodies under any applied force can be found by minimizing the potential energy of the system. Find the displacements of the bodies when forces of 1000 lb and 2000 lb are applied to bodies A and B, respectively, using Newton's method. Use the starting vector, $\mathbf{X}_1 = \begin{Bmatrix} 0 \\ 0 \end{Bmatrix}$. *Hint:*

Potential energy of the system = strain energy of springs $-$ potential of applied loads

where the strain energy of a spring of stiffness k and end displacements x_1 and x_2 is given by $\frac{1}{2}k(x_2 - x_1)^2$ and the potential of the applied force, F_i, is given by $x_i F_i$.

6.6   The potential energy of the two-bar truss shown in Fig. 6.22 under the applied load P is given by

$$f(x_1, x_2) = \frac{EA}{s}\left(\frac{l}{2s}\right)^2 x_1^2 + \frac{EA}{s}\left(\frac{h}{s}\right)^2 x_2^2 - Px_1 \cos\theta - Px_2 \sin\theta$$

where E is Young's modulus, A the cross-sectional area of each member, l the span of the truss, s the length of each member, h the depth of the truss, θ the angle at which load is applied, x_1 the horizontal displacement of free node, and x_2 the vertical displacement of the free node.

(a) Simplify the expression of f for the data $E = 207 \times 10^9$ Pa, $A = 10^{-5}$ m^2, $l = 1.5$ m, $h = 4$ m, $P = 10,000$ N, and $\theta = 30°$.

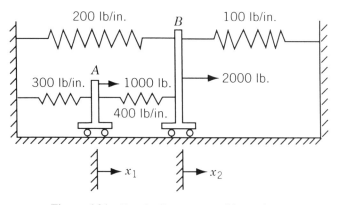

Figure 6.21   Two bodies connected by springs.

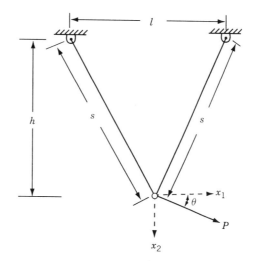

Figure 6.22   Two-bar truss.

(b) Find the steepest descent direction, $\mathbf{S}_1$, of f at the trial vector $\mathbf{X}_1 = \{ {}^0_0 \}$.

(c) Derive the one-dimensional minimization problem, $f(\lambda)$, at $\mathbf{X}_1$ along the direction $\mathbf{S}_1$.

(d) Find the optimal step length λ^* using the calculus method and find the new design vector $\mathbf{X}_2$.

6.7 Three carts, interconnected by springs, are subjected to the loads P_1, P_2, and P_3 as shown in Fig. 6.23. The displacements of the carts can be found by minimizing the potential energy of the system (f):

$$f(\mathbf{X}) = \tfrac{1}{2}\mathbf{X}^{\mathrm{T}}[K]\mathbf{X} - \mathbf{X}^{\mathrm{T}}\mathbf{P}$$

where

$$[K] = \begin{bmatrix} k_1 + k_4 + k_5 & -k_4 & -k_5 \\ -k_4 & k_2 + k_4 + k_6 & -k_6 \\ -k_5 & -k_6 & k_3 + k_5 + k_6 + k_7 + k_8 \end{bmatrix}$$

$$\mathbf{P} = \begin{Bmatrix} P_1 \\ P_2 \\ P_3 \end{Bmatrix} \quad \text{and} \quad \mathbf{X} = \begin{Bmatrix} x_1 \\ x_2 \\ x_3 \end{Bmatrix}$$

Derive the function $f(x_1, x_2, x_3)$ for the following data: $k_1 = 5000$ N/m , $k_2 = 1500$ N/m, $k_3 = 2000$ N/m, $k_4 = 1000$ N/m, $k_5 = 2500$ N/m, $k_6 = 500$ N/m, $k_7 = 3000$ N/m, $k_8 = 3500$ N/m, $P_1 = 1000$ N, $P_2 = 2000$ N, and $P_3 = 3000$ N. Complete one iteration of Newton's method and find the equilibrium configuration of the carts. Use $\mathbf{X}_1 = \{0\ 0\ 0\}^{\mathrm{T}}$.

6.8 Plot the contours of the following function over the region $(-5 \le x_1 \le 5, -3 \le x_2 \le 6)$ and identify the optimum point:

$$f(x_1, x_2) = (x_1 + 2x_2 - 7)^2 + (2x_1 + x_2 - 5)^2$$

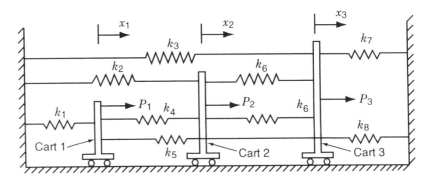

Figure 6.23   Three carts interconnected by springs.

6.9   Plot the contours of the following function in the two dimensional (x_1, x_2) space over the region $(-4 \le x_1 \le 4, -3 \le x_2 \le 6)$ and identify the optimum point:

$$f(x_1, x_2) = 2(x_2 - x_1^2)^2 + (1 - x_1)^2$$

6.10   Consider the problem

$$f(x_1, x_2) = 100(x_2 - x_1^2)^2 + (1 - x_1)^2$$

Plot the contours of f over the region $(-4 \le x_1 \le 4, -3 \le x_2 \le 6)$ and identify the optimum point.

6.11   It is required to find the solution of a system of linear algebraic equations given by $[A]\mathbf{X} = \mathbf{b}$, where $[A]$ is a known $n \times n$ symmetric positive-definite matrix and $\mathbf{b}$ is an n-component vector of known constants. Develop a scheme for solving the problem as an unconstrained minimization problem.

6.12   Solve the following equations using the steepest descent method (two iterations only) with the starting point, $\mathbf{X}_1 = \{0\ 0\ 0\}$:

$$2x_1 + x_2 = 4, \quad x_1 + 2x_2 + x_3 = 8, \quad x_2 + 3x_3 = 11$$

6.13   An electric power of 100 MW generated at a hydroelectric power plant is to be transmitted 400 km to a stepdown transformer for distribution at 11 kV. The power dissipated due to the resistance of conductors is $i^2 c^{-1}$, where i is the line current in amperes and c is the conductance in mhos. The resistance loss, based on the cost of power delivered, can be expressed as $0.15i^2 c^{-1}$ dollars. The power transmitted (k) is related to the transmission line voltage at the power plant (e) by the relation $k = \sqrt{3}ei$, where e is in kilovolts. The cost of conductors is given by $2c$ millions of dollars, and the investment in equipment needed to accommodate the voltage e is given by $500e$ dollars. Find the values of e and c to minimize the total cost of transmission using Newton's method (one iteration only).

6.14   Find a suitable transformation of variables to reduce the condition number of the Hessian matrix of the following function to one:

$$f = 2x_1^2 + 16x_2^2 - 2x_1 x_2 - x_1 - 6x_2 - 5$$

6.15 Find a suitable transformation or scaling of variables to reduce the condition number of the Hessian matrix of the following function to one:

$$f = 4x_1^2 + 3x_2^2 - 5x_1x_2 - 8x_1 + 10$$

6.16 Determine whether the following vectors serve as conjugate directions for minimizing the function $f = 2x_1^2 + 16x_2^2 - 2x_1x_2 - x_1 - 6x_2 - 5$.

(a) $\mathbf{S}_1 = \begin{Bmatrix} 15 \\ -1 \end{Bmatrix}, \quad \mathbf{S}_2 = \begin{Bmatrix} 1 \\ 1 \end{Bmatrix}$

(b) $\mathbf{S}_1 = \begin{Bmatrix} -1 \\ 15 \end{Bmatrix}, \quad \mathbf{S}_2 = \begin{Bmatrix} 1 \\ 1 \end{Bmatrix}$

6.17 Consider the problem:

$$\text{Minimize } f = x_1 - x_2 + 2x_1^2 + 2x_1x_2 + x_2^2$$

Find the solution of this problem in the range $-10 \le x_i \le 10$, $i = 1, 2$, using the random jumping method. Use a maximum of 10,000 function evaluations.

6.18 Consider the problem:

$$\text{Minimize } f = 6x_1^2 - 6x_1x_2 + 2x_2^2 - x_1 - 2x_2$$

Find the minimum of this function in the range $-5 \le x_i \le 5$, $i = 1, 2$, using the random walk method with direction exploitation.

6.19 Find the condition number of each matrix.

(a) $[A] = \begin{bmatrix} 1 & 2 \\ 1.0001 & 2 \end{bmatrix}$

(b) $[B] = \begin{bmatrix} 3.9 & 1.6 \\ 6.8 & 2.9 \end{bmatrix}$

6.20 Perform two iterations of the Newton's method to minimize the function

$$f(x_1, x_2) = 100(x_2 - x_1^2)^2 + (1 - x_1)^2$$

from the starting point $\begin{Bmatrix} -1.2 \\ 1.0 \end{Bmatrix}$.

6.21 Perform two iterations of univariate method to minimize the function given in Problem 6.20 from the stated starting vector.

6.22 Perform four iterations of Powell's method to minimize the function given in Problem 6.20 from the stated starting point.

6.23 Perform two iterations of the steepest descent method to minimize the function given in Problem 6.20 from the stated starting point.

6.24 Perform two iterations of the Fletcher–Reeves method to minimize the function given in Problem 6.20 from the stated starting point.

6.25 Perform two iterations of the DFP method to minimize the function given in Problem 6.20 from the stated starting vector.

6.26 Perform two iterations of the BFGS method to minimize the function given in Problem 6.20 from the indicated starting point.

6.27 Perform two iterations of the Marquardt's method to minimize the function given in Problem 6.20 from the stated starting point.

6.28 Prove that the search directions used in the Fletcher–Reeves method are $[A]$-conjugate while minimizing the function

$$f(x_1, x_2) = x_1^2 + 4x_2^2$$

6.29 Generate a regular simplex of size 4 in a two-dimensional space using each base point:

(a) $\begin{Bmatrix} 4 \\ -3 \end{Bmatrix}$ (b) $\begin{Bmatrix} 1 \\ 1 \end{Bmatrix}$ (c) $\begin{Bmatrix} -1 \\ -2 \end{Bmatrix}$

6.30 Find the coordinates of the vertices of a simplex in a three-dimensional space such that the distance between vertices is 0.3 and one vertex is given by $(2, -1, -8)$.

6.31 Generate a regular simplex of size 3 in a three-dimensional space using each base point.

(a) $\begin{Bmatrix} 0 \\ 0 \\ 0 \end{Bmatrix}$ (b) $\begin{Bmatrix} 4 \\ 3 \\ 2 \end{Bmatrix}$ (c) $\begin{Bmatrix} 1 \\ -2 \\ 3 \end{Bmatrix}$

6.32 Find a vector $\mathbf{S}_2$ that is conjugate to the vector

$$\mathbf{S}_1 = \begin{Bmatrix} 2 \\ -3 \\ 6 \end{Bmatrix}$$

with respect to the matrix:

$$[A] = \begin{bmatrix} 1 & 2 & 3 \\ 2 & 5 & 6 \\ 3 & 6 & 9 \end{bmatrix}$$

6.33 Compare the gradients of the function $f(\mathbf{X}) = 100(x_2 - x_1^2)^2 + (1 - x_1)^2$ at $\mathbf{X} = \begin{Bmatrix} 0.5 \\ 0.5 \end{Bmatrix}$ given by the following methods:

(a) Analytical differentiation
(b) Central difference method
(c) Forward difference method
(d) Backward difference method

Use a perturbation of 0.005 for x_1 and x_2 in the finite-difference methods.

6.34 It is required to evaluate the gradient of the function

$$f(x_1, x_2) = 100(x_2 - x_1^2)^2 + (1 - x_1)^2$$

at point $\mathbf{X} = \begin{Bmatrix} 0.5 \\ 0.5 \end{Bmatrix}$ using a finite-difference scheme. Determine the step size Δx to be used to limit the error in any of the components, $\partial f / \partial x_i$, to 1 % of the exact value, in the following methods:

(a) Central difference method
(b) Forward difference method
(c) Backward difference method

6.35 Consider the minimization of the function

$$f = \frac{1}{x_1^2 + x_2^2 + 2}$$

Perform one iteration of Newton's method from the starting point $\mathbf{X}_1 = \begin{Bmatrix} 4 \\ 0 \end{Bmatrix}$ using Eq. (6.86). How much improvement is achieved with $\mathbf{X}_2$?

6.36 Consider the problem:

$$\text{Minimize } f = 2(x_1 - x_1^2)^2 + (1 - x_1)^2$$

If a base simplex is defined by the vertices

$$\mathbf{X}_1 = \begin{Bmatrix} 0 \\ 0 \end{Bmatrix}, \quad \mathbf{X}_2 = \begin{Bmatrix} 1 \\ 0 \end{Bmatrix}, \quad \mathbf{X}_3 = \begin{Bmatrix} 0 \\ 1 \end{Bmatrix}$$

find a sequence of four improved vectors using reflection, expansion, and/or contraction.

6.37 Consider the problem:

$$\text{Minimize } f = (x_1 + 2x_2 - 7)^2 + (2x_1 + x_2 - 5)^2$$

If a base simplex is defined by the vertices

$$\mathbf{X}_1 = \begin{Bmatrix} -2 \\ -2 \end{Bmatrix}, \quad \mathbf{X}_2 = \begin{Bmatrix} -3 \\ 0 \end{Bmatrix}, \quad \mathbf{X}_3 = \begin{Bmatrix} -1 \\ -1 \end{Bmatrix}$$

find a sequence of four improved vectors using reflection, expansion, and/or contraction.

6.38 Consider the problem:

$$f = 100(x_2 - x_1^2)^2 + (1 - x_1)^2$$

Find the solution of the problem using grid search with a step size $\Delta x_i = 0.1$ in the range $-3 \le x_i \le 3, i = 1, 2$.

6.39 Show that the property of quadratic convergence of conjugate directions is independent of the order in which the one-dimensional minimizations are performed by considering the minimization of

$$f = 6x_1^2 + 2x_2^2 - 6x_1x_2 - x_1 - 2x_2$$

using the conjugate directions $\mathbf{S}_1 = \begin{Bmatrix} 1 \\ 2 \end{Bmatrix}$ and $\mathbf{S}_2 = \begin{Bmatrix} 1 \\ 0 \end{Bmatrix}$ and the starting point $\mathbf{X}_1 = \begin{Bmatrix} 0 \\ 0 \end{Bmatrix}$.

6.40 Show that the optimal step length λ_i^* that minimizes $f(\mathbf{X})$ along the search direction $\mathbf{S}_i = -\nabla f_i$ is given by Eq. (6.75).

6.41 Show that β_2 in Eq. (6.76) is given by Eq. (6.77).

6.42 Minimize $f = 2x_1^2 + x_2^2$ from the starting point $(1, 2)$ using the univariate method (two iterations only).

6.43 Minimize $f = 2x_1^2 + x_2^2$ by using the steepest descent method with the starting point $(1, 2)$ (two iterations only).

6.44 Minimize $f = x_1^2 + 3x_2^2 + 6x_3^2$ by the Newton's method using the starting point as $(2, -1, 1)$.

6.45 Minimize $f = 4x_1^2 + 3x_2^2 - 5x_1x_2 - 8x_1$ starting from point $(0, 0)$ using Powell's method. Perform four iterations.

6.46 Minimize $f(x_1, x_2) = x_1^4 - 2x_1^2x_2 + x_1^2 + x_2^2 + 2x_1 + 1$ by the simplex method. Perform two steps of reflection, expansion, and/or contraction.

6.47 Solve the following system of equations using Newton's method of unconstrained minimization with the starting point

$$\mathbf{X}_1 = \begin{Bmatrix} 0 \\ 0 \\ 0 \end{Bmatrix}$$

$$2x_1 - x_2 + x_3 = -1, \quad x_1 + 2x_2 = 0, \quad 3x_1 + x_2 + 2x_3 = 3$$

6.48 It is desired to solve the following set of equations using an unconstrained optimization method:

$$x^2 + y^2 = 2, \quad 10x^2 - 10y - 5x + 1 = 0$$

Formulate the corresponding problem and complete two iterations of optimization using the DFP method starting from $\mathbf{X}_1 = \begin{Bmatrix} 0 \\ 0 \end{Bmatrix}$.

6.49 Solve Problem 6.48 using the BFGS method (two iterations only).

6.50 The following nonlinear equations are to be solved using an unconstrained optimization method:

$$2xy = 3, \quad x^2 - y = 2$$

Complete two one-dimensional minimization steps using the univariate method starting from the origin.

6.51 Consider the two equations

$$7x^3 - 10x - y = 1, \quad 8y^3 - 11y + x = 1$$

Formulate the problem as an unconstrained optimization problem and complete two steps of the Fletcher–Reeves method starting from the origin.

6.52 Solve the equations $5x_1 + 3x_2 = 1$ and $4x_1 - 7x_2 = 76$ using the BFGS method with the starting point $(0, 0)$.

6.53 Indicate the number of one-dimensional steps required for the minimization of the function $f = x_1^2 + x_2^2 - 2x_1 - 4x_2 + 5$ according to each scheme:

(a) Steepest descent method
(b) Fletcher–Reeves method
(c) DFP method
(d) Newton's method
(e) Powell's method
(f) Random search method
(g) BFGS method
(h) Univariate method

6.54 Same as Problem 6.53 for the following function:

$$f = (x_2 - x_1^2)^2 + (1 - x_1)^2$$

6.55 Verify whether the following search directions are [A]-conjugate while minimizing the function

$$f = x_1 - x_2 + 2x_1^2 + 2x_1x_2 + x_2^2$$

(a) $\mathbf{S}_1 = \begin{Bmatrix} -1 \\ 1 \end{Bmatrix}$, $\mathbf{S}_2 = \begin{Bmatrix} 1 \\ 0 \end{Bmatrix}$

(b) $\mathbf{S}_1 = \begin{Bmatrix} -1 \\ 1 \end{Bmatrix}$, $\mathbf{S}_2 = \begin{Bmatrix} 0 \\ 1 \end{Bmatrix}$

6.56 Solve the equations $x_1 + 2x_2 + 3x_3 = 14$, $x_1 - x_2 + x_3 = 1$, and $3x_1 - 2x_2 + x_3 = 2$ using Marquardt's method of unconstrained minimization. Use the starting point $\mathbf{X}_1 = \{0, 0, 0\}^T$.

6.57 Apply the simplex method to minimize the function f given in Problem 6.20. Use the point $(-1.2, 1.0)$ as the base point to generate an initial regular simplex of size 2 and go through three steps of reflection, expansion, and/or contraction.

6.58 Write a computer program to implement Powell's method using the golden section method of one-dimensional search.

6.59 Write a computer program to implement the Davidon–Fletcher–Powell method using the cubic interpolation method of one-dimensional search. Use a finite-difference scheme to evaluate the gradient of the objective function.

6.60 Write a computer program to implement the BFGS method using the cubic interpolation method of one-dimensional minimization. Use a finite-difference scheme to evaluate the gradient of the objective function.

6.61 Write a computer program to implement the steepest descent method of unconstrained minimization with the direct root method of one-dimensional search.

6.62 Write a computer program to implement the Marquardt method coupled with the direct root method of one-dimensional search.

6.63 Find the minimum of the quadratic function given by Eq. (6.141) starting from the solution $\mathbf{X}_1 = \{0, 0\}^T$ using MATLAB.

6.64 Find the minimum of the Powell's quatic function given by Eq. (6.142) starting from the solution $\mathbf{X}_1 = \{3, -1, 0, 1\}^T$ using MATLAB.

6.65 Find the minimum of the Fletcher and Powell's helical valley function given by Eq. (6.143) starting from the solution $\mathbf{X}_1 = \{-1, 0, 0\}^T$ using MATLAB.

6.66 Find the minimum of the nonlinear function given by Eq. (6.144) starting from the solution $\mathbf{X}_1 = \{0, 1, 2\}^T$ using MATLAB.

6.67 Find the minimum of the Wood's function given by Eq. (6.149) starting from the solution $\mathbf{X}_1 = \{-3, -1, -3, -1\}^T$ using MATLAB.

7

Nonlinear Programming III: Constrained Optimization Techniques

7.1 INTRODUCTION

This chapter deals with techniques that are applicable to the solution of the constrained optimization problem:

$$\text{Find } \mathbf{X} \text{ which minimizes } f(\mathbf{X})$$

subject to

$$g_j(\mathbf{X}) \le 0, \quad j = 1, 2, \ldots, m$$
$$h_k(\mathbf{X}) = 0, \quad k = 1, 2, \ldots, p \tag{7.1}$$

There are many techniques available for the solution of a constrained nonlinear programming problem. All the methods can be classified into two broad categories: direct methods and indirect methods, as shown in Table 7.1. In the *direct methods*, the constraints are handled in an explicit manner, whereas in most of the *indirect methods*, the constrained problem is solved as a sequence of unconstrained minimization problems. We discuss in this chapter all the methods indicated in Table 7.1.

7.2 CHARACTERISTICS OF A CONSTRAINED PROBLEM

In the presence of constraints, an optimization problem may have the following features [7.1, 7.51]:

1. The constraints may have no effect on the optimum point; that is, the constrained minimum is the same as the unconstrained minimum as shown in Fig. 7.1. In this case the minimum point $\mathbf{X}^*$ can be found by making use of the necessary and sufficient conditions

$$\nabla f|_{\mathbf{X}^*} = \mathbf{0} \tag{7.2}$$

$$\mathbf{J}_{\mathbf{X}^*} = \left[\frac{\partial^2 f}{\partial x_i \partial x_j} \right]_{\mathbf{X}^*} = \text{ positive definite} \tag{7.3}$$

Table 7.1   Constrained Optimization Techniques

Direct methods	Indirect methods
Random search methods	Transformation of variables technique
Heuristic search methods	Sequential unconstrained minimization
Complex method	techniques
Objective and constraint approximation	Interior penalty function method
methods	Exterior penalty function method
Sequential linear programming method	Augmented Lagrange multiplier method
Sequential quadratic programming method	
Methods of feasible directions	
Zoutendijk's method	
Rosen's gradient projection method	
Generalized reduced gradient method	

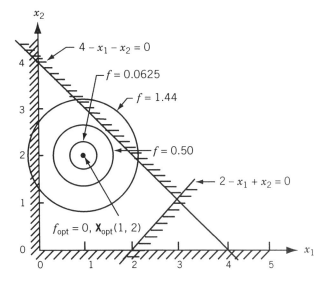

Figure 7.1   Constrained and unconstrained minima are the same (linear constraints).

However, to use these conditions, one must be certain that the constraints are not going to have any effect on the minimum. For simple optimization problems like the one shown in Fig. 7.1, it may be possible to determine beforehand whether or not the constraints have an influence on the minimum point. However, in most practical problems, even if we have a situation as shown in Fig. 7.1, it will be extremely difficult to identify it. Thus one has to proceed with the general assumption that the constraints have some influence on the optimum point.

2. The optimum (unique) solution occurs on a constraint boundary as shown in Fig. 7.2. In this case the Kuhn–Tucker necessary conditions indicate that the negative of the gradient must be expressible as a positive linear combination of the gradients of the active constraints.

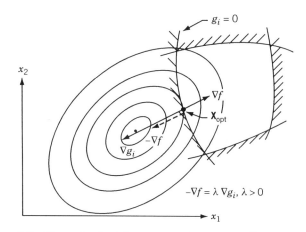

Figure 7.2   Constrained minimum occurring on a nonlinear constraint.

3. If the objective function has two or more unconstrained local minima, the constrained problem may have multiple minima as shown in Fig. 7.3.

4. In some cases, even if the objective function has a single unconstrained minimum, the constraints may introduce multiple local minima as shown in Fig. 7.4.

A constrained optimization technique must be able to locate the minimum in all the situations outlined above.

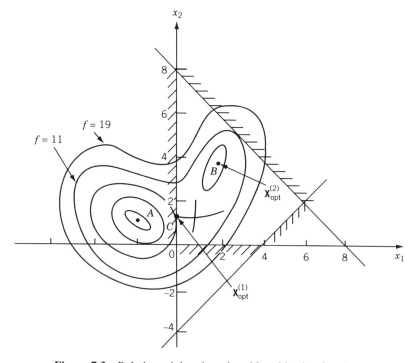

Figure 7.3   Relative minima introduced by objective function.

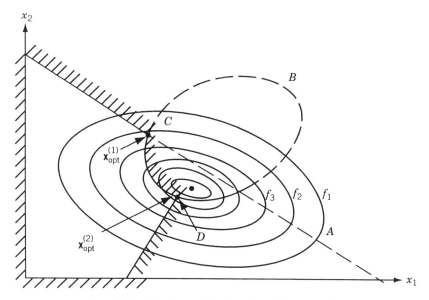

Figure 7.4   Relative minima introduced by constraints.

Direct Methods

7.3   RANDOM SEARCH METHODS

The random search methods described for unconstrained minimization (Section 6.2) can be used, with minor modifications, to solve a constrained optimization problem. The basic procedure can be described by the following steps:

1. Generate a trial design vector using one random number for each design variable.
2. Verify whether the constraints are satisfied at the trial design vector. Usually, the equality constraints are considered satisfied whenever their magnitudes lie within a specified tolerance. If any constraint is violated, continue generating new trial vectors until a trial vector that satisfies all the constraints is found.
3. If all the constraints are satisfied, retain the current trial vector as the best design if it gives a reduced objective function value compared to the previous best available design. Otherwise, discard the current feasible trial vector and proceed to step 1 to generate a new trial design vector.
4. The best design available at the end of generating a specified maximum number of trial design vectors is taken as the solution of the constrained optimization problem.

It can be seen that several modifications can be made to the basic procedure indicated above. For example, after finding a feasible trial design vector, a feasible direction can be generated (using random numbers) and a one-dimensional search can be conducted along the feasible direction to find an improved feasible design vector.

Another procedure involves constructing an unconstrained function, $F(\mathbf{X})$, by adding penalty for violating any constraint as (as described in Section 7.12):

$$F(\mathbf{X}) = f(\mathbf{X}) + a \sum_{j=1}^{m} [G_j(\mathbf{X})]^2 + b \sum_{k=1}^{p} [H_k(\mathbf{X})]^2 \qquad (7.4)$$

where

$$[G_j(\mathbf{X})]^2 = [\max(0, g_j(\mathbf{X}))]^2 \qquad (7.5)$$

$$[H_k(\mathbf{X})]^2 = h_k^2(\mathbf{X}) \qquad (7.6)$$

indicate the squares of violations of inequality and equality constraints, respectively, and a and b are constants. Equation (7.4) indicates that while minimizing the objective function $f(\mathbf{X})$, a positive penalty is added whenever a constraint is violated, the penalty being proportional to the square of the amount of violation. The values of the constants a and b can be adjusted to change the contributions of the penalty terms relative to the magnitude of the objective function.

Note that the random search methods are not efficient compared to the other methods described in this chapter. However, they are very simple to program and usually are reliable in finding a nearly optimal solution with a sufficiently large number of trial vectors. Also, these methods can find near global optimal solution even when the feasible region is nonconvex.

7.4 COMPLEX METHOD

In 1965, Box extended the simplex method of unconstrained minimization (discussed in Section 6.7) to solve constrained minimization problems of the type [7.2]:

$$\text{Minimize } f(\mathbf{X}) \qquad (7.7a)$$

subject to

$$g_j(\mathbf{X}) \leq 0, \qquad j = 1, 2, \dots, m \qquad (7.7b)$$

$$x_i^{(l)} \leq x_i \leq x_i^{(u)}, \qquad i = 1, 2, \dots, n \qquad (7.7c)$$

In general, the satisfaction of the side constraints (lower and upper bounds on the variables x_i) may not correspond to the satisfaction of the constraints $g_j(\mathbf{X}) \leq 0$. This method cannot handle nonlinear equality constraints. The formation of a sequence of geometric figures each having $k = n + 1$ vertices in an n-dimensional space (called the *simplex*) is the basic idea in the simplex method. In the complex method also, a sequence of geometric figures each having $k \geq n + 1$ vertices is formed to find the constrained minimum point. The method assumes that an initial feasible point $\mathbf{X}_1$ (which satisfies all the m constraints) is available.

Iterative Procedure

1. Find $k \geq n + 1$ points, each of which satisfies all m constraints. In actual practice, we start with only one feasible point $\mathbf{X}_1$, and the remaining $k - 1$ points

are found one at a time by the use of random numbers generated in the range 0 to 1, as

$$x_{i,j} = x_i^{(l)} + r_{i,j}(x_i^{(u)} - x_i^{(l)}), \quad i = 1, 2, \ldots, n, \ \ j = 2, 3, \ldots, k \qquad (7.8)$$

where $x_{i,j}$ is the ith component of the point $\mathbf{X}_j$, and $r_{i,j}$ is a random number lying in the interval $(0, 1)$. It is to be noted that the points $\mathbf{X}_2, \mathbf{X}_3, \ldots, \mathbf{X}_k$ generated according to Eq. (7.8) satisfy the side constraints, Eqs. (7.7c) but may not satisfy the constraints given by Eqs. (7.7b).

As soon as a new point $\mathbf{X}_j$ is generated ($j = 2, 3, \ldots, k$), we find whether it satisfies all the constraints, Eqs. (7.7b). If $\mathbf{X}_j$ violates any of the constraints stated in Eqs. (7.7b), the trial point $\mathbf{X}_j$ is moved halfway toward the centroid of the remaining, already accepted points (where the given initial point $\mathbf{X}_1$ is included). The centroid $\mathbf{X}_0$ of already accepted points is given by

$$\mathbf{X}_0 = \frac{1}{j-1} \sum_{l=1}^{j-1} \mathbf{X}_l \qquad (7.9)$$

If the trial point $\mathbf{X}_j$ so found still violates some of the constraints, Eqs. (7.7b), the process of moving halfway in toward the centroid $\mathbf{X}_0$ is continued until a feasible point $\mathbf{X}_j$ is found. Ultimately, we will be able to find a feasible point $\mathbf{X}_j$ by this procedure provided that the feasible region is convex. By proceeding in this way, we will ultimately be able to find the required feasible points $\mathbf{X}_2, \mathbf{X}_3, \ldots, \mathbf{X}_k$.

2. The objective function is evaluated at each of the k points (vertices). If the vertex $\mathbf{X}_h$ corresponds to the largest function value, the process of reflection is used to find a new point $\mathbf{X}_r$ as

$$\mathbf{X}_r = (1 + \alpha)\mathbf{X}_0 - \alpha\mathbf{X}_h \qquad (7.10)$$

where $\alpha \geq 1$ (to start with) and $\mathbf{X}_0$ is the centroid of all vertices except $\mathbf{X}_h$:

$$\mathbf{X}_0 = \frac{1}{k-1} \sum_{\substack{l=1 \\ l \neq k}}^{k} \mathbf{X}_l \qquad (7.11)$$

3. Since the problem is a constrained one, the point $\mathbf{X}_r$ has to be tested for feasibility. If the point $\mathbf{X}_r$ is feasible and $f(\mathbf{X}_r) < f(\mathbf{X}_h)$, the point $\mathbf{X}_h$ is replaced by $\mathbf{X}_r$, and we go to step 2. If $f(\mathbf{X}_r) \geq f(\mathbf{X}_h)$, a new trial point $\mathbf{X}_r$ is found by reducing the value of α in Eq. (7.10) by a factor of 2 and is tested for the satisfaction of the relation $f(\mathbf{X}_r) < f(\mathbf{X}_h)$. If $f(\mathbf{X}_r) \geq f(\mathbf{X}_h)$, the procedure of finding a new point $\mathbf{X}_r$ with a reduced value of α is repeated again. This procedure is repeated, if necessary, until the value of α becomes smaller than a prescribed small quantity ε, say, 10^{-6}. If an improved point $\mathbf{X}_r$, with $f(\mathbf{X}_r) < f(\mathbf{X}_h)$, cannot be obtained even with that small value of α, the point $\mathbf{X}_r$ is discarded and the entire procedure of reflection is restarted by using the point $\mathbf{X}_p$ (which has the second-highest function value) instead of $\mathbf{X}_h$.

4. If at any stage, the reflected point $\mathbf{X}_r$ (found in step 3) violates any of the constraints [Eqs. (7.7b)], it is moved halfway in toward the centroid until it becomes feasible, that is,

$$(\mathbf{X}_r)_{\text{new}} = \tfrac{1}{2}(\mathbf{X}_0 + \mathbf{X}_r) \qquad\qquad (7.12)$$

This method will progress toward the optimum point as long as the complex has not collapsed into its centroid.

5. Each time the worst point $\mathbf{X}_h$ of the current complex is replaced by a new point, the complex gets modified and we have to test for the convergence of the process. We assume convergence of the process whenever the following two conditions are satisfied:

(a) The complex shrinks to a specified small size (i.e., the distance between any two vertices among $\mathbf{X}_1, \mathbf{X}_2, \ldots, \mathbf{X}_k$ becomes smaller than a prescribed small quantity, ε_1.

(b) The standard deviation of the function value becomes sufficiently small (i.e., when

$$\left\{ \frac{1}{k} \sum_{j=1}^{k} [f(\mathbf{X}) - f(\mathbf{X}_j)]^2 \right\}^{1/2} \leq \varepsilon_2 \qquad\qquad (7.13)$$

where $\mathbf{X}$ is the centroid of all the k vertices of the current complex, and $\varepsilon_2 > 0$ is a specified small number).

Discussion.   This method does not require the derivatives of $f(\mathbf{X})$ and $g_j(\mathbf{X})$ to find the minimum point, and hence it is computationally very simple. The method is very simple from programming point of view and does not require a large computer storage.

1. A value of 1.3 for the initial value of α in Eq. (7.10) has been found to be satisfactory by Box.

2. Box recommended a value of $k \simeq 2n$ (although a lesser value can be chosen if n is greater than, say, 5). If k is not sufficiently large, the complex tends to collapse and flatten along the first constraint boundary encountered.

3. From the procedure above, it can be observed that the complex rolls over and over, normally expanding. However, if a boundary is encountered, the complex contracts and flattens itself. It can then roll along this constraint boundary and leave it if the contours change. The complex can also accommodate more than one boundary and can turn corners.

4. If the feasible region is nonconvex, there is no guarantee that the centroid of all feasible points is also feasible. If the centroid is not feasible, we cannot apply the procedure above to find the new points $\mathbf{X}_r$.

5. The method becomes inefficient rapidly as the number of variables increases.

6. It cannot be used to solve problems having equality constraints.

7. This method requires an initial point $\mathbf{X}_1$ that is feasible. This is not a major restriction. If an initial feasible point is not readily available, the method described in Section 7.13 can be used to find a feasible point $\mathbf{X}_1$.

7.5   SEQUENTIAL LINEAR PROGRAMMING

In the *sequential linear programming* (SLP) *method*, the solution of the original nonlinear programming problem is found by solving a series of linear programming problems. Each LP problem is generated by approximating the nonlinear objective and constraint functions using first-order Taylor series expansions about the current design vector, $\mathbf{X}_i$. The resulting LP problem is solved using the simplex method to find the new design vector $\mathbf{X}_{i+1}$. If $\mathbf{X}_{i+1}$ does not satisfy the stated convergence criteria, the problem is relinearized about the point $\mathbf{X}_{i+1}$ and the procedure is continued until the optimum solution $\mathbf{X}^*$ is found.

If the problem is a convex programming problem, the linearized constraints always lie entirely outside the feasible region. Hence the optimum solution of the approximating LP problem, which lies at a vertex of the new feasible region, will lie outside the original feasible region. However, by relinearizing the problem about the new point and repeating the process, we can achieve convergence to the solution of the original problem in few iterations. The SLP method, also known as the *cutting plane method*, was originally presented by Cheney and Goldstein [7.3] and Kelly [7.4].

Algorithm.   The SLP algorithm can be stated as follows:

1. Start with an initial point $\mathbf{X}_1$ and set the iteration number as $i = 1$. The point $\mathbf{X}_1$ need not be feasible.
2. Linearize the objective and constraint functions about the point $\mathbf{X}_i$ as

$$f(\mathbf{X}) \approx f(\mathbf{X}_i) + \nabla f(\mathbf{X}_i)^{\mathrm{T}}(\mathbf{X} - \mathbf{X}_i) \tag{7.14}$$

$$g_j(\mathbf{X}) \approx g_j(\mathbf{X}_i) + \nabla g_j(\mathbf{X}_i)^{\mathrm{T}}(\mathbf{X} - \mathbf{X}_i) \tag{7.15}$$

$$h_k(\mathbf{X}) \approx h_k(\mathbf{X}_i) + \nabla h_k(\mathbf{X}_i)^{\mathrm{T}}(\mathbf{X} - \mathbf{X}_i) \tag{7.16}$$

3. Formulate the approximating linear programming problem as[†]

$$\text{Minimize } f(\mathbf{X}_i) + \nabla f_i^{\mathrm{T}}(\mathbf{X} - \mathbf{X}_i)$$

subject to

$$g_j(\mathbf{X}_i) + \nabla g_j(\mathbf{X}_i)^T(\mathbf{X} - \mathbf{X}_i) \le 0, \quad j = 1, 2, \ldots, m$$

$$h_k(\mathbf{X}_i) + \nabla h_k(\mathbf{X}_i)^T(\mathbf{X} - \mathbf{X}_i) = 0, \quad k = 1, 2, \ldots, p \tag{7.17}$$

4. Solve the approximating LP problem to obtain the solution vector $\mathbf{X}_{i+1}$.
5. Evaluate the original constraints at $\mathbf{X}_{i+1}$; that is, find

$$g_j(\mathbf{X}_{i+1}), \quad j = 1, 2, \ldots, m \quad \text{and} \quad h_k(\mathbf{X}_{i+1}), \quad k = 1, 2, \ldots, p$$

[†]Notice that the LP problem stated in Eq. (7.17) may sometimes have an unbounded solution. This can be avoided by formulating the first approximating LP problem by considering only the following constraints:

$$l_i \le x_i \le u_i, \quad i = 1, 2, \ldots, n \tag{7.18}$$

In Eq. (7.18), l_i and u_i represent the lower and upper bounds on x_i, respectively. The values of l_i and u_i depend on the problem under consideration, and their values have to be chosen such that the optimum solution of the original problem does not fall outside the range indicated by Eq. (7.18).

If $g_j(\mathbf{X}_{i+1}) \leq \varepsilon$ for $j = 1, 2, \ldots, m$, and $|h_k(\mathbf{X}_{i+1})| \leq \varepsilon$, $k = 1, 2, \ldots, p$, where ε is a prescribed small positive tolerance, all the original constraints can be assumed to have been satisfied. Hence stop the procedure by taking

$$\mathbf{X}_{\mathrm{opt}} \simeq \mathbf{X}_{i+1}$$

If $g_j(\mathbf{X}_{i+1}) > \varepsilon$ for some j, or $|h_k(\mathbf{X}_{i+1})| > \varepsilon$ for some k, find the most violated constraint, for example, as

$$g_k(\mathbf{X}_{i+1}) = \max_j[g_j(\mathbf{X}_{i+1})] \tag{7.19}$$

Relinearize the constraint $g_k(\mathbf{X}) \leq 0$ about the point $\mathbf{X}_{i+1}$ as

$$g_k(\mathbf{X}) \simeq g_k(\mathbf{X}_{i+1}) + \nabla g_k(\mathbf{X}_{i+1})^{\mathrm{T}}(\mathbf{X} - \mathbf{X}_{i+1}) \leq 0 \tag{7.20}$$

and add this as the $(m + 1)$th inequality constraint to the previous LP problem.

6. Set the new iteration number as $i = i + 1$, the total number of constraints in the new approximating LP problem as $m + 1$ inequalities and p equalities, and go to step 4.

The sequential linear programming method has several advantages:

1. It is an efficient technique for solving convex programming problems with nearly linear objective and constraint functions.

2. Each of the approximating problems will be a LP problem and hence can be solved quite efficiently. Moreover, any two consecutive approximating LP problems differ by only one constraint, and hence the dual simplex method can be used to solve the sequence of approximating LP problems much more efficiently.[†]

3. The method can easily be extended to solve integer programming problems. In this case, one integer LP problem has to be solved in each stage.

Geometric Interpretation of the Method.   The SLP method can be illustrated with the help of a one-variable problem:

$$\text{Minimize } f(x) = c_1 x$$

subject to

$$g(x) \leq 0 \tag{7.21}$$

where c_1 is a constant and $g(x)$ is a nonlinear function of x. Let the feasible region and the contour of the objective function be as shown in Fig. 7.5. To avoid any possibility of unbounded solution, let us first take the constraints on x as $c \leq x \leq d$, where c and d represent the lower and upper bounds on x. With these constraints, we formulate the LP problem:

$$\text{Minimize } f(x) = c_1 x$$

[†]The dual simplex method was discussed in Section 4.3.

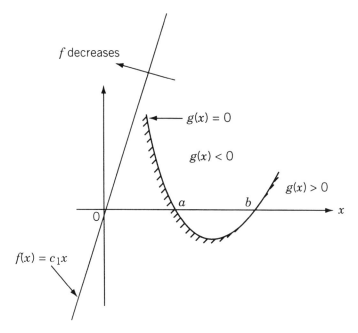

Figure 7.5   Graphical representation of the problem stated by Eq. (7.21).

subject to

$$c \leq x \leq d \tag{7.22}$$

The optimum solution of this approximating LP problem can be seen to be $x^* = c$. Next, we linearize the constraint $g(x)$ about point c and add it to the previous constraint set. Thus the new LP problem becomes

$$\text{Minimize } f(x) = c_1 x \tag{7.23a}$$

subject to

$$c \leq x \leq d \tag{7.23b}$$

$$g(c) + \frac{dg}{dx}(c)(x - c) \leq 0 \tag{7.23c}$$

The feasible region of x, according to the constraints (7.23b) and (7.23c), is given by $e \leq x \leq d$ (Fig. 7.6). The optimum solution of the approximating LP problem given by Eqs. (7.23) can be seen to be $x^* = e$. Next, we linearize the constraint $g(x) \leq 0$ about the current solution $x^* = e$ and add it to the previous constraint set to obtain the next approximating LP problem as

$$\text{Minimize } f(x) = c_1 x \tag{7.24a}$$

subject to

$$c \leq x \leq d \tag{7.24b}$$

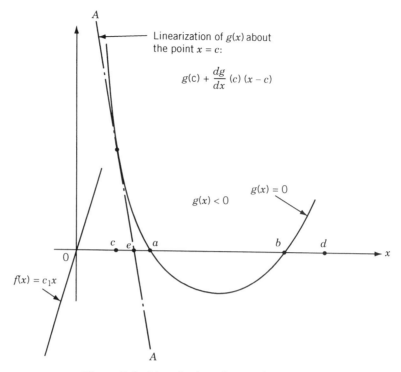

Figure 7.6   Linearization of constraint about c.

$$g(c) + \frac{dg}{dx}(c)(x - c) \leq 0 \tag{7.24c}$$

$$g(e) + \frac{dg}{dx}(e)(x - e) \leq 0 \tag{7.24d}$$

The permissible range of x, according to the constraints (7.24b), (7.24c), and (7.24d), can be seen to be $f \leq x \leq d$ from Fig. 7.7. The optimum solution of the LP problem of Eqs. (7.24) can be obtained as $x^* = f$.

We then linearize $g(x) \leq 0$ about the present point $x^* = f$ and add it to the previous constraint set [Eqs. (7.24)] to define a new approximating LP problem. This procedure has to be continued until the optimum solution is found to the desired level of accuracy. As can be seen from Figs. 7.6 and 7.7, the optimum of all the approximating LP problems (e.g., points $c, e, f, \ldots$) lie outside the feasible region and converge toward the true optimum point, $x = a$. The process is assumed to have converged whenever the solution of an approximating problem satisfies the original constraint within some specified tolerance level as

$$g(x_k^*) \leq \varepsilon$$

where ε is a small positive number and x_k^* is the optimum solution of the kth approximating LP problem. It can be seen that the lines (hyperplanes in a general problem) defined by $g(x_k^*) + dg/dx(x_k^*)(x - x_k^*)$ cut off a portion of the existing feasible region. Hence this method is called the *cutting plane method*.

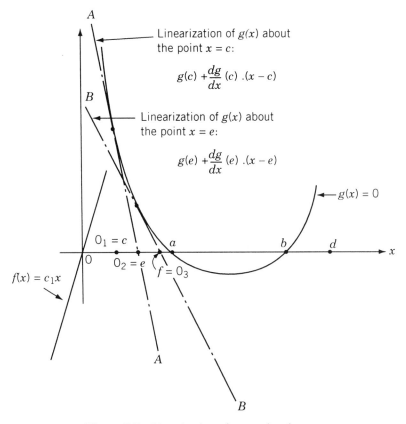

Figure 7.7   Linearization of constraint about e.

Example 7.1

$$\text{Minimize } f(x_1, x_2) = x_1 - x_2$$

subject to

$$g_1(x_1, x_2) = 3x_1^2 - 2x_1 x_2 + x_2^2 - 1 \le 0$$

using the cutting plane method. Take the convergence limit in step 5 as $\varepsilon = 0.02$.

Note: This example was originally given by Kelly [7.4]. Since the constraint boundary represents an ellipse, the problem is a convex programming problem. From graphical representation, the optimum solution of the problem can be identified as $x_1^* = 0$, $x_2^* = 1$, and $f_{\min} = -1$.

SOLUTION

Steps 1, 2, 3:   Although we can start the solution from any initial point $\mathbf{X}_1$, to avoid the possible unbounded solution, we first take the bounds on x_1 and x_2 as $-2 \le x_1 \le 2$ and $-2 \le x_2 \le 2$ and solve the following LP problem:

$$\text{Minimize } f = x_1 - x_2$$

subject to

$$-2 \leq x_1 \leq 2$$

$$-2 \leq x_2 \leq 2 \qquad (E_1)$$

The solution of this problem can be obtained as

$$\mathbf{X} = \begin{bmatrix} -2 \\ 2 \end{bmatrix} \text{ with } f(\mathbf{X}) = -4$$

Step 4: Since we have solved one LP problem, we can take

$$\mathbf{X}_{i+1} = \mathbf{X}_2 = \begin{Bmatrix} -2 \\ 2 \end{Bmatrix}$$

Step 5: Since $g_1(\mathbf{X}_2) = 23 > \varepsilon$, we linearize $g_1(\mathbf{X})$ about point $\mathbf{X}_2$ as

$$g_1(\mathbf{X}) \simeq g_1(\mathbf{X}_2) + \nabla g_1(\mathbf{X}_2)^{\mathrm{T}}(\mathbf{X} - \mathbf{X}_2) \leq 0 \qquad (E_2)$$

As

$$g_1(\mathbf{X}_2) = 23, \quad \frac{\partial g_1}{\partial x_1}\bigg|_{\mathbf{X}_2} = (6x_1 - 2x_2)|_{\mathbf{X}_2} = -16$$

$$\frac{\partial g_1}{\partial x_2}\bigg|_{\mathbf{X}_2} = (-2x_1 + 2x_2)|_{\mathbf{X}_2} = 8$$

Eq. (E_2) becomes

$$g_1(\mathbf{X}) \simeq -16x_1 + 8x_2 - 25 \leq 0$$

By adding this constraint to the previous LP problem, the new LP problem becomes

$$\text{Minimize } f = x_1 - x_2$$

subject to

$$-2 \leq x_1 \leq 2$$

$$-2 \leq x_2 \leq 2 \qquad (E_3)$$

$$-16x_1 + 8x_2 - 25 \leq 0$$

Step 6: Set the iteration number as $i = 2$ and go to step 4.

Step 4: Solve the approximating LP problem stated in Eqs. (E_3) and obtain the solution

$$\mathbf{X}_3 = \begin{Bmatrix} -0.5625 \\ 2.0 \end{Bmatrix} \text{ with } f_3 = f(\mathbf{X}_3) = -2.5625$$

This procedure is continued until the specified convergence criterion, $g_1(\mathbf{X}_i) \leq \varepsilon$, in step 5 is satisfied. The computational results are summarized in Table 7.2.

Table 7.2   Results for Example 7.1

Iteration number, i	New linearized constraint considered	Solution of the approximating LP problem $\mathbf{X}_{i+1}$	$f(\mathbf{X}_{i+1})$	$g_1(\mathbf{X}_{i+1})$
1	$-2 \leq x_1 \leq 2$ and $-2 \leq x_2 \leq 2$	$(-2.0, 2.0)$	-4.00000	23.00000
2	$-16.0x_1 + 8.0x_2 - 25.0 \leq 0$	$(-0.56250, 2.00000)$	-2.56250	6.19922
3	$-7.375x_1 + 5.125x_2$ $-8.19922 \leq 0$	$(0.27870, 2.00000)$	-1.72193	2.11978
4	$-2.33157x_1 + 3.44386x_2$ $-4.11958 \leq 0$	$(-0.52970, 0.83759)$	-1.36730	1.43067
5	$-4.85341x_1 + 2.73459x_2$ $-3.43067 \leq 0$	$(-0.05314, 1.16024)$	-1.21338	0.47793
6	$-2.63930x_1 + 2.42675x_2$ $-2.47792 \leq 0$	$(0.42655, 1.48490)$	-1.05845	0.48419
7	$-0.41071x_1 + 2.11690x_2$ $-2.48420 \leq 0$	$(0.17058, 1.20660)$	-1.03603	0.13154
8	$-1.38975x_1 + 2.07205x_2$ $-2.13155 \leq 0$	$(0.01829, 1.04098)$	-1.02269	0.04656
9	$-1.97223x_1 + 2.04538x_2$ $-2.04657 \leq 0$	$(-0.16626, 0.84027)$	-1.00653	0.06838
10	$-2.67809x_1 + 2.01305x_2$ $-2.06838 \leq 0$	$(-0.07348, 0.92972)$	-1.00321	0.01723

7.6   BASIC APPROACH IN THE METHODS OF FEASIBLE DIRECTIONS

In the methods of feasible directions, basically we choose a starting point satisfying all the constraints and move to a better point according to the iterative scheme

$$\mathbf{X}_{i+1} = \mathbf{X}_i + \lambda \mathbf{S}_i \tag{7.25}$$

where $\mathbf{X}_i$ is the starting point for the ith iteration, $\mathbf{S}_i$ the direction of movement, λ the distance of movement (step length), and $\mathbf{X}_{i+1}$ the final point obtained at the end of the ith iteration. The value of λ is always chosen so that $\mathbf{X}_{i+1}$ lies in the feasible region. The search direction $\mathbf{S}_i$ is found such that (1) a small move in that direction violates no constraint, and (2) the value of the objective function can be reduced in that direction. The new point $\mathbf{X}_{i+1}$ is taken as the starting point for the next iteration and the entire procedure is repeated several times until a point is obtained such that no direction satisfying both properties 1 and 2 can be found. In general, such a point denotes the constrained local minimum of the problem. This local minimum need not be a global one unless the problem is a convex programming problem. A direction satisfying property 1 is called *feasible* while a direction satisfying both properties 1 and 2 is called a *usable feasible direction*. This is the reason that these methods are known as *methods of feasible directions*. There are many ways of choosing usable feasible directions, and hence there are many different methods of feasible directions. As seen in Chapter 2, a direction $\mathbf{S}$ is feasible at a point $\mathbf{X}_i$ if it satisfies the relation

$$\frac{d}{d\lambda} g_j(\mathbf{X}_i + \lambda \mathbf{S})|_{\lambda=0} = \mathbf{S}^\mathrm{T} \nabla g_j(\mathbf{X}_i) \leq 0 \qquad (7.26)$$

where the equality sign holds true only if a constraint is linear or strictly concave, as shown in Fig. 2.8. A vector $\mathbf{S}$ will be a usable feasible direction if it satisfies the relations

$$\frac{d}{d\lambda} f(\mathbf{X}_i + \lambda \mathbf{S})|_{\lambda=0} = \mathbf{S}^\mathrm{T} \nabla f(\mathbf{X}_i) < 0 \qquad (7.27)$$

$$\frac{d}{d\lambda} g_j(\mathbf{X}_i + \lambda \mathbf{S})|_{\lambda=0} = \mathbf{S}^\mathrm{T} \nabla g_j(\mathbf{X}_i) \leq 0 \qquad (7.28)$$

It is possible to reduce the value of the objective function at least by a small amount by taking a step length $\lambda > 0$ along such a direction.

The detailed iterative procedure of the methods of feasible directions will be considered in terms of two well-known methods: Zoutendijk's method of feasible directions and Rosen's gradient projection method.

7.7 ZOUTENDIJK'S METHOD OF FEASIBLE DIRECTIONS

In *Zoutendijk's method of feasible directions*, the usable feasible direction is taken as the negative of the gradient direction if the initial point of the iteration lies in the interior (not on the boundary) of the feasible region. However, if the initial point lies on the boundary of the feasible region, some constraints will be active and the usable feasible direction is found so as to satisfy Eqs. (7.27) and (7.28). The iterative procedure of Zoutendijk's method can be stated as follows (only inequality constraints are considered in Eq. (7.1), for simplicity.

Algorithm

1. Start with an initial feasible point $\mathbf{X}_1$ and small numbers ε_1, ε_2, and ε_3 to test the convergence of the method. Evaluate $f(\mathbf{X}_1)$ and $g_j(\mathbf{X}_1)$, $j = 1, 2, \ldots, m$. Set the iteration number as $i = 1$.

2. If $g_j(\mathbf{X}_i) < 0$, $j = 1, 2, \ldots, m$ (i.e., $\mathbf{X}_i$ is an interior feasible point), set the current search direction as

$$\mathbf{S}_i = -\nabla f(\mathbf{X}_i) \qquad (7.29)$$

Normalize $\mathbf{S}_i$ in a suitable manner and go to step 5. If at least one $g_j(\mathbf{X}_i) = 0$, go to step 3.

3. Find a usable feasible direction $\mathbf{S}$ by solving the direction-finding problem:

$$\text{Minimize } -\alpha \qquad (7.30a)$$

subject to

$$\mathbf{S}^{\mathrm{T}}\nabla g_j(\mathbf{X}_i) + \theta_j\alpha \le 0, \quad j = 1, 2, \ldots, p \tag{7.30b}$$

$$\mathbf{S}^{\mathrm{T}}\nabla f + \alpha \le 0 \tag{7.30c}$$

$$-1 \le s_i \le 1, \quad i = 1, 2, \ldots, n \tag{7.30d}$$

where s_i is the ith component of $\mathbf{S}$, the first p constraints have been assumed to be active at the point $\mathbf{X}_i$ (the constraints can always be renumbered to satisfy this requirement), and the values of all θ_j can be taken as unity. Here α can be taken as an additional design variable.

4. If the value of α^* found in step 3 is very nearly equal to zero, that is, if $\alpha^* \le \varepsilon_1$, terminate the computation by taking $\mathbf{X}_{\mathrm{opt}} \simeq \mathbf{X}_i$. If $\alpha^* > \varepsilon_1$, go to step 5 by taking $\mathbf{S}_i = \mathbf{S}$.

5. Find a suitable step length λ_i along the direction $\mathbf{S}_i$ and obtain a new point $\mathbf{X}_{i+1}$ as

$$\mathbf{X}_{i+1} = \mathbf{X}_i + \lambda_i \mathbf{S}_i \tag{7.31}$$

The methods of finding the step length λ_i will be considered later.

6. Evaluate the objective function $f(\mathbf{X}_{i+1})$.

7. Test for the convergence of the method. If

$$\left| \frac{f(\mathbf{X}_i) - f(\mathbf{X}_{i+1})}{f(\mathbf{X}_i)} \right| \le \varepsilon_2 \quad \text{and} \quad \|\mathbf{X}_i - \mathbf{X}_{i+1}\| \le \varepsilon_3 \tag{7.32}$$

terminate the iteration by taking $\mathbf{X}_{\mathrm{opt}} \simeq \mathbf{X}_{i+1}$. Otherwise, go to step 8.

8. Set the new iteration number as $i = i + 1$, and repeat from step 2 onward.

There are several points to be considered in applying this algorithm. These are related to (1) finding an appropriate usable feasible direction (**S**), (2) finding a suitable step size along the direction **S**, and (3) speeding up the convergence of the process. All these aspects are discussed below.

7.7.1   Direction-Finding Problem

If the point $\mathbf{X}_i$ lies in the interior of the feasible region [i.e., $g_j(\mathbf{X}_i) < 0$ for $j = 1, 2, \ldots, m$], the usable feasible direction is taken as

$$\mathbf{S}_i = -\nabla f(\mathbf{X}_i) \tag{7.33}$$

The problem becomes complicated if one or more of the constraints are critically satisfied at $\mathbf{X}_i$, that is, when some of the $g_j(\mathbf{X}_i) = 0$. One simple way to find a usable feasible direction at a point $\mathbf{X}_i$ at which some of the constraints are active is to generate a random vector and verify whether it satisfies Eqs. (7.27) and (7.28). This approach is a crude one but is very simple and easy to program. The relations to be checked for each random vector are also simple, and hence it will not require much computer time. However, a more systematic procedure is generally adopted to find a usable feasible direction in practice. Since there will be, in general, several directions that satisfy

Eqs. (7.27) and (7.28), one would naturally be tempted to choose the "best" possible usable feasible direction at $\mathbf{X}_i$.

Thus we seek to find a feasible direction that, in addition to decreasing the value of f, also points away from the boundaries of the active nonlinear constraints. Such a direction can be found by solving the following optimization problem. Given the point $\mathbf{X}_i$, find the vector $\mathbf{S}$ and the scalar α that maximize α subject to the constraints

$$\mathbf{S}^{\mathrm{T}} \nabla g_j(\mathbf{X}_i) + \theta_j \alpha \leq 0, \quad j \in J \tag{7.34}$$

$$\mathbf{S}^{\mathrm{T}} \nabla f(\mathbf{X}_i) + \alpha \leq 0 \tag{7.35}$$

where J represents the set of active constraints and $\mathbf{S}$ is normalized by one of the following relations:

$$\mathbf{S}^{\mathrm{T}} \mathbf{S} = \sum_{i=1}^{n} s_i^2 = 1 \tag{7.36}$$

$$-1 \leq s_i \leq 1, \quad i = 1, 2, \ldots, n \tag{7.37}$$

$$\mathbf{S}^{\mathrm{T}} \nabla f(\mathbf{X}_i) \leq 1 \tag{7.38}$$

In this problem, θ_j are arbitrary positive scalar constants, and for simplicity, we can take all $\theta_j = 1$. Any solution of this problem with $\alpha > 0$ is a usable feasible direction. The maximum value of α gives the best direction ($\mathbf{S}$) that makes the value of $\mathbf{S}^{\mathrm{T}} \nabla f_i$ negative and the values of $\mathbf{S}^{\mathrm{T}} \nabla g_j(\mathbf{X}_i)$ as negative as possible simultaneously. In other words, the maximum value of α makes the direction $\mathbf{S}$ steer away from the active nonlinear constraint boundaries. It can easily be seen that by giving different values for different θ_j, we can give more importance to certain constraint boundaries compared to others. Equations (7.36) to (7.38) represent the normalization of the vector $\mathbf{S}$ so as to ensure that the maximum of α will be a finite quantity. If the normalization condition is not included, the maximum of α may be made to approach ∞ without violating the constraints [Eqs. (7.34) and (7.35)].

Notice that the objective function α, and the constraint equations (7.34) and (7.35) are linear in terms of the variables $s_1, s_2, \ldots, s_n, \alpha$. The normalization constraint will also be linear if we use either Eq. (7.37) or (7.38). However, if we use Eq. (7.36) for normalization, it will be a quadratic function. Thus the direction-finding problem can be posed as a linear programming problem by using either Eq. (7.37) or (7.38) for normalization. Even otherwise, the problem will be a LP problem except for one quadratic constraint. It was shown by Zoutendijk [7.5] that this problem can be handled by a modified version of linear programming. Thus the direction-finding problem can be solved with reasonable efficiency. We use Eq. (7.37) in our presentation. The direction-finding problem can be stated more explicitly as

$$\text{Minimize } -\alpha$$

subject to

$$s_1 \frac{\partial g_1}{\partial x_1} + s_2 \frac{\partial g_1}{\partial x_2} + \cdots + s_n \frac{\partial g_1}{\partial x_n} + \theta_1 \alpha \leq 0$$

$$s_1 \frac{\partial g_2}{\partial x_1} + s_2 \frac{\partial g_2}{\partial x_2} + \cdots + s_n \frac{\partial g_2}{\partial x_n} + \theta_2 \alpha \leq 0$$

$$\vdots$$

$$s_1 \frac{\partial g_p}{\partial x_1} + s_2 \frac{\partial g_p}{\partial x_2} + \cdots + s_n \frac{\partial g_p}{\partial x_n} + \theta_p \alpha \leq 0 \qquad (7.39)$$

$$s_1 \frac{\partial f}{\partial x_1} + s_2 \frac{\partial f}{\partial x_2} + \cdots + s_n \frac{\partial f}{\partial x_n} + \alpha \leq 0$$

$$s_1 - 1 \leq 0$$

$$s_2 - 1 \leq 0$$

$$\vdots$$

$$s_n - 1 \leq 0$$

$$-1 - s_1 \leq 0$$

$$-1 - s_2 \leq 0$$

$$\vdots$$

$$-1 - s_n \leq 0$$

where p is the number of active constraints and the partial derivatives $\partial g_1/\partial x_1$, $\partial g_1/\partial x_2$, ..., $\partial g_p/\partial x_n$, $\partial f/\partial x_1$, ..., $\partial f/\partial x_n$ have been evaluated at point $\mathbf{X}_i$. Since the components of the search direction, s_i, $i = 1$ to n, can take any value between -1 and 1, we define new variables t_i as $t_i = s_i + 1$, $i = 1$ to n, so that the variables will always be nonnegative. With this change of variables, the problem above can be restated as a standard linear programming problem as follows:

$$\text{Find } (t_1, t_2, \ldots, t_n, \alpha, y_1, y_2, \ldots, y_{p+n+1}) \text{ which}$$

$$\text{minimizes } - \alpha$$

subject to

$$t_1 \frac{\partial g_1}{\partial x_1} + t_2 \frac{\partial g_1}{\partial x_2} + \cdots + t_n \frac{\partial g_1}{\partial x_n} + \theta_1 \alpha + y_1 = \sum_{i=1}^{n} \frac{\partial g_1}{\partial x_i}$$

$$t_1 \frac{\partial g_2}{\partial x_1} + t_2 \frac{\partial g_2}{\partial x_2} + \cdots + t_n \frac{\partial g_2}{\partial x_n} + \theta_2 \alpha + y_2 = \sum_{i=1}^{n} \frac{\partial g_2}{\partial x_i}$$

$$\vdots$$

$$t_1 \frac{\partial g_p}{\partial x_1} + t_2 \frac{\partial g_p}{\partial x_2} + \cdots + t_n \frac{\partial g_p}{\partial x_n} + \theta_p \alpha + y_p = \sum_{i=1}^{n} \frac{\partial g_p}{\partial x_i} \qquad (7.40)$$

$$t_1 \frac{\partial f}{\partial x_1} + t_2 \frac{\partial f}{\partial x_2} + \cdots + t_n \frac{\partial f}{\partial x_n} + \alpha + y_{p+1} = \sum_{i=1}^{n} \frac{\partial f}{\partial x_i}$$

$$t_1 + y_{p+2} = 2$$

$$t_2 + y_{p+3} = 2$$

$$\vdots$$

$$t_n + y_{p+n+1} = 2$$

$$t_1 \geq 0$$

$$t_2 \geq 0$$

$$\vdots$$

$$t_n \geq 0$$

$$\alpha \geq 0$$

where $y_1, y_2, \ldots, y_{p+n+1}$ are the nonnegative slack variables. The simplex method discussed in Chapter 3 can be used to solve the direction-finding problem stated in Eqs. (7.40). This problem can also be solved by more sophisticated methods that treat the upper bounds on t_i in a special manner instead of treating them as constraints [7.6]. If the solution of the direction-finding problem gives a value of $\alpha^* > 0$, $f(\mathbf{X})$ can be improved by moving along the usable feasible direction

$$\mathbf{S} = \begin{Bmatrix} s_1 \\ s_2 \\ \vdots \\ s_n \end{Bmatrix} = \begin{Bmatrix} t_1^* - 1 \\ t_2^* - 1 \\ \vdots \\ t_n^* - 1 \end{Bmatrix}$$

If, however, $\alpha^* = 0$, it can be shown that the Kuhn–Tucker optimality conditions are satisfied at $\mathbf{X}_i$ and hence point $\mathbf{X}_i$ can be taken as the optimal solution.

7.7.2   Determination of Step Length

After finding a usable feasible direction $\mathbf{S}_i$ at any point $\mathbf{X}_i$, we have to determine a suitable step length λ_i to obtain the next point $\mathbf{X}_{i+1}$ as

$$\mathbf{X}_{i+1} = \mathbf{X}_i + \lambda_i \mathbf{S}_i \tag{7.41}$$

There are several ways of computing the step length. One of the methods is to determine an optimal step length (λ_i) that minimizes $f(\mathbf{X}_i + \lambda \mathbf{S}_i)$ such that the new point $\mathbf{X}_{i+1}$ given by Eq. (7.41) lies in the feasible region. Another method is to choose the step length (λ_i) by trial and error so that it satisfies the relations

$$f(\mathbf{X}_i + \lambda_i \mathbf{S}_i) \leq f(\mathbf{X}_i)$$

$$g_j(\mathbf{X}_i + \lambda_i \mathbf{S}_i) \leq 0, \quad j = 1, 2, \ldots, m \tag{7.42}$$

Method 1. The optimal step length, λ_i, can be found by any of the one-dimensional minimization methods described in Chapter 5. The only drawback with these methods is that the constraints will not be considered while finding λ_i. Thus the new point $\mathbf{X}_{i+1} = \mathbf{X}_i + \lambda_i \mathbf{S}_i$ may lie either in the interior of the feasible region (Fig. 7.8a), or on the boundary of the feasible region (Fig. 7.8b), or in the infeasible region (Fig. 7.8c).

If the point $\mathbf{X}_{i+1}$ lies in the interior of the feasible region, there are no active constraints and hence we proceed to the next iteration by setting the new usable feasible direction as $\mathbf{S}_{i+1} = -\nabla f(X_{i+1})$ (i.e., we go to step 2 of the algorithm). On the other hand, if $\mathbf{X}_{i+1}$ lies on the boundary of the feasible region, we generate a new usable feasible direction $\mathbf{S} = \mathbf{S}_{i+1}$ by solving a new direction-finding problem (i.e., we go to step 3 of the algorithm). One practical difficulty has to be noted at this stage. To detect that point $\mathbf{X}_{i+1}$ is lying on the constraint boundary, we have to find whether one or more $g_j(\mathbf{X}_{i+1})$ are zero. Since the computations are done numerically, will we say that

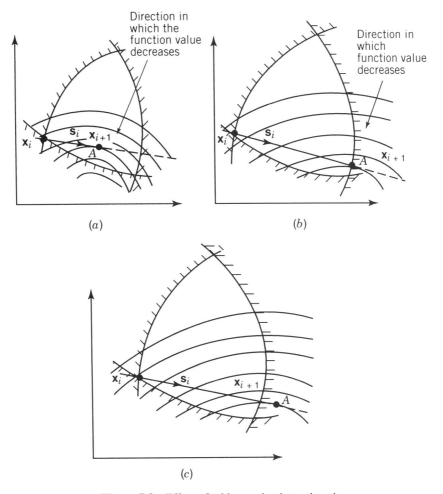

Figure 7.8 Effect of taking optimal step length.

the constraint g_j is active if $g_j(\mathbf{X}_{i+1}) = 10^{-2}, 10^{-3}, 10^{-8}$, and so on? We immediately notice that a small margin ε has to be specified to detect an active constraint. Thus we can accept a point $\mathbf{X}$ to be lying on the constraint boundary if $|g_j(\mathbf{X})| \leq \varepsilon$ where ε is a prescribed small number. If point $\mathbf{X}_{i+1}$ lies in the infeasible region, the step length has to be reduced (corrected) so that the resulting point lies in the feasible region only. It is to be noted that an initial trial step size (ε_1) has to be specified to initiate the one-dimensional minimization process.

Method 2.   Even if we do not want to find the optimal step length, some sort of a trial-and-error method has to be adopted to find the step length λ_i so as to satisfy the relations (7.42). One possible method is to choose an arbitrary step length ε and compute the values of

$$\tilde{f} = f(\mathbf{X}_i + \varepsilon \mathbf{S}_i) \quad \text{and} \quad \tilde{g}_j = g_j(\mathbf{X}_i + \varepsilon \mathbf{S}_i)$$

Depending on the values of $\tilde{f}$ and $\tilde{g}_j$, we may need to adjust the value of ε until we improve the objective function value without violating the constraints.

Initial Trial Step Length.   It can be seen that in whatever way we want to find the step size λ_i, we need to specify an initial trial step length ε. The value of ε can be chosen in several ways. Some of the possibilities are given below.

1. The average of the final step lengths λ_i obtained in the last few iterations can be used as the initial trial step length ε for the next step. Although this method is often satisfactory, it has a number of disadvantages:

   (a) This method cannot be adopted for the first iteration.
   (b) This method cannot take care of the rate of variation of $f(\mathbf{X})$ in different directions.
   (c) This method is quite insensitive to rapid changes in the step length that take place generally as the optimum point is approached.

2. At each stage, an initial step length ε is calculated so as to reduce the objective function value by a given percentage. For doing this, we can approximate the behavior of the function $f(\lambda)$ to be linear in λ. Thus if

$$f(\mathbf{X}_i) = f(\lambda = 0) = f_1 \tag{7.43}$$

$$\frac{df}{d\lambda}(\mathbf{X}_i) = \frac{df}{d\lambda}(\mathbf{X}_i + \lambda \mathbf{S}_i)\bigg|_{\lambda=0} = \mathbf{S}^T \nabla f_i = f_1' \tag{7.44}$$

are known to us, the linear approximation of $f(\lambda)$ is given by

$$f(\lambda) \simeq f_1 + f_1'\lambda$$

To obtain a reduction of $\delta\%$ in the objective function value compared to $|f_1|$, the step length $\lambda = \varepsilon$ is given by

$$f_1 + f_1'\varepsilon = f_1 - \frac{\delta}{100}|f_1|$$

that is,

$$\varepsilon = -\frac{\delta}{100}\frac{|f_1|}{f_1'} \tag{7.45}$$

It is to be noted that the value of ε will always be positive since f_1' given in Eq. (7.44) is always negative. This method yields good results if the percentage reduction (δ) is restricted to small values on the order of 1 to 5.

7.7.3   Termination Criteria

In steps 4 and 5 of the algorithm, the optimization procedure is assumed to have converged whenever the maximum value of $\alpha(\alpha^*)$ becomes approximately zero and the results of the current iteration satisfy the relations stated in Eq. (7.32). In addition, one can always test the Kuhn–Tucker necessary conditions before terminating the procedure.

However, we can show that if the Kuhn–Tucker conditions are satisfied, the value of α^* will become zero. The Kuhn–Tucker conditions are given by

$$\nabla f + \sum_{j=1}^{p}\lambda_j\nabla g_j = 0 \tag{7.46}$$

$$\lambda_j > 0, \quad = 1, 2, \ldots, p \tag{7.47}$$

where the first p constraints are assumed to be the active constraints. Equation (7.46) gives

$$\mathbf{S}^T\nabla f = -\sum_{j=1}^{p}\lambda_j\mathbf{S}^T\nabla g_j > 0 \tag{7.48}$$

if $\mathbf{S}$ is a usable feasible direction. Thus if the Kuhn–Tucker conditions are satisfied at a point $\mathbf{X}_i$, we will not be able to find any search direction $\mathbf{S}$ that satisfies the strict inequalities in the relations

$$\mathbf{S}^T\nabla g_j \le 0, \quad j = 1, 2, \ldots, p$$
$$\mathbf{S}^T\nabla f \le 0 \tag{7.49}$$

However, these relations can be satisfied with strict equality sign by taking the trivial solution $\mathbf{S} = \mathbf{0}$, which means that the value of α^* in the direction-finding problem, Eqs. (7.39), is zero. Some modifications and accelerating techniques have been suggested to improve the convergence of the algorithm presented in this section and the details can be found in Refs. [7.7] and [7.8].

Example 7.2

$$\text{Minimize } f(x_1, x_2) = x_1^2 + x_2^2 - 4x_1 - 4x_2 + 8$$

subject to

$$g_1(x_1, x_2) = x_1 + 2x_2 - 4 \le 0$$

with the starting point $\mathbf{X}_1 = \begin{Bmatrix} 0 \\ 0 \end{Bmatrix}$. Take $\varepsilon_1 = 0.001$, $\varepsilon_2 = 0.001$, and $\varepsilon_3 = 0.01$.

SOLUTION

Step 1: At $\mathbf{X}_1 = \{^0_0\}$:

$$f(\mathbf{X}_1) = 8 \quad \text{and} \quad g_1(\mathbf{X}_1) = -4$$

Iteration 1

Step 2: Since $g_1(\mathbf{X}_1) < 0$, we take the search direction as

$$\mathbf{S}_1 = -\nabla f(\mathbf{X}_1) = -\left\{ \begin{matrix} \partial f/\partial x_1 \\ \partial f/\partial x_2 \end{matrix} \right\}_{\mathbf{X}_1} = \left\{ \begin{matrix} 4 \\ 4 \end{matrix} \right\}$$

This can be normalized to obtain $\mathbf{S}_1 = \{^1_1\}$.

Step 5: To find the new point $\mathbf{X}_2$, we have to find a suitable step length along $\mathbf{S}_1$. For this, we choose to minimize $f(\mathbf{X}_1 + \lambda \mathbf{S}_1)$ with respect to λ. Here

$$f(\mathbf{X}_1 + \lambda \mathbf{S}_1) = f(0 + \lambda, 0 + \lambda) = 2\lambda^2 - 8\lambda + 8$$

$$\frac{df}{d\lambda} = 0 \quad \text{at} \quad \lambda = 2$$

Thus the new point is given by $\mathbf{X}_2 = \{^2_2\}$ and $g_1(\mathbf{X}_2) = 2$. As the constraint is violated, the step size has to be corrected.

As $g_1' = g_1|_{\lambda=0} = -4$ and $g_1'' = g_1|_{\lambda=2} = 2$, linear interpolation gives the new step length as

$$\tilde{\lambda} = -\frac{g_1'}{g_1'' - g_1'}\lambda = \frac{4}{3}$$

This gives $g_1|_{\lambda=\tilde{\lambda}} = 0$ and hence $\mathbf{X}_2 = \left\{ \begin{matrix} \frac{4}{3} \\ \frac{4}{3} \end{matrix} \right\}$.

Step 6: $f(\mathbf{X}_2) = \frac{8}{9}$.

Step 7: Here

$$\left| \frac{f(\mathbf{X}_1) - f(\mathbf{X}_2)}{f(\mathbf{X}_1)} \right| = \left| \frac{8 - \frac{8}{9}}{8} \right| = \frac{8}{9} > \varepsilon_2$$

$$\|\mathbf{X}_1 - \mathbf{X}_2\| = [(0 - \tfrac{4}{3})^2 + (0 - \tfrac{4}{3})^2]^{1/2} = 1.887 > \varepsilon_2$$

and hence the convergence criteria are not satisfied.

Iteration 2

Step 2: As $g_1 = 0$ at $\mathbf{X}_2$, we proceed to find a usable feasible direction.

Step 3: The direction-finding problem can be stated as [Eqs. (7.40)]:

Minimize $f = -\alpha$

subject to

$$t_1 + 2t_2 + \alpha + y_1 = 3$$

$$-\tfrac{4}{3}t_1 - \tfrac{4}{3}t_2 + \alpha + y_2 = -\tfrac{8}{3}$$

$$t_1 + y_3 = 2$$

$$t_2 + y_4 = 2$$

$$t_1 \geq 0$$

$$t_2 \geq 0$$

$$\alpha \geq 0$$

where y_1 to y_4 are the nonnegative slack variables. Since an initial basic feasible solution is not readily available, we introduce an artificial variable $y_5 \geq 0$ into the second constraint equation. By adding the infeasibility form $w = y_5$, the LP problem can be solved to obtain the solution:

$$t_1^* = 2, \quad t_2^* = \tfrac{3}{10}, \quad \alpha^* = \tfrac{4}{10}, \quad y_4^* = \tfrac{17}{10}, \quad y_1^* = y_2^* = y_3^* = 0$$

$$-f_{\min} = -\alpha^* = -\tfrac{4}{10}$$

As $\alpha^* > 0$, the usable feasible direction is given by

$$\mathbf{S} = \begin{Bmatrix} s_1 \\ s_2 \end{Bmatrix} = \begin{Bmatrix} t_1^* - 1 \\ t_2^* - 1 \end{Bmatrix} = \begin{Bmatrix} 1.0 \\ -0.7 \end{Bmatrix}$$

Step 4: Since $\alpha^* > \varepsilon_1$, we go to the next step.

Step 5: We have to move along the direction $\mathbf{S}_2 = \begin{Bmatrix} 1.0 \\ -0.7 \end{Bmatrix}$ from the point $\mathbf{X}_2 = \begin{Bmatrix} 1.333 \\ 1.333 \end{Bmatrix}$. To find the minimizing step length, we minimize

$$f(\mathbf{X}_2 + \lambda \mathbf{S}_2) = f(1.333 + \lambda, \ 1.333 - 0.7\lambda)$$

$$= 1.49\lambda^2 - 0.4\lambda + 0.889$$

As $df/d\lambda = 2.98\lambda - 0.4 = 0$ at $\lambda = 0.134$, the new point is given by

$$\mathbf{X}_3 = \mathbf{X}_2 + \lambda \mathbf{S}_2 = \begin{Bmatrix} 1.333 \\ 1.333 \end{Bmatrix} + 0.134 \begin{Bmatrix} 1.0 \\ -0.7 \end{Bmatrix} = \begin{Bmatrix} 1.467 \\ 1.239 \end{Bmatrix}$$

At this point, the constraint is satisfied since $g_1(\mathbf{X}_3) = -0.055$. Since point $\mathbf{X}_3$ lies in the interior of the feasible domain, we go to step 2.

The procedure is continued until the optimum point $\mathbf{X}^* = \begin{Bmatrix} 1.6 \\ 1.2 \end{Bmatrix}$ and $f_{\min} = 0.8$ are obtained.

7.8 ROSEN'S GRADIENT PROJECTION METHOD

The gradient projection method of Rosen [7.9, 7.10] does not require the solution of an auxiliary linear optimization problem to find the usable feasible direction. It uses the projection of the negative of the objective function gradient onto the constraints that are currently active. Although the method has been described by Rosen for a general nonlinear programming problem, its effectiveness is confined primarily to problems in which the constraints are all linear. Consider a problem with linear constraints:

$$\text{Minimize } f(\mathbf{X})$$

subject to

$$g_j(\mathbf{X}) = \sum_{i=1}^{n} a_{ij} x_i - b_j \leq 0, \quad j = 1, 2, \ldots, m \tag{7.50}$$

Let the indices of the active constraints at any point be $j_1, j_2, \ldots, j_p$. The gradients of the active constraints are given by

$$\nabla g_j(\mathbf{X}) = \begin{Bmatrix} a_{1j} \\ a_{2j} \\ \vdots \\ a_{nj} \end{Bmatrix}, \quad j = j_1, j_2, \ldots, j_p \tag{7.51}$$

By defining a matrix $\mathbf{N}$ of order $n \times p$ as

$$\mathbf{N} = [\nabla g_{j1} \nabla g_{j2} \ldots \nabla g_{jp}] \tag{7.52}$$

the direction-finding problem for obtaining a usable feasible direction $\mathbf{S}$ can be posed as follows.

$$\text{Find } \mathbf{S} \text{ which minimizes } \mathbf{S}^T \nabla f(\mathbf{X}) \tag{7.53}$$

subject to

$$\mathbf{N}^T \mathbf{S} = 0 \tag{7.54}$$

$$\mathbf{S}^T \mathbf{S} - 1 = 0 \tag{7.55}$$

where Eq. (7.55) denotes the normalization of the vector $\mathbf{S}$. To solve this equality-constrained problem, we construct the Lagrangian function as

$$L(\mathbf{S}, \boldsymbol{\lambda}, \beta) = \mathbf{S}^T \nabla f(\mathbf{X}) + \boldsymbol{\lambda}^T \mathbf{N}^T \mathbf{S} + \beta(\mathbf{S}^T \mathbf{S} - 1) \tag{7.56}$$

where

$$\boldsymbol{\lambda} = \begin{Bmatrix} \lambda_1 \\ \lambda_2 \\ \vdots \\ \lambda_p \end{Bmatrix}$$

is the vector of Lagrange multipliers associated with Eqs. (7.54) and β is the Lagrange multiplier associated with Eq. (7.55). The necessary conditions for the minimum are given by

$$\frac{\partial L}{\partial \mathbf{S}} = \nabla f(\mathbf{X}) + \mathbf{N}\boldsymbol{\lambda} + 2\beta\mathbf{S} = \mathbf{0} \tag{7.57}$$

$$\frac{\partial L}{\partial \boldsymbol{\lambda}} = \mathbf{N}^{\mathrm{T}}\mathbf{S} = \mathbf{0} \tag{7.58}$$

$$\frac{\partial L}{\partial \beta} = \mathbf{S}^{\mathrm{T}}\mathbf{S} - 1 = 0 \tag{7.59}$$

Equation (7.57) gives

$$\mathbf{S} = -\frac{1}{2\beta}(\nabla f + \mathbf{N}\boldsymbol{\lambda}) \tag{7.60}$$

Substitution of Eq. (7.60) into Eq. (7.58) gives

$$\mathbf{N}^{\mathrm{T}}\mathbf{S} = -\frac{1}{2\beta}(\mathbf{N}^{\mathrm{T}}\nabla f + \mathbf{N}^{\mathrm{T}}\mathbf{N}\boldsymbol{\lambda}) = \mathbf{0} \tag{7.61}$$

If $\mathbf{S}$ is normalized according to Eq. (7.59), β will not be zero, and hence Eq. (7.61) gives

$$\mathbf{N}^{\mathrm{T}}\nabla f + \mathbf{N}^{\mathrm{T}}\mathbf{N}\boldsymbol{\lambda} = \mathbf{0} \tag{7.62}$$

from which $\boldsymbol{\lambda}$ can be found as

$$\boldsymbol{\lambda} = -(\mathbf{N}^{\mathrm{T}}\mathbf{N})^{-1}\mathbf{N}^{\mathrm{T}}\nabla f \tag{7.63}$$

This equation, when substituted in Eq. (7.60), gives

$$\mathbf{S} = -\frac{1}{2\beta}(\mathbf{I} - \mathbf{N}(\mathbf{N}^{\mathrm{T}}\mathbf{N})^{-1}\mathbf{N}^{\mathrm{T}})\nabla f = -\frac{1}{2\beta}\mathbf{P}\nabla f \tag{7.64}$$

where

$$\mathbf{P} = \mathbf{I} - \mathbf{N}(\mathbf{N}^{\mathrm{T}}\mathbf{N})^{-1}\mathbf{N}^{\mathrm{T}} \tag{7.65}$$

is called the *projection matrix*. Disregarding the scaling constant 2β, we can say that the matrix $\mathbf{P}$ projects the vector $-\nabla f(\mathbf{X})$ onto the intersection of all the hyperplanes perpendicular to the vectors

$$\nabla g_j, \quad j = j_1, j_2, \ldots, j_p$$

We assume that the constraints $g_j(\mathbf{X})$ are independent so that the columns of the matrix $\mathbf{N}$ will be linearly independent, and hence $\mathbf{N}^{\mathrm{T}}\mathbf{N}$ will be nonsingular and can be inverted. The vector $\mathbf{S}$ can be normalized [without having to know the value of β in Eq. (7.64)] as

$$\mathbf{S} = -\frac{\mathbf{P}\nabla f}{\|\mathbf{P}\nabla f\|} \tag{7.66}$$

If $\mathbf{X}_i$ is the starting point for the ith iteration (at which $g_{j1}, g_{j2}, \ldots, g_{jp}$ are critically satisfied), we find $\mathbf{S}_i$ from Eq. (7.66) as

$$\mathbf{S}_i = -\frac{\mathbf{P}_i \nabla f(\mathbf{X}_i)}{\|\mathbf{P}_i \nabla f(\mathbf{X}_i)\|} \tag{7.67}$$

where $\mathbf{P}_i$ indicates the projection matrix $\mathbf{P}$ evaluated at the point $\mathbf{X}_i$. If $\mathbf{S}_i \neq \mathbf{0}$, we start from $\mathbf{X}_i$ and move along the direction $\mathbf{S}_i$ to find a new point $\mathbf{X}_{i+1}$ according to the familiar relation

$$\mathbf{X}_{i+1} = \mathbf{X}_i + \lambda_i \mathbf{S}_i \tag{7.68}$$

where λ_i is the step length along the search direction $\mathbf{S}_i$. The computational details for calculating λ_i will be considered later. However, if $\mathbf{S}_i = \mathbf{0}$, we have from Eqs. (7.64) and (7.63),

$$-\nabla f(\mathbf{X}_i) = \mathbf{N}\boldsymbol{\lambda} = \lambda_1 \nabla g_{j1} + \lambda_2 \nabla g_{j2} + \cdots + \lambda_p \nabla g_{jp} \tag{7.69}$$

where

$$\boldsymbol{\lambda} = -(\mathbf{N}^\mathrm{T}\mathbf{N})^{-1}\mathbf{N}^\mathrm{T}\nabla f(\mathbf{X}_i) \tag{7.70}$$

Equation (7.69) denotes that the negative of the gradient of the objective function is given by a linear combination of the gradients of the active constraints at $\mathbf{X}_i$. Further, if all λ_j, given by Eq. (7.63), are nonnegative, the Kuhn–Tucker conditions [Eqs. (7.46) and (7.47)] will be satisfied and hence the procedure can be terminated.

However, if some λ_j are negative and $\mathbf{S}_i = \mathbf{0}$, Eq. (7.69) indicates that some constraint normals ∇g_j make an obtuse angle with $-\nabla f$ at $\mathbf{X}_i$. This also means that the constraints g_j, for which λ_j are negative, are active at $\mathbf{X}_i$ but should not be considered in finding a new search direction $\mathbf{S}$ that will be both feasible and usable. (If we consider all of them, the search direction $\mathbf{S}$ comes out to be zero.) This is illustrated in Fig. 7.9, where the constraint normal $\nabla g_1(\mathbf{X}_i)$ should not be considered in finding a usable feasible direction $\mathbf{S}$ at point $\mathbf{X}_i$.

In actual practice we do not discard all the active constraints for which λ_j are negative in forming the matrix $\mathbf{N}$. Rather, we delete only one active constraint that corresponds to the most negative value of λ_j. That is, the new $\mathbf{N}$ matrix is taken as

$$\mathbf{N}_{\text{new}} = [\nabla g_{j1} \, \nabla g_{j2} \, \cdots \, \nabla g_{jq-1} \, \nabla g_{jq+1} \, \nabla g_{jq+2} \, \cdots \, \nabla g_{jp}] \tag{7.71}$$

where ∇g_{jq} is dropped from $\mathbf{N}$ by assuming that λ_q is most negative among λ_j obtained from Eq. (7.63). The new projection matrix is formed, by dropping the constraint g_{jq}, as

$$\mathbf{P}_{\text{new}} = (\mathbf{I} - \mathbf{N}_{\text{new}}(\mathbf{N}_{\text{new}}^\mathrm{T}\mathbf{N}_{\text{new}})^{-1}\mathbf{N}_{\text{new}}^\mathrm{T}) \tag{7.72}$$

and the new search direction $(\mathbf{S}_i)_{\text{new}}$ as

$$(\mathbf{S}_i)_{\text{new}} = -\frac{\mathbf{P}_{\text{new}} \nabla f(\mathbf{X_i})}{\|\mathbf{P}_{\text{new}} \nabla f(\mathbf{X_i})\|} \tag{7.73}$$

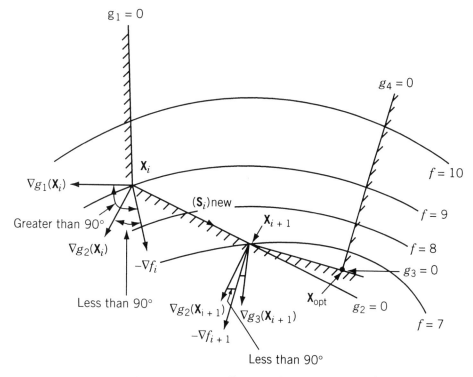

Figure 7.9   Situation when $\mathbf{S}_i = 0$ and some λ_j are negative.

and this vector will be a nonzero vector in view of the new computations we have made. The new approximation $\mathbf{X}_{i+1}$ is found as usual by using Eq. (7.68). At the new point $\mathbf{X}_{i+1}$, a new constraint may become active (in Fig. 7.9, the constraint g_3 becomes active at the new point $\mathbf{X}_{i+1}$). In such a case, the new active constraint has to be added to the set of active constraints to find the new projection matrix at $\mathbf{X}_{i+1}$.

We shall now consider the computational details for computing the step length λ_i in Eq. (7.68).

7.8.1   Determination of Step Length

The step length λ_i in Eq. (7.68) may be taken as the minimizing step length λ_i^* along the direction $\mathbf{S}_i$, that is,

$$f(\mathbf{X}_i + \lambda_i^* \mathbf{S}_i) = \min_{\lambda} f(\mathbf{X}_i + \lambda \mathbf{S}_i) \qquad (7.74)$$

However, this minimizing step length λ_i^* may give the point

$$\mathbf{X}_{i+1} = \mathbf{X}_i + \lambda_i^* \mathbf{S}_i$$

that lies outside the feasible region. Hence the following procedure is generally adopted to find a suitable step length λ_i. Since the constraints $g_j(\mathbf{X})$ are linear, we have

$$g_j(\lambda) = g_j(\mathbf{X}_i + \lambda \mathbf{S}_i) = \sum_{i=1}^{n} a_{ij}(x_i + \lambda s_i) - b_j$$

$$= \sum_{i=1}^{n} a_{ij}x_i - b_j + \lambda \sum_{i=1}^{n} a_{ij}s_i$$

$$= g_j(\mathbf{X}_i) + \lambda \sum_{i=1}^{n} a_{ij}s_i, \quad j = 1, 2, \ldots, m \qquad (7.75)$$

where

$$\mathbf{X}_i = \begin{Bmatrix} x_1 \\ x_2 \\ \vdots \\ x_n \end{Bmatrix} \quad \text{and} \quad \mathbf{S}_i = \begin{Bmatrix} s_1 \\ s_2 \\ \vdots \\ s_n \end{Bmatrix}.$$

This equation shows that $g_j(\lambda)$ will also be a linear function of λ. Thus if a particular constraint, say the kth, is not active at $\mathbf{X}_i$, it can be made to become active at the point $\mathbf{X}_i + \lambda_k \mathbf{S}_i$ by taking a step length λ_k where

$$g_k(\lambda_k) = g_k(\mathbf{X}_i) + \lambda_k \sum_{i=1}^{n} a_{ik}s_i = 0$$

that is,

$$\lambda_k = -\frac{g_k(\mathbf{X}_i)}{\sum_{i=1}^{n} a_{ik}s_i} \qquad (7.76)$$

Since the kth constraint is not active at $\mathbf{X}_i$, the value of $g_k(\mathbf{X}_i)$ will be negative and hence the sign of λ_k will be same as that of the quantity $\left(\sum_{i=1}^{n} a_{ik}s_i\right)$. From Eqs. (7.75) we have

$$\frac{dg_k}{d\lambda}(\lambda) = \sum_{i=1}^{n} a_{ik}s_i \qquad (7.77)$$

and hence the sign of λ_k depends on the rate of change of g_k with respect to λ. If this rate of change is negative, we will be moving away from the kth constraint in the positive direction of λ. However, if the rate of change $(dg_k/d\lambda)$ is positive, we will be violating the constraint g_k if we take any step length λ larger than λ_k. Thus to avoid violation of any constraint, we have to take the step length (λ_M) as

$$\lambda_M = \min_{\substack{\lambda_k > 0 \text{ and } k \\ \text{is any integer among} \\ 1 \text{ to } m \text{ other than} \\ j_1, j_2, \ldots, j_p}} (\lambda_k) \qquad (7.78)$$

In some cases, the function $f(\lambda)$ may have its minimum along the line $\mathbf{S}_i$ in between $\lambda = 0$ and $\lambda = \lambda_M$. Such a situation can be detected by calculating the value of

$$\frac{df}{d\lambda} = \mathbf{S}_i^\mathrm{T} \nabla f(\lambda) \quad \text{at} \quad \lambda = \lambda_M$$

If the minimum value of λ, λ_i^*, lies in between $\lambda = 0$ and $\lambda = \lambda_M$, the quantity $df/d\lambda(\lambda_M)$ will be positive. In such a case we can find the minimizing step length λ_i^* by interpolation or by using any of the techniques discussed in Chapter 5.

An important point to be noted is that if the step length is given by λ_i (not by λ_i^*), at least one more constraint will be active at $\mathbf{X}_{i+1}$ than at $\mathbf{X}_i$. These additional constraints will have to be considered in generating the projection matrix at $\mathbf{X}_{i+1}$. On the other hand, if the step length is given by λ_i^*, no new constraint will be active at $\mathbf{X}_{i+1}$, and hence the projection matrix at $\mathbf{X}_{i+1}$ involves only those constraints that were active at $\mathbf{X}_i$.

Algorithm. The procedure involved in the application of the gradient projection method can be described by the following steps:

1. Start with an initial point $\mathbf{X}_1$. The point $\mathbf{X}_1$ has to be feasible, that is,

$$g_j(\mathbf{X}_1) \le 0, \quad j = 1, 2, \ldots, m$$

2. Set the iteration number as $i = 1$.
3. If $\mathbf{X}_i$ is an interior feasible point [i.e., if $g_j(\mathbf{X}_i) < 0$ for $j = 1, 2, \ldots, m$], set the direction of search as $\mathbf{S}_i = -\nabla f(\mathbf{X}_i)$, normalize the search direction as

$$\mathbf{S}_i = \frac{-\nabla f(\mathbf{X}_i)}{\|\nabla f(\mathbf{X}_i)\|}$$

and go to step 5. However, if $g_j(\mathbf{X}_i) = 0$ for $j = j_1, j_2, \ldots, j_p$, go to step 4.
4. Calculate the projection matrix $\mathbf{P}_i$ as

$$\mathbf{P}_i = \mathbf{I} - \mathbf{N}_p(\mathbf{N}_p^\mathrm{T}\mathbf{N}_p)^{-1}\mathbf{N}_p^\mathrm{T}$$

where

$$\mathbf{N}_p = [\nabla g_{j1}(\mathbf{X}_i)\nabla g_{j2}(\mathbf{X}_i)\ldots\nabla g_{jp}(\mathbf{X}_i)]$$

and find the normalized search direction $\mathbf{S}_i$ as

$$\mathbf{S}_i = \frac{-\mathbf{P}_i\nabla f(\mathbf{X}_i)}{\|\mathbf{P}_i\nabla f(\mathbf{X}_i)\|}$$

5. Test whether or not $\mathbf{S}_i = \mathbf{0}$. If $\mathbf{S}_i \ne \mathbf{0}$, go to step 6. If $\mathbf{S}_i = \mathbf{0}$, compute the vector $\boldsymbol{\lambda}$ at $\mathbf{X}_i$ as

$$\boldsymbol{\lambda} = -(\mathbf{N}_p^\mathrm{T}\mathbf{N}_p)^{-1}\mathbf{N}_p^\mathrm{T}\nabla f(\mathbf{X}_i)$$

If all the components of the vector $\boldsymbol{\lambda}$ are nonnegative, take $\mathbf{X}_{\mathrm{opt}} = \mathbf{X}_i$ and stop the iterative procedure. If some of the components of $\boldsymbol{\lambda}$ are negative, find the component λ_q that has the most negative value and form the new matrix $\mathbf{N}_p$ as

$$\mathbf{N}_p = [\nabla g_{j1} \nabla g_{j2} \cdots \nabla g_{jq-1} \nabla g_{jq+1} \cdots \nabla g_{jp}]$$

and go to step 3.

6. If $\mathbf{S}_i \neq \mathbf{0}$, find the maximum step length λ_M that is permissible without violating any of the constraints as $\lambda_M = \min(\lambda_k)$, $\lambda_k > 0$ and k is any integer among 1 to m other than $j_1, j_2, \ldots, j_p$. Also find the value of $df/d\lambda(\lambda_M) = \mathbf{S}_i^T \nabla f(\mathbf{X}_i + \lambda_M \mathbf{S}_i)$. If $df/d\lambda(\lambda_M)$ is zero or negative, take the step length as $\lambda_i = \lambda_M$. On the other hand, if $df/d\lambda(\lambda_M)$ is positive, find the minimizing step length λ_i^* either by interpolation or by any of the methods discussed in Chapter 5, and take $\lambda_i = \lambda_i^*$.

7. Find the new approximation to the minimum as

$$\mathbf{X}_{i+1} = \mathbf{X}_i + \lambda_i \mathbf{S}_i$$

If $\lambda_i = \lambda_M$ or if $\lambda_M \leq \lambda_i^*$, some new constraints (one or more) become active at $\mathbf{X}_{i+1}$ and hence generate the new matrix $\mathbf{N}_p$ to include the gradients of all active constraints evaluated at $\mathbf{X}_{i+1}$. Set the new iteration number as $i = i + 1$, and go to step 4. If $\lambda_i = \lambda_i^*$ and $\lambda_i^* < \lambda_M$, no new constraint will be active at $\mathbf{X}_{i+1}$ and hence the matrix $\mathbf{N}_p$ remains unaltered. Set the new value of i as $i = i + 1$, and go to step 3.

Example 7.3

$$\text{Minimize } f(x_1, x_2) = x_1^2 + x_2^2 - 2x_1 - 4x_2$$

subject to

$$g_1(x_1, x_2) = x_1 + 4x_2 - 5 \leq 0$$

$$g_2(x_1, x_2) = 2x_1 + 3x_2 - 6 \leq 0$$

$$g_3(x_1, x_2) = -x_1 \leq 0$$

$$g_4(x_1, x_2) = -x_2 \leq 0$$

starting from the point $\mathbf{X}_1 = \left\{ {1.0 \atop 1.0} \right\}$.

SOLUTION

Iteration $i = 1$

Step 3: Since $g_j(\mathbf{X}_1) = 0$ for $j = 1$, we have $p = 1$ and $j_1 = 1$.

Step 4: As $\mathbf{N}_1 = [\nabla g_1(\mathbf{X}_1)] = \left[{1 \atop 4} \right]$, the projection matrix is given by

$$\mathbf{P}_1 = \begin{bmatrix} 1 & 0 \\ 0 & 1 \end{bmatrix} - \begin{bmatrix} 1 \\ 4 \end{bmatrix} \left[\begin{bmatrix} 1 & 4 \end{bmatrix} \begin{bmatrix} 1 \\ 4 \end{bmatrix} \right]^{-1} \begin{bmatrix} 1 & 4 \end{bmatrix}$$

$$= \frac{1}{17} \begin{bmatrix} 16 & -4 \\ -4 & 1 \end{bmatrix}$$

The search direction $\mathbf{S}_1$ is given by

$$\mathbf{S}_1 = -\frac{1}{17} \begin{bmatrix} 16 & -4 \\ -4 & 1 \end{bmatrix} \left\{ {0 \atop -2} \right\} = \left\{ {-\frac{8}{17} \atop \frac{2}{17}} \right\} = \left\{ {-0.4707 \atop 0.1177} \right\}$$

as

$$\nabla f(\mathbf{X}_1) = \begin{Bmatrix} 2x_1 - 2 \\ 2x_2 - 4 \end{Bmatrix}_{\mathbf{X}_1} = \begin{Bmatrix} 0 \\ -2 \end{Bmatrix}$$

The normalized search direction can be obtained as

$$\mathbf{S}_1 = \frac{1}{[(-0.4707)^2 + (0.1177)^2]^{1/2}} \begin{Bmatrix} -0.4707 \\ 0.1177 \end{Bmatrix} = \begin{Bmatrix} -0.9701 \\ 0.2425 \end{Bmatrix}$$

Step 5: Since $\mathbf{S}_1 \neq \mathbf{0}$, we go step 6.

Step 6: To find the step length λ_M, we set

$$\mathbf{X} = \begin{Bmatrix} x_1 \\ x_2 \end{Bmatrix} = \mathbf{X}_1 + \lambda \mathbf{S}$$

$$= \begin{Bmatrix} 1.0 - 0.9701\lambda \\ 1.0 + 0.2425\lambda \end{Bmatrix}$$

For $j = 2$:

$$g_2(\mathbf{X}) = (2.0 - 1.9402\lambda) + (3.0 + 0.7275\lambda) - 6.0 = 0 \quad \text{at} \quad \lambda = \lambda_2$$
$$= -0.8245$$

For $j = 3$:

$$g_3(\mathbf{X}) = -(1.0 - 0.9701\lambda) = 0 \quad \text{at} \quad \lambda = \lambda_3 = 1.03$$

For $j = 4$:

$$g_4(\mathbf{X}) = -(1.0 + 0.2425\lambda) = 0 \quad \text{at} \quad \lambda = \lambda_4 = -4.124$$

Therefore,

$$\lambda_M = \lambda_3 = 1.03$$

Also,

$$f(\mathbf{X}) = f(\lambda) = (1.0 - 0.9701\lambda)^2 + (1.0 + 0.2425\lambda)^2$$
$$- 2(1.0 - 0.9701\lambda) - 4(1.0 + 0.2425\lambda)$$
$$= 0.9998\lambda^2 - 0.4850\lambda - 4.0$$

$$\frac{df}{d\lambda} = 1.9996\lambda - 0.4850$$

$$\frac{df}{d\lambda}(\lambda_M) = 1.9996(1.03) - 0.4850 = 1.5746$$

As $df/d\lambda(\lambda_M) > 0$, we compute the minimizing step length λ_1^* by setting $df/d\lambda = 0$. This gives

$$\lambda_1 = \lambda_1^* = \frac{0.4850}{1.9996} = 0.2425$$

Step 7: We obtain the new point $\mathbf{X}_2$ as

$$\mathbf{X}_2 = \mathbf{X}_1 + \lambda_1 \mathbf{S}_1 = \begin{Bmatrix} 1.0 \\ 1.0 \end{Bmatrix} + 0.2425 \begin{Bmatrix} -0.9701 \\ 0.2425 \end{Bmatrix} = \begin{Bmatrix} 0.7647 \\ 1.0588 \end{Bmatrix}$$

Since $\lambda_1 = \lambda_1^*$ and $\lambda_1^* < \lambda_M$, no new constraint has become active at $\mathbf{X}_2$ and hence the matrix $\mathbf{N}_1$ remains unaltered.

Iteration i = 2

Step 3: Since $g_1(\mathbf{X}_2) = 0$, we set $p = 1$, $j_1 = 1$ and go to step 4.

Step 4:

$$\mathbf{N}_1 = \begin{bmatrix} 1 \\ 4 \end{bmatrix}$$

$$\mathbf{P}_2 = \frac{1}{17} \begin{bmatrix} 16 & -4 \\ -4 & 1 \end{bmatrix}$$

$$\Delta f(\mathbf{X}_2) = \begin{Bmatrix} 2x_1 - 2 \\ 2x_2 - 4 \end{Bmatrix}_{\mathbf{X}_2} = \begin{Bmatrix} 1.5294 - 2.0 \\ 2.1176 - 4.0 \end{Bmatrix} = \begin{Bmatrix} -0.4706 \\ -1.8824 \end{Bmatrix}$$

$$\mathbf{S}_2 = -\mathbf{P}_2 \nabla f(\mathbf{X}_2) = \frac{1}{17} \begin{bmatrix} 16 & -4 \\ -4 & 1 \end{bmatrix} \begin{Bmatrix} 0.4706 \\ 1.8824 \end{Bmatrix} = \begin{Bmatrix} 0.0 \\ 0.0 \end{Bmatrix}$$

Step 5: Since $\mathbf{S}_2 = 0$, we compute the vector $\boldsymbol{\lambda}$ at $\mathbf{X}_2$ as

$$\boldsymbol{\lambda} = -(\mathbf{N}_1^T \mathbf{N}_1)^{-1} \mathbf{N}_1^T \nabla f(\mathbf{X}_2)$$

$$= -\frac{1}{17}[1 \quad 4] \begin{Bmatrix} -0.4706 \\ -1.8824 \end{Bmatrix} = 0.4707 > 0$$

The nonnegative value of $\boldsymbol{\lambda}$ indicates that we have reached the optimum point and hence that

$$\mathbf{X}_{\text{opt}} = \mathbf{X}_2 = \begin{Bmatrix} 0.7647 \\ 1.0588 \end{Bmatrix} \quad \text{with } f_{\text{opt}} = -4.059$$

7.9   GENERALIZED REDUCED GRADIENT METHOD

The *generalized reduced gradient* (GRG) *method* is an extension of the reduced gradient method that was presented originally for solving problems with linear constraints only [7.11]. To see the details of the GRG method, consider the nonlinear programming problem:

$$\text{Minimize } f(\mathbf{X}) \tag{7.79}$$

subject to

$$h_j(\mathbf{X}) \leq 0, \quad j = 1, 2, \ldots, m \tag{7.80}$$

$$l_k(\mathbf{X}) = 0, \quad k = 1, 2, \ldots, l \tag{7.81}$$

$$x_i^{(l)} \leq x_i \leq x_i^{(u)}, \quad i = 1, 2, \ldots, n \tag{7.82}$$

By adding a nonnegative slack variable to each of the inequality constraints in Eq. (7.80), the problem can be stated as

$$\text{Minimize } f(\mathbf{X}) \tag{7.83}$$

subject to

$$h_j(\mathbf{X}) + x_{n+j} = 0, \quad j = 1, 2, \ldots, m \tag{7.84}$$

$$h_k(\mathbf{X}) = 0, \quad k = 1, 2, \ldots, l \tag{7.85}$$

$$x_i^{(l)} \leq x_i \leq x_i^{(u)}, \quad i = 1, 2, \ldots, n \tag{7.86}$$

$$x_{n+j} \geq 0, \quad j = 1, 2, \ldots, m \tag{7.87}$$

with $n + m$ variables $(x_1, x_2, \ldots, x_n, x_{n+1}, \ldots, x_{n+m})$. The problem can be rewritten in a general form as:

$$\text{Minimize } f(\mathbf{X}) \tag{7.88}$$

subject to

$$g_j(\mathbf{X}) = 0, \quad j = 1, 2, \ldots, m + l \tag{7.89}$$

$$x_i^{(l)} \leq x_i \leq x_i^{(u)}, \quad i = 1, 2, \ldots, n + m \tag{7.90}$$

where the lower and upper bounds on the slack variable, x_i, are taken as 0 and a large number (infinity), respectively ($i = n + 1, n + 2, \ldots, n + m$).

   The GRG method is based on the idea of elimination of variables using the equality constraints (see Section 2.4.1). Thus theoretically, one variable can be reduced from the set x_i ($i = 1, 2, \ldots, n + m$) for each of the $m + l$ equality constraints given by Eqs. (7.84) and (7.85). It is convenient to divide the $n + m$ design variables arbitrarily into two sets as

$$\mathbf{X} = \begin{Bmatrix} \mathbf{Y} \\ \mathbf{Z} \end{Bmatrix} \tag{7.91}$$

$$\mathbf{Y} = \begin{Bmatrix} y_1 \\ y_2 \\ \vdots \\ y_{n-l} \end{Bmatrix} = \text{design or independent variables} \tag{7.92}$$

$$\mathbf{Z} = \begin{Bmatrix} z_1 \\ z_2 \\ \vdots \\ z_{m+l} \end{Bmatrix} = \text{state or dependent variables} \tag{7.93}$$

and where the design variables are completely independent and the state variables are dependent on the design variables used to satisfy the constraints $g_j(\mathbf{X}) = 0$, $j = 1, 2, \ldots, m + l$.

Consider the first variations of the objective and constraint functions:

$$df(\mathbf{X}) = \sum_{i=1}^{n-l} \frac{\partial f}{\partial y_i} dy_i + \sum_{i=1}^{m+l} \frac{\partial f}{\partial z_i} dz_i = \nabla_{\mathbf{Y}}^{\mathrm{T}} f \, d\mathbf{Y} + \nabla_{\mathbf{Z}}^{\mathrm{T}} f \, d\mathbf{Z} \tag{7.94}$$

$$dg_i(\mathbf{X}) = \sum_{j=1}^{n-l} \frac{\partial g_i}{\partial y_j} dy_j + \sum_{j=1}^{m+l} \frac{\partial g_i}{\partial z_j} dz_j$$

or

$$d\mathbf{g} = [C] \, d\mathbf{Y} + [D] \, d\mathbf{Z} \tag{7.95}$$

where

$$\nabla_{\mathbf{Y}} f = \left\{ \begin{array}{c} \dfrac{\partial f}{\partial y_1} \\[2mm] \dfrac{\partial f}{\partial y_2} \\[2mm] \vdots \\[2mm] \dfrac{\partial f}{\partial y_{n-l}} \end{array} \right\} \tag{7.96}$$

$$\nabla_{\mathbf{Z}} f = \left\{ \begin{array}{c} \dfrac{\partial f}{\partial z_1} \\[2mm] \dfrac{\partial f}{\partial z_2} \\[2mm] \vdots \\[2mm] \dfrac{\partial f}{\partial z_{m+l}} \end{array} \right\} \tag{7.97}$$

$$[C] = \begin{bmatrix} \dfrac{\partial g_1}{\partial y_1} & \cdots & \dfrac{\partial g_1}{\partial y_{n-l}} \\[2mm] \vdots & & \vdots \\[2mm] \dfrac{\partial g_{m+l}}{\partial y_1} & \cdots & \dfrac{\partial g_{m+l}}{\partial y_{n-l}} \end{bmatrix} \tag{7.98}$$

$$[D] = \begin{bmatrix} \dfrac{\partial g_1}{\partial z_1} & \cdots & \dfrac{\partial g_1}{\partial z_{m+l}} \\[2mm] \vdots & & \vdots \\[2mm] \dfrac{\partial g_{m+l}}{\partial z_1} & \cdots & \dfrac{\partial g_{m+l}}{\partial z_{m+l}} \end{bmatrix} \tag{7.99}$$

$$dY = \begin{Bmatrix} dy_1 \\ dy_2 \\ \vdots \\ dy_{n-l} \end{Bmatrix} \qquad (7.100)$$

$$dZ = \begin{Bmatrix} dz_1 \\ dz_2 \\ \vdots \\ dz_{m+l} \end{Bmatrix} \qquad (7.101)$$

Assuming that the constraints are originally satisfied at the vector $\mathbf{X}$, $(\mathbf{g}(\mathbf{X}) = \mathbf{0})$, any change in the vector $d\mathbf{X}$ must correspond to $d\mathbf{g} = \mathbf{0}$ to maintain feasibility at $\mathbf{X} + d\mathbf{X}$. Equation (7.95) can be solved to express $d\mathbf{Z}$ as

$$d\mathbf{Z} = -[D]^{-1}[C]d\mathbf{Y} \qquad (7.102)$$

The change in the objective function due to the change in $\mathbf{X}$ is given by Eq. (7.94), which can be expressed, using Eq. (7.102), as

$$df(\mathbf{X}) = (\nabla_{\mathbf{Y}}^{T} f - \nabla_{\mathbf{Z}}^{T} f[D]^{-1}[C])d\mathbf{Y} \qquad (7.103)$$

or

$$\frac{df}{d\mathbf{Y}}(\mathbf{X}) = \mathbf{G}_R \qquad (7.104)$$

where

$$\mathbf{G}_R = \nabla_{\mathbf{Y}} f - ([D]^{-1}[C])^{T}\nabla_{\mathbf{Z}} f \qquad (7.105)$$

is called the *generalized reduced gradient*. Geometrically, the reduced gradient can be described as a projection of the original n-dimensional gradient onto the $(n - m)$-dimensional feasible region described by the design variables.

We know that a necessary condition for the existence of a minimum of an unconstrained function is that the components of the gradient vanish. Similarly, a constrained function assumes its minimum value when the appropriate components of the reduced gradient are zero. This condition can be verified to be same as the Kuhn–Tucker conditions to be satisfied at a relative minimum. In fact, the reduced gradient $\mathbf{G}_R$ can be used to generate a search direction $\mathbf{S}$ to reduce the value of the constrained objective function similar to the gradient ∇f that can be used to generate a search direction $\mathbf{S}$ for an unconstrained function. A suitable step length λ is to be chosen to minimize the value of f along the search direction $\mathbf{S}$. For any specific value of λ, the dependent variable vector $\mathbf{Z}$ is updated using Eq. (7.102). Noting that Eq. (7.102) is based on using a linear approximation to the original nonlinear problem, we find that the constraints may not be exactly equal to zero at λ, that is, $d\mathbf{g} \neq \mathbf{0}$. Hence when $\mathbf{Y}$ is held fixed, in order to have

$$g_i(\mathbf{X}) + dg_i(\mathbf{X}) = 0, \quad i = 1, 2, \ldots, m + l \qquad (7.106)$$

we must have

$$\mathbf{g}(\mathbf{X}) + d\mathbf{g}(\mathbf{X}) = \mathbf{0} \qquad (7.107)$$

Using Eq. (7.95) for $d\mathbf{g}$ in Eq. (7.107), we obtain

$$d\mathbf{Z} = [D]^{-1}(-\mathbf{g}(\mathbf{X}) - [C]d\mathbf{Y}) \qquad (7.108)$$

The value of $d\mathbf{Z}$ given by Eq. (7.108) is used to update the value of $\mathbf{Z}$ as

$$\mathbf{Z}_{\text{update}} = \mathbf{Z}_{\text{current}} + d\mathbf{Z} \qquad (7.109)$$

The constraints evaluated at the updated vector $\mathbf{X}$, and the procedure [of finding $d\mathbf{Z}$ using Eq. (7.108)] is repeated until $d\mathbf{Z}$ is sufficiently small. Note that Eq. (7.108) can be considered as Newton's method of solving simultaneous equations for $d\mathbf{Z}$.

Algorithm

1. *Specify the design and state variables*. Start with an initial trial vector $\mathbf{X}$. Identify the design and state variables ($\mathbf{Y}$ and $\mathbf{Z}$) for the problem using the following guidelines.

   (a) The state variables are to be selected to avoid singularity of the matrix, $[D]$.

   (b) Since the state variables are adjusted during the iterative process to maintain feasibility, any component of $\mathbf{X}$ that is equal to its lower or upper bound initially is to be designated a design variable.

   (c) Since the slack variables appear as linear terms in the (originally inequality) constraints, they should be designated as state variables. However, if the initial value of any state variable is zero (its lower bound value), it should be designated a design variable.

2. *Compute the generalized reduced gradient*. The GRG is determined using Eq. (7.105). The derivatives involved in Eq. (7.105) can be evaluated numerically, if necessary.

3. *Test for convergence*. If all the components of the GRG are close to zero, the method can be considered to have converged and the current vector $\mathbf{X}$ can be taken as the optimum solution of the problem. For this, the following test can be used:

$$\|\mathbf{G}_R\| \leq \varepsilon$$

   where ε is a small number. If this relation is not satisfied, we go to step 4.

4. *Determine the search direction*. The GRG can be used similar to a gradient of an unconstrained objective function to generate a suitable search direction, $\mathbf{S}$. The techniques such as steepest descent, Fletcher–Reeves, Davidon–Fletcher–Powell, or Broydon–Fletcher–Goldfarb–Shanno methods

can be used for this purpose. For example, if a steepest descent method is used, the vector $\mathbf{S}$ is determined as

$$\mathbf{S} = -\mathbf{G}_R \qquad (7.110)$$

5. *Find the minimum along the search direction.* Although any of the one -dimensional minimization procedures discussed in Chapter 5 can be used to find a local minimum of f along the search direction $\mathbf{S}$, the following procedure can be used conveniently.

(a) Find an estimate for λ as the distance to the nearest side constraint. When design variables are considered, we have

$$\lambda = \begin{cases} \dfrac{y_i^{(u)} - (y_i)_{\text{old}}}{s_i} & \text{if } s_i > 0 \\[2ex] \dfrac{y_i^{(l)} - (y_i)_{\text{old}}}{s_i} & \text{if } s_i < 0 \end{cases} \qquad (7.111)$$

where s_i is the ith component of $\mathbf{S}$. Similarly, when state variables are considered, we have, from Eq. (7.102),

$$d\mathbf{Z} = -[D]^{-1}[C]\,d\mathbf{Y} \qquad (7.112)$$

Using $d\mathbf{Y} = \lambda \mathbf{S}$, Eq. (7.112) gives the search direction for the variables $\mathbf{Z}$ as

$$\mathbf{T} = -[D]^{-1}[C]\mathbf{S} \qquad (7.113)$$

Thus

$$\lambda = \begin{cases} \dfrac{z_i^{(u)} - (z_i)_{\text{old}}}{t_i} & \text{if } t_i > 0 \\[2ex] \dfrac{z_i^{(l)} - (z_i)_{\text{old}}}{t_i} & \text{if } t_i < 0 \end{cases} \qquad (7.114)$$

where t_i is the ith component of $\mathbf{T}$.

(b) The minimum value of λ given by Eq. (7.111), λ_1, makes some design variable attain its lower or upper bound. Similarly, the minimum value of λ given by Eq. (7.114), λ_2, will make some state variable attain its lower or upper bound. The smaller of λ_1 or λ_2 can be used as an upper bound on the value of λ for initializing a suitable one-dimensional minimization procedure. The quadratic interpolation method can be used conveniently for finding the optimal step length λ^*.

(c) Find the new vector $\mathbf{X}_{\text{new}}$:

$$\mathbf{X}_{\text{new}} = \begin{Bmatrix} \mathbf{Y}_{\text{old}} + d\mathbf{Y} \\ \mathbf{Z}_{\text{old}} + d\mathbf{Z} \end{Bmatrix} = \begin{Bmatrix} \mathbf{Y}_{\text{old}} + \lambda^*\mathbf{S} \\ \mathbf{Z}_{\text{old}} + \lambda^*\mathbf{T} \end{Bmatrix} \qquad (7.115)$$

If the vector $\mathbf{X}_{\text{new}}$ corresponding to λ^* is found infeasible, then $\mathbf{Y}_{\text{new}}$ is held constant and $\mathbf{Z}_{\text{new}}$ is modified using Eq. (7.108) with $d\mathbf{Z} = \mathbf{Z}_{\text{new}} - \mathbf{Z}_{\text{old}}$. Finally, when convergence is achieved with Eq. (7.108), we find that

$$\mathbf{X}_{\text{new}} = \begin{Bmatrix} \mathbf{Y}_{\text{old}} + \Delta\mathbf{Y} \\ \mathbf{Z}_{\text{old}} + \Delta\mathbf{Z} \end{Bmatrix} \tag{7.116}$$

and go to step 1.

Example 7.4

$$\text{Minimize } f(x_1, x_2, x_3) = (x_1 - x_2)^2 + (x_2 - x_3)^4$$

subject to

$$g_1(\mathbf{X}) = x_1(1 + x_2^2) + x_3^4 - 3 = 0$$

$$-3 \le x_i \le 3, \quad i = 1, 2, 3$$

using the GRG method.

SOLUTION

Step 1: We choose arbitrarily the independent and dependent variables as

$$\mathbf{Y} = \begin{Bmatrix} y_1 \\ y_2 \end{Bmatrix} = \begin{Bmatrix} x_1 \\ x_2 \end{Bmatrix}, \quad \mathbf{Z} = \{z_1\} = \{x_3\}$$

Let the starting vector be

$$\mathbf{X}_1 = \begin{Bmatrix} -2.6 \\ 2 \\ 2 \end{Bmatrix}$$

with $f(\mathbf{X}_1) = 21.16$.

Step 2: Compute the GRG at $\mathbf{X}_1$. Noting that

$$\frac{\partial f}{\partial x_1} = 2(x_1 - x_2)$$

$$\frac{\partial f}{\partial x_2} = -2(x_1 - x_2) + 4(x_2 - x_3)^3$$

$$\frac{\partial f}{\partial x_3} = -4(x_2 - x_3)^3$$

$$\frac{\partial g_1}{\partial x_1} = 1 + x_2^2$$

$$\frac{\partial g_1}{\partial x_2} = 2x_1 x_2$$

$$\frac{\partial g_1}{\partial x_3} = 4x_3^3$$

we find, at $\mathbf{X}_1$,

$$\nabla_{\mathbf{Y}} f = \left\{ \begin{array}{c} \dfrac{\partial f}{\partial x_1} \\[2mm] \dfrac{\partial f}{\partial x_2} \end{array} \right\}_{\mathbf{X}_1} = \left\{ \begin{array}{c} 2(-2.6 - 2) \\ -2(-2.6 - 2) + 4(2 - 2)^3 \end{array} \right\} = \left\{ \begin{array}{c} -9.2 \\ 9.2 \end{array} \right\}$$

$$\nabla_{\mathbf{Z}} f = \left\{ \dfrac{\partial f}{\partial x_3} \right\}_{\mathbf{X}_1} = \{ -4(x_2 - x_3)^3 \}_{\mathbf{X}_1} = 0$$

$$[C] = \left[\dfrac{\partial g_1}{\partial x_1} \quad \dfrac{\partial g_1}{\partial x_2} \right]_{\mathbf{X}_1} = [5 \quad -10.4]$$

$$[D] = \left[\dfrac{\partial g_1}{\partial x_3} \right]_{\mathbf{X}_1} = [32]$$

$$D^{-1} = [\tfrac{1}{32}], \quad [D]^{-1}[C] = \tfrac{1}{32}[5 \ -10.4] = [0.15625 \ -0.325]$$

$$\mathbf{G}_R = \nabla_{\mathbf{Y}} f - [[D]^{-1}[C]]^{\mathrm{T}} \nabla_{\mathbf{Z}} f$$

$$= \left\{ \begin{array}{c} -9.2 \\ 9.2 \end{array} \right\} - \left\{ \begin{array}{c} 0.15625 \\ -0.325 \end{array} \right\} (0) = \left\{ \begin{array}{c} -9.2 \\ 9.2 \end{array} \right\}$$

Step 3: Since the components of $\mathbf{G}_R$ are not zero, the point $\mathbf{X}_1$ is not optimum, and hence we go to step 4.

Step 4: We use the steepest descent method and take the search direction as

$$\mathbf{S} = -\mathbf{G}_R = \left\{ \begin{array}{c} 9.2 \\ -9.2 \end{array} \right\}$$

Step 5: We find the optimal step length along $\mathbf{S}$.

(a) Considering the design variables, we use Eq. (7.111) to obtain For $y_1 = x_1$:

$$\lambda = \frac{3 - (-2.6)}{9.2} = 0.6087$$

For $y_2 = x_2$:

$$\lambda = \frac{-3 - (2)}{-9.2} = 0.5435$$

Thus the smaller value gives $\lambda_1 = 0.5435$. Equation (7.113) gives

$$\mathbf{T} = -([D]^{-1}[C])\mathbf{S} = -(0.15625 \ -0.325) \left\{ \begin{array}{c} 9.2 \\ -9.2 \end{array} \right\} = -4.4275$$

and hence Eq. (7.114) leads to

For $z_1 = x_3 : \lambda = \dfrac{-3 - (2)}{-4.4275} = 1.1293$

Thus $\lambda_2 = 1.1293$.

(b) The upper bound on λ is given by the smaller of λ_1 and λ_2, which is equal to 0.5435. By expressing

$$\mathbf{X} = \begin{Bmatrix} \mathbf{Y} + \lambda \mathbf{S} \\ \mathbf{Z} + \lambda \mathbf{T} \end{Bmatrix}$$

we obtain

$$\mathbf{X} = \begin{Bmatrix} x_1 \\ x_2 \\ x_3 \end{Bmatrix} = \begin{Bmatrix} -2.6 \\ 2 \\ 2 \end{Bmatrix} + \lambda \begin{Bmatrix} 9.2 \\ -9.2 \\ -4.4275 \end{Bmatrix} = \begin{Bmatrix} -2.6 + 9.2\lambda \\ 2 - 9.2\lambda \\ 2 - 4.4275\lambda \end{Bmatrix}$$

and hence

$$
\begin{aligned}
f(\lambda) = f(\mathbf{X}) &= (-2.6 + 9.2\lambda - 2 + 9.2\lambda)^2 \\
&\quad + (2 - 9.2\lambda - 2 + 4.4275\lambda)^4 \\
&= 518.7806\lambda^4 + 338.56\lambda^2 - 169.28\lambda + 21.16
\end{aligned}
$$

$df/d\lambda = 0$ gives

$$2075.1225\lambda^3 + 677.12\lambda - 169.28 = 0$$

from which we find the root as $\lambda^* \approx 0.22$. Since λ^* is less than the upper bound value 0.5435, we use λ^*.

(c) The new vector $\mathbf{X}_{\text{new}}$ is given by

$$
\begin{aligned}
\mathbf{X}_{\text{new}} &= \begin{Bmatrix} \mathbf{Y}_{\text{old}} + d\mathbf{Y} \\ \mathbf{Z}_{\text{old}} + d\mathbf{Z} \end{Bmatrix} \\
&= \begin{Bmatrix} \mathbf{Y}_{\text{old}} + \lambda^*\mathbf{S} \\ \mathbf{Z}_{\text{old}} + \lambda^*\mathbf{T} \end{Bmatrix} = \begin{Bmatrix} -2.6 + 0.22(9.2) \\ 2 + 0.22(-9.2) \\ 2 + 0.22(-4.4275) \end{Bmatrix} = \begin{Bmatrix} -0.576 \\ -0.024 \\ 1.02595 \end{Bmatrix}
\end{aligned}
$$

with

$$d\mathbf{Y} = \begin{Bmatrix} 2.024 \\ -2.024 \end{Bmatrix}, \quad d\mathbf{Z} = \{-0.97405\}$$

Now, we need to check whether this vector is feasible. Since

$$g_1(\mathbf{X}_{\text{new}}) = (-0.576)[1 + (-0.024)^2] + (1.02595)^4 - 3 = -2.4684 \neq 0$$

the vector $\mathbf{X}_{\text{new}}$ is infeasible. Hence we hold $\mathbf{Y}_{\text{new}}$ constant and modify $\mathbf{Z}_{\text{new}}$ using Newton's method [Eq. (7.108)] as

$$d\mathbf{Z} = [D]^{-1}[-\mathbf{g}(\mathbf{X}) - [C]d\mathbf{Y}]$$

Since

$$[D] = \left[\frac{\partial g_1}{\partial z_1} \right] = [4x_3^3] = [4(1.02595)^3] = [4.319551]$$

$$g_1(\mathbf{X}) = \{-2.4684\}$$

$$[C] = \left[\frac{\partial g_1}{\partial y_1} \frac{\partial g_1}{\partial y_2} \right] = \{[2(-0.576 + 0.024)][-2(-0.576 + 0.024)$$

$$+ 4(-0.024 - 1.02595)^3]\}$$

$$= [-1.104 \ \ -3.5258]$$

$$dZ = \frac{1}{4.319551} \left[2.4684 - \{-1.104 \ \ -3.5258\} \right.$$

$$\left. \times \begin{Bmatrix} 2.024 \\ -2.024 \end{Bmatrix} \right] = \{-0.5633\}$$

we have $\mathbf{Z}_{new} = \mathbf{Z}_{old} + d\mathbf{Z} = \{2 - 0.5633\} = \{1.4367\}$. The current $\mathbf{X}_{new}$ becomes

$$\mathbf{X}_{new} = \begin{Bmatrix} \mathbf{Y}_{old} + d\mathbf{Y} \\ \mathbf{Z}_{old} + d\mathbf{Z} \end{Bmatrix} = \begin{Bmatrix} -0.576 \\ -0.024 \\ 1.4367 \end{Bmatrix}$$

The constraint becomes

$$g_1 = (-0.576)(1 - (-0.024)^2) + (1.4367)^4 - 3 = 0.6842 \neq 0$$

Since this $\mathbf{X}_{new}$ is infeasible, we need to apply Newton's method [Eq. (7.108)] at the current $\mathbf{X}_{new}$. In the present case, instead of repeating Newton's iteration, we can find the value of $\mathbf{Z}_{new} = \{x_3\}_{new}$ by satisfying the constraint as

$$g_1(\mathbf{X}) = (-0.576)[1 + (-0.024)^2] + x_3^4 - 3 = 0$$

$$\text{or} \quad x_3 = (2.4237)^{0.25} = 1.2477$$

This gives

$$\mathbf{X}_{new} = \begin{Bmatrix} -0.576 \\ -0.024 \\ 1.2477 \end{Bmatrix} \quad \text{and}$$

$$f(\mathbf{X}_{new}) = (-0.576 + 0.024)^2 + (-0.024 - 1.2477)^4 = 2.9201$$

Next we go to step 1.

Step 1: We do not have to change the set of independent and dependent variables and hence we go to the next step.

Step 2: We compute the GRG at the current $\mathbf{X}$ using Eq. (7.105). Since

$$\nabla_{\mathbf{Y}} f = \left\{ \begin{array}{c} \dfrac{\partial f}{\partial x_1} \\ \dfrac{\partial f}{\partial x_2} \end{array} \right\} = \left\{ \begin{array}{c} 2(-0.576 + 0.024) \\ -2(-0.576 + 0.024) + 4(-0.024 - 1.2477)^3 \end{array} \right\}$$

$$= \left\{ \begin{array}{c} -1.104 \\ -7.1225 \end{array} \right\}$$

$$\nabla_{\mathbf{Z}} f = \left\{ \dfrac{\partial f}{\partial z_1} \right\} = \left\{ \dfrac{\partial f}{\partial x_3} \right\} = \{-4(-0.024 - 1.2477)^3\} = \{8.2265\}$$

$$[C] = \left[\dfrac{\partial g_1}{\partial x_1} \quad \dfrac{\partial g_1}{\partial x_2} \right] = [(1 + (-0.024)^2) \quad 2(-0.576)(-0.024)]$$

$$= [1.000576 \quad 0.027648]$$

$$[D] = \left[\dfrac{\partial g_1}{\partial x_3} \right] = [4x_3^3] = [4(1.2477)^3] = [7.7694]$$

$$[D]^{-1}[C] = \dfrac{1}{7.7694}[1.000576 \quad 0.027648] = [0.128784 \quad 0.003558]$$

$$\mathbf{G}_R = \nabla_{\mathbf{Y}} f - [[D]^{-1}[C]]^{\mathrm{T}} \nabla_{\mathbf{Z}} f$$

$$= \left\{ \begin{array}{c} -1.104 \\ -7.1225 \end{array} \right\} - \left\{ \begin{array}{c} 0.128784 \\ 0.003558 \end{array} \right\} (8.2265) = \left\{ \begin{array}{c} -2.1634 \\ -7.1518 \end{array} \right\}$$

Since $\mathbf{G}_R \neq 0$, we need to proceed to the next step.

Note: It can be seen that the value of the objective function reduced from an initial value of 21.16 to 2.9201 in one iteration.

7.10   SEQUENTIAL QUADRATIC PROGRAMMING

The sequential quadratic programming is one of the most recently developed and perhaps one of the best methods of optimization. The method has a theoretical basis that is related to (1) the solution of a set of nonlinear equations using Newton's method, and (2) the derivation of simultaneous nonlinear equations using Kuhn–Tucker conditions to the Lagrangian of the constrained optimization problem. In this section we present both the derivation of the equations and the solution procedure of the sequential quadratic programming approach.

7.10.1   Derivation

Consider a nonlinear optimization problem with only equality constraints:

Find $\mathbf{X}$ which minimizes $f(\mathbf{X})$

subject to

$$h_k(\mathbf{X}) = 0, \quad k = 1, 2, \ldots, p \tag{7.117}$$

The extension to include inequality constraints will be considered at a later stage. The Lagrange function, $L(\mathbf{X}, \boldsymbol{\lambda})$, corresponding to the problem of Eq. (7.117) is given by

$$L = f(\mathbf{X}) + \sum_{k=1}^{p} \lambda_k h_k(\mathbf{X}) \tag{7.118}$$

where λ_k is the Lagrange multiplier for the kth equality constraint. The Kuhn–Tucker necessary conditions can be stated as

$$\nabla L = \mathbf{0} \quad \text{or} \quad \nabla f + \sum_{k=1}^{p} \lambda_k \nabla h_k = \mathbf{0} \quad \text{or} \quad \nabla f + [A]^{\mathrm{T}} \boldsymbol{\lambda} = \mathbf{0} \tag{7.119}$$

$$h_k(\mathbf{X}) = 0, \quad k = 1, 2, \dots, p \tag{7.120}$$

where $[A]$ is an $n \times p$ matrix whose kth column denotes the gradient of the function h_k. Equations (7.119) and (7.120) represent a set of $n + p$ nonlinear equations in $n + p$ unknowns ($x_i, i = 1, \dots, n$ and $\lambda_k, k = 1, \dots, p$). These nonlinear equations can be solved using Newton's method. For convenience, we rewrite Eqs. (7.119) and (7.120) as

$$\mathbf{F}(\mathbf{Y}) = \mathbf{0} \tag{7.121}$$

where

$$\mathbf{F} = \left\{ \begin{matrix} \nabla L \\ \mathbf{h} \end{matrix} \right\}_{(n+p) \times 1}, \quad \mathbf{Y} = \left\{ \begin{matrix} \mathbf{X} \\ \boldsymbol{\lambda} \end{matrix} \right\}_{(n+p) \times 1}, \quad \mathbf{0} = \left\{ \begin{matrix} \mathbf{0} \\ \mathbf{0} \end{matrix} \right\}_{(n+p) \times 1} \tag{7.122}$$

According to Newton's method, the solution of Eqs. (7.121) can be found iteratively as (see Section 6.11)

$$\mathbf{Y}_{j+1} = \mathbf{Y}_j + \Delta \mathbf{Y}_j \tag{7.123}$$

with

$$[\nabla F]_j^{\mathrm{T}} \Delta \mathbf{Y}_j = -\mathbf{F}(\mathbf{Y}_j) \tag{7.124}$$

where $\mathbf{Y}_j$ is the solution at the start of jth iteration and $\Delta \mathbf{Y}_j$ is the change in $\mathbf{Y}_j$ necessary to generate the improved solution, $\mathbf{Y}_{j+1}$, and $[\nabla F]_j = [\nabla F(\mathbf{Y}_j)]$ is the $(n + p) \times (n + p)$ Jacobian matrix of the nonlinear equations whose ith column denotes the gradient of the function $F_i(\mathbf{Y})$ with respect to the vector $\mathbf{Y}$. By substituting Eqs. (7.121) and (7.122) into Eq. (7.124), we obtain

$$\begin{bmatrix} [\nabla^2 L] & [H] \\ [H]^{\mathrm{T}} & [0] \end{bmatrix}_j \left\{ \begin{matrix} \Delta \mathbf{X} \\ \Delta \boldsymbol{\lambda} \end{matrix} \right\}_j = - \left\{ \begin{matrix} \nabla L \\ \mathbf{h} \end{matrix} \right\}_j \tag{7.125}$$

$$\Delta \mathbf{X}_j = \mathbf{X}_{j+1} - \mathbf{X}_j \tag{7.126}$$

$$\Delta \boldsymbol{\lambda}_j = \boldsymbol{\lambda}_{j+1} - \boldsymbol{\lambda}_j \tag{7.127}$$

where $[\nabla^2 L]_{n \times n}$ denotes the Hessian matrix of the Lagrange function. The first set of equations in (7.125) can be written separately as

$$[\nabla^2 L]_j \Delta \mathbf{X}_j + [H]_j \Delta \boldsymbol{\lambda}_j = -\nabla L_j \qquad (7.128)$$

Using Eq. (7.127) for $\Delta \boldsymbol{\lambda}_j$ and Eq. (7.119) for ∇L_j, Eq. (7.128) can be expressed as

$$[\nabla^2 L]_j \Delta \mathbf{X}_j + [H]_j (\boldsymbol{\lambda}_{j+1} - \boldsymbol{\lambda}_j) = -\nabla f_j - [H]_j^T \boldsymbol{\lambda}_j \qquad (7.129)$$

which can be simplified to obtain

$$[\nabla^2 L]_j \Delta \mathbf{X}_j + [H]_j \boldsymbol{\lambda}_{j+1} = -\nabla f_j \qquad (7.130)$$

Equation (7.130) and the second set of equations in (7.125) can now be combined as

$$\begin{bmatrix} [\nabla^2 L] & [H] \\ [H]^T & [0] \end{bmatrix}_j \begin{Bmatrix} \Delta \mathbf{X}_j \\ \boldsymbol{\lambda}_{j+1} \end{Bmatrix} = - \begin{Bmatrix} \nabla f_j \\ \mathbf{h}_j \end{Bmatrix} \qquad (7.131)$$

Equations (7.131) can be solved to find the change in the design vector $\Delta \mathbf{X}_j$ and the new values of the Lagrange multipliers, $\boldsymbol{\lambda}_{j+1}$. The iterative process indicated by Eq. (7.131) can be continued until convergence is achieved.

Now consider the following quadratic programming problem:

Find $\Delta \mathbf{X}$ that minimizes the quadratic objective function

$$Q = \nabla f^T \Delta \mathbf{X} + \tfrac{1}{2} \Delta \mathbf{X}^T [\nabla^2 L] \Delta \mathbf{X}$$

subject to the linear equality constraints   (7.132)

$$h_k + \nabla h_k^T \Delta \mathbf{X} = 0, \quad k = 1, 2, \ldots, p \quad \text{or} \quad \mathbf{h} + [H]^T \Delta \mathbf{X} = \mathbf{0}$$

The lagrange function, $\tilde{L}$, corresponding to the problem of Eq. (7.132) is given by

$$\tilde{L} = \nabla f^T \Delta \mathbf{X} + \tfrac{1}{2} \Delta \mathbf{X}^T [\nabla^2 L] \Delta \mathbf{X} + \sum_{k=1}^{p} \lambda_k (h_k + \nabla h_k^T \Delta \mathbf{X}) \qquad (7.133)$$

where λ_k is the Lagrange multiplier associated with the kth equality constraint.

The Kuhn–Tucker necessary conditions can be stated as

$$\nabla f + [\nabla^2 L] \Delta \mathbf{X} + [H] \boldsymbol{\lambda} = \mathbf{0} \qquad (7.134)$$

$$h_k + \nabla h_k^T \Delta \mathbf{X} = 0, \quad k = 1, 2, \ldots, p \qquad (7.135)$$

Equations (7.134) and (7.135) can be identified to be same as Eq. (7.131) in matrix form. This shows that the original problem of Eq. (7.117) can be solved iteratively by solving the quadratic programming problem defined by Eq. (7.132). In fact, when inequality constraints are added to the original problem, the quadratic programming problem of Eq. (7.132) becomes

Find $\mathbf{X}$ which minimizes $Q = \nabla f^T \Delta \mathbf{X} + \tfrac{1}{2} \Delta \mathbf{X}^T [\nabla^2 L] \Delta \mathbf{X}$

subject to

$$g_j + \nabla g_j^{\mathrm{T}} \Delta \mathbf{X} \leq 0, \quad j = 1, 2, \ldots, m$$

$$h_k + \nabla h_k^{\mathrm{T}} \Delta \mathbf{X} = 0, \quad k = 1, 2, \ldots, p \tag{7.136}$$

with the Lagrange function given by

$$\tilde{L} = f(\mathbf{X}) + \sum_{j=1}^{m} \lambda_j g_j(\mathbf{X}) + \sum_{k=1}^{p} \lambda_{m+k} h_k(\mathbf{X}) \tag{7.137}$$

Since the minimum of the augmented Lagrange function is involved, the sequential quadratic programming method is also known as the *projected Lagrangian method*.

7.10.2   Solution Procedure

As in the case of Newton's method of unconstrained minimization, the solution vector $\Delta \mathbf{X}$ in Eq. (7.136) is treated as the search direction, $\mathbf{S}$, and the quadratic programming subproblem (in terms of the design vector $\mathbf{S}$) is restated as:

$$\text{Find } \mathbf{S} \text{ which minimizes } Q(\mathbf{S}) = \nabla f(\mathbf{X})^{\mathrm{T}} \mathbf{S} + \tfrac{1}{2} \mathbf{S}^{\mathrm{T}} [H] \mathbf{S}$$

subject to

$$\beta_j g_j(\mathbf{X}) + \nabla g_j(\mathbf{X})^{\mathrm{T}} \mathbf{S} \leq 0, \quad j = 1, 2, \ldots, m$$

$$\overline{\beta} h_k(\mathbf{X}) + \nabla h_k(\mathbf{X})^{\mathrm{T}} \mathbf{S} = 0, \quad k = 1, 2, \ldots, p \tag{7.138}$$

where $[H]$ is a positive definite matrix that is taken initially as the identity matrix and is updated in subsequent iterations so as to converge to the Hessian matrix of the Lagrange function of Eq. (7.137), and β_j and $\overline{\beta}$ are constants used to ensure that the linearized constraints do not cut off the feasible space completely. Typical values of these constants are given by

$$\overline{\beta} \approx 0.9; \quad \beta_j = \begin{cases} 1 & \text{if } g_j(\mathbf{X}) \leq 0 \\ \overline{\beta} & \text{if } g_j(\mathbf{X}) \geq 0 \end{cases} \tag{7.139}$$

The subproblem of Eq. (7.138) is a quadratic programming problem and hence the method described in Section 4.8 can be used for its solution. Alternatively, the problem can be solved by any of the methods described in this chapter since the gradients of the function involved can be evaluated easily. Since the Lagrange multipliers associated with the solution of the problem, Eq. (7.138), are needed, they can be evaluated using Eq. (7.263). Once the search direction, $\mathbf{S}$, is found by solving the problem in Eq. (7.138), the design vector is updated as

$$\mathbf{X}_{j+1} = \mathbf{X}_j + \alpha^* \mathbf{S} \tag{7.140}$$

where α^* is the optimal step length along the direction $\mathbf{S}$ found by minimizing the function (using an exterior penalty function approach):

$$\phi = f(\mathbf{X}) + \sum_{j=1}^{m} \lambda_j (\max[0, g_j(\mathbf{X})]) + \sum_{k=1}^{p} \lambda_{m+k} |h_k(\mathbf{X})| \tag{7.141}$$

with

$$\lambda_j = \begin{cases} |\lambda_j|, & j = 1, 2, \ldots, m + p \text{ in first iteration} \\ \max\{|\lambda_j|, \frac{1}{2}(\tilde{\lambda}_j, |\lambda_j|)\} \text{in subsequent iterations} \end{cases} \tag{7.142}$$

and $\tilde{\lambda}_j = \lambda_j$ of the previous iteration. The one-dimensional step length α^* can be found by any of the methods discussed in Chapter 5.

Once $\mathbf{X}_{j+1}$ is found from Eq. (7.140), for the next iteration the Hessian matrix $[H]$ is updated to improve the quadratic approximation in Eq. (7.138). Usually, a modified BFGS formula, given below, is used for this purpose [7.12]:

$$[H_{i+1}] = [H_i] - \frac{[H_i]\mathbf{P}_i\mathbf{P}_i^T[H_i]}{\mathbf{P}_i^T[H_i]\mathbf{P}_i} + \frac{\gamma\gamma^T}{\mathbf{P}_i^T\mathbf{P}_i} \tag{7.143}$$

$$\mathbf{P}_i = \mathbf{X}_{i+1} - \mathbf{X}_i \tag{7.144}$$

$$\gamma = \theta\mathbf{Q}_i + (1 - \theta)[H_i]\mathbf{P}_i \tag{7.145}$$

$$\mathbf{Q}_i = \nabla_x\tilde{L}(\mathbf{X}_{i+1}, \lambda_{i+1}) - \nabla_x\tilde{L}(\mathbf{X}_i, \lambda_i) \tag{7.146}$$

$$\theta = \begin{cases} 1.0 & \text{if } \mathbf{P}_i^T\mathbf{Q}_i \geq 0.2\mathbf{P}_i^T[H_i]\mathbf{P}_i \\ \dfrac{0.8\mathbf{P}_i^T[H_i]\mathbf{P}_i}{\mathbf{P}_i^T[H_i]\mathbf{P}_i - \mathbf{P}_i^T\mathbf{Q}_i} & \text{if } \mathbf{P}_i^T\mathbf{Q}_i < 0.2\mathbf{P}_i^T[H_i]\mathbf{P}_i \end{cases} \tag{7.147}$$

where $\tilde{L}$ is given by Eq. (7.137) and the constants 0.2 and 0.8 in Eq. (7.147) can be changed, based on numerical experience.

Example 7.5 Find the solution of the problem (see Problem 1.31):

$$\text{Minimize } f(\mathbf{X}) = 0.1x_1 + 0.05773x_2 \tag{E_1}$$

subject to

$$g_1(\mathbf{X}) = \frac{0.6}{x_1} + \frac{0.3464}{x_2} - 0.1 \leq 0 \tag{E_2}$$

$$g_2(\mathbf{X}) = 6 - x_1 \leq 0 \tag{E_3}$$

$$g_3(\mathbf{X}) = 7 - x_2 \leq 0 \tag{E_4}$$

using the sequential quadratic programming technique.

SOLUTION   Let the starting point be $\mathbf{X}_1 = (11.8765, 7.0)^T$ with $g_1(\mathbf{X}_1) = g_3(\mathbf{X}_1) = 0$, $g_2(\mathbf{X}_1) = -5.8765$, and $f(\mathbf{X}_1) = 1.5917$. The gradients of the objective and constraint functions at $\mathbf{X}_1$ are given by

$$\nabla f(\mathbf{X}_1) = \begin{Bmatrix} 0.1 \\ 0.05773 \end{Bmatrix}, \quad \nabla g_1(\mathbf{X}_1) = \begin{Bmatrix} \dfrac{-0.6}{x_1^2} \\ \dfrac{-0.3464}{x_2^2} \end{Bmatrix}_{\mathbf{X}_1} = \begin{Bmatrix} -0.004254 \\ -0.007069 \end{Bmatrix}$$

$$\nabla g_2(\mathbf{X}_1) = \begin{Bmatrix} -1 \\ 0 \end{Bmatrix}, \quad \nabla g_3(\mathbf{X}_1) = \begin{Bmatrix} 0 \\ -1 \end{Bmatrix}$$

We assume the matrix $[H_1]$ to be the identity matrix and hence the objective function of Eq. (7.138) becomes

$$Q(\mathbf{S}) = 0.1s_1 + 0.05773s_2 + 0.5s_1^2 + 0.5s_2^2 \qquad (E_5)$$

Equation (7.139) gives $\beta_1 = \beta_3 = 0$ since $g_1 = g_3 = 0$ and $\beta_2 = 1.0$ since $g_2 < 0$, and hence the constraints of Eq. (7.138) can be expressed as

$$\tilde{g}_1 = -0.004254s_1 - 0.007069s_2 \le 0 \qquad (E_6)$$

$$\tilde{g}_2 = -5.8765 - s_1 \le 0 \qquad (E_7)$$

$$\tilde{g}_3 = -s_2 \le 0 \qquad (E_8)$$

We solve this quadratic programming problem [Eqs. (E$_5$) to (E$_8$)] directly with the use of the Kuhn–Tucker conditions. The Kuhn–Tucker conditions are given by

$$\frac{\partial Q}{\partial s_1} + \sum_{j=1}^{3} \lambda_j \frac{\partial \tilde{g}_j}{\partial s_1} = 0 \qquad (E_9)$$

$$\frac{\partial Q}{\partial s_2} + \sum_{j=1}^{3} \lambda_j \frac{\partial \tilde{g}_j}{\partial s_2} = 0 \qquad (E_{10})$$

$$\lambda_j \tilde{g}_j = 0, \quad j = 1, 2, 3 \qquad (E_{11})$$

$$\tilde{g}_j \le 0, \quad j = 1, 2, 3 \qquad (E_{12})$$

$$\lambda_j \ge 0, \quad j = 1, 2, 3 \qquad (E_{13})$$

Equations (E$_9$) and (E$_{10}$) can be expressed, in this case, as

$$0.1 + s_1 - 0.004254\lambda_1 - \lambda_2 = 0 \qquad (E_{14})$$

$$0.05773 + s_2 - 0.007069\lambda_1 - \lambda_3 = 0 \qquad (E_{15})$$

By considering all possibilities of active constraints, we find that the optimum solution of the quadratic programming problem [Eqs. (E$_5$) to (E$_8$)] is given by

$$s_1^* = -0.04791, \quad s_2^* = 0.02883, \quad \lambda_1^* = 12.2450, \quad \lambda_2^* = 0, \quad \lambda_3^* = 0$$

The new design vector, $\mathbf{X}$, can be expressed as

$$\mathbf{X} = \mathbf{X}_1 + \alpha \mathbf{S} = \left\{ \begin{array}{c} 11.8765 - 0.04791\alpha \\ 7.0 + 0.02883\alpha \end{array} \right\}$$

where α can be found by minimizing the function ϕ in Eq. (7.141):

$$\phi = 0.1(11.8765 - 0.04791\alpha) + 0.05773(7.0 + 0.02883\alpha)$$

$$+ 12.2450 \left(\frac{0.6}{11.8765 - 0.04791\alpha} + \frac{0.3464}{7.0 + 0.02883\alpha} - 0.1 \right)$$

By using quadratic interpolation technique (unrestricted search method can also be used for simplicity), we find that ϕ attains its minimum value of 1.48 at $\alpha^* = 64.93$, which corresponds to the new design vector

$$\mathbf{X}_2 = \begin{Bmatrix} 8.7657 \\ 8.8719 \end{Bmatrix}$$

with $f(\mathbf{X}_2) = 1.38874$ and $g_1(\mathbf{X}_2) = +0.0074932$ (violated slightly). Next we update the matrix $[H]$ using Eq. (7.143) with

$$\tilde{L} = 0.1x_1 + 0.05773x_2 + 12.2450 \left(\frac{0.6}{x_1} + \frac{0.3464}{x_2} - 0.1 \right)$$

$$\nabla_x \tilde{L} = \begin{Bmatrix} \dfrac{\partial \tilde{L}}{\partial x_1} \\[2mm] \dfrac{\partial \tilde{L}}{\partial x_2} \end{Bmatrix} \quad \text{with} \quad \frac{\partial \tilde{L}}{\partial x_1} = 0.1 - \frac{7.3470}{x_1^2}$$

$$\text{and} \quad \frac{\partial \tilde{L}}{\partial x_2} = 0.05773 - \frac{4.2417}{x_2^2}$$

$$\mathbf{P}_1 = \mathbf{X}_2 - \mathbf{X}_1 = \begin{Bmatrix} -3.1108 \\ 1.8719 \end{Bmatrix}$$

$$\mathbf{Q}_1 = \nabla_x \tilde{L}(\mathbf{X}_2) - \nabla_x \tilde{L}(\mathbf{X}_1) = \begin{Bmatrix} 0.00438 \\ 0.00384 \end{Bmatrix} - \begin{Bmatrix} 0.04791 \\ -0.02883 \end{Bmatrix} = \begin{Bmatrix} -0.04353 \\ 0.03267 \end{Bmatrix}$$

$$\mathbf{P}_1^T[H_1]\mathbf{P}_1 = 13.1811, \quad \mathbf{P}_1^T\mathbf{Q}_1 = 0.19656$$

This indicates that $\mathbf{P}_1^T\mathbf{Q}_1 < 0.2\mathbf{P}_1^T[H_1]\mathbf{P}_1$, and hence θ is computed using Eq. (7.147) as

$$\theta = \frac{(0.8)(13.1811)}{13.1811 - 0.19656} = 0.81211$$

$$\gamma = \theta\mathbf{Q}_1 + (1 - \theta)[H_1]\mathbf{P}_1 = \begin{Bmatrix} 0.54914 \\ -0.32518 \end{Bmatrix}$$

Hence

$$[H_2] = \begin{bmatrix} 0.2887 & 0.4283 \\ 0.4283 & 0.7422 \end{bmatrix}$$

We can now start another iteration by defining a new quadratic programming problem using Eq. (7.138) and continue the procedure until the optimum solution is found. Note that the objective function reduced from a value of 1.5917 to 1.38874 in one iteration when $\mathbf{X}$ changed from $\mathbf{X}_1$ to $\mathbf{X}_2$.

Indirect Methods

7.11   TRANSFORMATION TECHNIQUES

If the constraints $g_j(\mathbf{X})$ are explicit functions of the variables x_i and have certain simple forms, it may be possible to make a transformation of the independent variables such that the constraints are satisfied automatically [7.13]. Thus it may be possible to convert a constrained optimization problem into an unconstrained one by making a change of variables. Some typical transformations are indicated below:

1. If lower and upper bounds on x_i are specified as

$$l_i \leq x_i \leq u_i \tag{7.148}$$

these can be satisfied by transforming the variable x_i as

$$x_i = l_i + (u_i - l_i)\sin^2 y_i \tag{7.149}$$

where y_i is the new variable, which can take any value.

2. If a variable x_i is restricted to lie in the interval $(0, 1)$, we can use the transformation:

$$x_i = \sin^2 y_i, \quad x_i = \cos^2 y_i$$

$$x_i = \frac{e^{y_i}}{e^{y_i} + e^{-y_i}} \quad \text{or} \quad x_i = \frac{y_i^2}{1 + y_i^2} \tag{7.150}$$

3. If the variable x_i is constrained to take only positive values, the transformation can be

$$x_i = \text{abs}(y_i), \quad x_i = y_i^2 \quad \text{or} \quad x_i = e^{y_i} \tag{7.151}$$

4. If the variable is restricted to take values lying only in between -1 and 1, the transformation can be

$$x_i = \sin y_i, \quad x_i = \cos y_i, \quad \text{or} \quad x_i = \frac{2y_i}{1 + y_i^2} \tag{7.152}$$

Note the following aspects of transformation techniques:

1. The constraints $g_j(\mathbf{X})$ have to be very simple functions of x_i.
2. For certain constraints it may not be possible to find the necessary transformation.
3. If it is not possible to eliminate all the constraints by making a change of variables, it may be better not to use the transformation at all. The partial transformation may sometimes produce a distorted objective function which might be more difficult to minimize than the original function.

To illustrate the method of transformation of variables, we consider the following problem.

Example 7.6   Find the dimensions of a rectangular prism-type box that has the largest volume when the sum of its length, width, and height is limited to a maximum value of 60 in. and its length is restricted to a maximum value of 36 in.

SOLUTION   Let x_1, x_2, and x_3 denote the length, width, and height of the box, respectively. The problem can be stated as follows:

$$\text{Maximize } f(x_1, x_2, x_3) = x_1 x_2 x_3 \tag{E_1}$$

subject to

$$x_1 + x_2 + x_3 \leq 60 \tag{E_2}$$

$$x_1 \leq 36 \tag{E_3}$$

$$x_i \geq 0, \quad i = 1, 2, 3 \tag{E_4}$$

By introducing new variables as

$$y_1 = x_1, \quad y_2 = x_2, \quad y_3 = x_1 + x_2 + x_3 \tag{E_5}$$

or

$$x_1 = y_1, \quad x_2 = y_2, \quad x_3 = y_3 - y_1 - y_2 \tag{E_6}$$

the constraints of Eqs. (E_2) to (E_4) can be restated as

$$0 \leq y_1 \leq 36, \quad 0 \leq y_2 \leq 60, \quad 0 \leq y_3 \leq 60 \tag{E_7}$$

where the upper bound, for example, on y_2 is obtained by setting $x_1 = x_3 = 0$ in Eq. (E_2). The constraints of Eq. (E_7) will be satisfied automatically if we define new variables z_i, $i = 1, 2, 3$, as

$$y_1 = 36 \sin^2 z_1, \quad y_2 = 60 \sin^2 z_2, \quad y_3 = 60 \sin^2 z_3 \tag{E_8}$$

Thus the problem can be stated as an unconstrained problem as follows:

Maximize $f(z_1, z_2, z_3)$

$$= y_1 y_2 (y_3 - y_1 - y_2) \tag{E_9}$$

$$= 2160 \sin^2 z_1 \sin^2 z_2 (60 \sin^2 z_3 - 36 \sin^2 z_1 - 60 \sin^2 z_2)$$

The necessary conditions of optimality yield the relations

$$\frac{\partial f}{\partial z_1} = 259{,}200 \sin z_1 \cos z_1 \sin^2 z_2 (\sin^2 z_3 - \tfrac{6}{5} \sin^2 z_1 - \sin^2 z_2) = 0 \tag{E_{10}}$$

$$\frac{\partial f}{\partial z_2} = 518{,}400 \sin^2 z_1 \sin z_2 \cos z_2 (\tfrac{1}{2} \sin^2 z_3 - \tfrac{3}{10} \sin^2 z_1 - \sin^2 z_2) = 0 \tag{E_{11}}$$

$$\frac{\partial f}{\partial z_3} = 259{,}200 \sin^2 z_1 \sin^2 z_2 \sin z_3 \cos z_3 = 0 \tag{E_{12}}$$

Equation (E_{12}) gives the nontrivial solution as $\cos z_3 = 0$ or $\sin^2 z_3 = 1$. Hence Eqs. (E_{10}) and (E_{11}) yield $\sin^2 z_1 = \tfrac{5}{9}$ and $\sin^2 z_2 = \tfrac{1}{3}$. Thus the optimum solution is given by $x_1^* = 20$ in., $x_2^* = 20$ in., $x_3^* = 20$ in., and the maximum volume $= 8000$ in^3.

7.12 BASIC APPROACH OF THE PENALTY FUNCTION METHOD

Penalty function methods transform the basic optimization problem into alternative formulations such that numerical solutions are sought by solving a sequence of unconstrained minimization problems. Let the basic optimization problem, with inequality constraints, be of the form:

$$\text{Find } \mathbf{X} \text{ which minimizes } f(\mathbf{X})$$

subject to

$$g_j(\mathbf{X}) \le 0, \quad j = 1, 2, \ldots, m \tag{7.153}$$

This problem is converted into an unconstrained minimization problem by constructing a function of the form

$$\phi_k = \phi(\mathbf{X}, r_k) = f(\mathbf{X}) + r_k \sum_{j=1}^{m} G_j[g_j(\mathbf{X})] \tag{7.154}$$

where G_j is some function of the constraint g_j, and r_k is a positive constant known as the *penalty parameter*. The significance of the second term on the right side of Eq. (7.154), called the *penalty term*, will be seen in Sections 7.13 and 7.15. If the unconstrained minimization of the ϕ function is repeated for a sequence of values of the penalty parameter $r_k (k = 1, 2, \ldots)$, the solution may be brought to converge to that of the original problem stated in Eq. (7.153). This is the reason why the penalty function methods are also known as *sequential unconstrained minimization techniques* (SUMTs).

The penalty function formulations for inequality constrained problems can be divided into two categories: interior and exterior methods. In the interior formulations, some popularly used forms of G_j are given by

$$G_j = -\frac{1}{g_j(\mathbf{X})} \tag{7.155}$$

$$G_j = \log[-g_j(\mathbf{X})] \tag{7.156}$$

Some commonly used forms of the function G_j in the case of exterior penalty function formulations are

$$G_j = \max[0, g_j(\mathbf{X})] \tag{7.157}$$

$$G_j = \{\max[0, g_i(\mathbf{X})]\}^2 \tag{7.158}$$

In the interior methods, the unconstrained minima of ϕ_k all lie in the feasible region and converge to the solution of Eq. (7.153) as r_k is varied in a particular manner. In the exterior methods, the unconstrained minima of ϕ_k all lie in the infeasible region and converge to the desired solution from the outside as r_k is changed in a specified manner. The convergence of the unconstrained minima of ϕ_k is illustrated in Fig. 7.10 for the simple problem

$$\text{Find } \mathbf{X} = \{x_1\} \text{ which minimizes } f(\mathbf{X}) = \alpha x_1$$

subject to $\tag{7.159}$

$$g_1(\mathbf{X}) = \beta - x_1 \le 0$$

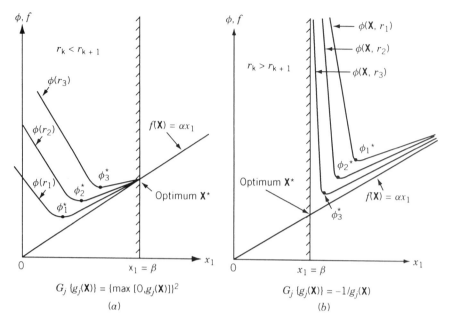

Figure 7.10   Penalty function methods: (*a*) exterior method; (*b*) interior method.

It can be seen from Fig. 7.10*a* that the unconstrained minima of $\phi(\mathbf{X}, r_k)$ converge to the optimum point $\mathbf{X}^*$ as the parameter r_k is increased sequentially. On the other hand, the interior method shown in Fig. 7.10*b* gives convergence as the parameter r_k is decreased sequentially.

There are several reasons for the appeal of the penalty function formulations. One main reason, which can be observed from Fig. 7.10, is that the sequential nature of the method allows a gradual or sequential approach to criticality of the constraints. In addition, the sequential process permits a graded approximation to be used in analysis of the system. This means that if the evaluation of f and g_j [and hence $\phi(\mathbf{X}, r_k)$] for any specified design vector $\mathbf{X}$ is computationally very difficult, we can use coarse approximations during the early stages of optimization (when the unconstrained minima of ϕ_k are far away from the optimum) and finer or more detailed analysis approximation during the final stages of optimization. Another reason is that the algorithms for the unconstrained minimization of rather arbitrary functions are well studied and generally are quite reliable. The algorithms of the interior and the exterior penalty function methods are given in Sections 7.13 and 7.15.

7.13   INTERIOR PENALTY FUNCTION METHOD

As indicated in Section 7.12, in the interior penalty function methods, a new function (ϕ function) is constructed by augmenting a penalty term to the objective function. The penalty term is chosen such that its value will be small at points away from the constraint boundaries and will tend to infinity as the constraint boundaries are approached. Hence the value of the ϕ function also "blows up" as the constraint boundaries are approached. This behavior can also be seen from Fig. 7.10*b*. Thus once the unconstrained minimization of $\phi(\mathbf{X}, r_k)$ is started from any feasible point $\mathbf{X}_1$, the subsequent points generated will always lie within the feasible domain since the constraint boundaries act as barriers during the minimization process. This is why the interior penalty function methods are also known as *barrier methods*. The ϕ function defined originally by Carroll [7.14] is

$$\phi(\mathbf{X}, r_k) = f(\mathbf{X}) - r_k \sum_{j=1}^{m} \frac{1}{g_j(\mathbf{X})} \qquad (7.160)$$

It can be seen that the value of the function ϕ will always be greater than f since $g_j(\mathbf{X})$ is negative for all feasible points $\mathbf{X}$. If any constraint $g_j(\mathbf{X})$ is satisfied critically (with equality sign), the value of ϕ tends to infinity. It is to be noted that the penalty term in Eq. (7.160) is not defined if $\mathbf{X}$ is infeasible. This introduces serious shortcoming while using the Eq. (7.160). Since this equation does not allow any constraint to be violated, it requires a feasible starting point for the search toward the optimum point. However, in many engineering problems, it may not be very difficult to find a point satisfying all the constraints, $g_j(\mathbf{X}) \leq 0$, at the expense of large values of the objective function, $f(\mathbf{X})$. If there is any difficulty in finding a feasible starting point, the method described in the latter part of this section can be used to find a feasible point. Since the initial point as well as each of the subsequent points generated in this method lies inside the acceptable region of the design space, the method is classified as an *interior penalty function formulation*. Since the constraint boundaries act as barriers, the method is also known as a barrier method. The iteration procedure of this method can be summarized as follows.

Iterative Process

1. Start with an initial feasible point $\mathbf{X}_1$ satisfying all the constraints with strict inequality sign, that is, $g_j(\mathbf{X}_1) < 0$ for $j = 1, 2, \ldots, m$, and an initial value of $r_1 > 0$. Set $k = 1$.
2. Minimize $\phi(\mathbf{X}, r_k)$ by using any of the unconstrained minimization methods and obtain the solution $\mathbf{X}_k^*$.
3. Test whether $\mathbf{X}_k^*$ is the optimum solution of the original problem. If $\mathbf{X}_k^*$ is found to be optimum, terminate the process. Otherwise, go to the next step.
4. Find the value of the next penalty parameter, r_{k+1}, as

$$r_{k+1} = cr_k$$

   where $c < 1$.
5. Set the new value of $k = k + 1$, take the new starting point as $\mathbf{X}_1 = \mathbf{X}_k^*$, and go to step 2.

Although the algorithm is straightforward, there are a number of points to be considered in implementing the method:

1. The starting feasible point $\mathbf{X}_1$ may not be readily available in some cases.
2. A suitable value of the initial penalty parameter (r_1) has to be found.
3. A proper value has to be selected for the multiplication factor, c.

4. Suitable convergence criteria have to be chosen to identify the optimum point.

5. The constraints have to be normalized so that each one of them vary between -1 and 0 only.

All these aspects are discussed in the following paragraphs.

Starting Feasible Point X_1.   In most engineering problems, it will not be very difficult to find an initial point $\mathbf{X}_1$ satisfying all the constraints, $g_j(\mathbf{X}_1) < 0$. As an example, consider the problem of minimum weight design of a beam whose deflection under a given loading condition has to remain less than or equal to a specified value. In this case one can always choose the cross section of the beam to be very large initially so that the constraint remains satisfied. The only problem is that the weight of the beam (objective) corresponding to this initial design will be very large. Thus in most of the practical problems, we will be able to find a feasible starting point at the expense of a large value of the objective function. However, there may be some situations where the feasible design points could not be found so easily. In such cases, the required feasible starting points can be found by using the interior penalty function method itself as follows:

1. Choose an arbitrary point $\mathbf{X}_1$ and evaluate the constraints $g_j(\mathbf{X})$ at the point $\mathbf{X}_1$. Since the point $\mathbf{X}_1$ is arbitrary, it may not satisfy all the constraints with strict inequality sign. If r out of a total of m constraints are violated, renumber the constraints such that the last r constraints will become the violated ones, that is,

$$g_j(\mathbf{X}_1) < 0, \quad j = 1, 2, \ldots, m - r$$
$$g_j(\mathbf{X}_1) \geq 0, \quad j = m - r + 1, m - r + 2, \ldots, m \qquad (7.161)$$

2. Identify the constraint that is violated most at the point $\mathbf{X}_1$, that is, find the integer k such that

$$g_k(\mathbf{X}_1) = \max[g_j(\mathbf{X}_1)]$$
$$\text{for } j = m - r + 1, m - r + 2, \ldots, m \qquad (7.162)$$

3. Now formulate a new optimization problem as

$$\text{Find } \mathbf{X} \text{ which minimizes } g_k(\mathbf{X})$$

subject to

$$g_j(\mathbf{X}) \leq 0, \quad j = 1, 2, \ldots, m - r$$
$$g_j(\mathbf{X}) - g_k(\mathbf{X}_1) \leq 0, \quad j = m - r + 1, m - r + 2, \ldots,$$
$$k - 1, k + 1, \ldots, m \qquad (7.163)$$

4. Solve the optimization problem formulated in step 3 by taking the point $\mathbf{X}_1$ as a feasible starting point using the interior penalty function method. Note that this optimization method can be terminated whenever the value of the objective function $g_k(\mathbf{X})$ drops below zero. Thus the solution obtained $\mathbf{X}_M$ will satisfy at least one more constraint than did the original point $\mathbf{X}_1$.

5. If all the constraints are not satisfied at the point $\mathbf{X}_M$, set the new starting point as $\mathbf{X}_1 = \mathbf{X}_M$, and renumber the constraints such that the last r constraints will be the unsatisfied ones (this value of r will be different from the previous value), and go to step 2.

This procedure is repeated until all the constraints are satisfied and a point $\mathbf{X}_1 = \mathbf{X}_M$ is obtained for which $g_j(\mathbf{X}_1) < 0,\ j = 1, 2, \ldots, m$.

If the constraints are consistent, it should be possible to obtain, by applying the procedure, a point $\mathbf{X}_1$ that satisfies all the constraints. However, there may exist situations in which the solution of the problem formulated in step 3 gives the unconstrained or constrained local minimum of $g_k(\mathbf{X})$ that is positive. In such cases one has to start afresh with a new point $\mathbf{X}_1$ from step 1 onward.

Initial Value of the Penalty Parameter (r_1).   Since the unconstrained minimization of $\phi(\mathbf{X}, r_k)$ is to be carried out for a decreasing sequence of r_k, it might appear that by choosing a very small value of r_1, we can avoid an excessive number of minimizations of the function ϕ. But from a computational point of view, it will be easier to minimize the unconstrained function $\phi(\mathbf{X}, r_k)$ if r_k is large. This can be seen qualitatively from Fig. 7.10b. As the value of r_k becomes smaller, the value of the function ϕ changes more rapidly in the vicinity of the minimum ϕ_k^*. Since it is easier to find the minimum of a function whose graph is smoother, the unconstrained minimization of ϕ will be easier if r_k is large. However, the minimum of ϕ_k, $\mathbf{X}_k^*$, will be farther away from the desired minimum $\mathbf{X}^*$ if r_k is large. Thus it requires an excessive number of unconstrained minimizations of $\phi(\mathbf{X}, r_k)$ (for several values of r_k) to reach the point $\mathbf{X}^*$ if r_1 is selected to be very large. Thus a moderate value has to be chosen for the initial penalty parameter (r_1). In practice, a value of r_1 that gives the value of $\phi(\mathbf{X}_1, r_1)$ approximately equal to 1.1 to 2.0 times the value of $f(\mathbf{X}_1)$ has been found to be quite satisfactory in achieving quick convergence of the process. Thus for any initial feasible starting point $\mathbf{X}_1$, the value of r_1 can be taken as

$$r_1 \simeq 0.1 \text{ to } 1.0 \frac{f(\mathbf{X}_1)}{-\sum_{j=1}^{m} 1/g_j(\mathbf{X}_1)} \tag{7.164}$$

Subsequent Values of the Penalty Parameter.   Once the initial value of r_k is chosen, the subsequent values of r_{k+1} have to be chosen such that

$$r_{k+1} < r_k \tag{7.165}$$

For convenience, the values of r_k are chosen according to the relation

$$r_{k+1} = c r_k \tag{7.166}$$

where $c < 1$. The value of c can be taken as 0.1, 0.2, or 0.5.

Convergence Criteria.   Since the unconstrained minimization of $\phi(\mathbf{X}, r_k)$ has to be carried out for a decreasing sequence of values r_k, it is necessary to use proper convergence criteria to identify the optimum point and to avoid an unnecessarily large number of unconstrained minimizations. The process can be terminated whenever the following conditions are satisfied.

1. The relative difference between the values of the objective function obtained at the end of any two consecutive unconstrained minimizations falls below a small number ε_1, that is,

$$\left| \frac{f(\mathbf{X}_k^*) - f(\mathbf{X}_{k-1}^*)}{f(\mathbf{X}_k^*)} \right| \leq \varepsilon_1 \tag{7.167}$$

2. The difference between the optimum points $\mathbf{X}_k^*$ and $\mathbf{X}_{k-1}^*$ becomes very small. This can be judged in several ways. Some of them are given below:

$$|(\Delta\mathbf{X})_i| \leq \varepsilon_2 \tag{7.168}$$

where $\Delta\mathbf{X} = \mathbf{X}_k^* - \mathbf{X}_{k-1}^*$, and $(\Delta\mathbf{X})_i$ is the ith component of the vector $\Delta\mathbf{X}$.

$$\max |(\Delta\mathbf{X})_i| \leq \varepsilon_3 \tag{7.169}$$

$$|\Delta\mathbf{X}| = [(\Delta\mathbf{X})_1^2 + (\Delta\mathbf{X})_2^2 + \cdots + (\Delta\mathbf{X})_n^2]^{1/2} \leq \varepsilon_4 \tag{7.170}$$

Note that the values of ε_1 to ε_4 have to be chosen depending on the characteristics of the problem at hand.

Normalization of Constraints. A structural optimization problem, for example, might be having constraints on the deflection (δ) and the stress (σ) as

$$g_1(\mathbf{X}) = \delta(\mathbf{X}) - \delta_{\max} \leq 0 \tag{7.171}$$

$$g_2(\mathbf{X}) = \sigma(\mathbf{X}) - \sigma_{\max} \leq 0 \tag{7.172}$$

where the maximum allowable values are given by $\delta_{\max} = 0.5$ in. and $\sigma_{\max} = 20{,}000$ psi. If a design vector $\mathbf{X}_1$ gives the values of g_1 and g_2 as -0.2 and $-10{,}000$, the contribution of g_1 will be much larger than that of g_2 (by an order of 10^4) in the formulation of the ϕ function given by Eq. (7.160). This will badly affect the convergence rate during the minimization of ϕ function. Thus it is advisable to normalize the constraints so that they vary between -1 and 0 as far as possible. For the constraints shown in Eqs. (7.171) and (7.172), the normalization can be done as

$$g_1'(\mathbf{X}) = \frac{g_1(\mathbf{X})}{\delta_{\max}} = \frac{\delta(\mathbf{X})}{\delta_{\max}} - 1 \leq 0 \tag{7.173}$$

$$g_2'(\mathbf{X}) = \frac{g_2(\mathbf{X})}{\sigma_{\max}} = \frac{\sigma(\mathbf{X})}{\sigma_{\max}} - 1 \leq 0 \tag{7.174}$$

If the constraints are not normalized as shown in Eqs. (7.173) and (7.174), the problem can still be solved effectively by defining different penalty parameters for different constraints as

$$\phi(\mathbf{X}, r_k) = f(\mathbf{X}) - r_k \sum_{j=1}^{m} \frac{R_j}{g_j(\mathbf{X})} \tag{7.175}$$

where $R_1, R_2, \ldots, R_m$ are selected such that the contributions of different $g_j(\mathbf{X})$ to the ϕ function will be approximately the same at the initial point $\mathbf{X}_1$. When the unconstrained minimization of $\phi(\mathbf{X}, r_k)$ is carried for a decreasing sequence of values of r_k, the values of $R_1, R_2, \ldots, R_m$ will not be altered; however, they are expected to be effective in reducing the disparities between the contributions of the various constraints to the ϕ function.

Example 7.7

$$\text{Minimize } f(x_1, x_2) = \tfrac{1}{3}(x_1 + 1)^3 + x_2$$

subject to

$$g_1(x_1, x_2) = -x_1 + 1 \leq 0$$

$$g_2(x_1, x_2) = -x_2 \leq 0$$

SOLUTION   To illustrate the interior penalty function method, we use the calculus method for solving the unconstrained minimization problem in this case. Hence there is no need to have an initial feasible point $\mathbf{X}_1$. The ϕ function is

$$\phi(\mathbf{X}, r) = \frac{1}{3}(x_1 + 1)^3 + x_2 - r\left(\frac{1}{-x_1 + 1} - \frac{1}{x_2}\right)$$

To find the unconstrained minimum of ϕ, we use the necessary conditions:

$$\frac{\partial \phi}{\partial x_1} = (x_1 + 1)^2 - \frac{r}{(1 - x_1)^2} = 0, \quad \text{that is,} \quad (x_1^2 - 1)^2 = r$$

$$\frac{\partial \phi}{\partial x_2} = 1 - \frac{r}{x_2^2} = 0, \quad \text{that is,} \quad x_2^2 = r$$

These equations give

$$x_1^*(r) = (r^{1/2} + 1)^{1/2}, \quad x_2^*(r) = r^{1/2}$$

$$\phi_{\min}(r) = \tfrac{1}{3}[(r^{1/2} + 1)^{1/2} + 1]^3 + 2r^{1/2} - \frac{1}{(1/r) - (1/r^{3/2} + 1/r^2)^{1/2}}$$

To obtain the solution of the original problem, we know that

$$f_{\min} = \lim_{r \to 0} \phi_{\min}(r)$$

$$x_1^* = \lim_{r \to 0} x_1^*(r)$$

$$x_2^* = \lim_{r \to 0} x_2^*(r)$$

The values of f, x_1^*, and x_2^* corresponding to a decreasing sequence of values of r are shown in Table 7.3.

Example 7.8

$$\text{Minimize } f(\mathbf{X}) = x_1^3 - 6x_1^2 + 11x_1 + x_3$$

subject to

$$x_1^2 + x_2^2 - x_3^2 \leq 0$$

$$4 - x_1^2 - x_2^2 - x_3^2 \leq 0$$

$$x_3 - 5 \leq 0$$

$$-x_i \leq 0, \quad i = 1, 2, 3$$

Table 7.3 Results for Example 7.7

Value of r	$x_1^*(r) = (r^{1/2} + 1)^{1/2}$	$x_2^*(r) = r^{1/2}$	$\phi_{\min}(r)$	$f(r)$
1000	5.71164	31.62278	376.2636	132.4003
100	3.31662	10.00000	89.9772	36.8109
10	2.04017	3.16228	25.3048	12.5286
1	1.41421	1.00000	9.1046	5.6904
0.1	1.14727	0.31623	4.6117	3.6164
0.01	1.04881	0.10000	3.2716	2.9667
0.001	1.01569	0.03162	2.8569	2.7615
0.0001	1.00499	0.01000	2.7267	2.6967
0.00001	1.00158	0.00316	2.6856	2.6762
0.000001	1.00050	0.00100	2.6727	2.6697
Exact solution 0	1	0	8/3	8/3

SOLUTION   The interior penalty function method, coupled with the Davidon–Fletcher–Powell method of unconstrained minimization and cubic interpolation method of one-dimensional search, is used to solve this problem. The necessary data are assumed as follows:

$$\text{Starting feasible point, } \mathbf{X}_1 = \begin{Bmatrix} 0.1 \\ 0.1 \\ 3.0 \end{Bmatrix}$$

$$r_1 = 1.0, \quad f(\mathbf{X}_1) = 4.041, \quad \phi(\mathbf{X}_1, r_1) = 25.1849$$

The optimum solution of this problem is known to be [7.15]

$$\mathbf{X} = \begin{Bmatrix} 0 \\ \sqrt{2} \\ \sqrt{2} \end{Bmatrix}, \quad f^* = \sqrt{2}$$

The results of numerical optimization are summarized in Table 7.4.

Convergence Proof.   The following theorem proves the convergence of the interior penalty function method.

Theorem 7.1 If the function

$$\phi(\mathbf{X}, r_k) = f(\mathbf{X}) - r_k \sum_{j=1}^{m} \frac{1}{g_j(\mathbf{X})} \tag{7.176}$$

is minimized for a decreasing sequence of values of r_k, the unconstrained minima $\mathbf{X}_k^*$ converge to the optimal solution of the constrained problem stated in Eq. (7.153) as $r_k \to 0$.

Table 7.4 Results for Example 7.8

k	Value of r_k	Starting point for minimizing ϕ_k	Number of iterations taken for minimizing ϕ_k	Optimum $\mathbf{X}_k^*$	ϕ_k^*	f_k^*
1	1.0×10^0	$\begin{bmatrix} 0.1 \\ 0.1 \\ 3.0 \end{bmatrix}$	9	$\begin{bmatrix} 0.37898 \\ 1.67965 \\ 2.34617 \end{bmatrix}$	10.36219	5.70766
2	1.0×10^{-1}	$\begin{bmatrix} 0.37898 \\ 1.67965 \\ 2.34617 \end{bmatrix}$	7	$\begin{bmatrix} 0.10088 \\ 1.41945 \\ 1.68302 \end{bmatrix}$	4.12440	2.73267
3	1.0×10^{-2}	$\begin{bmatrix} 0.10088 \\ 1.41945 \\ 1.68302 \end{bmatrix}$	5	$\begin{bmatrix} 0.03066 \\ 1.41411 \\ 1.49842 \end{bmatrix}$	2.25437	1.83012
4	1.0×10^{-3}	$\begin{bmatrix} 0.03066 \\ 1.41411 \\ 1.49842 \end{bmatrix}$	3	$\begin{bmatrix} 0.009576 \\ 1.41419 \\ 1.44081 \end{bmatrix}$	1.67805	1.54560
5	1.0×10^{-4}	$\begin{bmatrix} 0.009576 \\ 1.41419 \\ 1.44081 \end{bmatrix}$	7	$\begin{bmatrix} 0.003020 \\ 1.41421 \\ 1.42263 \end{bmatrix}$	1.49745	1.45579
6	1.0×10^{-5}	$\begin{bmatrix} 0.003020 \\ 1.41421 \\ 1.42263 \end{bmatrix}$	3	$\begin{bmatrix} 0.0009530 \\ 1.41421 \\ 1.41687 \end{bmatrix}$	1.44052	1.42735
7	1.0×10^{-6}	$\begin{bmatrix} 0.0009530 \\ 1.41421 \\ 1.41687 \end{bmatrix}$	3	$\begin{bmatrix} 0.0003013 \\ 1.41421 \\ 1.41505 \end{bmatrix}$	1.42253	1.41837
8	1.0×10^{-7}	$\begin{bmatrix} 0.0003013 \\ 1.41421 \\ 1.41505 \end{bmatrix}$	3	$\begin{bmatrix} 0.00009535 \\ 1.41421 \\ 1.41448 \end{bmatrix}$	1.41684	1.41553
9	1.0×10^{-8}	$\begin{bmatrix} 0.00009535 \\ 1.41421 \\ 1.41448 \end{bmatrix}$	5	$\begin{bmatrix} 0.00003019 \\ 1.41421 \\ 1.41430 \end{bmatrix}$	1.41505	1.41463
10	1.0×10^{-9}	$\begin{bmatrix} 0.00003019 \\ 1.41421 \\ 1.41430 \end{bmatrix}$	4	$\begin{bmatrix} 0.000009567 \\ 1.41421 \\ 1.41424 \end{bmatrix}$	1.41448	1.41435
11	1.0×10^{-10}	$\begin{bmatrix} 0.000009567 \\ 1.41421 \\ 1.41424 \end{bmatrix}$	3	$\begin{bmatrix} 0.00003011 \\ 1.41421 \\ 1.41422 \end{bmatrix}$	1.41430	1.41426
12	1.0×10^{-11}	$\begin{bmatrix} 0.000003011 \\ 1.41421 \\ 1.41422 \end{bmatrix}$	3	$\begin{bmatrix} 0.9562 \times 10^{-6} \\ 1.41421 \\ 1.41422 \end{bmatrix}$	1.41424	1.41423
13	1.0×10^{-12}	$\begin{bmatrix} 0.9562 \times 10^{-6} \\ 1.41421 \\ 1.41422 \end{bmatrix}$	4	$\begin{bmatrix} 0.3248 \times 10^{-6} \\ 1.41421 \\ 1.41421 \end{bmatrix}$	1.41422	1.41422

Proof: If $\mathbf{X}^*$ is the optimum solution of the constrained problem, we have to prove that

$$\lim_{r_k \to 0} [\min \phi(\mathbf{X}, r_k)] = \phi(\mathbf{X}_k^*, r_k) = f(\mathbf{X}^*) \qquad (7.177)$$

Since $f(\mathbf{X})$ is continous and $f(\mathbf{X}^*) \leq f(\mathbf{X})$ for all feasible points $\mathbf{X}$, we can choose feasible point $\underset{\sim}{\mathbf{X}}$ such that

$$f(\underset{\sim}{\mathbf{X}}) < f(\mathbf{X}^*) + \frac{\varepsilon}{2} \qquad (7.178)$$

for any value of $\varepsilon > 0$. Next select a suitable value of k, say K, such that

$$r_k \leq \left\{ \frac{\varepsilon}{2m} \Big/ \min_j \left[-\frac{1}{g_j(\underset{\sim}{\mathbf{X}})} \right] \right\} \qquad (7.179)$$

From the definition of the ϕ function, we have

$$f(\mathbf{X}^*) \leq \min \phi(\mathbf{X}, r_k) = \phi(\mathbf{X}_k^*, r_k) \qquad (7.180)$$

where $\mathbf{X}_k^*$ is the unconstrained minimum of $\phi(\mathbf{X}, r_k)$. Further,

$$\phi(\mathbf{X}_k^*, r_k) \leq \phi(\mathbf{X}_K^*, r_k) \qquad (7.181)$$

since $\mathbf{X}_k^*$ minimizes $\phi(\mathbf{X}, r_k)$ and any $\mathbf{X}$ other than $\mathbf{X}_k^*$ leads to a value of ϕ greater than or equal to $\phi(\mathbf{X}_k^*, r_k)$. Further, by choosing $r_k < r_K$, we obtain

$$\phi(\mathbf{X}_K^*, r_K) = f(\mathbf{X}_K^*) - r_K \sum_{j=1}^{m} \frac{1}{g_j(\mathbf{X}_K^*)}$$

$$> f(\mathbf{X}_K^*) - r_k \sum_{j=1}^{m} \frac{1}{g_j(\mathbf{X}_K^*)}$$

$$> \phi(\mathbf{X}_k^*, r_k) \qquad (7.182)$$

as $\mathbf{X}_k^*$ is the unconstrained minimum of $\phi(\mathbf{X}, r_k)$. Thus

$$f(\mathbf{X}^*) \leq \phi(\mathbf{X}_k^*, r_k) \leq \phi(\mathbf{X}_K^*, r_k) < \phi(\mathbf{X}_K^*, r_K) \qquad (7.183)$$

But

$$\phi(\mathbf{X}_K^*, r_K) \leq \phi(\underset{\sim}{\mathbf{X}}, r_K) = f(\underset{\sim}{\mathbf{X}}) - r_K \sum_{j=1}^{m} \frac{1}{g_j(\underset{\sim}{\mathbf{X}})} \qquad (7.184)$$

Combining the inequalities (7.183) and (7.184), we have

$$f(\mathbf{X}^*) \leq \phi(\mathbf{X}_k^*, r_k) \leq f(\underset{\sim}{\mathbf{X}}) - r_K \sum_{j=1}^{m} \frac{1}{g_j(\underset{\sim}{\mathbf{X}})} \qquad (7.185)$$

Inequality (7.179) gives

$$-r_K \sum_{j=1}^{m} \frac{1}{g_j(\underset{\sim}{\mathbf{X}})} < \frac{\varepsilon}{2} \qquad (7.186)$$

By using inequalities (7.178) and (7.186), inequality (7.185) becomes

$$f(\mathbf{X}^*) \le \phi(\mathbf{X}_k^*, r_k) < f(\mathbf{X}^*) + \frac{\varepsilon}{2} + \frac{\varepsilon}{2} = f(\mathbf{X}^*) + \varepsilon$$

or

$$\phi(\mathbf{X}_k^*, r_k) - f(\mathbf{X}^*) < \varepsilon \tag{7.187}$$

Given any $\varepsilon > 0$ (however small it may be), it is possible to choose a value of k so as to satisfy the inequality (7.187). Hence as $k \to \infty (r_k \to 0)$, we have

$$\lim_{r_k \to 0} \phi(\mathbf{X}_k^*, r_k) = f(\mathbf{X}^*)$$

This completes the proof of the theorem.

Additional Results.   From the proof above, it follows that as $r_k \to 0$,

$$\lim_{k \to \infty} f(\mathbf{X}_k^*) = f(\mathbf{X}^*) \tag{7.188}$$

$$\lim_{k \to \infty} r_k \left[-\sum_{j=1}^{m} \frac{1}{g_j(\mathbf{X}_k^*)} \right] = 0 \tag{7.189}$$

It can also be shown that if $r_1, r_2, \ldots$ is a strictly decreasing sequence of positive values, the sequence $f(\mathbf{X}_1^*), f(\mathbf{X}_2^*), \ldots$ will also be strictly decreasing. For this, consider two consecutive parameters, say, r_k and r_{k+1}, with

$$0 < r_{k+1} < r_k \tag{7.190}$$

Then we have

$$f(\mathbf{X}_{k+1}^*) - r_{k+1} \sum_{j=1}^{m} \frac{1}{g_j(\mathbf{X}_{k+1}^*)} < f(\mathbf{X}_k^*) - r_{k+1} \sum_{j=1}^{m} \frac{1}{g_j(\mathbf{X}_k^*)} \tag{7.191}$$

since $\mathbf{X}_{k+1}^*$ alone minimizes $\phi(\mathbf{X}, r_{k+1})$. Similarly,

$$f(\mathbf{X}_k^*) - r_k \sum_{j=1}^{m} \frac{1}{g_j(\mathbf{X}_k^*)} < f(\mathbf{X}_{k+1}^*) - r_k \sum_{j=1}^{m} \frac{1}{g_j(\mathbf{X}_{k+1}^*)} \tag{7.192}$$

Divide Eq. (7.191) by r_{k+1}, Eq. (7.192) by r_k, and add the resulting inequalities to obtain

$$\frac{1}{r_{k+1}} f(\mathbf{X}_{k+1}^*) - \sum_{j=1}^{m} \frac{1}{g_j(\mathbf{X}_{k+1}^*)} + \frac{1}{r_k} f(\mathbf{X}_k^*) - \sum_{j=1}^{m} \frac{1}{g_j(\mathbf{X}_k^*)}$$

$$< \frac{1}{r_{k+1}} f(\mathbf{X}_k^*) - \sum_{j=1}^{m} \frac{1}{g_j(\mathbf{X}_k^*)} + \frac{1}{r_k} f(\mathbf{X}_{k+1}^*) - \sum_{j=1}^{m} \frac{1}{g_j(\mathbf{X}_{k+1}^*)} \tag{7.193}$$

Canceling the common terms from both sides, we can write the inequality (7.193) as

$$f(\mathbf{X}_{k+1}^*)\left(\frac{1}{r_{k+1}} - \frac{1}{r_k}\right) < f(\mathbf{X}_k^*)\left(\frac{1}{r_{k+1}} - \frac{1}{r_k}\right) \tag{7.194}$$

since

$$\frac{1}{r_{k+1}} - \frac{1}{r_k} = \frac{r_k - r_{k+1}}{r_k r_{k+1}} > 0 \tag{7.195}$$

we obtain

$$f(\mathbf{X}_{k+1}^*) < f(\mathbf{X}_k^*) \tag{7.196}$$

7.14   CONVEX PROGRAMMING PROBLEM

In Section 7.13 we saw that the sequential minimization of

$$\phi(\mathbf{X}, r_k) = f(\mathbf{X}) - r_k \sum_{j=1}^{m} \frac{1}{g_j(\mathbf{X})}, \quad r_k > 0 \tag{7.197}$$

for a decreasing sequence of values of r_k gives the minima $\mathbf{X}_k^*$. As $k \to \infty$, these points $\mathbf{X}_k^*$ converge to the minimum of the constrained problem:

Minimize $f(\mathbf{X})$

subject to $\tag{7.198}$

$$g_j(\mathbf{X}) \le 0, \quad j = 1, 2, \ldots, m$$

To ensure the existence of a global minimum of $\phi(\mathbf{X}, r_k)$ for every positive value of r_k, ϕ has to be strictly convex function of $\mathbf{X}$. The following theorem gives the sufficient conditions for the ϕ function to be strictly convex. If ϕ is convex, for every $r_k > 0$ there exists a unique minimum of $\phi(\mathbf{X}, r_k)$.

Theorem 7.2 If $f(\mathbf{X})$ and $g_j(\mathbf{X})$ are convex and at least one of $f(\mathbf{X})$ and $g_j(\mathbf{X})$ is strictly convex, the function $\phi(\mathbf{X}, r_k)$ defined by Eq. (7.197) will be a strictly convex function of $\mathbf{X}$.

Proof: This theorem can be proved in two steps. In the first step we prove that if a function $g_k(\mathbf{X})$ is convex, $1/g_k(\mathbf{X})$ will be concave. In the second step, we prove that a positive combination of convex functions is convex, and strictly convex if at least one of the functions is strictly convex.

Thus Theorem A.3 of Appendix A guarantees that the sequential minimization of $\phi(\mathbf{X}, r_k)$ for a decreasing sequence of values of r_k leads to the global minimum of the original constrained problem. When the convexity conditions are not satisfied, or when the functions are so complex that we do not know beforehand whether the convexity conditions are satisfied, it will not be possible to prove that the minimum found by the

SUMT method is a global one. In such cases one has to satisfy with a local minimum only. However, one can always reapply the SUMT method from different feasible starting points and try to find a better local minimum point if the problem has several local minima. Of course, this procedure requires more computational effort.

7.15   EXTERIOR PENALTY FUNCTION METHOD

In the exterior penalty function method, the ϕ function is generally taken as

$$\phi(\mathbf{X}, r_k) = f(\mathbf{X}) + r_k \sum_{j=1}^{m} \langle g_j(\mathbf{X}) \rangle^q \qquad (7.199)$$

where r_k is a positive penalty parameter, the exponent q is a nonnegative constant, and the bracket function $\langle g_j(\mathbf{X}) \rangle$ is defined as

$$\langle g_j(\mathbf{X}) \rangle = \max\langle g_j(\mathbf{X}), 0 \rangle$$

$$= \begin{cases} g_j(\mathbf{X}) & \text{if } g_j(\mathbf{X}) > 0 \\ & \text{(constraint is violated)} \\ 0 & \text{if } g_j(\mathbf{X}) \leq 0 \\ & \text{(constraint is satisfied)} \end{cases} \qquad (7.200)$$

It can be seen from Eq. (7.199) that the effect of the second term on the right side is to increase $\phi(\mathbf{X}, r_k)$ in proportion to the qth power of the amount by which the constraints are violated. Thus there will be a penalty for violating the constraints, and the amount of penalty will increase at a faster rate than will the amount of violation of a constraint (for $q > 1$). This is the reason why the formulation is called the penalty function method. Usually, the function $\phi(\mathbf{X}, r_k)$ possesses a minimum as a function of $\mathbf{X}$ in the infeasible region. The unconstrained minima $\mathbf{X}_k^*$ converge to the optimal solution of the original problem as $k \to \infty$ and $r_k \to \infty$. Thus the unconstrained minima approach the feasible domain gradually, and as $k \to \infty$, the $\mathbf{X}_k^*$ eventually lies in the feasible region. Let us consider Eq. (7.199) for various values of q.

1. $q = 0$. Here the ϕ function is given by

$$\phi(\mathbf{X}, r_k) = f(\mathbf{X}) + r_k \sum_{j=1}^{m} \langle g_j(\mathbf{X}) \rangle^0$$

$$= \begin{cases} f(\mathbf{X}) + mr_k & \text{if   all } g_j(\mathbf{X}) > 0 \\ f(\mathbf{X}) & \text{if   all } g_j(\mathbf{X}) \leq 0 \end{cases} \qquad (7.201)$$

This function is discontinuous on the boundary of the acceptable region as shown in Fig. 7.11 and hence it would be very difficult to minimize this function.

2. $0 < q < 1$. Here the ϕ function will be continuous, but the penalty for violating a constraint may be too small. Also, the derivatives of the function are discontinuous along the boundary. Thus it will be difficult to minimize the ϕ function. Typical contours of the ϕ function are shown in Fig. 7.12.

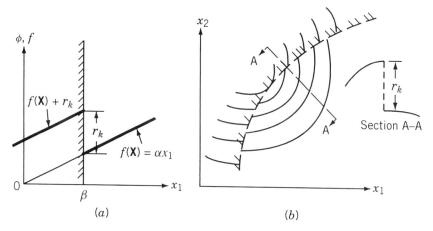

Figure 7.11 A ϕ function discontinuous for $q = 0$.

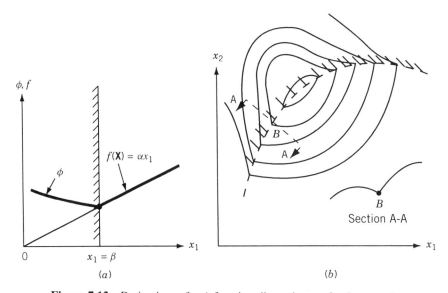

Figure 7.12 Derivatives of a ϕ function discontinuous for $0 < q < 1$.

3. $q = 1$. In this case, under certain restrictions, it has been shown by Zangwill [7.16] that there exists an r_0 so large that the minimum of $\phi(\mathbf{X}, r_k)$ is exactly the constrained minimum of the original problem for all $r_k > r_0$. However, the contours of the ϕ function look similar to those shown in Fig. 7.12 and possess discontinuous first derivatives along the boundary. Hence despite the convenience of choosing a single r_k that yields the constrained minimum in one unconstrained minimization, the method is not very attractive from computational point of view.

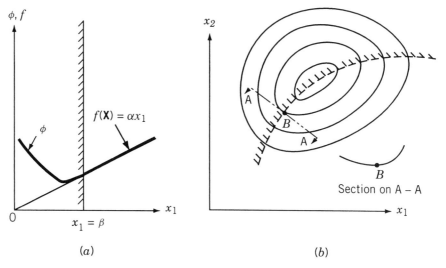

Figure 7.13   A ϕ function for $q > 1$.

4. $q > 1$. The ϕ function will have continuous first derivatives in this case as shown in Fig. 7.13. These derivatives are given by

$$\frac{\partial \phi}{\partial x_i} = \frac{\partial f}{\partial x_i} + r_k \sum_{j=1}^{m} q \langle g_j(\mathbf{X}) \rangle^{q-1} \frac{\partial g_j(\mathbf{X})}{\partial x_i} \tag{7.202}$$

Generally, the value of q is chosen as 2 in practical computation. We assume a value of $q > 1$ in subsequent discussion of this method.

Algorithm.   The exterior penalty function method can be stated by the following steps:

1. Start from any design $\mathbf{X}_1$ and a suitable value of r_1. Set $k = 1$.
2. Find the vector $\mathbf{X}_k^*$ that minimizes the function

$$\phi(\mathbf{X}, r_k) = f(\mathbf{X}) + r_k \sum_{j=1}^{m} \langle g_j(\mathbf{X}) \rangle^q$$

3. Test whether the point $\mathbf{X}_k^*$ satisfies all the constraints. If $\mathbf{X}_k^*$ is feasible, it is the desired optimum and hence terminate the procedure. Otherwise, go to step 4.
4. Choose the next value of the penalty parameter that satisfies the relation

$$r_{k+1} > r_k$$

and set the new value of k as original k plus 1 and go to step 2. Usually, the value of r_{k+1} is chosen according to the relation $r_{k+1} = cr_k$, where c is a constant greater than 1.

Example 7.9

$$\text{Minimize } f(x_1, x_2) = \tfrac{1}{3}(x_1 + 1)^3 + x_2$$

subject to

$$g_1(x_1, x_2) = 1 - x_1 \leq 0$$
$$g_2(x_1, x_2) = -x_2 \leq 0$$

SOLUTION   To illustrate the exterior penalty function method, we solve the unconstrained minimization problem by using differential calculus method. As such, it is not necessary to have an initial trial point $\mathbf{X}_1$. The ϕ function is

$$\phi(\mathbf{X}_1, r) = \tfrac{1}{3}(x_1 + 1)^3 + x_2 + r[\max(0, 1 - x_1)]^2 + r[\max(0, -x_2)]^2$$

The necessary conditions for the unconstrained minimum of $\phi(\mathbf{X}, r)$ are

$$\frac{\partial \phi}{\partial x_1} = (x_1 + 1)^2 - 2r[\max(0, 1 - x_1)] = 0$$

$$\frac{\partial \phi}{\partial x_2} = 1 - 2r[\max(0, -x_2)] = 0$$

These equations can be written as

$$\min[(x_1 + 1)^2, (x_1 + 1)^2 - 2r(1 - x_1)] = 0 \tag{E_1}$$
$$\min[1, 1 + 2rx_2] = 0 \tag{E_2}$$

In Eq. (E_1), if $(x_1 + 1)^2 = 0$, $x_1 = -1$ (this violates the first constraint), and if

$$(x_1 + 1)^2 - 2r(1 - x_1) = 0, \quad x_1 = -1 - r + \sqrt{r^2 + 4r}$$

In Eq. (E_2), the only possibility is that $1 + 2rx_2 = 0$ and hence $x_2 = -1/2r$. Thus the solution of the unconstrained minimization problem is given by

$$x_1^*(r) = -1 - r + r\left(1 + \frac{4}{r}\right)^{1/2} \tag{E_3}$$

$$x_2^*(r) = -\frac{1}{2r} \tag{E_4}$$

From this, the solution of the original constrained problem can be obtained as

$$x_1^* = \lim_{r \to \infty} x_1^*(r) = 1, \quad x_2^* = \lim_{r \to \infty} x_2^*(r) = 0$$

$$f_{\min} = \lim_{r \to \infty} \phi_{\min}(r) = \tfrac{8}{3}$$

The convergence of the method, as r increases gradually, can be seen from Table 7.5.

Table 7.5   Results for Example 7.9

Value of r	x_1^*	x_2^*	$\phi_{\min}(r)$	$f_{\min}(r)$
0.001	-0.93775	-500.00000	-249.9962	-500.0000
0.01	-0.80975	-50.00000	-24.9650	-49.9977
0.1	-0.45969	-5.00000	-2.2344	-4.9474
1	0.23607	-0.50000	0.9631	0.1295
10	0.83216	-0.05000	2.3068	2.0001
100	0.98039	-0.00500	2.6249	2.5840
1,000	0.99800	-0.00050	2.6624	2.6582
10,000	0.99963	-0.00005	2.6655	2.6652
∞	1	0	$\frac{8}{3}$	$\frac{8}{3}$

Convergence Proof.   To prove the convergence of the algorithm given above, we assume that f and g_j, $j = 1, 2, \ldots, m$, are continuous and that an optimum solution exists for the given problem. The following results are useful in proving the convergence of the exterior penalty function method.

Theorem 7.3 If

$$\phi(\mathbf{X}, r_k) = f(\mathbf{X}) + r_k G[\mathbf{g}(\mathbf{X})] = f(\mathbf{X}) + r_k \sum_{j=1}^{m} \langle g_j(\mathbf{X}) \rangle^q$$

the following relations will be valid for any $0 < r_k < r_{k+1}$:

1. $\phi(\mathbf{X}_k^*, r_k) \le \phi(\mathbf{X}_{k+1}^*, r_{k+1})$.
2. $f(\mathbf{X}_k^*) \le f(\mathbf{X}_{k+1}^*)$.
3. $G[\mathbf{g}(\mathbf{X}_k^*)] \ge G[\mathbf{g}(\mathbf{X}_{k+1}^*)]$.

Proof: The proof is similar to that of Theorem 7.1.

Theorem 7.4 If the function $\phi(\mathbf{X}, r_k)$ given by Eq. (7.199) is minimized for an increasing sequence of values of r_k, the unconstrained minima $\mathbf{X}_k^*$ converge to the optimum solution $(\mathbf{X}^*)$ of the constrained problem as $r_k \to \infty$.

Proof: The proof is similar to that of Theorem 7.1 (see Problem 7.46).

7.16   EXTRAPOLATION TECHNIQUES IN THE INTERIOR PENALTY FUNCTION METHOD

In the interior penalty function method, the ϕ function is minimized sequentially for a decreasing sequence of values $r_1 > r_2 > \cdots > r_k$ to find the unconstrained minima $\mathbf{X}_1^*, \mathbf{X}_2^*, \ldots, \mathbf{X}_k^*$, respectively. Let the values of the objective function corresponding to $\mathbf{X}_1^*, \mathbf{X}_2^*, \ldots, \mathbf{X}_k^*$ be $f_1^*, f_2^*, \ldots, f_k^*$, respectively. It has been proved that the sequence

$\mathbf{X}_1^*, \mathbf{X}_2^*, \ldots, \mathbf{X}_k^*$ converges to the minimum point $\mathbf{X}^*$, and the sequence $f_1^*, f_2^*, \ldots, f_k^*$ to the minimum value f^* of the original constrained problem stated in Eq. (7.153) as $r_k \to 0$. After carrying out a certain number of unconstrained minimizations of ϕ, the results obtained thus far can be used to estimate the minimum of the original constrained problem by a method known as the *extrapolation technique*. The extrapolations of the design vector and the objective function are considered in this section.

7.16.1   Extrapolation of the Design Vector X

Since different vectors $\mathbf{X}_i^*$, $i = 1, 2, \ldots, k$, are obtained as unconstrained minima of $\phi(\mathbf{X}, r_i)$ for different r_i, $i = 1, 2, \ldots, k$, the unconstrained minimum $\phi(\mathbf{X}, r)$ for any value of r, $\mathbf{X}^*(r)$, can be approximated by a polynomial in r as

$$\mathbf{X}^*(r) = \sum_{j=0}^{k-1} \mathbf{A}_j (r)^j = \mathbf{A}_0 + r\mathbf{A}_1 + r^2\mathbf{A}_2 + \cdots + r^{k-1}\mathbf{A}_{k-1} \qquad (7.203)$$

where $\mathbf{A}_j$ are n-component vectors. By substituting the known conditions

$$\mathbf{X}^*(r = r_i) = \mathbf{X}_i^*, \quad i = 1, 2, \ldots, k \qquad (7.204)$$

in Eq. (7.203), we can determine the vectors $\mathbf{A}_j$, $j = 0, 1, 2, \ldots, k - 1$ uniquely. Then $\mathbf{X}^*(r)$, given by Eq. (7.203), will be a good approximation for the unconstrained minimum of $\phi(\mathbf{X}, r)$ in the interval $(0, r_1)$. By setting $r = 0$ in Eq. (7.203), we can obtain an estimate to the true minimum, $\mathbf{X}^*$, as

$$\mathbf{X}^* = \mathbf{X}^*(r = 0) = \mathbf{A}_0 \qquad (7.205)$$

It is to be noted that it is not necessary to approximate $\mathbf{X}^*(r)$ by a $(k-1)$ st-order polynomial in r. In fact, any polynomial of order $1 \leq p \leq k - 1$ can be used to approximate $\mathbf{X}^*(r)$. In such a case we need only $p + 1$ points out of $\mathbf{X}_1^*, \mathbf{X}_2^*, \ldots, \mathbf{X}_k^*$ to define the polynomial completely.

As a simplest case, let us consider approximating $\mathbf{X}^*(r)$ by a first-order polynomial (linear equation) in r as

$$\mathbf{X}^*(r) = \mathbf{A}_0 + r\mathbf{A}_1 \qquad (7.206)$$

To evaluate the vectors $\mathbf{A}_0$ and $\mathbf{A}_1$, we need the data of two unconstrained minima. If the extrapolation is being done at the end of the kth unconstrained minimization, we generally use the latest information to find the constant vectors $\mathbf{A}_0$ and $\mathbf{A}_1$. Let $\mathbf{X}_{k-1}^*$ and $\mathbf{X}_k^*$ be the unconstrained minima corresponding to r_{k-1} and r_k, respectively. Since $r_k = cr_{k-1}$ $(c < 1)$, Eq. (7.206) gives

$$\mathbf{X}^*(r = r_{k-1}) = \mathbf{A}_0 + r_{k-1}\mathbf{A}_1 = \mathbf{X}_{k-1}^*$$
$$\mathbf{X}^*(r = r_k) = \mathbf{A}_0 + cr_{k-1}\mathbf{A}_1 = \mathbf{X}_k^* \qquad (7.207)$$

These equations give

$$\mathbf{A}_0 = \frac{\mathbf{X}_k^* - c\mathbf{X}_{k-1}^*}{1 - c}$$
$$\mathbf{A}_1 = \frac{\mathbf{X}_{k-1}^* - \mathbf{X}_k^*}{r_{k-1}(1 - c)} \qquad (7.208)$$

From Eqs. (7.206) and (7.208), the extrapolated value of the true minimum can be obtained as

$$\mathbf{X}^*(r = 0) = \mathbf{A}_0 = \frac{\mathbf{X}_k^* - c\mathbf{X}_{k-1}^*}{1 - c} \tag{7.209}$$

The extrapolation technique [Eq. (7.203)] has several advantages:

1. It can be used to find a good estimate to the optimum of the original problem with the help of Eq. (7.205).

2. It can be used to provide an additional convergence criterion to terminate the minimization process. The point obtained at the end of the kth iteration, $\mathbf{X}_k^*$, can be taken as the true minimum if the relation

$$|\mathbf{X}_k^* - \mathbf{X}^*(r = 0)| \leq \boldsymbol{\varepsilon} \tag{7.210}$$

is satisfied, where $\boldsymbol{\varepsilon}$ is the vector of prescribed small quantities.

3. This method can also be used to estimate the next minimum of the ϕ function after a number of minimizations have been completed. This estimate[†] can be used as a starting point for the $(k + 1)$st minimization of the ϕ function. The estimate of the $(k + 1)$st minimum, based on the information collected from the previous k minima, is given by Eq. (7.203) as

$$\mathbf{X}_{k+1}^* \simeq \mathbf{X}^*(r = r_{k+1} = r_1 c^k)$$
$$= \mathbf{A}_0 + (r_1 c^k)\mathbf{A}_1 + (r_1 c^k)^2 \mathbf{A}_2 + \cdots + \mathbf{A}_{k-1}(r_1 c^k)^{k-1} \tag{7.211}$$

If Eqs. (7.206) and (7.208) are used, this estimate becomes

$$\mathbf{X}_{k+1} \simeq \mathbf{X}^*(r = c^2 r_{k-1}) = \mathbf{A}_0 + c^2 r_{k-1}\mathbf{A}_1$$
$$= (1 + c)\mathbf{X}_k^* - c\mathbf{X}_{k-1}^* \tag{7.212}$$

Discussion.   It has been proved that under certain conditions, the difference between the true minimum $\mathbf{X}^*$ and the estimate $\mathbf{X}^*(r = 0) = \mathbf{A}_0$ will be of the order r_1^k [7.17]. Thus as $r_1 \to 0$, $\mathbf{A}_0 \to \mathbf{X}^*$. Moreover, if $r_1 < 1$, the estimates of $\mathbf{X}^*$ obtained by using k minima will be better than those using $(k - 1)$ minima, and so on. Hence as more minima are achieved, the estimate of $\mathbf{X}^*$ or $\mathbf{X}_{k+1}^*$ presumably gets better. This estimate can be used as the starting point for the $(k + 1)$st minimization of the ϕ function. This accelerates the entire process by substantially reducing the effort needed to minimize the successive ϕ functions. However, the computer storage requirements and accuracy considerations (such as numerical round-off errors that become important for higher-order estimates) limit the order of polynomial in Eq. (7.203). It has been found in practice that extrapolations with the help of even quadratic and cubic equations in r generally yield good estimates for $\mathbf{X}_{k+1}^*$ and $\mathbf{X}^*$. Note that the extrapolated points given by any of Eqs. (7.205), (7.209), (7.211), and (7.212) may sometimes violate the constraints. Hence we have to check any extrapolated point for feasibility before using it as a starting point for the next minimization of ϕ. If the extrapolated point is found infeasible, it has to be rejected.

[†]The estimate obtained for $\mathbf{X}^*$ can also be used as a starting point for the $(k + 1)$st minimization of the ϕ function.

7.16.2   Extrapolation of the Function f

As in the case of the design vector, it is possible to use extrapolation technique to estimate the optimum value of the original objective function, f^*. For this, let $f_1^*, f_2^*, \ldots, f_k^*$ be the values of the objective function corresponding to the vectors $\mathbf{X}_1^*, \mathbf{X}_2^*, \ldots, \mathbf{X}_k^*$. Since the points $\mathbf{X}_1^*, \mathbf{X}_2^*, \ldots, \mathbf{X}_k^*$ have been found to be the unconstrained minima of the ϕ function corresponding to $r_1, r_2, \ldots, r_k$, respectively, the objective function, f^*, can be assumed to be a function of r. By approximating f^* by a $(k-1)$st-order polynomial in r, we have

$$f^*(r) = \sum_{j=0}^{k-1} a_j(r)^j = a_0 + a_1 r + a_2 r^2 + \cdots + a_{k-1} r^{k-1} \qquad (7.213)$$

where the k constants a_j, $j = 0, 1, 2, \ldots, k-1$ can be evaluated by substituting the known conditions

$$f^*(r = r_i) = f_i^* = a_0 + a_1 r_i + a_2 r_i^2 + \cdots + a_{k-1} r_i^{k-1}, \quad i = 1, 2, \ldots, k \quad (7.214)$$

Since Eq. (7.213) is a good approximation for the true f^* in the interval $(0, r_1)$, we can obtain an estimate for the constrained minimum of f as

$$f^* \simeq f^*(r = 0) = a_0 \qquad (7.215)$$

As a particular case, a linear approximation can be made for f^* by using the last two data points. Thus if f_{k-1}^* and f_k^* are the function values corresponding to r_{k-1} and $r_k = c r_{k-1}$, we have

$$\begin{aligned} f_{k-1}^* &= a_0 + r_{k-1} a_1 \\ f_k^* &= a_0 + c r_{k-1} a_1 \end{aligned} \qquad (7.216)$$

These equations yield

$$a_0 = \frac{f_k^* - c f_{k-1}^*}{1 - c} \qquad (7.217)$$

$$a_1 = \frac{f_{k-1}^* - f_k^*}{r_{k-1}(1 - c)} \qquad (7.218)$$

$$f^*(r) = \frac{f_k^* - c f_{k-1}^*}{1 - c} + \frac{r}{r_{k-1}} \frac{f_{k-1}^* - f_k^*}{1 - c} \qquad (7.219)$$

Equation (7.219) gives an estimate of f^* as

$$f^* \simeq f^*(r = 0) = a_0 = \frac{f_k^* - c f_{k-1}^*}{1 - c} \qquad (7.220)$$

The extrapolated value a_0 can be used to provide an additional convergence criterion for terminating the interior penalty function method. The criterion is that whenever the value of f_k^* obtained at the end of kth unconstrained minimization of ϕ is sufficiently close to the extrapolated value a_0, that is, when

$$\left| \frac{f_k^* - a_0}{f_k^*} \right| \leq \varepsilon \qquad (7.221)$$

where ε is a specified small quantity, the process can be terminated.

Example 7.10   Find the extrapolated values of $\mathbf{X}$ and f in Example 7.8 using the results of minimization of $\phi(\mathbf{X}, r_1)$ and $\phi(\mathbf{X}, r_2)$.

SOLUTION   From the results of Example 7.8, we have for $r_1 = 1.0$,

$$\mathbf{X}_1^* = \begin{Bmatrix} 0.37898 \\ 1.67965 \\ 2.34617 \end{Bmatrix}, \quad f_1^* = 5.70766$$

and for $r_2 = 0.1$,

$$c = 0.1, \quad \mathbf{X}_2^* = \begin{Bmatrix} 0.10088 \\ 1.41945 \\ 1.68302 \end{Bmatrix}, \quad f_2^* = 2.73267$$

By using Eq. (7.206) for approximating $\mathbf{X}^*(r)$, the extrapolated vector $\mathbf{X}^*$ is given by Eq. (7.209) as

$$\mathbf{X}^* \simeq \mathbf{A}_0 = \frac{\mathbf{X}_2^* - c\mathbf{X}_1^*}{1 - c} = \frac{1}{0.9} \left[\begin{Bmatrix} 0.10088 \\ 1.41945 \\ 1.68302 \end{Bmatrix} - 0.1 \begin{Bmatrix} 0.37898 \\ 1.67865 \\ 2.34617 \end{Bmatrix} \right] \tag{E_1}$$

$$= \begin{Bmatrix} 0.06998 \\ 1.39053 \\ 1.60933 \end{Bmatrix} \tag{E_2}$$

Similarly, the linear resltionships $f^*(r) = a_0 + a_1 r$ leads to [from Eq. (7.220)]

$$f^* \simeq \frac{f_2^* - cf_1^*}{1 - c} = \frac{1}{0.9}[2.73267 - 0.1(5.707667)] = 2.40211 \tag{E_3}$$

It can be verified that the extrapolated design vector $\mathbf{X}^*$ is feasible and hence can be used as a better starting point for the subsequent minimization of the function ϕ.

7.17   EXTENDED INTERIOR PENALTY FUNCTION METHODS

In the interior penalty function approach, the ϕ function is defined within the feasible domain. As such, if any of the one-dimensional minimization methods discussed in Chapter 5 is used, the resulting optimal step lengths might lead to infeasible designs. Thus the one-dimensional minimization methods have to be modified to avoid this problem. An alternative method, known as the *extended interior penalty function method*, has been proposed in which the ϕ function is defined outside the feasible region. The extended interior penalty function method combines the best features of the interior and exterior methods for inequality constraints. Several types of extended interior penalty function formulations are described in this section.

7.17.1   Linear Extended Penalty Function Method

The linear extended penalty function method was originally proposed by Kavlie and Moe [7.18] and later improved by Cassis and Schmit [7.19]. In this method, the ϕ_k

function is constructed as follows:

$$\phi_k = \phi(\mathbf{X}, r_k) = f(\mathbf{X}) + r_k \sum_{j=1}^{m} \tilde{g}_j(\mathbf{X}) \tag{7.222}$$

where

$$\tilde{g}_j(\mathbf{X}) = \begin{cases} -\dfrac{1}{g_j(\mathbf{X})} & \text{if } g_j(\mathbf{X}) \le \varepsilon \\[2mm] -\dfrac{2\varepsilon - g_j(\mathbf{X})}{\varepsilon^2} & \text{if } g_j(\mathbf{X}) > \varepsilon \end{cases} \tag{7.223}$$

and ε is a small negative number that marks the transition from the interior penalty $[g_j(\mathbf{X}) \le \varepsilon]$ to the extended penalty $[g_j(\mathbf{X}) > \varepsilon]$. To produce a sequence of improved feasible designs, the value of ε is to be selected such that the function ϕ_k will have a positive slope at the constraint boundary. Usually, ε is chosen as

$$\varepsilon = -c(r_k)^a \tag{7.224}$$

where c and a are constants. The constant a is chosen such that $\frac{1}{3} \le a \le \frac{1}{2}$, where the value of $a = \frac{1}{3}$ guarantees that the penalty for violating the constraints increases as r_k goes to zero while the value of $a = \frac{1}{2}$ is required to help keep the minimum point $\mathbf{X}^*$ in the quadratic range of the penalty function. At the start of optimization, ε is selected in the range $-0.3 \le \varepsilon \le -0.1$. The value of r_1 is selected such that the values of $f(\mathbf{X})$ and $r_1 \sum_{j=1}^{m} \tilde{g}_j(\mathbf{X})$ are equal at the initial design vector $\mathbf{X}_1$. This defines the value of c in Eq. (7.224). The value of ε is computed at the beginning of each unconstrained minimization using the current value of r_k from Eq. (7.224) and is kept constant throughout that unconstrained minimization. A flowchart for implementing the linear extended penalty function method is given in Fig. 7.14.

7.17.2   Quadratic Extended Penalty Function Method

The ϕ_k function defined by Eq. (7.222) can be seen to be continuous with continuous first derivatives at $g_j(\mathbf{X}) = \varepsilon$. However, the second derivatives can be seen to be discontinuous at $g_j(\mathbf{X}) = \varepsilon$. Hence it is not possible to use a second-order method for unconstrained minimization [7.20]. The quadratic extended penalty function is defined so as to have continuous second derivatives at $g_j(\mathbf{X}) = \varepsilon$ as follows:

$$\phi_k = \phi(\mathbf{X}, r_k) = f(\mathbf{X}) + r_k \sum_{j=1}^{m} \tilde{g}_j(\mathbf{X}) \tag{7.225}$$

where

$$\tilde{g}_j(\mathbf{X}) = \begin{cases} -\dfrac{1}{g_j(\mathbf{X})} & \text{if } g_j(\mathbf{X}) \le \varepsilon \\[2mm] \left\{ -\dfrac{1}{\varepsilon}\left[\dfrac{g_j(\mathbf{X})}{\varepsilon}\right]^2 - 3\dfrac{g_j(\mathbf{X})}{\varepsilon} + 3 \right\} & \text{if } g_j(\mathbf{X}) > \varepsilon \end{cases} \tag{7.226}$$

With this definition, second-order methods can be used for the unconstrained minimization of ϕ_k. It is to be noted that the degree of nonlinearity of ϕ_k is increased in

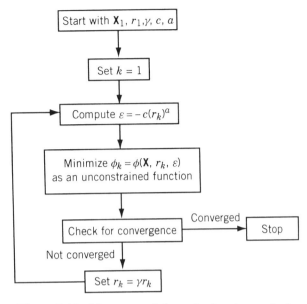

Figure 7.14   Linear extended penalty function method.

Eq. (7.225) compared to Eq. (7.222). The concept of extended interior penalty function approach can be generalized to define a variable penalty function method from which the linear and quadratic methods can be derived as special cases [7.24].

Example 7.11   Plot the contours of the ϕ_k function using the linear extended interior penalty function for the following problem:

$$\text{Minimize } f(x) = (x - 1)^2$$

subject to

$$g_1(x) = 2 - x \le 0$$
$$g_2(x) = x - 4 \le 0$$

SOLUTION   We choose $c = 0.2$ and $a = 0.5$ so that $\varepsilon = -0.2\sqrt{r_k}$. The ϕ_k function is defined by Eq. (7.222). By selecting the values of r_k as 10.0, 1.0, 0.1, and 0.01 sequentially, we can determine the values of ϕ_k for different values of x, which can then be plotted as shown in Fig. 7.15. The graph of $f(x)$ is also shown in Fig. 7.15 for comparison.

7.18   PENALTY FUNCTION METHOD FOR PROBLEMS WITH MIXED EQUALITY AND INEQUALITY CONSTRAINTS

The algorithms described in previous sections cannot be directly applied to solve problems involving strict equality constraints. In this section we consider some of the

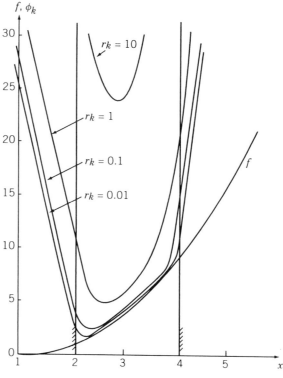

Figure 7.15   Graphs of ϕ_k.

methods that can be used to solve a general class of problems.

$$\text{Minimize } f(\mathbf{X})$$

subject to

$$
\begin{aligned}
g_j(\mathbf{X}) &\le 0, \quad j = 1, 2, \ldots, m \\
l_j(\mathbf{X}) &= 0, \quad j = 1, 2, \ldots, p
\end{aligned}
\tag{7.227}
$$

7.18.1   Interior Penalty Function Method

Similar to Eq. (7.154), the present problem can be converted into an unconstrained minimization problem by constructing a function of the form

$$
\phi_k = \phi(\mathbf{X}, r_k) = f(\mathbf{X}) + r_k \sum_{j=1}^{m} G_j[g_j(\mathbf{X})] + H(r_k) \sum_{j=1}^{p} l_j^2(\mathbf{X})
\tag{7.228}
$$

where G_j is some function of the constraint g_j tending to infinity as the constraint boundary is approached, and $H(r_k)$ is some function of the parameter r_k tending to infinity as r_k tends to zero. The motivation for the third term in Eq. (7.228) is that as

$H(r_k) \to \infty$, the quantity $\Sigma_{j=1}^p l_j^2(\mathbf{X})$ must tend to zero. If $\Sigma_{j=1}^p l_j^2(\mathbf{X})$ does not tend to zero, ϕ_k would tend to infinity, and this cannot happen in a sequential minimization process if the problem has a solution. Fiacco and McCormick [7.17, 7.21] used the following form of Eq. (7.228):

$$\phi_k = \phi(\mathbf{X}, r_k) = f(\mathbf{X}) - r_k \sum_{j=1}^m \frac{1}{g_j(\mathbf{X})} + \frac{1}{\sqrt{r_k}} \sum_{j=1}^p l_j^2(\mathbf{X}) \qquad (7.229)$$

If ϕ_k is minimized for a decreasing sequence of values r_k, the following theorem proves that the unconstrained minima $\mathbf{X}_k^*$ will converge to the solution $\mathbf{X}^*$ of the original problem stated in Eq. (7.227).

Theorem 7.5 If the problem posed in Eq. (7.227) has a solution, the unconstrained minima, $\mathbf{X}_k^*$, of $\phi(\mathbf{X}, r_k)$, defined by Eq. (7.229) for a sequence of values $r_1 > r_2 > \cdots > r_k$, converge to the optimal solution of the constrained problem [Eq. (7.227)] as $r_k \to 0$.

Proof: A proof similar to that of Theorem 7.1 can be given to prove this theorem. Further, the solution obtained at the end of sequential minimization of ϕ_k is guaranteed to be the global minimum of the problem, Eqs. (7.227), if the following conditions are satisfied:

    (i) $f(\mathbf{X})$ is convex.
    (ii) $g_j(\mathbf{X})$, $j = 1, 2, \ldots, m$ are convex.
    (iii) $\Sigma_{j=1}^p l_j^2(\mathbf{X})$ is convex in the interior feasible domain defined by the inequality constraints.
    (iv) One of the functions among $f(\mathbf{X})$, $g_1(\mathbf{X})$, $g_2(\mathbf{X})$, ..., $g_m(\mathbf{X})$, and $\Sigma_{j=1}^p l_j^2(\mathbf{X})$ is strictly convex.

Note:

1. To start the sequential unconstrained minimization process, we have to start from a point $\mathbf{X}_1$ at which the inequality constraints are satisfied and not necessarily the equality constraints.

2. Although this method has been applied to solve a variety of practical problems, it poses an extremely difficult minimization problem in many cases, mainly because of the scale disparities that arise between the penalty terms

$$-r_k \sum_{j=1}^m \frac{1}{g_j(\mathbf{X})} \quad \text{and} \quad \frac{1}{r_k^{1/2}} \sum_{j=1}^p l_j^2(\mathbf{X})$$

as the minimization process proceeds.

7.18.2 Exterior Penalty Function Method

To solve an optimization problem involving both equality and inequality constraints as stated in Eqs. (7.227), the following form of Eq. (7.228) has been proposed:

$$\phi_k = \phi(\mathbf{X}, r_k) = f(\mathbf{X}) + r_k \sum_{j=1}^m \langle g_j(\mathbf{X}) \rangle^2 + r_k \sum_{j=1}^p l_j^2(\mathbf{X}) \qquad (7.230)$$

As in the case of Eq. (7.199), this function has to be minimized for an increasing sequence of values of r_k. It can be proved that as $r_k \to \infty$, the unconstrained minima, $\mathbf{X}_k^*$, of $\phi(\mathbf{X}, r_k)$ converge to the minimum of the original constrained problem stated in Eq. (7.227).

7.19 PENALTY FUNCTION METHOD FOR PARAMETRIC CONSTRAINTS

7.19.1 Parametric Constraint

In some optimization problems, a particular constraint may have to be satisfied over a range of some parameter (θ) as

$$g_j(\mathbf{X}, \theta) \leq 0, \quad \theta_l \leq \theta \leq \theta_u \tag{7.231}$$

where θ_l and θ_u are lower and the upper limits on θ, respectively. These types of constraints are called *parametric constraints*. As an example, consider the design of a four-bar linkage shown in Fig. 7.16. The angular position of the output link ϕ will depend on the angular position of the input link, θ, and the lengths of the links, l_1, l_2, l_3, and l_4. If $l_i (i = 1 \text{ to } 4)$ are taken as the design variables $x_i (i = 1 \text{ to } 4)$, the angular position of the output link, $\phi(\mathbf{X}, \theta)$, for any fixed value of $\theta(\theta_i)$ can be changed by changing the design vector, $\mathbf{X}$. Thus if $\overline{\phi}(\theta)$ is the output desired, the output $\phi(\mathbf{X}, \theta)$ generated will, in general, be different from that of $\overline{\phi}(\theta)$, as shown in Fig. 7.17. If the linkage is used in some precision equipment, we would like to restrict the difference $|\overline{\phi}(\theta) - \phi(\mathbf{X}, \theta)|$ to be smaller than some permissible value, say, ε. Since this restriction has to be satisfied for all values of the parameter θ, the constraint can be stated as a parametric constraint as

$$|\overline{\phi}(\theta) - \phi(\mathbf{X}, \theta)| \leq \varepsilon, \quad 0° \leq \theta \leq 360° \tag{7.232}$$

Sometimes the number of parameters in a parametric constraint may be more than one. For example, consider the design of a rectangular plate acted on by an arbitrary load as shown in Fig. 7.18. If the magnitude of the stress induced under the given loading, $|\sigma(x, y)|$, is restricted to be smaller than the allowable value $\sigma_{\max}$, the constraint can

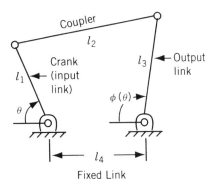

Figure 7.16   Four-bar linkage.

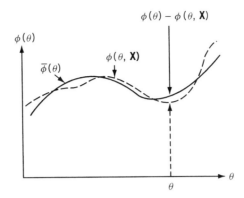

Figure 7.17 Output angles generated and desired.

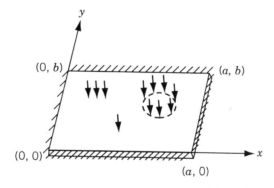

Figure 7.18 Rectangular plate under arbitrary load.

be stated as a parametric constraint as

$$|\sigma(x, y)| - \sigma_{\max} \leq 0, \quad 0 \leq x \leq a, \quad 0 \leq y \leq b \qquad (7.233)$$

Thus this constraint has to be satisfied at all the values of parameters x and y.

7.19.2 Handling Parametric Constraints

One method of handling a parametric constraint is to replace it by a number of ordinary constraints as

$$g_j(\mathbf{X}, \theta_i) \leq 0, \quad i = 1, 2, \ldots, r \qquad (7.234)$$

where $\theta_1, \theta_2, \ldots, \theta_r$ are discrete values taken in the range of θ. This method is not efficient, for the following reasons:

1. It results in a very large number of constraints in the optimization problem.
2. Even if all the r constraints stated in Eq. (7.234) are satisfied, the constraint may still be violated at some other value of θ [i.e., $g_j(\mathbf{X}, \theta) > 0$ where $\theta_k < \theta < \theta_{k+1}$ for some k].

Another method of handling the parametric constraints is to construct the ϕ function in a different manner as follows [7.1, 7.15].

Interior Penalty Function Method

$$\phi(\mathbf{X}, r_k) = f(\mathbf{X}) - r_k \sum_{j=1}^{m}\left[\int_{\theta_l}^{\theta_u}\frac{1}{g_j(\mathbf{X},\theta)}d\theta\right] \tag{7.235}$$

The idea behind using the integral in Eq. (7.235) for a parametric constraint is to make the integral tend to infinity as the value of the constraint $g_j(\mathbf{X}, \theta)$ tends to zero even at one value of θ in its range. If a gradient method is used for the unconstrained minimization of $\phi(\mathbf{X}, r_k)$, the derivatives of ϕ with respect to the design variables $x_i (i = 1, 2, \ldots, n)$ are needed. Equation (7.235) gives

$$\frac{\partial\phi}{\partial x_i}(\mathbf{X}, r_k) = \frac{\partial f}{\partial x_i}(\mathbf{X}) + r_k \sum_{j=1}^{m}\left[\int_{\theta_l}^{\theta_u}\frac{1}{g_j^2(\mathbf{X},\theta)}\frac{\partial g_j}{\partial x_i}(\mathbf{X},\theta)d\theta\right] \tag{7.236}$$

by assuming that the limits of integration, θ_l and θ_u, are indepdnent of the design variables x_i. Thus it can be noticed that the computation of $\phi(\mathbf{X}, r_k)$ or $\partial\phi(\mathbf{X}, r_k)/\partial x_i$ involves the evaluation of an integral. In most of the practical problems, no closed-form expression will be available for $g_j(\mathbf{X}, \theta)$, and hence we have to use some sort of a numerical integration process to evaluate ϕ or $\partial\phi/\partial x_i$. If trapezoidal rule [7.22] is used to evaluate the integral in Eq. (7.235), we obtain[†]

$$\phi(\mathbf{X}, r_k) = f(\mathbf{X}) - r_k \sum_{r=1}^{m}\left\{\frac{\Delta\theta}{2}\left[\frac{1}{g_j(\mathbf{X},\theta_l)} + \frac{1}{g_j(\mathbf{X},\theta_u)}\right]\right.$$
$$\left. +\Delta\theta\sum_{p=2}^{r-1}\frac{1}{g_j(\mathbf{X},\theta_p)}\right\} \tag{7.237}$$

[†]Let the interval of the parameter θ be divided into $r-1$ equal divisions so that

$$\theta_1 = \theta_l, \quad \theta_2 = \theta_1 + \Delta\theta, \quad \theta_3 = \theta_1 + 2.\,\Delta\theta, \ldots, \theta_r = \theta_1 + (r-1)\Delta\theta = \theta_u,$$

$$\Delta\theta = \frac{\theta_u - \theta_l}{r-1}$$

If the graph of the function $g_j(\mathbf{X}, \theta)$ looks as shown in Fig. 7.19, the integral of $1/g_j(\mathbf{X}, \theta)$ can be found approximately by adding the areas of all the trapeziums, like $ABCD$. This is the reason why the method is known as *trapezoidal rule*. The sum of all the areas is given by

$$\int_{\theta_l}^{\theta_u}\frac{d\theta}{g_j(\mathbf{X},\theta)} \approx \sum_{l=1}^{r-1}A_l = \sum_{p=1}^{r-1}\left[\frac{1}{g_j(\mathbf{X},\theta_p)} + \frac{1}{g_j(\mathbf{X},\theta_{p+1})}\right]\frac{\Delta\theta}{2}$$

$$= \frac{\Delta\theta}{2}\left[\frac{1}{g_j(\mathbf{X},\theta_l)} + \frac{1}{g_j(\mathbf{X},\theta_u)}\right] + \sum_{p=2}^{r-1}\frac{\Delta\theta}{g_j(\mathbf{X},\theta_p)}$$

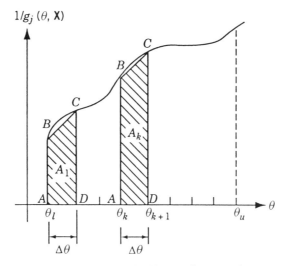

Figure 7.19   Numerical integration procedure.

where r is the number of discrete values of θ, and $\Delta\theta$ is the uniform spacing between the discrete values so that

$$\theta_1 = \theta_l, \quad \theta_2 = \theta_1 + \Delta\theta,$$

$$\theta_3 = \theta_1 + 2\Delta\theta, \dots, \theta_r = \theta_1 + (r-1)\Delta\theta = \theta_u$$

If $g_j(\mathbf{X}, \theta)$ cannot be expressed as a closed-form function of $\mathbf{X}$, the derivative $\partial g_j/\partial x_i$ occurring in Eq. (7.236) has to be evaluated by using some form of a finite-difference formula.

Exterior Penalty Function Method

$$\phi(\mathbf{X}, r_k) = f(\mathbf{X}) + r_k \sum_{j=1}^{m} \left[\int_{\theta_l}^{\theta_u} \langle g_j(\mathbf{X}, \theta) \rangle^2 d\theta \right] \tag{7.238}$$

The method of evaluating $\phi(\mathbf{X}, r_k)$ will be similar to that of the interior penalty function method.

7.20   AUGMENTED LAGRANGE MULTIPLIER METHOD

7.20.1   Equality-Constrained Problems

The augmented Lagrange multiplier (ALM) method combines the Lagrange multiplier and the penalty function methods. Consider the following equality-constrained problem:

$$\text{Minimize } f(\mathbf{X}) \tag{7.239}$$

subject to

$$h_j(\mathbf{X}) = 0, \quad j = 1, 2, \ldots, p, \quad p < n \tag{7.240}$$

The Lagrangian corresponding to Eqs. (7.239) and (7.240) is given by

$$L(\mathbf{X}, \boldsymbol{\lambda}) = f(\mathbf{X}) + \sum_{j=1}^{p} \lambda_j h_j(\mathbf{X}) \tag{7.241}$$

where λ_j, $j = 1, 2, \ldots, p$, are the Lagrange multipliers. The necessary conditions for a stationary point of $L(\mathbf{X}, \boldsymbol{\lambda})$ include the equality constraints, Eq. (7.240). The exterior penalty function approach is used to define the new objective function $A(\mathbf{X}, \boldsymbol{\lambda}, r_k)$, termed the *augmented Lagrangian function*, as

$$A(\mathbf{X}, \boldsymbol{\lambda}, r_k) = f(\mathbf{X}) + \sum_{j=1}^{p} \lambda_j h_j(\mathbf{X}) + r_k \sum_{j=1}^{p} h_j^2(\mathbf{X}) \tag{7.242}$$

where r_k is the penalty parameter. It can be noted that the function A reduces to the Lagrangian if $r_k = 0$ and to the ϕ function used in the classical penalty function method if all $\lambda_j = 0$. It can be shown that if the Lagrange multipliers are fixed at their optimum values λ_j^*, the minimization of $A(\mathbf{X}, \boldsymbol{\lambda}, r_k)$ gives the solution of the problem stated in Eqs. (7.239) and (7.240) in one step for any value of r_k. In such a case there is no need to minimize the function A for an increasing sequence of values of r_k. Since the values of λ_j^* are not known in advance, an iterative scheme is used to find the solution of the problem. In the first iteration ($k = 1$), the values of $\lambda_j^{(k)}$ are chosen as zero, the value of r_k is set equal to an arbitrary constant, and the function A is minimized with respect to $\mathbf{X}$ to find $\mathbf{X}^{*(k)}$. The values of $\lambda_j^{(k)}$ and r_k are then updated to start the next iteration. For this, the necessary conditions for the stationary point of L, given by Eq. (7.241), are written as

$$\frac{\partial L}{\partial x_i} = \frac{\partial f}{\partial x_i} + \sum_{j=1}^{p} \lambda_j^* \frac{\partial h_j}{\partial x_i} = 0, \quad i = 1, 2, \ldots, n \tag{7.243}$$

where λ_j^* denote the values of Lagrange multipliers at the stationary point of L. Similarly, the necessary conditions for the minimum of A can be expressed as

$$\frac{\partial A}{\partial x_i} = \frac{\partial f}{\partial x_i} + \sum_{j=1}^{p} (\lambda_j + 2 r_k h_j) \frac{\partial h_j}{\partial x_i} = 0, \quad i = 1, 2, \ldots, n \tag{7.244}$$

A comparison of the right-hand sides of Eqs. (7.243) and (7.244) yields

$$\lambda_j^* = \lambda_j + 2 r_k h_j, \quad j = 1, 2, \ldots, p \tag{7.245}$$

These equations are used to update the values of λ_j as

$$\lambda_j^{(k+1)} = \lambda_j^{(k)} + 2 r_k h_j(\mathbf{X}^{(k)}), \quad j = 1, 2, \ldots, p \tag{7.246}$$

where $\mathbf{X}^{(k)}$ denotes the starting vector used in the minimization of A. The value of r_k is updated as

$$r_{k+1} = cr_k, \quad c > 1 \tag{7.247}$$

The function A is then minimized with respect to $\mathbf{X}$ to find $\mathbf{X}^{*(k+1)}$ and the iterative process is continued until convergence is achieved for $\lambda_j^{(k)}$ or $\mathbf{X}^*$. If the value of r_{k+1} exceeds a prespecified maximum value r_{max}, it is set equal to r_{max}. The iterative process is indicated as a flow diagram in Fig. 7.20.

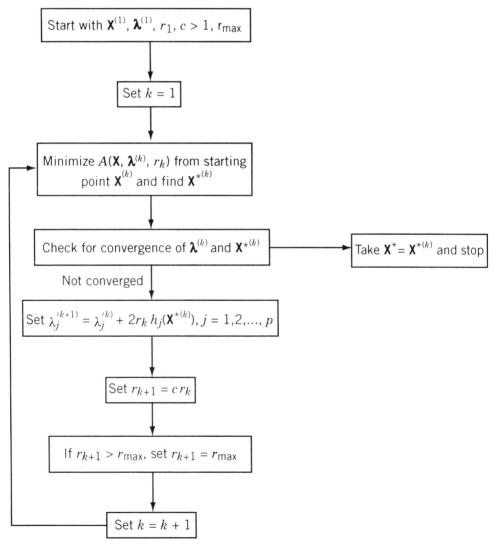

Figure 7.20   Flowchart of augmented Lagrange multiplier method.

7.20.2   Inequality-Constrained Problems

Consider the following inequality-constrained problem:

$$\text{Minimize } f(\mathbf{X}) \qquad (7.248)$$

subject to

$$g_j(\mathbf{X}) \le 0, \quad j = 1, 2, \ldots, m \qquad (7.249)$$

To apply the ALM method, the inequality constraints of Eq. (7.249) are first converted to equality constraints as

$$g_j(\mathbf{X}) + y_j^2 = 0, \quad j = 1, 2, \ldots, m \qquad (7.250)$$

where y_j^2 are the slack variables. Then the augmented Lagrangian function is constructed as

$$A(\mathbf{X}, \boldsymbol{\lambda}, \mathbf{Y}, r_k) = f(\mathbf{X}) + \sum_{j=1}^{m} \lambda_j [g_j(\mathbf{X}) + y_j^2] + \sum_{j=1}^{m} r_k [g_j(\mathbf{X}) + y_j^2]^2 \qquad (7.251)$$

where the vector of slack variables, $\mathbf{Y}$, is given by

$$\mathbf{Y} = \begin{Bmatrix} y_1 \\ y_2 \\ \vdots \\ y_m \end{Bmatrix}$$

If the slack variables y_j, $j = 1, 2, \ldots, m$, are considered as additional unknowns, the function A is to be minimized with respect to $\mathbf{X}$ and $\mathbf{Y}$ for specified values of λ_j and r_k. This increases the problem size. It can be shown [7.23] that the function A given by Eq. (7.251) is equivalent to

$$A(\mathbf{X}, \boldsymbol{\lambda}, r_k) = f(\mathbf{X}) + \sum_{j=1}^{m} \lambda_j \alpha_j + r_k \sum_{j=1}^{m} \alpha_j^2 \qquad (7.252)$$

where

$$\alpha_j = \max \left\{ g_j(\mathbf{X}), -\frac{\lambda_j}{2r_k} \right\} \qquad (7.253)$$

Thus the solution of the problem stated in Eqs. (7.248) and (7.249) can be obtained by minimizing the function A, given by Eq. (7.252), as in the case of equality-constrained problems using the update formula

$$\lambda_j^{(k+1)} = \lambda_j^{(k)} + 2r_k \alpha_j^{(k)}, \quad j = 1, 2, \ldots, m \qquad (7.254)$$

in place of Eq. (7.246). It is to be noted that the function A, given by Eq. (7.252), is continuous and has continuous first derivatives but has discontinuous second derivatives with respect to $\mathbf{X}$ at $g_j(\mathbf{X}) = -\lambda_j/2r_k$. Hence a second-order method cannot be used to minimize the function A.

7.20.3   Mixed Equality–Inequality-Constrained Problems

Consider the following general optimization problem:

$$\text{Minimize } f(\mathbf{X}) \tag{7.255}$$

subject to

$$g_j(\mathbf{X}) \le 0, \quad j = 1, 2, \ldots, m \tag{7.256}$$

$$h_j(\mathbf{X}) = 0, \quad j = 1, 2, \ldots, p \tag{7.257}$$

This problem can be solved by combining the procedures of the two preceding sections. The augmented Lagrangian function, in this case, is defined as

$$A(\mathbf{X}, \lambda, r_k) = f(\mathbf{X}) + \sum_{j=1}^{m} \lambda_j \alpha_j + \sum_{j=1}^{p} \lambda_{m+j} h_j(\mathbf{X})$$

$$+ r_k \sum_{j=1}^{m} \alpha_j^2 + r_k \sum_{j=1}^{p} h_j^2(\mathbf{X}) \tag{7.258}$$

where α_j is given by Eq. (7.253). The solution of the problem stated in Eqs. (7.255) to (7.257) can be found by minimizing the function A, defined by Eq. (7.258), as in the case of equality-constrained problems using the update formula

$$\lambda_j^{(k+1)} = \lambda_j^{(k)} + 2r_k \max\left\{ g_j(\mathbf{X}), -\frac{\lambda_j^{(k)}}{2r_k} \right\}, \quad j = 1, 2, \ldots, m \tag{7.259}$$

$$\lambda_{m+j}^{(k+1)} = \lambda_{m+j}^{(k)} + 2r_k h_j(\mathbf{X}), \quad j = 1, 2, \ldots, p \tag{7.260}$$

The ALM method has several advantages. As stated earlier, the value of r_k need not be increased to infinity for convergence. The starting design vector, $\mathbf{X}^{(1)}$, need not be feasible. Finally, it is possible to achieve $g_j(\mathbf{X}) = 0$ and $h_j(\mathbf{X}) = 0$ precisely and the nonzero values of the Lagrange multipliers ($\lambda_j \ne 0$) identify the active contraints automatically.

Example 7.12

$$\text{Minimize } f(\mathbf{X}) = 6x_1^2 + 4x_1 x_2 + 3x_2^2 \tag{E$_1$}$$

subject to

$$h(\mathbf{X}) = x_1 + x_2 - 5 = 0 \tag{E$_2$}$$

using the ALM method.

SOLUTION   The augmented Lagrangian function can be constructed as

$$A(\mathbf{X}, \lambda, r_k) = 6x_1^2 + 4x_1 x_2 + 3x_2^2 + \lambda(x_1 + x_2 - 5)$$

$$+ r_k(x_1 + x_2 - 5)^2 \tag{E$_3$}$$

Table 7.6 Results for Example 7.12

$\lambda^{(i)}$	r_k	$x_1^{*(i)}$	$x_2^{*(i)}$	Value of h
0.00000	1.00000	−0.23810	2.22222	−3.01587
−6.03175	1.00000	−0.38171	3.56261	−1.81910
−9.66994	1.00000	−0.46833	4.37110	−1.09723
−11.86441	1.00000	−0.52058	4.85876	−0.66182
−13.18806	1.00000	−0.55210	5.15290	−0.39919
−13.98645	1.00000	−0.57111	5.33032	−0.24078
−14.46801	1.00000	−0.58257	5.43734	−0.14524
−14.75848	1.00000	−0.58949	5.50189	−0.08760
−14.93369	1.00000	−0.59366	5.54082	−0.05284
−15.03937	1.00000	−0.59618	5.56430	−0.03187

For the stationary point of A, the necessary conditions, $\partial A/\partial x_i = 0$, $i = 1, 2$, yield

$$x_1(12 + 2r_k) + x_2(4 + 2r_k) = 10r_k - \lambda \tag{E_4}$$

$$x_1(4 + 2r_k) + x_2(6 + 2r_k) = 10r_k - \lambda \tag{E_5}$$

The solution of Eqs. (E$_4$) and (E$_5$) gives

$$x_1 = \frac{-90r_k^2 + 9r_k\lambda - 6\lambda + 60r_k}{(14 - 5r_k)(12 + 2r_k)} \tag{E_6}$$

$$x_2 = \frac{20r_k - 2\lambda}{14 - 5r_k} \tag{E_7}$$

Let the value of r_k be fixed at 1 and select a value of $\lambda^{(1)} = 0$. This gives

$$x_1^{*(1)} = -\tfrac{5}{21}, \quad x_2^{*(1)} = \tfrac{20}{9} \quad \text{with} \quad h = -\tfrac{5}{21} + \tfrac{20}{9} - 5 = -3.01587$$

For the next iteration,

$$\lambda^{(2)} = \lambda^{(1)} + 2r_k h(\mathbf{X}^{*(1)}) = 0 + 2(1)(-3.01587) = -6.03175$$

Substituting this value for λ along with $r_k = 1$ in Eqs. (E$_6$) and (E$_7$), we get

$$x_1^{*(2)} = -0.38171, \quad x_2^{*(2)} = 3.56261$$

$$\text{with} \quad h = -0.38171 + 3.56261 - 5 = -1.81910$$

This procedure can be continued until some specified convergence is satisfied. The results of the first ten iterations are given in Table 7.6.

7.21 CHECKING THE CONVERGENCE OF CONSTRAINED OPTIMIZATION PROBLEMS

In all the constrained optimization techniques described in this chapter, identification of the optimum solution is very important from the points of view of stopping the iterative process and using the solution with confidence. In addition to the convergence criteria discussed earlier, the following two methods can also be used to test the point for optimality.

7.21.1   Perturbing the Design Vector

Since the optimum point

$$\mathbf{X}^* = \begin{Bmatrix} x_1^* \\ x_2^* \\ \vdots \\ x_n^* \end{Bmatrix}$$

corresponds to the minimum function value subject to the satisfaction of the constraints $g_j(\mathbf{X}^*) \leq 0$, $j = 1, 2, \ldots, m$ (the equality constraints can also be included, if necessary), we perturb $\mathbf{X}^*$ by changing each of the design variables, one at a time, by a small amount, and evaluate the values of f and g_j, $j = 1, 2, \ldots, m$. Thus if

$$\mathbf{X}_i^+ = \mathbf{X}^* + \Delta\mathbf{X}_i$$
$$\mathbf{X}_i^- = \mathbf{X}^* - \Delta\mathbf{X}_i$$

where

$$\Delta\mathbf{X}_i = \begin{Bmatrix} 0 \\ \vdots \\ 0 \\ \Delta x_i \\ 0 \\ \vdots \\ 0 \end{Bmatrix} \leftarrow i\text{th row}$$

Δx_i is a small perturbation in x_i that can be taken as 0.1 to 2.0 % of x_i^*. Evaluate

$$f(\mathbf{X}_i^+); \quad f(\mathbf{X}_i^-); \quad g_j(\mathbf{X}_i^+)$$
$$g_j(\mathbf{X}_i^-), \quad j = 1, 2, \ldots, m \quad \text{for} \quad i = 1, 2, \ldots, n$$

If

$$f(\mathbf{X}_i^+) \geq f(\mathbf{X}^*); \quad g_j(\mathbf{X}_i^+) \leq 0, \quad j = 1, 2, \ldots, m$$
$$f(\mathbf{X}_i^-) \geq f(\mathbf{X}^*); \quad g_j(\mathbf{X}_i^-) \leq 0, \quad j = 1, 2, \ldots, m$$

for $i = 1, 2, \ldots, n$, $\mathbf{X}^*$ can be taken as the constrained optimum point of the original problem.

7.21.2   Testing the Kuhn–Tucker Conditions

Since the Kuhn–Tucker conditions, Eqs. (2.73) and (2.74), are necessarily to be satisfied[†] by the optimum point of any nonlinear programming problem, we can at least

---

[†]These may not be sufficient to guarantee a global minimum point for nonconvex programming problems.

test for the satisfaction of these conditions before taking a point $\mathbf{X}$ as optimum. Equations (2.73) can be written as

$$\sum_{j \in j_1} \lambda_j \frac{\partial g_j}{\partial x_i} = -\frac{\partial f}{\partial x_i}, \quad i = 1, 2, \ldots, n \tag{7.261}$$

where J_1 indicates the set of active constraints at the point $\mathbf{X}$. If $g_{j1}(\mathbf{X}) = g_{j2}(\mathbf{X}) = \cdots = g_{jp}(\mathbf{X}) = 0$, Eqs. (7.261) can be expressed as

$$\underset{n \times p \; p \times 1}{\mathbf{G} \; \lambda} = \underset{n \times 1}{\mathbf{F}} \tag{7.262}$$

where

$$\mathbf{G} = \begin{bmatrix} \dfrac{\partial g_{j1}}{\partial x_1} & \dfrac{\partial g_{j2}}{\partial x_1} & \cdots & \dfrac{\partial g_{jp}}{\partial x_1} \\[2mm] \dfrac{\partial g_{j1}}{\partial x_2} & \dfrac{\partial g_{j2}}{\partial x_2} & \cdots & \dfrac{\partial g_{jp}}{\partial x_2} \\[2mm] \vdots & & & \\[2mm] \dfrac{\partial g_{j1}}{\partial x_n} & \dfrac{\partial g_{j2}}{\partial x_n} & \cdots & \dfrac{\partial g_{jp}}{\partial x_n} \end{bmatrix}_{\mathbf{X}}$$

$$\lambda = \begin{Bmatrix} \lambda_{j1} \\ \lambda_{j2} \\ \vdots \\ \lambda_{jp} \end{Bmatrix} \quad \text{and} \quad \mathbf{F} = \begin{Bmatrix} -\dfrac{\partial f}{\partial x_1} \\[2mm] -\dfrac{\partial f}{\partial x_2} \\[2mm] \vdots \\[2mm] -\dfrac{\partial f}{\partial x_n} \end{Bmatrix}_{\mathbf{X}}$$

From Eqs. (7.262) we can obtain an expression for λ as

$$\lambda = (\mathbf{G}^T \mathbf{G})^{-1} \mathbf{G}^T \mathbf{F} \tag{7.263}$$

If all the components of λ, given by Eq. (7.263) are positive, the Kuhn–Tucker conditions will be satisfied. A major difficulty in applying Eq. (7.263) arises from the fact that it is very difficult to ascertain which constraints are active at the point $\mathbf{X}$. Since no constraint will have exactly the value of 0.0 at the point $\mathbf{X}$ while working on the computer, we have to take a constraint g_j to be active whenever it satisifes the relation

$$|g_j(\mathbf{X})| \leq \varepsilon \tag{7.264}$$

where ε is a small number on the order of 10^{-2} to 10^{-6}. Notice that Eq. (7.264) assumes that the constraints were originally normalized.

7.22   TEST PROBLEMS

As discussed in previous sections, a number of algorithms are available for solving a constrained nonlinear programming problem. In recent years, a variety of computer programs have been developed to solve engineering optimization problems. Many of these are complex and versatile and the user needs a good understanding of the algorithms/computer programs to be able to use them effectively. Before solving a new engineering design optimization problem, we usually test the behavior and convergence of the algorithm/computer program on simple test problems. Five test problems are given in this section. All these problems have appeared in the optimization literature and most of them have been solved using different techniques.

7.22.1   Design of a Three-Bar Truss

The optimal design of the three-bar truss shown in Fig. 7.21 is considered using two different objectives with the cross-sectional areas of members 1 (and 3) and 2 as design variables [7.38].

Design vector:

$$\mathbf{X} = \begin{Bmatrix} x_1 \\ x_2 \end{Bmatrix} = \begin{Bmatrix} A_1 \\ A_2 \end{Bmatrix}$$

Objective functions:

$$f_1(\mathbf{X}) = \text{weight} = 2\sqrt{2}x_1 + x_2$$

$$f_2(\mathbf{X}) = \text{vertical deflection of loaded joint} = \frac{PH}{E}\frac{1}{x_1 + \sqrt{2}x_2}$$

Constraints:

$$\sigma_1(\mathbf{X}) - \sigma^{(u)} \leq 0$$

$$\sigma_2(\mathbf{X}) - \sigma^{(u)} \leq 0$$

$$\sigma_3(\mathbf{X}) - \sigma^{(l)} \leq 0$$

$$x_i^{(l)} \leq x_i \leq x_i^{(u)}, \quad i = 1, 2$$

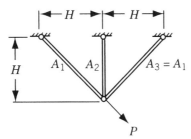

Figure 7.21   Three-bar truss [7.38].

where σ_i is the stress induced in member i, $\sigma^{(u)}$ the maximum permissible stress in tension, $\sigma^{(l)}$ the maximum permissible stress in compression, $x_i^{(l)}$ the lower bound on x_i, and $x_i^{(u)}$ the upper bound on x_i. The stresses are given by

$$\sigma_1(\mathbf{X}) = P\frac{x_2 + \sqrt{2}x_1}{\sqrt{2}x_1^2 + 2x_1x_2}$$

$$\sigma_2(\mathbf{X}) = P\frac{1}{x_1 + \sqrt{2}x_2}$$

$$\sigma_3(\mathbf{X}) = -P\frac{x_2}{\sqrt{2}x_1^2 + 2x_1x_2}$$

Data: $\sigma^{(u)} = 20$, $\sigma^{(l)} = -15$, $x_i^{(l)} = 0.1(i = 1, 2)$, $x_i^{(u)} = 5.0(i = 1, 2)$, $P = 20$, and $E = 1$.

Optimum design:

$$\mathbf{X}_1^* = \begin{Bmatrix} 0.78706 \\ 0.40735 \end{Bmatrix}, \quad f_1^* = \begin{matrix} 2.6335, & \text{stress constraint of} \\ & \text{member 1 is active at } \mathbf{X}_1^* \end{matrix}$$

$$\mathbf{X}_2^* = \begin{Bmatrix} 5.0 \\ 5.0 \end{Bmatrix}, \quad f_2^* = 1.6569$$

7.22.2 Design of a Twenty-Five-Bar Space Truss

The 25-bar space truss shown in Fig. 7.22 is required to support the two load conditions given in Table 7.7 and is to be designed with constraints on member stresses as well as Euler buckling [7.38]. A minimum allowable area is specified for each member. The allowable stresses for all members are specified as σ_{max} in both tension and compression. The Young's modulus and the material density are taken as $E = 10^7$ psi and $\rho = 0.1$ lb/in^3. The members are assumed to be tubular with a nominal diameter/thickness ratio of 100, so that the buckling stress in member i becomes

$$p_i = -\frac{100.01\pi E A_i}{8l_i^2}, \quad i = 1, 2, \ldots, 25$$

where A_i and l_i denote the cross-sectional area and length, respectively, of member i. The member areas are linked as follows:

$$A_1, \quad A_2 = A_3 = A_4 = A_5, \quad A_6 = A_7 = A_8 = A_9,$$

$$A_{10} = A_{11}, \quad A_{12} = A_{13}, \quad A_{14} = A_{15} = A_{16} = A_{17,}$$

$$A_{18} = A_{19} = A_{20} = A_{21}, \quad A_{22} = A_{23} = A_{24} = A_{25}$$

Thus there are eight independent area design variables in the problem. Three problems are solved using different objective functions:

$$f_1(\mathbf{X}) = \sum_{i=1}^{25} \rho A_i l_i = \text{weight}$$

$$f_2(\mathbf{X}) = (\delta_{1x}^2 + \delta_{1y}^2 + \delta_{1z}^2)^{1/2} + (\delta_{2x}^2 + \delta_{2y}^2 + \delta_{2z}^2)^{1/2}$$

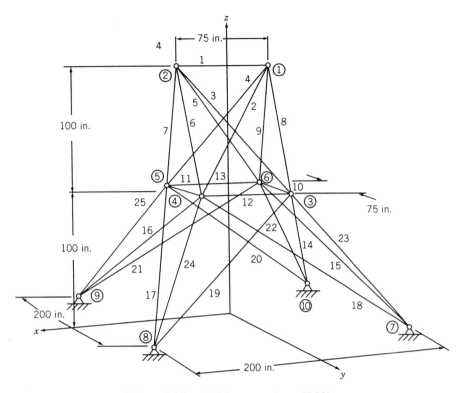

Figure 7.22   A 25-bar space truss [7.38].

$\quad$ = sum of deflections of nodes 1 and 2

$f_3(\mathbf{X}) = -\omega_1$ = negative of fundamental natural frequency of vibration

where δ_{ix} = deflection of node i along x direction.

Table 7.7   Loads Acting on the 25-Bar Truss

	Joint			
	1	2	3	6
	Load condition 1, loads in pounds			
F_x	0	0	0	0
F_y	20,000	−20,000	0	0
F_z	−5,000	−5,000	0	0
	Load condition 2, loads in pounds			
F_x	1,000	0	500	500
F_y	10,000	10,000	0	0
F_z	−5,000	−5,000	0	0

Constraints:

$$|\sigma_{ij}(\mathbf{X})| \le \sigma_{\max}, \qquad i = 1, 2, \ldots, 25, \qquad j = 1, 2$$

$$\sigma_{ij}(\mathbf{X}) \le p_i(\mathbf{X}), \qquad i = 1, 2, \ldots, 25, \qquad j = 1, 2$$

$$x_i^{(l)} \le x_i \le x_i^{(u)}, \qquad i = 1, 2, \ldots, 8$$

where σ_{ij} is the stress induced in member i under load condition j, $x_i^{(l)}$ the lower bound on x_i, and $x_i^{(u)}$ the upper bound on x_i.

Data: $\sigma_{\max} = 40{,}000\,\text{psi}$, $x_i^{(l)} = 0.1\,\text{in}^2$, $x_i^{(u)} = 5.0\,\text{in}^2$ for $i = 1, 2, \ldots, 25$.

Optimum solution: See Table 7.8.

7.22.3 Welded Beam Design

The welded beam shown in Fig. 7.23 is designed for minimum cost subject to constraints on shear stress in weld (τ), bending stress in the beam (σ), buckling load on the bar (P_c), end deflection of the beam (δ), and side constraints [7.39].

Design vector:

$$\begin{Bmatrix} x_1 \\ x_2 \\ x_3 \\ x_4 \end{Bmatrix} = \begin{Bmatrix} h \\ l \\ t \\ b \end{Bmatrix}$$

Table 7.8 Optimization Results of the 25-Bar Truss [7.38]

Quantity	Optimization problem		
	Minimization of weight	Minimization of deflection	Maximization of frequency
Design vector, $\mathbf{X}$	0.1[a]	3.7931	0.1[a]
	0.80228	5.0[a]	0.79769
	0.74789	5.0[a]	0.74605
	0.1[a]	3.3183	0.72817
	0.12452	5.0[a]	0.84836
	0.57117	5.0[a]	1.9944
	0.97851	5.0[a]	1.9176
	0.80247	5.0[a]	4.1119
Weight (lb)	233.07265	1619.3258	600.87891
Deflection (in.)	1.924989	0.30834	1.35503
Fundamental frequency (Hz)	73.25348	70.2082	108.6224
Number of active behavior constraints	9[b]	0	4[c]

[a] Active side constraint.

[b] Buckling stress in members, 2, 5, 7, 8, 19, and 20 in load condition 1 and in members 13, 16, and 24 in load condition 2.

[c] Buckling stress in members 2, 5, 7, and 8 in load condition 1.

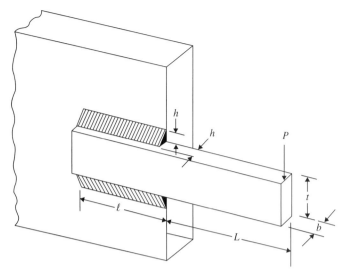

Figure 7.23   Welded beam [7.39].

Objective function: $f(\mathbf{X}) = 1.10471x_1^2 x_2 + 0.04811x_3 x_4 (14.0 + x_2)$
Constraints:

$$g_1(\mathbf{X}) = \tau(\mathbf{X}) - \tau_{max} \leq 0$$

$$g_2(\mathbf{X}) = \sigma(\mathbf{X}) - \sigma_{max} \leq 0$$

$$g_3(\mathbf{X}) = x_1 - x_4 \leq 0$$

$$g_4(\mathbf{X}) = 0.10471x_1^2 + 0.04811x_3 x_4 (14.0 + x_2) - 5.0 \leq 0$$

$$g_5(\mathbf{X}) = 0.125 - x_1 \leq 0$$

$$g_6(\mathbf{X}) = \delta(\mathbf{X}) - \delta_{max} \leq 0$$

$$g_7(\mathbf{X}) = P - P_c(\mathbf{X}) \leq 0$$

$$g_8(\mathbf{X}) \text{ to } g_{11}(\mathbf{X}): \ 0.1 \leq x_i \leq 2.0, \qquad i = 1, 4$$

$$g_{12}(\mathbf{X}) \text{ to } g_{15}(\mathbf{X}): \ 0.1 \leq x_i \leq 10.0, \quad i = 2, 3$$

where

$$\tau(\mathbf{X}) = \sqrt{(\tau')^2 + 2\tau'\tau''\frac{x_2}{2R} + (\tau'')^2}$$

$$\tau' = \frac{P}{\sqrt{2}x_1 x_2}, \quad \tau'' = \frac{MR}{J}, \quad M = P\left(L + \frac{x_2}{2}\right)$$

$$R = \sqrt{\frac{x_2^2}{4} + \left(\frac{x_1 + x_3}{2}\right)^2}$$

$$J = 2 \left\{ \frac{x_1 x_2}{\sqrt{2}} \left[\frac{x_2^2}{12} + \left(\frac{x_1 + x_3}{2} \right)^2 \right] \right\}$$

$$\sigma(\mathbf{X}) = \frac{6PL}{x_4 x_3^2}$$

$$\delta(\mathbf{X}) = \frac{4PL^3}{Ex_3^3 x_4}$$

$$P_c(\mathbf{X}) = \frac{4.013\sqrt{EG(x_3^2 x_4^6/36)}}{L^2} \left(1 - \frac{x_3}{2L}\sqrt{\frac{E}{4G}} \right)$$

Data: $P = 6000$ lb, $L = 14$ in., $E = 30 \times 10^6$ psi, $G = 12 \times 10^6$ psi, $\tau_{\text{max}} = 13{,}600$ psi, $\sigma_{\text{max}} = 30{,}000$ psi, and $\delta_{\text{max}} = 0.25$ in.

Starting and optimum solutions:

$$\mathbf{X}^{\text{start}} = \begin{Bmatrix} h \\ l \\ t \\ b \end{Bmatrix} = \begin{Bmatrix} 0.4 \\ 6.0 \\ 9.0 \\ 0.5 \end{Bmatrix} \text{ in.}, \quad f^{\text{start}} = \$5.3904, \quad \mathbf{X}^* = \begin{Bmatrix} h \\ l \\ t \\ b \end{Bmatrix}^* = \begin{Bmatrix} 0.2444 \\ 6.2177 \\ 8.2915 \\ 0.2444 \end{Bmatrix} \text{ in.},$$

$$f^* = \$2.3810$$

7.22.4   Speed Reducer (Gear Train) Design

The design of the speed reducer, shown in Fig. 7.24, is considered with the face width (b), module of teeth (m), number of teeth on pinion (z), length of shaft 1 between bearings (l_1), length of shaft 2 between bearings (l_2), diameter of shaft 1 (d_1), and diameter of shaft 2 (d_2) as design variables $x_1, x_2, \ldots, x_7$, respectively. The constraints include limitations on the bending stress of gear teeth, surface stress, transverse deflections of shafts 1 and 2 due to transmitted force, and stresses in shafts 1 and 2 [7.40, 7.41].

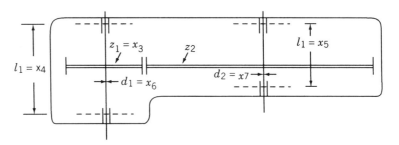

Figure 7.24   Speed reducer (gear pair) [7.40].

Objective (minimization of weight of speed reducer)*:*

$$f(\mathbf{X}) = 0.7854x_1x_2^2(3.3333x_3^2 + 14.9334x_3 - 43.0934) - 1.508x_1(x_6^2 + x_7^2)$$
$$+ 7.477(x_6^3 + x_7^3) + 0.7854(x_4x_6^2 + x_5x_7^2)$$

Constraints:

$$g_1(x) = 27x_1^{-1}x_2^{-2}x_3^{-1} \leq 1$$

$$g_2(x) = 397.5x_1^{-1}x_2^{-2}x_3^{-2} \leq 1$$

$$g_3(x) = 1.93x_2^{-1}x_3^{-1}x_4^3x_6^{-4} \leq 1$$

$$g_4(x) = 1.93x_2^{-1}x_3^{-1}x_5^3x_7^{-4} \leq 1$$

$$g_5(x) = \left[\left(\frac{745x_4}{x_2x_3}\right)^2 + (16.9)10^6\right]^{0.5} \Big/ 0.1x_6^3 \leq 1100$$

$$g_6(x) = \left[\left(\frac{745x_5}{x_2x_3}\right)^2 + (157.5)10^6\right]^{0.5} \Big/ 0.1x_7^3 \leq 850$$

$$g_7(x) = x_2x_3 \leq 40$$

$$g_8(x) : 5 \leq \frac{x_1}{x_2} \leq 12 : g_9(x)$$

$$g_{10}(x) : 2.6 \leq x_1 \leq 3.6 : g_{11}(x)$$

$$g_{12}(x) : 0.7 \leq x_2 \leq 0.8 : g_{13}(x)$$

$$g_{14}(x) : 17 \leq x_3 \leq 28 : g_{15}(x)$$

$$g_{16}(x) : 7.3 \leq x_4 \leq 8.3 : g_{17}(x)$$

$$g_{18}(x) : 7.3 \leq x_5 \leq 8.3 : g_{19}(x)$$

$$g_{20}(x) : 2.9 \leq x_6 \leq 3.9 : g_{21}(x)$$

$$g_{22}(x) : 5.0 \leq x_7 \leq 5.5 : g_{23}(x)$$

$$g_{24}(x) = (1.5x_6 + 1.9)x_4^{-1} \leq 1$$

$$g_{25}(x) = (1.1x_7 + 1.9)x_5^{-1} \leq 1$$

Optimum solution:

$$\mathbf{X}^* = \{3.5 \;\; 0.7 \;\; 17.0 \;\; 7.3 \;\; 7.3 \;\; 3.35 \;\; 5.29\}^{\mathrm{T}}, \qquad f^* = 2985.22$$

7.22.5   Heat Exchanger Design [7.42]

Objective function: Minimize $f(\mathbf{X}) = x_1 + x_2 + x_3$

Constraints:

$$g_1(\mathbf{X}) = 0.0025(x_4 + x_6) - 1 \leq 0$$

$$g_2(\mathbf{X}) = 0.0025(-x_4 + x_5 + x_7) - 1 \leq 0$$

$$g_3(\mathbf{X}) = 0.01(-x_5 + x_8) - 1 \leq 0$$

$$g_4(\mathbf{X}) = 100x_1 - x_1x_6 + 833.33252x_4 - 83{,}333.333 \leq 0$$

$$g_5(\mathbf{X}) = x_2x_4 - x_2x_7 - 1250x_4 + 1250x_5 \leq 0$$

$$g_6(\mathbf{X}) = x_3x_5 - x_3x_8 - 2500x_5 + 1{,}250{,}000 \leq 0$$

$$g_7 : 100 \leq x_1 \leq 10{,}000 : g_8$$

$$g_9 : 1000 \leq x_2 \leq 10{,}000 : g_{10}$$

$$g_{11} : 1000 \leq x_3 \leq 10{,}000 : g_{12}$$

$$g_{13} \text{ to } g_{22} : 10 \leq x_i \leq 1000, \quad i = 4, 5, \ldots, 8$$

Optimum   solution: $\mathbf{X}^* = \{567 \quad 1357 \quad 5125 \quad 181 \quad 295 \quad 219 \quad 286 \quad 395\}^T$, $f^* = 7049$

7.23   MATLAB SOLUTION OF CONSTRAINED OPTIMIZATION PROBLEMS

The solution of multivariable minimization problems, with inequality and equality constraints, using the MATLAB function `fmincon` is illustrated in this section.

Example 7.13   Find the solution of Example 7.8 starting from the initial point $\mathbf{X}_1 = \{0.1 \quad 0.1 \quad 3.0\}^T$

SOLUTION

Step 1:   Write an M-file `objfun.m` for the objective function.

```
function f= objfun (x)
f= x(1)^3-6*x(1)^2+11*x(1)+x(3);
```

Step 2:   Write an M-file `constraints.m` for the constraints.

```
function [c, ceq] = constraints (x)
% Nonlinear inequality constraints
c = [x(1)^2+x(2)^2-x(3)^2;4-x(1)^2-x(2)^2-x(3)^2;x(3)-5;
   -x(1);-x(2);-x(3)];
% Nonlinear equality constraints
ceq = [];
```

Step 3: Invoke constrained optimization program (write this in new MATLAB file).

```
clc
clear all
warning off
x0 = [.1,.1, 3.0]; % Starting guess
fprintf ('The values of function value and constraints at
  starting pointn');
f=objfun (x0)
[c, ceq] = constraints (x0)
options  = optimset ('LargeScale', 'off');
[x, fval]=fmincon (@objfun, x0, [], [], [], [], [], [],
  @constraints, options)
fprintf ('The values of constraints at optimum solutionn');
[c, ceq] = constraints (x)
  % Check the constraint values at x
```

This Produces the Solution or Ouput as follows:

```
The values of function value and constraints at
  starting point
f=
 4.0410
c=
 -8.9800
 -5.0200
 -2.0000
 -0.1000
 -0.1000
 -3.0000
ceq =
  []
Optimization terminated: first-order optimality measure
  less
 than options. TolFun and maximum constraint violation is
  less
 than options.TolCon.
Active inequalities (to within options.TolCon = 1e-006):
 lower upper ineqlin  ineqnonlin
               1
               2
               4
x=
    0 1.4142 1.4142
fval =
 1.4142
The values of constraints at optimum solution
c=
```

```
    -0.0000
    -0.0000
    -3.5858
        0
    -1.4142
    -1.4142
ceq =
[]
```

REFERENCES AND BIBLIOGRAPHY

7.1   R. L. Fox, *Optimization Methods for Engineering Design*, Addison-Wesley, Reading, MA, 1971.

7.2   M. J. Box, A new method of constrained optimization and a comparison with other methods, *Computer Journal*, Vol. 8, No. 1, pp. 42–52, 1965.

7.3   E. W. Cheney and A. A. Goldstein, Newton's method of convex programming and Tchebycheff approximation, *Numerische Mathematik*, Vol. 1, pp. 253–268, 1959.

7.4   J. E. Kelly, The cutting plane method for solving convex programs, *Journal of SIAM*, Vol. VIII, No. 4, pp. 703–712, 1960.

7.5   G. Zoutendijk, *Methods of Feasible Directions*, Elsevier, Amsterdam, 1960.

7.6   W.W. Garvin, *Introduction to Linear Programming*, McGraw-Hill, New York, 1960.

7.7   S. L. S. Jacoby, J. S. Kowalik, and J. T. Pizzo, *Iterative Methods for Nonlinear Optimization Problems*, Prentice Hall, Englewood Cliffs, NJ, 1972.

7.8   G. Zoutendijk, Nonlinear programming: a numerical survey, *SIAM Journal of Control Theory and Applications*, Vol. 4, No. 1, pp. 194–210, 1966.

7.9   J. B. Rosen, The gradient projection method of nonlinear programming, Part I: linear constraints, *SIAM Journal*, Vol. 8, pp. 181–217, 1960.

7.10 J. B. Rosen, The gradient projection method for nonlinear programming, Part II: nonlinear constraints, *SIAM Journal*, Vol. 9, pp. 414–432, 1961.

7.11 G. A. Gabriele and K. M. Ragsdell, The generalized reduced gradient method: a reliable tool for optimal design, *ASME Journal of Engineering for Industry*, Vol. 99, pp. 384–400, 1977.

7.12 M. J. D. Powell, A fast algorithm for nonlinearity constrained optimization calculations, in *Lecture Notes in Mathematics*, G. A. Watson et al., Eds., Springer-Verlag, Berlin, 1978.

7.13 M. J. Box, A comparison of several current optimization methods and the use of transformations in constrained problems, *Computer Journal*, Vol. 9, pp. 67–77, 1966.

7.14 C. W. Carroll, The created response surface technique for optimizing nonlinear restrained systems, *Operations Research*, Vol. 9, pp. 169–184, 1961.

7.15 A. V. Fiacco and G. P. McCormick, *Nonlinear Programming: Sequential Unconstrained Minimization Techniques*, Wiley, New York, 1968.

7.16 W. I. Zangwill, Nonlinear programming via penalty functions, *Management Science*, Vol. 13, No. 5, pp. 344–358, 1967.

7.17 A. V. Fiacco and G. P. McCormick, Extensions of SUMT for nonlinear programming: equality constraints and extrapolation, *Management Science*, Vol. 12, No. 11, pp. 816–828, July 1966.

7.18   D. Kavlie and J. Moe, Automated design of frame structure, *ASCE Journal of the Structural Division*, Vol. 97, No. ST1, pp. 33–62, Jan. 1971.

7.19   J. H. Cassis and L. A. Schmit, On implementation of the extended interior penalty function, *International Journal for Numerical Methods in Engineering*, Vol. 10, pp. 3–23, 1976.

7.20   R. T. Haftka and J. H. Starnes, Jr., Application of a quadratic extended interior penalty function for structural optimization, *AIAA Journal*, Vol. 14, pp. 718–728, 1976.

7.21   A. V. Fiacco and G. P. McCormick, *SUMT Without Parmaeters*, System Research Memorandum 121, Technical Institute, Northwestern University, Evanston, IL, 1965.

7.22   A. Ralston, *A First Course in Numerical Analysis*, McGraw-Hill, New York, 1965.

7.23   R. T. Rockafellar, The multiplier method of Hestenes and Powell applied to convex programming, *Journal of Optimization Theory and Applications*, Vol. 12, No. 6, pp. 555–562, 1973.

7.24   B. Prasad, A class of generalized variable penalty methods for nonlinear programming, *Journal of Optimization Theory and Applications*, Vol. 35, pp. 159–182, 1981.

7.25   L. A. Schmit and R. H. Mallett, Structural synthesis and design parameter hierarchy, *Journal of the Structural Division, Proceedings of ASCE*, Vol. 89, No. ST4, pp. 269–299, 1963.

7.26   J. Kowalik and M. R. Osborne, *Methods for Unconstrained Optimization Problems*, American Elsevier, New York, 1968.

7.27   N. Baba, Convergence of a random optimization method for constrained optimization problems, *Journal of Optimization Theory and Applications*, Vol. 33, pp. 451–461, 1981.

7.28   J. T. Betts, A gradient projection-multiplier method for nonlinear programming, *Journal of Optimization Theory and Applications*, Vol. 24, pp. 523–548, 1978.

7.29   J. T. Betts, An improved penalty function method for solving constrained parameter optimization problems, *Journal of Optimization Theory and Applications*, Vol. 16, pp. 1–24, 1975.

7.30   W. Hock and K. Schittkowski, Test examples for nonlinear programming codes, *Journal of Optimization Theory and Applications*, Vol. 30, pp. 127–129, 1980.

7.31   J. C. Geromel and L. F. B. Baptistella, Feasible direction method for large-scale nonconvex programs: decomposition approach, *Journal of Optimization Theory and Applications*, Vol. 35, pp. 231–249, 1981.

7.32   D. M. Topkis, A cutting-plane algorithm with linear and geometric rates of convergence, *Journal of Optimization Theory and Applications*, Vol. 36, pp. 1–22, 1982.

7.33   M. Avriel, *Nonlinear Programming: Analysis and Methods*, Prentice Hall, Englewood Cliffs, NJ, 1976.

7.34   H. W. Kuhn, Nonlinear programming: a historical view, in *Nonlinear Programming, SIAM-AMS Proceedings*, Vol. 9, American Mathematical Society, Providence, RI, 1976.

7.35   J. Elzinga and T. G. Moore, A central cutting plane algorithm for the convex programming problem, *Mathematical Programming*, Vol. 8, pp. 134–145, 1975.

7.36   V. B. Venkayya, V. A. Tischler, and S. M. Pitrof, Benchmarking in structural optimization, *Proceedings of the 4th AIAA/USAF/NASA/OAI Symposium on Multidisciplinary Analysis and Optimization*, Sept. 21–23, 1992, Cleveland, Ohio, AIAA Paper AIAA-92-4794.

7.37   W. Hock and K. Schittkowski, *Test Examples for Nonlinear Programming Codes*, Lecture Notes in Economics and Mathematical Systems, No. 187, Springer-Verlag, Berlin, 1981.

7.38   S. S. Rao, Multiobjective optimization of fuzzy structural systems, *International Journal for Numerical Methods in Engineering*, Vol. 24, pp. 1157–1171, 1987.

7.39   K. M. Ragsdell and D. T. Phillips, Optimal design of a class of welded structures using geometric programming, *ASME Journal of Engineering for Industry*, Vol. 98, pp. 1021–1025, 1976.

7.40   J. Golinski, An adaptive optimization system applied to machine synthesis, *Mechanism and Machine Synthesis*, Vol. 8, pp. 419–436, 1973.

7.41   H. L. Li and P. Papalambros, A production system for use of global optimization knowledge, *ASME Journal of Mechanisms, Transmissions, and Automation in Design*, Vol. 107, pp. 277–284, 1985.

7.42   M. Avriel and A. C. Williams, An extension of geometric programming with application in engineering optimization, *Journal of Engineering Mathematics*, Vol. 5, pp. 187–194, 1971.

7.43   G. A. Gabriele and K. M. Ragsdell, Large scale nonlinear programming using the generalized reduced gradient method, *ASME Journal of Mechanical Design*, Vol. 102, No. 3, pp. 566–573, 1980.

7.44   A. D. Belegundu and J. S. Arora, A recursive quadratic programming algorithm with active set strategy for optimal design, *International Journal for Numerical Methods in Engineering*, Vol. 20, No. 5, pp. 803–816, 1984.

7.45   G. A. Gabriele and T. J. Beltracchi, An investigation of Pschenichnyi's recursive quadratic programming method for engineering optimization, *ASME Journal of Mechanisms, Transmissions, and Automation in Design*, Vol. 109, pp. 248–253, 1987.

7.46   F. Moses, Optimum structural design using linear programming, *ASCE Journal of the Structural Division*, Vol. 90, No. ST6, pp. 89–104, 1964.

7.47   S. L. Lipson and L. B. Gwin, The complex method applied to optimal truss configuration, *Computers and Structures*, Vol. 7, pp. 461–468, 1977.

7.48   G. N. Vanderplaats, *Numerical Optimization Techniques for Engineering Design with Applications*, McGraw-Hill, New York, 1984.

7.49   T. F. Edgar and D. M. Himmelblau, *Optimization of Chemical Processes*, McGraw-Hill, New York, 1988.

7.50   A. Ravindran, K. M. Ragsdell, and G. V. Reklaitis, *Engineering Optimization Methods and Applications*, 2nd ed., Wiley, New York, 2006.

7.51   L. S. Lasdon, *Optimization Theory for Large Systems*, Macmillan, New York, 1970.

7.52   R. T. Haftka and Z. Gürdal, *Elements of Structural Optimization*, 3rd ed., Kluwer Academic, Dordrecht, The Netherlands, 1992.

REVIEW QUESTIONS

7.1   Answer true or false:

(a) The complex method is similar to the simplex method.

(b) The optimum solution of a constrained problem can be the same as the unconstrained optimum.

(c) The constraints can introduce local minima in the feasible space.

(d) The complex method can handle both equality and inequality constraints.

(e) The complex method can be used to solve both convex and nonconvex problems.

(f) The number of inequality constraints cannot exceed the number of design variables.

(g) The complex method requires a feasible starting point.

(**h**) The solutions of all LP problems in the SLP method lie in the infeasible domain of the original problem.

(**i**) The SLP method is applicable to both convex and nonconvex problems.

(**j**) The usable feasible directions can be generated using random numbers.

(**k**) The usable feasible direction makes an obtuse angle with the gradients of all the constraints.

(**l**) If the starting point is feasible, all subsequent unconstrained minima will be feasible in the exterior penalty function method.

(**m**) The interior penalty function method can be used to find a feasible starting point.

(**n**) The penalty parameter r_k approaches zero as k approaches infinity in the exterior penalty function method.

(**o**) The design vector found through extrapolation can be used as a starting point for the next unconstrained minimization in the interior penalty function method.

7.2 Why is the SLP method called the cutting plane method?

7.3 How is the direction-finding problem solved in Zoutendijk's method?

7.4 What is SUMT?

7.5 How is a parametric constraint handled in the interior penalty function method?

7.6 How can you identify an active constraint during numerical optimization?

7.7 Formulate the equivalent unconstrained objective function that can be used in random search methods.

7.8 How is the perturbation method used as a convergence check?

7.9 How can you compute Lagrange multipliers during numerical optimization?

7.10 What is the use of extrapolating the objective function in the penalty function approach?

7.11 Why is handling of equality constraints difficult in the penalty function methods?

7.12 What is the geometric interpretation of the reduced gradient?

7.13 Is the generalized reduced gradient zero at the optimum solution?

7.14 What is the relation between the sequential quadratic programming method and the Lagrangian function?

7.15 Approximate the nonlinear function $f(\mathbf{X})$ as a linear function at $\mathbf{X}_0$.

7.16 What is the limitation of the linear extended penalty function?

7.17 What is the difference between the interior and extended interior penalty function methods?

7.18 What is the basic principle used in the augmented Lagrangian method?

7.19 When can you use the steepest descent direction as a usable feasible direction in Zoutendijk's method?

7.20 Construct the augmented Lagrangian function for a constrained optimization problem.

7.21 Construct the ϕ_k function to be used for a mixed equality–inequality constrained problem in the interior penalty function approach.

7.22 What is a parametric constraint?

7.23 Match the following methods:

(a)	Zoutendijk method	Heuristic method
(b)	Cutting plane method	Barrier method
(c)	Complex method	Feasible directions method
(d)	Projected Lagrangian method	Sequential linear programming method
(e)	Penalty function method	Gradient projection method
(f)	Rosen's method	Sequential unconstrained minimization method
(g)	Interior penalty function method	Sequential quadratic programming method

7.24 Answer true or false:

(a) The Rosen's gradient projection method is a method of feasible directions.

(b) The starting vector can be infeasible in Rosen's gradient projection method.

(c) The transformation methods seek to convert a constrained problem into an unconstrained one.

(d) The ϕ_k function is defined over the entire design space in the interior penalty function method.

(e) The sequence of unconstrained minima generated by the interior penalty function method lies in the feasible space.

(f) The sequence of unconstrained minima generated by the exterior penalty function method lies in the feasible space.

(g) The random search methods are applicable to convex and nonconvex optimization problems.

(h) The GRG method is related to the method of elimination of variables.

(i) The sequential quadratic programming method can handle only equality constraints.

(j) The augmented Lagrangian method is based on the concepts of penalty function and Lagrange multiplier methods.

(k) The starting vector can be infeasible in the augmented Lagrangiam method.

PROBLEMS

7.1 Find the solution of the problem:

$$\text{Minimize } f(\mathbf{X}) = x_1^2 + 2x_2^2 - 2x_1x_2 - 14x_1 - 14x_2 + 10$$

subject to

$$4x_1^2 + x_2^2 - 25 \leq 0$$

using a graphical procedure.

7.2 Generate four feasible design vectors to the welded beam design problem (Section 7.22.3) using random numbers.

7.3 Generate four feasible design vectors to the three-bar truss design problem (Section 7.22.1) using random numbers.

7.4 Consider the tubular column described in Example 1.1. Starting from the design vector ($d = 8.0$ cm, $t = 0.4$ cm), complete two steps of reflection, expansion, and/or contraction of the complex method.

7.5 Consider the problem:

$$\text{Minimize } f(\mathbf{X}) = x_1 - x_2$$

subject to

$$3x_1^2 - 2x_1x_2 + x_2^2 - 1 \leq 0$$

(a) Generate the approximating LP problem at the vector, $\mathbf{X}_1 = \left\{ \begin{smallmatrix} -2 \\ 2 \end{smallmatrix} \right\}$.

(b) Solve the approximating LP problem using graphical method and find whether the resulting solution is feasible to the original problem.

7.6 Approximate the following optimization problem as (a) a quadratic programming problem, and (b) a linear programming problem at $\mathbf{X} = \left\{ \begin{smallmatrix} 1 \\ -2 \end{smallmatrix} \right\}$.

$$\text{Minimize } f(\mathbf{X}) = 2x_1^3 + 15x_2^2 - 8x_1x_2 + 15$$

subject to

$$x_1^2 + x_1x_2 + 1 = 0$$
$$4x_1 - x_2^2 \leq 4$$

7.7 The problem of minimum volume design subject to stress constraints of the three-bar truss shown in Fig. 7.21 can be stated as follows:

$$\text{Minimize } f(\mathbf{X}) = 282.8x_1 + 100.0x_2$$

subject to

$$\sigma_1 - \sigma_0 = \frac{20(x_2 + \sqrt{2}x_1)}{2x_1x_2 + \sqrt{2}x_1^2} - 20 \leq 0$$

$$-\sigma_3 - \sigma_0 = \frac{20x_2}{2x_1x_2 + \sqrt{2}x_1^2} - 20 \leq 0$$

$$0 \leq x_i \leq 0.3, \quad i = 1, 2$$

where σ_i is the stress induced in member i, $\sigma_0 = 20$ the permissible stress, x_1 the area of cross section of members 1 and 3, and x_2 the area of cross section of member 2. Approximate the problem as a LP problem at ($x_1 = 1, x_2 = 1$).

7.8 $$\text{Minimize } f(\mathbf{X}) = x_1^2 + x_2^2 - 6x_1 - 8x_2 + 10$$

subject to

$$4x_1^2 + x_2^2 \leq 16$$
$$3x_1 + 5x_2 \leq 15$$
$$x_i \geq 0, \quad i = 1, 2$$

with the starting point $\mathbf{X}_1 = \left\{ \begin{smallmatrix} 1 \\ 1 \end{smallmatrix} \right\}$. Using the cutting plane method, complete one step of the process.

7.9   Minimize $f(\mathbf{X}) = 9x_1^2 + 6x_2^2 + x_3^2 - 18x_1 - 12x_2 - 6x_3 - 8$
subject to

$$x_1 + 2x_2 + x_3 \leq 4$$

$$x_i \geq 0, \quad i = 1, 2, 3$$

Using the starting point $\mathbf{X}_1 = \{0, 0, 0\}^T$, complete one step of sequential linear programming method.

7.10   Complete one cycle of the sequential linear programming method for the truss of Section 7.22.1 using the starting point, $\mathbf{X}_1 = \{ {1 \atop 1} \}$.

7.11   A flywheel is a large mass that can store energy during coasting of an engine and feed it back to the drive when required. A solid disk-type flywheel is to be designed for an engine to store maximum possible energy with the following specifications: maximum permissible weight = 150 lb, maximum permissible diameter $(d) = 25$ in., maximum rotational speed = 3000 rpm, maximum allowable stress $(\sigma_{max}) = 20{,}000$ psi, unit weight $(\gamma) = 0.283$ lb/in³, and Poisson's ratio $(\nu) = 0.3$. The energy stored in the flywheel is given by $\frac{1}{2}I\omega^2$, where I is the mass moment of inertia and ω is the angular velocity, and the maximum tangential and radial stresses developed in the flywheel are given by

$$\sigma_t = \sigma_r = \frac{\gamma(3+\nu)\omega^2 d^2}{8g}$$

where g is the acceleration due to gravity and d the diameter of the flywheel. The distortion energy theory of failure is to be used, which leads to the stress constraint

$$\sigma_t^2 + \tau_r^2 - \sigma_t\sigma_r \leq \sigma_{max}^2$$

Considering the diameter (d) and the width (w) as design variables, formulate the optimization problem. Starting from $(d = 15$ in., $w = 2$ in.$)$, complete one iteration of the SLP method.

7.12   Derive the necessary conditions of optimality and find the solution for the following problem:
$$\text{Minimize } f(\mathbf{X}) = 5x_1x_2$$

subject to
$$25 - x_1^2 - x_2^2 \geq 0$$

7.13   Consider the following problem:
$$\text{Minimize } f = (x_1 - 5)^2 + (x_2 - 5)^2$$

subject to
$$x_1 + 2x_2 \leq 15$$

$$1 \leq x_i \leq 10, \quad i = 1, 2$$

Derive the conditions to be satisfied at the point $\mathbf{X} = \{ {1 \atop 7} \}$ by the search direction $\mathbf{S} = \{ {s_1 \atop s_2} \}$ if it is to be a usable feasible direction.

7.14 Consider the problem:

$$\text{Minimize } f = (x_1-1)^2 + (x_2-5)^2$$

subject to

$$g_1 = -x_1^2 + x_2 - 4 \le 0$$

$$g_2 = -(x_1 - 2)^2 + x_2 - 3 \le 0$$

Formulate the direction-finding problem at $\mathbf{X}_i = \left\{ {-1 \atop 5} \right\}$ as a linear programming problem (in Zoutendijk method).

7.15 Minimize $f(\mathbf{X}) = (x_1 - 1)^2 + (x_2 - 5)^2$
subject to

$$-x_1^2 + x_2 \le 4$$

$$-(x_1 - 2)^2 + x_2 \le 3$$

starting from the point $\mathbf{X}_1 = \left\{ {1 \atop 1} \right\}$ and using Zoutendijk's method. Complete two one-dimensional minimization steps.

7.16 Minimize $f(\mathbf{X}) = (x_1 - 1)^2 + (x_2 - 2)^2 - 4$
subject to

$$x_1 + 2x_2 \le 5$$

$$4x_1 + 3x_2 \le 10$$

$$6x_1 + x_2 \le 7$$

$$x_i \ge 0, \quad i = 1, 2$$

by using Zoutendijk's method from the starting point $\mathbf{X}_1 = \left\{ {1 \atop 1} \right\}$. Perform two one-dimensional minimization steps of the process.

7.17 Complete one iteration of Rosen's gradient projection method for the following problem:

$$\text{Minimize } f = (x_1 - 1)^2 + (x_2 - 2)^2 - 4$$

subject to

$$x_1 + 2x_2 \le 5$$

$$4x_1 + 3x_2 \le 10$$

$$6x_1 + x_2 \le 7$$

$$x_i \ge 0, \quad i = 1, 2$$

Use the starting point, $\mathbf{X}_1 = \left\{ {1 \atop 1} \right\}$.

7.18 Complete one iteration of the GRG method for the problem:

$$\text{Minimize } f = x_1^2 + x_2^2$$

subject to

$$x_1 x_2 - 9 = 0$$

starting from $\mathbf{X}_1 = \left\{ {2.0 \atop 4.5} \right\}$.

7.19 Approximate the following problem as a quadratic programming problem at ($x_1 = 1$, $x_2 = 1$):

$$\text{Minimize } f = x_1^2 + x_2^2 - 6x_1 - 8x_2 + 15$$

subject to

$$4x_1^2 + x_2^2 \leq 16$$

$$3x_1^2 + 5x_2^2 \leq 15$$

$$x_i \geq 0, \quad i = 1, 2$$

7.20 Consider the truss structure shown in Fig. 7.25. The minimum weight design of the truss subject to a constraint on the deflection of node S along with lower bounds on the cross sectional areas of members can be started as follows:

$$\text{Minimize } f = 0.1847x_1 + 0.1306x_2$$

subject to

$$\frac{26.1546}{x_1} + \frac{30.1546}{x_2} \leq 1.0$$

$$x_i \geq 25 \text{ mm}^2, \quad i = 1, 2$$

Complete one iteration of sequential quadratic programming method for this problem.

7.21 Find the dimensions of a rectangular prism type parcel that has the largest volume when each of its sides is limited to 42 in. and its depth plus girth is restricted to a maximum value of 72 in. Solve the problem as an unconstrained minimization problem using suitable transformations.

7.22 Transform the following constrained problem into an equivalent unconstrained problem:

$$\text{Maximize } f(x_1, x_2) = [9 - (x_1 - 3)^2]\frac{x_2^3}{27\sqrt{3}}$$

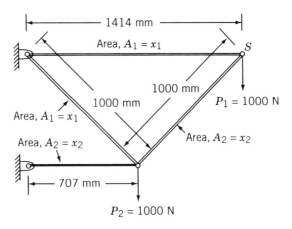

Figure 7.25   Four-bar truss.

subject to

$$0 \leq x_1$$

$$0 \leq x_2 \leq \frac{x_1}{\sqrt{3}}$$

$$0 \leq x_1 + \sqrt{3}x_2 \leq 6$$

7.23 Construct the ϕ_k function, according to **(a)** interior and **(b)** exterior penalty function methods and plot its contours for the following problem:

$$\text{Maximize } f = 2x$$

subject to

$$2 \leq x \leq 10$$

7.24 Construct the ϕ_k function according to the exterior penalty function approach and complete the minimization of ϕ_k for the following problem.

$$\text{Minimize } f(x) = (x - 1)^2$$

subject to

$$g_1(x) = 2 - x \leq 0, \quad g_2(x) = x - 4 \leq 0$$

7.25 Plot the contours of the ϕ_k function using the quadratic extended interior penalty function method for the following problem:

$$\text{Minimize } f(x) = (x - 1)^2$$

subject to

$$g_1(x) = 2 - x \leq 0, \quad g_2(x) = x - 4 \leq 0$$

7.26 Consider the problem:

$$\text{Minimize } f(x) = x^2 - 10x - 1$$

subject to

$$1 \leq x \leq 10$$

Plot the contours of the ϕ_k function using the linear extended interior penalty function method.

7.27 Consider the problem:

$$\text{Minimize } f(x_1, x_2) = (x_1 - 1)^2 + (x_2 - 2)^2$$

subject to

$$2x_1 - x_2 = 0 \quad \text{and} \quad x_1 \leq 5$$

Construct the ϕ_k function according to the interior penalty function approach and complete the minimization of ϕ_1.

7.28 Solve the following problem using an interior penalty function approach coupled with the calculus method of unconstrained minimization:

$$\text{Minimize } f = x^2 - 2x - 1$$

subject to

$$1 - x \geq 0$$

Note: Sequential minimization is not necessary.

7.29 Consider the problem:

$$\text{Minimize } f = x_1^2 + x_2^2 - 6x_1 - 8x_2 + 15$$

subject to

$$4x_1^2 + x_2^2 \geq 16, \quad 3x_1 + 5x_2 \leq 15$$

Normalize the constraints and find a suitable value of r_1 for use in the interior penalty function method at the starting point $(x_1, x_2) = (0, 0)$.

7.30 Determine whether the following optimization problem is convex, concave, or neither type:

$$\text{Minimize } f = -4x_1 + x_1^2 - 2x_1x_2 + 2x_2^2$$

subject to

$$2x_1 + x_2 \leq 6, \quad x_1 - 4x_2 \leq 0, \quad x_i \geq 0, \quad i = 1, 2$$

7.31 Find the solution of the following problem using an exterior penalty function method with classical method of unconstrained minimization:

$$\text{Minimize } f(x_1, x_2) = (2x_1 - x_2)^2 + (x_2 + 1)^2$$

subject to

$$x_1 + x_2 = 10$$

Consider the limiting case as $r_k \to \infty$ analytically.

7.32 Minimize $f = 3x_1^2 + 4x_2^2$ subject to $x_1 + 2x_2 = 8$ using an exterior penalty function method with the calculus method of unconstrained minimization.

7.33 A beam of uniform rectangular cross section is to be cut from a log having a circular cross section of diameter $2a$. The beam is to be used as a cantilever beam to carry a concentrated load at the free end. Find the cross-sectional dimensions of the beam which will have the maximum bending stress carrying capacity using an exterior penalty function approach with analytical unconstrained minimization.

7.34 Consider the problem:

$$\text{Minimize } f = \tfrac{1}{3}(x_1 + 1)^3 + x_2$$

subject to

$$1 - x_1 \leq 0, \quad x_2 \geq 0$$

The results obtained during the sequential minimization of this problem according to the exterior penalty function approach are given below:

Value of k	r_k	Starting point for minimization of $\phi(\mathbf{X}, r_k)$	Unconstrained minimum of $\phi(\mathbf{X}, r_k) = \mathbf{X}_k^*$	$f(\mathbf{X}_k^*) = f_k^*$
1	1	$(-0.4597, -5.0)$	$(0.2361, -0.5)$	0.1295
2	10	$(0.2361, -0.5)$	$(0.8322, -0.05)$	2.0001

Estimate the optimum solution, $\mathbf{X}^*$ and f^*, using a suitable extrapolation technique.

7.35 The results obtained in an exterior penalty function method of solution for the optimization problem stated in Problem 7.15 are given below:

$$r_1 = 0.01, \quad \mathbf{X}_1^* = \left\{ \begin{matrix} -0.80975 \\ -50.0 \end{matrix} \right\}, \quad \phi_1^* = -24.9650, \quad f_1^* = -49.9977$$

$$r_2 = 1.0, \quad \mathbf{X}_2^* = \left\{ \begin{matrix} 0.23607 \\ -0.5 \end{matrix} \right\}, \quad \phi_2^* = 0.9631, \quad f_2^* = 0.1295$$

Estimate the optimum design vector and optimum objective function using an extrapolation method.

7.36 The following results have been obtained during an exterior penalty function approach:

$$r_1 = 10^{-10}, \quad \mathbf{X}_1^* = \left\{ \begin{matrix} 0.66 \\ 28.6 \end{matrix} \right\}$$

$$r_2 = 10^{-9}, \quad \mathbf{X}_2^* = \left\{ \begin{matrix} 1.57 \\ 18.7 \end{matrix} \right\}$$

Find the optimum solution, $\mathbf{X}^*$, using an extrapolation technique.

7.37 The results obtained in a sequential unconstrained minimization technique (using an exterior penalty function approach) from the starting point $\mathbf{X}_1 = \left\{ \begin{matrix} 6.0 \\ 30.0 \end{matrix} \right\}$ are

$$r_1 = 10^{-10}, \quad \mathbf{X}_1^* = \left\{ \begin{matrix} 0.66 \\ 28.6 \end{matrix} \right\}; \quad r_2 = 10^{-9}, \quad \mathbf{X}_2^* = \left\{ \begin{matrix} 1.57 \\ 18.7 \end{matrix} \right\}$$

$$r_3 = 10^{-8}, \quad \mathbf{X}_3^* = \left\{ \begin{matrix} 1.86 \\ 18.8 \end{matrix} \right\}$$

Estimate the optimum solution using a suitable extrapolation technique.

7.38 The two-bar truss shown in Fig. 7.26 is acted on by a varying load whose magnitude is given by $P(\theta) = P_0 \cos 2\theta; \ 0° \le \theta \le 360°$. The bars have a tubular section with mean diameter d and wall thickness t. Using $P_0 = 50{,}000$ lb, $\sigma_{\text{yield}} = 30{,}000$ psi, and $E = 30 \times 10^6$ psi, formulate the problem as a parametric optimization problem for minimum volume design subject to buckling and yielding constraints. Assume the bars to be pin connected for the purpose of buckling analysis. Indicate the procedure that can be used for a graphical solution of the problem.

7.39   Minimize $f(\mathbf{X}) = (x_1 - 1)^2 + (x_2 - 2)^2$

subject to

$$x_1 + 2x_2 - 2 = 0$$

using the augmented Lagrange multiplier method with a fixed value of $r_p = 1$. Use a maximum of three iterations.

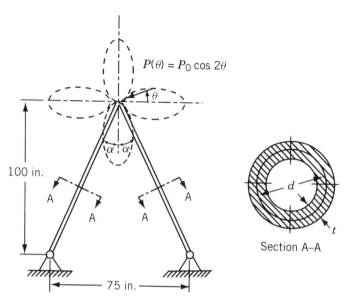

Figure 7.26   Two-bar truss subjected to a parametric load.

7.40 Solve the following optimization problem using the augmented Lagrange multiplier method keeping $r_p = 1$ throughout the iterative process and $\lambda^{(1)} = 0$:

$$\text{Minimize } f = (x_1 - 1)^2 + (x_2 - 2)^2$$

subject to

$$-x_1 + 2x_2 = 2$$

7.41 Consider the problem:

$$\text{Minimize } f = (x_1 - 1)^2 + (x_2 - 5)^2$$

subject to

$$x_1 + x_2 - 5 = 0$$

(a) Write the expression for the augmented Lagrange function with $r_p = 1$.
(b) Start with $\lambda_1^{(1)} = 0$ and perform two iterations.
(c) Find $\lambda_1^{(3)}$.

7.42 Consider the optimization problem:

$$\text{Minimize } f = x_1^3 - 6x_1^2 + 11x_1 + x_3$$

subject to

$$x_1^2 + x_2^2 - x_3^2 \le 0, \quad 4 - x_1^2 - x_2^2 - x_3^2 \le 0, \quad x_3 \le 5,$$

$$x_i \ge 0, \quad i = 1, 2, 3$$

Determine whether the solution

$$\mathbf{X} = \begin{Bmatrix} 0 \\ \sqrt{2} \\ \sqrt{2} \end{Bmatrix}$$

is optimum by finding the values of the Lagrange multipliers.

7.43 Determine whether the solution

$$\mathbf{X} = \begin{Bmatrix} 0 \\ \sqrt{2} \\ \sqrt{2} \end{Bmatrix}$$

is optimum for the problem considered in Example 7.8 using a perturbation method with $\Delta x_i = 0.001$, $i = 1, 2, 3$.

7.44 The following results are obtained during the minimization of

$$f(\mathbf{X}) = 9 - 8x_1 - 6x_2 - 4x_3 + 2x_1^2 + 2x_2^2 + x_3^2 + 2x_1x_2 + 2x_1x_3$$

subject to

$$x_1 + x_2 + 2x_3 \leq 3$$

$$x_i \geq 0, \quad i = 1, 2, 3$$

using the interior penalty function method:

Value of r_i	Starting point for minimization of $\phi(\mathbf{X}, r_i)$	Unconstrained minimum of $\phi(\mathbf{X}, r_i) = \mathbf{X}_i^*$	$f(\mathbf{X}_i^*) = f_i^*$
1	$\begin{Bmatrix} 0.1 \\ 0.1 \\ 0.1 \end{Bmatrix}$	$\begin{Bmatrix} 0.8884 \\ 0.7188 \\ 0.7260 \end{Bmatrix}$	0.7072
0.01	$\begin{Bmatrix} 0.8884 \\ 0.7188 \\ 0.7260 \end{Bmatrix}$	$\begin{Bmatrix} 1.3313 \\ 0.7539 \\ 0.3710 \end{Bmatrix}$	0.1564
0.0001	$\begin{Bmatrix} 1.3313 \\ 0.7539 \\ 0.3710 \end{Bmatrix}$	$\begin{Bmatrix} 1.3478 \\ 0.7720 \\ 0.4293 \end{Bmatrix}$	0.1158

Use an extrapolation technique to predict the optimum solution of the problem using the following relations:

(a) $\mathbf{X}(r) = \mathbf{A}_0 + r\mathbf{A}_1$; $f(r) = a_0 + ra_1$
(b) $\mathbf{X}(r) = \mathbf{A}_0 + r^{1/2}\mathbf{A}_1$; $f(r) = a_0 + r^{1/2}a_1$

Compare your results with the exact solution

$$\mathbf{X}^* = \begin{Bmatrix} \frac{12}{9} \\ \frac{7}{9} \\ \frac{4}{9} \end{Bmatrix}, \qquad f_{\min} = \frac{1}{9}$$

7.45   Find the extrapolated solution of Problem 7.44 by using quadratic relations for $\mathbf{X}(r)$ and $f(r)$.

7.46   Give a proof for the convergence of exterior penalty function method.

7.47   Write a computer program to implement the interior penalty function method with the DFP method of unconstrained minimization and the cubic interpolation method of one-dimensional search.

7.48   Write a computer program to implement the exterior penalty function method with the BFGS method of unconstrained minimization and the direct root method of one-dimensional search.

7.49   Write a computer program to implement the augmented Lagrange multiplier method with a suitable method of unconstrained minimization.

7.50   Write a computer program to implement the sequential linear programming method.

7.51   Find the solution of the welded beam design problem formulated in Section 7.22.3 using the MATLAB function fmincon with the starting point $\mathbf{X}_1 = \{0.4, 6.0, 9.0, 0.5\}^{\mathrm{T}}$

7.52   Find the solution of the following problem (known as Rosen–Suzuki problem) using the MATLAB function fmincon with the starting point $\mathbf{X}_1 = \{0, 0, 0, 0\}^{\mathrm{T}}$:

Minimize

$$f(\mathbf{X}) = x_1^2 + x_2^2 + 2x_3^2 - x_4^2 - 5x_1 - 5x_2 - 21x_3 + 7x_4 + 100$$

subject to

$$x_1^2 + x_2^2 + x_3^2 + x_4^2 + x_1 - x_2 + x_3 - x_4 - 100 \leq 0$$

$$x_1^2 + 2x_2^2 + x_3^2 + 2x_4^2 - x_1 - x_4 - 10 \leq 0$$

$$2x_1^2 + x_2^2 + x_3^2 + 2x_1 - x_2 - x_4 - 5 \leq 0$$

$$-100 \leq x_i \leq 100, \quad i = 1, 2, 3, 4$$

7.53   Find the solution of the following problem using the MATLAB function fmincon with the starting point $\mathbf{X}_1 = \{0.5, 1.0\}^{\mathrm{T}}$:

Minimize

$$f(\mathbf{X}) = x_1^2 + x_2^2 - 4x_1 - 6x_2$$

subject to

$$x_1 + x_2 \leq 2$$

$$2x_1 + 3x_2 \leq 12$$

$$x_i \geq 0, \quad i = 1, 2$$

7.54   Find the solution of the following problem using the MATLAB function fmincon with the starting point: $\mathbf{X}_1 = \{0.5, 1.0, 1.0\}$:

Minimize $f(\mathbf{X}) = x_1^2 + 3x_2^2 + x_3$
subject to

$$x_1^2 + x_2^2 + x_3^2 = 16$$

7.55 Find the solution of the following problem using the MATLAB function `fmincon` with the starting point: $\mathbf{X}_1 = \{1.0, \ 1.0\}^{\mathrm{T}}$:

Minimize $f(\mathbf{X}) = x_1^2 + x_2^2$
subject to

$$4 - x_1 - x_2^2 \leq 0$$

$$3x_2 - x_1 \leq 0$$

$$-3x_2 - x_1 \leq 0$$

8

Geometric Programming

8.1 INTRODUCTION

Geometric programming is a relatively new method of solving a class of nonlinear programming problems. It was developed by Duffin, Peterson, and Zener [8.1]. It is used to minimize functions that are in the form of posynomials subject to constraints of the same type. It differs from other optimization techniques in the emphasis it places on the relative magnitudes of the terms of the objective function rather than the variables. Instead of finding optimal values of the design variables first, geometric programming first finds the optimal value of the objective function. This feature is especially advantageous in situations where the optimal value of the objective function may be all that is of interest. In such cases, calculation of the optimum design vectors can be omitted. Another advantage of geometric programming is that it often reduces a complicated optimization problem to one involving a set of simultaneous linear algebraic equations. The major disadvantage of the method is that it requires the objective function and the constraints in the form of posynomials. We will first see the general form of a posynomial.

8.2 POSYNOMIAL

In an engineering design situation, frequently the objective function (e.g., the total cost) $f(\mathbf{X})$ is given by the sum of several component costs $U_i(\mathbf{X})$ as

$$f(\mathbf{X}) = U_1 + U_2 + \cdots + U_N \tag{8.1}$$

In many cases, the component costs U_i can be expressed as power functions of the type

$$U_i = c_i \; x_1^{a_{1i}} \; x_2^{a_{2i}} \; \cdots \; x_n^{a_{ni}} \tag{8.2}$$

where the coefficients c_i are positive constants, the exponents a_{ij} are real constants (positive, zero, or negative), and the design parameters $x_1, x_2, \ldots, x_n$ are taken to be positive variables. Functions like f, because of the positive coefficients and variables and real exponents, are called *posynomials*. For example,

$$f(x_1, x_2, x_3) = 6 + 3x_1 - 8x_2 + 7x_3 + 2x_1x_2$$
$$- 3x_1x_3 + \tfrac{4}{3}x_2x_3 + \tfrac{8}{7}x_1^2 - 9x_2^2 + x_3^2$$

is a second-degree polynomial in the variables, x_1, x_2, and x_3 (coefficients of the various terms are real) while

$$g(x_1, x_2, x_3) = x_1 x_2 x_3 + x_1^2 x_2 + 4x_3 + \frac{2}{x_1 x_2} + 5x_3^{-1/2}$$

is a posynomial. If the natural formulation of the optimization problem does not lead to posynomial functions, geometric programming techniques can still be applied to solve the problem by replacing the actual functions by a set of empirically fitted posynomials over a wide range of the parameters x_i.

8.3   UNCONSTRAINED MINIMIZATION PROBLEM

Consider the unconstrained minimization problem:

$$\text{Find } \mathbf{X} = \begin{Bmatrix} x_1 \\ x_2 \\ \vdots \\ x_n \end{Bmatrix}$$

that minimizes the objective function

$$f(\mathbf{X}) = \sum_{j=1}^{N} U_j(\mathbf{X}) = \sum_{j=1}^{N} \left(c_j \prod_{i=1}^{n} x_i^{a_{ij}} \right) = \sum_{j=1}^{N} (c_j \, x_1^{a_{1j}} \, x_2^{a_{2j}} \, \cdots \, x_n^{a_{nj}}) \tag{8.3}$$

where $c_j > 0$, $x_i > 0$, and the a_{ij} are real constants.

The solution of this problem can be obtained by various procedures. In the following sections, two approaches—one based on the differential calculus and the other based on the concept of geometric inequality—are presented for the solution of the problem stated in Eq. (8.3).

8.4   SOLUTION OF AN UNCONSTRAINED GEOMETRIC PROGRAMMING PROGRAM USING DIFFERENTIAL CALCULUS

According to the differential calculus methods presented in Chapter 2, the necessary conditions for the minimum of f are given by

$$\frac{\partial f}{\partial x_k} = \sum_{j=1}^{N} \frac{\partial U_j}{\partial x_k}$$

$$= \sum_{j=1}^{N} (c_j \, x_1^{a_{1j}} \, x_2^{a_{2j}} \, \cdots \, x_{k-1}^{a_{k-1,j}} \, a_{kj} x_k^{a_{kj}-1} \, a_{k+1}^{a_{k+1,j}} \, \cdots \, x_n^{a_{nj}}) = 0,$$

$$k = 1, 2, \ldots, n \tag{8.4}$$

By multiplying Eq. (8.4) by x_k, we can rewrite it as

$$x_k \frac{\partial f}{\partial x_k} = \sum_{j=1}^{N} a_{kj} (c_j \quad x_1^{a_{1j}} \quad x_2^{a_{2j}} \cdots x_{k-1}^{a_{k-1,j}} \quad x_k^{a_{kj}} \quad x_{k+1}^{a_{k+1,j}} \cdots x_n^{a_{nj}})$$

$$= \sum_{j=1}^{N} a_{kj} U_j (\mathbf{X}) = 0, \qquad k = 1, 2, \ldots, n \tag{8.5}$$

To find the minimizing vector

$$\mathbf{X}^* = \begin{Bmatrix} x_1^* \\ x_2^* \\ \vdots \\ x_n^* \end{Bmatrix}$$

we have to solve the n equations given by Eqs. (8.4), simultaneously. To ensure that the point $\mathbf{X}^*$ corresponds to the minimum of f (but not to the maximum or the stationary point of $\mathbf{X}$), the sufficiency condition must be satisfied. This condition states that the Hessian matrix of f is evaluated at $\mathbf{X}^*$:

$$\mathbf{J}_{\mathbf{X}^*} = \left[\frac{\partial^2 f}{\partial x_k \, \partial x_l} \right]_{\mathbf{X}^*}$$

must be positive definite. We will see this condition at a latter stage. Since the vector $\mathbf{X}^*$ satisfies Eqs. (8.5), we have

$$\sum_{j=1}^{N} a_{kj} U_j (\mathbf{X}^*) = 0, \qquad k = 1, 2, \ldots, n \tag{8.6}$$

After dividing by the minimum value of the objective function f^*, Eq. (8.6) becomes

$$\sum_{j=1}^{N} \Delta_j^* a_{kj} = 0, \qquad k = 1, 2, \ldots, n \tag{8.7}$$

where the quantities Δ_j^* are defined as

$$\Delta_j^* = \frac{U_j(\mathbf{X}^*)}{f^*} = \frac{U_j^*}{f^*} \tag{8.8}$$

and denote the relative contribution of jth term to the optimal objective function. From Eq. (8.8), we obtain

$$\sum_{j=1}^{N} \Delta_j^* = \Delta_1^* + \Delta_2^* + \cdots + \Delta_N^*$$

$$= \frac{1}{f^*} (U_1^* + U_2^* + \cdots + U_N^*) = 1 \tag{8.9}$$

Equations (8.7) are called the *orthogonality conditions* and Eq. (8.9) is called the *normality condition*. To obtain the minimum value of the objective function f^*, the following procedure can be adopted. Consider

$$f^* = (f^*)^1 = (f^*)^{\Sigma_{j=1}^N \Delta_j^*} = (f^*)^{\Delta_1^*}(f^*)^{\Delta_2^*}\cdots(f^*)^{\Delta_N^*} \tag{8.10}$$

Since

$$f^* = \frac{U_1^*}{\Delta_1^*} = \frac{U_2^*}{\Delta_2^*} = \cdots = \frac{U_N^*}{\Delta_N^*} \tag{8.11}$$

from Eq. (8.8), Eq. (8.10) can be rewritten as

$$f^* = \left(\frac{U_1^*}{\Delta_1^*}\right)^{\Delta_1^*}\left(\frac{U_2^*}{\Delta_2^*}\right)^{\Delta_2^*}\cdots\left(\frac{U_N^*}{\Delta_N^*}\right)^{\Delta_N^*} \tag{8.12}$$

By substituting the relation

$$U_j^* = c_j \prod_{i=1}^n (x_i^*)^{a_{ij}}, \qquad j = 1, 2, \ldots, N$$

Eq. (8.12) becomes

$$f^* = \left\{\left(\frac{c_1}{\Delta_1^*}\right)^{\Delta_1^*}\left[\prod_{i=1}^n (x_i^*)^{a_{i1}}\right]^{\Delta_1^*}\right\}\left\{\left(\frac{c_2}{\Delta_2^*}\right)^{\Delta_2^*}\left[\prod_{i=1}^n (x_i^*)^{a_{i2}}\right]^{\Delta_2^*}\right\}$$

$$\cdots\left\{\left(\frac{c_N}{\Delta_N^*}\right)^{\Delta_N^*}\left[\prod_{i=1}^n (x_i^*)^{a_{iN}}\right]^{\Delta_N^*}\right\}$$

$$= \left\{\prod_{j=1}^N \left(\frac{c_j}{\Delta_j^*}\right)^{\Delta_j^*}\right\}\left\{\prod_{j=1}^N \left[\prod_{i=1}^n (x_i^*)^{a_{ij}}\right]^{\Delta_j^*}\right\}$$

$$= \left\{\prod_{j=1}^N \left(\frac{c_j}{\Delta_j^*}\right)^{\Delta_j^*}\right\}\left[\prod_{i=1}^n (x_i^*)^{\sum_{j=1}^N a_{ij}\Delta_j^*}\right]$$

$$= \prod_{j=1}^N \left(\frac{c_j}{\Delta_j^*}\right)^{\Delta_j^*} \tag{8.13}$$

since

$$\sum_{j=1}^N a_{ij}\Delta_j^* = 0 \qquad \text{for any } i \quad \text{from Eq. (8.7)}$$

Thus the optimal objective function f^* can be found from Eq. (8.13) once Δ_j^* are determined. To determine Δ_j^* $(j = 1, 2, \ldots, N)$, Eqs. (8.7) and (8.9) can be used. It can be seen that there are $n+1$ equations in N unknowns. If $N = n+1$, there will be as many linear simultaneous equations as there are unknowns and we can find a unique solution.

Degree of Difficulty.   The quantity $N - n - 1$ is termed a *degree of difficulty* in geometric programming. In the case of a constrained geometric programming problem, N denotes the total number of terms in all the posynomials and n represents the number of design variables. If $N - n - 1 = 0$, the problem is said to have a zero degree of difficulty. In this case, the unknowns Δ_j^* ($j = 1, 2, \ldots, N$) can be determined uniquely from the orthogonality and normality conditions. If N is greater than $n + 1$, we have more number of variables (Δ_j^*s) than the equations, and the method of solution for this case will be discussed in subsequent sections. It is to be noted that geometric programming is not applicable to problems with negative degree of difficulty.

Sufficiency Condition.   We can see that Δ_j^* are found by solving Eqs. (8.7) and (8.9), which in turn are obtained by using the necessary conditions only. We can show that these conditions are also sufficient.

Finding the Optimal Values of Design Variables.   Since f^* and Δ_j^* ($j = 1, 2, \ldots, N$) are known, we can determine the optimal values of the design variables from the relations

$$U_j^* = \Delta_j^* f^* = c_j \prod_{i=1}^{n} (x_i^*)^{a_{ij}}, \quad j = 1, 2, \ldots, N \tag{8.14}$$

The simultaneous solution of these equations will yield the desired quantities x_i^* ($i = 1, 2, \ldots, n$). It can be seen that Eqs. (8.14) are nonlinear in terms of the variables $x_1^*, x_2^*, \ldots, x_n^*$, and hence their simultaneous solution is not easy if we want to solve them directly. To simplify the simultaneous solution of Eqs. (8.14), we rewrite them as

$$\frac{\Delta_j^* f^*}{c_j} = (x_1^*)^{a_{1j}} (x_2^*)^{a_{2j}} \cdots (x_n^*)^{a_{nj}}, \quad j = 1, 2, \ldots, N \tag{8.15}$$

By taking logarithms on both the sides of Eqs. (8.15), we obtain

$$\ln \frac{\Delta_j^* f^*}{c_j} = a_{1j} \ln x_1^* + a_{2j} \ln x_2^* + \cdots + a_{nj} \ln x_n^*,$$

$$j = 1, 2, \ldots, N \tag{8.16}$$

By letting

$$w_i = \ln x_i^*, \quad i = 1, 2, \ldots, n \tag{8.17}$$

Eqs. (8.16) can be written as

$$a_{11} w_1 + a_{21} w_2 + \cdots + a_{n1} w_n = \ln \frac{f^* \Delta_1^*}{c_1}$$

$$a_{12} w_1 + a_{22} w_2 + \cdots + a_{n2} w_n = \ln \frac{f^* \Delta_2^*}{c_2}$$

$$\vdots$$

$$a_{1N} w_1 + a_{2N} w_2 + \cdots + a_{nN} w_n = \ln \frac{f^* \Delta_N^*}{c_N}$$

$$\tag{8.18}$$

These equations, in the case of problems with a zero degree of difficulty, give a unique solution to $w_1, w_2, \ldots, w_n$. Once w_i are found, the desired solution can be obtained as

$$x_i^* = e^{wi}, \qquad i = 1, 2, \ldots, n \tag{8.19}$$

In a general geometric programming problem with a nonnegative degree of difficulty, $N \geq n + 1$, and hence Eqs. (8.18) denote N equations in n unknowns. By choosing any n linearly independent equations, we obtain a set of solutions w_i and hence x_i^*.

The solution of an unconstrained geometric programming problem is illustrated with the help of the following zero-degree-of-difficulty example [8.1].

Example 8.1   It has been decided to shift grain from a warehouse to a factory in an open rectangular box of length x_1 meters, width x_2 meters, and height x_3 meters. The bottom, sides, and the ends of the box cost, respectively, $80, $10, and $20/m^2. It costs $1 for each round trip of the box. Assuming that the box will have no salvage value, find the minimum cost of transporting 80 m^3 of grain.

SOLUTION   The total cost of transportation is given by

total cost = cost of box + cost of transportation

= (cost of sides + cost of bottom + cost of ends of the box)

+ (number of round trips required for transporting the grain

× cost of each round trip)

$$f(\mathbf{X}) = [(2x_1x_3)10 + (x_1x_2)80 + (2x_2x_3)20] + \left[\frac{80}{x_1x_2x_3}(1) \right]$$

$$= \$ \left(80x_1x_2 + 40x_2x_3 + 20x_1x_3 + \frac{80}{x_1x_2x_3} \right) \tag{E$_1$}$$

where x_1, x_2, and x_3 indicate the dimensions of the box, as shown in Fig. 8.1. By comparing Eq. (E$_1$) with the general posynomial of Eq. (8.1), we obtain

$$c_1 = 80, \quad c_2 = 40, \quad c_3 = 20, \quad c_4 = 80$$

$$\begin{pmatrix} a_{11} & a_{12} & a_{13} & a_{14} \\ a_{21} & a_{22} & a_{23} & a_{24} \\ a_{31} & a_{32} & a_{33} & a_{34} \end{pmatrix} = \begin{pmatrix} 1 & 0 & 1 & -1 \\ 1 & 1 & 0 & -1 \\ 0 & 1 & 1 & -1 \end{pmatrix}$$

The orthogonality and normality conditions are given by

$$\begin{bmatrix} 1 & 0 & 1 & -1 \\ 1 & 1 & 0 & -1 \\ 0 & 1 & 1 & -1 \\ 1 & 1 & 1 & 1 \end{bmatrix} \begin{Bmatrix} \Delta_1 \\ \Delta_2 \\ \Delta_3 \\ \Delta_4 \end{Bmatrix} = \begin{Bmatrix} 0 \\ 0 \\ 0 \\ 1 \end{Bmatrix}$$

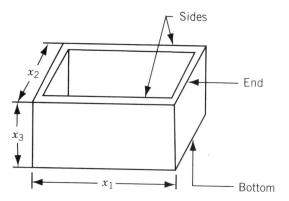

Figure 8.1   Open rectangular box.

that is,

$$\Delta_1 + \Delta_3 - \Delta_4 = 0 \tag{E$_2$}$$

$$\Delta_1 + \Delta_2 - \Delta_4 = 0 \tag{E$_3$}$$

$$\Delta_2 + \Delta_3 - \Delta_4 = 0 \tag{E$_4$}$$

$$\Delta_1 + \Delta_2 + \Delta_3 + \Delta_4 = 1 \tag{E$_5$}$$

From Eqs. (E$_2$) and (E$_3$), we obtain

$$\Delta_4 = \Delta_1 + \Delta_3 = \Delta_1 + \Delta_2 \quad \text{or} \quad \Delta_2 = \Delta_3 \tag{E$_6$}$$

Similarly, Eqs. (E$_3$) and (E$_4$) give us

$$\Delta_4 = \Delta_1 + \Delta_2 = \Delta_2 + \Delta_3 \quad \text{or} \quad \Delta_1 = \Delta_3 \tag{E$_7$}$$

Equations (E$_6$) and (E$_7$) yield

$$\Delta_1 = \Delta_2 = \Delta_3$$

while Eq. (E$_6$) gives

$$\Delta_4 = \Delta_1 + \Delta_3 = 2\Delta_1$$

Finally, Eq. (E$_5$) leads to the unique solution

$$\Delta_1^* = \Delta_2^* = \Delta_3^* = \tfrac{1}{5} \quad \text{and} \quad \Delta_4^* = \tfrac{2}{5}$$

Thus the optimal value of the objective function can be found from Eq. (8.13) as

$$f^* = \left(\frac{80}{1/5}\right)^{1/5} \left(\frac{40}{1/5}\right)^{1/5} \left(\frac{20}{1/5}\right)^{1/5} \left(\frac{80}{2/5}\right)^{2/5}$$

$$= (4 \times 10^2)^{1/5}(2 \times 10^2)^{1/5}(1 \times 10^2)^{1/5}(4 \times 10^4)^{1/5}$$

$$= (32 \times 10^{10})^{1/5} = \$200$$

It can be seen that the minimum total cost has been obtained before finding the optimal size of the box. To find the optimal values of the design variables, let us write Eqs. (8.14) as

$$U_1^* = 80x_1^* x_2^* = \Delta_1^* f^* = \frac{1}{5}(200) = 40 \tag{E_8}$$

$$U_2^* = 40x_2^* x_3^* = \Delta_2^* f^* = \frac{1}{5}(200) = 40 \tag{E_9}$$

$$U_3^* = 20x_1^* x_3^* = \Delta_3^* f^* = \frac{1}{5}(200) = 40 \tag{E_{10}}$$

$$U_4^* = \frac{80}{x_1^* x_2^* x_3^*} = \Delta_4^* f^* = \frac{2}{5}(200) = 80 \tag{E_{11}}$$

From these equations, we obtain

$$x_2^* = \frac{1}{2}\frac{1}{x_1^*} = \frac{1}{x_3^*}, \quad x_1^* = \frac{x_3^*}{2}, \quad x_2^* = \frac{1}{x_3^*}$$

$$\frac{1}{x_1^* x_2^* x_3^*} = 1 = \frac{2x_3^*}{x_3^* x_3^*}, \quad x_3^* = 2$$

Therefore,

$$x_1^* = 1 \text{ m}, \quad x_2^* = \frac{1}{2} \text{ m}, \quad x_3^* = 2 \text{ m} \tag{E_{12}}$$

It is to be noticed that there is one redundant equation among Eqs. (E_8) to (E_{11}), which is not needed for the solution of x_i^* $(i = 1 \text{ to } n)$.

The solution given in Eq. (E_{12}) can also be obtained using Eqs. (8.18). In the present case, Eqs. (8.18) lead to

$$1w_1 + 1w_2 + 0w_3 = \ln\frac{200 \times \frac{1}{5}}{80} = \ln\frac{1}{2} \tag{E_{13}}$$

$$0w_1 + 1w_2 + 1w_3 = \ln\frac{200 \times \frac{1}{5}}{40} = \ln 1 \tag{E_{14}}$$

$$1w_1 + 0w_2 + 1w_3 = \ln\frac{200 \times \frac{1}{5}}{20} = \ln 2 \tag{E_{15}}$$

$$-1w_1 - 1w_2 - 1w_3 = \ln\frac{200 \times \frac{2}{5}}{80} = \ln 1 \tag{E_{16}}$$

By adding Eqs. (E_{13}), (E_{14}), and (E_{16}), we obtain

$$w_2 = \ln\frac{1}{2} + \ln 1 + \ln 1 = \ln(\frac{1}{2} \cdot 1 \cdot 1) = \ln\frac{1}{2} = \ln x_2^*$$

or

$$x_2^* = \frac{1}{2}$$

Similarly, by adding Eqs. (E_{13}), (E_{15}), and (E_{16}), we get

$$w_1 = \ln\frac{1}{2} + \ln 2 + \ln 1 = \ln 1 = \ln x_1^*$$

or

$$x_1^* = 1$$

Finally, we can obtain x_3^* by adding Eqs. (E$_{14}$), (E$_{15}$), and (E$_{16}$) as

$$w_3 = \ln 1 + \ln 2 + \ln 1 = \ln 2 = \ln x_3^*$$

or

$$x_3^* = 2$$

It can be noticed that there are four equations, Eqs. (E$_{13}$) to (E$_{16}$) in three unknowns w_1, w_2, and w_3. However, not all of them are linearly independent. In this case, the first three equations only are linearly independent, and the fourth equation, (E$_{16}$), can be obtained by adding Eqs. (E$_{13}$), (E$_{14}$), and (E$_{15}$), and dividing the result by -2.

8.5 SOLUTION OF AN UNCONSTRAINED GEOMETRIC PROGRAMMING PROBLEM USING ARITHMETIC–GEOMETRIC INEQUALITY

The *arithmetic mean–geometric mean inequality* (also known as the *arithmetic–geometric inequality* or *Cauchy's inequality*) is given by [8.1]

$$\Delta_1 u_1 + \Delta_2 u_2 + \cdots + \Delta_N u_N \geq u_1^{\Delta_1} u_2^{\Delta_2} \cdots u_N^{\Delta_N} \tag{8.20}$$

with

$$\Delta_1 + \Delta_2 + \cdots + \Delta_N = 1 \tag{8.21}$$

This inequality is found to be very useful in solving geometric programming problems. Using the inequality of (8.20), the objective function of Eq. (8.3) can be written as (by setting $U_i = u_i \Delta_i$, $i = 1, 2, \ldots, N$)

$$U_1 + U_2 + \cdots + U_N \geq \left(\frac{U_1}{\Delta_1}\right)^{\Delta_1} \left(\frac{U_2}{\Delta_2}\right)^{\Delta_2} \cdots \left(\frac{U_N}{\Delta_N}\right)^{\Delta_N} \tag{8.22}$$

where $U_i = U_i(\mathbf{X})$, $i = 1, 2, \ldots, N$, and the weights $\Delta_1, \Delta_2, \ldots, \Delta_N$, satisfy Eq. (8.21). The left-hand side of the inequality (8.22) [i.e., the original function $f(\mathbf{X})$] is called the *primal function*. The right side of inequality (8.22) is called the *predual function*. By using the known relations

$$U_j = c_j \prod_{i=1}^{n} x_i^{a_{ij}}, \quad j = 1, 2, \ldots, N \tag{8.23}$$

the predual function can be expressed as

$$\left(\frac{U_1}{\Delta_1}\right)^{\Delta_1} \left(\frac{U_2}{\Delta_2}\right)^{\Delta_2} \cdots \left(\frac{U_N}{\Delta_N}\right)^{\Delta_N}$$

$$= \left(\frac{c_1 \prod_{i=1}^{n} x_i^{a_{i1}}}{\Delta_1}\right)^{\Delta_1} \left(\frac{c_2 \prod_{i=1}^{n} x_i^{a_{i2}}}{\Delta_2}\right)^{\Delta_2} \cdots \left(\frac{c_N \prod_{i=1}^{n} x_i^{a_{iN}}}{\Delta_N}\right)^{\Delta_N}$$

$$= \left(\frac{c_1}{\Delta_1}\right)^{\Delta_1} \left(\frac{c_2}{\Delta_2}\right)^{\Delta_2} \cdots \left(\frac{C_N}{\Delta_N}\right)^{\Delta_N} \left\{ \left(\prod_{i=1}^{n} x_i^{a_{i1}}\right)^{\Delta_1} \left(\prod_{i=1}^{n} x_i^{a_{i2}}\right)^{\Delta_2} \right.$$

$$\left. \cdots \left(\prod_{i=1}^{n} x_i^{a_{iN}}\right)^{\Delta_N} \right\}$$

$$= \left(\frac{c_1}{\Delta_1}\right)^{\Delta_1} \left(\frac{c_2}{\Delta_2}\right)^{\Delta_2} \cdots \left(\frac{c_N}{\Delta_N}\right)^{\Delta_N} \left\{ \left(x_1^{\sum_{j=1}^{N} a_{1j}\Delta_j} \right) \left(x_2^{\sum_{j=1}^{N} a_{2j}\Delta_j} \right) \right.$$

$$\left. \cdots \left(x_n^{\sum_{j=1}^{N} a_{nj}\Delta_j} \right) \right\} \tag{8.24}$$

If we select the weights Δ_j so as to satisfy the normalization condition, Eq. (8.21), and also the orthogonality relations

$$\sum_{j=1}^{N} a_{ij}\Delta_j = 0, \quad i = 1, 2, \ldots, n \tag{8.25}$$

Eq. (8.24) reduces to

$$\left(\frac{U_1}{\Delta_1}\right)^{\Delta_1} \left(\frac{U_2}{\Delta_2}\right)^{\Delta_2} \cdots \left(\frac{U_N}{\Delta_N}\right)^{\Delta_N} = \left(\frac{c_1}{\Delta_1}\right)^{\Delta_1} \left(\frac{c_2}{\Delta_2}\right)^{\Delta_2} \cdots \left(\frac{c_N}{\Delta_N}\right)^{\Delta_N} \tag{8.26}$$

Thus the inequality (8.22) becomes

$$U_1 + U_2 + \cdots + U_N \geq \left(\frac{c_1}{\Delta_1}\right)^{\Delta_1} \left(\frac{c_2}{\Delta_2}\right)^{\Delta_2} \cdots \left(\frac{c_N}{\Delta_N}\right)^{\Delta_N} \tag{8.27}$$

In this inequality, the right side is called the *dual function*, $v(\Delta_1, \Delta_2, \ldots, \Delta_N)$. The inequality (8.27) can be written simply as

$$f \geq v \tag{8.28}$$

A basic result is that the maximum of the dual function equals the minimum of the primal function. Proof of this theorem is given in the next section. The theorem enables us to accomplish the optimization by minimizing the primal or by maximizing the dual, whichever is easier. Also, the maximization of the dual function subject to the orthogonality and normality conditions is a sufficient condition for f, the primal function, to be a global minimum.

8.6 PRIMAL–DUAL RELATIONSHIP AND SUFFICIENCY CONDITIONS IN THE UNCONSTRAINED CASE

If f^* indicates the minimum of the primal function and v^* denotes the maximum of the dual function, Eq. (8.28) states that

$$f \geq f^* \geq v^* \geq v \tag{8.29}$$

In this section we prove that $f^* = v^*$ and also that f^* corresponds to the global minimum of $f(\mathbf{X})$. For convenience of notation, let us denote the objective function $f(\mathbf{X})$ by x_0 and make the exponential transformation

$$e^{w_i} = x_i \quad \text{or} \quad w_i = \ln x_i, \qquad i = 0, 1, 2, \ldots, n \tag{8.30}$$

where the variables w_i are unrestricted in sign. Define the new variables Δ_j, also termed *weights*, as

$$\Delta_j = \frac{U_j}{x_0} = \frac{c_j \prod_{i=1}^{n} x_i^{a_{ij}}}{x_0}, \qquad j = 1, 2, \ldots, N \tag{8.31}$$

which can be seen to be positive and satisfy the relation

$$\sum_{j=1}^{N} \Delta_j = 1 \tag{8.32}$$

By taking logarithms on both sides of Eq. (8.31), we obtain

$$\ln \Delta_j = \ln c_j + \sum_{i=1}^{n} a_{ij} \ln x_i - \ln x_0 \tag{8.33}$$

or

$$\ln \frac{\Delta_j}{c_j} = \sum_{i=1}^{n} a_{ij} w_i - w_0, \qquad j = 1, 2, \ldots, N \tag{8.34}$$

Thus the original problem of minimizing $f(\mathbf{X})$ with no constraints can be replaced by one of minimizing w_0 subject to the equality constraints given by Eqs. (8.32) and (8.34). The objective function x_0 is given by

$$x_0 = e^{w_0} = \sum_{j=1}^{N} c_j \prod_{i=1}^{n} e^{a_{ij} w_i}$$

$$= \sum_{j=1}^{N} c_j e^{\sum_{i=1}^{n} a_{ij} w_i} \tag{8.35}$$

Since the exponential function $(e^{a_{ij} w_i})$ is convex with respect to w_i, the objective function x_0, which is a positive combination of exponential functions, is also convex (see Problem 8.15). Hence there is only one stationary point for x_0 and it must be the global minimum. The global minimum point of w_0 can be obtained by constructing the following Lagrangian function and finding its stationary point:

$$L(\mathbf{w}, \mathbf{\Delta}, \lambda) = w_0 + \lambda_0 \left(\sum_{i=1}^{N} \Delta_i - 1 \right)$$

$$+ \sum_{j=1}^{N} \lambda_j \left(\sum_{i=1}^{n} a_{ij} w_i - w_0 - \ln \frac{\Delta_j}{c_j} \right) \tag{8.36}$$

where

$$\mathbf{w} = \begin{Bmatrix} w_0 \\ w_1 \\ \vdots \\ w_n \end{Bmatrix}, \qquad \boldsymbol{\Delta} = \begin{Bmatrix} \Delta_1 \\ \Delta_2 \\ \vdots \\ \Delta_N \end{Bmatrix}, \qquad \boldsymbol{\lambda} = \begin{Bmatrix} \lambda_0 \\ \lambda_1 \\ \vdots \\ \lambda_N \end{Bmatrix} \tag{8.37}$$

with $\boldsymbol{\lambda}$ denoting the vector of Lagrange multipliers. At the stationary point of L, we have

$$\frac{\partial L}{\partial w_i} = 0, \quad i = 0, 1, 2, \ldots, n$$

$$\frac{\partial L}{\partial \Delta_j} = 0, \quad j = 1, 2, \ldots, N \tag{8.38}$$

$$\frac{\partial L}{\partial \lambda_i} = 0, \quad i = 0, 1, 2, \ldots, N$$

These equations yield the following relations:

$$1 - \sum_{j=1}^{N} \lambda_j = 0 \quad \text{or} \quad \sum_{j=1}^{N} \lambda_j = 1 \tag{8.39}$$

$$\sum_{j=1}^{N} \lambda_j a_{ij} = 0, \qquad i = 1, 2, \ldots, n \tag{8.40}$$

$$\lambda_0 - \frac{\lambda_j}{\Delta_j} = 0 \quad \text{or} \quad \lambda_0 = \frac{\lambda_j}{\Delta_j}, \qquad j = 1, 2, \ldots, N \tag{8.41}$$

$$\sum_{j=1}^{N} \Delta_j - 1 = 0 \quad \text{or} \quad \sum_{j=1}^{N} \Delta_j = 1 \tag{8.42}$$

$$-\ln \frac{\Delta_j}{c_j} + \sum_{i=1}^{n} a_{ij} w_i - w_0 = 0, \qquad j = 1, 2, \ldots, N \tag{8.43}$$

Equations (8.39), (8.41), and (8.42) give the relation

$$\sum_{j=1}^{N} \lambda_j = 1 = \sum_{j=1}^{N} \lambda_0 \Delta_j = \lambda_0 \sum_{j=1}^{N} \Delta_j = \lambda_0 \tag{8.44}$$

Thus the values of the Lagrange multipliers are given by

$$\lambda_j = \begin{cases} 1 & \text{for} \quad j = 0 \\ \Delta_j & \text{for} \quad j = 1, 2, \ldots, N \end{cases} \tag{8.45}$$

By substituting Eq. (8.45) into Eq. (8.36), we obtain

$$L(\mathbf{\Delta}, \mathbf{w}) = -\sum_{j=1}^{N} \Delta_j \ln \frac{\Delta_j}{c_j} + (1 - w_0) \left(\sum_{j=1}^{N} \Delta_j - 1 \right) + \sum_{i=1}^{n} w_i \left(\sum_{j=1}^{N} a_{ij} \Delta_j \right)$$

(8.46)

The function given in Eq. (8.46) can be considered as the Lagrangian function corresponding to a new optimization problem whose objective function $\tilde{v}(\mathbf{\Delta})$ is given by

$$\tilde{v}(\mathbf{\Delta}) = -\sum_{j=1}^{N} \Delta_j \ln \frac{\Delta_j}{c_j} = \ln \left[\prod_{j=1}^{N} \left(\frac{c_j}{\Delta_j} \right)^{\Delta_j} \right]$$

(8.47)

and the constraints by

$$\sum_{j=1}^{N} \Delta_j - 1 = 0$$

(8.48)

$$\sum_{j=1}^{N} a_{ij} \Delta_j = 0, \qquad i = 1, 2, \ldots, n$$

(8.49)

This problem will be the dual for the original problem. The quantities $(1 - w_0)$, $w_1, w_2, \ldots, w_n$ can be regarded as the Lagrange multipliers for the constraints given by Eqs. (8.48) and (8.49).

Now it is evident that the vector $\mathbf{\Delta}$ which makes the Lagrangian of Eq. (8.46) stationary will automatically give a stationary point for that of Eq. (8.36). It can be proved that the function

$$\Delta_j \ln \frac{\Delta_j}{c_j}, \qquad j = 1, 2, \ldots, N$$

is convex (see Problem 8.16) since Δ_j is positive. Since the function $\tilde{v}(\mathbf{\Delta})$ is given by the negative of a sum of convex functions, it will be a concave function. Hence the function $\tilde{v}(\mathbf{\Delta})$ will have a unique stationary point that will be its global maximum point. Hence the minimum of the original primal function is same as the maximum of the function given by Eq. (8.47) subject to the normality and orthogonality conditions given by Eqs. (8.48) and (8.49) with the variables Δ_j constrained to be positive.

By substituting the optimal solution $\mathbf{\Delta}^*$, the optimal value of the objective function becomes

$$\tilde{v}^* = \tilde{v}(\mathbf{\Delta}^*) = L(\mathbf{w}^*, \mathbf{\Delta}^*) = w_0^* = L(\mathbf{w}^*, \mathbf{\Delta}^*, \mathbf{\lambda}^*)$$

$$= -\sum_{j=1}^{N} \Delta_j^* \ln \frac{\Delta_j^*}{c_j}$$

(8.50)

By taking the exponentials and using the transformation relation (8.30), we get

$$f^* = \prod_{j=1}^{N} \left(\frac{c_j}{\Delta_j^*} \right)^{\Delta_j^*}$$

(8.51)

Primal and Dual Problems.   We saw that geometric programming treats the problem of minimizing posynomials and maximizing product functions. The minimization problems are called *primal programs* and the maximization problems are called *dual programs*. Table 8.1 gives the primal and dual programs corresponding to an unconstrained minimization problem.

Computational Procedure.   To solve a given unconstrained minimization problem, we construct the dual function $v(\boldsymbol{\Delta})$ and maximize either $v(\boldsymbol{\Delta})$ or $\ln v(\boldsymbol{\Delta})$, whichever is convenient, subject to the constraints given by Eqs. (8.48) and (8.49). If the degree of difficulty of the problem is zero, there will be a unique solution for the Δ_j^*'s.

For problems with degree of difficulty greater than zero, there will be more variables Δ_j $(j = 1, 2, \ldots, N)$ than the number of equations $(n + 1)$. Sometimes it will be possible for us to express any $(n + 1)$ number of Δ_j's in terms of the remaining $(N - n - 1)$ number of Δ_j's. In such cases, our problem will be to maximize $v(\boldsymbol{\Delta})$ or $\ln v(\boldsymbol{\Delta})$ with respect to the $(N - n - 1)$ independent Δ_j's. This procedure is illustrated with the help of the following one-degree-of-difficulty example.

Example 8.2   In a certain reservoir pump installation, the first cost of the pipe is given by $(100D + 50D^2)$, where D is the diameter of the pipe in centimeters. The cost of the reservoir decreases with an increase in the quantity of fluid handled and is given by $20/Q$, where Q is the rate at which the fluid is handled (cubic meters per second).

Table 8.1   Primal and Dual Programs Corresponding to an Unconstrained Minimization Problem

Primal program	Dual program
Find $\mathbf{X} = \begin{Bmatrix} x_1 \\ x_2 \\ \vdots \\ x_n \end{Bmatrix}$	Find $\boldsymbol{\Delta} = \begin{Bmatrix} \Delta_1 \\ \Delta_2 \\ \vdots \\ \Delta_N \end{Bmatrix}$
so that	so that
$f(\mathbf{X}) = \sum_{j=1}^{N} c_j x_1^{a_{1j}} x_2^{a_{2j}} \cdots x_n^{a_{nj}}$ $\to$ minimum $x_1 > 0, x_2 > 0, \ldots, x_n > 0$	$v(\boldsymbol{\Delta}) = \prod_{j=1}^{N} \left(\dfrac{c_j}{\Delta_j}\right)^{\Delta_j}$ or $\ln v(\boldsymbol{\Delta}) = \ln \left[\prod_{j=1}^{N} \left(\dfrac{c_j}{\Delta_j}\right)^{\Delta_j}\right] \to$ maximum (8.47)

subject to the constraints
$$\sum_{j=1}^{N} \Delta_j = 1 \qquad (8.48)$$
$$\sum_{j=1}^{N} a_{ij}\Delta_j = 0, \quad i = 1, 2, \ldots, n \quad (8.49)$$

The pumping cost is given by $(300Q^2/D^5)$. Find the optimal size of the pipe and the amount of fluid handled for minimum overall cost.

SOLUTION

$$f(D, Q) = 100D^1 Q^0 + 50D^2 Q^0 + 20D^0 Q^{-1} + 300D^{-5}Q^2 \qquad (E_1)$$

Here we can see that

$$c_1 = 100, \quad c_2 = 50, \quad c_3 = 20, \quad c_4 = 300$$

$$\begin{pmatrix} a_{11} & a_{12} & a_{13} & a_{14} \\ a_{21} & a_{22} & a_{23} & a_{24} \end{pmatrix} = \begin{pmatrix} 1 & 2 & 0 & -5 \\ 0 & 0 & -1 & 2 \end{pmatrix}$$

The orthogonality and normality conditions are given by

$$\begin{pmatrix} 1 & 2 & 0 & -5 \\ 0 & 0 & -1 & 2 \\ 1 & 1 & 1 & 1 \end{pmatrix} \begin{Bmatrix} \Delta_1 \\ \Delta_2 \\ \Delta_3 \\ \Delta_4 \end{Bmatrix} = \begin{Bmatrix} 0 \\ 0 \\ 1 \end{Bmatrix}$$

Since $N > (n + 1)$, these equations do not yield the required Δ_j $(j = 1$ to $4)$ directly. But any three of the Δ_j's can be expressed in terms of the remaining one. Hence by solving for $\Delta_1, \Delta_2,$ and Δ_3 in terms of Δ_4, we obtain

$$\Delta_1 = 2 - 11\Delta_4$$
$$\Delta_2 = 8\Delta_4 - 1 \qquad (E_2)$$
$$\Delta_3 = 2\Delta_4$$

The dual problem can now be written as

Maximize $v(\Delta_1, \Delta_2, \Delta_3, \Delta_4)$

$$= \left(\frac{c_1}{\Delta_1} \right)^{\Delta_1} \left(\frac{c_2}{\Delta_2} \right)^{\Delta_2} \left(\frac{c_3}{\Delta_3} \right)^{\Delta_3} \left(\frac{c_4}{\Delta_4} \right)^{\Delta_4}$$

$$= \left(\frac{100}{2 - 11\Delta_4} \right)^{2-11\Delta_4} \left(\frac{50}{8\Delta_4 - 1} \right)^{8\Delta_4-1} \left(\frac{20}{2\Delta_4} \right)^{2\Delta_4} \left(\frac{300}{\Delta_4} \right)^{\Delta_4}$$

Since the maximization of v is equivalent to the maximization of $\ln v$, we will maximize $\ln v$ for convenience. Thus

$$\ln v = (2 - 11\Delta_4)[\ln 100 - \ln(2 - 11\Delta_4)] + (8\Delta_4 - 1)$$
$$\times [\ln 50 - \ln(8\Delta_4 - 1)] + 2\Delta_4[\ln 20 - \ln(2\Delta_4)]$$
$$+ \Delta_4[\ln 300 - \ln(\Delta_4)]$$

Since $\ln v$ is expressed as a function of Δ_4 alone, the value of Δ_4 that maximizes $\ln v$ must be unique (because the primal problem has a unique solution). The necessary condition for the maximum of $\ln v$ gives

$$\frac{\partial}{\partial \Delta_4}(\ln v) = -11[\ln 100 - \ln(2 - 11\Delta_4)] + (2 - 11\Delta_4)\frac{11}{2 - 11\Delta_4}$$

$$+ 8 \, [\ln 50 - \ln(8\Delta_4 - 1)] + (8\Delta_4 - 1)\left(-\frac{8}{8\Delta_4 - 1}\right)$$

$$+ 2 \, [\ln 20 - \ln(2\Delta_4)] + 2\Delta_4\left(-\frac{2}{2\Delta_4}\right)$$

$$+ 1 \, [\ln 300 - \ln(\Delta_4)] + \Delta_4\left(-\frac{1}{\Delta_4}\right) = 0$$

This gives after simplification

$$\ln \frac{(2 - 11\Delta_4)^{11}}{(8\Delta_4 - 1)^8(2\Delta_4)^2\Delta_4} - \ln \frac{(100)^{11}}{(50)^8(20)^2(300)} = 0$$

i.e.,

$$\frac{(2 - 11\Delta_4)^{11}}{(8\Delta_4 - 1)^8(2\Delta_4)^2\Delta_4} = \frac{(100)^{11}}{(50)^8(20)^2(300)} = 2130 \qquad (E_3)$$

from which the value of Δ_4^* can be obtained by using a trial-and-error process as follows:

Value of Δ_4^*	Value of left-hand side of Eq. (E_3)
$2/11 = 0.182$	0.0
0.15	$\dfrac{(0.35)^{11}}{(0.2)^8(0.3)^2(0.15)} \simeq 284$
0.147	$\dfrac{(0.385)^{11}}{(0.175)^8(0.294)^2(0.147)} \simeq 2210$
0.146	$\dfrac{(0.39)^{11}}{(0.169)^8(0.292)^2(0.146)} \simeq 4500$

Thus we find that $\Delta_4^* \simeq 0.147$, and Eqs. (E_2) give

$$\Delta_1^* = 2 - 11\Delta_4^* = 0.385$$

$$\Delta_2^* = 8\Delta_4^* - 1 = 0.175$$

$$\Delta_3^* = 2\Delta_4^* = 0.294$$

The optimal value of the objective function is given by

$$v^* = f^* = \left(\frac{100}{0.385}\right)^{0.385} \left(\frac{50}{0.175}\right)^{0.175} \left(\frac{20}{0.294}\right)^{0.294} \left(\frac{300}{0.147}\right)^{0.147}$$

$$= 8.5 \times 2.69 \times 3.46 \times 3.06 = 242$$

The optimum values of the design variables can be found from

$$U_1^* = \Delta_1^* f^* = (0.385)(242) = 92.2$$

$$U_2^* = \Delta_2^* f^* = (0.175)(242) = 42.4$$

$$U_3^* = \Delta_3^* f^* = (0.294)(242) = 71.1 \qquad (E_4)$$

$$U_4^* = \Delta_4^* f^* = (0.147)(242) = 35.6$$

From Eqs. (E$_1$) and (E$_4$), we have

$$U_1^* = 100 D^* = 92.2$$

$$U_2^* = 50 D^{*2} = 42.4$$

$$U_3^* = \frac{20}{Q^*} = 71.1$$

$$U_4^* = \frac{300 Q^{*2}}{D^{*5}} = 35.6$$

These equations can be solved to find the desired solution $D^* = 0.922\,\text{cm}$, $Q^* = 0.281\,\text{m}^3/\text{s}$.

8.7 CONSTRAINED MINIMIZATION

Most engineering optimization problems are subject to constraints. If the objective function and all the constraints are expressible in the form of posynomials, geometric programming can be used most conveniently to solve the optimization problem. Let the constrained minimization problem be stated as

$$\text{Find } \mathbf{X} = \begin{Bmatrix} x_1 \\ x_2 \\ \vdots \\ x_n \end{Bmatrix}$$

which minimizes the objective function

$$f(\mathbf{X}) = \sum_{j=1}^{N_0} c_{0j} \prod_{i=1}^{n} x_i^{a_{0ij}} \qquad (8.52)$$

and satisfies the constraints

$$g_k(\mathbf{X}) = \sum_{j=1}^{N_k} c_{kj} \prod_{i=1}^{n} x_i^{a_{kij}} \lessgtr 1, \quad k = 1, 2, \ldots, m \qquad (8.53)$$

where the coefficients c_{0j} $(j = 1, 2, \ldots, N_0)$ and c_{kj} $(k = 1, 2, \ldots, m; \ j = 1, 2, \ldots, N_k)$ are positive numbers, the exponents a_{0ij} $(i = 1, 2, \ldots, n; \ j = 1, 2, \ldots, N_0)$

and a_{kij} $(k = 1, 2, \ldots, m;\ i = 1, 2, \ldots, n;\ j = 1, 2, \ldots, N_k)$ are any real numbers, m indicates the total number of constraints, N_0 represents the number of terms in the objective function, and N_k denotes the number of terms in the kth constraint. The design variables $x_1, x_2, \ldots, x_n$ are assumed to take only positive values in Eqs. (8.52) and (8.53). The solution of the constrained minimization problem stated above is considered in the next section.

8.8   SOLUTION OF A CONSTRAINED GEOMETRIC PROGRAMMING PROBLEM

For simplicity of notation, let us denote the objective function as

$$x_0 = g_0(\mathbf{X}) = f(\mathbf{X}) = \sum_{i=1}^{N_0} c_{0j} \prod_{j=1}^{n} x_i^{a_{0ij}} \tag{8.54}$$

The constraints given in Eq. (8.53) can be rewritten as

$$f_k = \sigma_k[1 - g_k(\mathbf{X})] \geq 0, \qquad k = 1, 2, \ldots, m \tag{8.55}$$

where σ_k, the *signum function*, is introduced for the kth constraint so that it takes on the value $+1$ or -1, depending on whether $g_k(\mathbf{X})$ is ≤ 1 or ≥ 1, respectively. The problem is to minimize the objective function, Eq. (8.54), subject to the inequality constraints given by Eq. (8.55). This problem is called the *primal problem* and can be replaced by an equivalent problem (known as the *dual problem*) with linear constraints, which is often easier to solve. The dual problem involves the maximization of the dual function, $v(\boldsymbol{\lambda})$, given by

$$v(\boldsymbol{\lambda}) = \prod_{k=0}^{m} \prod_{j=1}^{N_k} \left(\frac{c_{kj}}{\lambda_{kj}} \sum_{l=1}^{N_k} \lambda_{kl} \right)^{\sigma_k \lambda_{kj}} \tag{8.56}$$

subject to the normality and orthogonality conditions

$$\sum_{j=1}^{N_0} \lambda_{0j} = 1 \tag{8.57}$$

$$\sum_{k=0}^{m} \sum_{j=1}^{N_k} \sigma_k a_{kij} \lambda_{kj} = 0, \quad i = 1, 2, \ldots, n \tag{8.58}$$

If the problem has zero degree of difficulty, the normality and orthogonality conditions [Eqs. (8.57) and (8.58)] yield a unique solution for $\boldsymbol{\lambda}^*$ from which the stationary value of the original objective function can be obtained as

$$f^* = x_0^* = v(\boldsymbol{\lambda}^*) = \prod_{k=0}^{m} \prod_{j=1}^{N_k} \left(\frac{c_{kj}}{\lambda_{kj}^*} \sum_{l=1}^{N_k} \lambda_{kl}^* \right)^{\sigma_k \lambda_{kj}^*} \tag{8.59}$$

If the function $f(\mathbf{X})$ is known to possess a minimum, the stationary value f^* given by Eq. (8.59) will be the global minimum of f since, in this case, there is a unique solution for $\boldsymbol{\lambda}^*$.

The degree of difficulty of the problem (D) is defined as

$$D = N - n - 1 \tag{8.60}$$

where N denotes the total number of posynomial terms in the problem:

$$N = \sum_{k=0}^{m} N_k \tag{8.61}$$

If the problem has a positive degree of difficulty, the linear Eqs. (8.57) and (8.58) can be used to express any $(n + 1)$ of the λ_{kj}'s in terms of the remaining D of the λ_{kj}'s. By using these relations, v can be expressed as a function of the D independent λ_{kj}'s. Now the stationary points of v can be found by using any of the unconstrained optimization techniques.

If calculus techniques are used, the first derivatives of the function v with respect to the independent dual variables are set equal to zero. This results in as many simultaneous nonlinear equations as there are degrees of difficulty (i.e., $N - n - 1$). The solution of these simultaneous nonlinear equations yields the best values of the dual variables, $\boldsymbol{\lambda}^*$. Hence this approach is occasionally impractical due to the computations required. However, if the set of nonlinear equations can be solved, geometric programming provides an elegant approach.

Optimum Design Variables. For problems with a zero degree of difficulty, the solution of $\boldsymbol{\lambda}^*$ is unique. Once the optimum values of λ_{kj} are obtained, the maximum of the dual function v^* can be obtained from Eq. (8.59), which is also the minimum of the primal function, f^*. Once the optimum value of the objective function $f^* = x_0^*$ is known, the next step is to determine the values of the design variables x_i^* $(i = 1, 2, \ldots, n)$. This can be achieved by solving simultaneously the following equations:

$$\Delta_{0j}^* = \lambda_{0j}^* \equiv \frac{c_{0j} \prod_{i=1}^{n} (x_i^*)^{a_{0ij}}}{x_0^*}, \qquad j = 1, 2, \ldots, N_0 \tag{8.62}$$

$$\Delta_{kj}^* = \frac{\lambda_{kj}^*}{\sum_{l=1}^{N_k} \lambda_{kl}^*} = c_{kj} \prod_{i=1}^{n} (x_i^*)^{a_{kij}}, \qquad \begin{aligned} & j = 1, 2, \ldots, N_k \\ & k = 1, 2, \ldots, m \end{aligned} \tag{8.63}$$

8.9 PRIMAL AND DUAL PROGRAMS IN THE CASE OF LESS-THAN INEQUALITIES

If the original problem has a zero degree of difficulty, the minimum of the primal problem can be obtained by maximizing the corresponding dual function. Unfortunately, this cannot be done in the general case where there are some greater than type of inequality constraints. However, if the problem has all the constraints in the form of

$g_k(\mathbf{X}) \leq 1$, the signum functions σ_k are all equal to $+1$, and the objective function $g_0(\mathbf{X})$ will be a strictly convex function of the transformed variables $w_1, w_2, \ldots, w_n$, where

$$x_i = e^{w_i}, \qquad i = 0, 1, 2, \ldots, n \tag{8.64}$$

In this case, the following primal–dual relationship can be shown to be valid:

$$f(\mathbf{X}) \geq f^* \equiv v^* \geq v(\boldsymbol{\lambda}) \tag{8.65}$$

Table 8.2 gives the primal and the corresponding dual programs. The following characteristics can be noted from this table:

1. The factors c_{kj} appearing in the dual function $v(\boldsymbol{\lambda})$ are the coefficients of the posynomials $g_k(\mathbf{X})$, $k = 0, 1, 2, \ldots, m$.
2. The number of components in the vector $\boldsymbol{\lambda}$ is equal to the number of terms involved in the posynomials $g_0, g_1, g_2, \ldots, g_m$. Associated with every term in $g_k(\mathbf{X})$, there is a corresponding Δ_{kj}.
3. Each factor $\left(\sum_{l=1}^{N_k} \lambda_{kl}\right)^{\lambda_{kj}}$ of $v(\boldsymbol{\lambda})$ comes from an inequality constraint $g_k(\mathbf{X}) \leq 1$. No such factor appears from the primal function $g_0(\mathbf{X})$ as the normality condition forces $\sum_{j=1}^{N_0} \lambda_{0j}$ to be unity.
4. The coefficient matrix $[a_{kij}]$ appearing in the orthogonality condition is same as the exponent matrix appearing in the posynomials of the primal program.

The following examples are considered to illustrate the method of solving geometric programming problems with less-than inequality constraints.

Example 8.3 Zero-degree-of-difficulty Problem Suppose that the problem considered in Example 8.1 is restated in the following manner. Minimize the cost of constructing the open rectangular box subject to the constraint that a maximum of 10 trips only are allowed for transporting the $80\,\mathrm{m}^3$ of grain.

SOLUTION   The optimization problem can be stated as

$$\text{Find } \mathbf{X} = \begin{Bmatrix} x_1 \\ x_2 \\ x_3 \end{Bmatrix} \text{ so as to minimize}$$

$$f(\mathbf{X}) = 20x_1x_2 + 40x_2x_3 + 80x_1x_2$$

subject to

$$\frac{80}{x_1x_2x_3} \leq 10 \quad \text{or} \quad \frac{8}{x_1x_2x_3} \leq 1$$

Since $n = 3$ and $N = 4$, this problem has zero degree of difficulty. As $N_0 = 3$, $N_1 = 1$, and $m = 1$, the dual problem can be stated as follows:

$$\text{Find } \boldsymbol{\lambda} = \begin{Bmatrix} \lambda_{01} \\ \lambda_{02} \\ \lambda_{03} \\ \lambda_{11} \end{Bmatrix} \text{ to maximize}$$

Table 8.2 Corresponding Primal and Dual Programs

Primal program	Dual program

Find $\mathbf{X} = \begin{Bmatrix} x_1 \\ x_2 \\ \vdots \\ x_n \end{Bmatrix}$

so that

$$g_0(\mathbf{X}) \equiv f(\mathbf{X}) \rightarrow \text{minimum}$$

subject to the constraints

$$x_1 > 0$$
$$x_2 > 0$$
$$\vdots$$
$$x_n > 0,$$
$$g_1(\mathbf{X}) \leq 1$$
$$g_2(\mathbf{X}) \leq 1$$
$$\vdots$$
$$g_m(\mathbf{X}) \leq 1,$$

Find $\boldsymbol{\lambda} = \begin{Bmatrix} \lambda_{01} \\ \lambda_{02} \\ \vdots \\ \lambda_{0N_0} \\ \cdots \\ \lambda_{11} \\ \lambda_{12} \\ \vdots \\ \lambda_{1N_1} \\ \cdots \\ \vdots \\ \cdots \\ \lambda_{m1} \\ \lambda_{m2} \\ \vdots \\ \lambda_{mN_m} \end{Bmatrix}$

so that

$$v(\boldsymbol{\lambda}) = \prod_{k=0}^{m} \prod_{j=1}^{N_k} \left(\frac{c_{kj}}{\lambda_{kj}} \sum_{l=1}^{N_k} \lambda_{kl} \right)^{\lambda_{kj}}$$
$$\rightarrow \text{maximum}$$

subject to the constraints

with

$$g_0(\mathbf{X}) = \sum_{j=1}^{N_0} c_{0j} x_1^{a_{01j}} x_2^{a_{02j}} \cdots x_n^{a_{0nj}}$$

$$g_1(\mathbf{X}) = \sum_{j=1}^{N_1} c_{1j} x_1^{a_{11j}} x_2^{a_{12j}} \cdots x_n^{a_{1nj}}$$

$$g_2(\mathbf{X}) = \sum_{j=1}^{N_2} c_{2j} x_1^{a_{21j}} x_2^{a_{22j}} \cdots x_n^{a_{2nj}}$$

$$\vdots$$

$$g_m(\mathbf{X}) = \sum_{j=1}^{N_m} c_{mj} x_1^{a_{m1j}} x_2^{a_{m2j}} \cdots x_n^{a_{mnj}}$$

$$\lambda_{01} \geq 0$$
$$\lambda_{02} \geq 0$$
$$\vdots$$
$$\lambda_{0N_0} \geq 0$$
$$\lambda_{11} \geq 0$$
$$\vdots$$
$$\lambda_{1N_1} \geq 0$$
$$\vdots$$
$$\lambda_{m1} \geq 0$$
$$\lambda_{m2} \geq 0$$
$$\vdots$$
$$\lambda_{mN_m} \geq 0$$

(continues)

Table 8.2   (*continued*)

Primal program	Dual program
the exponents a_{kij} are real numbers, and the coefficients c_{kj} are positive numbers.	$$\sum_{j=1}^{N_0} \lambda_{0j} = 1$$ $$\sum_{k=0}^{m} \sum_{j=1}^{N_k} a_{kij}\lambda_{kj} = 0, \quad i = 1, 2, \ldots, n$$ the factors c_{kj} are positive, and the coefficients a_{kij} are real numbers.

Terminology

$g_0 = f$ = primal function	v = dual function
$x_1, x_2, \ldots, x_n$ = primal variables	$\lambda_{01}, \lambda_{02}, \ldots, \lambda_{mNm}$ = dual variables
$g_k \leq 1$ are primal constraints	$\displaystyle\sum_{j=1}^{N_0} \lambda_{0j} = 1$ is the normality constraint
$(k = 1, 2, \ldots, m)$	$\displaystyle\sum_{k=0}^{m} \sum_{j=1}^{N_k} a_{kij}\lambda_{kj} = 0, \ i = 1, 2, \ldots, n$ are the orthogonality constraints
$x_i > 0, \ i = 1, 2, \ldots, n$ positive restrictions.	
n = number of primal variables	
m = number of primal constraints	$\lambda_{kj} \geq 0, j = 1, 2, \ldots, N_k;$
$N = N_0 + N_1 + \cdots + N_m$ = total number of terms in the posynomials	$k = 0, 1, 2, \ldots, m$
$N - n - 1$ = degree of difficulty of the problem	are nonnegativity restrictions
	$N = N_0 + N_1 + \cdots + N_m$ = number of dual variables
	$n + 1$ number of dual constraints

$$v(\lambda) = \prod_{k=0}^{1} \prod_{j=1}^{N_k} \left(\frac{c_{kj}}{\lambda_{kj}} \sum_{l=1}^{N_k} \lambda_{kl} \right)^{\lambda_{kj}}$$

$$= \prod_{j=1}^{N_0=3} \left(\frac{c_{0j}}{\lambda_{0j}} \sum_{l=1}^{N_0=3} \lambda_{0l} \right)^{\lambda_{0j}} \prod_{j=1}^{N_1=1} \left(\frac{c_{1j}}{\lambda_{1j}} \sum_{l=1}^{N_1=1} \lambda_{1l} \right)^{\lambda_{1j}}$$

$$= \left[\frac{c_{01}}{\lambda_{01}} (\lambda_{01} + \lambda_{02} + \lambda_{03}) \right]^{\lambda_{01}} \left[\frac{c_{02}}{\lambda_{02}} (\lambda_{01} + \lambda_{02} + \lambda_{03}) \right]^{\lambda_{02}}$$

$$\cdot \left[\frac{c_{03}}{\lambda_{03}} (\lambda_{01} + \lambda_{02} + \lambda_{03}) \right]^{\lambda_{03}} \left(\frac{c_{11}}{\lambda_{11}} \lambda_{11} \right)^{\lambda_{11}} \tag{E_1}$$

subject to the constraints

$$\lambda_{01} + \lambda_{02} + \lambda_{03} = 1$$

$$a_{011}\lambda_{01} + a_{012}\lambda_{02} + a_{013}\lambda_{03} + a_{111}\lambda_{11} = 0$$

$$a_{021}\lambda_{01} + a_{022}\lambda_{02} + a_{023}\lambda_{03} + a_{121}\lambda_{11} = 0$$

$$a_{031}\lambda_{01} + a_{032}\lambda_{02} + a_{033}\lambda_{03} + a_{131}\lambda_{11} = 0 \qquad (E_2)$$

$$\lambda_{0j} \geq 0, \quad j = 1, 2, 3$$

$$\lambda_{11} \geq 0$$

In this problem, $c_{01} = 20$, $c_{02} = 40$, $c_{03} = 80$, $c_{11} = 8$, $a_{011} = 1$, $a_{021} = 0$, $a_{031} = 1$, $a_{012} = 0$, $a_{022} = 1$, $a_{032} = 1$, $a_{013} = 1$, $a_{023} = 1$, $a_{033} = 0$, $a_{111} = -1$, $a_{121} = -1$, and $a_{131} = -1$. Hence Eqs. (E$_1$) and (E$_2$) become

$$v(\lambda) = \left[\frac{20}{\lambda_{01}}(\lambda_{01} + \lambda_{02} + \lambda_{03})\right]^{\lambda_{01}} \left[\frac{40}{\lambda_{02}}(\lambda_{01} + \lambda_{02} + \lambda_{03})\right]^{\lambda_{02}}$$

$$\times \left[\frac{80}{\lambda_{03}}(\lambda_{01} + \lambda_{02} + \lambda_{03})\right]^{\lambda_{03}} \left(\frac{8}{\lambda_{11}}\lambda_{11}\right)^{\lambda_{11}} \qquad (E_3)$$

subject to

$$\lambda_{01} + \lambda_{02} + \lambda_{03} = 1$$

$$\lambda_{01} + \lambda_{03} - \lambda_{11} = 0$$

$$\lambda_{02} + \lambda_{03} - \lambda_{11} = 0 \qquad (E_4)$$

$$\lambda_{01} + \lambda_{02} - \lambda_{11} = 0$$

$$\lambda_{01} \geq 0, \quad \lambda_{02} \geq 0, \quad \lambda_{03} \geq 0, \quad \lambda_{11} \geq 0$$

The four linear equations in Eq. (E$_4$) yield the unique solution

$$\lambda_{01}^* = \lambda_{02}^* = \lambda_{03}^* = \tfrac{1}{3}, \quad \lambda_{11}^* = \tfrac{2}{3}$$

Thus the maximum value of v or the minimum value of x_0 is given by

$$v^* = x_0^* = (60)^{1/3}(120)^{1/3}(240)^{1/3}(8)^{2/3}$$

$$= [(60)^3]^{1/3}(8)^{1/3}(8)^{2/3} = (60)(8) = 480$$

The values of the design variables can be obtained by applying Eqs. (8.62) and (8.63) as

$$\lambda_{01}^* = \frac{c_{01}(x_1^*)^{a_{011}}(x_2^*)^{a_{021}}(x_3^*)^{a_{031}}}{x_0^*}$$

$$\tfrac{1}{3} = \frac{20(x_1^*)(x_3^*)}{480} = \frac{x_1^* x_3^*}{24} \qquad (E_5)$$

$$\lambda_{02}^* = \frac{c_{02}(x_1^*)^{a_{012}}(x_2^*)^{a_{022}}(x_3^*)^{a_{032}}}{x_0^*}$$

$$\tfrac{1}{3} = \frac{40(x_2^*)(x_3^*)}{480} = \frac{x_2^* x_3^*}{12} \qquad (E_6)$$

$$\lambda_{03}^* = \frac{c_{03}(x_1^*)^{a013}(x_2^*)^{a023}(x_3^*)^{a033}}{x_0^*}$$

$$\frac{1}{3} = \frac{80(x_1^*)(x_2^*)}{480} = \frac{x_1^* x_2^*}{6} \tag{E_7}$$

$$\frac{\lambda_{11}^*}{\lambda_{11}^*} = c_{11}(x_1^*)^{a111}(x_2^*)^{a121}(x_3^*)^{a131}$$

$$1 = 8(x_1^*)^{-1}(x_2^*)^{-1}(x_3^*)^{-1} = \frac{8}{x_1^* x_2^* x_3^*} \tag{E_8}$$

Equations (E_5) to (E_8) give

$$x_1^* = 2, \quad x_2^* = 1, \quad x_3^* = 4$$

Example 8.4 *One-degree-of-difficulty Problem*

$$\text{Minimize } f = x_1 x_2^2 x_3^{-1} + 2x_1^{-1} x_2^{-3} x_4 + 10 x_1 x_3$$

subject to

$$3x_1^{-1} x_3 x_4^{-2} + 4 x_3 x_4 \leq 1$$

$$5 x_1 x_2 \leq 1$$

SOLUTION Here $N_0 = 3$, $N_1 = 2$, $N_2 = 1$, $N = 6$, $n = 4$, $m = 2$, and the degree of difficulty of this problem is $N - n - 1 = 1$. The dual problem can be stated as follows:

$$\text{Maximixze } v(\lambda) = \prod_{k=0}^{m} \prod_{j=1}^{N_k} \left(\frac{c_{kj}}{\lambda_{kj}} \sum_{l=1}^{N_k} \lambda_{kl} \right)^{\lambda_{kj}}$$

subject to

$$\sum_{j=1}^{N_0} \lambda_{0j} = 1$$

$$\sum_{k=0}^{m} \sum_{j=1}^{N_k} a_{kij} \lambda_{kj} = 0, \qquad i = 1, 2, \ldots, n \tag{E_1}$$

$$\sum_{j=1}^{N_k} \lambda_{kj} \geq 0, \qquad k = 1, 2, \ldots, m$$

As $c_{01} = 1$, $c_{02} = 2$, $c_{03} = 10$, $c_{11} = 3$, $c_{12} = 4$, $c_{21} = 5$, $a_{011} = 1$, $a_{021} = 2$, $a_{031} = -1$, $a_{041} = 0$, $a_{012} = -1$, $a_{022} = -3$, $a_{032} = 0$, $a_{042} = 1$, $a_{013} = 1$, $a_{023} = 0$, $a_{033} = 1$, $a_{043} = 0$, $a_{111} = -1$, $a_{121} = 0$, $a_{131} = 1$, $a_{141} = -2$, $a_{112} = 0$, $a_{122} = 0$, $a_{132} = 1$,

$a_{142} = 1$, $a_{211} = 1$, $a_{221} = 1$, $a_{231} = 0$, and $a_{241} = 0$, Eqs. (E$_1$) become

$$\text{Maximize } v(\boldsymbol{\lambda}) = \left[\frac{c_{01}}{\lambda_{01}}(\lambda_{01} + \lambda_{02} + \lambda_{03})\right]^{\lambda_{01}} \left[\frac{c_{02}}{\lambda_{02}}(\lambda_{01} + \lambda_{02} + \lambda_{03})\right]^{\lambda_{02}}$$

$$\times \left[\frac{c_{03}}{\lambda_{03}}(\lambda_{01} + \lambda_{02} + \lambda_{03})\right]^{\lambda_{03}} \left[\frac{c_{11}}{\lambda_{11}}(\lambda_{11} + \lambda_{12})\right]^{\lambda_{11}}$$

$$\times \left[\frac{c_{12}}{\lambda_{12}}(\lambda_{11} + \lambda_{12})\right]^{\lambda_{12}} \left(\frac{c_{21}}{\lambda_{21}}\lambda_{21}\right)^{\lambda_{21}}$$

subject to

$$\lambda_{01} + \lambda_{02} + \lambda_{03} = 1$$

$$a_{011}\lambda_{01} + a_{012}\lambda_{02} + a_{013}\lambda_{03} + a_{111}\lambda_{11} + a_{112}\lambda_{12} + a_{211}\lambda_{21} = 0$$

$$a_{021}\lambda_{01} + a_{022}\lambda_{02} + a_{023}\lambda_{03} + a_{121}\lambda_{11} + a_{122}\lambda_{12} + a_{221}\lambda_{21} = 0$$

$$a_{031}\lambda_{01} + a_{032}\lambda_{02} + a_{033}\lambda_{03} + a_{131}\lambda_{11} + a_{132}\lambda_{12} + a_{231}\lambda_{21} = 0$$

$$a_{041}\lambda_{01} + a_{042}\lambda_{02} + a_{043}\lambda_{03} + a_{141}\lambda_{11} + a_{142}\lambda_{12} + a_{241}\lambda_{21} = 0$$

$$\lambda_{11} + \lambda_{12} \geq 0$$

$$\lambda_{21} \geq 0$$

or

$$\text{Maximize } v(\boldsymbol{\lambda}) = \left(\frac{1}{\lambda_{01}}\right)^{\lambda_{01}} \left(\frac{2}{\lambda_{02}}\right)^{\lambda_{02}} \left(\frac{10}{\lambda_{03}}\right)^{\lambda_{03}} \left[\frac{3}{\lambda_{11}}(\lambda_{11} + \lambda_{12})\right]^{\lambda_{11}}$$

$$\times \left[\frac{4}{\lambda_{12}}(\lambda_{11} + \lambda_{12})\right]^{\lambda_{12}} (5)^{\lambda_{21}} \tag{E$_2$}$$

subject to

$$\lambda_{01} + \lambda_{02} + \lambda_{03} = 1$$

$$\lambda_{01} - \lambda_{02} + \lambda_{03} - \lambda_{11} + \lambda_{21} = 0$$

$$2\lambda_{01} - 3\lambda_{02} \qquad + \lambda_{21} = 0 \tag{E$_3$}$$

$$-\lambda_{01} + \lambda_{03} + \lambda_{11} + \lambda_{12} = 0$$

$$\lambda_{02} \qquad - 2\lambda_{11} + \lambda_{12} = 0$$

$$\lambda_{11} + \lambda_{12} \geq 0$$

$$\lambda_{21} \geq 0$$

Equations (E$_3$) can be used to express any five of the λ's in terms of the remaining one as follows: Equations (E$_3$) can be rewritten as

$$\lambda_{02} + \lambda_{03} = 1 - \lambda_{01} \tag{E$_4$}$$

$$\lambda_{02} - \lambda_{03} + \lambda_{11} - \lambda_{21} = \lambda_{01} \tag{E$_5$}$$

$$3\lambda_{02} - \lambda_{21} = 2\lambda_{01} \tag{E$_6$}$$

$$\lambda_{12} = \lambda_{01} - \lambda_{03} - \lambda_{11} \tag{E_7}$$

$$\lambda_{12} = 2\lambda_{11} - \lambda_{02} \tag{E_8}$$

From Eqs. (E$_7$) and (E$_8$), we have

$$\lambda_{12} = \lambda_{01} - \lambda_{03} - \lambda_{11} = 2\lambda_{11} - \lambda_{02}$$

$$3\lambda_{11} - \lambda_{02} + \lambda_{03} = \lambda_{01} \tag{E_9}$$

Adding Eqs. (E$_5$) and (E$_9$), we obtain

$$\lambda_{21} = 4\lambda_{11} - 2\lambda_{01} \tag{E_{10}}$$

$$= 3\lambda_{02} - 2\lambda_{01} \quad \text{from Eq. (E}_6\text{)}$$

$$\lambda_{11} = \tfrac{3}{4}\lambda_{02} \tag{E_{11}}$$

Substitution of Eq. (E$_{11}$) in Eq. (E$_8$) gives

$$\lambda_{12} = \tfrac{3}{2}\lambda_{02} - \lambda_{02} = \tfrac{1}{2}\lambda_{02} \tag{E_{12}}$$

Equations (E$_{11}$), (E$_{12}$), and (E$_7$) give

$$\lambda_{03} = \lambda_{01} - \lambda_{11} - \lambda_{12} = \lambda_{01} - \tfrac{3}{4}\lambda_{02} - \tfrac{1}{2}\lambda_{02} = \lambda_{01} - \tfrac{5}{4}\lambda_{02} \tag{E_{13}}$$

By substituting for λ_{03}, Eq. (E$_4$) gives

$$\lambda_{02} = 8\lambda_{01} - 4 \tag{E_{14}}$$

Using this relation for λ_{02}, the expressions for $\lambda_{03}, \lambda_{11}, \lambda_{12}$, and λ_{21} can be obtained as

$$\lambda_{03} = \lambda_{01} - \tfrac{5}{4}\lambda_{02} = -9\lambda_{01} + 5 \tag{E_{15}}$$

$$\lambda_{11} = \tfrac{3}{4}\lambda_{02} = 6\lambda_{01} - 3 \tag{E_{16}}$$

$$\lambda_{12} = \tfrac{1}{2}\lambda_{02} = 4\lambda_{01} - 2 \tag{E_{17}}$$

$$\lambda_{21} = 4\lambda_{11} - 2\lambda_{01} = 22\lambda_{01} - 12 \tag{E_{18}}$$

Thus the objective function in Eq. (E$_2$) can be stated in terms of λ_{01} as

$$v(\lambda_{01}) = \left(\frac{1}{\lambda_{01}}\right)^{\lambda_{01}} \left(\frac{2}{8\lambda_{01}-4}\right)^{8\lambda_{01}-4} \left(\frac{10}{5-9\lambda_{01}}\right)^{5-9\lambda_{01}}$$

$$\times \left(\frac{30\lambda_{01}-15}{6\lambda_{01}-3}\right)^{6\lambda_{01}-3} \left(\frac{40\lambda_{01}-20}{4\lambda_{01}-2}\right)^{4\lambda_{01}-2} (5)^{22\lambda_{01}-12}$$

$$= \left(\frac{1}{\lambda_{01}}\right)^{\lambda_{01}} \left(\frac{1}{4\lambda_{01}-2}\right)^{8\lambda_{01}-4} \left(\frac{10}{5-9\lambda_{01}}\right)^{5-9\lambda_{01}}$$

$$\times (5)^{6\lambda_{01}-3}(10)^{4\lambda_{01}-2}(5)^{22\lambda_{01}-12}$$

$$= \left(\frac{1}{\lambda_{01}}\right)^{\lambda_{01}} \left(\frac{1}{4\lambda_{01}-2}\right)^{8\lambda_{01}-4} \left(\frac{10}{5-9\lambda_{01}}\right)^{5-9\lambda_{01}} (5)^{32\lambda_{01}-17}(2)^{4\lambda_{01}-2}$$

To find the maximum of v, we set the derivative of v with respect to λ_{01} equal to zero. To simplify the calculations, we set $d\,(\ln v)/d\lambda_{01} = 0$ and find the value of λ_{01}^*. Then the values of $\lambda_{02}^*, \lambda_{03}^*, \lambda_{11}^*, \lambda_{12}^*$, and λ_{21}^* can be found from Eqs. (E$_{14}$) to (E$_{18}$). Once the dual variables (λ_{kj}^*) are known, Eqs. (8.62) and (8.63) can be used to find the optimum values of the design variables as in Example 8.3.

8.10   GEOMETRIC PROGRAMMING WITH MIXED INEQUALITY CONSTRAINTS

In this case the geometric programming problem contains at least one signum function with a value of $\sigma_k = -1$ among $k = 1, 2, \ldots, m$. (Note that $\sigma_0 = +1$ corresponds to the objective function.) Here no general statement can be made about the convexity or concavity of the constraint set. However, since the objective function is continuous and is bounded below by zero, it must have a constrained minimum provided that there exist points satisfying the constraints.

Example 8.5

$$\text{Minimize } f = x_1 x_2^2 x_3^{-1} + 2x_1^{-1} x_2^{-3} x_4 + 10 x_1 x_3$$

subject to

$$3x_1 x_3^{-1} x_4^2 + 4x_3^{-1} x_4^{-1} \geq 1$$

$$5x_1 x_2 \leq 1$$

SOLUTION   In this problem, $m = 2$, $N_0 = 3$, $N_1 = 2$, $N_2 = 1$, $N = 6$, $n = 4$, and the degree of difficulty is 1. The signum functions are $\sigma_0 = 1$, $\sigma_1 = -1$, and $\sigma_2 = 1$. The dual objective function can be stated, using Eq. (8.56), as follows:

$$\text{Maximize } v(\boldsymbol{\lambda}) = \prod_{k=0}^{2} \prod_{j=1}^{N_k} \left(\frac{c_{kj}}{\lambda_{kj}} \sum_{l=1}^{N_k} \lambda_{kl} \right)^{\sigma_k \lambda_{kj}}$$

$$= \left[\frac{c_{01}}{\lambda_{01}} (\lambda_{01} + \lambda_{02} + \lambda_{03}) \right]^{\lambda_{01}} \left[\frac{c_{02}}{\lambda_{02}} (\lambda_{01} + \lambda_{02} + \lambda_{03}) \right]^{\lambda_{02}}$$

$$\times \left[\frac{c_{03}}{\lambda_{03}} (\lambda_{01} + \lambda_{02} + \lambda_{03}) \right]^{\lambda_{03}}$$

$$\times \left[\frac{c_{11}}{\lambda_{11}} (\lambda_{11} + \lambda_{12}) \right]^{-\lambda_{11}} \left[\frac{c_{12}}{\lambda_{12}} (\lambda_{11} + \lambda_{12}) \right]^{-\lambda_{12}} \left(\frac{c_{21}}{\lambda_{21}} \lambda_{21} \right)^{\lambda_{21}}$$

$$= \left(\frac{1}{\lambda_{01}} \right)^{\lambda_{01}} \left(\frac{2}{\lambda_{02}} \right)^{\lambda_{02}} \left(\frac{10}{\lambda_{03}} \right)^{\lambda_{03}} \left[\frac{3(\lambda_{11} + \lambda_{12})}{\lambda_{11}} \right]^{-\lambda_{11}}$$

$$\times \left[\frac{4(\lambda_{11} + \lambda_{12})}{\lambda_{12}} \right]^{-\lambda_{12}} (5)^{\lambda_{21}} \tag{E$_1$}$$

The constraints are given by (see Table 8.2)

$$\sum_{j=1}^{N_0} \lambda_{0j} = 1$$

$$\sum_{k=0}^{m} \sum_{j=1}^{N_k} \sigma_k a_{kij} \lambda_{kj} = 0, \quad i = 1, 2, \ldots, n$$

$$\sum_{j=1}^{N_k} \lambda_{kj} \geq 0, \quad k = 1, 2, \ldots, m$$

that is,

$$\lambda_{01} + \lambda_{02} + \lambda_{03} = 1$$

$$\sigma_0 a_{011} \lambda_{01} + \sigma_0 a_{012} \lambda_{02} + \sigma_0 a_{013} \lambda_{03} + \sigma_1 a_{111} \lambda_{11} + \sigma_1 a_{112} \lambda_{12} + \sigma_2 a_{211} \lambda_{21} = 0$$

$$\sigma_0 a_{021} \lambda_{01} + \sigma_0 a_{022} \lambda_{02} + \sigma_0 a_{023} \lambda_{03} + \sigma_1 a_{121} \lambda_{11} + \sigma_1 a_{122} \lambda_{12} + \sigma_2 a_{221} \lambda_{21} = 0$$

$$\sigma_0 a_{031} \lambda_{01} + \sigma_0 a_{032} \lambda_{02} + \sigma_0 a_{033} \lambda_{03} + \sigma_1 a_{131} \lambda_{11} + \sigma_1 a_{132} \lambda_{12} + \sigma_2 a_{231} \lambda_{21} = 0$$

$$\sigma_0 a_{041} \lambda_{01} + \sigma_0 a_{042} \lambda_{02} + \sigma_0 a_{043} \lambda_{03} + \sigma_1 a_{141} \lambda_{11} + \sigma_1 a_{142} \lambda_{12} + \sigma_2 a_{241} \lambda_{21} = 0$$

$$\lambda_{11} + \lambda_{12} \geq 0$$

$$\lambda_{21} \geq 0$$

that is,

$$\lambda_{01} + \lambda_{02} + \lambda_{03} = 1$$

$$\lambda_{01} - \lambda_{02} + \lambda_{03} - \lambda_{11} + \lambda_{21} = 0$$

$$2\lambda_{01} - 3\lambda_{02} + \lambda_{21} = 0 \tag{E$_2$}$$

$$-\lambda_{01} + \lambda_{03} + \lambda_{11} + \lambda_{12} = 0$$

$$\lambda_{02} - 2\lambda_{11} + \lambda_{12} = 0$$

$$\lambda_{11} + \lambda_{12} \geq 0$$

$$\lambda_{21} \geq 0$$

Since Eqs. (E$_2$) are same as Eqs. (E$_3$) of the preceding example, the equality constraints can be used to express λ_{02}, λ_{03}, λ_{11}, λ_{12}, and λ_{21} in terms of λ_{01} as

$$\lambda_{02} = 8\lambda_{01} - 4$$

$$\lambda_{03} = -9\lambda_{01} + 5$$

$$\lambda_{11} = 6\lambda_{01} - 3 \tag{E$_3$}$$

$$\lambda_{12} = 4\lambda_{01} - 2$$

$$\lambda_{21} = 22\lambda_{01} - 12$$

By using Eqs. (E_3), the dual objective function of Eq. (E_1) can be expressed as

$$
v(\lambda_{01}) = \left(\frac{1}{\lambda_{01}}\right)^{\lambda_{01}} \left(\frac{2}{8\lambda_{01} - 4}\right)^{8\lambda_{01} - 4} \left(\frac{10}{-9\lambda_{01} + 5}\right)^{5 - 9\lambda_{01}}
$$

$$
\times \left[\frac{3(10\lambda_{01} - 5)}{6\lambda_{01} - 3}\right]^{-6\lambda_{01} + 3} \left[\frac{4(10\lambda_{01} - 5)}{4\lambda_{01} - 2}\right]^{-4\lambda_{01} + 2} (5)^{22\lambda_{01} - 12}
$$

$$
= \left(\frac{1}{\lambda_{01}}\right)^{\lambda_{01}} \left(\frac{1}{4\lambda_{01} - 2}\right)^{8\lambda_{01} - 4} \left(\frac{10}{5 - 9\lambda_{01}}\right)^{5 - 9\lambda_{01}} (5)^{3 - 6\lambda_{01}} (10)^{2 - 4\lambda_{01}}
$$

$$
\times (5)^{22\lambda_{01} - 12}
$$

$$
= \left(\frac{1}{\lambda_{01}}\right)^{\lambda_{01}} \left(\frac{1}{4\lambda_{01} - 2}\right)^{8\lambda_{01} - 4} \left(\frac{10}{5 - 9\lambda_{01}}\right)^{5 - 9\lambda_{01}} (5)^{12\lambda_{01} - 7} (2)^{2 - 4\lambda_{01}}
$$

To maximize v, set $d(\ln v)/d\lambda_{01} = 0$ and find λ_{01}^*. Once λ_{01}^* is known, λ_{kj}^* can be obtained from Eqs. (E_3) and the optimum design variables from Eqs. (8.62) and (8.63).

8.11 COMPLEMENTARY GEOMETRIC PROGRAMMING

Avriel and Williams [8.4] extended the method of geometric programming to include any rational function of posynomial terms and called the method *complementary geometric programming*.[†] The case in which some terms may be negative will then become a special case of complementary geometric programming. While geometric programming problems have the remarkable property that every constrained local minimum is also a global minimum, no such claim can generally be made for complementary geometric programming problems. However, in many practical situations, it is sufficient to find a local minimum.

The algorithm for solving complementary geometric programming problems consists of successively approximating rational functions of posynomial terms by posynomials. Thus solving a complementary geometric programming problem by this algorithm involves the solution of a sequence of ordinary geometric programming problems. It has been proved that the algorithm produces a sequence whose limit is a local minimum of the complementary geometric programming problem (except in some pathological cases).

Let the complementary geometric programming problem be stated as follows:

$$\text{Minimize } R_0(\mathbf{X})$$

subject to

$$R_k(\mathbf{X}) \leq 1, \quad k = 1, 2, \ldots, m$$

where

$$R_k(\mathbf{X}) = \frac{A_k(\mathbf{X}) - B_k(\mathbf{X})}{C_k(\mathbf{X}) - D_k(\mathbf{X})}, \quad k = 0, 1, 2, \ldots, m \tag{8.66}$$

[†]The application of geometric programming to problems involving generalized polynomial functions was presented by Passy and Wilde [8.2].

where $A_k(\mathbf{X})$, $B_k(\mathbf{X})$, $C_k(\mathbf{X})$, and $D_k(\mathbf{X})$ are posynomials in $\mathbf{X}$ and possibly some of them may be absent. We assume that $R_0(\mathbf{X}) > 0$ for all feasible $\mathbf{X}$. This assumption can always be satisfied by adding, if necessary, a sufficiently large constant to $R_0(\mathbf{X})$.

To solve the problem stated in Eq. (8.66), we introduce a new variable $x_0 > 0$, constrained to satisfy the relation $x_0 \geq R_0(\mathbf{X})$ [i.e., $R_0(\mathbf{X})/x_0 \leq 1$], so that the problem can be restated as

$$\text{Minimize } x_0 \tag{8.67}$$

subject to

$$\frac{A_k(\mathbf{X}) - B_k(\mathbf{X})}{C_k(\mathbf{X}) - D_k(\mathbf{X})} \leq 1, \quad k = 0, 1, 2, \ldots, m \tag{8.68}$$

where

$$A_0(\mathbf{X}) = R_0(\mathbf{X}), \quad C_0(\mathbf{X}) = x_0, \quad B_0(\mathbf{X}) = 0, \quad \text{and} \quad D_0(\mathbf{X}) = 0$$

It is to be noted that the constraints have meaning only if $C_k(\mathbf{X}) - D_k(\mathbf{X})$ has a constant sign throughout the feasible region. Thus if $C_k(\mathbf{X}) - D_k(\mathbf{X})$ is positive for some feasible $\mathbf{X}$, it must be positive for all other feasible $\mathbf{X}$. Depending on the positive or negative nature of the term $C_k(\mathbf{X}) - D_k(\mathbf{X})$, Eq. (8.68) can be rewritten as

$$\frac{A_k(\mathbf{X}) + D_k(\mathbf{X})}{B_k(\mathbf{X}) + C_k(\mathbf{X})} \leq 1$$

or

$$\tag{8.69}$$

$$\frac{B_k(\mathbf{X}) + C_k(\mathbf{X})}{A_k(\mathbf{X}) + D_k(\mathbf{X})} \leq 1$$

Thus any complementary geometric programming problem (CGP) can be stated in standard form as

$$\text{Minimize } x_0 \tag{8.70}$$

subject to

$$\frac{P_k(\mathbf{X})}{Q_k(\mathbf{X})} \leq 1, \quad k = 1, 2, \ldots, m \tag{8.71}$$

$$\mathbf{X} = \begin{Bmatrix} x_0 \\ x_1 \\ x_2 \\ \vdots \\ x_n \end{Bmatrix} > \mathbf{0} \tag{8.72}$$

where $P_k(\mathbf{X})$ and $Q_k(\mathbf{X})$ are posynomials of the form

$$P_k(\mathbf{X}) = \sum_j c_{kj} \prod_{i=0}^{n} (x_i)^{akij} = \sum_j p_{kj}(\mathbf{X}) \tag{8.73}$$

$$Q_k(\mathbf{X}) = \sum_j d_{kj} \prod_{i=0}^{n} (x_i)^{bkij} = \sum_j q_{kj}(\mathbf{X}) \tag{8.74}$$

Solution Procedure.

1. Approximate each of the posynomials $Q(\mathbf{X})^\dagger$ by a posynomial term. Then all the constraints in Eq. (8.71) can be expressed as a posynomial to be less than or equal to 1. This follows because a posynomial divided by a posynomial term is again a posynomial. Thus with this approximation, the problem reduces to an ordinary geometric programming problem. To approximate $Q(\mathbf{X})$ by a single-term posynomial, we choose any $\underset{\sim}{\mathbf{X}} > \mathbf{0}$ and let

$$U_j = q_j(\mathbf{X}) \tag{8.75}$$

$$\Delta_j = \frac{q_j(\underset{\sim}{\mathbf{X}})}{Q(\underset{\sim}{\mathbf{X}})} \tag{8.76}$$

where q_j denotes the jth term of the posynomial $Q(\mathbf{X})$. Thus we obtain, by using the arithmetic–geometric inequality, Eq. (8.22),

$$Q(\mathbf{X}) = \sum_j q_j(\mathbf{X}) \geq \prod_j \left[\frac{q_j(\mathbf{X})}{q_j(\underset{\sim}{\mathbf{X}})} Q(\underset{\sim}{\mathbf{X}}) \right]^{q_j(\underset{\sim}{\mathbf{X}})/Q(\underset{\sim}{\mathbf{X}})} \tag{8.77}$$

By using Eq. (8.74), the inequality (8.77) can be restated as

$$Q(\mathbf{X}) \geq \underset{\sim}{Q}(\mathbf{X}, \underset{\sim}{\mathbf{X}}) \equiv Q(\underset{\sim}{\mathbf{X}}) \prod_i \left(\frac{x_i}{\underset{\sim}{x_i}} \right)^{\sum_j [b_{ij} q_j(\underset{\sim}{\mathbf{X}})/Q(\underset{\sim}{\mathbf{X}})]} \tag{8.78}$$

where the equality sign holds true if $x_i = \underset{\sim}{x_i}$. We can take $\underset{\sim}{Q}(\mathbf{X}, \underset{\sim}{\mathbf{X}})$ as an approximation for $Q(\mathbf{X})$ at $\underset{\sim}{\mathbf{X}}$.

2. At any feasible point $\mathbf{X}^{(1)}$, replace $Q_k(\mathbf{X})$ in Eq. (8.71) by their approximations $\underset{\sim}{Q}_k(\mathbf{X}, \mathbf{X}^{(1)})$, and solve the resulting ordinary geometric programming problem to obtain the next point $\mathbf{X}^{(2)}$.

3. By continuing in this way, we generate a sequence $\{\mathbf{X}^{(\alpha)}\}$, where $\mathbf{X}^{(\alpha+1)}$ is an optimal solution for the αth ordinary geometric programming problem (OGP$_\alpha$):

$$\text{Minimize } x_0$$

subject to

$$\frac{P_k(\mathbf{X})}{\underset{\sim}{Q}_k(\mathbf{X}, \mathbf{X}^{(\alpha)})} \leq 1, \quad k = 1, 2, \ldots, m$$

$$\mathbf{X} = \begin{Bmatrix} x_0 \\ x_1 \\ x_2 \\ \vdots \\ x_n \end{Bmatrix} > \mathbf{0} \tag{8.79}$$

It has been proved [8.4] that under certain mild restrictions, the sequence of points $\{\mathbf{X}^{(\alpha)}\}$ converges to a local minimum of the complementary geometric programming problem.

$\dagger$The subscript k is removed for $Q(\mathbf{X})$ for simplicity.

Degree of Difficulty. The degree of difficulty of a complementary geometric programming problem (CGP) is also defined as

$$\text{degree of difficulty} = N - n - 1$$

where N indicates the total number of terms appearing in the numerators of Eq. (8.71). The relation between the degree of difficulty of a CGP and that of the OGP_α, the approximating ordinary geometric program, is important. The degree of difficulty of a CGP is always equal to that of the approximating OGP_α, solved at each iteration. Thus a CGP with zero degree of difficulty and an arbitrary number of negative terms can be solved by a series of solutions to square systems of linear equations. If the CGP has one degree of difficulty, at each iteration we solve an OGP with one degree of difficulty, and so on. The degree of difficulty is independent of the choice of $\mathbf{X}^{(\alpha)}$ and is fixed throughout the iterations. The following example is considered to illustrate the procedure of complementary geometric programming.

Example 8.6
$$\text{Minimize } x_1$$

subject to

$$-4x_1^2 + 4x_2 \leq 1$$

$$x_1 + x_2 \geq 1$$

$$x_1 > 0, \quad x_2 > 0$$

SOLUTION This problem can be stated as a complementary geometric programming problem as

$$\text{Minimize } x_1 \tag{E_1}$$

subject to

$$\frac{4x_2}{1 + 4x_1^2} \leq 1 \tag{E_2}$$

$$\frac{x_1^{-1}}{1 + x_1^{-1}x_2} \leq 1 \tag{E_3}$$

$$x_1 > 0 \tag{E_4}$$

$$x_2 > 0 \tag{E_5}$$

Since there are two variables (x_1 and x_2) and three posynomial terms [one term in the objective function and one term each in the numerators of the constraint Eqs. (E_2) and (E_3)], the degree of difficulty of the CGP is zero. If we denote the denominators of Eqs. (E_2) and (E_3) as

$$Q_1(\mathbf{X}) = 1 + 4x_1^2$$

$$Q_2(\mathbf{X}) = 1 + x_1^{-1}x_2$$

they can each be approximated by a single-term posynomial with the help of Eq. (8.78) as

$$Q_1(\mathbf{X}, \underset{\sim}{\mathbf{X}}) = (1 + 4\underset{\sim}{x_1^2}) \left(\frac{x_1}{x_2}\right)^{8\underset{\sim}{x_1^2}/(1+4\underset{\sim}{x_1^2})}$$

$$Q_2(\mathbf{X}, \underset{\sim}{\mathbf{X}}) = \left(1 + \frac{\underset{\sim}{x_2}}{\underset{\sim}{x_1}}\right) \left(\frac{x_1}{\underset{\sim}{x_1}}\right)^{-\underset{\sim}{x_2}/(\underset{\sim}{x_1}+\underset{\sim}{x_2})} \left(\frac{x_2}{\underset{\sim}{x_1}}\right)^{\underset{\sim}{x_2}/(\underset{\sim}{x_1}+\underset{\sim}{x_2})}$$

Let us start the iterative process from the point $\mathbf{X}^{(1)} = \{\begin{smallmatrix}1\\1\end{smallmatrix}\}$, which can be seen to be feasible. By taking $\underset{\sim}{\mathbf{X}} = \mathbf{X}^{(1)}$, we obtain

$$\mathbf{Q}_1(\mathbf{X}, \mathbf{X}^{(1)}) = 5x_1^{8/5}$$

$$Q_2(\mathbf{X}, \mathbf{X}^{(1)}) = 2x_1^{-1/2}x_2^{1/2}$$

and we formulate the first ordinary geometric programming problem (OGP$_1$) as

$$\text{Minimize } x_1$$

subject to

$$\tfrac{4}{5}x_1^{-8/5}x_2 \le 1$$

$$\tfrac{1}{2}x_1^{-1/2}x_2^{-1/2} \le 1$$

$$x_1 > 0$$

$$x_2 > 0$$

Since this (OGP$_1$) is a geometric programming problem with zero degree of difficulty, its solution can be found by solving a square system of linear equations, namely

$$\lambda_1 = 1$$

$$\lambda_1 - \tfrac{8}{5}\lambda_2 - \tfrac{1}{2}\lambda_3 = 0$$

$$\lambda_2 - \tfrac{1}{2}\lambda_3 = 0$$

The solution is $\lambda_1^* = 1$, $\lambda_2^* = \frac{5}{13}$, $\lambda_3^* = \frac{10}{13}$. By substituting this solution into the dual objective function, we obtain

$$v(\boldsymbol{\lambda}^*) = (\tfrac{4}{5})^{5/13} (\tfrac{1}{2})^{10/13} \simeq 0.5385$$

From the duality relations, we get

$$x_1 \simeq 0.5385 \quad \text{and} \quad x_2 = \tfrac{5}{4}(x_1)^{8/15} \simeq 0.4643$$

Thus the optimal solution of OGP$_1$ is given by

$$\mathbf{X}_{\text{opt}}^{(1)} = \begin{Bmatrix} 0.5385 \\ 0.4643 \end{Bmatrix}$$

Next we choose $\mathbf{X}^{(2)}$ to be the optimal solution of OGP_1 [i.e., $\mathbf{X}_{opt}^{(1)}$] and approximate Q_1 and Q_2 about this point, solve OGP_2, and so on. The sequence of optimal solutions of OGP_α as generated by the iterative procedure is shown below:

Iteration number, α	$\mathbf{X}_{opt}$	
	x_1	x_2
0	1.0	1.0
1	0.5385	0.4643
2	0.5019	0.5007
3	0.5000	0.5000

The optimal values of the variables for the CGP are $x_1^* = 0.5$ and $x_2^* = 0.5$. It can be seen that in three iterations, the solution of the approximating geometric programming problems OGP_α is correct to four significant figures.

8.12 APPLICATIONS OF GEOMETRIC PROGRAMMING

Example 8.7 Determination of Optimum Machining Conditions [8.9, 8.10] Geometric programming has been applied for the determination of optimum cutting speed and feed which minimize the unit cost of a turning operation.

Formulation as a Zero-degree-of-difficulty Problem
The total cost of turning per piece is given by

$$f_0(\mathbf{X}) = \text{machining cost} + \text{tooling cost} + \text{handling cost}$$

$$= K_m t_m + \frac{t_m}{T}(K_m t_c + K_t) + K_m t_h \tag{E_1}$$

where K_m is the cost of operating time (\$/min), K_t the tool cost (\$/cutting edge), t_m the machining time per piece (min) $= \pi DL/(12VF)$, T the tool life (min/cutting edge) $= (a/VF^b)^{1/c}$, t_c the tool changing time (minutes/workpiece), t_h the handling time (min/workpiece), D the diameter of the workpiece (in), L the axial length of the workpiece (in.), V the cutting speed (ft/min), F the feed (in./revolution), a, b, and c are constants in tool life equation, and

$$\mathbf{X} = \begin{Bmatrix} x_1 \\ x_2 \end{Bmatrix} = \begin{Bmatrix} V \\ F \end{Bmatrix}$$

Since the constant term will not affect the minimization, the objective function can be taken as

$$f(\mathbf{X}) = C_{01}V^{-1}F^{-1} + C_{02}V^{1/c-1}F^{b/c-1} \tag{E_2}$$

where

$$C_{01} = \frac{K_m \pi DL}{12} \quad \text{and} \quad C_{02} = \frac{\pi DL(K_m t_c + K_t)}{12\,a^{1/c}} \tag{E_3}$$

If the maximum feed allowable on the lathe is F_{max}, we have the constraint

$$C_{11}F \leq 1 \tag{E$_4$}$$

where

$$C_{11} = F_{max}^{-1} \tag{E$_5$}$$

Since the total number of terms is three and the number of variables is two, the degree of difficulty of the problem is zero. By using the data

$$K_m = 0.10, \quad K_t = 0.50, \quad t_c = 0.5, \quad t_h = 2.0, \quad D = 6.0,$$
$$L = 8.0, \quad a = 140.0, \quad b = 0.29, \quad c = 0.25, \quad F_{max} = 0.005$$

the solution of the problem [minimize f given in Eq. (E$_2$) subject to the constraint (E$_4$)] can be obtained as

$$f^* = \$1.03 \text{ per piece}, \quad V^* = 323 \text{ ft/min}, \quad F^* = 0.005 \text{ in./rev}$$

Formulation as a One-degree-of-difficulty Problem

If the maximum horsepower available on the lathe is given by P_{max}, the power required for machining should be less than P_{max}. Since the power required for machining can be expressed as $a_1 V^{b_1} F^{c_1}$, where a_1, b_1, and c_1 are constants, this constraint can be stated as follows:

$$C_{21} V^{b_1} F^{c_1} \leq 1 \tag{E$_6$}$$

where

$$C_{21} = a_1 P_{max}^{-1} \tag{E$_7$}$$

If the problem is to minimize f given by Eq. (E$_2$) subject to the constraints (E$_4$) and (E$_6$), it will have one degree of difficulty. By taking $P_{max} = 2.0$ and the values of a_1, b_1, and c_1 as 3.58, 0.91, and 0.78, respectively, in addition to the previous data, the following result can be obtained:

$$f^* = \$1.05 \text{ per piece}, \quad V^* = 290.0 \text{ ft/min}, \quad F^* = 0.005 \text{ in./rev}$$

Formulation as a Two-degree-of-difficulty Problem

If a constraint on the surface finish is included as

$$a_2 V^{b_2} F^{c_2} \leq S_{max}$$

where a_2, b_2, and c_2 are constants and S_{max} is the maximum permissible surface roughness in microinches, we can restate this restriction as

$$C_{31} V^{b_2} F^{c_2} \leq 1 \tag{E$_8$}$$

where

$$C_{31} = a_2 S_{max}^{-1} \tag{E_9}$$

If the constraint (E$_8$) is also included, the problem will have a degree of difficulty two. By taking $a_2 = 1.36 \times 10^8$, $b_2 = -1.52$, $c_2 = 1.004$, $S_{max} = 100\,\mu in.$, $F_{max} = 0.01$, and $P_{max} = 2.0$ in addition to the previous data, we obtain the following result:

$$f^* = \$1.11 \text{ per piece}, \quad V^* = 311\,ft/min, \quad F^* = 0.0046\,in./rev$$

Example 8.8 Design of a Hydraulic Cylinder [8.11] The minimum volume design of a hydraulic cylinder (subject to internal pressure) is considered by taking the piston diameter (d), force (f), hydraulic pressure (p), stress (s), and the cylinder wall thickness (t) as design variables. The following constraints are considered:

Minimum force required is F, that is,

$$f = p\frac{\pi d^2}{4} \geq F \tag{E_1}$$

Hoop stress induced should be less than S, that is,

$$s = \frac{pd}{2t} \leq S \tag{E_2}$$

Side constraints:

$$d + 2t \leq D \tag{E_3}$$

$$p \leq P \tag{E_4}$$

$$t \geq T \tag{E_5}$$

where D is the maximum outside diameter permissible, P the maximum pressure of the hydraulic system and T the minimum cylinder wall thickness required. Equations (E$_1$) to (E$_5$) can be stated in normalized form as

$$\frac{4}{\pi} F p^{-1} d^{-2} \leq 1$$

$$\tfrac{1}{2} S^{-1} p d t^{-1} \leq 1$$

$$D^{-1} d + 2 D^{-1} t \leq 1$$

$$P^{-1} p \leq 1$$

$$T t^{-1} \leq 1$$

The volume of the cylinder per unit length (objective) to be minimized is given by $\pi t(d + t)$.

Example 8.9 Design of a Cantilever Beam Formulate the problem of determining the cross-sectional dimensions of the cantilever beam shown in Fig. 8.2 for minimum weight. The maximum permissible bending stress is σ_y.

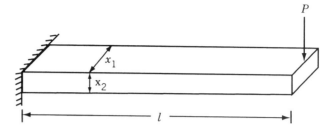

Figure 8.2 Cantilever beam of rectangular cross section.

SOLUTION   The width and depth of the beam are considered as design variables. The objective function (weight) is given by

$$f(\mathbf{X}) = \rho l x_1 x_2 \tag{E$_1$}$$

where ρ is the weight density and l is the length of the beam. The maximum stress induced at the fixed end is given by

$$\sigma = \frac{Mc}{I} = Pl\frac{x_2}{2}\frac{1}{\frac{1}{12}x_1 x_2^3} = \frac{6Pl}{x_1 x_2^2} \tag{E$_2$}$$

and the constraint becomes

$$\frac{6Pl}{\sigma_y}x_1^{-1}x_2^{-2} \leq 1 \tag{E$_3$}$$

Example 8.10 Design of a Cone Clutch [8.23]   Find the minimum volume design of the cone clutch shown in Fig.1.18 such that it can transmit a specified minimum torque.

SOLUTION   By selecting the outer and inner radii of the cone, R_1 and R_2, as design variables, the objective function can be expressed as

$$f(R_1, R_2) = \tfrac{1}{3}\pi h(R_1^2 + R_1 R_2 + R_2^2) \tag{E$_1$}$$

where the axial thickness, h, is given by

$$h = \frac{R_1 - R_2}{\tan \alpha} \tag{E$_2$}$$

Equations (E$_1$) and (E$_2$) yield

$$f(R_1, R_2) = k_1(R_1^3 - R_2^3) \tag{E$_3$}$$

where

$$k_1 = \frac{\pi}{3 \tan \alpha} \tag{E$_4$}$$

The axial force applied (F) and the torque developed (T) are given by [8.37]

$$F = \int p \, dA \sin\alpha = \int_{R_2}^{R_1} p \frac{2\pi r \, dr}{\sin\alpha} \sin\alpha = \pi p(R_1^2 - R_2^2) \tag{E5}$$

$$T = \int rfp \, dA = \int_{R_2}^{R_1} rfp \frac{2\pi r}{\sin\alpha} dr = \frac{2\pi fp}{3\sin\alpha}(R_1^3 - R_2^3) \tag{E6}$$

where p is the pressure, f the coefficient of friction, and A the area of contact. Substitution of p from Eq. (E5) into (E6) leads to

$$T = \frac{k_2(R_1^2 + R_1 R_2 + R_2^2)}{R_1 + R_2} \tag{E7}$$

where

$$k_2 = \frac{2Ff}{3\sin\alpha} \tag{E8}$$

Since k_1 is a constant, the objective function can be taken as $f = R_1^3 - R_2^3$. The minimum torque to be transmitted is assumed to be $5k_2$. In addition, the outer radius R_1 is assumed to be equal to at least twice the inner radius R_2. Thus the optimization problem becomes

$$\text{Minimize } f(R_1, R_2) = R_1^3 - R_2^3$$

subject to

$$\frac{R_1^2 + R_2 R_2 + R_2^2}{R_1 + R_2} \geq 5 \tag{E9}$$

$$\frac{R_1}{R_2} \geq 2$$

This problem has been solved using complementary geometric programming [8.23] and the solution was found iteratively as shown in Table 8.3. Thus the final solution is taken as $R_1^* = 4.2874$, $R_2^* = 2.1437$, and $f^* = 68.916$.

Example 8.11 Design of a Helical Spring   Formulate the problem of minimum weight design of a helical spring under axial load as a geometric programming problem. Consider constraints on the shear stress, natural frequency, and buckling of the spring.

SOLUTION   By selecting the mean diameter of the coil and the diameter of the wire as the design variables, the design vector is given by

$$\mathbf{X} = \begin{Bmatrix} x_1 \\ x_2 \end{Bmatrix} = \begin{Bmatrix} D \\ d \end{Bmatrix} \tag{E1}$$

The objective function (weight) of the helical spring can be expressed as

$$f(\mathbf{X}) = \frac{\pi d^2}{4}(\pi D)\rho(n + Q) \tag{E2}$$

Table 8.3 Results for Example 8.10

Iteration number	Starting design	Ordinary geometric programming problem	Solution of OGP
1	$x_1 = R_0 = 40$ $x_2 = R_1 = 3$ $x_3 = R_2 = 3$	Minimize $x_1^1 x_2^0 x_3^0$ subject to $0.507 x_1^{-0.597} x_2^3 x_3^{-1.21} \leq 1$ $1.667(x_2^{-1} + x_3^{-1}) \leq 1$	$x_1 = 162.5$ $x_2 = 5.0$ $x_3 = 2.5$
2	$x_1 = R_0 = 162.5$ $x_2 = R_1 = 5.0$ $x_3 = R_2 = 2.5$	Minimize $x_1^1 x_2^0 x_3^0$ subject to $0.744 x_1^{-0.912} x_2^3 x_3^{-0.2635} \leq 1$ $3.05(x_2^{-0.43} x_3^{-0.571} + x_2^{-1.43} x_3^{0.429}) \leq 1$ $2 x_2^{-1} x_3 \leq 1$	$x_1 = 82.2$ $x_2 = 4.53$ $x_3 = 2.265$
3	$x_1 = R_0 = 82.2$ $x_2 = R_1 = 4.53$ $x_3 = R_2 = 2.265$	Minimize $x_1^1 x_2^0 x_3^0$ subject to $0.687 x_1^{-0.876} x_2^3 x_3^{-0.372} \leq 1$ $1.924 x_1^0 x_2^{-0.429} x_3^{-0.571} +$ $1.924 x_1^0 x_2^{-1.492} x_3^{0.429} \leq 1$ $2 x_2^{-1} x_3 \leq 1$	$x_1 = 68.916$ $x_2 = 4.2874$ $x_3 = 2.1437$

where n is the number of active turns, Q the number of inactive turns, and ρ the weight density of the spring. If the deflection of the spring is δ, we have

$$\delta = \frac{8PC^3 n}{Gd} \quad \text{or} \quad n = \frac{Gd\delta}{8PC^3} \tag{E_3}$$

where G is the shear modulus, P the axial load on the spring, and C the spring index ($C = D/d$). Substitution of Eq. (E$_3$) into (E$_2$) gives

$$f(\mathbf{X}) = \frac{\pi^2 \rho G \delta}{32 P} \frac{d^6}{D^2} + \frac{\pi^2 \rho Q}{4} d^2 D \tag{E_4}$$

If the maximum shear stress in the spring (τ) is limited to τ_{max}, the stress constraint can be expressed as

$$\tau = \frac{8KPC}{\pi d^2} \leq \tau_{max} \quad \text{or} \quad \frac{8KPC}{\pi d^2 \tau_{max}} \leq 1 \tag{E_5}$$

where K denotes the stress concentration factor given by

$$K \approx \frac{2}{C^{0.25}} \tag{E_6}$$

The use of Eq. (E$_6$) in (E$_5$) results in

$$\frac{16P}{\pi \tau_{max}} \frac{D^{3/4}}{d^{11/4}} \leq 1 \tag{E_7}$$

To avoid fatigue failure, the natural frequency of the spring (f_n) is to be restricted to be greater than $(f_n)_{\min}$. The natural frequency of the spring is given by

$$f_n = \frac{2d}{\pi D^2 n}\left(\frac{Gg}{32\rho}\right)^{1/2} \tag{E_8}$$

where g is the acceleration due to gravity. Using $g = 9.81$ m/s^2, $G = 8.56 \times 10^{10}$ N/m^2, and $(f_n)_{\min} = 13$, Eq. (E$_8$) becomes

$$\frac{13(f_n)_{\min}\delta G}{288,800 P}\frac{d^3}{D} \leq 1 \tag{E_9}$$

Similarly, in order to avoid buckling, the free length of the spring is to be limited as

$$L \leq \frac{11.5(D/2)^2}{P/K^1} \tag{E_{10}}$$

Using the relations

$$K^1 = \frac{Gd^4}{8D^3 n} \tag{E_{11}}$$

$$L = nd(1 + Z) \tag{E_{12}}$$

and $Z = 0.4$, Eq. (E$_{10}$) can be expressed as

$$0.0527\left(\frac{G\delta^2}{P}\right)\frac{d^5}{D^5} \leq 1 \tag{E_{13}}$$

It can be seen that the problem given by the objective function of Eq. (E$_4$) and constraints of Eqs. (E$_7$), (E$_9$), and (E$_{13}$) is a geometric programming problem.

Example 8.12 Design of a Lightly Loaded Bearing [8.29] A lightly loaded bearing is to be designed to minimize a linear combination of frictional moment and angle of twist of the shaft while carrying a load of 1000 lb. The angular velocity of the shaft is to be greater than 100 rad/s.

SOLUTION

Formulation as a Zero-Degree-of-Difficulty Problem
The frictional moment of the bearing (M) and the angle of twist of the shaft (ϕ) are given by

$$M = \frac{8\pi}{\sqrt{1-n^2}}\frac{\mu\Omega}{c}R^2 L \tag{E_1}$$

$$\phi = \frac{S_e l}{GR} \tag{E_2}$$

where μ is the viscosity of the lubricant, n the eccentricity ratio $(= e/c)$, e the eccentricity of the journal relative to the bearing, c the radial clearance, Ω the angular velocity of the shaft, R the radius of the journal, L the half-length of the bearing, S_e the shear stress, l the length between the driving point and the rotating mass, and G the shear modulus. The load on each bearing (W) is given by

$$W = \frac{2\mu\Omega R L^2 n}{c^2(1-n^2)^2}[\pi^2(1-n^2)+16n^2]^{1/2} \tag{E_3}$$

For the data $W = 1000\,\text{lb}$, $c/R = 0.0015$, $n = 0.9$, $l = 10$ in., $S_e = 30{,}000\,\text{psi}$, $\mu = 10^{-6}$ lb-s/in^2, and $G = 12 \times 10^6$ psi, the objective function and the constraint reduce to

$$f(R,L) = aM + b\phi = 0.038\Omega R^2 L + 0.025 R^{-1} \tag{E_4}$$

$$\Omega R^{-1} L^3 = 11.6 \tag{E_5}$$

$$\Omega \geq 100 \tag{E_6}$$

where a and b are constants assumed to be $a = b = 1$. Using the solution of Eq. (E$_5$) gives

$$\Omega = 11.6 R L^{-3} \tag{E_7}$$

the optimization problem can be stated as

$$\text{Minimize } f(R,L) = 0.45 R^3 L^{-2} + 0.025 R^{-1} \tag{E_8}$$

subject to

$$8.62 R^{-1} L^3 \leq 1 \tag{E_9}$$

The solution of this zero-degree-of-difficulty problem can be determined as $R^* = 0.212$ in., $L^* = 0.291$ in., and $f^* = 0.17$.

Formulation as a One-Degree-of-Difficulty Problem

By considering the objective function as a linear combination of the frictional moment (M), the angle of twist of the shaft (ϕ), and the temperature rise of the oil (T), we have

$$f = aM + b\phi + cT \tag{E_{10}}$$

where a, b, and c are constants. The temperature rise of the oil in the bearing is given by

$$T = 0.045 \frac{\mu\Omega R^2}{c^2 n \sqrt{(1-n^2)}} \tag{E_{11}}$$

By assuming that 1 in.-lb of frictional moment in bearing is equal to 0.0025 rad of angle of twist, which, in turn, is equivalent to 1 °F rise in temperature, the constants a, b, and c can be determined. By using Eq. (E$_7$), the optimization problem can be stated as

$$\text{Minimize } f(R,L) = 0.44 R^3 L^{-2} + 10 R^{-1} + 0.592 R L^{-3} \tag{E_{12}}$$

subject to

$$8.62R^{-1}L^3 \leq 1 \tag{E$_{13}$}$$

The solution of this one-degree-of-difficulty problem can be found as $R^* = 1.29$, $L^* = 0.53$, and $f^* = 16.2$.

Example 8.13 Design of a Two-bar Truss [8.33]   The two-bar truss shown in Fig. 8.3 is subjected to a vertical load $2P$ and is to be designed for minimum weight. The members have a tubular section with mean diameter d and wall thickness t and the maximum permissible stress in each member (σ_0) is equal to 60,000 psi. Determine the values of h and d using geometric programming for the following data: $P = 33,000$ lb, $t = 0.1$ in., $b = 30$ in., $\sigma_0 = 60,000$ psi, and ρ (density) $= 0.3$ lb/in^3.

SOLUTION   The objective function is given by

$$\begin{aligned} f(d, h) &= 2\rho\pi dt\sqrt{b^2 + h^2} \\ &= 2(0.3)\pi d(0.1)\sqrt{900 + h^2} = 0.188d\sqrt{900 + h^2} \end{aligned} \tag{E$_1$}$$

The stress constraint can be expressed as

$$\sigma = \frac{P}{\pi dt}\frac{\sqrt{900 + h^2}}{h} \leq \sigma_0$$

or

$$\frac{33,000}{\pi d(0.1)}\frac{\sqrt{900 + h^2}}{h} \leq 60,000$$

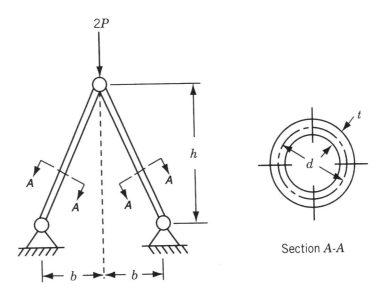

Figure 8.3   Two-bar truss under load.

or

$$1.75 \frac{\sqrt{900 + h^2}}{dh} \leq 1 \tag{E_2}$$

It can be seen that the functions in Eqs. (E$_1$) and (E$_2$) are not posynomials, due to the presence of the term $\sqrt{900 + h^2}$. The functions can be converted to posynomials by introducing a new variable y as

$$y = \sqrt{900 + h^2} \quad \text{or} \quad y^2 = 900 + h^2$$

and a new constraint as

$$\frac{900 + h^2}{y^2} \leq 1 \tag{E_3}$$

Thus the optimization problem can be stated, with $x_1 = y$, $x_2 = h$, and $x_3 = d$ as design variables, as

$$\text{Minimize } f = 0.188 yd \tag{E_4}$$

subject to

$$1.75 yh^{-1} d^{-1} \leq 1 \tag{E_5}$$

$$900 y^{-2} + y^{-2} h^2 \leq 1 \tag{E_6}$$

For this zero-degree-of-difficulty problem, the associated dual problem can be stated as

$$\text{Maximize } v(\lambda_{01}, \lambda_{11}, \lambda_{21}, \lambda_{22})$$

$$= \left(\frac{0.188}{\lambda_{01}}\right)^{\lambda_{01}} \left(\frac{1.75}{\lambda_{11}}\right)^{\lambda_{11}} \left(\frac{900}{\lambda_{21}}\right)^{\lambda_{21}} \left(\frac{1}{\lambda_{22}}\right)^{\lambda_{22}} (\lambda_{21} + \lambda_{22})^{\lambda_{21} + \lambda_{22}} \tag{E_7}$$

subject to

$$\lambda_{01} = 1 \tag{E_8}$$

$$\lambda_{01} + \lambda_{11} - 2\lambda_{21} - 2\lambda_{22} = 0 \tag{E_9}$$

$$-\lambda_{11} + 2\lambda_{22} = 0 \tag{E_{10}}$$

$$\lambda_{01} - \lambda_{11} = 0 \tag{E_{11}}$$

The solution of Eqs. (E$_8$) to (E$_{11}$) gives $\lambda_{01}^* = 1$, $\lambda_{11}^* = 1$, $\lambda_{21}^* = \frac{1}{2}$, and $\lambda_{22}^* = \frac{1}{2}$. Thus the maximum value of v and the minimum value of f is given by

$$v^* = \left(\frac{0.188}{1}\right)^1 (1.75)^1 \left(\frac{900}{0.5}\right)^{0.5} \left(\frac{1}{0.5}\right)^{0.5} (0.5 + 0.5)^{0.5 + 0.5} = 19.8 = f^*$$

The optimum values of x_i can be found from Eqs. (8.62) and (8.63):

$$1 = \frac{0.188 y^* d^*}{19.8}$$

$$1 = 1.75 y^* h^{*-1} d^{*-1}$$

$$\tfrac{1}{2} = 900 y^{*-2}$$

$$\tfrac{1}{2} = y^{*-2} h^{*2}$$

These equations give the solution: $y^* = 42.426$, $h^* = 30\,\text{in.}$, and $d^* = 2.475\,\text{in.}$

Example 8.14 Design of a Four-bar Mechanism [8.24]   Find the link lengths of the four-bar linkage shown in Fig. 8.4 for minimum structural error.

SOLUTION   Let a, b, c, and d denote the link lengths, θ the input angle, and ϕ the output angle of the mechanism. The loop closure equation of the linkage can be expressed as

$$2ad \cos\theta - 2cd \cos\phi + (a^2 - b^2 + c^2 + d^2)$$
$$- 2ac \cos(\theta - \phi) = 0 \tag{E_1}$$

In function-generating linkages, the value of ϕ generated by the mechanism is made equal to the desired value, ϕ_d, only at some values of θ. These are known as *precision points*. In general, for arbitrary values of the link lengths, the actual output angle (ϕ_i) generated for a particular input angle (θ_i) involves some error (ε_i) compared to the desired value (ϕ_{di}), so that

$$\phi_i = \phi_{di} + \varepsilon_i \tag{E_2}$$

where ε_i is called the *structural error* at θ_i. By substituting Eq. (E_2) into (E_1) and assuming that $\sin \varepsilon_i \approx \varepsilon_i$ and $\cos \varepsilon_i \approx 1$ for small values of ε_i, we obtain

$$\varepsilon_i = \frac{K + 2ad \cos\theta_i - 2cd \cos\theta_{di} - 2ac \cos\theta_i \cos(\phi_{di} - \theta_i)}{-2ac \sin(\phi_{di} - \theta_i) - 2cd \sin\phi_{di}} \tag{E_3}$$

where

$$K = a^2 - b^2 + c^2 + d^2 \tag{E_4}$$

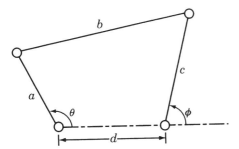

Figure 8.4   Four-bar linkage.

The objective function for minimization is taken as the sum of squares of structural error at a number of precision or design positions, so that

$$f = \sum_{i=1}^{n} \varepsilon_i^2 \tag{E_5}$$

where n denotes the total number of precision points considered. Note that the error ε_i is minimized when f is minimized (ε_i will not be zero, usually).

For simplicity, we assume that $a \ll d$ and that the error ε_i is zero at θ_0. Thus $\varepsilon_0 = 0$ at $\theta_i = \theta_0$, and Eq. (E_3) yields

$$K = 2cd \cos \phi_{di} + 2ac \cos \theta_0 \cos(\phi_{d0} - \theta_0) - 2ad \cos \theta_0 \tag{E_6}$$

In view of the assumption $a \ll d$, we impose the constraint as (for convenience)

$$\frac{3a}{d} \leq 1 \tag{E_7}$$

where any larger number can be used in place of 3. Thus the objective function for minimization can be expressed as

$$f = \sum_{i=1}^{n} \frac{a^2(\cos \theta_i - \cos \theta_0)^2 - 2ac(\cos \theta_i - \cos \theta_0)(\cos \phi_{di} - \cos \phi_{d0})}{c^2 \sin^2 \phi_{di}} \tag{E_8}$$

Usually, one of the link lengths is taken as unity. By selecting a and c as the design variables, the normality and orthogonality conditions can be written as

$$\Delta_1^* + \Delta_2^* = 1 \tag{E_9}$$

$$2\Delta_1^* + \Delta_2^* = 0 \tag{E_{10}}$$

$$2\Delta_1^* + 0.5\Delta_2^* + \Delta_3^* = 0 \tag{E_{11}}$$

These equations yield the solution $\Delta_1^* = -1$, $\Delta_2^* = 2$, and $\Delta_3^* = 1$, and the maximum value of the dual function is given by

$$v(\mathbf{\Delta}^*) = \left(\frac{c_1}{\Delta_1^*} \right)^{\Delta_1^*} \left(\frac{c_2}{\Delta_2^*} \right)^{\Delta_2^*} \left(\frac{c_3}{\Delta_3^*} \right)^{\Delta_3^*} \tag{E_{12}}$$

where c_1, c_2, and c_3 denote the coefficients of the posynomial terms in Eqs. (E_7) and (E_8).

For numerical computation, the following data are considered:

Precision point, i	1	2	3	4	5	6
Input, θ_i (deg)	0	10	20	30	40	45
Desired output, ϕ_{di} (deg)	30	38	47	58	71	86

If we select the precision point 4 as the point where the structural error is zero ($\theta_0 = 30°$, $\phi_{d0} = 58°$), Eq. (E_8) gives

$$f = 0.1563 \frac{a^2}{c^2} - \frac{0.76a}{c} \tag{E_{13}}$$

subject to

$$\frac{3a}{d} \leq 1$$

Noting that $c_1 = 0.1563$, $c_2 = 0.76$, and $c_3 = 3/d$, we see that Eq. (E_{12}) gives

$$v(\Delta) = \left(\frac{0.1563}{-1}\right)^{-1} \left(\frac{-0.76}{2}\right)^2 \left(\frac{3}{d}\right)^1 (1)^1 = \frac{2.772}{d}$$

Noting that

$$0.1563\frac{a^2}{c^2} = \left(-\frac{2.772}{d}\right)(-1) = \frac{2.772}{d}$$

$$-0.76\frac{a}{c} = -\frac{2.772}{d}(2) = -\frac{5.544}{d}$$

and using $a = 1$, we find that $c^* = 0.41$ and $d^* = 3.0$. In addition, Eqs. (E_6) and (E_4) yield

$$a^2 - b^2 + c^2 + d^2$$
$$= 2cd \cos \phi_{d0} + 2ac \cos \theta_0 \cos(\phi_{d0} - \theta_0) - 2ad \cos \theta_0$$

or $b^* = 3.662$. Thus the optimal link dimensions are given by $a^* = 1$, $b^* = 3.662$, $c^* = 0.41$, and $d^* = 3.0$.

REFERENCES AND BIBLIOGRAPHY

8.1 R. J. Duffin, E. Peterson, and C. Zener, *Geometric Programming*, Wiley, New York, 1967.

8.2 U. Passy and D. J. Wilde, Generalized polynomial optimization, *SIAM Journal of Applied Mathematics*, Vol. 15, No. 5, pp. 1344–1356, Sept. 1967.

8.3 D. J. Wilde and C. S. Beightler, *Foundations of Optimization*, Prentice-Hall, Englewood Cliffs, NJ, 1967.

8.4 M. Avriel and A. C. Williams, Complementary geometric programming, *SIAM Journal of Applied Mathematics*, Vol. 19, No. 1, pp. 125–141, July 1970.

8.5 R. M. Stark and R. L. Nicholls, *Mathematical Foundations for Design: Civil Engineering Systems*, McGraw-Hill, New York, 1972.

8.6 C. McMillan, Jr., *Mathematical Programming: An Introduction to the Design and Application of Optimal Decision Machines*, Wiley, New York, 1970.

8.7 C. Zener, A mathematical aid in optimizing engineering designs, *Proceedings of the National Academy of Science*, Vol. 47, p. 537, 1961.

8.8 C. Zener, *Engineering Design by Geometric Programming*, Wiley-Interscience, New York, 1971.

8.9 D. S. Ermer, Optimization of the constrained machining economics problem by geometric programming, *Journal of Engineering for Industry, Transactions of ASME*, Vol. 93, pp. 1067–1072, Nov. 1971.

8.10 S. K. Hati and S. S. Rao, Determination of optimum machining conditions: deterministic and probabilistic approaches, *Journal of Engineering for Industry, Transactions of ASME*, Vol. 98, pp. 354–359, May 1976.

8.11   D. Wilde, Monotonicity and dominance in optimal hydraulic cylinder design, *Journal of Engineering for Industry, Transactions of ASME*, Vol. 97, pp. 1390–1394, Nov. 1975.

8.12   A. B. Templeman, On the solution of geometric programs via separable programming, *Operations Research Quarterly*, Vol. 25, pp. 184–185, 1974.

8.13   J. J. Dinkel and G. A. Kochenberger, On a cofferdam design optimization, *Mathematical Programming*, Vol. 6, pp. 114–117, 1974.

8.14   A. J. Morris, A transformation for geometric programming applied to the minimum weight design of statically determinate structure, *International Journal of Mechanical Science*, Vol. 17, pp. 395–396, 1975.

8.15   A. B. Templeman, Optimum truss design by sequential geometric programming, *Journal of Structural Engineering*, Vol. 3, pp. 155–163, 1976.

8.16   L. J. Mancini and R. L. Piziali, Optimum design of helical springs by geometric programming, *Engineering Optimization*, Vol. 2, pp. 73–81, 1976.

8.17   K. M. Ragsdell and D. T. Phillips, Optimum design of a class of welded structures using geometric programming, *Journal of Engineering for Industry*, Vol. 98, pp. 1021–1025, 1976.

8.18   R. S. Dembo, A set of geometric programming test problems and their solutions, *Mathematical Programming*, Vol. 10, pp. 192–213, 1976.

8.19   M. J. Rijckaert and X. M. Martens, Comparison of generalized geometric programming algorithms, *Journal of Optimization Theory and Applications*, Vol. 26, pp. 205–242, 1978.

8.20   P.V.L.N. Sarma, X. M. Martens, G. V. Reklaitis, and M. J. Rijckaert, A comparison of computational strategies for geometric programs, *Journal of Optimization Theory and Applications*, Vol. 26, pp. 185–203, 1978.

8.21   R. S. Dembo, Current state of the art of algorithms and computer software for geometric programming, *Journal of Optimization Theory and Applications*, Vol. 26, pp. 149–183, 1978.

8.22   R. S. Dembo, Sensitivity analysis in geometric programming, *Journal of Optimization Theory and Applications*, Vol. 37, pp. 1–22, 1982.

8.23   S. S. Rao, Application of complementary geometric programming to mechanical design problems, *International Journal of Mechanical Engineering Education*, Vol. 13, No. 1, pp. 19–29, 1985.

8.24   A. C. Rao, Synthesis of 4-bar function generators using geometric programming, *Mechanism and Machine Theory*, Vol. 14, pp. 141–149, 1979.

8.25   M. Avriel and J. D. Barrett, Optimal design of pitched laminated wood beams, pp. 407–419 in *Advances in Geometric Programming*, M. Avriel, Ed., Plenum Press, New York, 1980.

8.26   P. Petropoulos, Optimal selection of machining rate variables by geometric programming, *International Journal of Production Research*, Vol. 11, pp. 305–314, 1973.

8.27   G. K. Agrawal, Helical torsion springs for minimum weight by geometric programming, *Journal of Optimization Theory and Applications*, Vol. 25, No. 2, pp. 307–310, 1978.

8.28   C. S. Beightler and D. T. Phillips, *Applied Geometric Programming*, Wiley, New York, 1976.

8.29   C. S. Beightler, T.-C. Lo, and H. G. Rylander, Optimal design by geometric programming, *ASME Journal of Engineering for Industry*, Vol. 92, No. 1, pp. 191–196, 1970.

8.30   Y. T. Sin and G. V. Reklaitis, On the computational utility of generalized geometric programming solution methods: Review and test procedure design, pp. 7–14, *Results and interpretation*, pp. 15–21, in *Progress in Engineering Optimization–1981*, R. W. Mayne and K. M. Ragsdell, Eds., ASME, New York, 1981.

8.31   M. Avriel, R. Dembo, and U. Passey, Solution of generalized geometric programs, *International Journal for Numerical Methods in Engineering*, Vol. 9, pp. 149–168, 1975.

8.32   Computational aspects of geometric programming: 1. Introduction and basic notation, pp. 115–120 (A. B. Templeman), 2. Polynomial programming, pp. 121–145 (J. Bradley), 3. Some primal and dual algorithms for posynomial and signomial geometric programs, pp. 147–160 (J. G. Ecker, W. Gochet, and Y. Smeers), 4. Computational experiments in geometric programming, pp. 161–173 (R. S. Dembo and M. J. Rijckaert), *Engineering Optimization*, Vol. 3, No. 3, 1978.

8.33   A. J. Morris, Structural optimization by geometric programming, *International Journal of Solids and Structures*, Vol. 8, pp. 847–864, 1972.

8.34   A. J. Morris, The optimisation of statically indeterminate structures by means of approximate geometric programming, pp. 6.1–6.17 *in Proceedings of the 2nd Symposium on Structural Optimization, AGARD Conference Proceedings 123*, Milan, 1973.

8.35   A. B. Templeman and S. K. Winterbottom, Structural design applications of geometric programming, pp. 5.1–5.15 *in Proceedings of the 2nd Symposium on Structural Optimization, AGARD Conference Proceedings 123*, Milan, 1973.

8.36   A. B. Templeman, Structural design for minimum cost using the method of geometric programming, *Proceedings of the Institute of Civil Engineers*, London, Vol. 46, pp. 459–472, 1970.

8.37   J. E. Shigley and C. R. Mischke, *Mechanical Engineering Design*, 5th ed., McGraw-Hill, New York, 1989.

REVIEW QUESTIONS

8.1   State whether each of the following functions is a polynomial, posynomial, or both.

(a) $f = 4 - x_1^2 + 6x_1x_2 + 3x_2^2$
(b) $f = 4 + 2x_1^2 + 5x_1x_2 + x_2^2$
(c) $f = 4 + 2x_1^2x_2^{-1} + 3x_2^{-4} + 5x_1^{-1}x_2^3$

8.2   Answer true or false:

(a) The optimum values of the design variables are to be known before finding the optimum value of the objective function in geometric programming.
(b) Δ_j^* denotes the relative contribution of the jth term to the optimum value of the objective function.
(c) There are as many orthogonality conditions as there are design variables in a geometric programming problem.
(d) If f is the primal and v is the dual, $f \leq v$.
(e) The degree of difficulty of a complementary geometric programming problem is given by $(N - n - 1)$, where n denotes the number of design variables and N represents the total number of terms appearing in the numerators of the rational functions involved.
(f) In a geometric programming problem, there are no restrictions on the number of design variables and the number of posynomial terms.

8.3   How is the degree of difficulty defined for a constrained geometric programming problem?

8.4   What is arithmetic–geometric inequality?

8.5 What is normality condition in a geometric programming problem?

8.6 Define a complementary geometric programming problem.

PROBLEMS

Using arithmetic mean–geometric mean inequality, obtain a lower bound v for each function $[f(x) \geq v$, where v is a constant] in Problems 8.1–8.3.

8.1 $f(x) = \dfrac{x^{-2}}{3} + \dfrac{2}{3}x^{-3} + \dfrac{4}{3}x^{3/2}$

8.2 $f(x) = 1 + x + \dfrac{1}{x} + \dfrac{1}{x^2}$

8.3 $f(x) = \frac{1}{2}x^{-3} + x^2 + 2x$

8.4 An open cylindrical vessel is to be constructed to transport $80\,\text{m}^3$ of grain from a warehouse to a factory. The sheet metal used for the bottom and sides cost $80 and $10 per square meter, respectively. If it costs $1 for each round trip of the vessel, find the dimensions of the vessel for minimizing the transportation cost. Assume that the vessel has no salvage upon completion of the operation.

8.5 Find the solution of the problem stated in Problem 8.4 by assuming that the sides cost $20 per square meter, instead of $10.

8.6 Solve the problem stated in Problem 8.4 if only 10 trips are allowed for transporting the $80\,\text{m}^3$ of grain.

8.7 An automobile manufacturer needs to allocate a maximum sum of 2.5×10^6 between the development of two different car models. The profit expected from both the models is given by $x_1^{1.5}x_2$, where x_i denotes the money allocated to model i ($i = 1, 2$). Since the success of each model helps the other, the amount allocated to the first model should not exceed four times the amount allocated to the second model. Determine the amounts to be allocated to the two models to maximize the profit expected. *Hint:* Minimize the inverse of the profit expected.

8.8 Write the dual of the heat exchanger design problem stated in Problem 1.12.

8.9 Minimize the following function:

$$f(\mathbf{X}) = x_1 x_2 x_3^{-2} + 2x_1^{-1}x_2^{-1}x_3 + 5x_2 + 3x_1 x_2^{-2}$$

8.10 Minimize the following function:

$$f(\mathbf{X}) = \frac{1}{2}x_1^2 + x_2 + \frac{3}{2}x_1^{-1}x_2^{-1}$$

8.11   Minimize $f(\mathbf{X}) = 20x_2 x_3 x_4^4 + 20x_1^2 x_3^{-1} + 5x_2 x_3^2$

subject to

$$5x_2^{-5}x_3^{-1} \leq 1$$

$$10x_1^{-1}x_2^3 x_4^{-1} \leq 1$$

$$x_i > 0, \quad i = 1 \text{ to } 4$$

8.12
$$\text{Minimize } f(\mathbf{X}) = x_1^{-2} + \tfrac{1}{4}x_2^2 x_3$$

subject to

$$\tfrac{3}{4}x_1^2 x_2^{-2} + \tfrac{3}{8}x_2 x_3^{-2} \le 1$$
$$x_i > 0, \quad i = 1, 2, 3$$

8.13
$$\text{Minimize } f(\mathbf{X}) = x_1^{-3}x_2 + x_1^{3/2}x_3^{-1}$$

subject to

$$x_1^2 x_2^{-1} + \tfrac{1}{2}x_1^{-2}x_3^3 \le 1$$
$$x_1 > 0, \quad x_2 > 0, \quad x_3 > 0$$

8.14
$$\text{Minimize } f = x_1^{-1}x_2^{-2}x_3^{-2}$$

subject to

$$x_1^3 + x_2^2 + x_3 \le 1$$
$$x_i > 0, \quad i = 1, 2, 3$$

8.15 Prove that the function $y = c_1 e^{a_1 x_1} + c_2 e^{a_2 x_2} + \cdots + c_n e^{a_n x_n}$, $c_i \ge 0$, $i = 1, 2, \ldots, n$, is a convex function with respect to $x_1, x_2, \ldots, x_n$.

8.16 Prove that $f = \ln x$ is a concave function for positive values of x.

8.17 The problem of minimum weight design of a helical torsional spring subject to a stress constraint can be expressed as [8.27]

$$\text{Minimize } f(d, D) = \frac{\pi^2 \rho E \phi}{14{,}680M}d^6 + \frac{\pi^2 \rho Q}{4}Dd^2$$

subject to

$$\frac{14.5M}{d^{2.885}D^{0.115}\sigma_{max}} \le 1$$

where d is the wire diameter, D the mean coil diameter, ρ the density, E is Young's modulus, ϕ the angular deflection in degrees, M the torsional moment, and Q the number of inactive turns. Solve this problem using geometric programming approach for the following data: $E = 20 \times 10^{10}$ Pa, $\sigma_{max} = 15 \times 10^7$ Pa, $\phi = 20°$, $Q = 2$, $M = 0.3$ N-m, and $\rho = 7.7 \times 10^4$ N/m^3.

8.18 Solve the machining economics problem given by Eqs. (E$_2$) and (E$_4$) of Example 8.7 for the given data.

8.19 Solve the machining economics problem given by Eqs. (E$_2$), (E$_4$), and (E$_6$) of Example 8.7 for the given data.

8.20 Determine the degree of difficulty of the problem stated in Example 8.8.

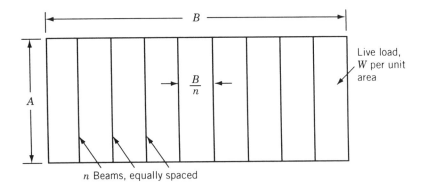

Figure 8.5   Floor consisting of a plate with supporting beams [8.36].

8.21   A rectangular area of dimensions A and B is to be covered by steel plates with supporting beams as shown in Fig. 8.5. The problem of minimum cost design of the floor subject to a constraint on the maximum deflection of the floor under a specified uniformly distributed live load can be stated as [8.36]

$$\text{Minimize } f(\mathbf{X}) = \text{cost of plates} + \text{cost of beams}$$
$$= k_f \gamma A B t + k_b \gamma A k_1 n Z^{2/3} \tag{1}$$

subject to

$$\frac{56.25 W B^4}{EA} t^{-3} n^{-4} + \left(\frac{4.69 W B A^3}{E k_2} \right) n^{-1} Z^{-4/3} \leq 1 \tag{2}$$

where W is the live load on the floor per unit area, k_f and k_b are the unit costs of plates and beams, respectively, γ the weight density of steel, t the thickness of plates, n the number of beams, $k_1 Z^{2/3}$ the cross-sectional area of each beam, $k_2 Z^{4/3}$ the area moment of inertia of each beam, k_1 and k_2 are constants, Z the section modulus of each beam, and E the elastic modulus of steel. The two terms on the left side of Eq. (2) denote the contributions of steel plates and beams to the deflection of the floor. By assuming the data as $A = 10$ m, $B = 50$ m, $W = 1000$ kgf/m^2, $k_b = \$0.05/$ kgf, $k_f = \$0.06/$ kgf, $\gamma = 7850$ kgf/m^3, $E = 2.1 \times 10^5$ MN/m^2, $k_1 = 0.78$, and $k_2 = 1.95$, determine the solution of the problem (i.e., the values of t^*, n^*, and Z^*).

8.22   Solve the zero-degree-of-difficulty bearing problem given by Eqs. (E$_8$) and (E$_9$) of Example 8.12.

8.23   Solve the one-degree-of-difficulty bearing problem given by Eqs. (E$_{12}$) and (E$_{13}$) of Example 8.12.

8.24   The problem of minimum volume design of a statically determinate truss consisting of n members (bars) with m unsupported nodes and subject to q load conditions can be stated as follows [8.14]:

$$\text{Minimize } f = \sum_{i=1}^{n} l_i x_i \tag{1}$$

subject to

$$\frac{F_i^{(k)}}{x_i \sigma_i^*} \leq 1, \quad i = 1, 2, \ldots, n, \quad k = 1, 2, \ldots, q \tag{2}$$

$$\sum_{i=1}^{n} \frac{F_i^{(k)} l_i}{x_i E \Delta_i^*} s_{ij} \leq 1, \quad j = 1, 2, \ldots, m, \quad k = 1, 2, \ldots, q \tag{3}$$

where $F_i^{(k)}$ is the tension in the ith member in the kth load condition, x_i the cross-sectional area of member i, l_i the length of member i, E is Young's modulus, σ_i^* the maximum permissible stress in member i, and Δ_j^* the maximum allowable displacement of node j. Develop a suitable transformation technique and express the problem of Eqs. (1) to (3) as a geometric programming problem in terms of the design variables x_i.

9

Dynamic Programming

9.1 INTRODUCTION

In most practical problems, decisions have to be made sequentially at different points in time, at different points in space, and at different levels, say, for a component, for a subsystem, and/or for a system. The problems in which the decisions are to be made sequentially are called *sequential decision problems*. Since these decisions are to be made at a number of stages, they are also referred to as *multistage decision problems*. Dynamic programming is a mathematical technique well suited for the optimization of multistage decision problems. This technique was developed by Richard Bellman in the early 1950s [9.2, 9.6].

The dynamic programming technique, when applicable, represents or decomposes a multistage decision problem as a sequence of single-stage decision problems. Thus an N-variable problem is represented as a sequence of N single-variable problems that are solved successively. In most cases, these N subproblems are easier to solve than the original problem. The decomposition to N subproblems is done in such a manner that the optimal solution of the original N-variable problem can be obtained from the optimal solutions of the N one-dimensional problems. It is important to note that the particular optimization technique used for the optimization of the N single-variable problems is irrelevant. It may range from a simple enumeration process to a differential calculus or a nonlinear programming technique.

Multistage decision problems can also be solved by direct application of the classical optimization techniques. However, this requires the number of variables to be small, the functions involved to be continuous and continuously differentiable, and the optimum points not to lie at the boundary points. Further, the problem has to be relatively simple so that the set of resultant equations can be solved either analytically or numerically. The nonlinear programming techniques can be used to solve slightly more complicated multistage decision problems. But their application requires the variables to be continuous and prior knowledge about the region of the global minimum or maximum. In all these cases, the introduction of stochastic variability makes the problem extremely complex and renders the problem unsolvable except by using some sort of an approximation such as chance constrained programming.[†] Dynamic programming, on the other hand, can deal with discrete variables, nonconvex, noncontinuous, and nondifferentiable functions. In general, it can also take into account the stochastic variability by a simple modification of the deterministic procedure. The dynamic programming

---

[†]The chance constrained programming is discussed in Chapter 11.

544

technique suffers from a major drawback, known as the *curse of dimensionality*. However, despite this disadvantage, it is very suitable for the solution of a wide range of complex problems in several areas of decision making.

9.2   MULTISTAGE DECISION PROCESSES

9.2.1   Definition and Examples

As applied to dynamic programming, a multistage decision process is one in which a number of single-stage processes are connected in series so that the output of one stage is the input of the succeeding stage. Strictly speaking, this type of process should be called a *serial multistage decision process* since the individual stages are connected head to tail with no recycle. Serial multistage decision problems arise in many types of practical problems. A few examples are given below and many others can be found in the literature.

Consider a chemical process consisting of a heater, a reactor, and a distillation tower connected in series. The objective is to find the optimal value of temperature in the heater, the reaction rate in the reactor, and the number of trays in the distillation tower such that the cost of the process is minimum while satisfying all the restrictions placed on the process. Figure 9.1 shows a missile resting on a launch pad that is expected to hit a moving aircraft (target) in a given time interval. The target will naturally take evasive action and attempts to avoid being hit. The problem is to generate a set of commands to the missile so that it can hit the target in the specified time interval. This can be done by observing the target and, from its actions, generate periodically a new direction and speed for the missile. Next, consider the minimum cost design of a water tank. The system consists of a tank, a set of columns, and a foundation. Here the tank supports the water, the columns support the weights of water and tank, and the foundation supports the weights of water, tank, and columns. The components can be

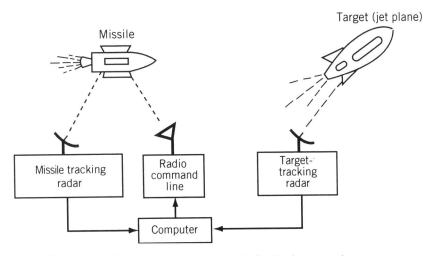

Figure 9.1   Ground-radar-controlled missile chasing a moving target.

seen to be in series and the system has to be treated as a multistage decision problem. Finally, consider the problem of loading a vessel with stocks of N items. Each unit of item i has a weight w_i and a monetary value c_i. The maximum permissible cargo weight is W. It is required to determine the cargo load that corresponds to maximum monetary value without exceeding the limitation of the total cargo weight. Although the multistage nature of this problem is not directly evident, it can be posed as a multistage decision problem by considering each item of the cargo as a separate stage.

9.2.2   Representation of a Multistage Decision Process

A single-stage decision process (which is a component of the multistage problem) can be represented as a rectangular block (Fig. 9.2). A decision process can be characterized by certain input parameters, $\mathbf{S}$ (or data), certain decision variables ($\mathbf{X}$), and certain output parameters ($\mathbf{T}$) representing the outcome obtained as a result of making the decision. The input parameters are called *input state variables*, and the output parameters are called *output state variables*. Finally, there is a return or objective function R, which measures the effectiveness of the decisions made and the output that results from these decisions. For a single-stage decision process shown in Fig. 9.2, the output is related to the input through a stage transformation function denoted by

$$\mathbf{T} = \mathbf{t}(\mathbf{X}, \mathbf{S}) \tag{9.1}$$

Since the input state of the system influences the decisions we make, the return function can be represented as

$$R = r(\mathbf{X}, \mathbf{S}) \tag{9.2}$$

A serial multistage decision process can be represented schematically as shown in Fig. 9.3. Because of some convenience, which will be seen later, the stages n, $n - 1, \ldots, i, \ldots, 2, 1$ are labeled in decreasing order. For the ith stage, the input state vector is denoted by $\mathbf{s}_{i+1}$ and the output state vector as $\mathbf{s}_i$. Since the system is a serial one, the output from stage $i + 1$ must be equal to the input to stage i. Hence the state transformation and return functions can be represented as

$$\mathbf{s}_i = \mathbf{t}_i(\mathbf{s}_{i+1}, \mathbf{x}_i) \tag{9.3}$$

$$R_i = r_i(\mathbf{s}_{i+1}, \mathbf{x}_i) \tag{9.4}$$

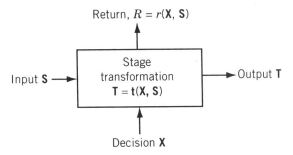

Figure 9.2   Single-stage decision problem.

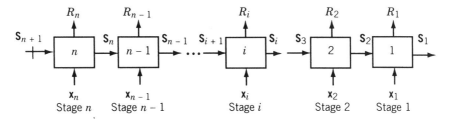

Figure 9.3 Multistage decision problem (initial value problem).

where $\mathbf{x}_i$ denotes the vector of decision variables at stage i. The state transformation equations (9.3) are also called *design equations*.

The objective of a multistage decision problem is to find $\mathbf{x}_1, \mathbf{x}_2, \ldots, \mathbf{x}_n$ so as to optimize some function of the individual statge returns, say, $f(R_1, R_2, \ldots, R_n)$ and satisfy Eqs. (9.3) and (9.4). The nature of the n-stage return function, f, determines whether a given multistage problem can be solved by dynamic programming. Since the method works as a decomposition technique, it requires the separability and monotonicity of the objective function. To have separability of the objective function, we must be able to represent the objective function as the composition of the individual stage returns. This requirement is satisfied for additive objective functions:

$$f = \sum_{i=1}^{n} R_i = \sum_{i=1}^{n} R_i(\mathbf{x}_i, \mathbf{s}_{i+1}) \tag{9.5}$$

where $\mathbf{x}_i$ are real, and for multiplicative objective functions,

$$f = \prod_{i=1}^{n} R_i = \prod_{i=1}^{n} R_i(\mathbf{x}_i, \mathbf{s}_{i+1}) \tag{9.6}$$

where $\mathbf{x}_i$ are real and nonnegative. On the other hand, the following objective function is not separable:

$$f = [R_1(\mathbf{x}_1, \mathbf{s}_2) + R_2(\mathbf{x}_2, \mathbf{s}_3)][R_3(\mathbf{x}_3, \mathbf{s}_4) + R_4(\mathbf{x}_4, \mathbf{s}_5)] \tag{9.7}$$

Fortunately, there are many practical problems that satisfy the separability condition. The objective function is said to be *monotonic* if for all values of $\mathbf{a}$ and $\mathbf{b}$ that make

$$R_i(\mathbf{x}_i = \mathbf{a}, \mathbf{s}_{i+1}) \geq R_i(\mathbf{x}_i = \mathbf{b}, \mathbf{s}_{i+1})$$

the following inequality is satisfied:

$$f(\mathbf{x}_n, \mathbf{x}_{n-1}, \ldots, \mathbf{x}_{i+1}, \mathbf{x}_i = \mathbf{a}, \mathbf{x}_{i-1}, \ldots, \mathbf{x}_1, \mathbf{s}_{n+1})$$
$$\geq f(\mathbf{x}_n, \mathbf{x}_{n-1}, \ldots, \mathbf{x}_{i+1}, \mathbf{x}_i = \mathbf{b}, \mathbf{x}_{i-1}, \ldots, \mathbf{x}_1, \mathbf{s}_{n+1}), \quad i = 1, 2, \ldots, n \tag{9.8}$$

9.2.3 Conversion of a Nonserial System to a Serial System

According to the definition, a serial system is one whose components (stages) are connected in such a way that the output of any component is the input of the succeeding component. As an example of a nonserial system, consider a steam power plant consisting of a pump, a feedwater heater, a boiler, a superheater, a steam turbine, and an electric generator, as shown in Fig. 9.4. If we assume that some steam is taken from the turbine to heat the feedwater, a loop will be formed as shown in Fig. 9.4*a*. This nonserial system can be converted to an equivalent serial system by regrouping the components so that a loop is redefined as a single element as shown in Fig. 9.4*b* and *c*. Thus the new serial multistage system consists of only three components: the pump, the boiler and turbine system, and the electric generator. This procedure can easily be extended to convert multistage systems with more than one loop to equivalent serial systems.

9.2.4 Types of Multistage Decision Problems

The serial multistage decision problems can be classified into three categories as follows.

1. *Initial value problem.* If the value of the initial state variable, s_{n+1}, is prescribed, the problem is called an *initial value problem*.
2. *Final value problem.* If the value of the final state variable, s_1 is prescribed, the problem is called a *final value problem*. Notice that a final value problem can be transformed into an initial value problem by reversing the directions of s_i, $i = 1, 2, \ldots, n + 1$. The details of this are given in Section 9.7.

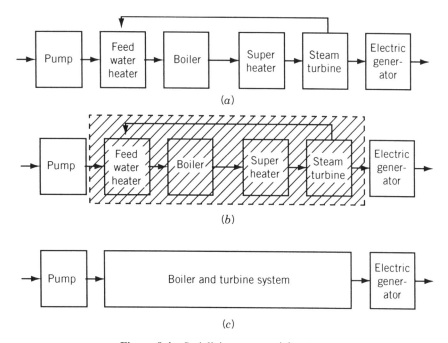

(a)

(b)

(c)

Figure 9.4 Serializing a nonserial system.

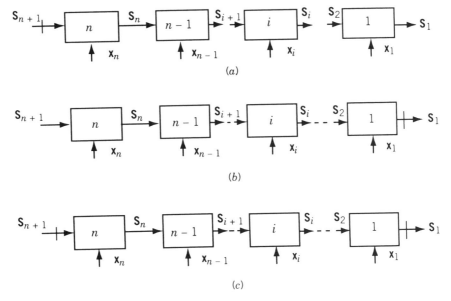

Figure 9.5   Types of multistage problems: (*a*) initial value problem; (*b*) final value problem; (*c*) boundary value problem.

3. *Boundary value problem.* If the values of both the input and output variables are specified, the problem is called a *boundary value problem*. The three types of problems are shown schematically in Fig. 9.5, where the symbol ⊢ is used to indicate a prescribed state variable.

9.3   CONCEPT OF SUBOPTIMIZATION AND PRINCIPLE OF OPTIMALITY

A dynamic programming problem can be stated as follows.[†] Find $x_1, x_2, \ldots, x_n$, which optimizes

$$f(x_1, x_2, \ldots, x_n) = \sum_{i=1}^{n} R_i = \sum_{i=1}^{n} r_i (s_{i+1}, x_i)$$

and satisfies the design equations

$$s_i = t_i (s_{i+1}, x_i), \quad i = 1, 2, \ldots, n$$

The dynamic programming makes use of the concept of suboptimization and the principle of optimality in solving this problem. The concept of suboptimization and the principle of optimality will be explained through the following example of an initial value problem.

[†]In the subsequent discussion, the design variables x_i and state variables s_i are denoted as scalars for simplicity, although the theory is equally applicable even if they are vectors.

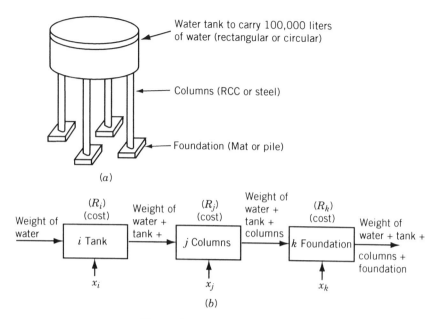

Figure 9.6 Water tank system.

Example 9.1 Explain the concept of suboptimization in the context of the design of the water tank shown in Fig. 9.6*a*. The tank is required to have a capacity of 100,000 liters of water and is to be designed for minimum cost [9.10].

SOLUTION Instead of trying to optimize the complete system as a single unit, it would be desirable to *break* the system into components which could be optimized more or less individually. For this breaking and component suboptimization, a logical procedure is to be used; otherwise, the procedure might result in a poor solution. This concept can be seen by breaking the system into three components: component i (tank), component j (columns), and component k (foundation). Consider the suboptimization of component j (columns) without a consideration of the other components. If the cost of steel is very high, the minimum cost design of component j may correspond to heavy concrete columns without reinforcement. Although this design may be acceptable for columns, the entire weight of the columns has to be carried by the foundation. This may result in a foundation that is prohibitively expensive. This shows that the suboptimization of component j has adversely influenced the design of the following component k. This example shows that the design of any interior component affects the designs of all the subsequent (downstream) components. As such, it cannot be suboptimized without considering its effect on the downstream components. The following mode of suboptimization can be adopted as a rational optimization strategy. Since the last component in a serial system influences no other component, it can be suboptimized independently. Then the last two components can be considered together as a single (larger) component and can be suboptimized without adversely influencing any of the downstream components. This process can be continued to group any number of end components as a single (larger) end component and suboptimize them. This process of

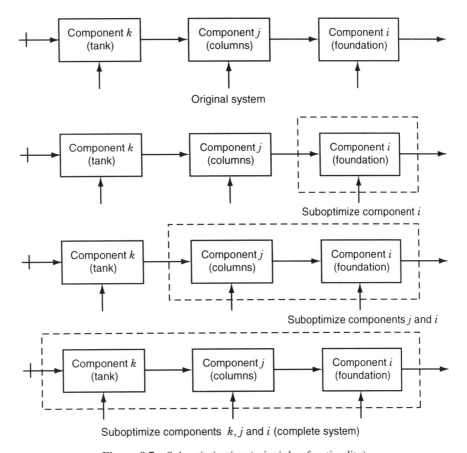

Figure 9.7   Suboptimization (principle of optimality).

suboptimization is shown in Fig. 9.7. Since the suboptimizations are to be done in the reverse order, the components of the system are also numbered in the same manner for convenience (see Fig. 9.3).

The process of suboptimization was stated by Bellman [9.2] as the principle of optimality:

> An optimal policy (or a set of decisions) has the property that whatever the initial state and initial decision are, the remaining decisions must constitute an optimal policy with regard to the state resulting from the first decision.

Recurrence Relationship.   Suppose that the desired objective is to minimize the n-stage objective function f, which is given by the sum of the individual stage returns:

$$\text{Minimize } f = R_n(x_n, s_{n+1}) + R_{n-1}(x_{n-1}, s_n) + \cdots + R_1(x_1, s_2) \tag{9.9}$$

where the state and decision variables are related as

$$s_i = t_i(s_{i+1}, x_i), \quad i = 1, 2, \ldots, n \tag{9.10}$$

Consider the first subproblem by starting at the final stage, $i = 1$. If the input to this stage s_2 is specified, then according to the principle of optimality, x_1 must be selected to optimize R_1. Irrespective of what happens to the other stages, x_1 must be selected such that $R_1(x_1, s_2)$ is an optimum for the input s_2. If the optimum is denoted as f_1^*, we have

$$f_1^*(s_2) = \underset{x_1}{\text{opt}}[R_1(x_1, s_2)] \tag{9.11}$$

This is called a *one-stage policy* since once the input state s_2 is specified, the optimal values of R_1, x_1, and s_1 are completely defined. Thus Eq. (9.11) is a parametric equation giving the optimum f_1^* as a function of the input parameter s_2.

Next, consider the second subproblem by grouping the last two stages together. If f_2^* denotes the optimum objective value of the second subproblem for a specified value of the input s_3, we have

$$f_2^*(s_3) = \underset{x_1,x_2}{\text{opt}}\,[R_2(x_2, s_3) + R_1(x_1, s_2)] \tag{9.12}$$

The principle of optimality requires that x_1 be selected so as to optimize R_1 for a given s_2. Since s_2 can be obtained once x_2 and s_3 are specified, Eq. (9.12) can be written as

$$f_2^*(s_3) = \underset{x_2}{\text{opt}}[R_2(x_2, s_3) + f_1^*(s_2)] \tag{9.13}$$

Thus f_2^* represents the optimal policy for the *two-stage subproblem*. It can be seen that the principle of optimality reduced the dimensionality of the problem from two [in Eq. (9.12)] to one [in Eq. (9.13)]. This can be seen more clearly by rewriting Eq. (9.13) using Eq. (9.10) as

$$f_2^*(s_3) = \underset{x_2}{\text{opt}}[R_2(x_2, s_3) + f_1^*\{t_2(x_2, s_3)\}] \tag{9.14}$$

In this form it can be seen that for a specified input s_3, the optimum is determined solely by a suitable choice of the decision variable x_2. Thus the optimization problem stated in Eq. (9.12), in which both x_2 and x_1 are to be simultaneously varied to produce the optimum f_2^*, is reduced to two subproblems defined by Eqs. (9.11) and (9.13). Since the optimization of each of these subproblems involves only a single decision variable, the optimization is, in general, much simpler.

This idea can be generalized and the ith subproblem defined by

$$f_i^*(s_{i+1}) = \underset{x_i,x_{i-1},\ldots,x_1}{\text{opt}}\,[R_i(x_i, s_{i+1}) + R_{i-1}(x_{i-1}, s_i) + \cdots + R_1(x_1, s_2)] \tag{9.15}$$

which can be written as

$$f_i^*(s_{i+1}) = \underset{x_i}{\text{opt}}[R_i(x_i, s_{i+1}) + f_{i-1}^*(s_i)] \tag{9.16}$$

where f_{i-1}^* denotes the optimal value of the objective function corresponding to the last $i - 1$ stages, and s_i is the input to the stage $i - 1$. The original problem in Eq. (9.15) requires the simultaneous variation of i decision variables, $x_1, x_2, \ldots, x_i$, to determine the optimum value of $f_i = \sum_{k=1}^{i} R_k$ for any specified value of the input s_{i+1}. This problem, by using the principle of optimality, has been decomposed into i separate problems, each involving only one decision variable. Equation (9.16) is the desired recurrence relationship valid for $i = 2, 3, \ldots, n$.

9.4 COMPUTATIONAL PROCEDURE IN DYNAMIC PROGRAMMING

The use of the recurrence relationship derived in Section 9.3 in actual computations is discussed in this section [9.10]. As stated, dynamic programming begins by suboptimizing the last component, numbered 1. This involves the determination of

$$f_1^*(s_2) = \underset{x_1}{\text{opt}}[R_1(x_1, s_2)] \tag{9.17}$$

The best value of the decision variable x_1, denoted as x_1^*, is that which makes the return (or objective) function R_1 assume its optimum value, denoted by f_1^*. Both x_1^* and f_1^* depend on the condition of the input or feed that the component 1 receives from the upstream, that is, on s_2. Since the particular value s_2 will assume after the upstream components are optimized is not known at this time, this last-stage suboptimization problem is solved for a "range" of possible values of s_2 and the results are entered into a graph or a table. This graph or table contains a complete summary of the results of suboptimization of stage 1. In some cases, it may be possible to express f_1^* as a function of s_2. If the calculations are to be performed on a computer, the results of suboptimization have to be stored in the form of a table in the computer. Figure 9.8 shows a typical table in which the results obtained from the suboptimization of stage 1 are entered.

Next we move up the serial system to include the last two components. In this two-stage suboptimization, we have to determine

$$f_2^*(s_3) = \underset{x_2, x_1}{\text{opt}}[R_2(x_2, s_3) + R_1(x_1, s_2)] \tag{9.18}$$

Since all the information about component 1 has already been encoded in the table corresponding to f_1^*, this information can then be substituted for R_1 in Eq. (9.18) to

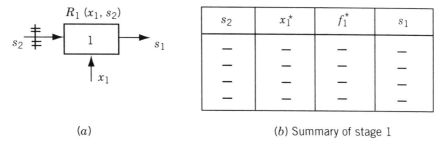

(a)   (b) Summary of stage 1

Figure 9.8 Suboptimization of component 1 for various settings of the input state variable s_2.

get the following simplified statement:

$$f_2^*(s_3) = \underset{x_2}{\text{opt}}[R_2(x_2, s_3) + f_1^*(s_2)] \tag{9.19}$$

Thus the number of variables to be considered has been reduced from two (x_1 and x_2) to one (x_2). A range of possible values of s_3 must be considered and for each one, x_2^* must be found so as to optimize $[R_2 + f_1^*(s_2)]$. The results (x_2^* and f_2^* for different s_3) of this suboptimization are entered in a table as shown in Fig. 9.9.

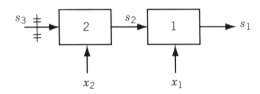

$$\underset{x_2}{\text{Opt}} \{R_2 + f_1^*(s_2)\} = f_1^*(s_3)$$

(a)

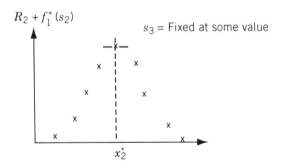

For each setting of s_3, draw
a graph as shown above to
obtain the following:

s_3	$x_2{}^*$	$f_2{}^*$	s_2
—	—	—	—
—	—	—	—
—	—	—	—
—	—	—	—

(b) Summary of stages 2 and 1

Figure 9.9 Suboptimization of components 1 and 2 for various settings of the input state variable s_3.

Assuming that the suboptimization sequence has been carried on to include $i - 1$ of the end components, the next step will be to suboptimize the i end components. This requires the solution of

$$f_i^*(s_{i+1}) = \underset{x_i, x_{i-1}, \ldots, x_1}{\text{opt}} \; [R_i + R_{i-1} + \cdots + R_1] \qquad (9.20)$$

However, again, all the information regarding the suboptimization of $i - 1$ end components is known and has been entered in the table corresponding to f_{i-1}^*. Hence this information can be substituted in Eq. (9.20) to obtain

$$f_i^*(s_{i+1}) = \underset{x_i}{\text{opt}}[R_i(x_i, s_{i+1}) + f_{i-1}^*(s_i)] \qquad (9.21)$$

Thus the dimensionality of the i-stage suboptimization has been reduced to 1, and the equation $s_i = t_i(s_{i+1}, x_i)$ provides the functional relation between x_i and s_i. As before, a range of values of s_{i+1} are to be considered, and for each one, x_i^* is to be found so as to optimize $[R_i + f_{i-1}^*]$. A table showing the values of x_i^* and f_i^* for each of the values of s_{i+1} is made as shown in Fig. 9.10.

The suboptimization procedure above is continued until stage n is reached. At this stage only one value of s_{n+1} needs to be considered (for initial value problems), and the optimization of the n components completes the solution of the problem.

The final thing needed is to retrace the steps through the tables generated, to gather the complete set of x_i^* $(i = 1, 2, \ldots, n)$ for the system. This can be done as follows. The nth suboptimization gives the values of x_n^* and f_n^* for the specified value of s_{n+1} (for initial value problem). The known design equation $s_n = t_n(s_{n+1}, x_n^*)$ can be used to find the input, s_n^*, to the $(n - 1)$th stage. From the tabulated results for $f_{n-1}^*(s_n)$, the optimum values f_{n-1}^* and x_{n-1}^* corresponding to s_n^* can readily be obtained. Again the known design equation $s_{n-1} = t_{n-1}(s_n, x_{n-1}^*)$ can be used to find the input, s_{n-1}^*, to the $(n - 2)$th stage. As before, from the tabulated results of $f_{n-2}^*(s_{n-1})$, the optimal values x_{n-2}^* and f_{n-2}^* corresponding to s_{n-1}^* can be found. This procedure is continued until the values x_1^* and f_1^* corresponding to s_2^* are obtained. Then the optimum solution vector of the original problem is given by $(x_1^*, x_2^*, \ldots, x_n^*)$ and the optimum value of the objective function by f_n^*.

9.5   EXAMPLE ILLUSTRATING THE CALCULUS METHOD OF SOLUTION

Example 9.2   The four-bar truss shown in Fig. 9.11 is subjected to a vertical load of 2×10^5 lb at joint A as shown. Determine the cross-sectional areas of the members (bars) such that the total weight of the truss is minimum and the vertical deflection of joint A is equal to 0.5 in. Assume the unit weight as 0.01 lb/in^3 and the Young's modulus as 20×10^6 psi.

SOLUTION   Let x_i denote the area of cross section of member $i(i = 1, 2, 3, 4)$. The lengths of members are given by $l_1 = l_3 = 100$ in., $l_2 = 120$ in., and $l_4 = 60$ in. The

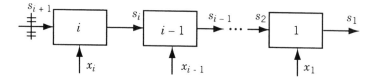

$$\underset{x_i}{\text{Opt}} \left[R_i + f^*_{i-1}(s_i) \right] = f^*_i(s_{i+1})$$

For each setting of s_{i+1}, consider a graph as shown below:

(a)

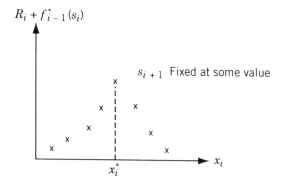

And obtain the following

s_{i+1}	x^*_i	f^*_i	s_i
—	—	—	—
—	—	—	—
—	—	—	—

(b) Summary of stages i, i-1, ...2, and 1

Figure 9.10   Suboptimization of components $1, 2, \dots, i$ for various settings of the input state variable s_{i+1}.

weight of the truss is given by

$$f(x_1, x_2, x_3, x_4) = 0.01(100x_1 + 120x_2 + 100x_3 + 60x_4)$$

$$= x_1 + 1.2x_2 + x_3 + 0.6x_4 \tag{E_1}$$

From structural analysis [9.5], the force developed in member i due to a unit load acting at joint $A(p_i)$, the deformation of member i (d_i), and the contribution of member i to the vertical deflection of A $(\delta_i = p_i d_i)$ can be determined as follows:

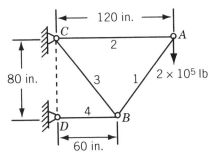

Figure 9.11   Four-bar truss.

Member i	p_i	$d_i = \dfrac{(\text{stress}_i)l_i}{E} = \dfrac{Pp_il_i}{x_iE}$ (in.)	$\delta_i = p_id_i$ (in.)
1	-1.25	$-1.25/x_1$	$1.5625/x_1$
2	0.75	$0.9/x_2$	$0.6750/x_2$
3	1.25	$1.25/x_3$	$1.5625/x_3$
4	-1.50	$-0.9/x_4$	$1.3500/x_4$

The vertical deflection of joint A is given by

$$d_A = \sum_{i=1}^{4} \delta_i = \frac{1.5625}{x_1} + \frac{0.6750}{x_2} + \frac{1.5625}{x_3} + \frac{1.3500}{x_4} \tag{E$_2$}$$

Thus the optimization problem can be stated as

$$\text{Minimize } f(\mathbf{X}) = x_1 + 1.2x_2 + x_3 + 0.6x_4$$

subject to

$$\frac{1.5625}{x_1} + \frac{0.6750}{x_2} + \frac{1.5625}{x_3} + \frac{1.3500}{x_4} = 0.5 \tag{E$_3$}$$

$$x_1 \geq 0, \quad x_2 \geq 0, \quad x_3 \geq 0, \quad x_4 \geq 0$$

Since the deflection of joint A is the sum of contributions of the various members, we can consider the 0.5 in. deflection as a resource to be allocated to the various activities x_i and the problem can be posed as a multistage decision problem as shown in Fig. 9.12. Let s_2 be the displacement (resource) available for allocation to the first member (stage 1), δ_1 the displacement contribution due to the first member, and $f_1^*(s_2)$ the minimum weight of the first member. Then

$$f_1^*(s_2) = \min[R_1 = x_1] = \frac{1.5625}{s_2} \tag{E$_4$}$$

such that

$$\delta_1 = \frac{1.5625}{x_1} \quad \text{and} \quad x_1 \geq 0$$

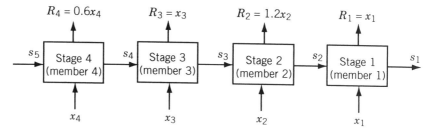

Figure 9.12   Example 9.2 as a four-stage decision problem.

since $\delta_1 = s_2$, and

$$x_1^* = \frac{1.5625}{s_2} \tag{E_5}$$

Let s_3 be the displacement available for allocation to the first two members, δ_2 the displacement contribution due to the second member, and $f_2^*(s_3)$ the minimum weight of the first two members. Then we have, from the recurrence relationship of Eq. (9.16),

$$f_2^*(s_3) = \min_{x_2 \geq 0} [R_2 + f_1^*(s_2)] \tag{E_6}$$

where s_2 represents the resource available after allocation to stage 2 and is given by

$$s_2 = s_3 - \delta_2 = s_3 - \frac{0.6750}{x_2}$$

Hence from Eq. (E_4), we have

$$f_1^*(s_2) = f_1^* \left(s_3 - \frac{0.6750}{x_2} \right) = \left[1.5625 \Bigg/ \left(s_3 - \frac{0.6750}{x_2} \right) \right] \tag{E_7}$$

Thus Eq. (E_6) becomes

$$f_2^*(s_3) = \min_{x_2 \geq 0} \left[1.2x_2 + \frac{1.5625}{s_3 - 0.6750/x_2} \right] \tag{E_8}$$

Let

$$F(s_3, x_2) = 1.2x_2 + \frac{1.5625}{s_3 - 0.6750/x_2} = 1.2x_2 + \frac{1.5625x_2}{s_3x_2 - 0.6750}$$

For any specified value of s_3, the minimum of F is given by

$$\frac{\partial F}{\partial x_2} = 1.2 - \frac{(1.5625)(0.6750)}{(s_3x_2 - 0.6750)^2} = 0 \quad \text{or} \quad x_2^* = \frac{1.6124}{s_3} \tag{E_9}$$

$$f_2^*(s_3) = 1.2x_2^* + \frac{1.5625}{s_3 - 0.6750/x_2^*} = \frac{1.9349}{s_3} + \frac{2.6820}{s_3} = \frac{4.6169}{s_3} \tag{E_{10}}$$

Let s_4 be the displacement available for allocation to the first three members. Let δ_3 be the displacement contribution due to the third member and $f_3^*(s_4)$ the minimum weight of the first three members. Then

$$f_3^*(s_4) = \min_{x_3 \geq 0} [x_3 + f_2^*(s_3)] \tag{E_{11}}$$

where s_3 is the resource available after allocation to stage 3 and is given by

$$s_3 = s_4 - \delta_3 = s_4 - \frac{1.5625}{x_3}$$

From Eq. (E_{10}) we have

$$f_2^*(s_3) = \frac{4.6169}{s_4 - 1.5625/x_3} \tag{E_{12}}$$

and Eq. (E_{11}) can be written as

$$f_3^*(s_4) = \min_{x_3 \geq 0} \left[x_3 + \frac{4.6169 x_3}{s_4 x_3 - 1.5625} \right] \tag{E_{13}}$$

As before, by letting

$$F(s_4, x_3) = x_3 + \frac{4.6169 x_3}{s_4 x_3 - 1.5625} \tag{E_{14}}$$

the minimum of F, for any specified value of s_4, can be obtained as

$$\frac{\partial F}{\partial x_3} = 1.0 - \frac{(4.6169)(1.5625)}{(s_4 x_3 - 1.5625)^2} = 0 \quad \text{or} \quad x_3^* = \frac{4.2445}{s_4} \tag{E_{15}}$$

$$f_3^*(s_4) = x_3^* + \frac{4.6169 x_3^*}{s_4 x_3^* - 1.5625} = \frac{4.2445}{s_4} + \frac{7.3151}{s_4} = \frac{11.5596}{s_4} \tag{E_{16}}$$

Finally, let s_5 denote the displacement available for allocation to the first four members. If δ_4 denotes the displacement contribution due to the fourth member, and $f_4^*(s_5)$ the minimum weight of the first four members, then

$$f_4^*(s_5) = \min_{x_4 \geq 0} [0.6 x_4 + f_3^*(s_4)] \tag{E_{17}}$$

where the resource available after allocation to the fourth member (s_4) is given by

$$s_4 = s_5 - \delta_4 = s_5 - \frac{1.3500}{x_4} \tag{E_{18}}$$

From Eqs. (E_{16}), (E_{17}), and (E_{18}), we obtain

$$f_4^*(s_5) = \min_{x_4 \geq 0} \left[0.6 x_4 + \frac{11.5596}{s_5 - 1.3500/x_4} \right] \tag{E_{19}}$$

By setting

$$F(s_5, x_4) = 0.6 x_4 + \frac{11.5596}{s_5 - 1.3500/x_4}$$

the minimum of $F(s_5, x_4)$, for any specified value of s_5, is given by

$$\frac{\partial F}{\partial x_4} = 0.6 - \frac{(11.5596)(1.3500)}{(s_5 x_4 - 1.3500)^2} = 0 \quad \text{or} \quad x_4^* = \frac{6.44}{s_5} \tag{E$_{20}$}$$

$$f_4^*(s_5) = 0.6 x_4^* + \frac{11.5596}{s_5 - 1.3500/x_4^*} = \frac{3.864}{s_5} + \frac{16.492}{s_5} = \frac{20.356}{s_5} \tag{E$_{21}$}$$

Since the value of s_5 is specified as 0.5 in., the minimum weight of the structure can be calculated from Eq. (E$_{21}$) as

$$f_4^*(s_5 = 0.5) = \frac{20.356}{0.5} = 40.712 \, \text{lb} \tag{E$_{22}$}$$

Once the optimum value of the objective function is found, the optimum values of the design variables can be found with the help of Eqs. (E$_{20}$), (E$_{15}$), (E$_9$), and (E$_5$) as

$$x_4^* = 12.88 \, \text{in}^2$$

$$s_4 = s_5 - \frac{1.3500}{x_4^*} = 0.5 - 0.105 = 0.395 \, \text{in.}$$

$$x_3^* = \frac{4.2445}{s_4} = 10.73 \, \text{in}^2$$

$$s_3 = s_4 - \frac{1.5625}{x_3^*} = 0.3950 - 0.1456 = 0.2494 \, \text{in.}$$

$$x_2^* = \frac{1.6124}{s_3} = 6.47 \, \text{in}^2$$

$$s_2 = s_3 - \frac{0.6750}{x_2^*} = 0.2494 - 0.1042 = 0.1452 \, \text{in.}$$

$$x_1^* = \frac{1.5625}{s_2} = 10.76 \, \text{in}^2$$

9.6   EXAMPLE ILLUSTRATING THE TABULAR METHOD OF SOLUTION

Example 9.3   Design the most economical reinforced cement concrete (RCC) water tank (Fig. 9.6a) to store 100,000 liters of water. The structural system consists of a tank, four columns each 10 m high, and a foundation to transfer all loads safely to the ground [9.10]. The design involves the selection of the most appropriate types of tank, columns, and foundation among the seven types of tanks, three types of columns, and three types of foundations available. The data on the various types of tanks, columns, and foundations are given in Tables 9.1, 9.2, and 9.3, respectively.

SOLUTION   The structural system can be represented as a multistage decision process as shown in Fig. 9.13. The decision variables x_1, x_2, and x_3 represent the type of

Table 9.1   Component 3 (Tank)

Type of tank	Load acting on the tank, s_4 (kgf)	R_3 cost ($)	Self-weight of the component (kgf)	$s_3 = s_4 +$ self-weight (kgf)
(a) Cylindrical RCC tank	100,000	5,000	45,000	145,000
(b) Spherical RCC tank	100,000	8,000	30,000	130,000
(c) Rectangular RCC tank	100,000	6,000	25,000	125,000
(d) Cylindrical steel tank	100,000	9,000	15,000	115,000
(e) Spherical steel tank	100,000	15,000	5,000	105,000
(f) Rectangular steel tank	100,000	12,000	10,000	110,000
(g) Cylindrical RCC tank with hemispherical RCC dome	100,000	10,000	15,000	115,000

Table 9.2   Component 2 (Columns)

Type of columns	s_3 (kgf)	R_2 cost ($)	Self-weight (kgf)	$s_2 = s_3 +$ self-weight (kgf)
(a) RCC columns	150,000	6,000	70,000	220,000
	130,000	5,000	50,000	180,000
	110,000	4,000	40,000	150,000
	100,000	3,000	40,000	140,000
(b) Concrete columns	150,000	8,000	60,000	210,000
	130,000	6,000	50,000	180,000
	110,000	4,000	30,000	140,000
	100,000	3,000	15,000	115,000
(c) Steel columns	150,000	15,000	30,000	180,000
	130,000	10,000	20,000	150,000
	110,000	9,000	15,000	125,000
	100,000	8,000	10,000	110,000

foundation, columns, and the tank used in the system, respectively. Thus the variable x_1 can take three discrete values, each corresponding to a particular type of foundation (among mat, concrete pile, and steel pile types). Similarly the variable x_2 is assumed to take three discrete values, each corresponding to one of the columns (out of RCC columns, concrete columns, and steel columns). Finally, the variable x_3 can take seven discrete values, each corresponding to a particular type of tank listed in Table 9.1.

Since the input load, that is, the weight of water, is known to be 100,000 kgf, s_4 is fixed and the problem can be considered as an initial value problem. We assume that the theories of structural analysis and design in the various materials provide the design equations

$$s_i = t_i(x_i, s_{i+1})$$

Table 9.3   Component 1 (Foundation)

Type of foundation	s_2 (kg$_f$)	R_1 cost (\$)	Self-weight (kg$_f$)	$s_1 = s_2 +$ self-weight (kg$_f$)
(a) Mat foundation	220,000	5,000	60,000	280,000
	200,000	4,000	45,000	245,000
	180,000	3,000	35,000	215,000
	140,000	2,500	25,000	165,000
	100,000	500	20,000	120,000
(b) Concrete pile foundation	220,000	3,500	55,000	275,000
	200,000	3,000	40,000	240,000
	180,000	2,500	30,000	210,000
	140,000	1,500	20,000	160,000
	100,000	1,000	15,000	115,000
(c) Steel pile foundation	220,000	3,000	10,000	230,000
	200,000	2,500	9,000	209,000
	180,000	2,000	8,000	188,000
	140,000	2,000	6,000	146,000
	100,000	1,500	5,000	105,000

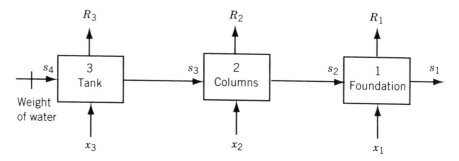

Figure 9.13   Example 9.3 as a three-stage decision problem.

which yield information for the various system components as shown in Tables 9.1 to 9.3 (these values are given only for illustrative purpose).

Suboptimization of Stage 1 (Component 1)

For the suboptimization of stage 1, we isolate component 1 as shown in Fig. 9.14a and minimize its cost $R_1(x_1, s_2)$ for any specified value of the input state s_2 to obtain $f_1^*(s_2)$ as

$$f_1^*(s_2) = \min_{x_1} [R_1(x_1, s_2)]$$

Since five settings of the input state variable s_2 are given in Table 9.3, we obtain f_1^* for each of these values as shown below:

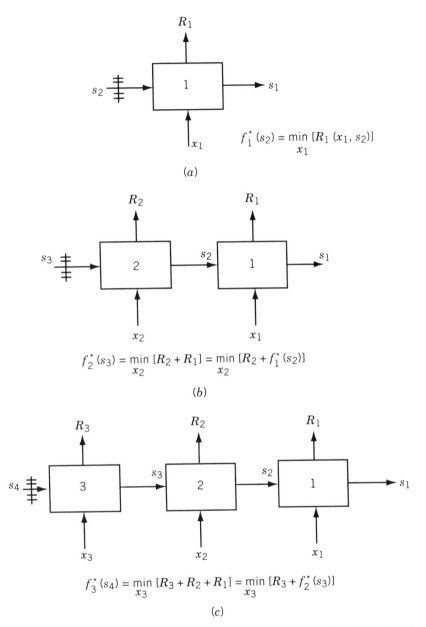

Figure 9.14   Various stages of suboptimization of Example 9.3: (*a*) suboptimization of component 1; (*b*) suboptimization of components 1 and 2; (*c*) suboptimization of components 1, 2, and 3.

Specific value of s_2 (kg$_f$)	x_1^* (type of foundation for minimum cost)	f_1^* ($)	Corresponding value of s_1 (kg$_f$)
220,000	(c)	3,000	230,000
200,000	(c)	2,500	209,000
180,000	(c)	2,000	188,000
140,000	(b)	1,500	160,000
100,000	(a)	500	120,000

Suboptimization of Stages 2 and 1 (Components 2 and 1)

Here we combine components 2 and 1 as shown in Fig. 9.14b and minimize the cost $(R_2 + R_1)$ for any specified value s_3 to obtain $f_2^*(s_3)$ as

$$f_2^*(s_3) = \min_{x_2, x_1} [R_2(x_2, s_3) + R_1(x_1, s_2)] = \min_{x_2} [R_2(x_2, s_3) + f_1^*(s_2)]$$

Since four settings of the input state variable s_3 are given in Table 9.2, we can find f_2^* for each of these four values. Since this number of settings for s_3 is small, the values of the output state variable s_2 that result will not necessarily coincide with the values of s_2 tabulated in Table 9.3. Hence we interpolate linearly the values of s_2 (if it becomes necessary) for the purpose of present computation. However, if the computations are done on a computer, more settings, more closely spaced, can be considered without much difficulty. The suboptimization of stages 2 and 1 gives the following results:

Specific value of s_3 (kg$_f$)	Value of x_2 (type of columns)	Cost of columns, R_2 ($)	Value of the output state variable s_2 (kg$_f$)	x_1^* (Type of foundation)	f_1^* ($)	$R_2 + f_1^*$ ($)
150,000	(a)	6,000	220,000	(c)	3,000	9,000
	(b)	8,000	210,000	(c)	2,750**	10,750
	(c)	15,000	180,000	(c)	2,000	17,000
130,000	(a)	5,000	180,000	(c)	2,000	7,000
	(b)	6,000	180,000	(c)	2,000	8,000
	(c)	10,000	150,000	(b)	1,625**	11,625
110,000	(a)	4,000	150,000	(b)	1,625**	5,625
	(b)	4,000	140,000	(b)	1,500	5,500
	(c)	9,000	125,000	(b)	1,125**	10,125
100,000	(a)	3,000	140,000	(b)	1,500	4,500
	(b)	3,000	115,000	(a)	875**	3,875
	(c)	8,000	110,000	(a)	750**	8,750

Notice that the double-starred quantities indicate interpolated values and the boxed quantities the minimum cost solution for the specified value of s_3. Now the desired quantities (i.e., f_2^* and x_2^*) corresponding to the various discrete values of s_3 can be summarized as follows:

Specified value of s_3 (kg$_f$)	Type of columns corresponding to minimum cost of stages 2 and 1, (x_2^*)	Minimum cost of stages 2 and 1, f_2^* ($)	Value of the corresponding state variable, s_2 (kg$_f$)
150,000	(a)	9,000	220,000
130,000	(a)	7,000	180,000
110,000	(b)	5,500	140,000
100,000	(b)	3,875	115,000

Suboptimization of Stages 3, 2, and 1 (Components 3, 2, and 1)

For the suboptimization of stages 3, 2, and 1, we consider all three components together as shown in Fig. 9.14c and minimize the cost $(R_3 + R_2 + R_1)$ for any specified value of s_4 to obtain $f_3^*(s_4)$. However, since there is only one value of s_4 (initial value problem) to be considered, we obtain the following results by using the information given in Table 9.1:

$$f_3^*(s_4) = \min_{x_3, x_2, x_1} \; [R_3(x_3, s_4) + R_2(x_2, s_3) + R_1(x_1, s_2)]$$

$$= \min_{x_3} \; [R_3(x_3, s_4) + f_2^*(s_3)]$$

Specific value of s_4 (kg$_f$)	Type of tank (x_3)	Cost of tank R_3 ($)	Corresponding output state, s_3 (kg$_f$)	x_2^* (type of columns for minimum cost)	f_2^* ($)	$R_3 + f_2^*$ ($)
100,000	(a)	5,000	145,000	(a)	8,500**	13,500
	(b)	8,000	130,000	(a)	7,000	15,000
	(c)	6,000	125,000	(a)	6,625**	12,625
	(d)	9,000	115,000	(b)	5,875**	14,875
	(e)	15,000	105,000	(b)	$4,687\frac{1}{2}$**	$19,687\frac{1}{2}$
	(f)	12,000	110,000	(b)	5,500	17,500
	(g)	10,000	115,000	(b)	5,875**	15,875

Here also the double-starred quantities indicate the interpolated values and the boxed quantity the minimum cost solution. From the results above, the minimum cost solution is given by

$$s_4 = 100,000 \text{ kg}_f$$

$$x_3^* = \text{type (c) tank}$$

$$f_3^*(s_4 = 100,000) = \$12,625$$

$$s_3 = 125,000 \text{ kg}_f$$

Now, we retrace the steps to collect the optimum values of x_3^*, x_2^*, and x_1^* and obtain

$$x_3^* = \text{type (c) tank,} \qquad s_3 = 125,000 \text{ kg}_f$$

$$x_2^* = \text{type (a) columns,} \qquad s_2 = 170,000 \text{ kg}_f$$

$$x_1^* = \text{type (c) foundation,} \quad s_1 = 181,000 \text{ kg}_f$$

and the total minimum cost of the water tank is \$12,625. Thus the minimum cost water tank consists of a rectangular RCC tank, RCC columns, and a steel pile foundation.

9.7   CONVERSION OF A FINAL VALUE PROBLEM INTO AN INITIAL VALUE PROBLEM

In previous sections the dynamic programming technique has been described with reference to an initial value problem. If the problem is a final value problem as shown in Fig. 9.15a, it can be solved by converting it into an equivalent initial value problem. Let the stage transformation (design) equation be given by

$$s_i = t_i(s_{i+1}, x_i), \quad i = 1, 2, \ldots, n \tag{9.22}$$

Assuming that the inverse relations exist, we can write Eqs. (9.22) as

$$s_{i+1} = \bar{t}_i(s_i, x_i), \quad i = 1, 2, \ldots, n \tag{9.23}$$

where the input state to stage i is expressed as a function of its output state and the decision variable. It can be noticed that the roles of input and output state variables are interchanged in Eqs. (9.22) and (9.23). The procedure of obtaining Eq. (9.23) from Eq. (9.22) is called *state inversion*. If the return (objective) function of stage i is originally expressed as

$$R_i = r_i(s_{i+1}, x_i), \quad i = 1, 2, \ldots, n \tag{9.24}$$

Eq. (9.23) can be used to express it in terms of the output state and the decision variable as

$$R_i = r_i[\bar{t}_i(s_i, x_i), x_i] = \bar{r}_i(s_i, x_i), \quad i = 1, 2, \ldots, n \tag{9.25}$$

The optimization problem can now be stated as follows:

Find $x_1, x_2, \ldots, x_n$ so that

$$f(x_1, x_2, \ldots, x_n) = \sum_{i=1}^{n} R_i = \sum_{i=1}^{n} \bar{r}_i(s_i, x_i) \tag{9.26}$$

will be optimum where the s_i are related by Eq. (9.23).

The use of Eq. (9.23) amounts to reversing the direction of the flow of information through the state variables. Thus the optimization process can be started at stage n and stages $n - 1, n - 2, \ldots, 1$ can be reached in a sequential manner. Since s_1 is specified (fixed) in the original problem, the problem stated in Eq. (9.26) will be equivalent to

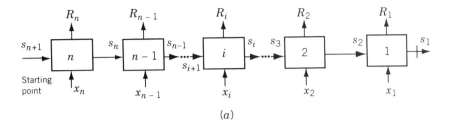

(a)

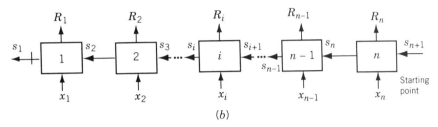

(b)

Figure 9.15 Conversion of a final value problem to an initial value problem: (a) final value problem; (b) initial value problem.

an initial value problem as shown in Fig. 9.15b. This initial value problem is identical to the one considered in Fig. 9.3 except for the stage numbers. If the stage numbers $1, 2, \ldots, n$ are reversed to $n, n - 1, \ldots, 1$, Fig. 9.15b will become identical to Fig. 9.3. Once this is done, the solution technique described earlier can be applied for solving the final value problem shown in Fig. 9.15a.

Example 9.4 A small machine tool manufacturing company entered into a contract to supply 80 drilling machines at the end of the first month and 120 at the end of the second month. The unit cost of manufacturing a drilling machine in any month is given by $\$(50x + 0.2x^2)$, where x denotes the number of drilling machines manufactured in that month. If the company manufactures more units than needed in the first month, there is an inventory carrying cost of $\$8$ for each unit carried to the next month. Find the number of drilling machines to be manufactured in each month to minimize the total cost. Assume that the company has enough facilities to manufacture up to 200 drilling machines per month and that there is no initial inventory. Solve the problem as a final value problem.

SOLUTION The problem can be stated as follows:

$$\text{Minimize } f(x_1, x_2) = (50x_1 + 0.2x_1^2) + (50x_2 + 0.2x_2^2) + 8(x_1 - 80)$$

subject to

$$x_1 \geq 80$$

$$x_1 + x_2 = 200$$

$$x_1 \geq 0, \quad x_2 \geq 0$$

where x_1 and x_2 indicate the number of drilling machines manufactured in the first month and the second month, respectively. To solve this problem as a final value problem, we start from the second month and go backward. If I_2 is the inventory at the beginning of the second month, the optimum number of drilling machines to be manufactured in the second month is given by

$$x_2^* = 120 - I_2 \tag{E$_1$}$$

and the cost incurred in the second month by

$$R_2(x_2^*, I_2) = 8I_2 + 50x_2^* + 0.2x_2^{*2}$$

By using Eq. (E$_1$), R_2 can be expressed as

$$R_2(I_2) = 8I_2 + 50(120 - I_2) + 0.2(120 - I_2)^2 = 0.2I_2^2 - 90I_2 + 8880 \tag{E$_2$}$$

Since the inventory at the beginning of the first month is zero, the cost involved in the first month is given by

$$R_1(x_1) = 50x_1 + 0.2x_1^2$$

Thus the total cost involved is given by

$$f_2(I_2, x_1) = (50x_1 + 0.2x_1^2) + (0.2I_2^2 - 90I_2 + 8880) \tag{E$_3$}$$

But the inventory at the beginning of the second month is related to x_1 as

$$I_2 = x_1 - 80 \tag{E$_4$}$$

Equations (E$_3$) and (E$_4$) lead to

$$f = f_2(I_2) = (50x_1 + 0.2x_1^2) + 0.2(x_1 - 80)^2 - 90(x_1 - 80) + 8880$$

$$= 0.4x_1^2 - 72x_1 + 17,360 \tag{E$_5$}$$

Since f is a function of x_1 only, the optimum value of x_1 can be obtained as

$$\frac{df}{dx_1} = 0.8x_1 - 72 = 0 \quad \text{or} \quad x_1^* = 90$$

As $d^2 f(x_1^*)/dx_1^2 = 0.8 > 0$, the value of x_1^* corresponds to the minimum of f. Thus the optimum solution is given by

$$f_{\min} = f(x_1^*) = \$14,120$$

$$x_1^* = 90 \quad \text{and} \quad x_2^* = 110$$

9.8   LINEAR PROGRAMMING AS A CASE OF DYNAMIC PROGRAMMING

A linear programming problem with n decision variables and m constraints can be considered as an n-stage dynamic programming problem with m state variables. In fact, a linear programming problem can be formulated as a dynamic programming problem. To illustrate the conversion of a linear programming problem into a dynamic programming problem, consider the following linear programming problem:

$$\text{Maximize } f(x_1, x_2, \ldots, x_n) = \sum_{j=1}^{n} c_j x_j$$

subject to

$$\sum_{j=1}^{n} a_{ij} x_j \leq b_i, \quad i = 1, 2, \ldots, m$$

$$x_j \geq 0, \quad j = 1, 2, \ldots, n$$

(9.27)

This problem can be considered as an n-stage decision problem where the value of the decision variable x_j must be determined at stage j. The right-hand sides of the constraints, b_i, $i = 1, 2, \ldots, m$, can be treated as m types of resources to be allocated among different kinds of activities x_j. For example, b_1 may represent the available machines, b_2 the available time, and so on, in a workshop. The variable x_1 may denote the number of castings produced, x_2 the number of forgings produced, x_3 the number of machined components produced, and so on, in the workshop. The constant c_j may represent the profit per unit of x_j. The coefficients a_{ij} represent the amount of ith resource b_i needed for 1 unit of jth activity x_j (e.g., the amount of material required to produce one casting). Hence when the value of the decision variable x_j at the jth stage is determined, $a_{1j} x_j$ units of resource 1, $a_{2j} x_j$ units of resource 2, $\ldots, a_{mj} x_j$ units of resource m will be allocated to jth activity if sufficient unused resources exist. Thus the amounts of the available resources must be determined before allocating them to any particular activity. For example, when the value of the first activity x_1 is determined at stage 1, there must be sufficient amounts of resources b_i for allocation to activity 1. The resources remaining after allocation to activity 1 must be determined before the value of x_2 is found at stage 2, and so on. In other words, the state of the system (i.e., the amounts of resources remaining for allocation) must be known before making a decision (about allocation) at any stage of the n-stage system. In this problem there are m state parameters constituting the state vector.

By denoting the optimal value of the composite objective function over n stages as f_n^*, we can state the problem as

Find

$$f_n^* = f_n^*(b_1, b_2, \ldots, b_m) = \max_{x_1, x_2, \ldots, x_n} \left(\sum_{j=1}^{n} c_j x_j \right)$$

(9.28)

such that

$$\sum_{j=1}^{n} a_{ij} x_j \le b_i, \qquad i = 1, 2, \ldots, m \tag{9.29}$$

$$x_j \ge 0, \qquad j = 1, 2, \ldots, n \tag{9.30}$$

The recurrence relationship (9.16), when applied to this problem yields

$$f_1^*(\beta_1, \beta_2, \ldots, \beta_m) = \max_{0 \le x_i \le \overline{\beta}} [c_i x_i + f_{i-1}^*(\beta_1 - a_{1i} x_i,$$

$$\beta_2 - a_{2i} x_i, \ldots, \beta_m - a_{mi} x_i)], \quad i = 2, 3, \ldots, n \tag{9.31}$$

where $\beta_1, \beta_2, \ldots, \beta_m$ are the resources available for allocation at stage i; $a_{1i} x_i, \ldots,$ $a_{mi} x_i$ are the resources allocated to the activity x_i, $\beta_1 - a_{1i} x_i, \beta_2 - a_{2i} x_i, \ldots, \beta_m - a_{mi} x_i$ are the resources available for allocation to the activity $i - 1$, and $\overline{\beta}$ indicates the maximum value that x_i can take without violating any of the constraints stated in Eqs. (9.29). The value of $\overline{\beta}$ is given by

$$\overline{\beta} = \min \left(\frac{\beta_1}{a_{1i}}, \frac{\beta_2}{a_{2i}}, \ldots, \frac{\beta_m}{a_{mi}} \right) \tag{9.32}$$

since any value larger than $\overline{\beta}$ would violate at least one constraint. Thus at the ith stage, the optimal values x_i^* and f_i^* can be determined as functions of $\beta_1, \beta_2, \ldots, \beta_m$.

Finally, at the nth stage, since the values of β_1, β_2, ..., β_m are known to be $b_1, b_2, \ldots, b_m$, respectively, we can determine x_n^* and f_n^*. Once x_n^* is known, the remaining values, $x_{n-1}^*, x_{n-2}^*, \ldots, x_1^*$ can be determined by retracing the suboptimization steps.

Example 9.5[†]

$$\text{Maximize } f(x_1, x_2) = 50x_1 + 100x_2$$

subject to

$$10x_1 + 5x_2 \le 2500$$

$$4x_1 + 10x_2 \le 2000$$

$$x_1 + 1.5x_2 \le 450$$

$$x_1 \ge 0, \quad x_2 \ge 0$$

SOLUTION Since $n = 2$ and $m = 3$, this problem can be considered as a two-stage dynamic programming problem with three state parameters. The first-stage problem is to find the maximum value of f_1:

$$\max f_1(\beta_1, \beta_2, \beta_3, x_1) = \max_{0 \le x_1 \le \overline{\beta}} (50x_1)$$

[†]This problem is the same as the one stated in Example 3.2.

where β_1, β_2, and β_3 are the resources available for allocation at stage 1, and x_1 is a nonnegative value that satisfies the side constraints $10x_1 \leq \beta_1$, $4x_1 \leq \beta_2$, and $x_1 \leq \beta_3$. Here $\beta_1 = 2500 - 5x_2$, $\beta_2 = 2000 - 10x_2$, and $\beta_3 = 450 - 1.5x_2$, and hence the maximum value β that x_1 can assume is given by

$$\overline{\beta} = x_1^* = \min \left[\frac{2500 - 5x_2}{10}, \frac{2000 - 10x_2}{4}, 450 - 1.5x_2 \right] \qquad (E_1)$$

Thus

$$f_1^* \left(\frac{2500 - 5x_2}{10}, \frac{2000 - 10x_2}{4}, 450 - 1.5x_2 \right) = 50x_1^*$$

$$= 50 \min \left(\frac{2500 - 5x_2}{10}, \frac{2000 - 10x_2}{4}, 450 - 1.5x_2 \right)$$

The second-stage problem is to find the maximum value of f_2:

$$\max f_2(\beta_1, \beta_2, \beta_3) = \max_{0 \leq x_2 \leq \overline{\beta}} \left[100x_2 + f_1^* \left(\frac{2500 - 5x_2}{10}, \right. \right.$$

$$\left. \left. \frac{2000 - 10x_2}{4}, 450 - 1.5x_2 \right) \right] \qquad (E_2)$$

where β_1, β_2, and β_3 are the resources available for allocation at stage 2, which are equal to 2500, 2000, and 450, respectively. The maximum value that x_2 can assume without violating any constraint is given by

$$\overline{\beta} = \min \left(\frac{2500}{5}, \frac{2000}{10}, \frac{450}{1.5} \right) = 200$$

Thus the recurrence relation, Eq. (E_2), can be restated as

$$\max \; f_2(2500, 2000, 450)$$

$$= \max_{0 \leq x_2 \leq 200} \left\{ 100x_2 + 50 \min \left(\frac{2500 - 5x_2}{10}, \frac{2000 - 10x_2}{4}, 450 - 1.5x_2 \right) \right\}$$

Since

$$\min \left(\frac{2500 - 5x_2}{10}, \frac{2000 - 10x_2}{4}, 450 - 1.5x_2 \right)$$

$$= \begin{cases} \dfrac{2500 - 5x_2}{10} & \text{if} \quad 0 \leq x_2 \leq 125 \\[2mm] \dfrac{2000 - 10x_2}{4} & \text{if} \quad 125 \leq x_2 \leq 200 \end{cases}$$

we obtain

$$\max_{0 \le x_2 \le 200} \left[100x_2 + 50 \ \min \left(\frac{2500 - 5x_2}{10}, \frac{2000 - 10x_2}{4}, 450 - 1.5x_2 \right) \right]$$

$$= \max \begin{cases} 100x_2 + 50 \left(\dfrac{2500 - 5x_2}{10} \right) & \text{if} \quad 0 \le x_2 \le 125 \\[3mm] 100x_2 + 50 \left(\dfrac{2000 - 10x_2}{4} \right) & \text{if} \quad 125 \le x_2 \le 200 \end{cases}$$

$$= \max \begin{cases} 75x_2 + 12{,}500 & \text{if} \quad 0 \le x_2 \le 125 \\[2mm] 25{,}000 - 25x_2 & \text{if} \quad 125 \le x_2 \le 200 \end{cases}$$

Now,

$$\max(75x_2 + 12{,}500) = 21{,}875 \text{ at } x_2 = 125$$

$$\max(25{,}000 - 25x_2) = 21{,}875 \text{ at } x_2 = 125$$

Hence

$$f_2^*(2500, 2000, 450) = 21{,}875 \quad \text{at} \quad x_2^* = 125.0$$

From Eq. (E_1) we have

$$x_1^* = \min \left(\frac{2500 - 5x_2^*}{10}, \frac{2000 - 10x_2^*}{4}, 450 - 1.5x_2^* \right)$$

$$= \min(187.5, 187.5, 262.5) = 187.5$$

Thus the optimum solution of the problem is given by $x_1^* = 187.5$, $x_2^* = 125.0$, and $f_{\max} = 21{,}875.0$, which can be seen to be identical with the one obtained earlier.

Problem of Dimensionality in Dynamic Programming.   The application of dynamic programming for the solution of a linear programming problem has a serious limitation due to the dimensionality restriction. The number of calculations needed will increase very rapidly as the number of decision variables and state parameters increases. As an example, consider a linear programming problem with 100 constraints. This means that there are 100 state variables. By the procedure outlined in Section 9.4, if a table of f_i^* is to be constructed in which 100 discrete values (settings) are given to each parameter, the table contains 100^{100} entries. This is a gigantic number, and if the calculations are to be performed on a high-speed digital computer, it would require 100^{96} seconds or about 100^{92} years[†] merely to compute one table of f_i^*. Like this, 100 tables have to be prepared, one for each decision variable. Thus it is totally out of the question to solve a general linear programming problem of any reasonable size[‡] by dynamic programming.

[†]The computer is assumed to be capable of computing 10^8 values of f_i^* per second.
[‡]As stated in Section 4.7, LP problems with 150,000 variables and 12,000 constraints have been solved in a matter of a few hours using some special techniques.

These comments are equally applicable for all dynamic programming problems involving many state variables, since the computations have to be performed for different possible values of each of the state variables. Thus this problem causes not only an increase in the computational time, but also requires a large computer memory. This problem is known as the *problem of dimensionality* or the *curse of dimensionality*, as termed by Bellman. This presents a serious obstacle in solving medium- and large-size dynamic programming problems.

9.9 CONTINUOUS DYNAMIC PROGRAMMING

If the number of stages in a multistage decision problem tends to infinity, the problem becomes an infinite stage or continuous problem and dynamic programming can still be used to solve the problem. According to this notion, the trajectory optimization problems, defined in Section 1.5, can also be considered as *infinite-stage* or *continuous problems*.

An infinite-stage or continuous decision problem may arise in several practical problems. For example, consider the problem of a missile hitting a target in a specified (finite) time interval. Theoretically, the target has to be observed and commands to the missile for changing its direction and speed have to be given continuously. Thus an infinite number of decisions have to be made in a finite time interval. Since a stage has been defined as a point where decisions are made, this problem will be an infinite-stage or continuous problem. Another example where an infinite-stage or continuous decision problem arises is in planning problems. Since large industries are assumed to function for an indefinite amount of time, they have to do their planning on this basis. They make their decisions at discrete points in time by anticipating a maximum profit in the long run (essentially over an infinite period of time). In this section we consider the application of continuous decision problems.

We have seen that the objective function in dynamic programming formulation is given by the sum of individual stage returns. If the number of stages tends to infinity, the objective function will be given by the sum of infinite terms, which amounts to having the objective function in the form of an integral. The following examples illustrate the formulation of continuous dynamic programming problems.

Example 9.6 Consider a manufacturing firm that produces a certain product. The rate of demand of this product (p) is known to be $p = p[x(t), t]$, where t is the time of the year and $x(t)$ is the amount of money spent on advertisement at time t. Assume that the rate of production is exactly equal to the rate of demand. The production cost, c, is known to be a function of the amount of production (p) and the production rate (dp/dt) as $c = c(p, dp/dt)$. The problem is to find the advertisement strategy, $x(t)$, so as to maximize the profit between t_1 and t_2. The unit selling price (s) of the product is known to be a function of the amount of production as $s = s(p) = a + b/p$, where a and b are known positive constants.

SOLUTION Since the profit is given by the difference between the income from sales and the expenditure incurred for production and advertisement, the total profit over the period t_1 to t_2 is given by

$$f = \int_{t_1}^{t_2} \left[p\left(a + \frac{b}{p}\right) - c\left(p, \frac{dp}{dt}, t\right) - x(t) \right] dt \tag{E_1}$$

where $p = p\{x(t), t\}$. Thus the optimization problem can be stated as follows: Find $x(t)$, $t_1 \le t \le t_2$, which maximizes the total profit, f given by Eq. (E_1).

Example 9.7   Consider the problem of determining the optimal temperature distribution in a plug-flow tubular reactor [9.1]. Let the reactions carried in this type of reactor be shown as follows:

$$X_1 \underset{k_2}{\overset{k_1}{\rightleftharpoons}} X_2 \xrightarrow{k_3} X_3$$

where X_1 is the reactant, X_2 the desired product, and X_3 the undesired product, and k_1, k_2, and k_3 are called rate constants. Let x_1 and x_2 denote the concentrations of the products X_1 and X_2, respectively. The equations governing the rate of change of the concentrations can be expressed as

$$\frac{dx_1}{dy} + k_1 x_1 = k_2 x_2 \tag{E_1}$$

$$\frac{dx_2}{dy} + k_2 x_2 + k_3 x_2 = k_1 x_1 \tag{E_2}$$

with the initial conditions $x_1(y = 0) = c_1$ and $x_2(y = 0) = c_2$, where y is the normalized reactor length such that $0 \le y \le 1$. In general, the rate constants depend on the temperature (t) and are given by

$$k_i = a_i e^{-(b_i/t)}, \quad i = 1, 2, 3 \tag{E_3}$$

where a_i and b_i are constants.

If the objective is to determine the temperature distribution $t(y)$, $0 \le y \le 1$, to maximize the yield of the product X_2, the optimization problem can be stated as follows:

Find $t(y)$, $0 \le y \le 1$, which maximizes

$$x_2(1) - x_2(0) = \int_{y=0}^{1} dx_2 = \int_{0}^{1} (k_1 x_1 - k_2 x_2 - k_3 x_2)\, dy$$

where $x_1(y)$ and $x_2(y)$ have to satisfy Eqs. (E_1) and (E_2). Here it is assumed that the desired temperature can be produced by some external heating device.

The classical method of approach to continuous decision problems is by the calculus of variations.[†] However, the analytical solutions, using calculus of variations, cannot be obtained except for very simple problems. The dynamic programming approach, on the other hand, provides a very efficient numerical approximation procedure for solving continuous decision problems. To illustrate the application of dynamic programming

---

[†]See Section 12.2 for additional examples of continuous decision problems and the solution techniques using calculus of variations.

to the solution of continuous decision problems, consider the following simple (unconstrained) problem. Find the function $y(x)$ that minimizes the integral

$$f = \int_{x=a}^{b} R\left(\frac{dy}{dx}, y, x\right) dx \tag{9.33}$$

subject to the known end conditions $y(x = a) = \alpha$, and $y(x = b) = \beta$. We shall see how dynamic programming can be used to determine $y(x)$ numerically. This approach will not yield an analytical expression for $y(x)$ but yields the value of $y(x)$ at a finite number of points in the interval $a \le x \le b$. To start with, the interval (a, b) is divided into n segments each of length Δx (all the segments are assumed to be of equal length only for convenience). The grid points defining the various segments are given by

$$x_1 = a, x_2 = a + \Delta x, \ldots,$$
$$x_i = a + (i - 1)\Delta x, \ldots, x_{n+1} = a + n\Delta x = b$$

If Δx is small, the derivative dy/dx at x_i can be approximated by a forward difference formula as

$$\frac{dy}{dx}(x_i) \simeq \frac{y_{i+1} - y_i}{\Delta x} \tag{9.34}$$

where $y_i = y(x_i)$, $i = 1, 2, \ldots, n + 1$. The integral in Eq. (9.33) can be approximated as

$$f \simeq \sum_{i=1}^{n} R\left[\frac{dy}{dx}(x_i), y(x_i), x_i\right] \Delta x \tag{9.35}$$

Thus the problem can be restated as

Find $y(x_2), y(x_3), \ldots, y(x_n)$, which minimizes

$$f \simeq \Delta x \sum_{i=1}^{n} R\left\{\frac{y_{i+1} - y_i}{\Delta x}, y_i, x_i\right\} \tag{9.36}$$

subject to the known conditions $y_1 = \alpha$ and $y_{n+1} = \beta$.

This problem can be solved as a final value problem. Let

$$f_i^*(\theta) = \min_{y_{i+1}, y_{i+2}, \ldots, y_n} \left\{ \sum_{k=1}^{n} R\left(\frac{y_{k+1} - y_k}{\Delta x}, y_k, x_k\right) \Delta x \right\} \tag{9.37}$$

where θ is a parameter representing the various values taken by y_i. Then $f_i^*(\theta)$ can also be written as

$$f_i^*(\theta) = \min_{y_{i+1}} \left[R\left\{\frac{y_{i+1} - \theta}{\Delta x}, \theta, x_i\right\} \Delta x + f_{i+1}^*(y_{i+1}) \right] \tag{9.38}$$

This relation is valid for $i = 1, 2, \ldots, n - 1$, and

$$f_n^*(\theta) = R\left(\frac{\beta - \theta}{\Delta x}, \theta, x_n\right) \Delta x \tag{9.39}$$

Finally the desired minimum value is given by $f_0^*(\theta = \alpha)$.

In Eqs. (9.37) to (9.39), θ or y_i is a continuous variable. However, for simplicity, we treat θ or y_i as a discrete variable. Hence for each value of i, we find a set of discrete values that θ or y_i can assume and find the value of $f_i^*(\theta)$ for each discrete value of θ or y_i. Thus $f_i^*(\theta)$ will be tabulated for only those discrete values that θ can take. At the final stage, we find the values of $f_0^*(\alpha)$ and y_1^*. Once y_1^* is known, the optimal values of $y_2, y_3, \ldots, y_n$ can easily be found without any difficulty, as outlined in the previous sections.

It can be seen that the solution of a continuous decision problem by dynamic programming involves the determination of a whole family of extremal trajectories as we move from b toward a. In the last step we find the particular extremal trajectory that passes through both points (a, α) and (b, β). This process is illustrated in Fig. 9.16. In this figure, $f_i^*(\theta)$ is found by knowing which of the extremal trajectories that terminate at x_{i+1} pass through the point (x_i, θ). If this procedure is followed, the solution of a continuous decision problem poses no additional difficulties. Although the simplest type of continuous decision problem is considered in this section, the same procedure can be adopted to solve any general continuous decision problem involving the determination of several functions, $y_1(x), y_2(x), \ldots, y_N(x)$ subject to m constraints $(m < N)$ in the form of differential equations [9.3].

9.10   ADDITIONAL APPLICATIONS

Dynamic programming has been applied to solve several types of engineering problems. Some representative applications are given in this section.

9.10.1   Design of Continuous Beams

Consider a continuous beam that rests on n rigid supports and carries a set of prescribed loads $P_1, P_2, \ldots, P_n$ as shown in Fig. 9.17 [9.11]. The locations of the supports are assumed to be known and the simple plastic theory of beams is assumed to

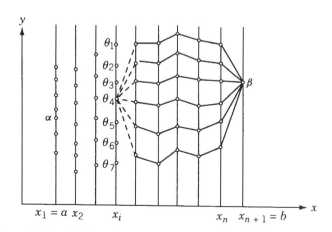

Figure 9.16   Solution of a continuous dynamic programming problem.

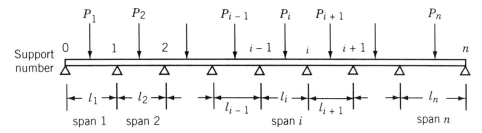

Figure 9.17   Continuous beam on rigid supports.

be applicable. Accordingly, the complete bending moment distribution can be determined once the reactant support moments $m_1, m_2, \ldots, m_n$ are known. Once the support moments are known (chosen), the plastic limit moment necessary for each span can be determined and the span can be designed. The bending moment at the center of the ith span is given by $-P_i l_i / 4$ and the largest bending moment in the ith span, M_i, can be computed as

$$M_i = \max \left\{ |m_{i-1}|, |m_i|, \left| \frac{m_{i-1} + m_i}{2} - \frac{P_i l_i}{4} \right| \right\}, \quad i = 1, 2, \ldots, n \qquad (9.40)$$

If the beam is uniform in each span, the limit moment for the ith span should be greater than or equal to M_i. The cross section of the beam should be selected so that it has the required limit moment. Thus the cost of the beam depends on the limit moment it needs to carry. The optimization problem becomes

$$\text{Find } \mathbf{X} = \{m_1, m_2, \ldots, m_n\}^{\mathrm{T}} \text{ which minimizes } \sum_{i=1}^{n} R_i(\mathbf{X})$$

while satisfying the constraints $m_i \geq M_i$, $i = 1, 2, \ldots, n$, where R_i denotes the cost of the beam in the ith span. This problem has a serial structure and hence can be solved using dynamic programming.

9.10.2   Optimal Layout (Geometry) of a Truss

Consider the planar, multibay, pin-jointed cantilever truss shown in Fig. 9.18 [9.11, 9.12, 9.22]. The configuration of the truss is defined by the x and y coordinates of the nodes. By assuming the lengths of the bays to be known (assumed to be unity in Fig. 9.18) and the truss to be symmetric about the x axis, the coordinates $y_1, y_2, \ldots, y_n$ define the layout (geometry) of the truss. The truss is subjected to a load (assumed to be unity in Fig. 9.18) at the left end. The truss is statically determinate and hence the forces in the bars belonging to bay i depend only on y_{i-1} and y_i and not on other coordinates $y_1, y_2, \ldots, y_{i-2}, y_{i+1}, \ldots, y_n$. Once the length of the bar and the force developed in it are known, its cross-sectional area can be determined. This, in turn, dictates the weight/cost of the bar. The problem of optimal layout of the truss can be formulated and solved as a dynamic programming problem.

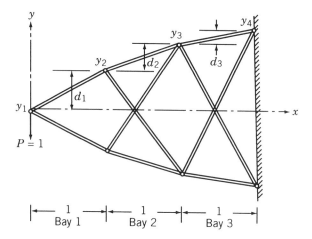

Figure 9.18   Multibay cantilever truss.

For specificness, consider a three-bay truss for which the following relationships are valid (see Fig. 9.18):

$$y_{i+1} = y_i + d_i, \quad i = 1, 2, 3 \tag{9.41}$$

Since the value of y_1 is fixed, the problem can be treated as an initial value problem. If the y coordinate of each node is limited to a finite number of alternatives that can take one of the four values 0.25, 0.5, 0.75, 1 (arbitrary units are used), there will be 64 possible designs, as shown in Fig. 9.19. If the cost of each bay is denoted by R_i, the resulting multistage decision problem can be represented as shown in Fig. 9.5a.

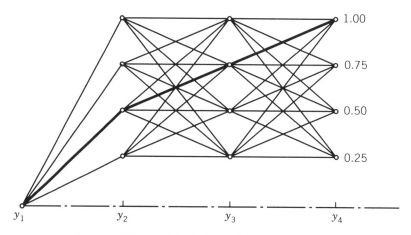

Figure 9.19   Possible designs of the cantilever truss.

9.10.3   Optimal Design of a Gear Train

Consider the gear train shown in Fig. 9.20, in which the gear pairs are numbered from 1 to n. The pitch diameters (or the number of teeth) of the gears are assumed to be known and the face widths of the gear pairs are treated as design variables [9.19, 9.20]. The minimization of the total weight of the gear train is considered as the objective. When the gear train transmits power at any particular speed, bending and surface wear stresses will be developed in the gears. These stresses should not exceed the respective permissible values for a safe design. The optimization problem can be stated as

$$\text{Find } \mathbf{X} = \{x_1, x_2, \ldots, x_n\}^{\mathrm{T}} \text{ which minimizes } \sum_{i=1}^{n} R_i(\mathbf{X}) \qquad (9.42)$$

subject to

$$\sigma_{bi}(\mathbf{X}) \leq \sigma_{b\,\max}, \quad \sigma_{wi}(\mathbf{X}) \leq \sigma_{w\,\max}, \quad i = 1, 2, \ldots, n$$

where x_i is the face width of gear pair i, R_i the weight of gear pair i, σ_{bi} (σ_{wi}) the bending (surface wear) stress induced in gear pair i, and $\sigma_{b\,\max}$ ($\sigma_{w\,\max}$) the maximum permissible bending (surface wear) stress. This problem can be considered as a multistage decision problem and can be solved using dynamic programming.

9.10.4   Design of a Minimum-Cost Drainage System

Underground drainage systems for stormwater or foul waste can be designed efficiently for minimum construction cost by dynamic programming [9.14]. Typically, a drainage system forms a treelike network in plan as shown in Fig. 9.21. The network slopes downward toward the outfall, using gravity to convey the wastewater to the outfall. Manholes are provided for cleaning and maintenance purposes at all pipe junctions. A representative three-element pipe segment is shown in Fig. 9.22. The design of an

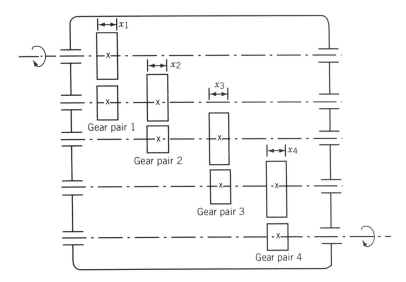

Figure 9.20   Gear train.

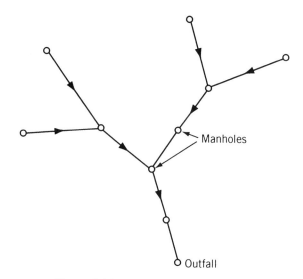

Figure 9.21   Typical drainage network.

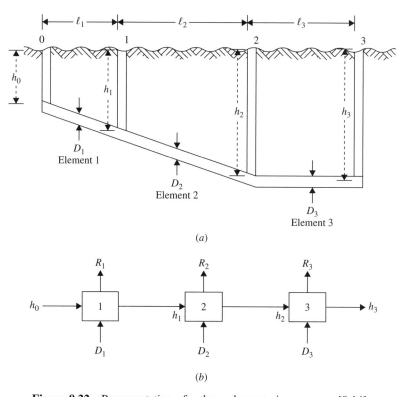

Figure 9.22   Representation of a three-element pipe segment [9.14].

element consists of selecting values for the diameter of the pipe, the slope of the pipe, and the mean depth of the pipe (D_i, h_{i-1}, and h_i). The construction cost of an element, R_i, includes cost of the pipe, cost of the upstream manhole, and earthwork related to excavation, backfilling, and compaction. Some of the constraints can be stated as follows:

1. The pipe must be able to discharge the specified flow.
2. The flow velocity must be sufficiently large.
3. The pipe slope must be greater than a specified minimum value.
4. The depth of the pipe must be sufficient to prevent damage from surface activities.

The optimum design problem can be formulated and solved as a dynamic programming problem.

REFERENCES AND BIBLIOGRAPHY

9.1 R. S. Schechter, *The Variational Method in Engineering*, McGraw-Hill, New York, 1967.

9.2 R. E. Bellman, *Dynamic Programming*, Princeton University Press, Princeton, NJ, 1957.

9.3 G. Hadley, *Nonlinear and Dynamic Programming*, Addison-Wesley, Reading, MA, 1964.

9.4 L. S. Lasdon, *Optimization Theory for Large Systems*, Macmillan, New York, 1970.

9.5 B. G. Neal, *Structural Theorems and Their Applications*, Pergamon Press, Oxford, UK, 1964.

9.6 R. E. Bellman and S. E. Dreyfus, *Applied Dynamic Programming*, Princeton University Press, Princeton, NJ, 1962.

9.7 G. L. Nemhauser, *Introduction to Dynamic Programming*, Wiley, New York, 1966.

9.8 S. Vajda, *Mathematical Programming*, Addison-Wesley, Reading, MA, 1961.

9.9 O.L.R. Jacobs, *An Introduction to Dynamic Programming*, Chapman & Hall, London, 1967.

9.10 R. J. Aguilar, *Systems Analysis and Design in Engineering, Architecture, Construction and Planning*, Prentice-Hall, Englewood Cliffs, NJ, 1973.

9.11 A. C. Palmer, Optimal structure design by dynamic programming, *ASCE Journal of the Structural Division*, Vol. 94, No. ST8, pp. 1887–1906, 1968.

9.12 D. J. Sheppard and A. C. Palmer, Optimal design of transmission towers by dynamic programming, *Computers and Structures*, Vol. 2, pp. 455–468, 1972.

9.13 J. A. S. Ferreira and R. V. V. Vidal, Optimization of a pump–pipe system by dynamic programming, *Engineering Optimization*, Vol. 7, pp. 241–251, 1984.

9.14 G. A. Walters and A. B. Templeman, Non-optimal dynamic programming algorithms in the design of minimum cost drainage systems, *Engineering Optimization*, Vol. 4, pp. 139–148, 1979.

9.15 J. S. Gero, P. J. Sheehan, and J. M. Becker, Building design using feedforward nonserial dynamic programming, *Engineering Optimization*, Vol. 3, pp. 183–192, 1978.

9.16 J. S. Gero and A. D. Radford, A dynamic programming approach to the optimum lighting problem, *Engineering Optimization*, Vol. 3, pp. 71–82, 1978.

9.17 W. S. Duff, Minimum cost solar thermal electric power systems: a dynamic programming based approach, *Engineering Optimization*, Vol. 2, pp. 83–95, 1976.

9.18 M. J. Harley and T. R. E. Chidley, Deterministic dynamic programming for long term reservoir operating policies, *Engineering Optimization*, Vol. 3, pp. 63–70, 1978.

9.19 S. G. Dhande, *Reliability Based Design of Gear Trains: A Dynamic Programming Approach*, Design Technology Transfer, ASME, New York, pp. 413–422, 1974.

9.20 S. S. Rao and G. Das, Reliability based optimum design of gear trains, *ASME Journal of Mechanisms, Transmissions, and Automation in Design*, Vol. 106, pp. 17–22, 1984.

9.21 A. C. Palmer and D. J. Sheppard, Optimizing the shape of pin-jointed structures, *Proceedings of the Institution of Civil Engineers*, Vol. 47, pp. 363–376, 1970.

9.22 U. Kirsch, *Optimum Structural Design: Concepts, Methods, and Applications*, McGraw-Hill, New York, 1981.

9.23 A. Borkowski and S. Jendo, *Structural Optimization*, Vol. 2—*Mathematical Programming*, M. Save and W. Prager (eds.), Plenum Press, New York, 1990.

9.24 L. Cooper and M. W. Cooper, *Introduction to Dynamic Programming*, Pergamon Press, Oxford, UK, 1981.

9.25 R. E. Larson and J. L. Casti, *Principles of Dynamic Programming, Part I—Basic Analytic and Computational Methods*, Marcel Dekker, New York, 1978.

9.26 D. K. Smith, *Dynamic Programming: A Practical Introduction*, Ellis Horwood, Chichester, UK, 1991.

9.27 W. F. Stoecker, *Design of Thermal Systems*, 3rd ed., McGraw-Hill, New York, 1989.

REVIEW QUESTIONS

9.1 What is a multistage decision problem?

9.2 What is the curse of dimensionality?

9.3 State two engineering examples of serial systems that can be solved by dynamic programming.

9.4 What is a return function?

9.5 What is the difference between an initial value problem and a final value problem?

9.6 How many state variables are to be considered if an LP problem with n variables and m constraints is to be solved as a dynamic programming problem?

9.7 How can you solve a trajectory optimization problem using dynamic programming?

9.8 Why are the components numbered in reverse order in dynamic programming?

9.9 Define the following terms:

   (a) Principle of optimality
   (b) Boundary value problem
   (c) Monotonic function
   (d) Separable function

9.10 Answer true or false:

   (a) Dynamic programming can be used to solve nonconvex problems.
   (b) Dynamic programming works as a decomposition technique.

(c) The objective function, $f = (R_1 + R_2)R_3$, is separable.

(d) A nonserial system can always be converted to an equivalent serial system by regrouping the components.

(e) Both the input and the output variables are specified in a boundary value problem.

(f) The state transformation equations are same as the design equations.

(g) The principle of optimality and the concept of suboptimization are same.

(h) A final value problem can always be converted into an initial value problem.

PROBLEMS

9.1 Four types of machine tools are to be installed (purchased) in a production shop. The costs of the various machine tools and the number of jobs that can be performed on each are given below.

Machine tool type	Cost of machine tool (\$)	Number of jobs that can be performed
1	3500	9
2	2500	4
3	2000	3
4	1000	2

If the total amount available is \$10,000, determine the number of machine tools of various types to be purchased to maximize the number of jobs performed. *Note:* The number of machine tools purchased must be integers.

9.2 The routes of an airline, which connects 16 cities ($A, B, \ldots, P$), are shown in Fig. 9.23. Journey from one city to another is possible only along the lines (routes) shown, with the associated costs indicated on the path segments. If a person wants to travel from city A to city P with minimum cost, without any backtracking, determine the optimal path (route) using dynamic programming.

9.3 A system consists of three subsystems in series, with each subsystem consisting of several components in parallel, as shown in Fig. 9.24. The weights and reliabilities of the various components are given below:

Subsystem, i	Weight of each component, w_i (lb)	Reliability of each component, r_i
1	4	0.96
2	2	0.92
3	6	0.98

The reliability of subsystem i is given by $R_i = 1 - (1 - r_i)^{n_i}$, $i = 1, 2, 3$, where n_i is the number of components connected in parallel in subsystem i, and the overall reliability of the system is given by $R_0 = R_1 R_2 R_3$. It was decided to use at least one and not more than three components in any subsystem. The system is to be transported into space by a space shuttle. If the total payload is restricted to 20 lb, find the number of components to be used in the three subsystems to maximize the overall reliability of the system.

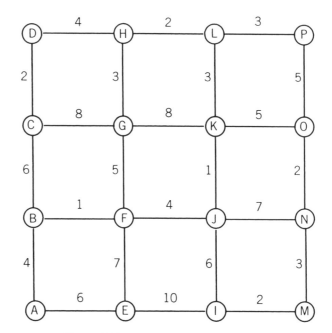

Figure 9.23   Possible paths from A to P.

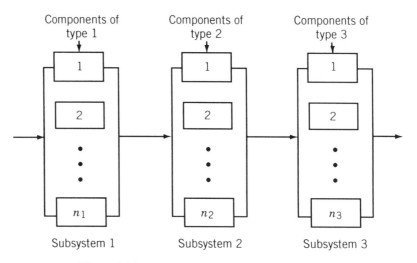

Figure 9.24   Three subsystems connected in series.

9.4   The altitude of an airplane flying between two cities A and F, separated by a distance of 2000 miles, can be changed at points B, C, D, and E (Fig. 9.25). The fuel cost involved in changing from one altitude to another between any two consecutive points is given in the following table. Determine the altitudes of the airplane at the intermediate points for minimum fuel cost.

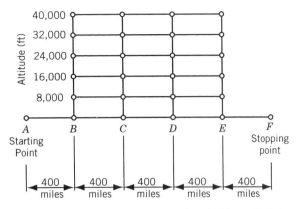

Figure 9.25 Altitudes of the airplane in Example 9.4.

From altitude (ft):	To altitude (ft):					
	0	8,000	16,000	24,000	32,000	40,000
0	—	4000	4800	5520	6160	6720
8,000	800	1600	2680	4000	4720	6080
16,000	320	480	800	2240	3120	4640
24,000	0	160	320	560	1600	3040
32,000	0	0	80	240	480	1600
40,000	0	0	0	0	160	240

9.5 Determine the path (route) corresponding to minimum cost in Problem 9.2 if a person wants to travel from city D to city M.

9.6 Each of the n lathes available in a machine shop can be used to produce two types of parts. If z lathes are used to produce the first part, the expected profit is $3z$ and if z of them are used to produce the second part, the expected profit is $2.5z$. The lathes are subject to attrition so that after completing the first part, only $z/3$ out of z remain available for further work. Similarly, after completing the second part, only $2z/3$ out of z remain available for further work. The process is repeated with the remaining lathes for two more stages. Find the number of lathes to be allocated to each part at each stage to maximize the total expected profit. Assume that any nonnegative real number of lathes can be assigned at each stage.

9.7 A minimum-cost pipeline is to be laid between points (towns) A and E. The pipeline is required to pass through one node out of B_1, B_2, and B_3, one out of C_1, C_2, and C_3, and one out of D_1, D_2, and D_3 (see Fig. 9.26). The costs associated with the various segments of the pipeline are given below:

For the segment starting at A		For the segment ending at E	
$A-B_1$	10	D_1-E	9
$A-B_2$	15	D_2-E	6
$A-B_3$	12	D_3-E	12

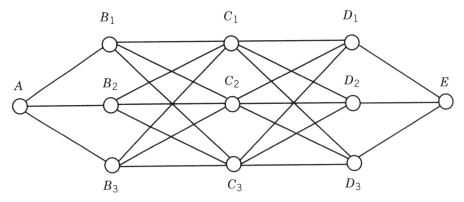

Figure 9.26 Pipe network.

For the segments B_i to C_j and C_i to D_j

From node i	To node j		
	1	2	3
1	8	12	19
2	9	11	13
3	7	15	14

Find the solution using dynamic programming.

9.8 Consider the problem of controlling a chemical reactor. The desired concentration of material leaving the reactor is 0.8 and the initial concentration is 0.2. The concentration at any time t, $x(t)$, is given by

$$\frac{dx}{dt} = \frac{1-x}{1+x}u(t)$$

where $u(t)$ is a design variable (control function).
Find $u(t)$ which minimizes

$$f = \int_0^T \{[x(t) - 0.8]^2 + u^2(t)\}\, dt$$

subject to

$$0 \le u(t) \le 1$$

Choose a grid and solve $u(t)$ numerically using dynamic programming.

9.9 It is proposed to build thermal stations at three different sites. The total budget available is 3 units (1 unit = \$10 million) and the feasible levels of investment on any thermal station are 0, 1, 2, or 3 units. The electric power obtainable (return function) for different investments is given below:

Return function, $R_i(x)$	Thermal Station, i		
	1	2	3
$R_i(0)$	0	0	0
$R_i(1)$	2	1	3
$R_i(2)$	4	5	5
$R_i(3)$	6	6	6

Find the investment policy for maximizing the total electric power generated.

9.10 Solve the following LP problem by dynamic programming:

$$\text{Maximize } f(x_1, x_2) = 10x_1 + 8x_2$$

subject to

$$2x_1 + x_2 \leq 25$$
$$3x_1 + 2x_2 \leq 45$$
$$x_2 \leq 10$$
$$x_1 \geq 0, \quad x_2 \geq 0$$

Verify your solution by solving it graphically.

9.11 A fertilizer company needs to supply 50 tons of fertilizer at the end of the first month, 70 tons at the end of second month, and 90 tons at the end of third month. The cost of producing x tons of fertilizer in any month is given by $\$(4500x + 20x^2)$. It can produce more fertilizer in any month and supply it in the next month. However, there is an inventory carrying cost of $400 per ton per month. Find the optimal level of production in each of the three periods and the total cost involved by solving it as an initial value problem.

9.12 Solve Problem 9.11 as a final value problem.

9.13 Solve the following problem by dynamic programming:

$$\underset{d_i \geq 0}{\text{Maximize}} \sum_{i=1}^{3} d_i^2$$

subject to

$$d_i = x_{i+1} - x_i, \qquad i = 1, 2, 3$$
$$x_i = 0, 1, 2, \ldots, 5, \quad i = 1, 2$$
$$x_3 = 5, \quad x_4 = 0$$

10

Integer Programming

10.1 INTRODUCTION

In all the optimization techniques considered so far, the design variables are assumed to be continuous, which can take any real value. In many situations it is entirely appropriate and possible to have fractional solutions. For example, it is possible to use a plate of thickness 2.60 mm in the construction of a boiler shell, 3.34 hours of labor time in a project, and 1.78 lb of nitrate to produce a fertilizer. Also, in many engineering systems, certain design variables can only have discrete values. For example, pipes carrying water in a heat exchanger may be available only in diameter increments of $\frac{1}{8}$ in. However, there are practical problems in which the fractional values of the design variables are neither practical nor physically meaningful. For example, it is not possible to use 1.6 boilers in a thermal power station, 1.9 workers in a project, and 2.76 lathes in a machine shop. If an integer solution is desired, it is possible to use any of the techniques described in previous chapters and round off the optimum values of the design variables to the nearest integer values. However, in many cases, it is very difficult to round off the solution without violating any of the constraints. Frequently, the rounding of certain variables requires substantial changes in the values of some other variables to satisfy all the constraints. Further, the round-off solution may give a value of the objective function that is very far from the original optimum value. All these difficulties can be avoided if the optimization problem is posed and solved as an integer programming problem.

When all the variables are constrained to take only integer values in an optimization problem, it is called an *all-integer programming problem*. When the variables are restricted to take only discrete values, the problem is called a *discrete programming problem*. When some variables only are restricted to take integer (discrete) values, the optimization problem is called a *mixed-integer (discrete) programming problem*. When all the design variables of an optimization problem are allowed to take on values of either zero or 1, the problem is called a *zero–one programming problem*. Among the several techniques available for solving the all-integer and mixed-integer linear programming problems, the cutting plane algorithm of Gomory [10.7] and the branch-and-bound algorithm of Land and Doig [10.8] have been quite popular. Although the zero–one linear programming problems can be solved by the general cutting plane or the branch-and-bound algorithms, Balas [10.9] developed an efficient enumerative algorithm for solving those problems. Very little work has been done in the field of integer nonlinear programming. The generalized penalty function method and the sequential linear integer (discrete) programming method can be used to

Table 10.1 Integer Programming Methods

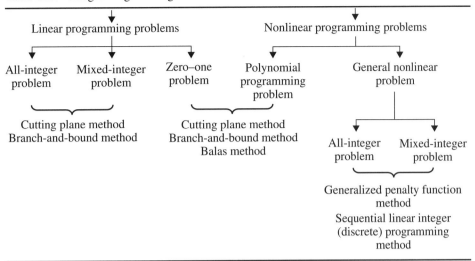

solve all integer and mixed-integer nonlinear programming problems. The various solution techniques of solving integer programming problems are summarized in Table 10.1. All these techniques are discussed briefly in this chapter.

Integer Linear Programming

10.2 GRAPHICAL REPRESENTATION

Consider the following integer programming problem:

$$\text{Maximize } f(\mathbf{X}) = 3x_1 + 4x_2$$

subject to

$$
\begin{aligned}
3x_1 - x_2 &\leq 12 \\
3x_1 + 11x_2 &\leq 66 \\
x_1 &\geq 0 \\
x_2 &\geq 0
\end{aligned}
\tag{10.1}
$$

$$x_1 \text{ and } x_2 \text{ are integers}$$

The graphical solution of this problem, by ignoring the integer requirements, is shown in Fig. 10.1. It can be seen that the solution is $x_1 = 5\frac{1}{2}$, $x_2 = 4\frac{1}{2}$ with a value of $f = 34\frac{1}{2}$. Since this is a noninteger solution, we truncate the fractional parts and obtain the new solution as $x_1 = 5$, $x_2 = 4$, and $f = 31$. By comparing this solution with all other integer feasible solutions (shown by dots in Fig. 10.1), we find that this solution is optimum for the integer LP problem stated in Eqs. (10.1).

It is to be noted that truncation of the fractional part of a LP problem will not always give the solution of the corresponding integer LP problem. This can be illustrated

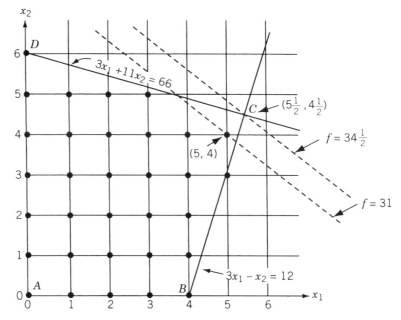

Figure 10.1   Graphical solution of the problem stated in Eqs. (10.1).

by changing the constraint $3x_1 + 11x_2 \leq 66$ to $7x_1 + 11x_2 \leq 88$ in Eqs. (10.1). With this altered constraint, the feasible region and the solution of the LP problem, without considering the integer requirement, are shown in Fig. 10.2. The optimum solution of this problem is identical with that of the preceding problem: namely, $x_1 = 5\frac{1}{2}$,

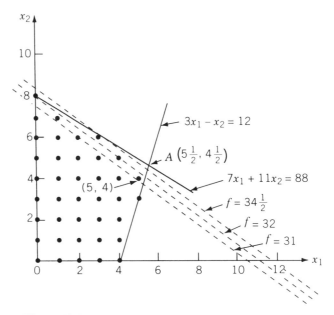

Figure 10.2   Graphical solution with modified constraint.

$x_2 = 4\frac{1}{2}$, and $f = 34\frac{1}{2}$. The truncation of the fractional part of this solution gives $x_1 = 5$, $x_2 = 4$, and $f = 31$. Although this truncated solution happened to be optimum to the corresponding integer problem in the earlier case, it is not so in the present case. In this case the optimum solution of the integer programming problem is given by $x_1^* = 0$, $x_2^* = 8$, and $f^* = 32$.

10.3   GOMORY'S CUTTING PLANE METHOD

10.3.1   Concept of a Cutting Plane

Gomory's method is based on the idea of generating a cutting plane. To illustrate the concept of a cutting plane, we again consider the problem stated in Eqs. (10.1). The feasible region of the problem is denoted by $ABCD$ in Fig. 10.1. The optimal solution of the problem, without considering the integer requirement, is given by point C. This point corresponds to $x_1 = 5\frac{1}{2}$, $x_2 = 4\frac{1}{2}$, and $f = 34\frac{1}{2}$, which is not optimal to the integer programming problem since the values of x_1 and x_2 are not integers. The feasible integer solutions of the problem are denoted by dots in Fig. 10.1. These points are called the *integer lattice points*.

In Fig. 10.3, the original feasible region is reduced to a new feasible region $ABEFGD$ by including the additional (arbitrarily selected) constraints. The idea behind

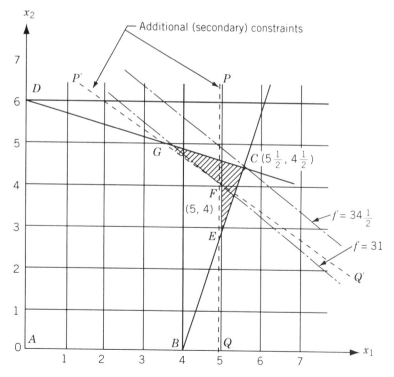

Figure 10.3   Effect of additional constraints.

adding these additional constraints is to reduce the original feasible convex region *ABCD* to a new feasible convex region (such as *ABEFGD*) such that an extreme point of the new feasible region becomes an integer optimal solution to the integer programming problem. There are two main considerations to be taken while selecting the additional constraints: (1) the new feasible region should also be a convex set, and (2) the part of the original feasible region that is sliced off because of the additional constraints should not include any feasible integer solutions of the original problem.

In Fig. 10.3, the inclusion of the two arbitrarily selected additional constraints *PQ* and $P'Q'$ gives the extreme point $F(x_1 = 5, x_2 = 4, f = 31)$ as the optimal solution of the integer programming problem stated in Eqs. (10.1). Gomory's method is one in which the additional constraints are developed in a systematic manner.

10.3.2 Gomory's Method for All-Integer Programming Problems

In this method the given problem [Eqs. (10.1)] is first solved as an ordinary LP problem by neglecting the integer requirement. If the optimum values of the variables of the problem happen to be integers, there is nothing more to be done since the integer solution is already obtained. On the other hand, if one or more of the basic variables have fractional values, some additional constraints, known as *Gomory constraints*, that will force the solution toward an all-integer point will have to be introduced. To see how the Gomory constraints are generated, let the tableau corresponding to the optimum (noninteger) solution of the ordinary LP problem be as shown in Table 10.2. Here it is assumed that there are a total of $m + n$ variables (n original variables plus m slack variables). At the optimal solution, the basic variables are represented as $x_i (i = 1, 2, \ldots, m)$ and the nonbasic variables as $y_j (j = 1, 2, \ldots, n)$ for convenience.

Gomory's Constraint.   From Table 10.2, choose the basic variable with the largest fractional value. Let this basic variable be x_i. When there is a tie in the fractional values of the basic variables, any of them can be taken as x_i. This variable can be

Table 10.2   Optimum Noninteger Solution of Ordinary LP Problem

Basic variables	Coefficient corresponding to:								Objective function	Constants
	x_1	$x_2 \ldots x_i \ldots x_m$			y_1	y_2	$\ldots\ y_j\ \ldots$	y_n		
x_1	1	0	0	0	$\bar{a}_{11}$	$\bar{a}_{12}$	$\bar{a}_{1j}$	$\bar{a}_{1n}$	0	$\bar{b}_1$
x_2	0	1	0	0	$\bar{a}_{21}$	$\bar{a}_{22}$	$\bar{a}_{2j}$	$\bar{a}_{2n}$	0	$\bar{b}_2$
$\vdots$										
x_i	0	0	1	0	$\bar{a}_{i1}$	$\bar{a}_{i2}$	$\bar{a}_{ij}$	$\bar{a}_{in}$	0	$\bar{b}_i$
$\vdots$										
x_m	0	0	0	1	$\bar{a}_{m1}$	$\bar{a}_{m2}$	$\bar{a}_{mj}$	$\bar{a}_{mn}$	0	$\bar{b}_m$
f	0	$0 \ldots 0 \ldots 0$			$\bar{c}_1$	$\bar{c}_2$	$\bar{c}_j$	$\bar{c}_n$	1	$\bar{f}$

expressed, from the ith equation of Table 10.2, as

$$x_i = \bar{b}_i - \sum_{j=1}^{n} \bar{a}_{ij} y_j \qquad (10.2)$$

where $\bar{b}_i$ is a noninteger. Let us write

$$\bar{b}_i = \hat{b}_i + \beta_i \qquad (10.3)$$

$$\bar{a}_{ij} = \hat{a}_{ij} + \alpha_{ij} \qquad (10.4)$$

where $\hat{b}_i$ and $\hat{a}_{ij}$ denote the integers obtained by truncating the fractional parts from $\bar{b}_i$ and $\bar{a}_{ij}$, respectively. Thus β_i will be a strictly positive fraction ($0 < \beta_i < 1$) and α_{ij} will be a nonnegative fraction ($0 \leq \alpha_{ij} < 1$). With the help of Eqs. (10.3) and (10.4), Eq. (10.2) can be rewritten as

$$\beta_i - \sum_{j=1}^{n} \alpha_{ij} y_j = x_i - \hat{b}_i + \sum_{j=1}^{n} \hat{a}_{ij} y_j \qquad (10.5)$$

Since all the variables x_i and y_j must be integers at an optimal integer solution, the right-hand side of Eq. (10.5) must be an integer. Thus we obtain

$$\beta_i - \sum_{j=1}^{n} \alpha_{ij} y_j = \text{integer} \qquad (10.6)$$

Notice that α_{ij} are nonnegative fractions and y_j are nonnegative integers. Hence the quantity $\sum_{j=1}^{n} \alpha_{ij} y_j$ will always be a nonnegative number. Since β_i is a strictly positive fraction, we have

$$\left(\beta_i - \sum_{j=1}^{n} \alpha_{ij} y_j \right) \leq \beta_i < 1 \qquad (10.7)$$

As the quantity $\left(\beta_i - \sum_{j=1}^{n} \alpha_{ij} y_j \right)$ has to be an integer [from Eq. (10.6)], it can be either a zero or a negative integer. Hence we obtain the desired constraint as

$$+\beta_i - \sum_{j=1}^{n} \alpha_{ij} y_j \leq 0 \qquad (10.8)$$

By adding a nonnegative slack variable s_i, the Gomory constraint equation becomes

$$s_i - \sum_{j=1}^{n} \alpha_{ij} y_j = -\beta_i \qquad (10.9)$$

where s_i must also be an integer by definition.

Table 10.3   Optimal Solution with Gomory Constraint

Basic variables	Coefficient corresponding to:											
	x_1	$x_2 \ldots x_i \ldots x_m$			y_1	y_2	$\cdots$	y_j	$\cdots$	y_n	f	s_i	Constants
x_1	1	0	0	0	$\bar{a}_{11}$	$\bar{a}_{12}$		$\bar{a}_{1j}$		$\bar{a}_{1n}$	0	0	$\bar{b}_1$
x_2	0	1	0	0	$\bar{a}_{21}$	$\bar{a}_{22}$		$\bar{a}_{2j}$		$\bar{a}_{2n}$	0	0	$\bar{b}_2$
$\vdots$													
x_i	0	0	1	0	$\bar{a}_{i1}$	$\bar{a}_{i2}$		$\bar{a}_{ij}$		$\bar{a}_{in}$	0	0	$\bar{b}_i$
$\vdots$													
x_m	0	0	0	1	$\bar{a}_{m1}$	$\bar{a}_{m2}$		$\bar{a}_{mj}$		$\bar{a}_{mn}$	0	0	$\bar{b}_m$
f	0	0	0	0	$\bar{c}_1$	$\bar{c}_2$		$\bar{c}_j$		$\bar{c}_n$	1	0	$\bar{f}$
s_i	0	0	0	0	$-\alpha_{i1}$	$-\alpha_{i2}$		$-\alpha_{ij}$		$-\alpha_{in}$	0	1	$-\beta_i$

Computational Procedure.   Once the Gomory constraint is derived, the coefficients of this constraint are inserted in a new row of the final tableau of the ordinary LP problem (i.e., Table 10.2). Since all $y_j = 0$ in Table 10.2, the Gomory constraint equation (10.9), becomes

$$s_i = -\beta_i = \text{negative}$$

which is infeasible. This means that the original optimal solution is not satisfying this new constraint. To obtain a new optimal solution that satisfies the new constraint, Eq. (10.9), the dual simplex method discussed in Chapter 4 can be used. The new tableau, after adding the Gomory constraint, is as shown in Table 10.3.

After finding the new optimum solution by applying the dual simplex method, test whether the new solution is all-integer or not. If the new optimum solution is all-integer, the process ends. On the other hand, if any of the basic variables in the new solution take on fractional values, a new Gomory constraint is derived from the new simplex tableau and the dual simplex method is applied again. This procedure is continued until either an optimal integer solution is obtained or the dual simplex method indicates that the problem has no feasible integer solution.

Remarks:

1. If there is no feasible integer solution to the given (primal) problem, this can be detected by noting an unbounded condition for the dual problem.

2. The application of the dual simplex method to remove the infeasibility of Eq. (10.9) is equivalent to cutting off the original feasible solution toward the optimal integer solution.

3. This method has a serious drawback. This is associated with the round-off errors that arise during numerical computations. Due to these round-off errors, we may ultimately get a wrong optimal integer solution. This can be rectified by storing the numbers as fractions instead of as decimal quantities. However, the magnitudes of the numerators and denominators of the fractional numbers, after some calculations, may exceed the capacity of the computer. This difficulty can

be avoided by using the all-integer integer programming algorithm developed by Gomory [10.10].

4. For obtaining the optimal solution of an ordinary LP problem, we start from a basic feasible solution (at the start of phase II) and find a sequence of improved basic feasible solutions until the optimum basic feasible solution is found. During this process, if the computations have to be terminated at any stage (for some reason), the current basic feasible solution can be taken as an approximation to the optimum solution. However, this cannot be done if we apply Gomory's method for solving an integer programming problem. This is due to the fact that the problem remains infeasible in the sense that no integer solution can be obtained until the whole problem is solved. Thus we will not be having any good integer solution that can be taken as an approximate optimum solution in case the computations have to be terminated in the middle of the process.

5. From the description given above, the number of Gomory constraints to be generated might appear to be very large, especially if the solution converges slowly. If the number of constraints really becomes very large, the size of the problem also grows without bound since one (slack) variable and one constraint are added with the addition of each Gomory constraint. However, it can be observed that the total number of constraints in the modified tableau will not exceed the number of variables in the original problem, namely, $n + m$. The original problem has m equality constraints in $n + m$ variables and we observe that there are n nonbasic variables. When a Gomory constraint is added, the number of constraints and the number of variables will each be increased by one, but the number of nonbasic variables will remain n. Hence at most n slack variables of Gomory constraints can be nonbasic at any time, and any additional Gomory constraint must be redundant. In other words, at most n Gomory constraints can be binding at a time. If at all a $(n + 1)$th constraint is there (with its slack variable as a basic and positive variable), it must be implied by the remaining constraints. Hence we drop any Gomory constraint once its slack variable becomes basic in a feasible solution.

Example 10.1

$$\text{Minimize } f = -3x_1 - 4x_2$$

subject to

$$3x_1 - x_2 + x_3 = 12$$

$$3x_1 + 11x_2 + x_4 = 66$$

$$x_i \geq 0, \quad i = 1 \text{ to } 4$$

$$\text{all } x_i \text{ are integers}$$

This problem can be seen to be same as the one stated in Eqs. (10.1) with the addition of slack variables x_3 and x_4.

SOLUTION

Step 1: Solve the LP problem by neglecting the integer requirement of the variables x_i, $i = 1$ to 4, using the regular simplex method as shown below:

Basic variables	Coefficients of variables				$-f$	$\bar{b}_i$	$\bar{b}_i / \bar{a}_{is}$ for $\bar{a}_{is} > 0$
	x_1	x_2	x_3	x_4			
x_3	3	-1	1	0	0	12	
x_4	3	$\boxed{11}$ Pivot element	0	1	0	66	$6 \leftarrow$
$-f$	-3	-4	0	0	1	0	

$\uparrow$
Most negative $\bar{c}_j$

Result of pivoting:

	x_1	x_2	x_3	x_4	$-f$	$\bar{b}_i$	
x_3	$\boxed{\frac{36}{11}}$ Pivot element	0	1	$\frac{1}{11}$	0	18	$\frac{11}{2} \leftarrow$ Smaller one
x_2	$\frac{3}{11}$	1	0	$\frac{1}{11}$	0	6	22
$-f$	$-\frac{21}{11}$	0	0	$\frac{4}{11}$	1	24	

$\uparrow$
Most negative $\bar{c}_j$

Result of pivoting:

	x_1	x_2	x_3	x_4	$-f$	$\bar{b}_i$
x_1	1	0	$\frac{11}{36}$	$\frac{1}{36}$	0	$\frac{11}{2}$
x_2	0	1	$-\frac{1}{12}$	$\frac{1}{12}$	0	$\frac{9}{2}$
$-f$	0	0	$\frac{7}{12}$	$\frac{5}{12}$	1	$\frac{69}{2}$

Since all the cost coefficients are nonnegative, the last tableau gives the optimum solution as

$$x_1 = \tfrac{11}{2}, \quad x_2 = \tfrac{9}{2}, \quad x_3 = 0, \quad x_4 = 0, \quad f_{\min} = -\tfrac{69}{2}$$

which can be seen to be identical to the graphical solution obtained in Section 10.2.

Step 2: *Generate a Gomory constraint.* Since the solution above is noninteger, a Gomory constraint has to be added to the last tableau. Since there is a tie between x_1 and x_2, let us select x_1 as the basic variable having the largest fractional value. From the row corresponding to x_1 in the last tableau, we can write

$$x_1 = \tfrac{11}{2} - \tfrac{11}{36}y_1 - \tfrac{1}{36}y_2 \qquad (E_1)$$

where y_1 and y_2 are used in place of x_3 and x_4 to denote the nonbasic variables. By comparing Eq. (E_1) with Eq. (10.2), we find that

$$i = 1, \quad \bar{b}_1 = \tfrac{11}{2}, \quad \hat{b}_1 = 5, \quad \beta_1 = \tfrac{1}{2}, \quad \bar{a}_{11} = \tfrac{11}{36},$$

$$\hat{a}_{11} = 0, \quad \alpha_{11} = \tfrac{11}{36}, \quad \bar{a}_{12} = \tfrac{1}{36}, \quad \hat{a}_{12} = 0, \quad \text{and} \quad \alpha_{12} = \tfrac{1}{36}$$

From Eq. (10.9), the Gomory constraint can be expressed as

$$s_1 - \alpha_{11}y_1 - \alpha_{12}y_2 = -\beta_1 \qquad (E_2)$$

where s_1 is a new nonnegative (integer) slack variable. Equation (E_2) can be written as

$$s_1 - \tfrac{11}{36}y_1 - \tfrac{1}{36}y_2 = -\tfrac{1}{2} \qquad (E_3)$$

By introducing this constraint, Eq. (E_3), into the previous optimum tableau, we obtain the new tableau shown below:

Basic variables	Coefficients of variables						$\bar{b}_i$	$\bar{b}_i/\bar{a}_{is}$ for $\bar{a}_{is} > 0$
	x_1	x_2	y_1	y_2	$-f$	s_1		
x_1	1	0	$\tfrac{11}{36}$	$\tfrac{1}{36}$	0	0	$\tfrac{11}{2}$	
x_2	0	1	$-\tfrac{1}{12}$	$\tfrac{1}{12}$	0	0	$\tfrac{9}{2}$	
$-f$	0	0	$\tfrac{7}{12}$	$\tfrac{5}{12}$	1	0	$\tfrac{69}{2}$	
s_1	0	0	$-\tfrac{11}{36}$	$-\tfrac{1}{36}$	0	1	$-\tfrac{1}{2}$	

Step 3: Apply the dual simplex method to find a new optimum solution. For this, we select the pivotal row r such that $\bar{b}_r = \min(\bar{b}_i < 0) = -\tfrac{1}{2}$ corresponding to s_1 in this case. The first column s is selected such that

$$\frac{\bar{c}_s}{-\bar{a}_{rs}} = \min_{\bar{a}_{rj} < 0} \left(\frac{\bar{c}_j}{-\bar{a}_{rj}} \right)$$

Here

$$\frac{\overline{c}_j}{-\overline{a}_{rj}} = \frac{7}{12} \times \frac{36}{11} = \frac{21}{11} \quad \text{for column } y_1$$

$$= \frac{5}{12} \times \frac{36}{1} = 15 \quad \text{for column } y_2.$$

Since $\frac{21}{11}$ is minimum out of $\frac{21}{11}$ and 15, the pivot element will be $-\frac{11}{36}$. The result of pivot operation is given in the following tableau:

Basic variables	\multicolumn{5}{c}{Coefficients of variables}		$\overline{b}_i$	$\overline{b}_i/\overline{a}_{is}$ for $\overline{a}_{is} > 0$
	x_1	x_2	y_1	y_2	$-f$	s_1		
x_1	1	0	0	0	0	1	5	
x_2	0	1	0	$\frac{1}{11}$	0	$-\frac{3}{11}$	$\frac{51}{11}$	
$-f$	0	0	0	$\frac{4}{11}$	1	$\frac{21}{11}$	$\frac{369}{11}$	
y_1	0	0	1	$\frac{1}{11}$	0	$-\frac{36}{11}$	$\frac{18}{11}$	

The solution given by the present tableau is $x_1 = 5$, $x_2 = 4\frac{7}{11}$, $y_1 = 1\frac{7}{11}$, and $f = -33\frac{6}{11}$, in which some variables are still nonintegers.

Step 4: Generate a new Gomory constraint. To generate the new Gomory constraint, we arbitrarily select x_2 as the variable having the largest fractional value (since there is a tie between x_2 and y_1). The row corresponding to x_2 gives

$$x_2 = \frac{51}{11} - \frac{1}{11}y_2 + \frac{3}{11}s_1$$

From this equation, the Gomory constraint [Eq. (10.9)] can be written as

$$s_2 - \frac{1}{11}y_2 + \frac{3}{11}s_1 = -\frac{7}{11}$$

When this constraint is added to the previous tableau, we obtain the following tableau:

Basic variables	\multicolumn{4}{c}{Coefficients of variables}				$\overline{b}_i$
	x_1	x_2	y_1	y_2	$-f$	s_1	s_2	
x_1	1	0	0	0	0	1	0	5
x_2	0	1	0	$\frac{1}{11}$	0	$-\frac{3}{11}$	0	$\frac{51}{11}$
y_1	0	0	1	$\frac{1}{11}$	0	$-\frac{36}{11}$	0	$\frac{18}{11}$
$-f$	0	0	0	$\frac{4}{11}$	1	$\frac{21}{11}$	0	$\frac{369}{11}$
s_2	0	0	0	$-\frac{1}{11}$	0	$\frac{3}{11}$	1	$-\frac{7}{11}$

Step 5: Apply the dual simplex method to find a new optimum solution. To carry the pivot operation, the pivot row is selected to correspond to the most negative value of $\overline{b}_i$. This is the s_2 row in this case.

Since only $\overline{a}_{rj}$ corresponding to column y_2 is negative, the pivot element will be $-\frac{1}{11}$ in the s_2 row. The pivot operation on this element leads to the following tableau:

Basic variables	Coefficients of variables							$\overline{b}_i$
	x_1	x_2	y_1	y_2	$-f$	s_1	s_2	
x_1	1	0	0	0	0	1	0	5
x_2	0	1	0	0	0	0	1	4
y_1	0	0	1	0	0	-3	1	1
$-f$	0	0	0	0	1	3	4	31
y_2	0	0	0	1	0	-3	-11	7

The solution given by this tableau is $x_1 = 5$, $x_2 = 4$, $y_1 = 1$, $y_2 = 7$, and $f = -31$, which can be seen to satisfy the integer requirement. Hence this is the desired solution.

10.3.3 Gomory's Method for Mixed-Integer Programming Problems

The method discussed in Section 10.3.2 is applicable to solve all integer programming problems where both the decision and slack variables are restricted to integer values in the optimal solution. In the mixed-integer programming problems, only a subset of the decision and slack variables are restricted to integer values. The procedure for solving mixed-integer programming problems is similar to that of all-integer programming problems in many respects.

Solution Procedure. As in the case of an all-integer programming problem, the first step involved in the solution of a mixed-integer programming problem is to obtain an optimal solution of the ordinary LP problem without considering the integer restrictions. If the values of the basic variables, which were restricted to integer values, happen to be integers in this optimal solution, there is nothing more to be done. Otherwise, a Gomory constraint is formulated by taking the integer-restricted basic variable, which has the largest fractional value in the optimal solution of the ordinary LP problem.

Let x_i be the basic variable that has the largest fractional value in the optimal solution (as shown in Table 10.2), although it is restricted to take on only integer values. If the nonbasic variables are denoted as y_j, $j = 1, 2, \ldots, n$, the basic variable x_i can be expressed as (from Table 10.2)

$$x_i = \overline{b}_i - \sum_{j=1}^{n} \overline{a}_{ij} y_j \tag{10.2}$$

We can write

$$\overline{b}_i = \hat{b}_i + \beta_i \tag{10.3}$$

where $\hat{b}_i$ is the integer obtained by truncating the fractional part of $\overline{b}_i$ and β_i is the fractional part of $\overline{b}_i$. By defining

$$\overline{a}_{ij} = a_{ij}^+ + a_{ij}^- \tag{10.10}$$

where

$$a_{ij}^+ = \begin{cases} \overline{a}_{ij} & \text{if } \overline{a}_{ij} \geq 0 \\ 0 & \text{if } \overline{a}_{ij} < 0 \end{cases} \qquad (10.11)$$

$$a_{ij}^- = \begin{cases} 0 & \text{if } \overline{a}_{ij} \geq 0 \\ \overline{a}_{ij} & \text{if } \overline{a}_{ij} < 0 \end{cases} \qquad (10.12)$$

Eq. (10.2) can be rewritten as

$$\sum_{j=1}^{n} (a_{ij}^+ + a_{ij}^-) y_j = \beta_i + (\hat{b}_i - x_i) \qquad (10.13)$$

Here, by assumption, x_i is restricted to integer values while $\overline{b}_i$ is not an integer. Since $0 < \beta_i < 1$ and $\hat{b}_i$ is an integer, we can have the value of $\beta_i + (\hat{b}_i - x_i)$ either ≥ 0 or < 0. First, we consider the case where

$$\beta_i + (\hat{b}_i - x_i) \geq 0 \qquad (10.14)$$

In this case, in order for x_i to be an integer, we must have

$$\beta_i + (\hat{b}_i - x_i) = \beta_i \quad \text{or} \quad \beta_i + 1 \quad \text{or} \quad \beta_i + 2, \ldots \qquad (10.15)$$

Thus Eq. (10.13) gives

$$\sum_{j=1}^{n} (a_{ij}^+ + a_{ij}^-) y_j \geq \beta_i \qquad (10.16)$$

Since $\overline{a}_{ij}$ are nonpositive and y_j are nonnegative by definition, we have

$$\sum_{j=1}^{n} a_{ij}^+ y_j \geq \sum_{j=1}^{n} (a_{ij}^+ - a_{ij}^-) y_j \qquad (10.17)$$

and hence

$$\sum_{j=1}^{n} a_{ij}^+ y_j \geq \beta_i \qquad (10.18)$$

Next, we consider the case where

$$\beta_i + (\hat{b}_i - x_i) < 0 \qquad (10.19)$$

For x_i to be an integer, we must have (since $0 < \beta_i < 1$)

$$\beta_i + (\hat{b}_i - x_i) = -1 + \beta_i \quad \text{or} \quad -2 + \beta_i \quad \text{or} \quad -3 + \beta_i, \ldots \qquad (10.20)$$

Thus Eq. (10.13) yields

$$\sum_{j=1}^{n} (a_{ij}^{+} + a_{ij}^{-}) y_j \le \beta_i - 1 \tag{10.21}$$

Since

$$\sum_{j=1}^{n} a_{ij}^{-} y_j \le \sum_{j=1}^{n} (a_{ij}^{+} + a_{ij}^{-}) y_j$$

we obtain

$$\sum_{j=1}^{n} a_{ij}^{-} y_j \le \beta_i - 1 \tag{10.22}$$

Upon dividing this inequality by the negative quantity $(\beta_i - 1)$, we obtain

$$\frac{1}{\beta_i - 1} \sum_{j=1}^{n} a_{ij}^{-} y_j \ge 1 \tag{10.23}$$

Multiplying both sides of this inequality by $\beta_i > 0$, we can write the inequality (10.23) as

$$\frac{\beta_i}{\beta_i - 1} \sum_{j=1}^{n} a_{ij}^{-} y_j \ge \beta_i \tag{10.24}$$

Since one of the inequalities in (10.18) and (10.24) must be satisfied, the following inequality must hold true:

$$\sum_{j=1}^{n} a_{ij}^{+} y_j + \frac{\beta_i}{\beta_i - 1} \sum_{j=1}^{n} (a_{ij}^{-}) y_j \ge \beta_i \tag{10.25}$$

By introducing a slack variable s_i, we obtain the desired Gomory constraint as

$$s_i = \sum_{j=1}^{n} a_{ij}^{+} y_j + \frac{\beta_i}{\beta_i - 1} \sum_{j=1}^{n} \bar{a}_{ij} y_j - \beta_i \tag{10.26}$$

This constraint must be satisfied before the variable x_i becomes an integer. The slack variable s_i is not required to be an integer. At the optimal solution of the ordinary LP problem (given by Table 10.2), all $y_j = 0$ and hence Eq. (10.26) becomes

$$s_i = -\beta_i = \text{negative}$$

which can be seen to be infeasible. Hence the constraint Eq. (10.26) is added at the end of Table 10.2, and the dual simplex method applied. This procedure is repeated the required number of times until the optimal mixed integer solution is found.

Discussion. In the derivation of the Gomory constraint, Eq. (10.26), we have not made use of the fact that some of the variables (y_j) might be integer variables. We notice that any integer value can be added to or subtracted from the coefficient of $\bar{a}_{ik}(= a_{ik}^+ + a_{ik}^-)$ of an integer variable y_k provided that we subtract or add, respectively, the same value to x_i in Eq. (10.13), that is,

$$\sum_{\substack{j=1 \\ j \neq k}}^{n} \bar{a}_{ij} y_j + (\bar{a}_{ik} \pm \delta) y_k = \beta_i + \hat{b}_i - (x_i \mp \delta) \tag{10.27}$$

From Eq. (10.27), the same logic as was used in the derivation of Eqs. (10.18) and (10.24) can be used to obtain the same final equation, Eq. (10.26). Of course, the coefficients of integer variables y_k will be altered by integer amounts in Eq. (10.26). It has been established that to cut the feasible region as much as possible (through the Gomory constraint), we have to make the coefficients of integer variables y_k as small as possible. We can see that the smallest positive coefficient we can have for y_j in Eq. (10.13) is

$$\alpha_{ij} = \bar{a}_{ij} - \hat{a}_{ij}$$

and the largest negative coefficient as

$$1 - \alpha_{ij} = 1 - \bar{a}_{ij} + \hat{a}_{ij}$$

where $\hat{a}_{ij}$ is the integer obtained by truncating the fractional part of $\bar{a}_{ij}$ and α_{ij} is the fractional part. Thus we have a choice of two expressions, $(\bar{a}_{ij} - \hat{a}_{ij})$ and $(1 - \bar{a}_{ij} + \hat{a}_{ij})$, for the coefficients of y_j in Eq. (10.26). We choose the smaller one out of the two to make the Gomory constraint, Eq. (10.26), cut deeper into the original feasible space. Thus Eq. (10.26) can be rewritten as

$$s_i = \underbrace{\sum_j a_{ij}^+ y_j + \frac{\beta_i}{\beta_i - 1} \sum_j (+a_{ij}^-) y_j}_{\text{for noninterger variables } y_j} + \underbrace{\sum_j (\bar{a}_{ij} - \hat{a}_{ij}) y_j}_{\substack{\text{for integer variables } y_j \\ \text{and for } \bar{a}_{ij} - \hat{a}_{ij} \leq \beta_i}}$$

$$+ \underbrace{\frac{\beta_i}{\beta_i - 1} \sum_j (1 - \bar{a}_{ij} + \hat{a}_{ij}) y_j}_{\substack{\text{for integer variables } y_j \\ \text{and for } \bar{a}_{ij} - \hat{a}_{ij} > \beta_i}} - \beta_i$$

where the slack variable s_i is not restricted to be an integer.

Example 10.2 Solve the problem of Example 10.1 with x_2 only restricted to take integer values.

SOLUTION

Step 1: Solve the LP problem by simplex method by neglecting the integer requirement. This gives the following optimal tableau:

Basic variables	Coefficients of variables					$\bar{b}_i$
	x_1	x_2	y_1	y_2	$-f$	
x_1	1	0	$\frac{11}{36}$	$\frac{1}{36}$	0	$\frac{11}{2}$
x_2	0	1	$-\frac{1}{12}$	$\frac{1}{12}$	0	$\frac{9}{2}$
$-f$	0	0	$\frac{7}{12}$	$\frac{5}{12}$	1	$\frac{69}{2}$

The noninteger solution given by this tableau is

$$x_1 = 5\tfrac{1}{2}, \quad x_2 = 4\tfrac{1}{2}, \quad y_1 = y_2 = 0, \text{ and } f_{\min} = -34\tfrac{1}{2}.$$

Step 2: *Formulate a Gomory constraint.* Since x_2 is the only variable that is restricted to take integer values, we construct the Gomory constraint for x_2. From the tableau of step 1, we obtain

$$x_2 = \bar{b}_2 - \bar{a}_{21} y_1 - \bar{a}_{22} y_2$$

where

$$\bar{b}_2 = \tfrac{9}{2}, \quad \bar{a}_{21} = -\tfrac{1}{12}, \quad \text{and} \quad \bar{a}_{22} = \tfrac{1}{12}$$

According to Eq. (10.3), we write $\bar{b}_2$ as $\bar{b}_2 = \hat{b}_2 + \beta_2$ where $\hat{b}_2 = 4$ and $\beta_2 = \tfrac{1}{2}$. Similarly, we write from Eq. (10.10)

$$\bar{a}_{21} = a_{21}^+ + a_{21}^-$$
$$\bar{a}_{22} = a_{22}^+ + a_{22}^-$$

where

$$a_{21}^+ = 0, \quad a_{21}^- = -\tfrac{1}{12} \text{ (since } \bar{a}_{21} \text{ is negative)}$$
$$a_{22}^+ = \tfrac{1}{12}, \quad a_{22}^- = 0 \text{ (since } \bar{a}_{22} \text{ is nonnegative)}$$

The Gomory constraint can be expressed as [from Eq. (10.26)]:

$$s_2 - \sum_{j=1}^{2} a_{2j}^+ y_j + \frac{\beta_2}{\beta_2 - 1} \sum_{j=1}^{2} a_{2j}^- y_j = -\beta_2$$

where s_2 is a slack variable that is not required to take integer values. By substituting the values of a_{ij}^+, a_{ij}^-, and β_i, this constraint can be written as

$$s_2 + \tfrac{1}{12} y_1 - \tfrac{1}{12} y_2 = -\tfrac{1}{2}$$

When this constraint is added to the tableau above, we obtain the following:

Basic variables	Coefficients of variables				$-f$	s_2	$\bar{b}_i$
	x_1	x_2	y_1	y_2			
x_1	1	0	$\frac{11}{36}$	$\frac{1}{36}$	0	0	$\frac{11}{2}$
x_2	0	1	$-\frac{1}{12}$	$\frac{1}{12}$	0	0	$\frac{9}{2}$
$-f$	0	0	$\frac{7}{12}$	$\frac{5}{12}$	1	0	$\frac{69}{2}$
s_2	0	0	$\frac{1}{12}$	$-\frac{1}{12}$	0	1	$-\frac{1}{2}$

Step 3: Apply the dual simplex method to find a new optimum solution. Since $-\frac{1}{2}$ is the only negative $\bar{b}_i$ term, the pivot operation has to be done in s_2 row. Further, $\bar{a}_{ij}$ corresponding to y_2 column is the only negative coefficient in s_2 row and hence pivoting has to be done on this element, $-\frac{1}{12}$. The result of pivot operation is shown in the following tableau:

Basic variables	Coefficients of variables				$-f$	s_2	$\bar{b}_i$
	x_1	x_2	y_1	y_2			
x_1	1	0	$\frac{1}{3}$	0	0	$\frac{1}{3}$	$\frac{16}{3}$
x_2	0	1	0	0	0	1	4
$-f$	0	0	1	0	1	5	32
y_2	0	0	-1	1	0	-12	6

This tableau gives the desired integer solution as

$$x_1 = 5\tfrac{1}{2}, \quad x_2 = 4, \quad y_2 = 6, \quad y_1 = 0, \quad s_2 = 0, \quad \text{and} \quad f_{\min} = -32$$

10.4 BALAS' ALGORITHM FOR ZERO–ONE PROGRAMMING PROBLEMS

When all the variables of a LP problem are constrained to take values of 0 or 1 only, we have a zero–one (or binary) LP problem. A study of the various techniques available for solving zero–one programming problems is important for the following reasons:

1. As we shall see later in this chapter (Section 10.5), a certain class of integer nonlinear programming problems can be converted into equivalent zero–one LP problems,

2. A wide variety of industrial, management, and engineering problems can be formulated as zero–one problems. For example, in structural control, the problem of selecting optimal locations of actuators (or dampers) can be formulated as a zero–one problem. In this case, if a variable is zero or 1, it indicates the absence or presence of the actuator, respectively, at a particular location [10.31].

The zero–one LP problems can be solved by using any of the general integer LP techniques like Gomory's cutting plane method and Land and Doig's branch-and-bound method by introducing the additional constraint that all the variables must be less than or equal to 1. This additional constraint will restrict each of the variables to take a value of either zero (0) or one (1). Since the cutting plane and the branch-and-bound algorithms were developed primarily to solve a general integer LP problem, they do not take advantage of the special features of zero–one LP problems. Thus several methods have been proposed to solve zero–one LP problems more efficiently. In this section we present an algorithm developed by Balas (in 1965) for solving LP problems with binary variables only [10.9].

If there are n binary variables in a problem, an explicit enumeration process will involve testing 2^n possible solutions against the stated constraints and the objective function. In Balas method, all the 2^n possible solutions are enumerated, explicitly or implicitly. The efficiency of the method arises out of the clever strategy it adopts in selecting only a few solutions for explicit enumeration.

The method starts by setting all the n variables equal to zero and consists of a systematic procedure of successively assigning to certain variables the value 1, in such a way that after trying a (small) part of all the 2^n possible combinations, one obtains either an optimal solution or evidence of the fact that no feasible solution exists. The only operations required in the computation are additions and subtractions, and hence the round-off errors will not be there. For this reason the method is some times referred to as *additive algorithm*.

Standard Form of the Problem.   To describe the algorithm, consider the following form of the LP problem with zero–one variables:

$$\text{Find } \mathbf{X} = \begin{Bmatrix} x_1 \\ x_2 \\ \vdots \\ x_n \end{Bmatrix} \text{ such that } f(\mathbf{X}) = \mathbf{C}^T \mathbf{X} \rightarrow \text{minimum}$$

subject to

$$\mathbf{AX} + \mathbf{Y} = \mathbf{B}$$
$$x_i = 0 \quad \text{or} \quad 1$$
$$\mathbf{Y} \geq \mathbf{0}$$

(10.28)

where

$$\mathbf{C} = \begin{Bmatrix} c_1 \\ c_2 \\ \vdots \\ c_n \end{Bmatrix} \geq \mathbf{0}, \quad \mathbf{Y} = \begin{Bmatrix} y_1 \\ y_2 \\ \vdots \\ y_m \end{Bmatrix}, \quad \mathbf{B} = \begin{Bmatrix} b_1 \\ b_2 \\ \vdots \\ b_m \end{Bmatrix}$$

$$\mathbf{A} = \begin{bmatrix} a_{11} & a_{12} & \cdots & a_{1n} \\ a_{21} & a_{22} & \cdots & a_{2n} \\ \vdots & & & \\ a_{m1} & a_{m2} & \cdots & a_{mn} \end{bmatrix}$$

where $\mathbf{Y}$ is the vector of slack variables and c_i and a_{ij} need not be integers.

Initial Solution. An initial solution for the problem stated in Eqs. (10.28) can be taken as

$$f_0 = 0$$
$$x_i = 0, \quad i = 1, 2, \ldots, n \quad (10.29)$$
$$\mathbf{Y}^{(0)} = \mathbf{B}$$

If $\mathbf{B} \geq \mathbf{0}$, this solution will be feasible and optimal since $\mathbf{C} \geq \mathbf{0}$ in Eqs. (10.28). In this case there is nothing more to be done as the starting solution itself happens to be optimal. On the other hand, if some of the components b_j are negative, the solution given by Eqs. (10.29) will be optimal (since $\mathbf{C} \geq \mathbf{0}$) but infeasible. Thus the method starts with an optimal (actually better than optimal) and infeasible solution. The algorithm forces this solution toward feasibility while keeping it optimal all the time. This is the reason why Balas called his method the *pseudo dual simplex method*. The word *pseudo* has been used since the method is similar to the dual simplex method only as far as the starting solution is concerned and the subsequent procedure has no similarity at all with the dual simplex method. The details can be found in Ref. [10.9].

Integer Nonlinear Programming

10.5 INTEGER POLYNOMIAL PROGRAMMING

Watters [10.2] has developed a procedure for converting integer polynomial programming problems to zero–one LP problems. The resulting zero–one LP problem can be solved conveniently by the Balas method discussed in Section 10.4. Consider the optimization problem:

$$\text{Find } \mathbf{X} = \begin{Bmatrix} x_1 \\ x_2 \\ \vdots \\ x_n \end{Bmatrix} \text{ which minimizes } f(\mathbf{X})$$

subject to the constraints $\qquad\qquad\qquad\qquad\qquad\qquad\qquad\qquad\qquad (10.30)$

$$g_j(\mathbf{X}) \leq 0, \quad j = 1, 2, \ldots, m$$
$$x_i = \text{integer}, \quad i = 1, 2, \ldots, n$$

where f and g_j, $j = 1, 2, \ldots, m$, are polynomials in the variables $x_1, x_2, \ldots, x_n$. A typical term in the polynomials can be represented as

$$c_k \prod_{l=1}^{n_k} (x_l)^{a_{kl}} \quad (10.31)$$

where c_k is a constant, a_{kl} a nonnegative constant exponent, and n_k the number of variables appearing in the kth term. We shall convert the integer polynomial programming problem stated in Eq. (10.30) into an equivalent zero–one LP problem in two stages. In the first stage we see how an integer variable, x_i, can be represented by an equivalent system of zero–one (binary) variables. We consider the conversion of a zero–one polynomial programming problem into a zero–one LP problem in the second stage.

10.5.1   Representation of an Integer Variable by an Equivalent System of Binary Variables

Let x_i be any integer variable whose upper bound is given by u_i so that

$$x_i \leq u_i < \infty \tag{10.32}$$

We assume that the value of the upper bound u_i can be determined from the constraints of the given problem.

We know that in the decimal number system, an integer p is represented as

$$p = p_0 + 10^1 p_1 + 10^2 p_2 + \cdots, 0 \leq p_i \leq (10 - 1 = 9)$$

$$\text{for} \quad i = 0, 1, 2, \ldots$$

and written as $p = \cdots p_2 p_1 p_0$ by neglecting the zeros to the left. For example, we write the number $p = 008076$ as 8076 to represent $p = 6 + (10^1)7 + (10^2)(0) + (10^3)8 + (10^4)0 + (10^5)0$. In a similar manner, the integer p can also be represented in binary number system as

$$p = q_0 + 2^1 q_1 + 2^2 q_2 + 2^3 q_3 + \cdots$$

where $0 \leq q_i \leq (2 - 1 = 1)$ for $i = 0, 1, 2, \ldots$.

In general, if $y_i^{(0)}, y_i^{(1)}, y_i^{(2)}, \ldots$ denote binary numbers (which can take a value of 0 or 1), the variable x_i can be expressed as

$$x_i = \sum_{k=0}^{N_i} 2^k y_i^{(k)} \tag{10.33}$$

where N_i is the smallest integer such that

$$\frac{u_i + 1}{2} \leq 2^{N_i} \tag{10.34}$$

Thus the value of N_i can be selected for any integer variable x_i once its upper bound u_i is known. For example, for the number 97, we can take $u_i = 97$ and hence the relation

$$\frac{u_i + 1}{2} = \frac{98}{2} = 49 \leq 2^{N_i}$$

is satisfied for $N_i \geq 6$. Hence by taking $N_i = 6$, we can represent u_i as

$$97 = q_0 + 2^1 q_1 + 2^2 q_2 + 2^3 q_3 + 2^4 q_4 + 2^5 q_5 + 2^6 q_6$$

where $q_0 = 1$, $q_1 = q_2 = q_3 = q_4 = 0$, and $q_5 = q_6 = 1$. A systematic method of finding the values of $q_0, q_1, q_2, \ldots$ is given below.

Method of Finding $q_0, q_1, q_2, \ldots$.   Let M be the given positive integer. To find its binary representation $q_n q_{n-1} \ldots q_1 q_0$, we compute the following recursively:

$$b_0 = M \tag{10.35}$$

$$b_1 = \frac{b_0 - q_0}{2}$$

$$b_2 = \frac{b_1 - q_1}{2}$$

$$\vdots$$

$$b_k = \frac{b_{k-1} - q_{k-1}}{2}$$

where $q_k = 1$ if b_k is odd and $q_k = 0$ if b_k is even. The procedure terminates when $b_k = 0$.

Equation (10.33) guarantees that x_i can take any feasible integer value less than or equal to u_i. The use of Eq. (10.33) in the problem stated in Eq. (10.30) will convert the integer programming problem into a binary one automatically. The only difference is that the binary problem will have $N_1 + N_2 + \cdots + N_n$ zero–one variables instead of the n original integer variables.

10.5.2   Conversion of a Zero–One Polynomial Programming Problem into a Zero–One LP Problem

The conversion of a polynomial programming problem into a LP problem is based on the fact that

$$x_i^{a_{ki}} \equiv x_i \tag{10.36}$$

if x_i is a binary variable (0 or 1) and a_{ki} is a positive exponent. If $a_{ki} = 0$, then obviously the variable x_i will not be present in the kth term. The use of Eq. (10.36) permits us to write the kth term of the polynomial, Eq. (10.31), as

$$c_k \prod_{l=1}^{n_k} (x_l)^{a_{kl}} = c_k \prod_{l=1}^{n_k} x_l = c_k(x_1, x_2, \ldots, x_{n_k}) \tag{10.37}$$

Since each of the variables $x_1, x_2, \ldots$ can take a value of either 0 or 1, the product $(x_1 x_2 \cdots x_{nk})$ also will take a value of 0 or 1. Hence by defining a binary variable y_k as

$$y_k = x_1 x_2 \cdots x_{nk} = \prod_{l=1}^{n_k} x_l \tag{10.38}$$

the kth term of the polynomial simply becomes $c_k y_k$. However, we need to add the following constraints to ensure that $y_k = 1$ when all $x_i = 1$ and zero otherwise:

$$y_k \geq \left(\sum_{i=1}^{n_k} x_i \right) - (n_k - 1) \tag{10.39}$$

$$y_k \leq \frac{1}{n_k} \left(\sum_{i=1}^{n_k} x_i \right) \tag{10.40}$$

It can be seen that if all $x_i = 1$, $\sum_{i=1}^{n_k} x_i = n_k$, and Eqs. (10.39) and (10.40) yield

$$y_k \geq 1 \tag{10.41}$$

$$y_k \leq 1 \tag{10.42}$$

which can be satisfied only if $y_k = 1$. If at least one $x_i = 0$, we have $\sum_{i=1}^{n_k} x_i < n_k$, and Eqs. (10.39) and (10.40) give

$$y_k \geq -(n_k - 1) \tag{10.43}$$

$$y_k < 1 \tag{10.44}$$

Since n_k is a positive integer, the only way to satisfy Eqs. (10.43) and (10.44) under all circumstances is to have $y_k = 0$.

   This procedure of converting an integer polynomial programming problem into an equivalent zero–one LP problem can always be applied, at least in theory.

10.6   BRANCH-AND-BOUND METHOD

The branch-and-bound method is very effective in solving mixed-integer linear and nonlinear programming problems. The method was originally developed by Land and Doig [10.8] to solve integer linear programming problems and was later modified by Dakin [10.23]. Subsequently, the method has been extended to solve nonlinear mixed-integer programming problems. To see the basic solution procedure, consider the following nonlinear mixed-integer programming problem:

$$\text{Minimize } f(\mathbf{X}) \tag{10.45}$$

subject to

$$g_j(\mathbf{X}) \geq 0, \quad j = 1, 2, \ldots, m \tag{10.46}$$

$$h_k(\mathbf{X}) = 0, \quad k = 1, 2, \ldots, p \tag{10.47}$$

$$x_j = \text{integer}, \quad j = 1, 2, \ldots, n_0 \ (n_0 \leq n) \tag{10.48}$$

where $\mathbf{X} = \{x_1, x_2, \ldots, x_n\}^{\mathrm{T}}$. Note that in the design vector $\mathbf{X}$, the first n_0 variables are identified as the integer variables. If $n_0 = n$, the problem becomes an all-integer programming problem. A design vector $\mathbf{X}$ is called a *continuous feasible solution* if

X satisfies constraints (10.46) and (10.47). A design vector **X** that satisfies all the constraints, Eqs. (10.46) to (10.48), is called an *integer feasible solution*.

The simplest method of solving an integer optimization problem involves enumerating all integer points, discarding infeasible ones, evaluating the objective function at all integer feasible points, and identifying the point that has the best objective function value. Although such an exhaustive search in the solution space is simple to implement, it will be computationally expensive even for moderate-size problems. The branch-and-bound method can be considered as a refined enumeration method in which most of the nonpromising integer points are discarded without testing them. Also note that the process of complete enumeration can be used only if the problem is an all-integer programming problem. For mixed-integer problems in which one or more variables may assume continuous values, the process of complete enumeration cannot be used.

In the branch-and-bound method, the integer problem is not directly solved. Rather, the method first solves a continuous problem obtained by relaxing the integer restrictions on the variables. If the solution of the continuous problem happens to be an integer solution, it represents the optimum solution of the integer problem. Otherwise, at least one of the integer variables, say x_i, must assume a nonintegral value. If x_i is not an integer, we can always find an integer $[x_i]$ such that

$$[x_i] < x_i < [x_i] + 1 \tag{10.49}$$

Then two subproblems are formulated, one with the additional upper bound constraint

$$x_i \leq [x_i] \tag{10.50}$$

and another with the lower bound constraint

$$x_i \geq [x_i] + 1 \tag{10.51}$$

The process of finding these subproblems is called *branching*.

The branching process eliminates some portion of the continuous space that is not feasible for the integer problem, while ensuring that none of the integer feasible solutions are eliminated. Each of these two subproblems are solved again as a continuous problem. It can be seen that the solution of a continuous problem forms a *node* and from each node two branches may originate.

The process of branching and solving a sequence of continuous problems discussed above is continued until an integer feasible solution is found for one of the two continuous problems. When such a feasible integer solution is found, the corresponding value of the objective function becomes an upper bound on the minimum value of the objective function. At this stage we can eliminate from further consideration all the continuous solutions (nodes) whose objective function values are larger than the upper bound. The nodes that are eliminated are said to have been *fathomed* because it is not possible to find a better integer solution from these nodes (solution spaces) than what we have now. The value of the upper bound on the objective function is updated whenever a better bound is obtained.

It can be seen that a node can be fathomed if any of the following conditions are true:

1. The continuous solution is an integer feasible solution.
2. The problem does not have a continuous feasible solution.
3. The optimal value of the continuous problem is larger than the current upper bound.

The algorithm continues to select a node for further branching until all the nodes have been fathomed. At that stage, the particular fathomed node that has the integer feasible solution with the lowest value of the objective function gives the optimum solution of the original nonlinear integer programming problem.

Example 10.3 Solve the following LP problem using the branch-and-bound method:

$$\text{Maximize } f = 3x_1 + 4x_2$$

subject to (E$_1$)

$$7x_1 + 11x_2 \le 88, \quad 3x_1 - x_2 \le 12, \quad x_1 \ge 0, \quad x_2 \ge 0$$

$$x_i = \text{integer}, \quad i = 1, 2 \tag{E$_2$}$$

SOLUTION The various steps of the procedure are illustrated using graphical method.

Step 1: First the problem is solved as a continuous variable problem [without Eq. (E$_2$)] to obtain:

$$\text{Problem (E}_1) : \text{Fig. 10.2;} \quad (x_1^* = 5.5, x_2^* = 4.5, f^* = 34.5)$$

Step 2: The branching process, with integer bounds on x_1, yields the problems:

$$\text{Maximize } f = 3x_1 + 4x_2$$

subject to (E$_3$)

$$7x_1 + 11x_2 \le 88, \quad 3x_1 - x_2 \le 12, \quad x_1 \le 5, \quad x_2 \ge 0$$

and

$$\text{Maximize } f = 3x_1 + 4x_2$$

subject to (E$_4$)

$$7x_1 + 11x_2 \le 88, \quad 3x_1 - x_2 \le 12, \quad x_1 \ge 6, \quad x_2 \ge 0$$

The solutions of problems (E$_3$) and (E$_4$) are given by

$$\text{Problem (E}_3) : \text{Fig. 10.4;} \quad (x_1^* = 5, x_2^* = 4.8182, f^* = 34.2727)$$

$$\text{Problem (E}_4) : \text{Fig. 10.5;} \quad \text{no feasible solution exists.}$$

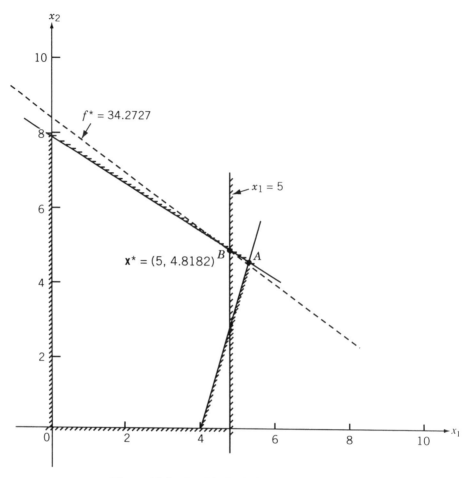

Figure 10.4   Graphical solution of problem (E_3).

Step 3: The next branching process, with integer bounds on x_2, leads to the following problems:

$$\text{Maximize } f = 3x_1 + 4x_2$$

subject to (E_5)

$$7x_1 + 11x_2 \le 88, \quad 3x_1 - x_2 \le 12, \quad x_1 \le 5, \quad x_2 \le 4$$

and

$$\text{Maximize } f = 3x_1 + 4x_2$$

subject to (E_6)

$$7x_1 + 11x_2 \le 88, \quad 3x_1 - x_2 \le 12, \quad x_1 \le 5, \quad x_2 \ge 5$$

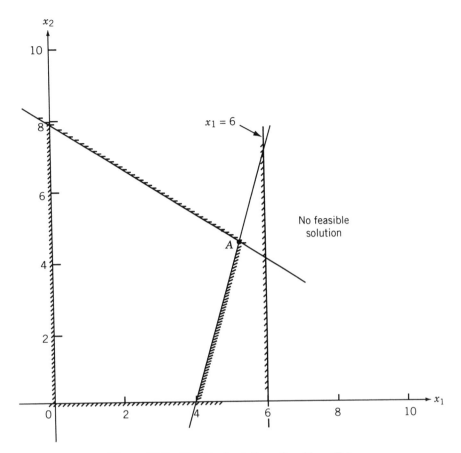

Figure 10.5 Graphical solution of problem (E$_4$).

The solutions of problems (E$_5$) and (E$_6$) are given by

$$\text{Problem (E}_5) : \text{Fig. 10.6;} \quad (x_1^* = 5, x_2^* = 4, f^* = 31)$$

$$\text{Problem (E}_6) : \text{Fig. 10.7;} \quad (x_1^* = 0, x_2^* = 8, f^* = 32)$$

Since both the variables assumed integer values, the optimum solution of the integer LP problem, Eqs. (E$_1$) and (E$_2$), is given by ($x_1^* = 0, x_2^* = 8, f^* = 32$).

Example 10.4 Find the solution of the welded beam problem of Section 7.22.3 by treating it as a mixed-integer nonlinear programming problem by requiring x_3 and x_4 to take integer values.

SOLUTION   The solution of this problem using the branch-and-bound method was reported in Ref. [10.25]. The optimum solution of the continuous variable nonlinear programming problem is given by

$$\mathbf{X}^* = \{0.24, 6.22, 8.29, 0.24\}^{\text{T}}, \quad f^* = 2.38$$

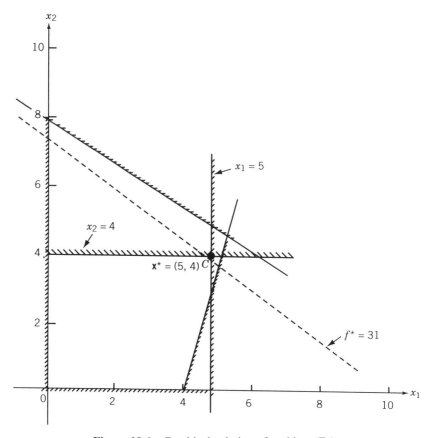

Figure 10.6   Graphical solution of problem (E_5).

Next, the branching problems, with integer bounds on x_3, are solved and the procedure is continued until the desired optimum solution is found. The results are shown in Fig. 10.8.

10.7 SEQUENTIAL LINEAR DISCRETE PROGRAMMING

Let the nonlinear programming problem with discrete variables be stated as follows:

$$\text{Minimize } f(\mathbf{X}) \tag{10.52}$$

subject to

$$g_j(\mathbf{X}) \leq 0, \quad j = 1, 2, \ldots, m \tag{10.53}$$

$$h_k(\mathbf{X}) = 0, \quad k = 1, 2, \ldots, p \tag{10.54}$$

$$x_i \in \{d_{i1}, d_{i2}, \ldots, d_{iq}\}, \quad i = 1, 2, \ldots, n_0 \tag{10.55}$$

$$x_i^{(l)} \leq x_i \leq x_i^{(u)}, \quad i = n_0 + 1, \quad n_0 + 2, \ldots, n \tag{10.56}$$

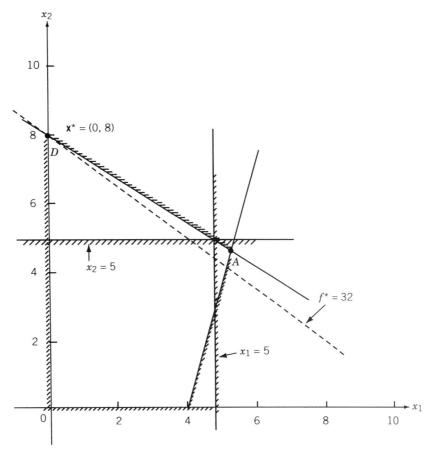

Figure 10.7   Graphical solution of problem (E₆).

where the first n_0 design variables are assumed to be discrete, d_{ij} is the jth discrete value for the variable i, and $\mathbf{X} = \{x_1, x_2, \ldots, x_n\}^{\mathrm{T}}$. It is possible to find the solution of this problem by solving a series of mixed-integer linear programming problems.

The nonlinear expressions in Eqs. (10.52) to (10.54) are linearized about a point $\mathbf{X}^0$ using a first-order Taylor's series expansion and the problem is stated as

$$\text{Minimize } f(\mathbf{X}) \approx f(\mathbf{X}^0) + \nabla f(\mathbf{X}^0)\delta\mathbf{X} \tag{10.57}$$

subject to

$$g_j(\mathbf{X}) \approx g_j(\mathbf{X}^0) + \nabla g_j(\mathbf{X}^0)\delta\mathbf{X} \le 0, \quad j = 1, 2, \ldots, m \tag{10.58}$$

$$h_k(\mathbf{X}) \approx h_k(\mathbf{X}^0) + \nabla h_k(\mathbf{X}^0)\delta\mathbf{X} = 0, \quad k = 1, 2, \ldots, p \tag{10.59}$$

$$x_i^0 + \delta x_i \in \{d_{i1}, d_{i2}, \ldots, d_{iq}\}, \quad i = 1, 2, \ldots, n_0 \tag{10.60}$$

$$x_i^{(l)} \le x_i^0 + \delta x_i \le x_i^{(u)}, \quad i = n_0 + 1, n_0 + 2, \ldots, n \tag{10.61}$$

$$\delta\mathbf{X} = \mathbf{X} - \mathbf{X}^0 \tag{10.62}$$

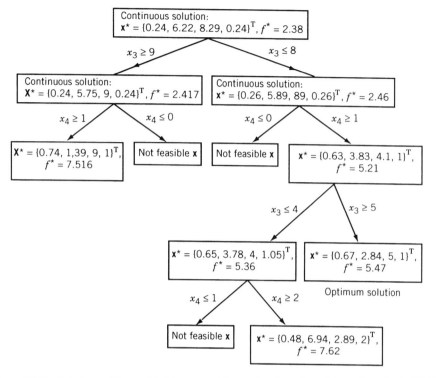

Figure 10.8   Solution of the welded beam problem using branch-and-bound method. [10.25]

The problem stated in Eqs. (10.57) to (10.62) cannot be solved using mixed-integer linear programming techniques since some of the design variables are discrete and noninteger. The discrete variables are redefined as [10.26]

$$x_i = y_{i1}d_{i1} + y_{i2}d_{i2} + \cdots + y_{iq}d_{iq} = \sum_{j=1}^{q} y_{ij}d_{ij}, \quad i = 1, 2, \ldots, n_0 \qquad (10.63)$$

with

$$y_{i1} + y_{i2} + \cdots + y_{iq} = \sum_{j=1}^{q} y_{ij} = 1 \qquad (10.64)$$

$$y_{ij} = 0 \text{ or } 1, \quad i = 1, 2, \ldots, n_0, \quad j = 1, 2, \ldots, q \qquad (10.65)$$

Using Eqs. (10.63) to (10.65) in Eqs. (10.57) to (10.62), we obtain

$$\text{Minimize } f(\mathbf{X}) \approx f(\mathbf{X}^0) + \sum_{i=1}^{n_0} \frac{\partial f}{\partial x_i} \left(\sum_{j=1}^{q} y_{ij}d_{ij} - x_i^0 \right)$$

$$+ \sum_{i=n_0+1}^{n} \frac{\partial f}{\partial x_i} (x_i - x_i^0) \qquad (10.66)$$

subject to

$$g_j(\mathbf{X}) \approx g_j(\mathbf{X}^0) + \sum_{i=1}^{n_0} \frac{\partial g_i}{\partial x_i} \left(\sum_{l=1}^{n_0} y_{il} d_{il} - x_i^0 \right) + \sum_{i=n_0+1}^{n} \frac{\partial g_j}{\partial x_i} (x_i - x_i^0) \leq 0,$$

$$j = 1, 2, \ldots, m \tag{10.67}$$

$$h_k(\mathbf{X}) \approx h_k(\mathbf{X}^0) + \sum_{i=1}^{n_0} \frac{\partial h_k}{\partial x_i} \left(\sum_{l=1}^{n_0} y_{il} d_{il} - x_i^0 \right) + \sum_{i=n_0+1}^{n} \frac{\partial h_k}{\partial x_i} (x_i - x_i^0) = 0,$$

$$k = 1, 2, \ldots, p \tag{10.68}$$

$$\sum_{j=1}^{q} y_{ij} = 1, \quad i = 1, 2, \ldots, n_0 \tag{10.69}$$

$$y_{ij} = 0 \text{ or } 1, \quad i = 1, 2, \ldots, n_0, \quad j = 1, 2, \ldots, q \tag{10.70}$$

$$x_i^{(l)} \leq x_i^0 + \delta x_i \leq x_i^{(u)}, \quad i = n_0 + 1, n_0 + 2, \ldots, n \tag{10.71}$$

The problem stated in Eqs. (10.66) to (10.71) can now be solved as a mixed-integer LP problem by treating both $y_{ij}(i = 1, 2, \ldots, n_0, j = 1, 2, \ldots, q)$ and $x_i(i = n_0 + 1, n_0 + 2, \ldots, n)$ as unknowns.

In practical implementation, the initial linearization point $\mathbf{X}^0$ is to be selected carefully. In many cases the solution of the discrete problem is expected to lie in the vicinity of the continuous optimum. Hence the original problem can be solved as a continuous nonlinear programming problem (by ignoring the discrete nature of the variables) using any of the standard nonlinear programming techniques. If the resulting continuous optimum solution happens to be a feasible discrete solution, it can be used as $\mathbf{X}^0$. Otherwise, the values of x_i from the continuous optimum solution are rounded (in a direction away from constraint violation) to obtain an initial feasible discrete solution $\mathbf{X}^0$. Once the first linearized discrete problem is solved, the subsequent linearizations can be made using the result of the previous optimization problem.

Example 10.5 **[10.26]**

$$\text{Minimize } f(\mathbf{X}) = 2x_1^2 + 3x_2^2$$

subject to

$$g(\mathbf{X}) = \frac{1}{x_1} + \frac{1}{x_2} - 4 \leq 0$$

$$x_1 \in \{0.3, 0.7, 0.8, 1.2, 1.5, 1.8\}$$

$$x_2 \in \{0.4, 0.8, 1.1, 1.4, 1.6\}$$

SOLUTION In this example, the set of discrete values of each variable is truncated by allowing only three values—its current value, the adjacent higher value, and the adjacent lower value—for simplifying the computations. Using $\mathbf{X}^0 = \left\{ {1.2 \atop 1.1} \right\}$, we have

$$f(\mathbf{X}^0) = 6.51, \quad g(\mathbf{X}^0) = -2.26$$

$$\nabla f(\mathbf{X}^0) = \left\{ {4x_1 \atop 6x_2} \right\}_{\mathbf{X}^0} = \left\{ {4.8 \atop 6.6} \right\}, \quad \nabla g(\mathbf{X}^0) = \left\{ {-\dfrac{1}{x_1^2} \atop -\dfrac{1}{x_2^2}} \right\}_{\mathbf{X}^0} = \left\{ {-0.69 \atop -0.83} \right\}$$

Now

$$x_1 = y_{11}(0.8) + y_{12}(1.2) + y_{13}(1.5)$$

$$x_2 = y_{21}(0.8) + y_{22}(1.1) + y_{23}(1.4)$$

$$\delta x_1 = y_{11}(0.8 - 1.2) + y_{12}(1.2 - 1.2) + y_{13}(1.5 - 1.2)$$

$$\delta x_2 = y_{21}(0.8 - 1.1) + y_{22}(1.1 - 1.1) + y_{23}(1.4 - 1.1)$$

$$f \approx 6.51 + \{4.8 \ 6.6\} \left\{ {-0.4y_{11} + 0.3y_{13} \atop -0.3y_{21} + 0.3y_{23}} \right\}$$

$$g \approx -2.26 + \{-0.69 - 0.83\} \left\{ {-0.4y_{11} + 0.3y_{13} \atop -0.3y_{21} + 0.3y_{23}} \right\}$$

Thus the first approximate problem becomes (in terms of the unknowns y_{11}, y_{12}, y_{13}, y_{21}, y_{22}, and y_{23}):

$$\text{Minimize } f = 6.51 - 1.92y_{11} + 1.44y_{13} - 1.98y_{21} + 1.98y_{23}$$

subject to

$$-2.26 + 0.28y_{11} + 0.21y_{13} + 0.25y_{21} - 0.25y_{23} \leq 0$$

$$y_{11} + y_{12} + y_{13} = 1$$

$$y_{21} + y_{22} + y_{23} = 1$$

$$y_{ij} = 0 \text{ or } 1, \quad i = 1, 2, \quad j = 1, 2, 3$$

In this problem, there are only nine possible solutions and hence they can all be enumerated and the optimum solution can be found as

$$y_{11} = 1, \quad y_{12} = 0, \quad y_{13} = 0, \quad y_{21} = 1, \quad y_{22} = 0, \quad y_{23} = 0$$

Thus the solution of the first approximate problem, in terms of original variables, is given by

$$x_1 = 0.8, \quad x_2 = 0.8, \quad f(\mathbf{X}) = 2.61, \quad \text{and} \quad g(\mathbf{X}) = -1.5$$

This point can be used to generate a second approximate problem and the process can be repeated until the final optimum solution is found.

10.8   GENERALIZED PENALTY FUNCTION METHOD

The solution of an integer nonlinear programming problem, based on the concept of penalty functions, was originally suggested by Gellatly and Marcal in 1967 [10.5]. This approach was later applied by Gisvold and Moe [10.4] and Shin et al. [10.24] to solve some design problems that have been formulated as nonlinear mixed-integer programming problems. The method can be considered as an extension of the interior penalty function approach considered in Section 7.13. To see the details of the approach, let the problem be stated as follows:

$$\text{Find} \mathbf{X} = \begin{Bmatrix} x_1 \\ x_2 \\ \vdots \\ x_n \end{Bmatrix} = \begin{Bmatrix} \mathbf{X}_d \\ \mathbf{X}_c \end{Bmatrix} \text{ which minimizes } f(\mathbf{X})$$

subject to the constraints   (10.72)

$$g_j(\mathbf{X}) \geq 0, \quad j = 1, 2, \ldots, m$$

$$\mathbf{X}_c \in S_c \quad \text{and} \quad \mathbf{X}_d \in S_d,$$

where the vector of variables ($\mathbf{X}$) is composed of two vectors $\mathbf{X}_d$ and $\mathbf{X}_c$, with $\mathbf{X}_d$ representing the set of integer variables and $\mathbf{X}_c$ representing the set of continuous variables. Notice that $\mathbf{X}_c$ will not be there if all the variables are constrained to take only integer values and $\mathbf{X}_d$ will not be there if none of the variables is restricted to take only integer values. The sets S_c and S_d denote the feasible sets of continuous and integer variables, respectively. To extend the interior penalty function approach to solve the present problem, given by Eq. (10.72), we first define the following transformed problem:

$$\text{Minimize } \phi_k(\mathbf{X}, r_k, s_k)$$

where

$$\phi_k(\mathbf{X}, r_k, s_k) = f(\mathbf{X}) + r_k \sum_{j=1}^{m} G_j[g_j(\mathbf{X})] + s_k Q_k(\mathbf{X}_d) \tag{10.73}$$

In this equation, r_k is a weighing factor (penalty parameter) and

$$r_k \sum_{j=1}^{m} G_j[g_j(\mathbf{X})]$$

is the contribution of the constraints to the ϕ_k function, which can be taken as

$$r_k \sum_{j=1}^{m} G_j[g_j(\mathbf{X})] = +r_k \sum_{j=1}^{m} \frac{1}{g_j(\mathbf{X})} \tag{10.74}$$

It can be noted that this term is positive for all $\mathbf{X}$ satisfying the relations $g_j(\mathbf{X}) > 0$ and tends to $+\infty$ if any one particular constraint tends to have a value of zero. This property ensures that once the minimization of the ϕ_k function is started from a feasible point, the point always remains in the feasible region. The term $s_k Q_k(\mathbf{X}_d)$ can be considered as a penalty term with s_k playing the role of a weighing factor (penalty parameter). The function $Q_k(\mathbf{X}_d)$ is constructed so as to give a penalty whenever some of the variables in $\mathbf{X}_d$ take values other than integer values. Thus the function $Q_k(\mathbf{X}_d)$ has the property that

$$Q_k(\mathbf{X}_d) = \begin{cases} 0 & \text{if } \mathbf{X}_d \in S_d \\ \mu > 0 & \text{if } \mathbf{X}_d \notin S_d \end{cases} \tag{10.75}$$

We can take, for example,

$$Q_k(\mathbf{X}_d) = \sum_{x_i \in \mathbf{X}_d} \left\{ 4 \left(\frac{x_i - y_i}{z_i - y_i} \right) \left(1 - \frac{x_i - y_i}{z_i - y_i} \right) \right\}^{\beta_k} \tag{10.76}$$

where $y_i \leq x_i$, $z_i \geq x_i$, and $\beta_k \geq 1$ is a constant. Here y_i and z_i are the two neighboring integer values for the value x_i. The function $Q_k(\mathbf{X}_d)$ is a normalized, symmetric beta function integrand. The variation of each of the terms under summation sign in Eq. (10.76) for different values of β_k is shown in Fig. 10.9. The value of β_k has to be greater than or equal to 1 if the function Q_k is to be continuous in its first derivative over the discretization or integer points.

The use of the penalty term defined by Eq. (10.76) makes it possible to change the shape of the ϕ_k function by changing β_k, while the amplitude can be controlled by the weighting factor s_k. The ϕ_k function given in Eq. (10.73) is now minimized for a sequence of values of r_k and s_k such that for $k \to \infty$, we obtain

$$\begin{aligned} \text{Min } \phi_k(\mathbf{X}, r_k, s_k) &\to \text{Min } f(\mathbf{X}) \\ g_j(\mathbf{X}) &\geq 0, \quad j = 1, 2, \ldots, m \\ Q_k(\mathbf{X}_d) &\to 0 \end{aligned} \tag{10.77}$$

In most of the practical problems, one can obtain a reasonably good solution by carrying out the minimization of ϕ_k even for 5 to 10 values of k. The method is illustrated in Fig. 10.10 in the case of a single-variable problem. It can be noticed from Fig. 10.10 that the shape of the ϕ function (also called the response function) depends strongly on the numerical values of r_k, s_k, and β_k.

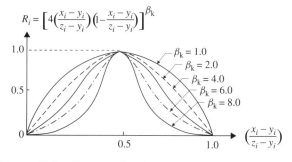

Figure 10.9   Contour of typical term in Eq. (10.62) [10.4].

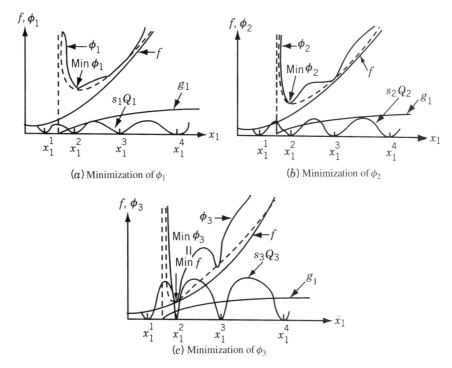

Figure 10.10   Solution of a single-variable integer problem by penalty function method. x_1, discrete variable; x_1^j, jth value of x_1 [10.4].

Choice of the Initial Values of r_k, s_k, and β_k.   The numerical values of r_k, s_k, and β_k have to be chosen carefully to achieve fast convergence. If these values are chosen such that they give the response surfaces of ϕ function as shown in Fig. 10.10c, several local minima will be introduced and the risk in finding the global minimum point will be more. Hence the initial value of s_k (namely, s_1) is to be chosen sufficiently small to yield a unimodal response surface. This can be achieved by setting

$$s_k Q_k' \ll P_k' \tag{10.78}$$

where Q_k' is an estimate of the maximum magnitude of the gradient to the Q_k surface and P_k' is a measure of the gradient of the function P_k defined by

$$P_k = f(\mathbf{X}) + r_k \sum_{j=1}^{m} G_j[g_j(\mathbf{X})] \tag{10.79}$$

Gisvold and Moe [10.4] have taken the values of Q_k' and P_k' as

$$Q_k' = \tfrac{1}{2} \cdot 4^{\beta_k} \beta_k (\beta_k - 1)^{\beta_k - 1} (2\beta_k - 1)^{1/2 - \beta_k} \tag{10.80}$$

$$P_k' = \left(\frac{\nabla P_k^T \nabla P_k}{n} \right)^{1/2} \tag{10.81}$$

where

$$\nabla P_k = \begin{Bmatrix} \partial P_k/\partial x_1 \\ \partial P_k/\partial x_2 \\ \vdots \\ \partial P_k/\partial x_n \end{Bmatrix} \tag{10.82}$$

The initial value of s_1, according to the requirement of Eq. (10.78), is given by

$$s_1 = c_1 \frac{P_1'(\mathbf{X}_1, r_1)}{Q_1'(\mathbf{X}_1^{(d)}, \beta_1)} \tag{10.83}$$

where $\mathbf{X}_1$ is the initial starting point for the minimization of ϕ_1, $\mathbf{X}_1^{(d)}$ the set of starting values of integer-restricted variables, and c_1 a constant whose value is generally taken in the range 0.001 and 0.1.

To choose the weighting factor r_1, the same consideration as discussed in Section 7.13 are to be taken into account. Accordingly, the value of r_1 is chosen as

$$r_1 = c_2 \frac{f(\mathbf{X}_1)}{+\sum_{j=1}^m 1/g_j(\mathbf{X}_1)} \tag{10.84}$$

with the value of c_2 ranging between 0.1 and 1.0. Finally, the parameter β_k must be taken greater than 1 to maintain the continuity of the first derivative of the function ϕ_k over the discretization points. Although no systematic study has been conducted to find the effect of choosing different values for β_k, the value of $\beta_1 \simeq 2.2$ has been found to give satisfactory convergence in some of the design problems.

Once the initial values of r_k, s_k, and β_k (for $k = 1$) are chosen, the subsequent values also have to be chosen carefully based on the numerical results obtained on similar formulations. The sequence of values r_k are usually determined by using the relation

$$r_{k+1} = c_3 r_k, \quad k = 1, 2, \ldots \tag{10.85}$$

where $c_3 < 1$. Generally, the value of c_3 is taken in the range 0.05 to 0.5. To select the values of s_k, we first notice that the effect of the term $Q_k(\mathbf{X}_d)$ is somewhat similar to that of an equality constraint. Hence the method used in finding the weighting factors for equality constraints can be used to find the factor s_{k+1}. For equality constraints, we use

$$\frac{s_{k+1}}{s_k} = \frac{r_k^{1/2}}{r_{k+1}^{1/2}} \tag{10.86}$$

From Eqs. (10.85) and (10.86), we can take

$$s_{k+1} = c_4 s_k \tag{10.87}$$

with c_4 approximately lying in the range $\sqrt{1/0.5}$ and $\sqrt{1/0.05}$ (i.e., 1.4 and 4.5). The values of β_k can be selected according to the relation

$$\beta_{k+1} = c_5 \beta_k \tag{10.88}$$

with c_5 lying in the range 0.7 to 0.9.

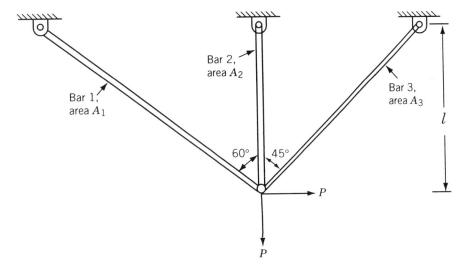

Figure 10.11   Three-bar truss.

A general convergence proof of the penalty function method, including the integer programming problems, was given by Fiacco [10.6]. Hence the present method is guaranteed to converge at least to a local minimum if the recovery procedure is applied the required number of times.

Example 10.6 [10.24] Find the minimum weight design of the three-bar truss shown in Fig. 10.11 with constraints on the stresses induced in the members. Treat the areas of cross section of the members as discrete variables with permissible values of the parameter $A_i \sigma_{\max}/P$ given by 0.1, 0.2, 0.3, 0.5, 0.8, 1.0, and 1.2.

SOLUTION   By defining the nondimensional quantities f and x_i as

$$f = \frac{W \sigma_{\max}}{P \rho l}, \quad x_i = \frac{A_i \sigma_{\max}}{P}, \quad i = 1, 2, 3$$

where W is the weight of the truss, $\sigma_{\max}$ the permissible (absolute) value of stress, P the load, ρ the density, l the depth, and A_i the area of cross section of member $i (i = 1, 2, 3)$, the discrete optimization problem can be stated as follows:

$$\text{Minimize } f = 2x_1 + x_2 + \sqrt{2} x_3$$

subject to

$$g_1(\mathbf{X}) = 1 - \frac{\sqrt{3} x_2 + 1.932 x_3}{1.5 x_1 x_2 + \sqrt{2} x_2 x_3 + 1.319 x_1 x_3} \geq 0$$

$$g_2(\mathbf{X}) = 1 - \frac{0.634 x_1 + 2.828 x_3}{1.5 x_1 x_2 + \sqrt{2} x_2 x_3 + 1.319 x_1 x_3} \geq 0$$

$$g_3(\mathbf{X}) = 1 - \frac{0.5x_1 - 2x_2}{1.5x_1x_2 + \sqrt{2}x_2x_3 + 1.319x_1x_3} \geq 0$$

$$g_4(\mathbf{X}) = 1 + \frac{0.5x_1 - 2x_2}{1.5x_1x_2 + \sqrt{2}x_2x_3 + 1.319x_1x_3} \geq 0$$

$$x_i \in \{0.1, 0.2, 0.3, 0.5, 0.8, 1.0, 1.2\}, \quad i = 1, 2, 3$$

The optimum solution of the continuous variable problem is given by $f^* = 2.7336$, $x_1^* = 1.1549$, $x_2^* = 0.4232$, and $x_3^* = 0.0004$. The optimum solution of the discrete variable problem is given by $f^* = 3.0414$, $x_1^* = 1.2$, $x_2^* = 0.5$, and $x_3^* = 0.1$.

10.9   SOLUTION OF BINARY PROGRAMMING PROBLEMS USING MATLAB

The MATLAB function `bintprog` can be used to solve a binary (or zero–one) programming problem. The following example illustrates the procedure.

Example 10.7   Find the solution of the following binary programming problem using the MATLAB function `bintprog`:

$$\text{Minimize } f(\mathbf{X}) = -5x_1 - 5x_2 - 8x_3 + 4x_4 + 4x_5$$

subject to

$$3x_1 - 6x_2 + 7x_3 - 9x_4 - 9x_5 \leq -10, \quad x_1 + 2x_2 - x_4 - 3x_5 \leq 0$$

$$x_i \text{ binary}; i = 1, 2, 3, 4, 5$$

SOLUTION

Step 1: State the problem in the form required by the program `bintprog`:

$$\text{Minimize } f(x) = f^{\mathrm{T}}x \text{ subject to } Ax \leq b \text{ and } A_{eq}x = b_{eq}$$

Here

$$f^{\mathrm{T}} = \{-5 \ \ -5 \ \ -8 \ \ 2 \ \ 4\}, \ \ x = \{x_1 \ \ x_2 \ \ x_3 \ \ x_4 \ \ x_5\}^{\mathrm{T}}$$

$$A = \begin{bmatrix} 3 & -6 & 7 & -9 & -9 \\ 1 & 2 & 0 & -1 & -3 \end{bmatrix}, \ b = \begin{Bmatrix} -10 \\ 0 \end{Bmatrix}$$

Step 2: The input is directly typed on the MATLAB command window and the program `bintprog` is called as indicated below:

```
clear; clc;
f = [-5 -5 -8 2 4]';
A = [3 -6 7 -9 -9; 1 2 0 -1 -3];
b = [-10 0]';
x = bintprog (f, A, b,[])
```

Step 3: The output of the program is shown below:

```
Optimization terminated.
x =
   1
   1
   1
   1
   1
```

REFERENCES AND BIBLIOGRAPHY

10.1   M. L. Balinski, Integer programming: methods, uses, computation, *Management Science*, Vol. 12, pp. 253–313, 1965.

10.2   L. J. Watters, Reduction of integer polynomial programming problems to zero–one linear programming problems, *Operations Research*, Vol. 15, No. 6, pp. 1171–1174, 1967.

10.3   S. Retter and D. B. Rice, Discrete optimizing solution procedures for linear and nonlinear integer programming problems, *Management Science*, Vol. 12, pp. 829–850, 1966.

10.4   K. M. Gisvold and J. Moe, A method for nonlinear mixed-integer programming and its application to design problems, *Journal of Engineering for Industry, Transactions of ASME*, Vol. 94, pp. 353–364, May 1972.

10.5   R. A. Gellatly and P. B. Marcal, *Investigation of Advanced Spacecraft Structural Design Technology*, NASA Report 2356–950001, 1967.

10.6   A. V. Fiacco, Penalty methods for mathematical programming in E^n with general constraints sets, *Journal of Optimization Theory and Applications*, Vol. 6, pp. 252–268, 1970.

10.7   R. E. Gomory, *An Algorithm for the Mixed Integer Problem*, Rand Report R.M. 25797, July 1960.

10.8   A. H. Land and A. Doig, An automatic method of solving discrete programming problems, *Econometrica*, Vol. 28, pp. 497–520, 1960.

10.9   E. Balas, An additive algorithm for solving linear programs with zero–one variables, *Operations Research*, Vol. 13, pp. 517–546, 1965.

10.10 R. E. Gomory, An all-integer integer programming algorithm, Chapter 13 in *Industrial Scheduling*, J. F. Muth and G. L. Thompson, Eds., Prentice-Hall, Englewood Cliffs, NJ, 1963.

10.11 H. A. Taha, *Operations Research: An Introduction*, Macmillan, New York, 1971.

10.12 S. Zionts, *Linear and Integer Programming*, Prentice-Hall, Englewood Cliffs, NJ, 1974.

10.13 C. McMillan, Jr., *Mathematical Programming: An Introduction to the Design and Application of Optimal Decision Machines*, Wiley, New York, 1970.

10.14 N. K. Kwak, *Mathematical Programming with Business Applications*, McGraw-Hill, New York, 1973.

10.15 H. Greenberg, *Integer Programming*, Academic Press, New York, 1971.

10.16 G. B. Dantzig and A. F. Veinott, Jr., Eds., *Mathematics of the Decision Sciences*, Part 1, American Mathematical Society, Providence, R I, 1968.

10.17 R. S. Garfinkel and G. L. Nemhauser, *Integer Programming*, Wiley, New York, 1972.

10.18 C. A. Trauth, Jr., and R. E. Woolsey, Integer linear programming: a study in computational efficiency, *Management Science*, Vol. 15, No. 9, pp. 481–493, 1969.

10.19 E. L. Lawler and M. D. Bell, A method for solving discrete optimization problems, *Operations Research*, Vol. 14, pp. 1098–1112, 1966.

10.20 P. Hansen, Quadratic zero–one programming by implicit enumeration, pp. 265–278 in *Numerical Methods for Nonlinear Optimization*, F. A. Lootsma, Ed., Academic Press, London, 1972.

10.21 R. R. Meyer, Integer and mixed-integer programming models: general properties, *Journal of Optimization Theory and Applications*, Vol. 16, pp. 191–206, 1975.

10.22 R. Karni, Integer linear programming formulation of the material requirements planning problem, *Journal of Optimization Theory and Applications*, Vol. 35, pp. 217–230, 1981.

10.23 R. J. Dakin, A tree-search algorithm for mixed integer programming problems, *Computer Journal*, Vol. 8, No. 3, pp. 250–255, 1965.

10.24 D. K. Shin, Z. Gürdal, and O. H. Griffin, Jr., A penalty approach for nonlinear optimization with discrete design variables, *Engineering Optimization*, Vol. 16, pp. 29–42, 1990.

10.25 O. K. Gupta and A. Ravindran, Nonlinear mixed integer programming and discrete optimization, pp. 27–32 in *Progress in Engineering Optimization—1981*, R. W. Mayne and K. M. Ragsdell, Eds., ASME, New York, 1981.

10.26 G. R. Olsen and G. N. Vanderplaats, Method for nonlinear optimization with discrete variables, *AIAA Journal*, Vol. 27, No. 11, pp. 1584–1589, 1989.

10.27 K. V. John, C. V. Ramakrishnan, and K. G. Sharma, Optimum design of trusses from available sections: use of sequential linear programming with branch and bound algorithm, *Engineering Optimization*, Vol. 13, pp. 119–145, 1988.

10.28 M. W. Cooper, A survey of methods for pure nonlinear integer programming, *Management Science*, Vol. 27, No. 3, pp. 353–361, 1981.

10.29 A. Glankwahmdee, J. S. Liebman, and G. L. Hogg, Unconstrained discrete nonlinear programming, *Engineering Optimization*, Vol. 4, pp. 95–107, 1979.

10.30 K. Hager and R. Balling, New approach for discrete structural optimization, *ASCE Journal of the Structural Division*, Vol. 114, No. ST5, pp. 1120–1134, 1988.

10.31 S. S. Rao, T. S. Pan, and V. B. Venkayya, Optimal placement of actuators in actively controlled structures using genetic algorithms, *AIAA Journal*, Vol. 29, No. 6, pp. 942–943, 1991.

10.32 R. G. Parker and R. L. Rardin, *Discrete Optimization*, Academic Press, Boston, 1988.

REVIEW QUESTIONS

10.1 Answer true or false:

(a) The integer and discrete programming problems are one and the same.

(b) Gomory's cutting plane method is applicable to mixed-integer programming problems.

(c) The Balas method was developed for the solution of all-integer programming problems.

(d) The branch-and-bound method can be used to solve zero–one programming problems.

(e) The branch-and-bound method is applicable to nonlinear integer programming problems.

10.2 Define the following terms:

(a) Cutting plane
(b) Gomory's constraint
(c) Mixed-integer programming problem
(d) Additive algorithm

10.3 Give two engineering examples of a discrete programming problem.

10.4 Name two engineering systems for which zero–one programming is applicable.

10.5 What are the disadvantages of truncating the fractional part of a continuous solution for an integer problem?

10.6 How can you solve an integer nonlinear programming problem?

10.7 What is a branch-and-bound method?

10.8 Match the following methods:

(a) Land and Doig   Cutting plane method
(b) Gomory   Zero–one programming method
(c) Balas   Generalized penalty function method
(d) Gisvold and Moe   Branch-and-bound method
(e) Reiter and Rice   Generalized quadratic programming method

PROBLEMS

Find the solution for Problems 10.1–10.5 using a graphical procedure.

10.1 $$\text{Minimize } f = 4x_1 + 5x_2$$

subject to

$$3x_1 + x_2 \geq 2$$
$$x_1 + 4x_2 \geq 5$$
$$3x_1 + 2x_2 \geq 7$$
$$x_1, x_2 \geq 0, \text{ integers}$$

10.2 $$\text{Maximize } f = 4x_1 + 8x_2$$

subject to

$$4x_1 + 5x_2 \leq 40$$
$$x_1 + 2x_2 \leq 12$$
$$x_1, x_2 \geq 0, \text{ integers}$$

10.3   Maximize $f = 4x_1 + 3x_2$

subject to

$$3x_1 + 2x_2 \le 18$$
$$x_1, x_2 \ge 0, \text{ integers}$$

10.4   Maximize $f = 3x_1 - x_2$

subject to

$$3x_1 - 2x_2 \le 3$$
$$-5x_1 - 4x_2 \le -10$$
$$x_1, x_2 \ge 0, \text{ integers}$$

10.5   Maximize $f = 2x_1 + x_2$

subject to

$$8x_1 + 5x_2 \le 15$$
$$x_1, x_2 \ge 0, \text{ integers}$$

10.6   Solve the following problem using Gomory's cutting plane method:

$$\text{Maximize } f = 6x_1 + 7x_2$$

subject to

$$7x_1 + 6x_2 \le 42$$
$$5x_1 + 9x_2 \le 45$$
$$x_1 - x_2 \le 4$$
$$x_i \ge 0 \text{ and integer, } \quad i = 1, 2$$

10.7   Solve the following problem using Gomory's cutting plane method:

$$\text{Maximize } f = x_1 + 2x_2$$

subject to

$$x_1 + x_2 \le 7$$
$$2x_1 \le 11, \quad 2x_2 \le 7$$
$$x_i \ge 0 \text{ and integer, } \quad i = 1, 2$$

10.8   Express 187 in binary form.

10.9   Three cities A, B, and C are to be connected by a pipeline. The distances between A and B, B and C, and C and A are 5, 3, and 4 units, respectively. The following restrictions are to be satisfied by the pipeline:

1. The pipes leading out of A should have a total capacity of at least 3.

2. The pipes leading out of B or of C should have total capacities of either 2 or 3.

3. No pipe between any two cities must have a capacity exceeding 2.

Only pipes of an integer number of capacity units are available and the cost of a pipe is proportional to its capacity and to its length. Determine the capacities of the pipe lines to minimize the total cost.

10.10 Convert the following integer quadratic problem into a zero–one linear programming problem:

$$\text{Minimize } f = 2x_1^2 + 3x_2^2 + 4x_1x_2 - 6x_1 - 3x_2$$

subject to

$$x_1 + x_2 \leq 1$$

$$2x_1 + 3x_2 \leq 4$$

$$x_1, x_2 \geq 0, \text{ integers}$$

10.11 Convert the following integer programming problem into an equivalent zero–one programming problem:

$$\text{Minimize } f = 6x_1 - x_2$$

subject to

$$3x_1 - x_2 \geq 4$$

$$2x_1 + x_2 \geq 3$$

$$-x_1 - x_2 \geq -3$$

$$x_1, \ x_2 \text{ nonnegative integers}$$

10.12 Solve the following zero–one programming problem using an exhaustive enumeration procedure:

$$\text{Maximize } f = -10x_1 - 5x_2 - 3x_3$$

subject to

$$x_1 + 2x_2 + x_3 \geq 4$$

$$2x_1 + x_2 + x_3 \leq 6$$

$$x_i = 0 \text{ or } 1, \quad i = 1, 2, 3$$

10.13 Solve the following binary programming problem using an exhaustive enumeration procedure:

$$\text{Minimize } f = -5x_1 + 7x_2 + 10x_3 - 3x_4 + x_5$$

subject to

$$x_1 + 3x_2 - 5x_3 + x_4 + 4x_5 \leq 0$$

$$2x_1 + 6x_2 - 3x_3 + 2x_4 + 2x_5 \geq 4$$

$$x_2 - 2x_3 - x_4 + x_5 \leq -2$$

$$x_i = 0 \text{ or } 1, \quad i = 1, 2, \ldots, 5$$

10.14   Find the solution of Problem 10.1 using the branch-and-bound method coupled with the graphical method of solution for the branching problems.

10.15   Find the solution of the following problem using the branch-and-bound method coupled with the graphical method of solution for the branching problems:

$$\text{Maximize } f = x_1 - 4x_2$$

subject to

$$x_1 - x_2 \geq -4, \quad 4x_1 + 5x_2 \leq 45$$
$$5x_1 - 2x_2 \leq 20, \quad 5x_1 + 2x_2 \geq 10$$
$$x_i \geq 0 \text{ and integer}, \quad i = 1, 2$$

10.16   Solve the following mixed integer programming problem using a graphical method:

$$\text{Minimize } f = 4x_1 + 5x_2$$

subject to

$$10x_1 + x_2 \geq 10, \quad 5x_1 + 4x_2 \geq 20$$
$$3x_1 + 7x_2 \geq 21, \quad x_2 + 12x_2 \geq 12$$
$$x_1 \geq 0 \text{ and integer}, \quad x_2 \geq 0$$

10.17   Solve Problem 10.16 using the branch-and-bound method coupled with a graphical method for the solution of the branching problems.

10.18   Convert the following problem into an equivalent zero–one LP problem:

$$\text{Maximize } f = x_1 x_2$$

subject to
$$x_1^2 + x_2^2 \leq 25, \quad x_i \geq 0 \text{ and integer}, \quad i = 1, 2$$

10.19   Consider the discrete variable problem:

$$\text{Maximize } f = x_1 x_2$$

subject to

$$x_1^2 + x_2^2 \leq 4$$
$$x_1 \in \{0.1, 0.5, 1.1, 1.6, 2.0\}$$
$$x_2 \in \{0.4, 0.8, 1.5, 2.0\}$$

Approximate this problem as a zero–one LP problem at the vector, $\mathbf{X}^0 = \{ {1.1 \atop 0.8} \}$.

10.20 Find the solution of the following problem using a graphical method based on the generalized penalty function approach:

$$\text{Minimize } f = x$$

subject to

$$x - 1 \geq 0 \quad \text{with} \quad x = \{1, 2, 3, \ldots\}$$

Select suitable values of r_k and s_k to construct the ϕ_k function.

10.21 Find the solution of the following binary programming problem using the MATLAB function `bintprog`:

$$\text{Minimize } f^{\mathrm{T}} x \text{ subject to } Ax \leq b \text{ and } Aeq\ x = beq$$

where

$$A = \begin{bmatrix} -1 & 1 & 0 & 0 & 0 & 0 & 0 & 0 & 0 \\ 0 & -1 & 1 & 0 & 0 & 0 & 0 & 0 & 0 \\ 0 & 0 & 0 & -1 & 1 & 0 & 0 & 0 & 0 \\ 0 & 0 & 0 & 0 & -1 & 1 & 0 & 0 & 0 \\ 0 & 0 & 0 & 0 & 0 & 0 & -1 & 1 & 0 \\ 0 & 0 & 0 & 0 & 0 & 0 & 0 & -1 & 1 \end{bmatrix}, \ b = \begin{Bmatrix} 0 \\ 0 \\ 0 \\ 0 \\ 0 \\ 0 \end{Bmatrix}$$

$$Aeq = [1\ 1\ 1\ 1\ 1\ 1\ 1\ 1\ 1] \text{ and } beq = \{5\}$$

10.22 Find the solution of the following binary programming problem using the MATLAB function `bintprog`:

$$\text{Minimize } f^{\mathrm{T}} x \text{ subject to } Ax \leq b$$

where

$$f^{\mathrm{T}} = \{-2 \ -3 \ -1 \ -4 \ -3 \ -2 \ -2 \ -1 \ -3\}$$

$$x = \{x_1 \ x_2 \ x_3 \ x_4 \ x_5 \ x_6 \ x_7 \ x_8 \ x_9\}^{\mathrm{T}}$$

$$A = \begin{bmatrix} 0 & -3 & 0 & -1 & -1 & 0 & 0 & 0 & 0 \\ 1 & 1 & 0 & 0 & 0 & 0 & 0 & 0 & 0 \\ 0 & 1 & 0 & 1 & -1 & -1 & 0 & 0 & 0 \\ 0 & -1 & 0 & 0 & 0 & -2 & -3 & -1 & -2 \\ 0 & 0 & -1 & 0 & 2 & 1 & 2 & -2 & 1 \end{bmatrix}, \ b = \begin{Bmatrix} -3 \\ 1 \\ -1 \\ -4 \\ 5 \end{Bmatrix}$$

11

Stochastic Programming

11.1 INTRODUCTION

Stochastic or *probabilistic programming* deals with situations where some or all of the parameters of the optimization problem are described by stochastic (or random or probabilistic) variables rather than by deterministic quantities. The sources of random variables may be several, depending on the nature and the type of problem. For instance, in the design of concrete structures, the strength of concrete is a random variable since the compressive strength of concrete varies considerably from sample to sample. In the design of mechanical systems, the actual dimension of any machined part is a random variable since the dimension may lie anywhere within a specified (permissible) tolerance band. Similarly, in the design of aircraft and rockets the actual loads acting on the vehicle depend on the atmospheric conditions prevailing at the time of the flight, which cannot be predicted precisely in advance. Hence the loads are to be treated as random variables in the design of such flight vehicles.

Depending on the nature of equations involved (in terms of random variables) in the problem, a stochastic optimization problem is called a *stochastic linear, geometric, dynamic*, or *nonlinear programming problem*. The basic idea used in stochastic programming is to convert the stochastic problem into an equivalent deterministic problem. The resulting deterministic problem is then solved by using familiar techniques such as linear, geometric, dynamic, and nonlinear programming. A review of the basic concepts of probability theory that are necessary for understanding the techniques of stochastic programming is given in Section 11.2. The stochastic linear, nonlinear, and geometric programming techniques are discussed in subsequent sections.

11.2 BASIC CONCEPTS OF PROBABILITY THEORY

The material of this section is by no means exhaustive of probability theory. Rather, it provides the basic background necessary for the continuity of presentation of this chapter. The reader interested in further details should consult Parzen [11.1], Ang and Tang [11.2], or Rao [11.3].

11.2.1 Definition of Probability

Every phenomenon in real life has a certain element of uncertainty. For example, the wind velocity at a particular locality, the number of vehicles crossing a bridge, the strength of a beam, and the life of a machine cannot be predicted exactly. These phenomena are chance dependent and one has to resort to probability theory to describe the characteristics of such phenomena.

Before introducing the concept of probability, it is necessary to define certain terms such as experiment and event. An *experiment* denotes the act of performing something the outcome of which is subject to uncertainty and is not known exactly. For example, tossing a coin, rolling a die, and measuring the yield strength of steel can be called experiments. The number of possible outcomes in an experiment may be finite or infinite, depending on the nature of the experiment. The outcome is a head or a tail in the case of tossing a coin, and any one of the numbers 1, 2, 3, 4, 5, and 6 in the case of rolling a die. On the other hand, the outcome may be any positive real number in the case of measuring the yield strength of steel. An *event* represents the outcome of a single experiment. For example, realizing a head on tossing a coin, getting the number 3 or 5 on rolling a die, and observing the yield strength of steel to be greater than 20,000 psi in measurement can be called events.

The *probability* is defined in terms of the likelihood of a specific event. If E denotes an event, the probability of occurrence of the event E is usually denoted by $P(E)$. The probability of occurrence depends on the number of observations or trials. It is given by

$$P(E) = \lim_{n \to \infty} \frac{m}{n} \tag{11.1}$$

where m is the number of successful occurrences of the event E and n is the total number of trials. From Eq. (11.1) we can see that probability is a nonnegative number and

$$0 \le P(E) \le 1.0 \tag{11.2}$$

where $P(E) = 0$ denotes that the event is impossible to realize while $P(E) = 1.0$ signifies that it is certain to realize that event. For example, the probability associated with the event of realizing both the head and the tail on tossing a coin is zero (impossible event), while the probability of the event that a rolled die will show up any number between 1 and 6 is 1 (certain event).

Independent Events.   If the occurrence of an event E_1 in no way affects the probability of occurrence of another event E_2, the events E_1 and E_2 are said to be *statistically independent*. In this case the probability of simultaneous occurrence of both the events is given by

$$P(E_1 E_2) = P(E_1)P(E_2) \tag{11.3}$$

For example, if $P(E_1) = P$(raining at a particular location) $= 0.4$ and $P(E_2) = P$(realizing the head on tossing a coin) $= 0.7$, obviously E_1 and E_2 are statistically independent and

$$P(E_1 E_2) = P(E_1)P(E_2) = 0.28$$

11.2.2   Random Variables and Probability Density Functions

An event has been defined as a possible outcome of an experiment. Let us assume that a random event is the measurement of a quantity X, which takes on various values in the range $-\infty$ to ∞. Such a quantity (like X) is called a *random variable*. We denote a random variable by a capital letter and the particular value taken by it by a lowercase letter. Random variables are of two types: (1) discrete and (2) continuous. If the random variable is allowed to take only discrete values $x_1, x_2, \ldots, x_n$, it is called a *discrete* random variable. On the other hand, if the random variable is permitted to take any real value in a specified range, it is called a *continuous* random variable. For example, the number of vehicles crossing a bridge in a day is a discrete random variable, whereas the yield strength of steel can be treated as a continuous random variable.

Probability Mass Function (for Discrete Random Variables). Corresponding to each x_i that a discrete random variable X can take, we can associate a probability of occurrence $P(x_i)$. We can describe the probabilities associated with the random variable X by a table of values, but it will be easier to write a general formula that permits one to calculate $P(x_i)$ by substituting the appropriate value of x_i. Such a formula is called the *probability mass function* of the random variable X and is usually denoted as $f_X(x_i)$, or simply as $f(x_i)$. Thus the function that gives the probability of realizing the random variable $X = x_i$ is called the probability mass function $f_X(x_i)$. Therefore,

$$f(x_i) = f_X(x_i) = P(X = x_i) \tag{11.4}$$

Cumulative Distribution Function (Discrete Case). Although a random variable X is described completely by the probability mass function, it is often convenient to deal with another, related function known as the *probability distribution function*. The probability that the value of the random variable X is less than or equal to some number x is defined as the *cumulative distribution function* $F_X(x)$.

$$F_X(x) = P(X \leq x) = \sum_i f_X(x_i) \tag{11.5}$$

where summation expends over those values of i such that $x_i \leq x$. Since the distribution function is a cumulative probability, it is also called the cumulative distribution function.

Example 11.1 Find the probability mass and distribution functions for the number realized when a fair die is thrown.

SOLUTION   Since each face is equally likely to show up, the probability of realizing any number between 1 and 6 is $\frac{1}{6}$.

$$P(X = 1) = P(X = 2) = \cdots = P(X = 6) = \tfrac{1}{6}$$
$$f_X(1) = f_X(2) = \cdots = f_X(6) = \tfrac{1}{6}$$

The analytical form of $F_X(x)$ is

$$F_X(x) = \frac{x}{6} \quad \text{for} \quad 1 \leq x \leq 6$$

It can be seen that for any discrete random variable, the distribution function will be a step function. If the least possible value of a variable X is S and the greatest possible value is T, then

$$F_X(x) = 0 \quad \text{for all } x < S \quad \text{and} \quad F_X(x) = 1 \quad \text{for all } x > T$$

Probability Density Function (Continuous Case).   The probability density function of a random variable is defined by

$$f_X(x)\,dx = P(x \leq X \leq x + dx) \tag{11.6}$$

which is equal to the probability of detecting X in the infinitesimal interval $(x, x + dx)$. The distribution function of X is defined as the probability of detecting X less than or equal to x, that is,

$$F_X(x) = \int_{-\infty}^{x} f_X(x')\,dx' \tag{11.7}$$

where the condition $F_X(-\infty) = 0$ has been used. As the upper limit of the integral goes to infinity, we have

$$\int_{-\infty}^{\infty} f_X(x)\,dx = F_X(\infty) = 1 \tag{11.8}$$

This is called the *normalization condition*. A typical probability density function and the corresponding distribution functions are shown in Fig. 11.1.

11.2.3   Mean and Standard Deviation

The probability density or distribution function of a random variable contains all the information about the variable. However, in many cases we require only the gross properties, not entire information about the random variable. In such cases one computes only the mean and the variation about the mean of the random variable as the salient features of the variable.

Mean.   The *mean value* (also termed the *expected value* or *average*) is used to describe the central tendency of a random variable.

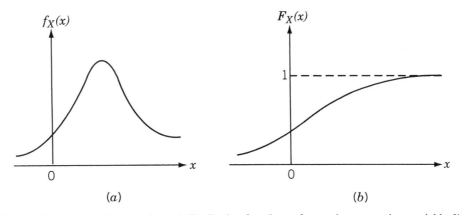

Figure 11.1   Probability density and distribution functions of a continuous random variable X: (*a*) density function; (*b*) distribution function.

Discrete Case. Let us assume that there are n trials in which the random variable X is observed to take on the value x_1 (n_1 times), x_2 (n_2 times), and so on, and $n_1 + n_2 + \cdots + n_m = n$. Then the arithmetic mean of X, denoted as $\overline{X}$, is given by

$$\overline{X} = \frac{\sum_{k=1}^{m} x_k n_k}{n} = \sum_{k=1}^{m} x_k \frac{n_k}{n} = \sum_{k=1}^{m} x_k f_X(x_k) \qquad (11.9)$$

where n_k/n is the relative frequency of occurrence of x_k and is same as the probability mass function $f_X(x_k)$. Hence in general, the expected value, $E(X)$, of a discrete random variable can be expressed as

$$\overline{X} = E(X) = \sum_{i} x_i f_X(x_i), \quad \text{sum over all } i \qquad (11.10)$$

Continuous Case. If $f_X(x)$ is the density function of a continuous random variable, X, the mean is given by

$$\overline{X} = \mu_x = E(X) = \int_{-\infty}^{\infty} x f_X(x) \, dx \qquad (11.11)$$

Standard Deviation.   The expected value or mean is a measure of the central tendency, indicating the location of a distribution on some coordinate axis. A measure of the variability of the random variable is usually given by a quantity known as the *standard deviation*. The mean-square deviation or variance of a random variable X is defined as

$$\begin{aligned} \sigma_X^2 = \operatorname{Var}(X) &= E[(X - \mu_X)^2] \\ &= E[X^2 - 2X\mu_X + \mu_X^2] \\ &= E(X^2) - 2\mu_X E(X) + E(\mu_X^2) \\ &= E(X^2) - \mu_X^2 \end{aligned} \qquad (11.12)$$

and the standard deviation as

$$\sigma_X = +\sqrt{\operatorname{Var}(X)} = \sqrt{E(X^2) - \mu_X^2} \qquad (11.13)$$

The coefficient of variation (a measure of dispersion in nondimensional form) is defined as

$$\text{coefficient of variation of } X = \gamma_X = \frac{\text{standard deviation}}{\text{mean}} = \frac{\sigma_X}{\mu_X} \qquad (11.14)$$

Figure 11.2 shows two density functions with the same mean μ_X but with different variances. As can be seen, the variance measures the *breadth* of a density function.

Example 11.2   The number of airplane landings at an airport in a minute (X) and their probabilities are given by

x_i	0	1	2	3	4	5	6
$p_X(x_i)$	0.02	0.15	0.22	0.26	0.17	0.14	0.04

Find the mean and standard deviation of X.

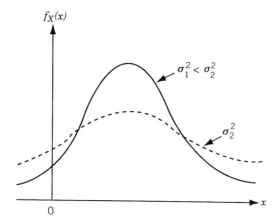

Figure 11.2 Two density functions with same mean.

SOLUTION

$$\overline{X} = \sum_{i=0}^{6} x_i \, p_X(x_i) = 0(0.02) + 1(0.15) + 2(0.22) + 3(0.26)$$

$$+ 4(0.17) + 5(0.14) + 6(0.04)$$

$$= 2.99$$

$$\overline{X^2} = \sum_{i=0}^{6} x_i^2 \, p_X(x_i) = 0(0.02) + 1(0.15) + 4(0.22) + 9(0.26)$$

$$+ 16(0.17) + 25(0.14) + 36(0.04)$$

$$= 11.03$$

Thus

$$\sigma_X^2 = \overline{X^2} - (\overline{X})^2 = 11.03 - (2.99)^2 = 2.0899 \quad \text{or} \quad \sigma_X = 1.4456$$

Example 11.3 The force applied on an engine brake (X) is given by

$$f_X(x) = \begin{cases} \dfrac{x}{48}, & 0 \le x \le 8 \, \text{lb} \\[2mm] \dfrac{12 - x}{24}, & 8 \le x \le 12 \, \text{lb} \end{cases}$$

Determine the mean and standard deviation of the force applied on the brake.

SOLUTION

$$\mu_X = E[X] = \int_{-\infty}^{\infty} x f_X(x) \, dx = \int_0^8 x \frac{x}{48} \, dx + \int_8^{12} x \frac{12 - x}{24} \, dx = 6.6667$$

$$E[X^2] = \int_{-\infty}^{\infty} x^2 f_X(x) \, dx = \int_0^8 x^2 \frac{x}{48} \, dx + \int_8^{12} x^2 \frac{12 - x}{24} \, dx$$

$$= 21.3333 + 29.3333 = 50.6666$$

$$\sigma_X^2 = E[X^2] - (E[X])^2 = 50.6666 - (6.6667)^2$$

$$= 6.2222 \quad \text{or} \quad \sigma_X = 2.4944$$

11.2.4 Function of a Random Variable

If X is a random variable, any other variable Y defined as a function of X will also be a random variable. If $f_X(x)$ and $F_X(x)$ denote, respectively, the probability density and distribution functions of X, the problem is to find the density function $f_Y(y)$ and the distribution function $F_Y(y)$ of the random variable Y. Let the functional relation be

$$Y = g(X) \tag{11.15}$$

By definition, the distribution function of Y is the probability of realizing Y less than or equal to y:

$$F_Y(y) = P(Y \le y) = P(g \le y)$$

$$= \int_{g(x) \le y} f_X(x) \, dx \tag{11.16}$$

where the integration is to be done over all values of x for which $g(x) \le y$.

For example, if the functional relation between y and x is as shown in Fig. 11.3, the range of integration is shown as $\Delta x_1 + \Delta x_2 + \Delta x_3 + \cdots$. The probability density function of Y is given by

$$f_Y(y) = \frac{\partial}{\partial y}[F_Y(y)] \tag{11.17}$$

If $Y = g(X)$, the mean and variance of Y are defined, respectively, by

$$E(Y) = \int_{-\infty}^{\infty} g(x) f_X(x) \, dx \tag{11.18}$$

$$\text{Var}[Y] = \int_{-\infty}^{\infty} [g(x) - E(Y)]^2 f_X(x) \, dx \tag{11.19}$$

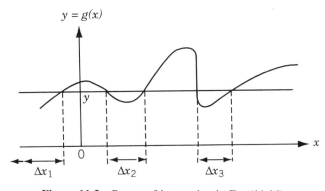

Figure 11.3 Range of integration in Eq. (11.16).

11.2.5   Jointly Distributed Random Variables

When two or more random variables are being considered simultaneously, their joint behavior is determined by a joint probability distribution function. The probability distributions of single random variables are called *univariate distributions* and the distributions that involve two random variables are called *bivariate distributions*. In general, if a distribution involves more than one random variable, it is called a *multivariate distribution*.

Joint Density and Distribution Functions.   We can define the *joint density function* of n continuous random variables $X_1, X_2, \ldots, X_n$ as

$$f_{X_1,\ldots,X_n}(x_1, \ldots, x_n)\, dx_1 \cdots dx_n = P(x_1 \leq X_1 \leq x_1 + dx_1,$$

$$x_2 \leq X_2 \leq x_2 + dx_2, \ldots, x_n \leq X_n \leq x_n + dx_n) \tag{11.20}$$

If the random variables are independent, the joint density function is given by the product of individual or marginal density functions as

$$f_{X_1,\ldots,X_n}(x_1, \ldots, x_n) = f_{X_1}(x_1) \cdots f_{X_n}(x_n) \tag{11.21}$$

The joint distribution function

$$F_{X_1,X_2,\ldots,X_n}(x_1, x_2, \ldots, x_n)$$

associated with the density function of Eq. (11.20) is given by

$$F_{X_1,\ldots,X_n}(x_1, \ldots, x_n)$$

$$= P[X_1 \leq x_1, \ldots, X_n \leq x_n]$$

$$= \int_{-\infty}^{x_1} \cdots \int_{-\infty}^{x_n} f_{X_1,\ldots,X_n}(x_1', x_2', \ldots, x_n')\, dx_1'\, dx_2' \cdots dx_n' \tag{11.22}$$

If $X_1, X_2, \ldots, X_n$ are independent random variables, we have

$$F_{X_1,\ldots,X_n}(x_1, \ldots, x_n) = F_{X_1}(x_1)\, F_{X_2}(x_2) \cdots F_{X_n}(x_n) \tag{11.23}$$

It can be seen that the joint density function can be obtained by differentiating the joint distribution function as

$$f_{X_1,\ldots,X_n}(x_1, \ldots, x_n) = \frac{\partial^n}{\partial x_1\, \partial x_2 \cdots \partial x_n} F_{X_1,\ldots,X_n}(x_1, \ldots, x_n) \tag{11.24}$$

Obtaining the Marginal or Individual Density Function from the Joint Density Function.   Let the joint density function of two random variables X and Y be denoted by $f(x, y)$ and the marginal density functions of X and Y by $f_X(x)$ and $f_Y(y)$, respectively. Take the infinitesimal rectangle with corners located at the points (x, y), $(x + dx, y)$, $(x, y + dy)$, and $(x + dx, y + dy)$. The probability of a random point (x', y') falling in this rectangle is $f_{X,Y}(x, y)\, dx\, dy$. The integral of such probability elements with respect to y (for a fixed value of x) is the sum of the probabilities of all the mutually exclusive ways of obtaining the points lying between x and $x + dx$. Let the lower and upper limits of y be $a_1(x)$ and $b_1(x)$. Then

$$P[x \leq x' \leq x + dx] = \left[\int_{a_1(x)}^{b_1(x)} f_{X,Y}(x, y)\, dy \right] dx = f_X(x)\, dx$$

$$f_X(x) = \int_{y_1=a_1(x)}^{y_2=b_1(x)} f_{X,Y}(x, y)\, dy \tag{11.25}$$

Similarly, we can show that

$$f_Y(y) = \int_{x_1=a_2(y)}^{x_2=b_2(y)} f_{X,Y}(x, y)\, dx \tag{11.26}$$

11.2.6   Covariance and Correlation

If X and Y are two jointly distributed random variables, the variances of X and Y are defined as

$$E[(X - \overline{X})^2] = \text{Var}[X] = \int_{-\infty}^{\infty} (x - \overline{X})^2 f_X(x)\, dx \tag{11.27}$$

$$E[(Y - \overline{Y})^2] = \text{Var}[Y] = \int_{-\infty}^{\infty} (y - \overline{Y})^2 f_Y(y)\, dy \tag{11.28}$$

and the covariance of X and Y as

$$E[(X - \overline{X})(Y - \overline{Y})] = \text{Cov}(X, Y)$$

$$= \int_{-\infty}^{\infty} \int_{-\infty}^{\infty} (x - \overline{X})(y - \overline{Y}) f_{X,Y}(x, y)\, dx\, dy$$

$$= \sigma_{X,Y} \tag{11.29}$$

The correlation coefficient, $\rho_{X,Y}$, for the random variables is defined as

$$\rho_{X,Y} = \frac{\text{Cov}(X, Y)}{\sigma_X \sigma_Y} \tag{11.30}$$

and it can be proved that $-1 \leq \rho_{X,Y} \leq 1$.

11.2.7   Functions of Several Random Variables

If Y is a function of several random variables $X_1, X_2, \ldots, X_n$, the distribution and density functions of Y can be found in terms of the joint density function of $X_1, X_2, \ldots, X_n$ as follows:
   Let

$$Y = g(X_1, X_2, \ldots, X_n) \tag{11.31}$$

Then the joint distribution function $F_Y(y)$, by definition, is given by

$$F_Y(y) = P(Y \leq y)$$

$$= \int_{x_1} \int_{x_2} \cdots \int_{x_n} f_{X_1, X_2, \ldots, X_n}(x_1, x_2, \ldots, x_n) dx_1\, dx_2 \cdots dx_n$$

$$g(x_1, x_2, \ldots, x_n) \leq y \qquad (11.32)$$

where the integration is to be done over the domain of the n-dimensional $(X_1, X_2, \ldots, X_n)$ space in which the inequality $g(x_1, x_2, \ldots, x_n) \leq y$ is satisfied. By differentiating Eq. (11.32), we can get the density function of y, $f_Y(y)$.

As in the case of a function of a single random variable, the mean and variance of a function of several random variables are given by

$$E(Y) = E[g(X_1, X_2, \ldots, X_n)] = \int_{-\infty}^{\infty} \cdots \int_{-\infty}^{\infty} g(x_1, x_2, \ldots, x_n) f_{X_1, X_2, \ldots, X_n}$$

$$\times (x_1, x_2, \ldots, x_n) dx_1\, dx_2 \cdots dx_n \qquad (11.33)$$

and

$$\text{Var}(Y) = \int_{-\infty}^{\infty} \cdots \int_{-\infty}^{\infty} [g(x_1, x_2, \ldots, x_n) - \overline{Y}]^2$$

$$\times f_{X_1, X_2 \ldots X_n}(x_1, x_2, \ldots, x_n) dx_1\, dx_2 \cdots dx_n \qquad (11.34)$$

In particular, if Y is a linear function of two random variables X_1 and X_2, we have

$$Y = a_1 X_1 + a_2 X_2$$

where a_1 and a_2 are constants. In this case

$$E(Y) = \int_{-\infty}^{\infty} \int_{-\infty}^{\infty} (a_1 x_1 + a_2 x_2) f_{X_1, X_2}(x_1, x_2) dx_1\, dx_2$$

$$= a_1 \int_{-\infty}^{\infty} x_1 f_{X_1}(x_1) dx_1 + a_2 \int_{-\infty}^{\infty} x_2 f_{X_2}(x_2) dx_2$$

$$= a_1 E(X_1) + a_2 E(X_2) \qquad (11.35)$$

Thus the expected value of a sum is given by the sum of the expected values. The variance of Y can be obtained as

$$\text{Var}(Y) = E[(a_1 X_1 + a_2 X_2) - (a_1 \overline{X} + a_2 \overline{X}_2)]^2$$

$$= E[a_1(X_1 - \overline{X}_1) + a_2(X_2 - \overline{X}_2)]^2$$

$$= E[a_1^2(X_1 - \overline{X}_1)^2 + 2a_1 a_2(X_1 - \overline{X}_1)(X_2 - \overline{X}_2) + a_2^2(X_2 - \overline{X}_2)^2] \qquad (11.36)$$

Noting that the expected values of the first and the third terms are variances, whereas that the middle term is a covariance, we obtain

$$\text{Var}(Y) = a_1^2 \text{Var}(X_1) + a_2^2 \text{Var}(X_2) + 2a_1 a_2 \text{Cov}(X_1, X_2) \qquad (11.37)$$

These results can be generalized to the case when Y is a linear function of several random variables. Thus if

$$Y = \sum_{i=1}^{n} a_i X_i \qquad (11.38)$$

then

$$E(Y) = \sum_{i=1}^{n} a_i E(X_i) \qquad (11.39)$$

$$\text{Var}(Y) = \sum_{i=1}^{n} a_i^2 \,\text{Var}(X_i) + \sum_{i=1}^{n}\sum_{j=1}^{n} a_i a_j \,\text{Cov}(X_i, X_j), \quad i \neq j \qquad (11.40)$$

Approximate Mean and Variance of a Function of Several Random Variables.
If $Y = g(X_1, \ldots, X_n)$, the approximate mean and variance of Y can be obtained as follows. Expand the function g in a Taylor series about the mean values $\overline{X}_1, \overline{X}_2, \ldots, \overline{X}_n$ to obtain

$$Y = g(\overline{X}_1, \overline{X}_2, \ldots, \overline{X}_n) + \sum_{i=1}^{n} (X_i - \overline{X}_i)\frac{\partial g}{\partial X_i}$$

$$+ \frac{1}{2}\sum_{i=1}^{n}\sum_{j=1}^{n}(X_i - \overline{X}_i)(X_j - \overline{X}_j)\frac{\partial^2 g}{\partial X_i \partial X_j} + \cdots \qquad (11.41)$$

where the derivatives are evaluated at $(\overline{X}_1, \overline{X}_2, \ldots, \overline{X}_n)$. By truncating the series at the linear terms, we obtain the first-order approximation to Y as

$$Y \simeq g(\overline{X}_1, \overline{X}_2, \ldots, \overline{X}_n) + \sum_{i=1}^{n}(X_i - \overline{X}_i)\left.\frac{\partial g}{\partial X_i}\right|_{(\overline{X}_1, \overline{X}_2, \ldots, \overline{X}_n)} \qquad (11.42)$$

The mean and variance of Y given by Eq. (11.42) can now be expressed as [using Eqs. (11.39) and (11.40)]

$$E(Y) \simeq g(\overline{X}_1, \overline{X}_2, \ldots, \overline{X}_n) \qquad (11.43)$$

$$\text{Var}(Y) \simeq \sum_{i=1}^{n} c_i^2 \,\text{Var}(X_i) + \sum_{i=1}^{n}\sum_{j=1}^{n} c_i c_j \,\text{Cov}(X_i, X_j), \quad i \neq j \qquad (11.44)$$

where c_i and c_j are the values of the partial derivatives $\partial g/\partial X_i$ and $\partial g/\partial X_j$, respectively, evaluated at $(\overline{X}_1, \overline{X}_2, \ldots, \overline{X}_n)$.

It is worth noting at this stage that the approximation given by Eq. (11.42) is frequently used in most of the practical problems to simplify the computations involved.

11.2.8 Probability Distributions

There are several types of probability distributions (analytical models) for describing various types of discrete and continuous random variables. Some of the common distributions are given below:

Discrete case	Continuous case
Discrete uniform distribution	Uniform distribution
Binomial	Normal or Gaussian
Geometric	Gamma
Multinomial	Exponential
Poisson	Beta
Hypergeometric	Rayleigh
Negative binomial (or Pascal's)	Weibull

In any physical problem, one chooses a particular type of probability distribution depending on (1) the nature of the problem, (2) the underlying assumptions associated with the distribution, (3) the shape of the graph between $f(x)$ or $F(x)$ and x obtained after plotting the available data, and (4) the convenience and simplicity afforded by the distribution.

Normal Distribution. The best known and most widely used probability distribution is the Gaussian or normal distribution. The normal distribution has a probability density function given by

$$f_X(x) = \frac{1}{\sqrt{2\pi}\sigma_X} e^{-1/2[(x-\mu_X)/\sigma_X]^2}, \quad -\infty < x < \infty \tag{11.45}$$

where μ_X and σ_X are the parameters of the distribution, which are also the mean and standard deviation of X, respectively. The normal distribution is often identified as $N(\mu_X, \sigma_X)$.

Standard Normal Distribution. A normal distribution with parameters $\mu_X = 0$ and $\sigma_X = 1$, called the *standard normal distribution*, is denoted as $N(0, 1)$. Thus the density function of a standard normal variable (Z) is given by

$$f_Z(z) = \frac{1}{\sqrt{2\pi}} e^{-(z^2/2)}, \quad -\infty < z < \infty \tag{11.46}$$

The distribution function of the standard normal variable (Z) is often designated as $\phi(z)$ so that, with reference to Fig. 11.4,

$$\phi(z_1) = p \quad \text{and} \quad z_1 = \phi^{-1}(p) \tag{11.47}$$

where p is the cumulative probability. The distribution function $N(0, 1)$ [i.e., $\phi(z)$] is tabulated widely as *standard normal tables*. For example, Table 11.1, gives the values of z, $f(z)$, and $\phi(z)$ for positive values of z. This is because the density function is symmetric about the mean value $(z = 0)$ and hence

$$f(-z) = f(z) \tag{11.48}$$

$$\phi(-z) = 1 - \phi(z) \tag{11.49}$$

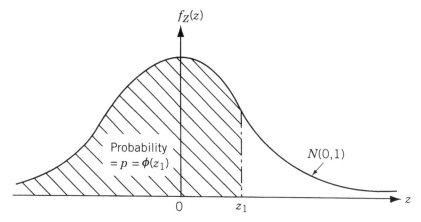

Figure 11.4   Standard normal density function.

By the same token, the values of z corresponding to $p < 0.5$ can be obtained as

$$z = \phi^{-1}(p) = -\phi^{-1}(1 - p) \tag{11.50}$$

Notice that any normally distributed variable (X) can be reduced to a standard normal variable by using the transformation

$$z = \frac{x - \mu_X}{\sigma_X} \tag{11.51}$$

For example, if $P(a < X \le b)$ is required, we have

$$P(a < X \le b) = \frac{1}{\sigma_X \sqrt{2\pi}} \int_a^b e^{-(1/2)[(x-\mu_X)/\sigma_X]^2} dx \tag{11.52}$$

By using Eq. (11.51) and $dx = \sigma_X \, dz$, Eq. (11.52) can be rewritten as

$$P(a < X \le b) = \frac{1}{\sqrt{2\pi}} \int_{(a-\mu_X)/\sigma_X}^{(b-\mu_X)/\sigma_X} e^{-z^2/2} dz \tag{11.53}$$

This integral can be recognized to be the area under the standard normal density curve between $(a - \mu_X)/\sigma_X$ and $(b - \mu_X)/\sigma_X$ and hence

$$P(a < X \le b) = \phi\left(\frac{b - \mu_X}{\sigma_X}\right) - \phi\left(\frac{a - \mu_X}{\sigma_X}\right) \tag{11.54}$$

Example 11.4   The width of a slot on a duralumin forging is normally distributed. The specification of the slot width is 0.900 ± 0.005. The parameters $\mu = 0.9$ and $\sigma = 0.003$ are known from past experience in production process. What is the percent of scrap forgings?

SOLUTION   If X denotes the width of the slot on the forging, the usable region is given by

$$0.895 \le x \le 0.905$$

Table 11.1 Standard Normal Distribution Table

z	$f(z)$	$\phi(z)$
0.0	0.398942	0.500000
0.1	0.396952	0.539828
0.2	0.391043	0.579260
0.3	0.381388	0.617912
0.4	0.368270	0.655422
0.5	0.352065	0.691463
0.6	0.333225	0.725747
0.7	0.312254	0.758036
0.8	0.289692	0.788145
0.9	0.266085	0.815940
1.0	0.241971	0.841345
1.1	0.217852	0.864334
1.2	0.194186	0.884930
1.3	0.171369	0.903199
1.4	0.149727	0.919243
1.5	0.129518	0.933193
1.6	0.110921	0.945201
1.7	0.094049	0.955435
1.8	0.078950	0.964070
1.9	0.065616	0.971284
2.0	0.053991	0.977250
2.1	0.043984	0.982136
2.2	0.035475	0.986097
2.3	0.028327	0.989276
2.4	0.022395	0.991802
2.5	0.017528	0.993790
2.6	0.013583	0.995339
2.7	0.010421	0.996533
2.8	0.007915	0.997445
2.9	0.005952	0.998134
3.0	0.004432	0.998650
3.5	0.000873	0.999767
4.0	0.000134	0.999968
4.5	0.000016	0.999996
5.0	0.0000015	0.9999997

and the amount of scrap is given by

$$\text{scrap} = P(x \leq 0.895) + P(x \geq 0.905)$$

In terms of the standardized normal variable,

$$\text{scrap} = P\left(Z \leq \frac{-0.9 + 0.895}{0.003}\right) + P\left(Z \geq \frac{-0.9 + 0.905}{0.003}\right)$$

$$= P(Z \leq -1.667) + P(Z \geq +1.667)$$

$$= [1 - P(Z \le 1.667)] + [1 - P(Z \le 1.667)]$$

$$= 2.0 - 2P(Z \le 1.667)$$

$$= 2.0 - 2(0.9525) = 0.095$$

$$= 9.5\%$$

Joint Normal Density Function.   If $X_1, X_2, \ldots, X_n$ follow normal distribution, any linear function, $Y = a_1 X_1 + a_2 X_2 + \cdots + a_n X_n$, also follows normal distribution with mean

$$\overline{Y} = a_1 \overline{X}_1 + a_2 \overline{X}_2 + \cdots + a_n \overline{X}_n \tag{11.55}$$

and variance

$$\text{Var}(Y) = a_1^2 \, \text{Var}(X_1) + a_2^2 \, \text{Var}(X_2) + \cdots + a_n^2 \, \text{Var}(X_n) \tag{11.56}$$

if $X_1, X_2, \ldots, X_n$ are independent. In general, the joint normal density function for n-independent random variables is given by

$$f_{X_1, X_2, \ldots, X_n}(x_1, x_2, \ldots, x_n) = \frac{1}{\sqrt{(2\pi)^n} \sigma_1 \sigma_2 \cdots \sigma_n} \exp\left[-\frac{1}{2} \sum_{k=1}^{n} \left(\frac{x_k - \overline{X}_k}{\sigma_k} \right)^2 \right]$$

$$= f_{X_1}(x_1) f_{X_2}(x_2) \cdots f_{X_n}(x_n) \tag{11.57}$$

where $\sigma_i = \sigma_{Xi}$. If the correlation between the random variables X_k and X_j is not zero, the joint density function is given by

$$f_{X_1, X_2, \ldots, X_n}(x_1, x_2, \ldots, x_n)$$

$$= \frac{1}{\sqrt{(2\pi)^n |\mathbf{K}|}} \exp\left[-\frac{1}{2} \sum_{j=1}^{n} \sum_{k=1}^{n} \{\mathbf{K}^{-1}\}_{jk} (x_j - \overline{X}_j)(x_k - \overline{X}_k) \right] \tag{11.58}$$

where

$$K_{X_j X_k} = K_{jk} = E[(x_j - \overline{X}_j)(x_k - \overline{X}_k)]$$

$$= \int_{-\infty}^{\infty} \int_{-\infty}^{\infty} (x_j - \overline{X}_j)(x_k - \overline{X}_k) f_{X_j, X_k}(x_j, x_k) dx_j dx_k$$

$$= \text{convariance between } X_j \text{ and } X_k$$

$$\mathbf{K} = \text{correlation matrix} = \begin{bmatrix} K_{11} & K_{12} & \cdots & K_{1n} \\ K_{21} & K_{22} & \cdots & K_{2n} \\ \vdots & & & \\ K_{n1} & K_{n2} & \cdots & K_{nn} \end{bmatrix} \tag{11.59}$$

and $\{\mathbf{K}^{-1}\}_{jk} = jk$th element of $\mathbf{K}^{-1}$. It is to be noted that $K_{X_j X_k} = 0$ for $j \ne k$ and $= \sigma_{X_j}^2$ for $j = k$ in case there is no correlation between X_j and X_k.

11.2.9   Central Limit Theorem

If $X_1, X_2, \ldots, X_n$ are n mutually independent random variables with finite mean and variance (they may follow different distributions), the sum

$$S_n = \sum_{i=1}^{n} X_i \tag{11.60}$$

tends to a normal variable if no single variable contributes significantly to the sum as n tends to infinity. Because of this theorem, we can approximate most of the physical phenomena as normal random variables. Physically, S_n may represent, for example, the tensile strength of a fiber-reinforced material, in which case the total tensile strength is given by the sum of the tensile strengths of individual fibers. In this case the tensile strength of the material may be represented as a normally distributed random variable.

11.3   STOCHASTIC LINEAR PROGRAMMING

A stochastic linear programming problem can be stated as follows:

$$\text{Minimize } f(\mathbf{X}) = \mathbf{C}^T \mathbf{X} = \sum_{j=1}^{n} c_j x_j \tag{11.61}$$

subject to

$$\mathbf{A}_i^T \mathbf{X} = \sum_{j=1}^{n} a_{ij} x_j \leq b_i, \quad i = 1, 2, \ldots, m \tag{11.62}$$

$$x_j \geq 0, \quad j = 1, 2, \ldots, n \tag{11.63}$$

where c_j, a_{ij}, and b_i are random variables (the decision variables x_j are assumed to be deterministic for simplicity) with known probability distributions. Several methods are available for solving the problem stated in Eqs. (11.61) to (11.63). We consider a method known as the *chance-constrained programming technique*, in this section.

As the name indicates, the chance-constrained programming technique can be used to solve problems involving chance constraints, that is, constraints having finite probability of being violated. This technique was originally developed by Charnes and Cooper [11.5]. In this method the stochastic programming problem is stated as follows:

$$\text{Minimize } f(\mathbf{X}) = \sum_{j=1}^{n} c_j x_j \tag{11.64}$$

subject to

$$P\left[\sum_{j=1}^{n} a_{ij} x_j \leq b_i \right] \geq p_i, \quad i = 1, 2, \ldots, m \tag{11.65}$$

$$x_j \geq 0, \quad j = 1, 2, \ldots, n \tag{11.66}$$

where c_j, a_{ij}, and b_i are random variables and p_i are specified probabilities. Notice that Eqs. (11.65) indicate that the ith constraint,

$$\sum_{j=1}^{n} a_{ij} x_j \leq b_i$$

has to be satisfied with a probability of at least p_i where $0 \leq p_i \leq 1$. For simplicity, we assume that the design variables x_j are deterministic and c_j, a_{ij}, and b_i are random variables. We shall further assume that all the random variables are normally distributed with known mean and standard deviations.

Since c_j are normally distributed random variables, the objective function $f(\mathbf{X})$ will also be a normally distributed random variable. The mean and variance of f are given by

$$\overline{f} = \sum_{j=1}^{n} \overline{c}_j x_j \tag{11.67}$$

$$\text{Var}(f) = \mathbf{X}^T \mathbf{V} \mathbf{X} \tag{11.68}$$

where $\overline{c}_j$ is the mean value of c_j and the matrix $\mathbf{V}$ is the covariance matrix of c_j defined as

$$\mathbf{V} = \begin{bmatrix} \text{Var}(c_1) & \text{Cov}(c_1, c_2) & \cdots & \text{Cov}(c_1, c_n) \\ \text{Cov}(c_2, c_1) & \text{Var}(c_2) & \cdots & \text{Cov}(c_2, c_n) \\ \vdots & & & \\ \text{Cov}(c_n, c_1) & \text{Cov}(c_n, c_2) & \cdots & \text{Var}(c_n) \end{bmatrix} \tag{11.69}$$

with $\text{Var}(c_j)$ and $\text{Cov}(c_i, c_j)$ denoting the variance of c_j and covariance between c_i and c_j, respectively. A new deterministic objective function for minimization can be formulated as

$$F(\mathbf{X}) = k_1 \overline{f} + k_2 \sqrt{\text{Var}(f)} \tag{11.70}$$

where k_1 and k_2 are nonnegative constants whose values indicate the relative importance of $\overline{f}$ and standard deviation of f for minimization. Thus $k_2 = 0$ indicates that the expected value of f is to be minimized without caring for the standard deviation of f. On the other hand, if $k_1 = 0$, it indicates that we are interested in minimizing the variability of f about its mean value without bothering about what happens to the mean value of f. Similarly, if $k_1 = k_2 = 1$, it indicates that we are giving equal importance to the minimization of the mean as well as the standard deviation of f. Notice that the new objective function stated in Eq. (11.70) is a nonlinear function in $\mathbf{X}$ in view of the expression for the variance of f.

The constraints of Eq. (11.65) can be expressed as

$$P[h_i \leq 0] \geq p_i, \quad i = 1, 2, \ldots, m \tag{11.71}$$

where h_i is a new random variable defined as

$$h_i = \sum_{j=1}^{n} a_{ij} x_j - b_i = \sum_{k=1}^{n+1} q_{ik} y_k \qquad (11.72)$$

where

$$q_{ik} = a_{ik}, \quad k = 1, 2, \ldots, n \quad q_{i,n+1} = b_i$$

$$y_k = x_k, \quad k = 1, 2, \ldots, n, \quad y_{n+1} = -1$$

Notice that the constant y_{n+1} is introduced for convenience. Since h_i is given by a linear combination of the normally distributed random variables q_{ik}, it will also follow normal distribution. The mean and the variance of h_i are given by

$$\overline{h}_i = \sum_{k=1}^{n+1} \overline{q}_{ik} y_k = \sum_{j=1}^{n} \overline{a}_{ij} x_j - \overline{b}_i \qquad (11.73)$$

$$\mathrm{Var}(h_i) = \mathbf{Y}^T \mathbf{V}_i \mathbf{Y} \qquad (11.74)$$

where

$$\mathbf{Y} = \begin{Bmatrix} y_1 \\ y_2 \\ \vdots \\ y_{n+1} \end{Bmatrix} \qquad (11.75)$$

$$\mathbf{V}_i = \begin{bmatrix} \mathrm{Var}(q_{i1}) & \mathrm{Cov}(q_{i1}, q_{i2}) & \cdots & \mathrm{Cov}(q_{i1}, q_{i,n+1}) \\ \mathrm{Cov}(q_{i2}, q_{i1}) & \mathrm{Var}(q_{i2}) & \cdots & \mathrm{Cov}(q_{i2}, q_{i,n+1}) \\ \vdots & & & \\ \mathrm{Cov}(q_{i,n+1}, q_{i1}) & \mathrm{Cov}(q_{i,n+1}, q_{i2}) & \cdots & \mathrm{Var}(q_{i,n+1}) \end{bmatrix} \qquad (11.76)$$

This can be written more explicitly as

$$\mathrm{Var}(h_i) = \sum_{k=1}^{n+1} \left[y_k^2 \, \mathrm{Var}(q_{ik}) + 2 \sum_{l=k+1}^{n+1} y_k y_l \, \mathrm{Cov}(q_{ik}, q_{il}) \right]$$

$$= \sum_{k=1}^{n} \left[y_k^2 \, \mathrm{Var}(q_{ik}) + 2 \sum_{l=k+1}^{n} y_k y_l \, \mathrm{Cov}(q_{ik}, q_{il}) \right]$$

$$+ y_{n+1}^2 \, \mathrm{Var}(q_{i,n+1}) + 2y_{n+1}^2 \, \mathrm{Cov}(q_{i,n+1}, q_{i,n+1})$$

$$+ \sum_{k=1}^{n} [2y_k y_{n+1} \, \mathrm{Cov}(q_{ik}, q_{i,n+1})]$$

$$= \sum_{k=1}^{n} \left[x_k^2 \, \text{Var}(a_{ik}) + 2 \sum_{l=k+1}^{n} x_k x_l \, \text{Cov}(a_{ik}, a_{il}) \right]$$

$$+ \text{Var}(b_i) - 2 \sum_{k=1}^{n} x_k \, \text{Cov}(a_{ik}, b_i) \tag{11.77}$$

Thus the constraints in Eqs. (11.71) can be restated as

$$P\left[\frac{h_i - \overline{h}_i}{\sqrt{\text{Var}(h_i)}} \leq \frac{-\overline{h}_i}{\sqrt{\text{Var}(h_i)}} \right] \geq p_i, \quad i = 1, 2, \ldots, m \tag{11.78}$$

where $[(h_i - \overline{h}_i)]/\sqrt{\text{Var}(h_i)}$ represents a standard normal variable with a mean value of zero and a variance of 1.

Thus if s_i denotes the value of the standard normal variable at which

$$\phi(s_i) = p_i \tag{11.79}$$

the constraints of Eq. (11.78) can be stated as

$$\phi\left(\frac{-\overline{h}_i}{\sqrt{\text{Var}(h_i)}} \right) \geq \phi(s_i), \quad i = 1, 2, \ldots, m \tag{11.80}$$

These inequalities will be satisfied only if the following deterministic nonlinear inequalities are satisfied:

$$\frac{-\overline{h}_i}{\sqrt{\text{Var}(h_i)}} \geq s_i, \quad i = 1, 2, \ldots, m$$

or

$$\overline{h}_i + s_i \sqrt{\text{Var}(h_i)} \leq 0, \quad i = 1, 2, \ldots, m \tag{11.81}$$

Thus the stochastic linear programming problem of Eqs. (11.64) to (11.66) can be stated as an equivalent deterministic nonlinear programming problem as

$$\text{Minimize } F(\mathbf{X}) = k_1 \sum_{j=1}^{n} \overline{c}_j x_j + k_2 \sqrt{\mathbf{X}^T \mathbf{V} \mathbf{X}}, \quad k_1 \geq 0, \quad k_2 \geq 0,$$

subject to

$$\overline{h}_i + s_i \sqrt{\text{Var}(h_i)} \leq 0, \quad i = 1, 2, \ldots, m$$

$$x_j \geq 0, \quad j = 1, 2, \ldots, n \tag{11.82}$$

Example 11.5 A manufacturing firm produces two machine parts using lathes, milling machines, and grinding machines. If the machining times required, maximum times available, and the unit profits are all assumed to be normally distributed random variables with the following data, find the number of parts to be manufactured per week to maximize the profit. The constraints have to be satisfied with a probability of at least 0.99.

| Type of machine | Machining time required per unit (min) | | | | Maximum time available per week (min) | |
| | Part I | | Part II | | | |
	Mean	Standard deviation	Mean	Standard deviation	Mean	Standard deviation
Lathes	$\bar{a}_{11} = 10$	$\sigma_{a_{11}} = 6$	$\bar{a}_{12} = 5$	$\sigma_{a_{12}} = 4$	$\bar{b}_1 = 2500$	$\sigma_{b_1} = 500$
Milling machines	$\bar{a}_{21} = 4$	$\sigma_{a_{21}} = 4$	$\bar{a}_{22} = 10$	$\sigma_{a_{22}} = 7$	$\bar{b}_2 = 2000$	$\sigma_{b_2} = 400$
Grinding machines	$\bar{a}_{31} = 1$	$\sigma_{a_{31}} = 2$	$\bar{a}_{32} = 1.5$	$\sigma_{a_{32}} = 3$	$\bar{b}_3 = 450$	$\sigma_{b_3} = 50$
Profit per unit	$\bar{c}_1 = 50$	$\sigma_{c_1} = 20$	$\bar{c}_2 = 100$	$\sigma_{c_2} = 50$		

SOLUTION   By defining new random variables h_i as

$$h_i = \sum_{j=1}^{n} a_{ij}x_j - b_i,$$

we find that h_i are also normally distributed. By assuming that there is no correlation between a_{ij}'s and b_i's, the means and variances of h_i can be obtained from Eqs. (11.73) and (11.77) as

$$\bar{h}_1 = \bar{a}_{11}x_1 + \bar{a}_{12}x_2 - \bar{b}_1 = 10x_1 + 5x_2 - 2500$$

$$\bar{h}_2 = \bar{a}_{21}x_1 + \bar{a}_{22}x_2 - \bar{b}_2 = 4x_1 + 10x_2 - 2000$$

$$\bar{h}_3 = \bar{a}_{31}x_1 + \bar{a}_{32}x_2 - \bar{b}_3 = x_1 + 1.5x_2 - 450$$

$$\sigma_{h_1}^2 = x_1^2\sigma_{a_{11}}^2 + x_2^2\sigma_{a_{12}}^2 + \sigma_{b_1}^2 = 36x_1^2 + 16x_2^2 + 250{,}000$$

$$\sigma_{h_2}^2 = x_1^2\sigma_{a_{21}}^2 + x_2^2\sigma_{a_{22}}^2 + \sigma_{b_2}^2 = 16x_1^2 + 49x_2^2 + 160{,}000$$

$$\sigma_{h_3}^2 = x_1^2\sigma_{a_{31}}^2 + x_2^2\sigma_{a_{32}}^2 + \sigma_{b_3}^2 = 4x_1^2 + 9x_2^2 + 2500$$

Assuming that the profits are independent random variables, the covariance matrix of c_j is given by

$$\mathbf{V} = \begin{bmatrix} \text{Var}(c_1) & 0 \\ 0 & \text{Var}(c_2) \end{bmatrix} = \begin{bmatrix} 400 & 0 \\ 0 & 2500 \end{bmatrix}$$

and the variance of the objective function by

$$\text{Var}(f) = \mathbf{X}^T\mathbf{V}\mathbf{X} = 400x_1^2 + 2500x_2^2$$

Thus the objective function can be taken as

$$F = k_1(50x_1 + 100x_2) + k_2\sqrt{400x_1^2 + 2500x_2^2}$$

The constraints can be stated as

$$P[h_i \leq 0] \geq p_i = 0.99, \quad i = 1, 2, 3$$

As the value of the standard normal variate (s_i) corresponding to the probability 0.99 is 2.33 (obtained from Table 11.1), we can state the equivalent deterministic nonlinear optimization problem as follows:

$$\text{Minimize } F = k_1(50x_1 + 100x_2) + k_2\sqrt{400x_1^2 + 2500x_2^2}$$

subject to

$$10x_1 + 5x_2 + 2.33\sqrt{36x_1^2 + 16x_2^2 + 250{,}000} - 2500 \leq 0$$

$$4x_1 + 10x_2 + 2.33\sqrt{16x_1^2 + 49x_2^2 + 160{,}000} - 2000 \leq 0$$

$$x_1 + 1.5x_2 + 2.33\sqrt{4x_1^2 + 9x_2^2 + 2500} - 450 \leq 0$$

$$x_1 \geq 0, \quad x_2 \geq 0$$

This problem can be solved by any of the nonlinear programming techniques once the values of k_1 and k_2 are specified.

11.4   STOCHASTIC NONLINEAR PROGRAMMING

When some of the parameters involved in the objective function and constraints vary about their mean values, a general optimization problem has to be formulated as a stochastic nonlinear programming problem. For the present purpose we assume that all the random variables are independent and follow normal distribution. A stochastic nonlinear programming problem can be stated in standard form as

$$\text{Find } \mathbf{X} \text{ which minimizes } f(\mathbf{Y}) \tag{11.83}$$

subject to

$$P[g_j(\mathbf{Y}) \geq 0] \geq p_j, \quad j = 1, 2, \ldots, m \tag{11.84}$$

where $\mathbf{Y}$ is the vector of N random variables $y_1, y_2, \ldots, y_N$ and it includes the decision variables $x_1, x_2, \ldots, x_n$. The case when $\mathbf{X}$ is deterministic can be obtained as a special case of the present formulation. Equations (11.84) denote that the probability of realizing $g_j(\mathbf{Y})$ greater than or equal to zero must be greater than or equal to the specified probability p_j. The problem stated in Eqs. (11.83) and (11.84) can be converted into an equivalent deterministic nonlinear programming problem by applying the chance constrained programming technique as follows.

11.4.1   Objective Function

The objective function $f(\mathbf{Y})$ can be expanded about the mean values of y_i, $\overline{y}_i$, as

$$f(\mathbf{Y}) = f(\overline{\mathbf{Y}}) + \sum_{i=1}^{N}\left(\frac{\partial f}{\partial y_i}\bigg|_{\overline{\mathbf{Y}}}\right)(y_i - \overline{y}_i) + \text{higher-order derivative terms} \tag{11.85}$$

If the standard deviations of y_i, σ_{yi}, are small, $f(\mathbf{Y})$ can be approximated by the first two terms of Eq. (11.85):

$$f(\mathbf{Y}) \simeq (\overline{\mathbf{Y}}) - \sum_{i=1}^{N} \left(\frac{\partial f}{\partial y_i} \bigg|_{\overline{\mathbf{Y}}} \right) \overline{y}_i + \sum_{i=1}^{N} \left(\frac{\partial f}{\partial y_i} \bigg|_{\overline{\mathbf{Y}}} \right) y_i = \psi(\mathbf{Y}) \qquad (11.86)$$

If all y_i $(i = 1, 2, \ldots, N)$ follow normal distribution, $\psi(\mathbf{Y})$, which is a linear function of $\mathbf{Y}$, also follows normal distribution. The mean and the variance of ψ are given by

$$\overline{\psi} = \psi(\overline{\mathbf{Y}}) \qquad (11.87)$$

$$\text{Var}(\psi) = \sigma_{\psi}^2 = \sum_{i=1}^{N} \left(\frac{\partial f}{\partial y_i} \bigg|_{\overline{\mathbf{Y}}} \right)^2 \sigma_{y_i}^2 \qquad (11.88)$$

since all y_i are independent. For the purpose of optimization, a new objective function $F(\mathbf{Y})$ can be constructed as

$$F(\mathbf{Y}) = k_1 \overline{\psi} + k_2 \sigma_{\psi} \qquad (11.89)$$

where $k_1 \geq 0$ and $k_2 \geq 0$, and their numerical values indicate the relative importance of $\overline{\psi}$ and σ_{ψ} for minimization. Another way of dealing with the standard deviation of ψ is to minimize $\overline{\psi}$ subject to the constraint $\sigma_{\psi} \leq k_3 \overline{\psi}$, where k_3 is a constant, along with the other constraints.

11.4.2 Constraints

If some parameters are random in nature, the constraints will also be probabilistic and one would like to have the probability that a given constraint is satisfied to be greater than a certain value. This is precisely what is stated in Eqs. (11.84) also. The constraint inequality (11.84) can be written as

$$\int_0^{\infty} f_{gj}(g_j) dg_j \geq p_j \qquad (11.90)$$

where $f_{gj}(g_j)$ is the probability density function of the random variable g_j (a function of several random variables is also a random variable) whose range is assumed to be $-\infty$ to ∞. The constraint function $g_j(\mathbf{Y})$ can be expanded around the vector of mean values of the random variables, $\overline{\mathbf{Y}}$, as

$$g_j(\mathbf{Y}) \simeq g_j(\overline{\mathbf{Y}}) + \sum_{i-1}^{N} \left(\frac{\partial g_j}{\partial y_i} \bigg|_{\overline{\mathbf{Y}}} \right) (y_i - \overline{y}_i) \qquad (11.91)$$

From this equation, the mean value, $\overline{g}_j$, and the standard deviation, σ_{gj}, of g_j can be obtained as

$$\overline{g}_j = g_j(\overline{\mathbf{Y}}) \qquad (11.92)$$

$$\sigma_{gj} = \left\{ \sum_{i=1}^{N} \left(\frac{\partial g_j}{\partial y_i} \bigg|_{\overline{\mathbf{Y}}} \right)^2 \sigma_{y_i}^2 \right\}^{1/2} \qquad (11.93)$$

By introducing the new variable

$$\theta = \frac{g_j - \overline{g}_j}{\sigma_{gj}} \tag{11.94}$$

and noting that

$$\int_{-\infty}^{\infty} \frac{1}{\sqrt{2\pi}} e^{-t^2/2} dt = 1 \tag{11.95}$$

Eq. (11.90) can be expressed as

$$\int_{-(\overline{g}_j/\sigma_{gj})}^{\infty} \frac{1}{\sqrt{2\pi}} e^{-\theta^2/2} d\theta \geq \int_{-\phi_j(p_j)}^{\infty} \frac{1}{\sqrt{2\pi}} e^{-t^2/2} dt \tag{11.96}$$

where $\phi_j(p_j)$ is the value of the standard normal variate corresponding to the probability p_j. Thus

$$-\frac{\overline{g}_j}{\sigma_{gj}} \leq -\phi_j(p_j)$$

or

$$-\overline{g}_j + \sigma_{gj}\phi_j(p_j) \leq 0 \tag{11.97}$$

Equation (11.97) can be rewritten as

$$\overline{g}_j - \phi_j(p_j) \left[\sum_{i=1}^{N} \left(\frac{\partial g_j}{\partial y_i} \bigg|_{\overline{\mathbf{Y}}} \right)^2 \sigma_{yi}^2 \right]^{1/2} \geq 0, \quad j = 1, 2, \ldots, m \tag{11.98}$$

Thus the optimization problem of Eqs. (11.83) and (11.84) can be stated in its equivalent deterministic form as: minimize $F(\mathbf{Y})$ given by Eq. (11.89) subject to the m constraints given by Eq. (11.98).

Example 11.6   Design a uniform column of tubular section shown in Fig. 11.5 to carry a compressive load P for minimum cost. The column is made up of a material that has a modulus of elasticity E and density ρ. The length of the column is l. The stress induced in the column should be less than the buckling stress as well as the yield stress. The mean diameter is restricted to lie between 2.0 and 14.0 cm, and columns with thickness outside the range 0.2 to 0.8 cm are not available in the market. The cost of the column includes material costs and construction costs and can be taken as $5W + 2d$, where W is the weight and d is the mean diameter of the column. The constraints have to be satisfied with a probability of at least 0.95.

The following quantities are probabilistic and follow normal distribution with mean and standard deviations as indicated:

Compressive load $= (\overline{P}, \sigma_P) = (2500, 500)\,\text{kg}$
Young's modulus $= (\overline{E}, \sigma_E) = (0.85 \times 10^6, 0.085 \times 10^6)\,\text{kg}_\text{f}/\text{cm}^2$
Density $= (\overline{\rho}, \sigma_\rho) = (0.0025, 0.00025)\,\text{kg}_\text{f}/\text{cm}^3$
Yield stress $= (\overline{f}_y, \sigma_{f_y}) = (500, 50)\,\text{kg}_\text{f}/\text{cm}^2$

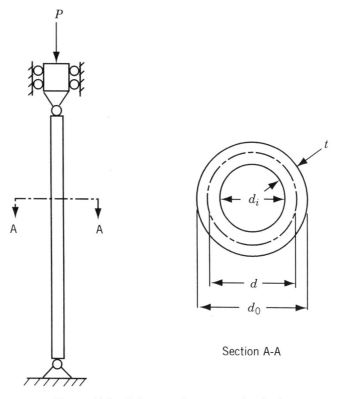

Figure 11.5   Column under compressive load.

Mean diameter of the section $= (\overline{d}, \sigma_d) = (\overline{d}, 0.01\overline{d})$
Column length $= (l, \sigma_l) = (250, 2.5)$ cm

SOLUTION   This problem, by neglecting standard deviations of the various quantities, can be seen to be identical to the one considered in Example 1.1. We will take the design variables as the mean tubular diameter $(\overline{d})$ and the tube thickness (t):

$$\mathbf{X} = \begin{Bmatrix} x_1 \\ x_2 \end{Bmatrix} = \begin{Bmatrix} \overline{d} \\ t \end{Bmatrix}$$

Notice that one of the design variables (d) is probabilistic in this case and we assume that $\overline{d}$ is unknown since σ_d is given in term of $(\overline{d})$. By denoting the vector of random variables as

$$\mathbf{Y} = \begin{Bmatrix} y_1 \\ y_2 \\ y_3 \\ y_4 \\ y_5 \\ y_6 \end{Bmatrix} = \begin{Bmatrix} P \\ E \\ \rho \\ f_y \\ l \\ d \end{Bmatrix}$$

the objective function can be expressed as $f(\mathbf{Y}) = 5W + 2d = 5\rho l \pi\, dt + 2d$. Since

$$\overline{\mathbf{Y}} = \begin{Bmatrix} \overline{P} \\ \overline{E} \\ \overline{\rho} \\ \overline{f}_y \\ \overline{l} \\ \overline{d} \end{Bmatrix} = \begin{Bmatrix} 2500 \\ 0.85 \times 10^6 \\ 0.0025 \\ 500 \\ 250 \\ \overline{d} \end{Bmatrix}$$

$$f(\overline{\mathbf{Y}}) = 5\overline{\rho}\,\overline{l}\pi\,\overline{d}t + 2\overline{d} = 9.8175\overline{d}t + 2\overline{d}$$

$$\left.\frac{\partial f}{\partial y_1}\right|_{\overline{\mathbf{Y}}} = \left.\frac{\partial f}{\partial y_2}\right|_{\overline{\mathbf{Y}}} = \left.\frac{\partial f}{\partial y_4}\right|_{\overline{\mathbf{Y}}} = 0$$

$$\left.\frac{\partial f}{\partial y_3}\right|_{\overline{\mathbf{Y}}} = 5\pi\overline{l}\,\overline{d}t = 3927.0\overline{d}t$$

$$\left.\frac{\partial f}{\partial y_5}\right|_{\overline{\mathbf{Y}}} = 5\pi\overline{\rho}\,\overline{d}t = 0.03927\overline{d}t$$

$$\left.\frac{\partial f}{\partial y_6}\right|_{\overline{\mathbf{Y}}} = 5\pi\overline{\rho}\,\overline{l}t + 2 = 9.8175t + 2.0$$

Equations (11.87) and (11.88) give

$$\psi(\overline{\mathbf{Y}}) = 9.8175\overline{d}t + 2\overline{d} \tag{E_1}$$

$$\sigma_\psi^2 = (3927.0\overline{d}t)^2\sigma_\rho^2 + (0.03927\overline{d}t)^2\sigma_l^2 + (9.8175t + 2.0)^2\sigma_d^2$$
$$= 0.9835\overline{d}^2 t^2 + 0.0004\overline{d}^2 + 0.003927\overline{d}^2 t \tag{E_2}$$

Thus the new objective function for minimization can be expressed as

$$F(\overline{d}, t) = k_1\overline{\psi} + k_2\sigma_\psi$$
$$= k_1(9.8175\overline{d}t + 2\overline{d}) + k_2(0.9835\overline{d}^2 t^2 + 0.0004\overline{d}^2 + 0.003927\overline{d}^2 t)^{1/2} \tag{E_3}$$

where $k_1 \geq 0$ and $k_2 \geq 0$ indicate the relative importances of $\overline{\psi}$ and σ_ψ for minimization. By using the expressions derived in Example 1.1, the constraints can be expressed as

$$P[g_1(\mathbf{Y}) \leq 0] = P\left(\frac{P}{\pi dt} - f_y \leq 0\right) \geq 0.95 \tag{E_4}$$

$$P[g_2(\mathbf{Y}) \leq 0] = P\left[\frac{P}{\pi dt} - \frac{\pi^2 E}{8l^2}(d^2 + t^2) \leq 0\right] \geq 0.95 \tag{E_5}$$

$$P[g_3(\mathbf{Y}) \leq 0] = P[-d + 2.0 \leq 0] \geq 0.95 \tag{E_6}$$

$$P[g_4(\mathbf{Y}) \leq 0] = P[d - 14.0 \leq 0] \geq 0.95 \tag{E_7}$$

$$P[g_5(\mathbf{Y}) \leq 0] = P[-t + 0.2 \leq 0] \geq 0.95 \tag{E_8}$$

$$P[g_6(\mathbf{Y}) \leq 0] = P[t - 0.8 \leq 0] \geq 0.95 \tag{E_9}$$

The mean values of the constraint functions are given by Eq. (11.92) as

$$\bar{g}_1 = \frac{\bar{P}}{\pi \bar{d}t} - \bar{f}_y = \frac{2500}{\pi \bar{d}t} - 500$$

$$\bar{g}_2 = \frac{\bar{P}}{\pi \bar{d}t} - \frac{\pi^2 \bar{E}(\bar{d}^2 + t^2)}{8\bar{l}^2} = \frac{2500}{\pi \bar{d}t} - \frac{\pi^2 (0.85 \times 10^6)(\bar{d}^2 + t^2)}{8(250)^2}$$

$$\bar{g}_3 = -\bar{d} + 2.0$$

$$\bar{g}_4 = \bar{d} - 14.0$$

$$\bar{g}_5 = -t + 0.2$$

$$\bar{g}_6 = t - 0.8$$

The partial derivatives of the constraint functions can be computed as follows:

$$\frac{\partial g_1}{\partial y_2}\bigg|_{\bar{\mathbf{Y}}} = \frac{\partial g_1}{\partial y_3}\bigg|_{\bar{\mathbf{Y}}} = \frac{\partial g_1}{\partial y_5}\bigg|_{\bar{\mathbf{Y}}} = 0$$

$$\frac{\partial g_1}{\partial y_1}\bigg|_{\bar{\mathbf{Y}}} = \frac{1}{\pi \bar{d}t}$$

$$\frac{\partial g_1}{\partial y_4}\bigg|_{\bar{\mathbf{Y}}} = -1$$

$$\frac{\partial g_1}{\partial y_6}\bigg|_{\bar{\mathbf{Y}}} = -\frac{\bar{P}}{\pi \bar{d}^2 t} = -\frac{2500}{\pi \bar{d}^2 t}$$

$$\frac{\partial g_2}{\partial y_3}\bigg|_{\bar{\mathbf{Y}}} = \frac{\partial g_2}{\partial y_4}\bigg|_{\bar{\mathbf{Y}}} = 0$$

$$\frac{\partial g_2}{\partial y_1}\bigg|_{\bar{\mathbf{Y}}} = \frac{1}{\pi \bar{d}t}$$

$$\frac{\partial g_2}{\partial y_2}\bigg|_{\bar{\mathbf{Y}}} = -\frac{\pi^2 (\bar{d}^2 + t^2)}{8\bar{l}^2} = -\frac{\pi^2 (\bar{d}^2 + t^2)}{500,000}$$

$$\frac{\partial g_2}{\partial y_5}\bigg|_{\bar{\mathbf{Y}}} = \frac{\pi^2 \bar{E}(\bar{d}^2 + t^2)}{4\bar{l}^3} = 0.0136\pi^2(\bar{d}^2 + t^2)$$

$$\frac{\partial g_2}{\partial y_6}\bigg|_{\bar{\mathbf{Y}}} = -\frac{\bar{P}}{\pi \bar{d}^2 t} - \frac{\pi^2 \bar{E}(2\bar{d})}{8\bar{l}^2} = -\frac{2500}{\pi \bar{d}^2 t} - \pi^2(3.4)\bar{d}$$

$$\frac{\partial g_3}{\partial y_i}\bigg|_{\bar{\mathbf{Y}}} = 0 \quad \text{for } i = 1 \text{ to } 5$$

$$\frac{\partial g_3}{\partial y_6}\bigg|_{\bar{\mathbf{Y}}} = -1.0$$

$$\frac{\partial g_4}{\partial y_i}\bigg|_{\bar{\mathbf{Y}}} = 0 \quad \text{for } i = 1 \text{ to } 5$$

$$\frac{\partial g_4}{\partial y_6}\bigg|_{\overline{\mathbf{Y}}} = 1.0$$

$$\frac{\partial g_5}{\partial y_i}\bigg|_{\overline{\mathbf{Y}}} = \frac{\partial g_6}{\partial y_i}\bigg|_{\overline{\mathbf{Y}}} = 0 \quad \text{for } i = 1 \text{ to } 6$$

Since the value of the standard normal variate $\phi_j(p_j)$ corresponding to the probability $p_j = 0.95$ is 1.645 (obtained from Table 11.1), the constraints in Eq. (11.98) can be expressed as follows.

For $j = 1^\dagger$:

$$\frac{2500}{\pi \overline{d} t} - 500 - 1.645 \left[\frac{\sigma_P^2}{\pi^2 \overline{d}^2 t^2} + \sigma_{f_y}^2 + \frac{(2500)^2}{\pi^2 \overline{d}^4 t^2} \sigma_d^2 \right]^{1/2} \leq 0$$

$$\frac{795}{\overline{d} t} - 500 - 1.645 \left(\frac{25,320}{\overline{d}^2 t^2} + 2500 + \frac{63.3}{\overline{d}^2 t^2} \right)^{1/2} \leq 0 \tag{E$_{10}$}$$

For $j = 2$:

$$\frac{2500}{\pi \overline{d} t} - 16.78(\overline{d}^2 + t^2) - 1.645 \left[\frac{\sigma_P^2}{\pi^2 \overline{d}^2 t^2} + \frac{\pi^4 (\overline{d}^2 + t^2)^2 \sigma_E^2}{25 \times 10^{10}} \right.$$

$$\left. + (0.0136\pi^2)^2 (\overline{d}^2 + t^2)^2 \sigma_l^2 + \left(\frac{2500}{\pi \overline{d}^2 t} + 3.4\pi^2 \overline{d} \right)^2 \sigma_d^2 \right]^{1/2} \leq 0$$

$$\frac{795}{\overline{d} t} - 16.78(\overline{d}^2 + t^2) - 1.645 \left[\frac{25,320}{\overline{d}^2 t^2} + 2.82(\overline{d}^2 + t^2)^2 \right.$$

$$\left. + 0.113(\overline{d}^2 + t^2)^2 + \frac{63.20}{\overline{d}^2 t^2} + 0.1126\overline{d}^4 + \frac{5.34\overline{d}}{t} \right]^{1/2} \leq 0 \tag{E$_{11}$}$$

For $j = 3$:

$$-\overline{d} + 2.0 - 1.645[(10^{-4})\overline{d}^2]^{1/2} \leq 0$$

$$-1.01645\overline{d} + 2.0 \leq 0 \tag{E$_{12}$}$$

For $j = 4$:

$$\overline{d} - 14.0 - 1.645[(10^{-4})\overline{d}^2]^{1/2} \leq 0$$

$$0.98335\overline{d} - 14.0 \leq 0 \tag{E$_{13}$}$$

For $j = 5$:

$$-t + 0.2 \leq 0 \tag{E$_{14}$}$$

For $j = 6$:

$$t - 0.8 \leq 0 \tag{E$_{15}$}$$

---

†The inequality sign is different from that of Eq. (11.98) due to the fact that the constraints are stated as $P[g_j(\mathbf{Y}) \leq 0] \geq p_j$.

Thus the equivalent deterministic optimization problem can be stated as follows: Minimize $F(\bar{d}, t)$ given by Eq. (E_3) subject to the constraints given by Eqs. (E_{10}) to (E_{15}). The solution of the problem can be found by applying any of the standard nonlinear programming techniques discussed in Chapter 7. In the present case, since the number of design variables is only two, a graphical method can also be used to find the solution.

11.5 STOCHASTIC GEOMETRIC PROGRAMMING

The deterministic geometric programming problem has been considered in Chapter 8. If the constants involved in the posynomials are random variables, the chance-constrained programming methods discussed in Sections 11.3 and 11.4 can be applied to this problem. The probabilistic geometric programming problem can be stated as follows:

Find $\mathbf{X} = \{x_1 x_2 \cdots x_n\}^{\mathrm{T}}$ which minimizes $f(\mathbf{Y})$

subject to (11.99)

$$P[g_j(\mathbf{Y}) > 0] \geq p_j, \quad j = 1, 2, \ldots, m$$

where $\mathbf{Y} = \{y_1, y_2, \ldots, y_N\}^{\mathrm{T}}$ is the vector of N random variables (may include the variables $x_1, x_2, \ldots, x_n$), and $f(\mathbf{Y})$ and $g_j(\mathbf{Y})$, $j = 1, 2, \ldots, m$, are posynomials. By expanding the objective function about the mean values of the random variables y_i, $\bar{y}_i$, and retaining only the first two terms, we can express the mean and variance of $f(\mathbf{Y})$ as in Eqs. (11.87) and (11.88). Thus the new objective function, $F(\mathbf{Y})$, can be expressed as in Eq. (11.89):

$$F(\mathbf{Y}) = k_1 \bar{\psi} + k_2 \sigma_\psi \tag{11.100}$$

The probabilistic constraints of Eq. (11.99) can be converted into deterministic form as in Section 11.4:

$$\bar{g}_j - \phi_j(p_j) \left[\sum_{i=1}^{N} \left(\frac{\partial g_j}{\partial y_i} \bigg|_{\overline{\mathbf{Y}}} \right)^2 \sigma_{y_i}^2 \right]^{1/2} \geq 0, \quad j = 1, 2, \ldots, m \tag{11.101}$$

Thus the optimization problem of Eq. (11.99) can be stated equivalently as follows: Find $\mathbf{Y}$ which minimizes $F(\mathbf{Y})$ given by Eq. (11.100) subject to the constraints of Eq. (11.101). The procedure is illustrated through the following example.

Example 11.7 Design a helical spring for minimum weight subject to a constraint on the shear stress (τ) induced in the spring under a compressive load P.

SOLUTION By selecting the coil diameter (D) and wire diameter (d) of the spring as design variables, we have $x_1 = D$ and $x_2 = d$. The objective function can be stated in deterministic form as [11.14, 11.15]:

$$f(\mathbf{X}) = \frac{\pi^2 d^2 D}{4} (N_c + Q)\rho \tag{E_1}$$

where N_c is the number of active turns, Q the number of inactive turns, and ρ the weight density. Noting that the deflection of the spring $(\bar{\delta})$ is given by

$$\bar{\delta} = \frac{8PC^3 N_c}{Gd} \tag{E_2}$$

where P is the load, $C = D/d$, and G is the shear modulus. By substituting the expression of N_c given by Eq. (E_2) into Eq. (E_1), the objective function can be expressed as

$$f(\mathbf{X}) = \frac{\pi^2 \rho G \bar{\delta}}{32 P} \frac{d^6}{D^2} + \frac{\pi^2 \rho Q}{4} d^2 D \tag{E_3}$$

The yield constraint can be expressed, in deterministic form, as

$$\tau = \frac{8KPC}{\pi d^2} \leq \tau_{\max} \tag{E_4}$$

where $\tau_{\max}$ is the maximum permissible value of shear stress and K the shear stress concentration factor given by (for $2 \leq C \leq 12$):

$$K = \frac{2}{C^{0.25}} \tag{E_5}$$

Using Eq. (E_5), the constraint of Eq. (E_4) can be rewritten as

$$\frac{16P}{\pi \tau_{\max}} \frac{D^{0.75}}{d^{2.75}} < 1 \tag{E_6}$$

By considering the design variables to be normally distributed with $(\bar{d}, \sigma_d) = \bar{d}(1, 0.05)$ and $(\bar{D}, \sigma_D) = \bar{D}(1, 0.05)$, $k_1 = 1$ and $k_2 = 0$ in Eq. (11.100) and using $p_j = 0.95$, the problem [Eqs. (11.100) and (11.101)] can be stated as follows:

$$\text{Minimize } F(\mathbf{Y}) = \frac{0.041 \pi^2 \rho \bar{\delta} G}{P} \frac{\bar{d}^6}{\bar{D}^2} + 0.278 \pi^2 \rho Q \bar{d}^2 \bar{D} \tag{E_7}$$

subject to

$$\frac{12.24P}{\pi \tau_{\max}} \frac{\bar{D}^{0.75}}{\bar{d}^{2.75}} \leq 1 \tag{E_8}$$

The data are assumed as $P = 510N$, $\rho = 78{,}000 \text{ N}/m^3$, $\bar{\delta} = 0.02$ m, $\tau_{\max} = 0.306 \times 10^9 Pa$, and $Q = 2$. The degree of difficulty of the problem can be seen to be zero and the normality and orthogonality conditions yield

$$\delta_1 + \delta_2 = 1$$
$$6\delta_1 + 2\delta_2 - 2.75\delta_3 = 0 \tag{E_9}$$
$$-2\delta_1 + \delta_2 + 0.75\delta_3 = 0$$

The solution of Eqs. (E_9) gives $\delta_1 = 0.81$, $\delta_2 = 0.19$, and $\delta_3 = 1.9$, which corresponds to $\bar{d} = 0.0053$ m, $\bar{D} = 0.0358$ m, and $f_{\min} = 2.266$ N.

REFERENCES AND BIBLIOGRAPHY

11.1   E. Parzen, *Modern Probability Theory and Its Applications*, Wiley, New York, 1960.

11.2   A. H. S. Ang and W. H. Tang, *Probability Concepts in Engineering Planning and Design*, Vol. I, *Basic Principles*, Wiley, New York, 1975.

11.3   S. S. Rao, *Reliability-based Design*, McGraw-Hill, New York, 1992.

11.4   G. B. Dantzig, Linear programming under uncertainty, *Management Science*, Vol. 1, pp. 197–207, 1955.

11.5   A. Charnes and W.W. Cooper, Chance constrained programming, *Management Science*, Vol. 6, pp. 73–79, 1959.

11.6   J. K. Sengupta and K. A. Fox, *Economic Analysis and Operations Research: Optimization Techniques in Quantitative Economic Models*, North-Holland, Amsterdam, 1971.

11.7   R. J. Aguilar, *Systems Analysis and Design in Engineering, Architecture, Construction and Planning*, Prentice-Hall, Englewood Cliffs, NJ, 1973.

11.8   G. L. Nemhauser, *Introduction to Dynamic Programming*, Wiley, New York, 1966.

11.9   A. Kaufmann and R. Cruon, *Dynamic Programming: Sequential Scientific Management*, tr. by H. C. Sneyd, Academic Press, New York, 1967.

11.10   G. E. Thompson, *Linear Programming: An Elementary Introduction*, Macmillan, New York, 1971.

11.11   M. Avriel and D. J. Wilde, Stochastic geometric programming, pp. 73–91 in *Proceedings of the Princeton Symposium on Mathematical Programming*, H. W. Kuhn, Ed., Princeton University Press, Princeton, NJ, 1970.

11.12   M. J. L. Kirby, The current state of chance constrained programming, pp. 93–111 in *Proceedings of the Princeton Symposium on Mathematical Programming*, H. W. Kuhn, Ed., Princeton University Press, Princeton, NJ, 1970.

11.13   W. K. Grassmann, *Stochastic Systems for Management*, North-Holland, New York, 1981.

11.14   J. E. Shigley and C. R. Mischke, *Mechanical Engineering Design*, 5th ed., McGraw-Hill, New York, 1989.

11.15   S. B. L. Beohar and A. C. Rao, Optimum design of helical springs using stochastic geometric programming, pp. 147–151 in *Progress in Engineering Optimization—1981*, R. W. Mayne and K. M. Ragsdell, Eds., ASME, New York, 1981.

11.16   L. L. Howell, S. S. Rao, and A. Midha, Reliability-based optimal design of a bistable compliant mechanism, *ASME Journal of Mechanical Design*, Vol. 116, pp. 1115–1121, 1995.

11.17   R. H. Crawford and S. S. Rao, Probabilistic analysis of function generating mechanisms, *ASME Journal of Mechanisms, Transmissions, and Automation in Design*, Vol. 111, pp. 479–481, 1989.

11.18   S. K. Hati and S. S. Rao, Determination of optimum machining conditions: deterministic and probabilistic approaches, *ASME Journal of Engineering for Industry*, Vol. 98, pp. 354–359, 1976.

11.19   S. S. Rao, Multiobjective optimization in structural design in the presence of uncertain parameters and stochastic process, *AIAA Journal*, Vol. 22, pp. 1670–1678, 1984.

11.20   S. S. Rao, Automated optimum design of wing structures: a probabilistic approach, *Computers and Structures*, Vol. 24, No. 5, pp. 799–808, 1986.

11.21   S. S. Rao, Reliability-based optimization under random vibration environment, *Computers and Structures*, Vol. 14, pp. 345–355, 1981.

11.22 S. S. Rao and C. P. Reddy, Mechanism design by chance constrained programming, *Mechanism and Machine Theory*, Vol. 14, pp. 413–424, 1979.

REVIEW QUESTIONS

11.1 Define the following terms:

(a) Mean

(b) Variance

(c) Standard deviation

(d) Probability

(e) Independent events

(f) Joint density function

(g) Covariance

(h) Central limit theorem

(i) Chance constrained programming

11.2 Match the following terms and descriptions:

(a) Marginal density function	Describes sum of several random variables
(b) Bivariate distribution	Described by probability density function
(c) Normal distribution	Describes one random variable
(d) Discrete distribution	Describes two random variables
(e) Continuous distribution	Described by probability mass function

11.3 Answer true or false:

(a) The uniform distribution can be used to describe only continuous random variables.

(b) The area under the probability density function can have any positive value.

(c) The standard normal variate has zero mean and unit variance.

(d) The magnitude of the correlation coefficient is bounded by one.

(e) Chance constrained programming method can be used to solve only stochastic LP problems.

(f) Chance constrained programming permits violation of constraints to some extent.

(g) Chance constrained programming assumes the random variables to be normally distributed.

(h) The design variables need not be random in a stochastic programming problem.

(i) Chance constrained programming always gives rise to a two-part objective function.

(j) Chance constrained programming converts a stochastic LP problem into a determinstic LP problem.

(k) Chance constrained programming converts a stochastic geometric programming problem into a deterministic geometric programming problem.

(l) The introduction of random variables increases the number of state variables in stochastic dynamic programming.

11.4 Explain the notation $N(\mu, \sigma)$.

11.5   What is a random variable?

11.6   Give two examples of random design parameters.

11.7   What is the difference between probability density and probability distribution functions?

11.8   What is the difference between discrete and continuous random variables?

11.9   How does correlation coefficient relate two random variables?

11.10   Identify possible random variables in a LP problem.

11.11   How do you find the mean and standard deviation of a sum of several random variables?

PROBLEMS

11.1   A contractor plans to use four tractors to work on a project in a remote area. The probability of a tractor functioning for a year without a break-down is known to be 80 %. If X denotes the number of tractors operating at the end of a year, determine the probability mass and distribution functions of X.

11.2   The absolute value of the velocity of a molecule in a perfect gas (V) obeys the Maxwell distribution

$$f_V(v) = \frac{4h^3}{\sqrt{\pi}} v^2 e^{-h^2 v^2}, \quad v \geq 0$$

where $h^2 = (m/2kT)$ is a constant (m is the mass of the molecule, k is Boltzmann's constant, and T is the absolute temperature). Find the mean and the standard deviation of the velocity of a molecule.

11.3   Find the expected value and the standard deviation of the number of tractors operating at the end of one year in Problem 11.1.

11.4   Mass-produced items always show random variation in their dimensions due to small unpredictable and uncontrollable disturbing influences. Suppose that the diameter, X, of the bolts manufactured in a production shop follow the distribution

$$f_X(x) = a(x - 0.9)(1.1 - x) \quad \text{for } 0.9 \leq x \leq 1.1$$
$$0 \qquad\qquad\qquad\qquad \text{elsewhere}$$

Find the values of a, μ_X and σ_X^2.

11.5   **(a)** The voltage V across a constant resistance R is known to fluctuate between 0 and 2 volts. If V follows uniform distribution, what is the distribution of the power expended in the resistance?

   (b) Find the distribution of the instantaneous voltage (V) given by $V = A \cos(\omega t + \phi)$, where A is a constant, ω the frequency, t the time, and ϕ the random phase angle uniformly distributed from 0 to 2π radians.

11.6   The hydraulic head loss (H) in a pipe due to friction is given by the Darcy–Weisbach equation,

$$H = f \frac{L}{2gD} V^2$$

where f is the friction factor, L the length of pipe, V the velocity of flow in pipe, g the acceleration due to gravity, and D the diameter of the pipe. If V follows exponential distribution,

$$f_V(v) = \begin{cases} \dfrac{1}{V_0} e^{-(v/V_0)} & \text{for } v \geq 0 \\ 0 & \text{for } v < 0 \end{cases}$$

where V_0 is the mean velocity, derive the density function for the head loss H.

11.7 The joint density function of two random variables X and Y is given by

$$f_{X,Y}(x, y) = \begin{cases} 3x^2 y + 3y^2 x & \text{for } 0 \leq x \leq 1, \quad 0 \leq y \leq 1 \\ 0 & \text{elsewhere} \end{cases}$$

Find the marginal density functions of X and Y.

11.8 Steel rods, manufactured with a nominal diameter of 3 cm, are considered acceptable if the diameter falls within the limits of 2.99 and 3.01 cm. It is observed that about 5 % are rejected oversize and 5 % are rejected undersize. Assuming that the diameters are normally distributed, find the standard deviation of the distribution. Compute the proportion of rejects if the permissible limits are changed to 2.985 and 3.015 cm.

11.9 Determine whether the random variables X and Y are dependent or independent when their joint density function is given by

$$f_{X,Y}(x, y) = \begin{cases} 4xy & \text{for } 0 \leq x \leq 1, \quad 0 \leq y \leq 1 \\ 0 & \text{elsewhere} \end{cases}$$

11.10 Determine whether the random variables X and Y are dependent or independent when their joint density function is given by

$$f_{X,Y}(x, y) = \begin{cases} \dfrac{1}{4\pi^2}[1 - \sin(x + y)] & \text{for } -\pi \leq x \leq \pi, \quad -\pi \leq y \leq \pi \\ 0 & \text{elsewhere} \end{cases}$$

11.11 The stress level at which steel yields (X) has been found to follow normal distribution. For a particular batch of steel, the mean and standard deviation of X are found to be 4000 and 300 kg$_f$/cm^2, respectively. Find

(a) The probability that a steel bar taken from this batch will have a yield stress between 3000 and 5000 kg$_f$/cm^2
(b) The probability that the yield stress will exceed 4500 kg$_f$/cm^2
(c) The value of X at which the distribution function has a value of 0.10

11.12 An automobile body is assembled using a large number of spot welds. The number of defective welds (X) closely follows the distribution

$$P(X = d) = \frac{e^{-2} 2^d}{d!}, \quad d = 0, 1, 2, \ldots$$

Find the probability that the number of defective welds is less than or equal to 2.

11.13 The range (R) of a projectile is given by

$$R = \frac{V_0^2}{g} \sin 2\phi$$

where V_0 is the initial velocity of the projectile, g the acceleration due to gravity, and ϕ the angle from the horizontal as shown in Fig. 11.6. If the mean and standard deviations of V_0 and ϕ are given by $\overline{V}_0 = 100$ ft/s, $\sigma_{V_0} = 10$ ft/s, $\overline{\phi} = 30°$, and $\sigma_\phi = 3°$, find the first-order mean and standard deviation of the range R, assuming that V_0 and ϕ are statistically independent. Evaluate also the second-order mean range. Assume that $g = 32.2$ ft/s^2.

11.14   Maximize $f = 4x_1 + 2x_2 + 3x_3 + c_4 x_4$

subject to

$$x_1 + x_3 + x_4 \leq 24$$

$$3x_1 + x_2 + 2x_3 + 4x_4 \leq 48$$

$$2x_1 + 2x_2 + 3x_3 + 2x_4 \leq 36$$

$$x_i \geq 0, \quad i = 1 \text{ to } 4$$

where c_4 is a discrete random variable that can take values of 4, 5, 6, or 7 with probabilities of 0.1, 0.2, 0.3, and 0.4, respectively. Using the simplex method, find the solution that maximizes the expected value of f.

11.15 Find the solution of Problem 11.14 if the objective is to maximize the variance of f.

11.16 A manufacturing firm can produce 1, 2, or 3 units of a product in a month, but the demand is uncertain. The demand is a discrete random variable that can take a value of 1, 2, or 3 with probabilities 0.2, 0.2, and 0.6, respectively. If the unit cost of production is \$400, unit revenue is \$1000, and unit cost of unfulfilled demand is \$0, determine the output that maximizes the expected total profit.

11.17 A factory manufactures products A, B, and C. Each of these products is processed through three different production stages. The times required to manufacture 1 unit of each of the three products at different stages and the daily capacity of the stages are probabilistic with means and standard deviations as indicated below.

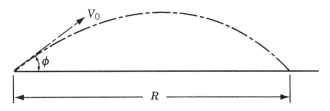

Figure 11.6 Range of a projectile.

Stage	Time per unit (min) for product:						Stage capacity (mins/day)	
	A		B		C			
	Mean	Standard deviation	Mean	Standard deviation	Mean	Standard deviation	Mean	Standard deviation
1	4	1	8	3	4	4	1720	172
2	12	2	0	0	8	2	1840	276
3	4	2	16	4	0	0	1680	336

The profit per unit is also a random variable with the following data:

	Profit($)	
Product	Mean	Standard deviation
A	6	2
B	4	1
C	10	3

Assuming that all amounts produced are absorbed by the market, determine the daily number of units to be manufactured of each product for the following cases.

(a) The objective is to maximize the expected profit.

(b) The objective is to maximize the standard deviation of the profit.

(c) The objective is to maximize the sum of expected profit and the standard deviation of the profit.

Assume that all the random variables follow normal distribution and the constraints have to be satisfied with a probability of 0.95.

11.18 In a belt-and-pulley drive, the belt embraces the shorter pulley $165°$ and runs over it at a mean speed of 1700 m/min with a standard deviation of 51 m/min. The density of the belt has a mean value of 1 g/cm^3 and a standard deviation of 0.05 g/cm^3. The mean and standard deviations of the permissible stress in the belt are 25 and 2.5 kg$_f$/cm^2, respectively. The coefficient of friction (μ) between the belt and the pulley is given by $\bar{\mu} = 0.25$ and $\sigma_\mu = 0.05$. Assuming a coefficient of variation of 0.02 for the belt dimensions, find the width and thickness of the belt to maximize the mean horsepower transmitted. The minimum permissible values for the width and the thickness of the belt are 10.0 and 0.5 cm, respectively. Assume that all the random variables follow normal distribution and the constraints have to be satisfied with a minimum probability of 0.95. *Hint:* Horsepower transmitted = $(T_1 - T_2)v/75$, where T_1 and T_2 are the tensions on the tight side and slack sides of the belt in kg$_f$ and v is the linear velocity of the belt in m/s:

$$T_1 = T_{\max} - T_c = T_{\max} - \frac{wv^2}{g} \text{ and } \frac{T_1}{T_2} = e^{\mu\theta}$$

where $T_{\max}$ is the maximum permissible tension, T_c the centrifugal tension, w the weight of the belt per meter length, g the acceleration due to gravity in m/s, and θ the angle of contact between the belt and the pulley.

11.19 An article is to be restocked every three months in a year. The quarterly demand U is random and its law of probability in any of the quarters is as given below:

U	Probability mass function, $P_U(u)$
0	0.2
1	0.3
2	0.4
3	0.1
> 3	0.0

The cost of stocking an article for a unit of time is 4, and when the stock is exhausted, there is a scarcity charge of 12. The orders that are not satisfied are lost, in other words, are not carried forward to the next period. Further, the stock cannot exceed three articles, owing to the restrictions on space. Find the annual policy of restocking the article so as to minimize the expected value of the sum of the cost of stocking and of the scarcity charge.

11.20 A close-coiled helical spring, made up of a circular wire of diameter d, is to be designed to carry a compressive load P. The permissible shear stress is σ_{max} and the permissible deflection is δ_{max}. The number of active turns of the spring is n and the solid height of the spring has to be greater than h. Formulate the problem of minimizing the volume of the material so as to satisfy the constraints with a minimum probability of p. Take the mean diameter of the coils (D) and the diameter of the wire (d) as design variables. Assume d, D, P, σ_{max}, δ_{max}, h, and the shear modulus of the material, G, to be normally distributed random variables. The coefficient of variation of d and D is k. The maximum shear stress, σ, induced in the spring is given by

$$\sigma = \frac{8PDK}{\pi d^3}$$

where K is the Wahl's stress factor defined by

$$K = \frac{4D - d}{4(D - d)} + \frac{0.615d}{D}$$

and the deflection (δ) by

$$\delta = \frac{8PD^3n}{Gd^4}$$

Formulate the optimization problem for the following data:

$$G = N(840,000, 84,000) \text{ kg}_f/\text{cm}^2, \quad \delta_{max} = N(2, 0.1) \text{ cm},$$

$$\sigma_{max} = N(3000, 150) \text{ kg}_f/\text{cm}^2,$$

$$P = N(12, 3) \text{ kg}_f, \quad n = 8, \quad h = N(2.0, 0.4) \text{ cm}, \quad k = 0.05,$$

$$p = 0.99$$

11.21 Solve Problem 11.20 using a graphical technique.

12

Optimal Control and Optimality Criteria Methods

12.1 INTRODUCTION

In this chapter we give a brief introduction to the following techniques of optimization:

1. Calculus of variations
2. Optimal control theory
3. Optimality criteria methods

If an optimization problem involves the minimization (or maximization) of a functional subject to the constraints of the same type, the decision variable will not be a number, but it will be a function. The calculus of variations can be used to solve this type of optimization problems. An optimization problem that is closely related to the calculus of variations problem is the optimal control problem. An optimal control problem involves two types of variables: the control and state variables, which are related to each other by a set of differential equations. Optimal control theory can be used for solving such problems. In some optimization problems, especially those related to structural design, the necessary conditions of optimality, for specialized design conditions, are used to develop efficient iterative techniques to find the optimum solution. Such techniques are known as *optimality criteria methods*.

12.2 CALCULUS OF VARIATIONS

12.2.1 Introduction

The calculus of variations is concerned with the determination of extrema (maxima and minima) or stationary values of functionals. A *functional* can be defined as a function of several other functions. Hence the calculus of variations can be used to solve trajectory optimization problems.[†] The subject of calculus of variations is almost as old as the calculus itself. The foundations of this subject were laid down by Bernoulli brothers and later important contributions were made by Euler, Lagrange, Weirstrass, Hamilton, and Bolzane. The calculus of variations is a powerful method for the solution of problems in several fields, such as statics and dynamics of rigid bodies, general elasticity, vibrations, optics, and optimization of orbits and controls. We shall see some of the fundamental concepts of calculus of variations in this section.

[†]See Section 1.5 for the definition of a trajectory optimization problem.

12.2.2 Problem of Calculus of Variations

A simple problem in the theory of the calculus of variations with no constraints can be stated as follows:

Find a function $u(x)$ that minimizes the functional (integral)

$$A = \int_{x_1}^{x_2} F(x, u, u', u'') \, dx \tag{12.1}$$

where A and F can be called functionals (functions of other functions). Here x is the independent variable,

$$u = u(x), \quad u' = \frac{du(x)}{dx}, \quad \text{and} \quad u'' = \frac{d^2 u(x)}{dx^2}$$

In mechanics, the functional usually possesses a clear physical meaning. For example, in the mechanics of deformable solids, the potential energy (π) plays the role of the functional (π is a function of the displacement components u, v, and w, which, in turn, are functions of the coordinates x, y, and z).

The integral in Eq. (12.1) is defined in the region or domain $[x_1, x_2]$. Let the values of u be prescribed on the boundaries as $u(x_1) = u_1$ and $u(x_2) = u_2$. These are called the *boundary conditions* of the problem. One of the procedures that can be used to solve the problem in Eq. (12.1) will be as follows:

1. Select a series of trial or tentative solutions $u(x)$ for the given problem and express the functional A in terms of each of the tentative solutions.
2. Compare the values of A given by the different tentative solutions.
3. Find the correct solution to the problem as that particular tentative solution which makes the functional A assume an extreme or stationary value.

The mathematical procedure used to select the correct solution from a number of tentative solutions is called the calculus of variations.

Stationary Values of Functionals. Any tentative solution $\bar{u}(x)$ in the neighborhood of the exact solution $u(x)$ may be represented as (Fig. 12.1)

$$\underset{\substack{\text{tentative} \\ \text{solution}}}{\bar{u}(x)} = \underset{\substack{\text{exact} \\ \text{solution}}}{u(x)} + \underset{\substack{\text{variation} \\ \text{of } u}}{\delta u(x)} \tag{12.2}$$

The variation in u (i.e., δu) is defined as an infinitesimal, arbitrary change in u for a fixed value of the variable x (i.e., for $\delta x = 0$). Here δ is called the *variational operator* (similar to the differential operator d). The operation of variation is commutative with both integration and differentiation, that is,

$$\delta \left(\int F \, dx \right) = \int (\delta F) \, dx \tag{12.3}$$

$$\delta \left(\frac{du}{dx} \right) = \frac{d}{dx} (\delta u) \tag{12.4}$$

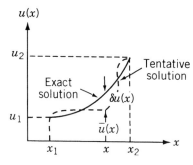

Figure 12.1   Tentative and exact solutions.

Also, we define the variation of a function of several variables or a functional in a manner similar to the calculus definition of a total differential:

$$\delta F = \frac{\partial F}{\partial u}\delta u + \frac{\partial F}{\partial u'}\delta u' + \frac{\partial F}{\partial u''}\delta u'' + \frac{\partial F}{\partial x}\underset{\underset{0}{\uparrow}}{\delta x} \tag{12.5}$$

(since we are finding variation of F for a fixed value of x, i.e., $\delta x = 0$).

Now, let us consider the variation in $A(\delta A)$ corresponding to variations in the solution (δu). If we want the condition for the stationariness of A, we take the necessary condition as the vanishing of first derivative of A (similar to maximization or minimization of simple functions in ordinary calculus).

$$\delta A = \int_{x_1}^{x_2}\left(\frac{\partial F}{\partial x}\delta u + \frac{\partial F}{\partial u'}\delta u' + \frac{\partial F}{\partial u''}\delta u''\right)dx = \int_{x_1}^{x_2}\delta F\ dx = 0 \tag{12.6}$$

Integrate the second and third terms by parts to obtain

$$\int_{x_1}^{x_2}\frac{\partial F}{\partial u'}\delta u'\ dx = \int_{x_1}^{x_2}\frac{\partial F}{\partial u'}\delta\left(\frac{\partial u}{\partial x}\right)dx = \int_{x_1}^{x_2}\frac{\partial F}{\partial u'}\frac{\partial}{\partial x}(\delta u)\ dx$$

$$= \frac{\partial F}{\partial u'}\delta u\Big|_{x_1}^{x_2} - \int_{x_1}^{x_2}\frac{d}{dx}\left(\frac{\partial F}{\partial u'}\right)\delta u\ dx \tag{12.7}$$

$$\int_{x_1}^{x_2}\frac{\partial F}{\partial u''}\delta u''\ dx = \int_{x_1}^{x_2}\frac{\partial F}{\partial u''}\frac{\partial}{\partial x}(\delta u')dx = \frac{\partial F}{\partial u''}\delta u'\Big|_{x_1}^{x_2}$$

$$- \int_{x_1}^{x_2}\frac{d}{dx}\left(\frac{\partial F}{\partial u''}\right)\delta u'\ dx$$

$$= \frac{\partial F}{\partial u''}\delta u'\Big|_{x_1}^{x_2} - \frac{d}{dx}\left(\frac{\partial F}{\partial u''}\right)\delta u\Big|_{x_1}^{x_2}$$

$$+ \int_{x_1}^{x_2}\frac{d^2}{dx^2}\left(\frac{\partial F}{\partial u''}\right)\delta u\ dx \tag{12.8}$$

Thus

$$\delta A = \int_{x_1}^{x_2} \left[\frac{\partial F}{\partial u} - \frac{d}{dx}\left(\frac{\partial F}{\partial u'}\right) + \frac{d^2}{dx^2}\left(\frac{\partial F}{\partial u''}\right) \right] \delta u \ dx$$

$$+ \left[\frac{\partial F}{\partial u'} - \frac{d}{dx}\left(\frac{\partial F}{\partial u''}\right) \right] \delta u \Big|_{x_1}^{x_2} + \left[\left(\frac{\partial F}{\partial u''}\right) \delta u' \right] \Big|_{x_1}^{x_2} = 0 \qquad (12.9)$$

Since δu is arbitrary, each term must vanish individually:

$$\frac{\partial F}{\partial u} - \frac{d}{dx}\left(\frac{\partial F}{\partial u'}\right) + \frac{d^2}{dx^2}\left(\frac{\partial F}{\partial u''}\right) = 0 \qquad (12.10)$$

$$\left[\frac{\partial F}{\partial u'} - \frac{d}{dx}\left(\frac{\partial F}{\partial u''}\right) \right] \delta u \Big|_{x_1}^{x_2} = 0 \qquad (12.11)$$

$$\frac{\partial F}{\partial u''} \delta u' \Big|_{x_1}^{x_2} = 0 \qquad (12.12)$$

Equation (12.10) will be the governing differential equation for the given problem and is called Euler equation or Euler–Lagrange equation. Equations (12.11) and (12.12) give the boundary conditions.

The conditions

$$\left[\frac{\partial F}{\partial u'} - \frac{d}{dx}\left(\frac{\partial F}{\partial u''}\right) \right] \Big|_{x_1}^{x_2} = 0 \qquad (12.13)$$

$$\frac{\partial F}{\partial u''} \Big|_{x_1}^{x_2} = 0 \qquad (12.14)$$

are called *natural* boundary conditions (if they are satisfied, they are called *free* boundary conditions). If the natural boundary conditions are not satisfied, we should have

$$\delta u(x_1) = 0, \qquad \delta u(x_2) = 0 \qquad (12.15)$$

$$\delta u'(x_1) = 0, \qquad \delta u'(x_2) = 0 \qquad (12.16)$$

in order to satisfy Eqs. (12.11) and (12.12). These are called *geometric* or *forced* boundary conditions.

Example 12.1 Brachistochrone Problem In June 1696, Johann Bernoulli set the following problem before the scholars of his time. "Given two points A and B in a vertical plane, find the path from A to B along which a particle of mass m will slide under the force of gravity, without friction, in the shortest time" (Fig. 12.2). The term *brachistochrone* derives from the Greek *brachistos* (shortest) and *chronos* (time).

If s is the distance along the path and v the velocity, we have

$$v = \frac{ds}{dt} = \frac{(dx^2 + dy^2)^{1/2}}{dt} = \frac{[1 + (y')^2]^{1/2}}{dt} dx$$

$$dt = \frac{1}{v}[1 + (y')^2]^{1/2} \ dx$$

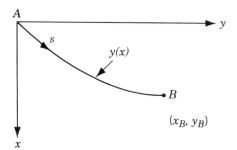

Figure 12.2 Curve of minimum time of descent.

Since potential energy is converted to kinetic energy as the particle moves down the path, we can write

$$\tfrac{1}{2}mv^2 = mgx$$

Hence

$$dt = \left[\frac{1 + (y')^2}{2gx}\right]^{1/2} dx \tag{E_1}$$

and the integral to be stationary is

$$t = \int_0^{x_B} \left[\frac{1 + (y')^2}{2gx}\right]^{1/2} dx \tag{E_2}$$

The integrand is a function of x and y' and so is a special case of Eq. (12.1). Using the Euler–Lagrange equation,

$$\frac{d}{dx}\left(\frac{\partial F}{\partial y'}\right) - \frac{\partial F}{\partial y} = 0 \quad \text{with } F = \left[\frac{1 + (y')^2}{2gx}\right]^{1/2}$$

we obtain

$$\frac{d}{dx}\left(\frac{y'}{\{x[1 + (y')^2]\}^{1/2}}\right) = 0$$

Integrating yields

$$y' = \frac{dy}{dx} = \left(\frac{C_1 x}{1 - C_1 x}\right)^{1/2} \tag{E_3}$$

where C_1 is a constant of integration. The ordinary differential equation (E_3) yields on integration the solution to the problem as

$$y(x) = C_1 \sin^{-1}(x/C_1) - (2C_1 x - x^2)^{1/2} + C_2 \tag{E_4}$$

Example 12.2 Design of a Solid Body of Revolution for Minimum Drag Next we consider the problem of determining the shape of a solid body of revolution for minimum drag. In the general case, the forces exerted on a solid body translating in a fluid depend on the shape of the body and the relative velocity in a very complex manner. However, if the density of the fluid is sufficiently small, the normal pressure (p) acting on the solid body can be approximately taken as [12.3]

$$p = 2\rho v^2 \, \sin^2 \theta \tag{E_1}$$

where ρ is the density of the fluid, v the velocity of the fluid relative to the solid body, and θ the angle between the direction of the velocity of the fluid and the tangent to the surface as shown in Fig. 12.3.

Since the pressure (p) acts normal to the surface, the x-component of the force acting on the surface of a slice of length dx and radius $y(x)$ shown in Fig. 12.4 can be written as

$$dP = (\text{normal pressure})(\text{surface area}) \sin\theta$$

$$= (2\rho v^2 \, \sin^2\theta)(2\pi y \, \sqrt{1 + (y')^2} \, dx) \, \sin\theta \tag{E_2}$$

where $y' = dy/dx$. The total drag force, P, is given by the integral of Eq. (E$_2$) as

$$P = \int_0^L 4\pi\rho v^2 y \, \sin^3\theta \sqrt{1 + (y')^2} \, dx \tag{E_3}$$

where L is the length of the body. To simplify the calculations, we assume that $y' \ll 1$ so that

$$\sin\,\theta = \frac{y'}{\sqrt{1 + (y')^2}} \simeq y' \tag{E_4}$$

Thus Eq. (E$_3$) can be approximated as

$$P = 4\pi\rho v^2 \int_0^L (y')^3 y \, dx \tag{E_5}$$

Now the minimum drag problem can be stated as follows.

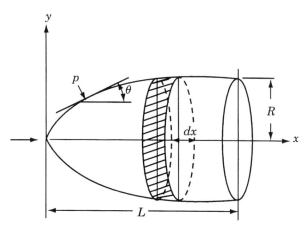

Figure 12.3   Solid body of revolution translating in a fluid medium.

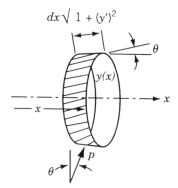

$dx\sqrt{1+(y')^2}$

θ

$y(x)$

x

x

p

θ

Figure 12.4 Element of surface area acted on by the pressure p.

Find $y(x)$ which minimizes the drag P given by Eq. (E$_5$) subject to the condition that $y(x)$ satisfies the end conditions

$$y(x = 0) = 0 \quad \text{and} \quad y(x = L) = R \tag{E$_6$}$$

By comparing the functional P of Eq. (E$_5$) with A of Eq. (12.1), we find that

$$F(x, y, y', y'') = 4\pi\rho v^2 (y')^3 y \tag{E$_7$}$$

The Euler–Lagrange equation, Eq. (12.10), corresponding to this functional can be obtained as

$$(y')^3 - 3\frac{d}{dx}[y(y')^2] = 0 \tag{E$_8$}$$

The boundary conditions, Eqs. (12.11) and (12.12), reduce to

$$[3y(y')^2]\delta y\Big|_{x_1=0}^{x_2=L} = 0 \tag{E$_9$}$$

Equation (E$_8$) can be written as

$$(y')^3 - 3[y'(y')^2 + y(2)y'y''] = 0$$

or

$$(y')^3 + 3yy'y'' = 0 \tag{E$_{10}$}$$

This equation, when integrated once, gives

$$y(y')^3 = k_1^3 \tag{E$_{11}$}$$

where k_1^3 is a constant of integration. Integrating Eq. (E$_{11}$), we obtain

$$y(x) = (k_1 x + k_2)^{3/4} \tag{E$_{12}$}$$

The application of the boundary conditions, Eqs. (E$_6$), gives the values of the constants as

$$k_1 = \frac{R^{4/3}}{L} \quad \text{and} \quad k_2 = 0$$

Hence the shape of the solid body having minimum drag is given by the equation

$$y(x) = R \left(\frac{x}{L} \right)^{3/4}$$

12.2.3   Lagrange Multipliers and Constraints

If the variable x is not completely independent but has to satisfy some condition(s) of constraint, the problem can be stated as follows:

Find the function $y(x)$ such that the integral

$$A = \int_{x_1}^{x_2} F \left(x, y, \frac{dy}{dx} \right) dx \rightarrow \text{minimum}$$

subject to the constraint   (12.17)

$$g \left(x, y, \frac{dy}{dx} \right) = 0$$

where g may be an integral function. The stationary value of a constrained calculus of variations problem can be found by the use of Lagrange multipliers. To illustrate the method, let us consider a problem known as isoperimetric problem given below.

Example 12.3   Optimum Design of a Cooling Fin   Cooling fins are used on radiators to increase the rate of heat transfer from a hot surface (wall) to the surrounding fluid. Often, we will be interested in finding the optimum tapering of a fin (of rectangular cross section) of specified total mass which transfers the maximum heat energy.

The configuration of the fin is shown in Fig. 12.5. If T_0 and T_∞ denote the wall and the ambient temperatures, respectively, the temperature of the fin at any point, $T(x)$, can be nondimensionalized as

$$t(x) = \frac{T(x) - T_\infty}{T_0 - T_\infty}$$ (E_1)

so that $t(0) = 1$ and $t(\infty) = 0$.

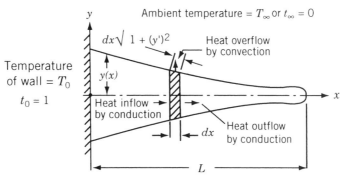

Figure 12.5   Geometry of a cooling fin.

To formulate the problem, we first write the heat balance equation for an elemental length, dx, of the fin:

heat inflow by conduction $=$ heat outflow by conduction and convection

that is,

$$\left(-kA\frac{dt}{dx}\right)_x = \left(-kA\frac{dt}{dx}\right)_{x+dx} + hS\,(t - t_\infty) \tag{E_2}$$

where k is the thermal conductivity, A the cross-sectional area of the fin $= 2y(x)$ per unit width of the fin, h the heat transfer coefficient, S the surface area of the fin element $= 2\sqrt{1 + (y')^2}\,dx$ per unit width, and $2y(x)$ the depth of the fin at any section x. By writing

$$\left(-kA\frac{dt}{dx}\right)_{x+dx} = \left(-kA\frac{dt}{dx}\right)_x + \frac{d}{dx}\left(-kA\frac{dt}{dx}\right)dx \tag{E_3}$$

and noting that $t_\infty = 0$, we can simplify Eq. (E$_2$) as

$$\frac{d}{dx}\left(ky\frac{dt}{dx}\right) = ht\,\sqrt{1 + (y')^2} \tag{E_4}$$

Assuming that $y' \ll 1$ for simplicity, this equation can be written as

$$k\frac{d}{dx}\left(y\frac{dt}{dx}\right) = ht \tag{E_5}$$

The amount of heat dissipated from the fin to the surroundings per unit time is given by

$$H = 2\int_0^L ht\,dx \tag{E_6}$$

by assuming that the heat flow from the free end of the fin is zero. Since the mass of the fin is specified as m, we have

$$2\int_0^L \rho y\,dx - m = 0 \tag{E_7}$$

where ρ is the density of fin.

Now the problem can be stated as follows: Find $t(x)$ that maximizes the integral in Eq. (E$_6$) subject to the constraint equation (E$_7$). Since $y(x)$ in Eq. (E$_7$) is also not known, it can be expressed in terms of $t(x)$ using the heat balance equation (E$_5$). By integrating Eq. (E$_5$) between the limits x and L, we obtain

$$-ky(x)\frac{dt}{dx}(x) = h\int_x^L t(x)\,dx \tag{E_8}$$

by assuming the heat flow from the free end to be zero. Equation (E$_8$) gives

$$y(x) = -\frac{h}{k}\frac{1}{dt/dx}\int_x^L t(x)\,dx \tag{E_9}$$

By substituting Eq. (E$_9$) in (E$_7$), the variational problem can be restated as

Find $y(x)$ which maximizes

$$H = 2h \int_0^L t(x)\, dx \qquad (E_{10})$$

subject to the constraint

$$g(x, t, t') = 2\rho \frac{h}{k} \int_0^L \frac{1}{dt/dx} \left[\int_x^L t(x)\, dx \right] dx + m = 0 \qquad (E_{11})$$

This problem can be solved by using the Lagrange multiplier method. The functional I to be extremized is given by

$$I = \int_0^L (H + \lambda g)\, dx = 2h \int_0^L \left[t(x) + \frac{\lambda \rho}{k} \frac{1}{dt/dx} \int_x^L t(x)\, dx \right] dx \qquad (E_{12})$$

where λ is the Lagrange multiplier.

By comparing Eq. (E$_{12}$) with Eq. (12.1) we find that

$$F(x, t, t') = 2ht + \frac{2h\lambda\rho}{k} \frac{1}{t'} \int_x^L t(x)\, dx \qquad (E_{13})$$

The Euler–Lagrange equation, Eq. (12.10), gives

$$h - \frac{\lambda h \rho}{k} \left[\frac{2t''}{(t')^3} \int_x^L t(x)\, dx + \frac{t(x)}{(t')^2} - \int_0^x \frac{dx}{t'} \right] = 0 \qquad (E_{14})$$

This integrodifferential equation has to be solved to find the solution $t(x)$. In this case we can verify that

$$t(x) = 1 - x \left(\frac{\lambda\rho}{k} \right)^{1/2} \qquad (E_{15})$$

satisfies Eq. (E$_{14}$). The thickness profile of the fin can be obtained from Eq. (E$_9$) as

$$
\begin{aligned}
y(x) &= -\frac{h}{k} \frac{1}{t'} \int_x^L t(x)\, dx = \frac{h}{k} \left(\frac{k}{\lambda\rho} \right)^{1/2} \int_x^L \left[1 - \left(\frac{\lambda\rho}{k} \right)^{1/2} x \right] dx \\
&= \frac{h}{(k\lambda\rho)^{1/2}} \left[L - \left(\frac{\lambda\rho}{k} \right)^{1/2} \frac{L^2}{2} - x + \left(\frac{\lambda\rho}{k} \right)^{1/2} \frac{x^2}{2} \right] \\
&= c_1 + c_2 x + c_3 x^2 \qquad (E_{16})
\end{aligned}
$$

where

$$c_1 = \frac{h}{(k\lambda\rho)^{1/2}} \left[L - \left(\frac{\lambda\rho}{k} \right)^{1/2} \frac{L^2}{2} \right] \qquad (E_{17})$$

$$c_2 = -\frac{h}{(k\lambda\rho)^{1/2}} \qquad (E_{18})$$

$$c_3 = \frac{h}{2(k\lambda\rho)^{1/2}} \left(\frac{\lambda\rho}{k} \right)^{1/2} = \frac{h}{2k} \qquad (E_{19})$$

The value of the unknown constant λ can be found by using Eq. (E$_7$) as

$$m = 2\rho \int_0^L y(x) \, dx = 2\rho \left(c_1 L + c_2 \frac{L^2}{2} + c_3 \frac{L^3}{3} \right)$$

that is,

$$\frac{m}{2\rho L} = c_1 + c_2 \frac{L}{2} + c_3 \frac{L^2}{3} = \frac{hL}{2(k\rho\lambda)^{1/2}} - \frac{1}{3} \frac{hL^2}{k} \tag{E$_{20}$}$$

Equation (E$_{20}$) gives

$$\lambda^{1/2} = \frac{hL}{(k\rho)^{1/2}} \frac{1}{(m/\rho L) + \frac{2}{3}(hL^2/k)} \tag{E$_{21}$}$$

Hence the desired solution can be obtained by substituting Eq. (E$_{21}$) in Eq. (E$_{16}$).

12.2.4   Generalization

The concept of including constraints can be generalized as follows. Let the problem be to find the functions $u_1(x, y, z), u_2(x, y, z), \ldots, u_n(x, y, z)$ that make the functional

$$\int_V f\left(x, y, z, u_1, u_2, \ldots, u_n, \frac{\partial u_1}{\partial x}, \ldots\right) dV \tag{12.18}$$

stationary subject to the m constraints

$$g_1\left(x, y, z, u_1, u_2, \ldots, u_n, \frac{\partial u_1}{\partial x}, \ldots\right) = 0$$

$$\vdots \tag{12.19}$$

$$g_m\left(x, y, z, u_1, u_2, \ldots, u_n, \frac{\partial u_1}{\partial x}, \ldots\right) = 0$$

The Lagrange multiplier method consists in taking variations in the functional

$$A = \int_V (f + \lambda_1 g_1 + \lambda_2 g_2 + \cdots + \lambda_m g_m) \, dV \tag{12.20}$$

where λ_i are now functions of position. In the special case where one or more of the g_i are integral conditions, the associated λ_i are constants.

12.3   OPTIMAL CONTROL THEORY

The basic optimal control problem can be stated as follows:

$$\text{Find the control vector } \mathbf{u} = \begin{Bmatrix} u_1 \\ u_2 \\ \vdots \\ u_m \end{Bmatrix}$$

which minimizes the functional, called the *performance index*,

$$J = \int_0^T f_0(\mathbf{x}, \mathbf{u}, t) \, dt \tag{12.21}$$

where

$$\mathbf{x} = \begin{Bmatrix} x_1 \\ x_2 \\ \vdots \\ x_n \end{Bmatrix}$$

is called the *state vector*, t the time parameter, T the terminal time, and f_0 is a function of $\mathbf{x}$, $\mathbf{u}$, and t. The state variables x_i and the control variables u_i are related as

$$\frac{dx_i}{dt} = f_i(x_1, x_2, \ldots, x_n; u_1, u_2, \ldots, u_m; t), \quad i = 1, 2, \ldots, n$$

or

$$\dot{\mathbf{x}} = \mathbf{f}(\mathbf{x}, \mathbf{u}, t) \tag{12.22}$$

In many problems, the system is linear and Eq. (12.22) can be stated as

$$\dot{\mathbf{x}} = [A]\mathbf{x} + [B]\mathbf{u} \tag{12.23}$$

where $[A]$ is an $n \times n$ matrix and $[B]$ is an $n \times m$ matrix. Further, while finding the control vector $\mathbf{u}$, the state vector $\mathbf{x}$ is to be transferred from a known initial vector $\mathbf{x}_0$ at $t = 0$ to a terminal vector $\mathbf{x}_T$ at $t = T$, where some (or all or none) of the state variables are specified.

12.3.1   Necessary Conditions for Optimal Control

To derive the necessary conditions for the optimal control, we consider the following simple problem:

$$\text{Find } u \text{ which minimizes } J = \int_0^T f_0(x, u, t) \, dt \tag{12.24}$$

subject to

$$\dot{x} = f(x, u, t) \tag{12.25}$$

with the boundary condition $x(0) = k_1$. To solve this optimal control problem, we introduce a Lagrange multiplier λ and define an augmented functional J^* as

$$J^* = \int_0^T \{f_0(x, u, t) + \lambda[f(x, u, t) - \dot{x}]\} \, dt \tag{12.26}$$

Since the integrand

$$F = f_0 + \lambda \, (f - \dot{x}) \tag{12.27}$$

is a function of the two variables x and u, we can write the Euler–Lagrange equations [with $u_1 = x$, $u_1' = \partial x/\partial t = \dot{x}$, $u_2 = u$ and $u_2' = \partial u/\partial t = \dot{u}$ in Eq. (12.10)] as

$$\frac{\partial F}{\partial x} - \frac{d}{dt}\left(\frac{\partial F}{\partial \dot{x}}\right) = 0 \tag{12.28}$$

$$\frac{\partial F}{\partial u} - \frac{d}{dt}\left(\frac{\partial F}{\partial \dot{u}}\right) = 0 \tag{12.29}$$

In view of relation (12.27), Eqs. (12.28) and (12.29) can be expressed as

$$\frac{\partial f_0}{\partial x} + \lambda\frac{\partial f}{\partial x} + \dot{\lambda} = 0 \tag{12.30}$$

$$\frac{\partial f}{\partial u} + \lambda\frac{\partial f}{\partial u} = 0 \tag{12.31}$$

A new functional H, called the *Hamiltonian*, is defined as

$$H = f_0 + \lambda f \tag{12.32}$$

and Eqs. (12.30) and (12.31) can be rewritten as

$$-\frac{\partial H}{\partial x} = \dot{\lambda} \tag{12.33}$$

$$\frac{\partial H}{\partial u} = 0 \tag{12.34}$$

Equations (12.33) and (12.34) represent two first-order differential equations. The integration of these equations leads to two constants whose values can be found from the known boundary conditions of the problem. If two boundary conditions are specified as $x(0) = k_1$ and $x(T) = k_2$, the two integration constants can be evaluated without any difficulty. On the other hand, if only one boundary condition is specified as, say, $x(0) = k_1$, the free-end condition is used as $\partial F/\partial \dot{x} = 0$ or $\lambda = 0$ at $t = T$.

Example 12.4 Find the optimal control u that makes the functional

$$J = \int_0^1 (x^2 + u^2)\, dt \tag{E_1}$$

stationary with

$$\dot{x} = u \tag{E_2}$$

and $x(0) = 1$. Note that the value of x is not specified at $t = 1$.

SOLUTION   The Hamiltonian can be expressed as

$$H = f_0 + \lambda u = x^2 + u^2 + \lambda u \tag{E_3}$$

and Eqs. (12.33) and (12.34) give

$$-2x = \dot{\lambda} \tag{E_4}$$

$$2u + \lambda = 0 \tag{E_5}$$

Differentiation of Eq. (E$_5$) leads to

$$2\dot{u} + \dot{\lambda} = 0 \qquad (E_6)$$

Equations (E$_4$) and (E$_6$) yield

$$\dot{u} = x \qquad (E_7)$$

Since $\dot{x} = u$ [Eq. (E$_2$)], we obtain

$$\ddot{x} = \dot{u} = x$$

that is,

$$\ddot{x} - x = 0 \qquad (E_8)$$

The solution of Eq. (E$_8$) is given by

$$x(t) = c_1 \sinh t + c_2 \cosh t \qquad (E_9)$$

where c_1 and c_2 are constants. By using the initial condition $x(0) = 1$, we obtain $c_2 = 1$. Since x is not fixed at the terminal point $t = T = 1$, we use the condition $\lambda = 0$ at $t = 1$ in Eq. (E$_5$) and obtain $u(t = 1) = 0$. But

$$u = \dot{x} = c_1 \cosh t + \sinh t \qquad (E_{10})$$

Thus

$$u(1) = 0 = c_1 \cosh 1 + \sinh 1$$

or

$$c_1 = \frac{-\sinh 1}{\cosh 1} \qquad (E_{11})$$

and hence the optimal control is

$$
\begin{aligned}
u(t) &= \frac{-\sinh 1}{\cosh 1} \cdot \cosh t + \sinh t \\
&= \frac{-\sinh 1 \cdot \cosh t + \cosh 1 \cdot \sinh t}{\cosh 1} = \frac{-\sinh(1-t)}{\cosh 1}
\end{aligned} \qquad (E_{12})
$$

The corresponding state trajectory is given by

$$x(t) = \dot{u} = \frac{\cosh(1-t)}{\cosh 1} \qquad (E_{13})$$

12.3.2   Necessary Conditions for a General Problem

We shall now consider the basic optimal control problem stated earlier:
Find the optimal control vector $\mathbf{u}$ that minimizes

$$J = \int_0^T f_0(\mathbf{x}, \mathbf{u}, t)\, dt \qquad (12.35)$$

subject to

$$\dot{x}_i = f_i(\mathbf{x}, \mathbf{u}, t), \, i = 1, 2, \ldots, n \qquad (12.36)$$

Now we introduce a *Lagrange multiplier* p_i, also known as the *adjoint variable*, for the ith constraint equation in (12.36) and form an augmented functional J^* as

$$J^* = \int_0^T \left[f_0 + \sum_{i=1}^n p_i(f_i - \dot{x}_i) \right] dt \qquad (12.37)$$

The Hamiltonian functional, H, is defined as

$$H = f_0 + \sum_{i=1}^n p_i f_i \qquad (12.38)$$

such that

$$J^* = \int_0^T \left(H - \sum_{i=1}^n p_i \dot{x}_i \right) dt \qquad (12.39)$$

Since the integrand

$$F = H - \sum_{i=1}^n p_i \dot{x}_i \qquad (12.40)$$

depends on $\mathbf{x}$, $\mathbf{u}$, and t, there are $n + m$ dependent variables ($\mathbf{x}$ and $\mathbf{u}$) and hence the Euler–Lagrange equations become

$$\frac{\partial F}{\partial x_i} - \frac{d}{dt}\left(\frac{\partial F}{\partial \dot{x}_i}\right) = 0, \quad i = 1, 2, \dots, n \qquad (12.41)$$

$$\frac{\partial F}{\partial u_j} - \frac{d}{dt}\left(\frac{\partial F}{\partial \dot{u}_j}\right) = 0, \quad j = 1, 2, \dots, m \qquad (12.42)$$

In view of relation (12.40), Eqs. (12.41) and (12.42) can be rewritten as

$$-\frac{\partial H}{\partial x_i} = p_i, \quad i = 1, 2, \dots, n \qquad (12.43)$$

$$\frac{\partial H}{\partial u_i} = 0, \quad j = 1, 2, \dots, m \qquad (12.44)$$

Equations (12.43) are knowns as *adjoint equations*.

The optimum solutions for $\mathbf{x}$, $\mathbf{u}$, and $\mathbf{p}$ can be obtained by solving Eqs. (12.36), (12.43), and (12.44). There are totally $2n + m$ equations with nx_i's, np_i's, and mu_j's as unknowns. If we know the initial conditions $x_i(0)$, $i = 1, 2, \dots, n$, and the terminal conditions $x_j(T)$, $j = 1, 2, \dots, l$, with $l < n$, we will have the terminal values of the remaining variables, namely $x_j(T)$, $j = l + 1, l + 2, \dots, n$, free. Hence we will have to use the free end conditions

$$p_j(T) = 0, \quad j = l + 1, l + 2, \dots, n \qquad (12.45)$$

Equations (12.45) are called the *transversality conditions*.

12.4   OPTIMALITY CRITERIA METHODS

The optimality criteria methods are based on the derivation of an appropriate criteria for specialized design conditions and developing an iterative procedure to find the optimum design. The optimality criteria methods were originally developed by Prager and his associates for distributed (continuous) systems [12.6] and extended by Venkayya, Khot, and Berke for discrete systems [12.7–12.10]. The methods were first presented for linear elastic structures with stress and displacement constraints and later extended to problems with other types of constraints. We will present the basic approach using only displacement constraints.

12.4.1   Optimality Criteria with a Single Displacement Constraint

Let the optimization problem be stated as follows:

$$\text{Find } \mathbf{X} \text{ which minimizes } f(\mathbf{X}) = \sum_{i=1}^{n} c_i x_i \qquad (12.46)$$

subject to

$$\sum_{i=1}^{n} \frac{a_i}{x_i} = y_{\max} \qquad (12.47)$$

where c_i are constants, $y_{\max}$ is the maximum permissible displacement, and a_i depends on the force induced in member i due to the applied loads, length of member i, and Young's modulus of member i. The Lagrangian function can be defined as

$$L(\mathbf{X}, \lambda) = \sum_{i=1}^{n} c_i x_i + \lambda \left(\sum_{i=1}^{n} \frac{a_i}{x_i} - y_{\max} \right) \qquad (12.48)$$

At the optimum solution, we have

$$\frac{\partial L}{\partial x_k} = c_k - \lambda \frac{a_k}{x_k^2} + \lambda \sum_{i=1}^{n} \frac{1}{x_i} \frac{\partial a_i}{\partial x_k} = 0, \quad k = 1, 2, \ldots, n \qquad (12.49)$$

It can be shown that the last term in Eq. (12.49) is zero for statically determinate as well as indeterminate structures [12.8] so that Eq. (12.49) reduces to

$$c_k - \lambda \frac{a_k}{x_k^2} = 0, \quad k = 1, 2, \ldots, n \qquad (12.50)$$

or

$$\lambda = \frac{c_k x_k^2}{a_k} \qquad (12.51)$$

Equation (12.51) indicates that the quantity $c_k x_k^2 / a_k$ is the same for all the design variables. If all the design variables are to be changed, this relation can be used. However, in practice, only a subset of design variables are involved in Eq. (12.49). Thus it is convenient to divide the design variables into two sets: active variables [those determined by the displacement constraint of Eq. (12.51)] and passive variables (those determined by other considerations). Assuming that the first $\bar{n}$ variables denote the active variables, we can rewrite Eqs. (12.46) and (12.47) as

$$f = \overline{f} + \sum_{i=1}^{\bar{n}} c_i x_i \qquad (12.52)$$

$$\sum_{i=1}^{\bar{n}} \frac{a_i}{x_i} = y_{\max} - \overline{y} = y^* \qquad (12.53)$$

where $\overline{f}$ and $\overline{y}$ denote the contribution of the passive variables to f and y, respectively. Equation (12.51) now gives

$$x_k = \sqrt{\lambda} \sqrt{\frac{a_k}{c_k}}, \quad k = 1, 2, \ldots, \bar{n} \qquad (12.54)$$

Substituting Eq. (12.54) into Eq. (12.53), and solving for λ, we obtain

$$\sqrt{\lambda} = \frac{1}{y^*} \sum_{k=1}^{\bar{n}} \sqrt{a_k c_k} \qquad (12.55)$$

Using Eq. (12.55) in Eq. (12.54) results in

$$x_k = \frac{1}{y^*} \sqrt{\frac{a_k}{c_k}} \sum_{i=1}^{\bar{n}} \sqrt{a_i c_i}, \quad k = 1, 2, \ldots, \bar{n} \qquad (12.56)$$

Equation (12.56) is the optimality criteria that must be satisfied at the optimum solution of the problem stated by Eqs. (12.46) and (12.47). This equation can be used to iteratively update the design variables x_k as

$$x_k^{(j+1)} = \left(\frac{1}{y^*} \sqrt{\frac{a_k}{c_k}} \sum_{i=1}^{\bar{n}} \sqrt{a_i c_i} \right)^{(j)}, \quad k = 1, 2, \ldots, \bar{n} \qquad (12.57)$$

where the superscript j denotes the iteration cycle. In each iteration, the components a_k and c_k are assumed to be constants (in general, they depend on the design vector).

12.4.2   Optimality Criteria with Multiple Displacement Constraints

When multiple displacement constraints are included, as in the case of a structure subjected to multiple-load conditions, the optimization problem can be stated as follows:

Find a set of active variables $\overline{\mathbf{X}} = \{x_1 \quad x_2 \ldots x_{\bar{n}}\}^{\mathrm{T}}$ which minimizes

$$f(\overline{X}) = f_0 + \sum_{i=1}^{\bar{n}} c_i x_i \qquad (12.58)$$

subject to

$$y_j = \sum_{i=1}^{\bar{n}} \frac{a_{ji}}{x_i} = y_j^*, \quad j = 1, 2, \ldots, J \qquad (12.59)$$

where J denotes the number of displacement (equality) constraints, y_j^* the maximum permissible value of the displacement y_j, and a_{ji} is a parameter that depends on the force induced in member i due to the applied loads, length of member i, and Young's modulus of member i. The Lagrangian function corresponding to Eqs. (12.58) and (12.59) can be expressed as

$$L(\overline{\mathbf{X}}, \lambda_1, \ldots, \lambda_J) = f_0 + \sum_{i=1}^{\overline{n}} c_i x_i + \sum_{j=1}^{J} \lambda_j \left(\sum_{i=1}^{\overline{n}} \frac{a_{ji}}{x_i} - y_j^* \right) \tag{12.60}$$

and the necessary conditions of optimality are given by

$$\frac{\partial L}{\partial x_k} = c_k - \sum_{j=1}^{J} \lambda_j \frac{a_{ji}}{x_k^2} = 0, \quad k = 1, 2, \ldots, \overline{n} \tag{12.61}$$

Equations (12.61) can be rewritten as

$$x_k = \left[\sum_{j=1}^{J} \left(\lambda_j \frac{a_{ji}}{c_k} \right) \right]^{1/2}, \quad k = 1, 2, \ldots, \overline{n} \tag{12.62}$$

Note that Eq. (12.62) can be used to iteratively update the variable x_k as

$$x_k^{(j+1)} = \left\{ \left[\sum_{j=1}^{J} \left(\lambda_j \frac{a_{ji}}{c_k} \right) \right]^{1/2} \right\}^{(j)}, \quad k = 1, 2, \ldots, \overline{n} \tag{12.63}$$

where the values of the Lagrange multipliers λ_j are also not known at the beginning. Several computational methods can be used to solve Eqs. (12.63) [12.7, 12.8].

12.4.3   Reciprocal Approximations

In some structural optimization problems, it is convenient and useful to consider the reciprocals of member cross-sectional areas $(1/A_i)$ as the new design variables (z_i). If the problem deals with the minimization of weight of a statically determinate structure subject to displacement or stress constraints, the objective function and its gradient can be expressed as explicit functions of the variables z_i and the constraints can be expressed as linear functions of the variables z_i. If the structure is statically indeterminate, the objective function remains a simple function of z_i but the constraints may not be linear in terms of z_i; however, a first-order Taylor series (linear) approximation of the constraints denote a very high-quality approximation of these constraints. With reciprocal variables, the optimization problem with a single displacement constraint can be stated as follows:

$$\text{Find } \mathbf{Z} = \{z_1 \ z_2 \ \ldots \ z_n\}^{\mathrm{T}} \text{ which minimizes } f(\mathbf{Z}) \tag{12.64}$$

subject to

$$g(\mathbf{Z}) = 0 \tag{12.65}$$

The necessary condition of optimality can be expressed as

$$\frac{\partial f}{\partial z_i} + \lambda \frac{\partial g}{\partial z_i} = 0, \quad i = 1, 2, \ldots, n \tag{12.66}$$

Assuming f to be linear in terms of the areas of cross section (original variables, $x_i = A_i$) and g to be linear in terms of z_i, we have

$$\frac{\partial f}{\partial z_i} = \frac{\partial f}{\partial x_i} \frac{\partial x_i}{\partial z_i} = -\frac{1}{z_i^2} \frac{\partial f}{\partial x_i} \tag{12.67}$$

and Eqs. (12.66) and (12.67) yield

$$x_i = \left(\lambda \frac{\partial g / \partial z_i}{\partial f / \partial x_i} \right)^{1/2}, \quad i = 1, 2, \ldots, n \tag{12.68}$$

To find λ we first find the linear approximation of g at a reference point (trial design) $\mathbf{Z}_0$ (or $\mathbf{X}_0$) as

$$g(\mathbf{Z}) \approx g(\mathbf{Z}_0) + \sum_{i=1}^{n} \left. \frac{\partial g}{\partial z_i} \right|_{\mathbf{Z}_0} (z_i - z_{0i}) \approx g_0 + \sum_{i=1}^{n} \left. \frac{\partial g}{\partial z_i} \right|_{\mathbf{Z}_0} z_i \tag{12.69}$$

where

$$g_0 = g(\mathbf{Z}_0) - \sum_{i=1}^{n} \left. \frac{\partial g}{\partial z_i} \right|_{\mathbf{Z}_0} z_{0i} = g(\mathbf{X}_0) + \sum_{i=1}^{n} \left. \frac{\partial g}{\partial x_i} \right|_{\mathbf{X}_0} x_{0i} \tag{12.70}$$

and z_{0i} is the ith component of $\mathbf{Z}_0$ with $x_{0i} = 1/z_{0i}$. By setting Eq. (12.69) equal to zero and substituting Eq. (12.68) for x_i, we obtain

$$\lambda = \left[\frac{1}{g_0} \sum_{i=1}^{n} \left(\frac{\partial f}{\partial x_i} \frac{\partial g}{\partial z_i} \right)^{1/2} \right]^2 \tag{12.71}$$

Equations (12.71) and (12.68) can now be used iteratively to find the optimal solution of the problem. The procedure is explained through the following example.

Example 12.5 The problem of minimum weight design subject to a constraint on the vertical displacement of node $S(U_1)$ of the three-bar truss shown in Fig. 12.6 can be stated as follows:

Find $\mathbf{X} = \begin{Bmatrix} x_1 \\ x_2 \end{Bmatrix}$ which minimizes

$$f(\mathbf{X}) = \rho(2\sqrt{2}\, l)x_1 + \rho l x_2 = 80.0445 x_1 + 28.3 x_2 \tag{E$_1$}$$

subject to

$$\frac{U_1}{U_{\text{max}}} - 1 \leq 0$$

or

$$g(\mathbf{X}) = \frac{1}{x_1 + \sqrt{2}\, x_2} - 1 \leq 0 \tag{E$_2$}$$

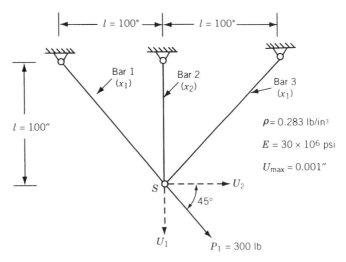

Figure 12.6 Three-bar truss.

where ρ is the weight density, E is Young's modulus, U_{max} the maximum permissible displacement, x_1 the area of cross section of bars 1 and 3, x_2 the area of cross section of bar 2, and the vertical displacement of node S is given by

$$U_1 = \frac{P_1 l}{E} \frac{1}{x_1 + \sqrt{2}\, x_2} \qquad (E_3)$$

Find the solution of the problem using the optimality criteria method.

SOLUTION   The partial derivatives of f and g required by Eqs. (12.68) and (12.71) can be computed as

$$\frac{\partial f}{\partial x_1} = 80.0445, \quad \frac{\partial f}{\partial x_2} = 28.3$$

$$\frac{\partial g}{\partial z_i} = \frac{\partial g}{\partial x_i} \frac{\partial x_i}{\partial z_i} = \frac{\partial g}{\partial x_i}(-x_i^2), \quad i = 1, 2$$

$$\frac{\partial g}{\partial x_1} = \frac{-1}{(x_1 + \sqrt{2}\, x_2)^2}, \quad \frac{\partial g}{\partial x_2} = \frac{-\sqrt{2}}{(x_1 + \sqrt{2}\, x_2)^2}$$

At any design $\mathbf{X}_i$, Eq. (12.70) gives

$$g_0 = g(\mathbf{X}_i) + \left.\frac{\partial g}{\partial x_1}\right|_{\mathbf{X}_i} x_{i1} + \left.\frac{\partial g}{\partial x_2}\right|_{\mathbf{X}_i} x_{i2}$$

$$= \frac{1}{x_{i1} + \sqrt{2}\, x_{i2}} - 1 - \frac{x_{i1}}{(x_{i1} + \sqrt{2}\, x_{i2})^2} - \frac{\sqrt{2}\, x_{i2}}{(x_{i1} + \sqrt{2}\, x_{i2})^2}$$

Thus the values of λ and (x_1, x_2) can be determined iteratively using Eqs. (12.71) and (12.68). Starting from the initial design $(x_1, x_2) = (2.0, 2.0)$ in², the results obtained are shown in Table 12.1.

Table 12.1 Results for Example 12.5[a]

Starting values		g_0 [Eq. (12.70)]	λ [Eq. (12.71)]	Solution from Eq. (12.68)	
x_1	x_2			x_1	x_2
0.20000E+01	0.20000E+01	-0.10000E+01	0.40022E+02	0.29289E+00	0.58579E+00
0.29289E+00	0.58579E+00	-0.10000E+01	0.31830E+02	0.16472E+00	0.65886E+00
0.16472E+00	0.65886E+00	-0.10000E+01	0.26475E+02	0.86394E-01	0.69115E+00
0.10000E+00	0.69115E+00	-0.10000E+01	0.23898E+02	0.50714E-01	0.70102E+00
0.10000E+00	0.70102E+00	-0.10000E+01	0.23846E+02	0.50011E-01	0.70117E+00
0.10000E+00	0.70117E+00	-0.10000E+01	0.23845E+02	0.50000E-01	0.70117E+00
0.10000E+00	0.70117E+00	-0.10000E+01	0.23845E+02	0.50000E-01	0.70117E+00

[a]With lower bounds on x_1 and x_2 as 0.1.

REFERENCES AND BIBLIOGRAPHY

12.1 R. S. Schechter, *The Variational Method in Engineering*, McGraw-Hill, New York, 1967.

12.2 M. M. Denn, *Optimization by Variational Methods*, McGraw-Hill, New York, 1969.

12.3 M. J. Forray, *Variational Calculus in Science and Engineering*, McGraw-Hill, New York, 1968.

12.4 A. E. Bryson and Y. C. Ho, *Applied Optimal Control*, Wiley, New York, 1975.

12.5 A. Shamaly, G. S. Christensen, and M. E. El-Hawary, Optimal control of a large turboalternator, *Journal of Optimization Theory and Applications*, Vol. 34, pp. 83–97, 1981.

12.6 W. Prager, Optimality criteria in structural design, *Proceedings of the National Academy of Science*, Vol. 61, No. 3, pp. 794–796, 1968.

12.7 V. B. Venkayya, N. S. Khot, and V. S. Reddy, *Energy Distribution in Optimum Structural Design*, AFFDL-TR-68-156, 1968.

12.8 L. Berke, *Convergence Behavior of Optimality Criteria Based Iterative Procedures*, AFFDL-TM-72-1-FBR, Jan. 1972.

12.9 V. B. Venkayya, Structural optimization using optimality criteria: a review and some recommendations, *International Journal for Numerical Methods in Engineering*, Vol. 13, pp. 203–228, 1978.

12.10 N. S. Khot, Algorithms based on optimality criteria to design minimum weight structures, *Engineering Optimization*, Vol. 5, pp. 73–90, 1981.

12.11 S. K. Hati and S. S. Rao, Determination of optimum profile of one-dimensional cooling fins, *ASME Journal of Vibration, Acoustics, Stress and Reliability in Design*, Vol. 105, pp. 317–320, 1983.

REVIEW QUESTIONS

12.1 Answer true or false:

(a) Design variables of an optimal control problem include both state and control variables.

(b) Reciprocal approximations consider reciprocals of member areas as design variables.

(c) Optimality criteria methods can be used for the optimization of nonlinear structures with displacement constraints.

(d) A variational operator is similar to a differential operator.

(e) Calculus of variations can be used only for finding the extrema of functionals with no constraints.

(f) Optimality criteria methods can be used to solve any optimization problem.

12.2 Define the following terms:

(a) Brachistochrone

(b) State vector

(c) Performance index

(d) Adjoint equations

(e) Transversality condition

(f) Optimality criteria methods

(g) Functional

(h) Hamiltonian

12.3   Match the following terms and descriptions:

(a) Adjoint variables	Linear elastic structures
(b) Optimality criteria methods	Lagrange multipliers
(c) Calculus of variations	Necessary conditions of optimality
(d) Optimal control theory	Optimization of functionals
(e) Governing equations	Hamiltonian used

12.4   What are the characteristics of a variational operator?

12.5   What are Euler–Lagrange equations?

12.6   Which method can be used to solve a trajectory optimization problem?

12.7   What is an optimality criteria method?

12.8   What is the basis of optimality criteria methods?

12.9   What are the advantages of using reciprocal approximations in structural optimization?

12.10   What is the difference between free and forced boundary conditions?

12.11   What type of problems require introduction of Lagrange multipliers?

12.12   Where are reciprocal approximations used? Why?

PROBLEMS

12.1   Find the curve connecting two points $A(0, 0)$ and $B(2, 0)$ such that the length of the line is a minimum and the area under the curve is $\pi/2$.

12.2   Prove that the shortest distance between two points is a straight line. Show that the necessary conditions yield a minimum and not a maximum.

12.3   Find the function $x(t)$ that minimizes the functional

$$A = \int_0^T \left[x^2 + 2xt + \left(\frac{dx}{dt}\right)^2 \right] dt$$

with the condition that $x(0) = 2$.

12.4   Find the closed plane curve of length L that encloses a maximum area.

12.5   The potential energy of an elastic circular annular plate of radii r_1 and r_2 shown in Fig. 12.7 is given by

$$\pi_0 = \pi D \int_{r_1}^{r_2} \left[r \left(\frac{d^2 w}{dr^2}\right)^2 + \frac{1}{r}\left(\frac{dw}{dr}\right)^2 + 2v \frac{dw}{dr}\frac{d^2 w}{dr^2} \right] dr$$

$$- 2\pi \int_{r_1}^{r_2} qrw \, dr + 2\pi \left[rM\frac{dw}{dr} - rQw \right]_{r=r_2}$$

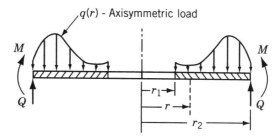

Figure 12.7   Circular annular plate under load.

where D is the flexural rigidity of the plate, w the transverse deflection of the plate, ν the Poisson's ratio, M the radial bending moment per unit of circumferential length, and Q the radial shear force per unit of circumferential length. Find the differential equation and the boundary conditions to be satisfied by minimizing π_0.

12.6   Consider the two-bar truss shown in Fig. 12.8. For the minimum-weight design of the truss with a bound on the horizontal displacement of node S, we need to solve the following problem:

Find $\mathbf{X} = \{x_1\ x_2\}^{\mathrm{T}}$ which minimizes

$$f(\mathbf{X}) = \sqrt{2}\,l(x_1 + x_2) = \sqrt{2}\,60(x_1 + x_2)$$

subject to

$$g(\mathbf{X}) = \frac{Pl}{2E}\left(\frac{1}{x_1} + \frac{1}{x_2}\right) - U_{\max}$$

$$= 10^{-3}\left(\frac{1}{x_1} + \frac{1}{x_2}\right) - 10^{-2} \leq 0$$

$$0.1\ \text{in.}^2 \leq x_i \leq 1.0\ \text{in.}^2, \quad i = 1, 2$$

Find the solution of the problem using the optimality criteria method.

12.7   In the three-bar truss considered in Example 12.5 (Fig. 12.6), if the constraint is placed on the resultant displacement of node S, the optimization problem can be stated as

Find $\mathbf{X} = \begin{Bmatrix} x_1 \\ x_2 \end{Bmatrix}$ which minimizes

$$f(\mathbf{X}) = 80.0445x_1 + 28.3x_2$$

subject to

$$\sqrt{U_1^2 + U_2^2} = \frac{P_1 l}{E}\left[\frac{1}{x_1^2} + \frac{1}{(x_1 + \sqrt{2}\,x_2)^2}\right]^{1/2} \leq U_{\max}$$

or

$$g(\mathbf{X}) = \left[\frac{1}{x_1^2} + \frac{1}{(x_1 + \sqrt{2}\,x_2)^2}\right]^{1/2} \leq U_{\max}$$

where the vertical and horizontal displacements of node S are given by

$$U_1 = \frac{Pl}{E}\,\frac{1}{x_1 + \sqrt{2}\,x_2} \quad \text{and} \quad U_2 = \frac{Pl}{E}\,\frac{1}{x_1}$$

Find the solution of the problem using the optimality criteria method.

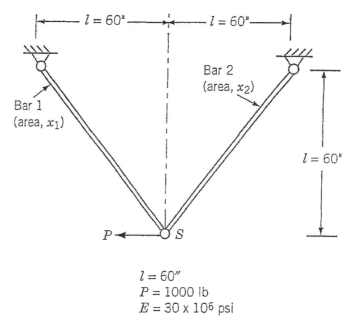

$$l = 60''$$
$$P = 1000 \text{ lb}$$
$$E = 30 \times 10^6 \text{ psi}$$

Figure 12.8   Two-bar truss subjected to horizontal load.

12.8   The problem of the minimum-weight design of the four-bar truss shown in Fig. 1.32 (Problem 1.31) subject to a constraint on the vertical displacement of joint A and limitations on design variables can be stated as follows:

Find $\mathbf{X} = \{x_1 \ x_2\}^T$ which minimizes

$$f(\mathbf{X}) = 0.1x_1 + 0.05773x_2$$

subject to

$$g(\mathbf{X}) = \frac{0.6}{x_1} + \frac{0.3464}{x_2} - 0.1 \le 0$$

$$x_i \ge 4, \quad i = 1, 2$$

where the maximum permissible vertical displacement of joint A is assumed to be 0.01 in. Solve the problem using the optimality criteria method.

13

Modern Methods of Optimization

13.1 INTRODUCTION

In recent years, some optimization methods that are conceptually different from the traditional mathematical programming techniques have been developed. These methods are labeled as modern or nontraditional methods of optimization. Most of these methods are based on certain characteristics and behavior of biological, molecular, swarm of insects, and neurobiological systems. The following methods are described in this chapter:

1. Genetic algorithms
2. Simulated annealing
3. Particle swarm optimization
4. Ant colony optimization
5. Fuzzy optimization
6. Neural-network-based methods

Most of these methods have been developed only in recent years and are emerging as popular methods for the solution of complex engineering problems. Most require only the function values (and not the derivatives). The *genetic algorithms* are based on the principles of natural genetics and natural selection. *Simulated annealing* is based on the simulation of thermal annealing of critically heated solids. Both genetic algorithms and simulated annealing are stochastic methods that can find the global minimum with a high probability and are naturally applicable for the solution of discrete optimization problems. The *particle swarm optimization* is based on the behavior of a colony of living things, such as a swarm of insects, a flock of birds, or a school of fish. *Ant colony optimization* is based on the cooperative behavior of real ant colonies, which are able to find the shortest path from their nest to a food source. In many practical systems, the objective function, constraints, and the design data are known only in vague and linguistic terms. *Fuzzy optimization methods* have been developed for solving such problems. In *neural-network-based methods,* the problem is modeled as a network consisting of several neurons, and the network is trained suitably to solve the optimization problem efficiently.

13.2   GENETIC ALGORITHMS

13.2.1   Introduction

Many practical optimum design problems are characterized by mixed continuous–discrete variables, and discontinuous and nonconvex design spaces. If standard nonlinear programming techniques are used for this type of problem they will be inefficient, computationally expensive, and, in most cases, find a relative optimum that is closest to the starting point. Genetic algorithms (GAs) are well suited for solving such problems, and in most cases they can find the global optimum solution with a high probability. Although GAs were first presented systematically by Holland [13.1], the basic ideas of analysis and design based on the concepts of biological evolution can be found in the work of Rechenberg [13.2]. Philosophically, GAs are based on Darwin's theory of survival of the fittest.

Genetic algorithms are based on the principles of natural genetics and natural selection. The basic elements of natural genetics—reproduction, crossover, and mutation—are used in the genetic search procedure. GAs differ from the traditional methods of optimization in the following respects:

1. A population of points (trial design vectors) is used for starting the procedure instead of a single design point. If the number of design variables is n, usually the size of the population is taken as $2n$ to $4n$. Since several points are used as candidate solutions, GAs are less likely to get trapped at a local optimum.

2. GAs use only the values of the objective function. The derivatives are not used in the search procedure.

3. In GAs the design variables are represented as strings of binary variables that correspond to the chromosomes in natural genetics. Thus the search method is naturally applicable for solving discrete and integer programming problems. For continuous design variables, the string length can be varied to achieve any desired resolution.

4. The objective function value corresponding to a design vector plays the role of fitness in natural genetics.

5. In every new generation, a new set of strings is produced by using randomized parents selection and crossover from the old generation (old set of strings). Although randomized, GAs are not simple random search techniques. They efficiently explore the new combinations with the available knowledge to find a new generation with better fitness or objective function value.

13.2.2   Representation of Design Variables

In GAs, the design variables are represented as strings of binary numbers, 0 and 1. For example, if a design variable x_i is denoted by a string of length four (or a four-bit string) as 0 1 0 1, its integer (decimal equivalent) value will be (1) 2^0 + (0) 2^1 + (1) 2^2 + (0) $2^3 = 1 + 0 + 4 + 0 = 5$. If each design variable x_i, $i = 1, 2, \ldots, n$ is coded in a string of length q, a design vector is represented using a string of total length nq. For example, if a string of length 5 is used to represent each variable, a total string of length 20 describes a design vector with $n = 4$. The following string of 20 binary digits denote the vector ($x_1 = 18$, $x_2 = 3$, $x_3 = 1$, $x_4 = 4$):

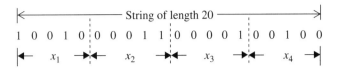

In general, if a binary number is given by $b_q b_{q-1} \cdots b_2 b_1 b_0$, where $b_k = 0$ or 1, $k = 0, 1, 2, \ldots, q$, then its equivalent decimal number y (integer) is given by

$$y = \sum_{k=0}^{q} 2^k b_k \tag{13.1}$$

This indicates that a continuous design variable x can only be represented by a set of discrete values if binary representation is used. If a variable x (whose bounds are given by $x^{(l)}$ and $x^{(u)}$) is represented by a string of q binary numbers, as shown in Eq. (13.1), its decimal value can be computed as

$$x = x^{(l)} + \frac{x^{(u)} - x^{(l)}}{2^q - 1} \sum_{k=0}^{q} 2^k b_k \tag{13.2}$$

Thus if a continuous variable is to be represented with high accuracy, we need to use a large value of q in its binary representation. In fact, the number of binary digits needed (q) to represent a continuous variable in steps (accuracy) of Δx can be computed from the relation

$$2^q \geq \frac{x^{(u)} - x^{(l)}}{\Delta x} + 1 \tag{13.3}$$

For example, if a continuous variable x with bounds 1 and 5 is to be represented with an accuracy of 0.01, we need to use a binary representation with q digits where

$$2^q \geq \frac{5 - 1}{0.01} + 1 = 401 \quad \text{or} \quad q = 9 \tag{13.4}$$

Equation (13.2) shows why GAs are naturally suited for solving discrete optimization problems.

Example 13.1 Steel plates are available in thicknesses (in inches) of

$$\frac{1}{32}, \frac{1}{16}, \frac{3}{32}, \frac{1}{8}, \frac{5}{32}, \frac{3}{16}, \frac{7}{32}, \frac{1}{4}, \frac{9}{32}, \frac{5}{16}, \frac{11}{32}, \frac{3}{8}, \frac{13}{32}, \frac{7}{16}, \frac{15}{32}, \frac{1}{2}$$

from a manufacturer. If the thickness of the steel plate, to be used in the construction of a pressure vessel, is considered as a discrete design variable, determine the size of the binary string to be used to select a thickness from the available values.

SOLUTION The lower and upper bounds on the steel plate (design variable, x) are given by $\frac{1}{32}$ and $\frac{1}{2}$ in., respectively, and the resolution or difference between any two adjacent thicknesses is $\frac{1}{32}$ in. Equation (13.3) gives

$$2^q \geq \frac{x^{(u)} - x^{(l)}}{\Delta x} + 1 = \frac{\frac{1}{2} \text{ in.} - \frac{1}{32} \text{ in.}}{\frac{1}{32} \text{ in.}} + 1 = 15$$

from which the size of the binary string to be used can be obtained as $q = 4$.

13.2.3   Representation of Objective Function and Constraints

Because genetic algorithms are based on the survival-of-the-fittest principle of nature, they try to maximize a function called the fitness function. Thus GAs are naturally suitable for solving unconstrained maximization problems. The fitness function, $F(\mathbf{X})$, can be taken to be same as the objective function $f(\mathbf{X})$ of an unconstrained maximization problem so that $F(\mathbf{X}) = f(\mathbf{X})$. A minimization problem can be transformed into a maximization problem before applying the GAs. Usually the fitness function is chosen to be nonnegative. The commonly used transformation to convert an unconstrained minimization problem to a fitness function is given by

$$F(\mathbf{X}) = \frac{1}{1 + f(\mathbf{X})} \tag{13.5}$$

It can be seen that Eq. (13.5) does not alter the location of the minimum of $f(\mathbf{X})$ but converts the minimization problem into an equivalent maximization problem.

A general constrained minimization problem can be stated as

$$\text{Minimize } f(\mathbf{X})$$

subject to

$$g_i(\mathbf{X}) \leq 0, \quad i = 1, 2, \ldots, m \tag{13.6}$$

and

$$h_j(\mathbf{X}) = 0, \quad j = 1, 2, \ldots, p$$

This problem can be converted into an equivalent unconstrained minimization problem by using the concept of penalty function as

$$\text{Minimize } \phi(\mathbf{X}) = f(\mathbf{X}) + \sum_{i=1}^{m} r_i \langle g_i(\mathbf{X}) \rangle^2 + \sum_{j=1}^{p} R_j \left(h_j(\mathbf{X}) \right)^2 \tag{13.7}$$

where r_i and R_j are the penalty parameters associated with the constraints $g_i(\mathbf{X})$ and $h_j(\mathbf{X})$, whose values are usually kept constant throughout the solution process. In Eq. (13.7), the function $\langle g_i(\mathbf{X}) \rangle$, called the bracket function, is defined as

$$\langle g_i(\mathbf{X}) \rangle = \begin{cases} g_i(\mathbf{X}) & \text{if } g_i(\mathbf{X}) > 0 \\ 0 & \text{if } g_i(\mathbf{X}) \leq 0 \end{cases} \tag{13.8}$$

In most cases, the penalty parameters associated with all the inequality and equality constraints are assumed to be the same constants as

$$r_i = r, \ i = 1, 2, \ldots, m \quad \text{and} \quad R_j = R, \ j = 1, 2, \ldots, p \tag{13.9}$$

where r and R are constants. The fitness function, $F(\mathbf{X})$, to be maximized in the GAs can be obtained, similar to Eq. (13.5), as

$$F(\mathbf{X}) = \frac{1}{1 + \phi(\mathbf{X})} \tag{13.10}$$

Equations (13.7) and (13.8) show that the penalty will be proportional to the square of the amount of violation of the inequality and equality constraints at the design vector **X**, while there will be no penalty added to $f(\mathbf{X})$ if all the constraints are satisfied at the design vector **X**.

13.2.4   Genetic Operators

The solution of an optimization problem by GAs starts with a population of random strings denoting several (population of) design vectors. The population size in GAs (n) is usually fixed. Each string (or design vector) is evaluated to find its fitness value. The population (of designs) is operated by three operators—reproduction, crossover, and mutation—to produce a new population of points (designs). The new population is further evaluated to find the fitness values and tested for the convergence of the process. One cycle of reproduction, crossover, and mutation and the evaluation of the fitness values is known as a generation in GAs. If the convergence criterion is not satisfied, the population is iteratively operated by the three operators and the resulting new population is evaluated for the fitness values. The procedure is continued through several generations until the convergence criterion is satisfied and the process is terminated. The details of the three operations of GAs are given below.

Reproduction. Reproduction is the first operation applied to the population to select good strings (designs) of the population to form a mating pool. The reproduction operator is also called the selection operator because it selects good strings of the population. The reproduction operator is used to pick above-average strings from the current population and insert their multiple copies in the mating pool based on a probabilistic procedure. In a commonly used reproduction operator, a string is selected from the mating pool with a probability proportional to its fitness. Thus if F_i denotes the fitness of the ith string in the population of size n, the probability for selecting the ith string for the mating pool (p_i) is given by

$$p_i = \frac{F_i}{\sum\limits_{j=1}^{n} F_j}; \quad i = 1, 2, \ldots, n \tag{13.11}$$

Note that Eq. (13.11) implies that the sum of the probabilities of the strings of the population being selected for the mating pool is one. The implementation of the selection process given by Eq. (13.11) can be understood by imagining a roulette wheel with its circumference divided into segments, one for each string of the population, with the segment lengths proportional to the fitness of the strings as shown in Fig. 13.1. By spinning the roulette wheel n times (n being the population size) and selecting, each time, the string chosen by the roulette-wheel pointer, we obtain a mating pool of size n. Since the segments of the circumference of the wheel are marked according to the fitness of the various strings of the original population, the roulette-wheel process is expected to select $F_i/\overline{F}$ copies of the ith string for the mating pool, where $\overline{F}$ denotes the average fitness of the population:

$$\overline{F} = \frac{1}{n} \sum_{j=1}^{n} F_j \tag{13.12}$$

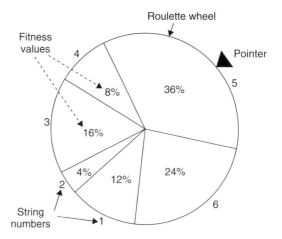

Figure 13.1 Roulette-wheel selection scheme.

In Fig. 13.1, the population size is assumed to be 6 with fitness values of the strings 1, 2, 3, 4, 5, and 6 given by 12, 4, 16, 8, 36, and 24, respectively. Since the fifth string (individual) has the highest value, it is expected to be selected most of the time (36% of the time, probabilistically) when the roulette wheel is spun n times ($n = 6$ in Fig. 13.1). The selection scheme, based on the spinning of the roulette wheel, can be implemented numerically during computations as follows.

The probabilities of selecting different strings based on their fitness values are calculated using Eq. (13.11). These probabilities are used to determine the cumulative probability of string i being copied to the mating pool, P_i, by adding the individual probabilities of strings 1 through i as

$$P_i = \sum_{j=1}^{i} p_j \tag{13.13}$$

Thus the roulette-wheel selection process can be implemented by associating the cumulative probability range $(P_{i-1} - P_i)$ to the ith string. To generate the mating pool of

Table 13.1 Roulette-Wheel Selection Process for Obtaining the Mating Pool

String number i	Fitness value F_i	Probability of selecting string i for the mating pool, p_i	Cumulative probability value of string i, $P_i = \sum_{j=1}^{i} p_j$	Range of cumulative probability of string i, (P_{i-1}, P_i)
1	12	0.12	0.12	0.00–0.12
2	4	0.04	0.16	0.12–0.16
3	16	0.16	0.32	0.16–0.32
4	8	0.08	0.40	0.32–0.40
5	36	0.36	0.76	0.40–0.76
6	24	0.24	1.00	0.76–1.00

size n during numerical computations, n random numbers, each in the range of zero to one, are generated (or chosen). By treating each random number as the cumulative probability of the string to be copied to the mating pool, n strings corresponding to the n random numbers are selected as members of the mating pool. By this process, the string with a higher (lower) fitness value will be selected more (less) frequently to the mating pool because it has a larger (smaller) range of cumulative probability. Thus strings with high fitness values in the population, probabilistically, get more copies in the mating pool. It is to be noted that no new strings are formed in the reproduction stage; only the existing strings in the population get copied to the mating pool. The reproduction stage ensures that highly fit individuals (strings) live and reproduce, and less fit individuals (strings) die. Thus the GAs simulate the principle of "survival-of-the-fittest" of nature.

Example 13.2   Consider six strings with fitness values 12, 4, 16, 8, 36, and 24 with the corresponding roulette wheel as shown in Fig. 13.1. Find the levels of contribution of the various strings to the mating pool using the roulette-wheel selection process with the following 12 random numbers: 0.41, 0.65, 0.42, 0.80, 0.67, 0.39, 0.63, 0.53, 0.86, 0.88, 0.75, 0.55.

Note: (1) These random numbers are taken from Ref. [13.20]. (2) Although the original population consists of only 6 strings, the mating pool is assumed to be composed of 12 strings to illustrate the roulette-wheel selection process.

SOLUTION   If the given random numbers are assumed to represent cumulative probabilities, the string numbers to be copied to the mating pool can be determined from the cumulative probability ranges listed in the last column of Table 13.1 as follows:

Random number (cumulative probability of the string to be copied)	0.41	0.65	0.42	0.80	0.67	0.39	0.63	0.53	0.86	0.88	0.75	0.55
String number to be copied to the mating pool	5	5	5	6	5	4	5	5	6	6	5	5

This indicates that the mating pool consists of 1 copy of string 4, 8 copies of string 5, and 3 copies of string 6. This shows that less fit individuals (strings 1, 2, and 3) did not contribute to the next generation (or died) because they could not contribute to the mating pool. String 4, although has a small fitness value, contributed 1 copy to the mating pool based on the random selection process used.

Crossover.   After reproduction, the crossover operator is implemented. The purpose of crossover is to create new strings by exchanging information among strings of the mating pool. Many crossover operators have been used in the literature of GAs. In most crossover operators, two individual strings (designs) are picked (or selected) at random from the mating pool generated by the reproduction operator and some portions of the strings are exchanged between the strings. In the commonly used process, known as a single-point crossover operator, a crossover site is selected at random along the string length, and the binary digits (alleles) lying on the right side of the crossover site are swapped (exchanged) between the two strings. The two strings selected for participation in the crossover operators are known as parent strings and the strings generated by the crossover operator are known as child strings.

For example, if two design vectors (parents), each with a string length of 10, are given by

$$\begin{array}{ll} \text{(Parent 1)} & \mathbf{X}_1 = \{0 \quad 1 \quad 0 \mid 1 \quad 0 \quad 1 \quad 1 \quad 0 \quad 1 \quad 1\} \\ \text{(Parent 2)} & \mathbf{X}_2 = \{1 \quad 0 \quad 0 \mid 0 \quad 1 \quad 1 \quad 1 \quad 1 \quad 0 \quad 0\} \end{array}$$

the result of crossover, when the crossover site is 3, is given by

$$\begin{array}{ll} \text{(Offspring 1)} & \mathbf{X}_3 = \{0 \quad 1 \quad 0 \mid 0 \quad 1 \quad 1 \quad 1 \quad 1 \quad 0 \quad 0\} \\ \text{(Offspring 2)} & \mathbf{X}_4 = \{1 \quad 0 \quad 0 \mid 1 \quad 0 \quad 1 \quad 1 \quad 0 \quad 1 \quad 1\} \end{array}$$

Since the crossover operator combines substrings from parent strings (which have good fitness values), the resulting child strings created are expected to have better fitness values provided an appropriate (suitable) crossover site is selected. However, the suitable or appropriate crossover site is not known before hand. Hence the crossover site is usually chosen randomly. The child strings generated using a random crossover site may or may not be as good or better than their parent strings in terms of their fitness values. If they are good or better than their parents, they will contribute to a faster improvement of the average fitness value of the new population. On the other hand, if the child strings created are worse than their parent strings, it should not be of much concern to the success of the GAs because the bad child strings will not survive very long as they are less likely to be selected in the next reproduction stage (because of the survival-of-the-fittest strategy used).

As indicated above, the effect of crossover may be useful or detrimental. Hence it is desirable not to use all the strings of the mating pool in crossover but to preserve some of the good strings of the mating pool as part of the population in the next generation. In practice, a crossover probability, p_c, is used in selecting the parents for crossover. Thus only 100 p_c percent of the strings in the mating pool will be used in the crossover operator while 100 $(1 - p_c)$ percent of the strings will be retained as they are in the new generation (of population).

Mutation. The crossover is the main operator by which new strings with better fitness values are created for the new generations. The mutation operator is applied to the new strings with a specific small mutation probability, p_m. The mutation operator changes the binary digit (allele's value) 1 to 0 and vice versa. Several methods can be used for implementing the mutation operator. In the single-point mutation, a mutation site is selected at random along the string length and the binary digit at that site is then changed from 1 to 0 or 0 to 1 with a probability of p_m. In the bit-wise mutation, each bit (binary digit) in the string is considered one at a time in sequence, and the digit is changed from 1 to 0 or 0 to 1 with a probability p_m. Numerically, the process can be implemented as follows. A random number between 0 and 1 is generated/chosen. If the random number is smaller than p_m, then the binary digit is changed. Otherwise, the binary digit is not changed.

The purpose of mutation is (1) to generate a string (design point) in the neighborhood of the current string, thereby accomplishing a local search around the current solution, (2) to safeguard against a premature loss of important genetic material at a particular position, and (3) to maintain diversity in the population.

As an example, consider the following population of size $n = 5$ with a string length 10:

$$1\ 0\ 0\ 0\ 1\quad 0\ 0\ 0\ 1\ 1$$

$$1\ 0\ 1\ 1\ 1\quad 1\ 0\ 1\ 0\ 0$$

$$1\ 1\ 0\ 0\ 0\quad 0\ 1\ 1\ 0\ 1$$

$$1\ 0\ 1\ 1\ 0\quad 1\ 0\ 0\ 1\ 0$$

$$1\ 1\ 1\ 0\ 0\quad 0\ 1\ 0\ 0\ 1$$

Here all the five strings have a 1 in the position of the first bit. The true optimum solution of the problem requires a 0 as the first bit. The required 0 cannot be created by either the reproduction or the crossover operators. However, when the mutation operator is used, the binary number will be changed from 1 to 0 in the location of the first bit with a probability of np_m.

Note that the three operators—reproduction, crossover, and mutation—are simple to implement. The reproduction operator selects good strings for the mating pool, the crossover operator recombines the substrings of good strings of the mating pool to create strings (next generation of population), and the mutation operator alters the string locally. The use of these three operators successively yields new generations with improved values of average fitness of the population. Although, the improvement of the fitness of the strings in successive generations cannot be proved mathematically, the process has been found to converge to the optimum fitness value of the objective function. Note that if any bad strings are created at any stage in the process, they will be eliminated by the reproduction operator in the next generation. The GAs have been successfully used to solve a variety of optimization problems in the literature.

13.2.5   Algorithm

The computational procedure involved in maximizing the fitness function $F(x_1, x_2, x_3, \ldots, x_n)$ in the genetic algorithm can be described by the following steps.

1. Choose a suitable string length $l = nq$ to represent the n design variables of the design vector **X**. Assume suitable values for the following parameters: population size m, crossover probability p_c, mutation probability p_m, permissible value of standard deviation of fitness values of the population $(s_f)_{\max}$ to use as a convergence criterion, and maximum number of generations $(i_{\max})$ to be used an a second convergence criterion.

2. Generate a random population of size m, each consisting of a string of length $l = nq$. Evaluate the fitness values F_i, $i = 1, 2, \ldots, m$, of the m strings.

3. Carry out the reproduction process.

4. Carry out the crossover operation using the crossover probability p_c.

5. Carry out the mutation operation using the mutation probability p_m to find the new generation of m strings.

6. Evaluate the fitness values F_i, $i = 1, 2, \ldots, m$, of the m strings of the new population. Find the standard deviation of the m fitness values.

7. Test for the convergence of the algorithm or process. If $s_f \leq (s_f)_{max}$, the convergence criterion is satisfied and hence the process may be stopped. Otherwise, go to step 8.

8. Test for the generation number. If $i \geq i_{max}$, the computations have been performed for the maximum permissible number of generations and hence the process may be stopped. Otherwise, set the generation number as $i = i + 1$ and go to step 3.

13.2.6   Numerical Results

The welded beam problem described in Section 7.22.3 (Fig. 7.23) was considered by Deb [13.20] with the following data: population size = 100, total string length = 40, substring length for each design variable = 10, probability of crossover = 0.9, and probability of mutation = 0.01. Different penalty parameters were considered for different constraints in order to have the contribution of each constraint violation to the objective function be approximately the same. Nearly optimal solutions were obtained after only about 15 generations with approximately $0.9 \times 100 \times 15 = 1350$ function evaluations. The optimum solution was found to be $x_1^* = 0.2489$, $x_2^* = 6.1730$, $x_3^* = 8.1789$, $x_4^* = 0.2533$, and $f^* = 2.43$, which can be compared with the solution obtained from geometric programming, $x_1^* = 0.2455$, $x_2^* = 6.1960$, $x_3^* = 8.2730$, $x_4^* = 0.2455$, and $f^* = 2.39$ [13.21]. Although the optimum solution given by the GAs corresponds to a slightly larger value of f^*, it satisfies all the constraints (the solution obtained from geometric programming violates three constraints slightly).

13.3   SIMULATED ANNEALING

13.3.1   Introduction

The simulated annealing method is based on the simulation of thermal annealing of critically heated solids. When a solid (metal) is brought into a molten state by heating it to a high temperature, the atoms in the molten metal move freely with respect to each other. However, the movements of atoms get restricted as the temperature is reduced. As the temperature reduces, the atoms tend to get ordered and finally form crystals having the minimum possible internal energy. The process of formation of crystals essentially depends on the cooling rate. When the temperature of the molten metal is reduced at a very fast rate, it may not be able to achieve the crystalline state; instead, it may attain a polycrystalline state having a higher energy state compared to that of the crystalline state. In engineering applications, rapid cooling may introduce defects inside the material. Thus the temperature of the heated solid (molten metal) needs to be reduced at a slow and controlled rate to ensure proper solidification with a highly ordered crystalline state that corresponds to the lowest energy state (internal energy). This process of cooling at a slow rate is known as annealing.

13.3.2   Procedure

The simulated annealing method simulates the process of slow cooling of molten metal to achieve the minimum function value in a minimization problem. The cooling phenomenon of the molten metal is simulated by introducing a temperature-like parameter and controlling it using the concept of Boltzmann's probability distribution. The Boltzmann's probability distribution implies that the energy (E) of a system in thermal equilibrium at temperature T is distributed probabilistically according to the relation

$$P(E) = e^{-E/kT} \tag{13.14}$$

where $P(E)$ denotes the probability of achieving the energy level E, and k is called the Boltzmann's constant. Equation (13.14) shows that at high temperatures the system has nearly a uniform probability of being at any energy state; however, at low temperatures, the system has a small probability of being at a high-energy state. This indicates that when the search process is assumed to follow Boltzmann's probability distribution, the convergence of the simulated annealing algorithm can be controlled by controlling the temperature T. The method of implementing the Boltzmann's probability distribution in simulated thermodynamic systems, suggested by Metropolis et al. [13.37], can also be used in the context of minimization of functions.

In the case of function minimization, let the current design point (state) be $\mathbf{X}_i$, with the corresponding value of the objective function given by $f_i = f(\mathbf{X}_i)$. Similar to the energy state of a thermodynamic system, the energy E_i at state $\mathbf{X}_i$ is given by

$$E_i = f_i = f(\mathbf{X}_i) \tag{13.15}$$

Then, according to the Metropolis criterion, the probability of the next design point (state) $\mathbf{X}_{i+1}$ depends on the difference in the energy state or function values at the two design points (states) given by

$$\Delta E = E_{i+1} - E_i = \Delta f = f_{i+1} - f_i \equiv f(\mathbf{X}_{i+1}) - f(\mathbf{X}_i) \tag{13.16}$$

The new state or design point $\mathbf{X}_{i+1}$ can be found using the Boltzmann's probability distribution:

$$P[E_{i+1}] = \min\left\{1, e^{-\Delta E/kT}\right\} \tag{13.17}$$

The Boltzmann's constant serves as a scaling factor in simulated annealing and, as such, can be chosen as 1 for simplicity. Note that if $\Delta E \le 0$, Eq. (13.17) gives $P[E_{i+1}] = 1$ and hence the point $\mathbf{X}_{i+1}$ is always accepted. This is a logical choice in the context of minimization of a function because the function value at $\mathbf{X}_{i+1}$, f_{i+1}, is better (smaller) than at $\mathbf{X}_i$, f_i, and hence the design vector $\mathbf{X}_{i+1}$ must be accepted. On the other hand, when $\Delta E > 0$, the function value f_{i+1} at $\mathbf{X}_{i+1}$ is worse (larger) than the one at $\mathbf{X}_i$. According to most conventional optimization procedures, the point $\mathbf{X}_{i+1}$ cannot be accepted as the next point in the iterative process. However, the probability of accepting the point $\mathbf{X}_{i+1}$, in spite of its being worse than $\mathbf{X}_i$ in terms of the objective function value, is finite (although it may be small) according to the Metropolis criterion. Note that the probability of accepting the point $\mathbf{X}_{i+1}$

$$P[E_{i+1}] = \left\{e^{-\Delta E/kT}\right\} \tag{13.18}$$

is not same in all situations. As can be seen from Eq. (13.18), this probability depends on the values of ΔE and T. If the temperature T is large, the probability will be high for design points $\mathbf{X}_{i+1}$ with larger function values (with larger values of $\Delta E = \Delta f$). Thus at high temperatures, even worse design points $\mathbf{X}_{i+1}$ are likely to be accepted because of larger probabilities. However, if the temperature T is small, the probability of accepting worse design points $\mathbf{X}_{i+1}$ (with larger values of $\Delta E = \Delta f$) will be small. Thus as the temperature values get smaller (that is, as the process gets closer to the optimum solution), the design points $\mathbf{X}_{i+1}$ with larger function values compared to the one at $\mathbf{X}_i$ are less likely to be accepted.

13.3.3   Algorithm

The SA algorithm can be summarized as follows. Start with an initial design vector $\mathbf{X}_1$ (iteration number $i = 1$) and a high value of temperature T. Generate a new design point randomly in the vicinity of the current design point and find the difference in function values:

$$\Delta E = \Delta f = f_{i+1} - f_i \equiv f(\mathbf{X}_{i+1}) - f(\mathbf{X}_i) \tag{13.19}$$

If f_{i+1} is smaller than f_i (with a negative value of Δf), accept the point $\mathbf{X}_{i+1}$ as the next design point. Otherwise, when Δf is positive, accept the point $\mathbf{X}_{i+1}$ as the next design point only with a probability $e^{-\Delta E/kT}$. This means that if the value of a randomly generated number is larger than $e^{-\Delta E/kT}$, accept the point $\mathbf{X}_{i+1}$; otherwise, reject the point $\mathbf{X}_{i+1}$. This completes one iteration of the SA algorithm. If the point $\mathbf{X}_{i+1}$ is rejected, then the process of generating a new design point $\mathbf{X}_{i+1}$ randomly in the vicinity of the current design point, evaluating the corresponding objective function value f_{i+1}, and deciding to accept $\mathbf{X}_{i+1}$ as the new design point, based on the use of the Metropolis criterion, Eq. (13.18), is continued. To simulate the attainment of thermal equilibrium at every temperature, a predetermined number (n) of new points $\mathbf{X}_{i+1}$ are tested at any specific value of the temperature T.

Once the number of new design points $\mathbf{X}_{i+1}$ tested at any temperature T exceeds the value of n, the temperature T is reduced by a prespecified fractional value c ($0 < c < 1$) and the whole process is repeated. The procedure is assumed to have converged when the current value of temperature T is sufficiently small or when changes in the function values (Δf) are observed to be sufficiently small.

The choices of the initial temperature T, the number of iterations n before reducing the temperature, and the temperature reduction factor c play important roles in the successful convergence of the SA algorithm. For example, if the initial temperature T is too large, it requires a larger number of temperature reductions for convergence. On the other hand, if the initial temperature is chosen to be too small, the search process may be incomplete in the sense that it might fail to thoroughly investigate the design space in locating the global minimum before convergence. The temperature reduction factor c has a similar effect. Too large a value of c (such as 0.8 or 0.9) requires too much computational effort for convergence. On the other hand, too small a value of c (such as 0.1 or 0.2) may result in a faster reduction in temperature that might not permit a thorough exploration of the design space for locating the global minimum solution. Similarly, a large value of the number of iterations n will help in achieving quasiequilibrium state at each temperature but will result in a larger computational effort. A smaller value of n, on the other hand, might result either in a premature convergence or convergence to a local minimum (due to inadequate exploration of the design space for the global minimum). Unfortunately, no unique set of values are available for T, n, and c that will work well for every problem. However, certain guidelines can be given for selecting these values. The initial temperature T can be chosen as the average value of the objective function computed at a number of randomly selected points in the design space. The number of iterations n can be chosen between 50 and 100 based on the computing resources and the desired accuracy of solution. The temperature reduction factor c can be chosen between 0.4 and 0.6 for a reasonable temperature reduction strategy (also termed the cooling schedule). More complex cooling schedules, based on the expected mathematical convergence rates, have been used in the literature for the solution of complex practical optimization problems [13.19]. In spite of all the research being done on SA algorithms, the choice of the initial temperature T, the number of iterations n at any specific temperature, and the temperature reduction factor (or cooling rate) c still remain an art and generally require a trial-and-error process to find suitable values for solving any particular type of optimization problems. The SA procedure is shown as a flowchart in Fig. 13.2.

13.3.4   Features of the Method

Some of the features of simulated annealing are as follows:

1. The quality of the final solution is not affected by the initial guesses, except that the computational effort may increase with worse starting designs.
2. Because of the discrete nature of the function and constraint evaluations, the convergence or transition characteristics are not affected by the continuity or differentiability of the functions.
3. The convergence is also not influenced by the convexity status of the feasible space.
4. The design variables need not be positive.
5. The method can be used to solve mixed-integer, discrete, or continuous problems.
6. For problems involving behavior constraints (in addition to lower and upper bounds on the design variables), an equivalent unconstrained function is to be formulated as in the case of genetic algorithms.

13.3.5   Numerical Results

The welded beam problem of Section 7.22.3 (Fig. 7.23) is solved using simulated annealing. The solution is given by $x_1^* = 0.2471$, $x_2^* = 6.1451$, $x_3^* = 8.2721$, $x_4^* = 0.2495$, and $f^* = 2.4148$. This solution can be compared with the solutions obtained by genetic algorithms ($x_1^* = 0.2489$, $x_2 = 6.1730$, $x_3^* = 8.1789$, $x_4^* = 0.2533$, and $f^* = 2.4331$) and geometric programming ($x_1^* = 0.2536$, $x_2^* = 7.1410$, $x_3^* = 7.1044$, $x_4^* = 0.2536$, and $f^* = 2.3398$). Notice that the solution given by geometric programming [13.21] violated three constraints slightly, while the solutions given by the genetic algorithms [13.20] and simulated annealing satisfied all the constraints.

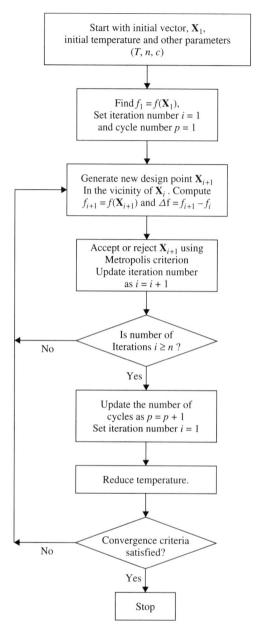

Figure 13.2   Simulated annealing procedure.

Example 13.3   Find the minimum of the following function using simulated annealing:

$$f(\mathbf{X}) = 500 - 20x_1 - 26x_2 - 4x_1x_2 + 4x_1^2 + 3x_2^2$$

SOLUTION   We follow the procedure indicated in the flowchart of Fig. 13.2.

Step 1: Choose the parameters of the SA method. The initial temperature is taken as the average value of f evaluated at four randomly selected points in the design space. By selecting the random points as $\mathbf{X}^{(1)} = \{{}^2_0\}$, $\mathbf{X}^{(2)} = \{{}^5_{10}\}$, $\mathbf{X}^{(3)} = \{{}^8_5\}$, $\mathbf{X}^{(4)} = \{{}^{10}_{10}\}$, we find the corresponding values of the objective function as $f^{(1)} = 476$, $f^{(2)} = 340$, $f^{(3)} = 381$, $f^{(4)} = 340$, respectively. Noting that the average value of the objective functions $f^{(1)}$, $f^{(2)}$, $f^{(3)}$, and $f^{(4)}$ is 384.25, we assume the initial temperature to be $T = 384.25$. The temperature reduction factor is chosen as $c = 0.5$. To make the computations brief, we choose the maximum permissible number of iterations (at any specific value of temperature) as $n = 2$. We select the initial design point as $\mathbf{X}_1 = \{{}^4_5\}$.

Step 2: Evaluate the objective function value at $\mathbf{X}_1$ as $f_1 = 349.0$ and set the iteration number as $i = 1$.

Step 3: Generate a new design point in the vicinity of the current design point. For this, we select two uniformly distributed random numbers u_1 and u_2; u_1 for x_1 in the vicinity of 4 and u_2 for x_2 in the vicinity of 5. The numbers u_1 and u_2 are chosen as 0.31 and 0.57, respectively. By choosing the ranges of x_1 and x_2 as $(-2, 10)$ and $(-1, 11)$, which represent ranges of ± 6 about their respective current values, the uniformly distributed random numbers r_1 and r_2 in the ranges of x_1 and x_2, corresponding to u_1 and u_2, can be found as

$$r_1 = -2 + u_1\{10 - (-2)\} = -2 + 0.31(12) = 1.72$$

$$r_2 = -1 + u_2\{11 - (-1)\} = -1 + 0.57(12) = 5.84$$

which gives $\mathbf{X}_2 = \begin{Bmatrix} r_1 \\ r_2 \end{Bmatrix} = \begin{Bmatrix} 1.72 \\ 5.84 \end{Bmatrix}$.

Since the objective function value $f_2 = f(\mathbf{X}_2) = 387.7312$, the value of Δf is given by

$$\Delta f = f_2 - f_1 = 387.7312 - 349.0 = 38.7312$$

Step 4: Since the value of Δf is positive, we use the Metropolis criterion to decide whether to accept or reject the current point. For this we choose a random number in the range $(0, 1)$ as $r = 0.83$. Equation (13.18) gives the probability of accepting the new design point $\mathbf{X}_2$ as

$$P[\mathbf{X}_2] = e^{-\Delta f/kT} \tag{E_1}$$

By assuming the value of the Boltzmann's constant k to be 1 for simplicity in Eq. (E_1), we obtain

$$P[\mathbf{X}_2] = e^{-\Delta f/kT} = e^{-38.7312/384.25} = 0.9041$$

Since $r = 0.83$ is smaller than 0.9041, we accept the point $\mathbf{X}_2 = \{{}^{1.72}_{5.84}\}$ as the next design point. Note that, although the objective function value f_2 is larger than f_1, we accept $\mathbf{X}_2$ because this is an early stage of simulation and the current temperature is high.

Step 3: Update the iteration number as $i = 2$. Since the iteration number i is less than or equal to n, we proceed to step 3.

Step 3: Generate a new design point in the vicinity of the current design point $\mathbf{X}_2 = \{ {}^{1.72}_{5.84} \}$. For this, we choose the range of each design variable as ± 6 about its current value so that the ranges are given by $(-6 + 1.72, 6 + 1.72) = (-4.28, 7.72)$ for x_1 and $(-6 + 5.84, 6 + 5.84) = (-0.16, 11.84)$ for x_2. By selecting two uniformly distributed random numbers in the range (0, 1) as $u_1 = 0.92$ and $u_2 = 0.73$, the corresponding uniformly distributed random numbers in the ranges of x_1 and x_2 become

$$r_1 = -4.28 + u_1 \{7.72 - (-4.28)\} = -4.28 + 0.92(12) = 6.76$$

$$r_2 = -0.16 + u_2 \{11.84 - (-0.16)\} = -0.16 + 0.73(12) = 8.60$$

which gives $\mathbf{X}_3 = \{ {}^{r_1}_{r_2} \} = \{ {}^{6.76}_{8.60} \}$ with a function value of $f_3 = 313.3264$. We note that the function value f_3 is better than f_2 with $\Delta f = f_3 - f_2 = 313.3264 - 387.7312 = -74.4048$.

Step 4: Since $\Delta f < 0$, we accept the current point as $\mathbf{X}_3$ and increase the iteration number to $i = 3$. Since $i > n$, we go to step 5.

Step 5: Since a cycle of iterations with the current value of temperature is completed, we reduce the temperature to a new value of $T = 0.5 (384.25) = 192.125$. Reset the current iteration number as $i = 1$ and go to step 3.

Step 3: Generate a new design point in the vicinity of the current design point $\mathbf{X}_3$ and continue the procedure until the temperature is reduced to a small value (until convergence).

13.4 PARTICLE SWARM OPTIMIZATION

13.4.1 Introduction

Particle swarm optimization, abbreviated as PSO, is based on the behavior of a colony or swarm of insects, such as ants, termites, bees, and wasps; a flock of birds; or a school of fish. The particle swarm optimization algorithm mimics the behavior of these social organisms. The word *particle* denotes, for example, a *bee* in a colony or a *bird* in a flock. Each individual or particle in a swarm behaves in a distributed way using its own intelligence and the collective or group intelligence of the swarm. As such, if one particle discovers a good path to *food,* the rest of the swarm will also be able to follow the good path instantly even if their location is far away in the swarm. Optimization methods based on swarm intelligence are called behaviorally inspired algorithms as opposed to the genetic algorithms, which are called evolution-based procedures. The PSO algorithm was originally proposed by Kennedy and Eberhart in 1995 [13.34].

In the context of multivariable optimization, the swarm is assumed to be of specified or fixed size with each particle located initially at random locations in the multidimensional design space. Each particle is assumed to have two characteristics: a *position* and a *velocity*. Each particle wanders around in the design space and remembers the best position (in terms of the food source or objective function value) it has discovered. The particles communicate information or good positions to each other and adjust their individual positions and velocities based on the information received on the good positions.

As an example, consider the behavior of birds in a flock. Although each bird has a limited intelligence by itself, it follows the following simple rules:

1. It tries not to come too close to other birds.
2. It steers toward the average direction of other birds.
3. It tries to fit the "average position" between other birds with no wide gaps in the flock.

Thus the behavior of the flock or swarm is based on a combination of three simple factors:

1. Cohesion—stick together.
2. Separation—don't come too close.
3. Alignment—follow the general heading of the flock.

The PSO is developed based on the following model:

1. When one bird locates a target or food (or maximum of the objective function), it instantaneously transmits the information to all other birds.
2. All other birds gravitate to the target or food (or maximum of the objective function), but not directly.
3. There is a component of each bird's own independent thinking as well as its past *memory*.

Thus the model simulates a random search in the design space for the maximum value of the objective function. As such, gradually over many iterations, the birds go to the target (or maximum of the objective function).

13.4.2   Computational Implementation of PSO

Consider an unconstrained maximization problem:

Maximize $f(\mathbf{X})$

with $\mathbf{X}^{(l)} \leq \mathbf{X} \leq \mathbf{X}^{(u)}$ (13.20)

where $\mathbf{X}^{(l)}$ and $\mathbf{X}^{(u)}$ denote the lower and upper bounds on $\mathbf{X}$, respectively. The PSO procedure can be implemented through the following steps.

1. Assume the size of the swarm (number of particles) is N. To reduce the total number of function evaluations needed to find a solution, we must assume a smaller size of the swarm. But with too small a swarm size it is likely to take us longer to find a solution or, in some cases, we may not be able to find a solution at all. Usually a size of 20 to 30 particles is assumed for the swarm as a compromise.
2. Generate the initial population of $\mathbf{X}$ in the range $\mathbf{X}^{(l)}$ and $\mathbf{X}^{(u)}$ randomly as $\mathbf{X}_1, \mathbf{X}_2, \ldots, \mathbf{X}_N$. Hereafter, for convenience, the particle (position of) j and its velocity in iteration i are denoted as $\mathbf{X}_j^{(i)}$ and $\mathbf{V}_j^{(i)}$, respectively. Thus the particles generated initially are denoted $\mathbf{X}_1(0)$, $\mathbf{X}_2(0), \ldots, \mathbf{X}_N(0)$. The vectors $\mathbf{X}_j(0)(j = 1, 2, \ldots, N)$ are called particles or vectors of coordinates of particles

(similar to chromosomes in genetic algorithms). Evaluate the objective function values corresponding to the particles as $f[\mathbf{X}_1(0)]$, $f[\mathbf{X}_2(0)]$, ..., $f[\mathbf{X}_N(0)]$.

3. Find the velocities of particles. All particles will be moving to the optimal point with a velocity. Initially, all particle velocities are assumed to be zero. Set the iteration number as $i = 1$.

4. In the ith iteration, find the following two important parameters used by a typical particle j:

(a) The historical best value of $\mathbf{X}_j(i)$ (coordinates of jth particle in the current iteration i), $\mathbf{P}_{\text{best},j}$, with the highest value of the objective function, $f[\mathbf{X}_j(i)]$, encountered by particle j in all the previous iterations.

The historical best value of $\mathbf{X}_j(i)$ (coordinates of all particles up to that iteration), $\mathbf{G}_{\text{best}}$, with the highest value of the objective function $f[\mathbf{X}_j(i)]$, encountered in all the previous iterations by any of the N particles.

(b) Find the velocity of particle j in the ith iteration as follows:

$$\mathbf{V}_j(i) = \mathbf{V}_j(i-1) + c_1 r_1 [\mathbf{P}_{\text{best},j} - \mathbf{X}_j(i-1)]$$
$$+ c_2 r_2 [\mathbf{G}_{\text{best}} - \mathbf{X}_j(i-1)]; \quad j = 1, 2, \ldots, N \qquad (13.21)$$

where c_1 and c_2 are the cognitive (individual) and social (group) learning rates, respectively, and r_1 and r_2 are uniformly distributed random numbers in the range 0 and 1. The parameters c_1 and c_2 denote the relative importance of the memory (position) of the particle itself to the memory (position) of the swarm. The values of c_1 and c_2 are usually assumed to be 2 so that $c_1 r_1$ and $c_2 r_2$ ensure that the particles would overfly the target about half the time.

(c) Find the position or coordinate of the jth particle in ith iteration as

$$\mathbf{X}_j(i) = \mathbf{X}_j(i-1) + \mathbf{V}_j(i); \quad j = 1, 2, \ldots, N \qquad (13.22)$$

where a time step of unity is assumed in the velocity term in Eq. (13.22). Evaluate the objective function values corresponding to the particles as $f[\mathbf{X}_1(i)]$, $F[\mathbf{X}_2(i)]$, ..., $F[\mathbf{X}_N(i)]$.

5. Check the convergence of the current solution. If the positions of all particles converge to the same set of values, the method is assumed to have converged. If the convergence criterion is not satisfied, step 4 is repeated by updating the iteration number as $i = i + 1$, and by computing the new values of $\mathbf{P}_{\text{best},j}$ and $\mathbf{G}_{\text{best}}$. The iterative process is continued until all particles converge to the same optimum solution.

13.4.3 Improvement to the Particle Swarm Optimization Method

It is found that usually the particle velocities build up too fast and the maximum of the objective function is skipped. Hence an inertia term, θ, is added to reduce the velocity. Usually, the value of θ is assumed to vary linearly from 0.9 to 0.4 as the iterative process progresses. The velocity of the jth particle, with the inertia term, is assumed as

$$\mathbf{V}_j(i) = \theta \mathbf{V}_j(i-1) + c_1 r_1 [\mathbf{P}_{\text{best},j} - \mathbf{X}_j(i-1)]$$
$$+ c_2 r_2 [\mathbf{G}_{\text{best}} - \mathbf{X}_j(i-1)]; \quad j = 1, 2, \ldots, N \tag{13.23}$$

The inertia weight θ was originally introduced by Shi and Eberhart in 1999 [13.36] to dampen the velocities over time (or iterations), enabling the swarm to converge more accurately and efficiently compared to the original PSO algorithm with Eq. (13.21). Equation (13.23) denotes an adapting velocity formulation, which improves its fine tuning ability in solution search. Equation (13.23) shows that a larger value of θ promotes global exploration and a smaller value promotes a local search. Thus a large value of θ makes the algorithm constantly explore new areas without much local search and hence fails to find the true optimum. To achieve a balance between global and local exploration to speed up convergence to the true optimum, an inertia weight whose value decreases linearly with the iteration number has been used:

$$\theta(i) = \theta_{\text{max}} - \left(\frac{\theta_{\text{max}} - \theta_{\text{min}}}{i_{\text{max}}} \right) i \tag{13.24}$$

where θ_{max} and θ_{min} are the initial and final values of the inertia weight, respectively, and i_{max} is the maximum number of iterations used in PSO. The values of $\theta_{\text{max}} = 0.9$ and $\theta_{\text{min}} = 0.4$ are commonly used.

13.4.4   Solution of the Constrained Optimization Problem

Let the constrained optimization problem be given by

$$\text{Maximize } f(\mathbf{X})$$

subject to   (13.25)

$$g_j(\mathbf{X}) \leq 0; \quad j = 1, 2, \ldots, m$$

An equivalent unconstrained function, $F(\mathbf{X})$, is constructed by using a penalty function for the constraints. Two types of penalty functions can be used in defining the function $F(\mathbf{X})$. The first type, known as the stationary penalty function, uses fixed penalty parameters throughout the minimization and the penalty value depends only on the degree of violation of the constraints. The second type, known as nonstationary penalty function, uses penalty parameters whose values change dynamically with the iteration number during optimization. The results obtained with the nonstationary penalty functions have been found to be superior to those obtained with stationary penalty functions in the numerical studies reported in the literature. As such, the nonstationary penalty function is to be used in practical computations.

According to the nonstationary penalty function approach, the function $F(\mathbf{X})$ is defined as

$$F(\mathbf{X}) = f(\mathbf{X}) + C(i)H(\mathbf{X}) \tag{13.26}$$

where $C(i)$ denotes a dynamically modified penalty parameter that varies with the iteration number i, and $H(\mathbf{X})$ represents the penalty factor associated with the constraints:

$$C(i) = (ci)^{\alpha} \tag{13.27}$$

$$H(\mathbf{X}) = \sum_{j=1}^{m} \left\{ \varphi[g_j(\mathbf{X})][q_j(\mathbf{X})]^{\gamma[q_i(\mathbf{X})]} \right\} \tag{13.28}$$

$$\varphi[q_j(\mathbf{X})] = a\left(1 - \frac{1}{e^{q_j(\mathbf{X})}}\right) + b \tag{13.29}$$

$$q_j(\mathbf{X}) = \max\left\{0, g_j(\mathbf{X})\right\}; \quad j = 1, 2, \ldots, m \tag{13.30}$$

where c, α, a, and b are constants. Note that the function $q_j(\mathbf{X})$ denotes the magnitude of violation of the jth constraint, $\varphi[q_j(\mathbf{X})]$ indicates a continuous assignment function, assumed to be of exponential form, as shown in Eq. (13.29), and $\gamma[q_i(\mathbf{X})]$ represents the power of the violated function. The values of $c = 0.5$, $\alpha = 2$, $a = 150$, and $b = 10$ along with

$$\gamma[q_j(\mathbf{X})] = \begin{cases} 1 & \text{if} \quad q_j(\mathbf{X}) \le 1 \\ 2 & \text{if} \quad q_j(\mathbf{X}) > 1 \end{cases} \tag{13.31}$$

were used by Liu and Lin [13.35].

Example 13.4 Find the maximum of the function

$$f(x) = -x^2 + 2x + 11$$

in the range $-2 \le x \le 2$ using the PSO method. Use 4 particles ($N = 4$) with the initial positions $x_1 = -1.5$, $x_2 = 0.0$, $x_3 = 0.5$, and $x_4 = 1.25$. Show the detailed computations for iterations 1 and 2.

SOLUTION

1. Choose the number of particles N as 4.
2. The initial population, chosen randomly (given as data), can be represented as $x_1(0) = -1.5$, $x_2(0) = 0.0$, $x_3(0) = 0.5$, and $x_4(0) = 1.25$. Evaluate the objective function values at current $x_j(0)$, $j = 1, 2, 3, 4$ as $f_1 = f[x_1(0)] = f(-1.5) = 5.75$, $f_2 = f[x_2(0)] = f(0.0) = 11.0$, $f_3 = f[x_3(0)] = f(0.5) = 11.75$, and $f_4 = f[x_4(0)] = f(1.25) = 11.9375$.
3. Set the initial velocities of each particle to zero:

$$v_1(0) = v_2(0) = v_3(0) = v_4(0) = 0$$

Set the iteration number as $i = 1$ and go to step 4.

4. (a) Find $P_{\text{best},1} = -1.5$, $P_{\text{best},2} = 0.0$, $P_{\text{best},3} = 0.5$, $P_{\text{best},4} = 1.25$, and $G_{\text{best}} = 1.25$.
   (b) Find the velocities of the particles as (by assuming $c_1 = c_2 = 1$ and using the random numbers in the range (0, 1) as $r_1 = 0.3294$ and $r_2 = 0.9542$):

$$v_j(i) = v_j(i-1) + r_1[P_{\text{best},j} - x_j(i-1)]$$
$$+ r_2[G_{\text{best}} - x_j(i-1)]; \quad j = 1, 2, 3, 4$$

so that

$$v_1(1) = 0 + 0.3294(-1.5 + 1.5) + 0.9542(1.25 + 1.5) = 2.6241$$

$$v_2(1) = 0 + 0.3294(0.0 - 0.0) + 0.9542(1.25 - 0.0) = 1.1927$$

$$v_3(1) = 0 + 0.3294(0.5 - 0.5) + 0.9542(1.25 - 0.5) = 0.7156$$

$$v_4(1) = 0 + 0.3294(1.25 - 1.25) + 0.9542(1.25 - 1.25) = 0.0$$

(c) Find the new values of $x_j(1)$, $j = 1, 2, 3, 4$, as $x_j(i) = x_j(i-1) + v_j(i)$:

$$x_1(1) = -1.5 + 2.6241 = 1.1241$$

$$x_2(1) = 0.0 + 1.1927 = 1.1927$$

$$x_3(1) = 0.5 + 0.7156 = 1.2156$$

$$x_4(1) = 1.25 + 0.0 = 1.25$$

5. Evaluate the objective function values at the current $x_j(i)$:

$$f[x_1(1)] = 11.9846, \quad f[x_2(1)] = 11.9629, \quad f[x_3(1)] = 11.9535,$$

$$f[x_4(1)] = 11.9375$$

Check the convergence of the current solution. Since the values of $x_j(i)$ did not converge, we increment the iteration number as $i = 2$ and go to step 4.

4. (a) Find $P_{best,1} = 1.1241$, $P_{best,2} = 1.1927$, $P_{best,3} = 1.2156$, $P_{best,4} = 1.25$, and $G_{best} = 1.1241$.

(b) Compute the new velocities of particles (by assuming $c_1 = c_2 = 1$ and using the random numbers in the range (0, 1) as $r_1 = 0.1482$ and $r_2 = 0.4867$):

$$v_j(i) = v_j(i-1) + r_1(P_{best,j} - x_j(i)) + r_2(G_{best} - x_j(i)); \quad j = 1, 2, 3, 4$$

so that

$$v_1(2) = 2.6240 + 0.1482(1.1241 - 1.1241) + 0.4867(1.1241 - 1.1241) = 2.6240$$

$$v_2(2) = 1.1927 + 0.1482(1.1927 - 1.1927) + 0.4867(1.1241 - 1.1927) = 1.1593$$

$$v_3(2) = 0.7156 + 0.1482(1.2156 - 1.2156) + 0.4867(1.1241 - 1.2156) = 0.6711$$

$$v_4(2) = 0.0 + 0.1482(1.25 - 1.25) + 0.4867(1.1241 - 1.25) = -0.0613$$

(c) Compute the current values of $x_j(i)$ as $x_j(i) = x_j(i-1) + v_j(i)$, $j = 1, 2, 3, 4$:

$$x_1(2) = 1.1241 + 2.6240 = 3.7481$$

$$x_2(2) = 1.1927 + 1.1593 = 2.3520$$

$$x_3(2) = 1.2156 + 0.6711 = 1.8867$$

$$x_4(2) = 1.25 - 0.0613 = 1.1887$$

6. Find the objective function values at the current $x_j(i)$:

$$f[x_1(2)] = 4.4480, \quad f[x_2(2)] = 10.1721, \quad f[x_3(2)] = 11.2138,$$

$$f[x_4(2)] = 11.9644$$

Check the convergence of the process. Since the values of $x_j(i)$ did not converge, we increment the iteration number as $i = 3$ and go to step 4. Repeat step 4 until the convergence of the process is achieved.

13.5 ANT COLONY OPTIMIZATION

13.5.1 Basic Concept

Ant colony optimization (ACO) is based on the cooperative behavior of real ant colonies, which are able to find the shortest path from their nest to a food source. The method was developed by Dorigo and his associates in the early 1990s [13.31, 13.32]. The ant colony optimization process can be explained by representing the optimization problem as a multilayered graph as shown in Fig. 13.3, where the number of

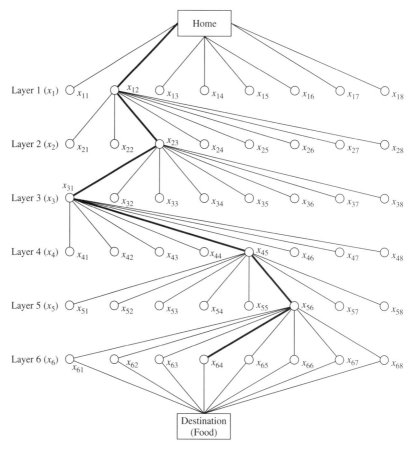

Figure 13.3 Graphical representation of the ACO process in the form of a multi-layered network.

layers is equal to the number of design variables and the number of nodes in a particular layer is equal to the number of discrete values permitted for the corresponding design variable. Thus each node is associated with a permissible discrete value of a design variable. Figure 13.3 denotes a problem with six design variables with eight permissible discrete values for each design variable.

The ACO process can be explained as follows. Let the colony consist of N ants. The ants start at the home node, travel through the various layers from the first layer to the last or final layer, and end at the destination node in each cycle or iteration. Each ant can select only one node in each layer in accordance with the state transition rule given by Eq. (13.32). The nodes selected along the path visited by an ant represent a candidate solution. For example, a typical path visited by an ant is shown by thick lines in Fig. 13.3. This path represents the solution $(x_{12}, x_{23}, x_{31}, x_{45}, x_{56}, x_{64})$. Once the path is complete, the ant deposits some pheromone on the path based on the local updating rule given by Eq. (13.33). When all the ants complete their paths, the pheromones on the globally best path are updated using the global updating rule described by Eqs. (13.32) and (13.33).

In the beginning of the optimization process (i.e., in iteration 1), all the edges or rays are initialized with an equal amount of pheromone. As such, in iteration 1, all the ants start from the home node and end at the destination node by randomly selecting a node in each layer. The optimization process is terminated if either the prespecified maximum number of iterations is reached or no better solution is found in a prespecified number of successive cycles or iterations. The values of the design variables denoted by the nodes on the path with largest amount of pheromone are considered as the components of the optimum solution vector. In general, at the optimum solution, all ants travel along the same best (converged) path.

13.5.2 Ant Searching Behavior

An ant k, when located at node i, uses the pheromone trail τ_{ij} to compute the probability of choosing j as the next node:

$$
p_{ij}^{(k)} = \begin{cases} \dfrac{\tau_{ij}^{\alpha}}{\sum\limits_{j \in N_i^{(k)}} \tau_{ij}^{\alpha}} & if \ j \in N_i^{(k)} \\[4mm] 0 & if \ j \notin N_i^{(k)} \end{cases} \tag{13.32}
$$

where α denotes the degree of importance of the pheromones and $N_i^{(k)}$ indicates the set of neighborhood nodes of ant k when located at node i. The neighborhood of node i contains all the nodes directly connected to node i except the predecessor node (i.e., the last node visited before i). This will prevent the ant from returning to the same node visited immediately before node i. An ant travels from node to node until it reaches the destination (food) node.

13.5.3 Path Retracing and Pheromone Updating

Before returning to the home node (backward node), the kth ant deposits $\Delta \tau^{(k)}$ of pheromone on arcs it has visited. The pheromone value τ_{ij} on the arc (i, j) traversed is updated as follows:

$$\tau_{ij} \leftarrow \tau_{ij} + \Delta \tau^{(k)} \tag{13.33}$$

Because of the increase in the pheromone, the probability of this arc being selected by the forthcoming ants will increase.

13.5.4 Pheromone Trail Evaporation

When an ant k moves to the next node, the pheromone evaporates from all the arcs ij according to the relation

$$\tau_{ij} \leftarrow (1 - p)\tau_{ij}; \quad \forall (i, j) \in A \tag{13.34}$$

where $p \in (0, 1]$ is a parameter and A denotes the segments or arcs traveled by ant k in its path from home to destination. The decrease in pheromone intensity favors the exploration of different paths during the search process. This favors the elimination of poor choices made in the path selection. This also helps in bounding the maximum value attained by the pheromone trails. An iteration is a complete cycle involving ant's movement, pheromone evaporation and pheromone deposit.

After all the ants return to the home node (nest), the pheromone information is updated according to the relation

$$\tau_{ij} = (1 - \rho)\tau_{ij} + \sum_{k=1}^{N} \Delta \tau_{ij}^{(k)} \tag{13.35}$$

where $\rho \in (0, 1]$ is the evaporation rate (also known as the pheromone decay factor) and $\Delta \tau_{ij}^{(k)}$ is the amount of pheromone deposited on arc ij by the best ant k. The goal of pheromone update is to increase the pheromone value associated with good or promising paths. The pheromone deposited on arc ij by the best ant is taken as

$$\Delta \tau_{ij}^{(k)} = \frac{Q}{L_k} \tag{13.36}$$

where Q is a constant and L_k is the length of the path traveled by the kth ant (in the case of the travel from one city to another in a traveling salesman problem). Equation (13.36) can be implemented as

$$\Delta \tau_{ij}^{(k)} = \begin{cases} \dfrac{\varsigma f_{\text{best}}}{f_{\text{worst}}}; & \text{if } (i, j) \in \text{global best tour} \\ 0; & \text{otherwise} \end{cases} \tag{13.37}$$

where f_{worst} is the worst value and f_{best} is the best value of the objective function among the paths taken by the N ants, and ζ is a parameter used to control the scale of the global updating of the pheromone. The larger the value of ζ, the more pheromone deposited on the global best path, and the better the exploitation ability. The aim of Eq. (13.37) is to provide a greater amount of pheromone to the tours (solutions) with better objective function values.

13.5.5 Algorithm

The step-by-step procedure of ACO algorithm for solving a minimization problem can be summarized as follows:

Step 1: Assume a suitable number of ants in the colony (N). Assume a set of permissible discrete values for each of the n design variables. Denote the permissible discrete values of the design variable x_i as $x_{i1}, x_{i2}, \ldots, x_{ip}$ ($i = 1, 2, \ldots, n$). Assume equal amounts of pheromone $\tau_{ij}^{(1)}$ initially along all the arcs or rays (discrete values of design variables) of the multilayered graph shown in Fig. 13.3. The superscript to τ_{ij} denotes the iteration number. For simplicity, $\tau_{ij}^{(1)} = 1$ can be assumed for all arcs ij. Set the iteration number $l = 1$.

Step 2:

(a) Compute the probability (p_{ij}) of selecting the arc or ray (or the discrete value) x_{ij} as

$$p_{ij} = \frac{\tau_{ij}^{(l)}}{\displaystyle\sum_{m=1}^{p} \tau_{im}^{(l)}}; \quad i = 1, 2, \ldots, n; \quad j = 1, 2, \ldots, p \qquad (13.38)$$

which can be seen to be same as Eq. (13.32) with $\alpha = 1$. A larger value can also be used for α.

(b) The specific path (or discrete values) chosen by the kth ant can be determined using random numbers generated in the range (0, 1). For this, we find the cumulative probability ranges associated with different paths of Fig. 13.3 based on the probabilities given by Eq. (13.38). The specific path chosen by ant k will be determined using the roulette-wheel selection process in step 3(a).

Step 3:

(a) Generate N random numbers $r_1, r_2, \ldots, r_N$ in the range (0, 1), one for each ant. Determine the discrete value or path assumed by ant k for variable i as the one for which the cumulative probability range [found in step 2(b)] includes the value r_i.

(b) Repeat step 3(a) for all design variables $i = 1, 2, \ldots, n$.

(c) Evaluate the objective function values corresponding to the complete paths (design vectors $\mathbf{X}^{(k)}$ or values of x_{ij} chosen for all design variables $i = 1, 2, \ldots, n$ by ant k, $k = 1, 2, \ldots, N$):

$$f_k = f(\mathbf{X}^{(k)}); \quad k = 1, 2, \ldots, N \qquad (13.39)$$

Determine the best and worst paths among the N paths chosen by different ants:

$$f_{\text{best}} = \mathop{\text{min}}_{k = 1, 2, \ldots, N} \{f_k\} \qquad (13.40)$$

$$f_{\text{worst}} = \mathop{\text{max}}_{k = 1, 2, \ldots, N} \{f_k\} \qquad (13.41)$$

Step 4: Test for the convergence of the process. The process is assumed to have converged if all N ants take the same best path. If convergence is not achieved, assume that all the ants return home and start again in search of food. Set the new iteration number as $l = l + 1$, and update the pheromones on different arcs (or discrete values of design variables) as

$$\tau_{ij}^{(l)} = \tau_{ij}^{(\text{old})} + \sum_k \Delta \tau_{ij}^{(k)} \tag{13.42}$$

where $\tau_{ij}^{(\text{old})}$ denotes the pheromone amount of the previous iteration left after evaporation, which is taken as

$$\tau_{ij}^{(\text{old})} = (1 - \rho) \tau_{ij}^{(l-1)} \tag{13.43}$$

and $\Delta \tau_{ij}^{(k)}$ is the pheromone deposited by the best ant k on its path and the summation extends over all the best ants k (if multiple ants take the same best path). Note that the best path involves only one arc ij (out of p possible arcs) for the design variable i. The evaporation rate or pheromone decay factor ρ is assumed to be in the range 0.5 to 0.8 and the pheromone deposited $\Delta \tau_{ij}^{(k)}$ is computed using Eq. (13.37).

With the new values of $\tau_{ij}^{(l)}$, go to step 2. Steps 2, 3, and 4 are repeated until the process converges, that is, until all the ants choose the same best path. In some cases, the iterative process is stopped after completing a prespecified maximum number of iterations (l_{max}).

Example 13.5   Find the minimum of the function $f(x) = x^2 - 2x - 11$ in the range $(0, 3)$ using the ant colony optimization method.

SOLUTION

Step 1: Assume the number of ants is $N = 4$. Note that there is only one design variable in this example ($n = 1$). The permissible discrete values of $x = x_1$ are assumed, within the range of x_1, as ($p = 7$):

$$x_{11} = 0.0, \ x_{12} = 0.5, \ x_{13} = 1.0, \ x_{14} = 1.5, \ x_{15} = 2.0, \ x_{16} = 2.5, \ x_{17} = 3.0$$

Each ant can choose any of the discrete values (paths or arcs) x_{1j}, $j = 1, 2, \ldots, 7$ shown in Fig. 13.4. Assume equal amounts of pheromone along each of the paths or arcs (τ_{1j}) shown in Fig. 13.4. For simplicity, $\tau_{1j} = 1$ is assumed for $j = 1, 2, \ldots, 7$.

Set the iteration number as $l = 1$.

Step 2: For any ant k, the probability of selecting path (or discrete variable) x_{1j} is given by

$$p_{1j} = \frac{\tau_{1j}}{\sum\limits_{p=1}^{7} \tau_{1p}} = \frac{1}{7}$$

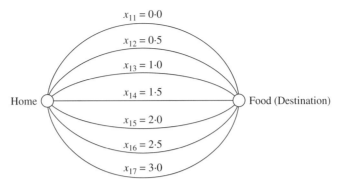

Figure 13.4   Possible paths for an ant (possible discrete values of $x \equiv x_1$).

To select the specific path (or discrete variable) chosen by an ant using a random number generated in the range $(0, 1)$, cumulative probability ranges are associated with different paths of Fig. 13.4 as (using roulette-wheel selection process in step 3):

$$x_{11} = \left(0, \tfrac{1}{7}\right) = (0.0, 0.1428), \quad x_{12} = \left(\tfrac{1}{7}, \tfrac{2}{7}\right) = (0.1428, 0.2857),$$

$$x_{13} = \left(\tfrac{2}{7}, \tfrac{3}{7}\right) = (0.2857, 0.4286),$$

$$x_{14} = \left(\tfrac{3}{7}, \tfrac{4}{7}\right) = (0.4286, 0.5714), \quad x_{15} = \left(\tfrac{4}{7}, \tfrac{5}{7}\right) = (0.5714, 0.7143),$$

$$x_{16} = \left(\tfrac{5}{7}, \tfrac{6}{7}\right) = (0.7143, 0.8571),$$

$$x_{17} = \left(\tfrac{6}{7}, 1\right) = (0.8571, 1.0)$$

Step 3: Generate four random numbers $r_i (i = 1, 2, 3, 4)$ in the range $(0, 1)$, one for each ant as $r_1 = 0.3122$, $r_2 = 0.8701$, $r_3 = 0.4729$, and $r_4 = 0.6190$. Using the cumulative probability range (given in step 2) in which the value of r_i falls, the discrete value assumed (or the path selected in Fig. 13.4) by different ants can be seen to be

ant 1 : $x_{13} = 1.0$; ant 2 : $x_{17} = 3.0$; ant 3 : $x_{14} = 1.5$; ant 4 : $x_{15} = 2.0$

The objective function values corresponding to the paths chosen by different ants are given by

ant 1 : $f_1 = f(x_{13}) = f(1.0) = -12.0$; ant 2 : $f_2 = f(x_{17}) = f(3.0) = -8.0$;

ant 3 : $f_3 = f(x_{14}) = f(1.5) = -11.75$; ant 4 : $f_4 = f(x_{15}) = f(2.0) = -11.0$

It can be seen that the path taken by ant 1 is the best one (with minimum value of the objective function): $x_{\text{best}} = x_{13} = 1.0$, $f_{\text{best}} = f_1 = -12.0$; and the path taken by ant 2 is the worst one (with maximum value of the objective function): $x_{\text{worst}} = x_{17} = 3.0$, $f_{\text{worst}} = f_2 = -8.0$.

Step 4: Assuming that the ants return home and start again in search of food, we set the iteration number as $l = 2$. We need to update the pheromone array as

$$\tau_{1j}^{(2)} = \tau_{1j}^{(\text{old})} + \sum_k \Delta\tau^{(k)} \tag{E$_1$}$$

where $\sum_k \Delta\tau^{(k)}$ is the pheromone deposited by the best ant k and the summation extends over all the best ants k (if multiple ants take the best path). In the present case, there is only one best ant, $k = 1$, which used the path x_{13}. Thus the value of $\sum_k \Delta\tau^{(k)}$ can be determined in this case as

$$\sum_k \Delta\tau^{(k)} = \Delta\tau^{(k=1)} = \frac{\varsigma f_{\text{best}}}{f_{\text{worst}}} = \frac{(2)(-12.0)}{(-8.0)} = 3.0$$

where the scaling parameter ς is assumed to be 2. Using a pheromone decay factor of $\rho = 0.5$ in Eq. (13.43), $\tau_{1j}^{(\text{old})}$ can be computed as

$$\tau_{1j}^{(\text{old})} = (1 - 0.5)\tau_{1j}^{(1)} = 0.5(1.0) = 0.5; \quad j = 1, 2, 4, 5, 6, 7$$

Thus Eq. (E$_1$) gives

$$\tau_{1j}^{(2)} = 1.0 + 3.0 = 4.0 \text{ for } j = 3 \text{ and } \tau_{1j}^{(2)} = 0.5 \text{ for } j = 1, 2, 4, 5, 6, 7$$

With this, we go to step 5.

Step 2: For any ant k, the probability of selecting path x_{1j} in Fig. 13.4 is given by

$$p_{1j} = \frac{\tau_{1j}}{\sum_{p=1}^{7} \tau_{1p}}; \quad j = 1, 2, \ldots, 7$$

where $\tau_{1j} = 0.5$; $j = 1, 2, 4, 5, 6, 7$ and $\tau_{13} = 4$. This gives

$$p_{1j} = \frac{0.5}{7.0} = 0.0714; \quad j = 1, 2, 4, 5, 6, 7; \quad p_{13} = \frac{4.0}{7.0} = 0.5714$$

To determine the discrete value or path selected by ant using a random number selected in the range (0, 1), cumulative probabilities are associated with different paths as (roulette wheel selection process):

$$x_{11} = (0, 0.0714), \quad x_{12} = (0.0714, 0.1429), \quad x_{13} = (0.1429, 0.7143),$$

$$x_{14} = (0.7143, 0.7857), \quad x_{15} = (0.7857, 0.8571), \quad x_{16} = (0.8571, 0.9286),$$

$$x_{17} = (0.9286, 1.0)$$

Step 3: Generate four random numbers in the range (0, 1), one for each of the ants as $r_1 = 0.3688, r_2 = 0.8577, r_3 = 0.0776, r_4 = 0.5791$. Using the cumulative probability range (given in step 2) in which the value of r_i falls, the discrete value assumed (or the path selected in Fig. 13.4) by different ants can be seen to be

$$\text{ant } 1 : x_{13} = 1.0; \text{ ant } 2 : x_{16} = 2.5; \text{ ant } 3 : x_{11} = 0.0; \text{ ant } 4 : x_{13} = 1.0$$

This shows that two ants (probabilistically) selected the path x_{13} due to higher pheromone left on the best path (x_{13}) found in the previous iteration. The objective function values corresponding to the paths chosen by different ants are given by

$$\text{ant } 1 : f_1 = f(x_{13}) = f(1.0) = -12.0; \text{ ant } 2 : f_2 = f(x_{16}) = f(2.5) = -9.75;$$

$$\text{ant } 3 : f_3 = f(x_{11}) = f(0.0) = -11.0; \text{ ant } 4 : f_4 = f(x_{13}) = f(1.0) = -12.0$$

It can be seen that the path taken by ants 1 and 4 is the best one with

$$x_{\text{best}} = x_{13} = 1.0 \text{ and } f_{\text{best}} = f_1 = f_4 = -12.0$$

and the path taken by ant 2 is the worst one with

$$x_{\text{worst}} = x_{16} = 2.5 \text{ and } f_{\text{worst}} = f_2 = -9.75$$

Now we go to step 4 to update the pheromone values on the various paths.

Step 4: Assuming that the ants return home and start again in search of food, we set the iteration number as $l = 3$. We need to update the pheromone array as

$$\tau_{1j}^{(3)} = \tau_{1j}^{(\text{old})} + \sum_k \Delta \tau^{(k)} \tag{E\textsubscript{2}}$$

where $\sum_k \Delta \tau^{(k)}$ is the pheromone deposited by the best ant k and the summation extends over all the best ants k (if multiple ants take the best path). In the present case, there are two best ants, $k = 1$ and 4, which used the path x_{13}. Thus the value of $\sum_k \Delta \tau^{(k)}$ can be determined in this case as

$$\sum_k \Delta \tau^{(k)} = \Delta \tau^{(k=1)} + \Delta \tau^{(k=4)} = \frac{2\varsigma f_{\text{best}}}{f_{\text{worst}}} = \frac{(2)(2)(-12.0)}{(-9.75)} = 4.9231$$

where the scaling parameter ς is assumed to be 2. Using a pheromone decay factor of $\rho = 0.5$ in Eq. (13.43), $\tau_{1j}^{(\text{old})}$ can be computed as

$$\tau_{1j}^{(\text{old})} = (1.0 - 0.5)\tau_{1j}^{(2)} = 0.5(0.5) = 0.25; \quad j = 1, 2, 4, 5, 6, 7$$

Thus Eq. (E\textsubscript{2}) gives

$$\tau_{1j}^{(3)} = 4.0 + 4.9231 = 8.9231 \text{ for } j = 3 \text{ and}$$

$$\tau_{1j}^{(3)} = 0.25 \text{ for } j = 1, 2, 4, 5, 6, 7$$

With this, we go to step 2.

Step 2: For any ant k, the probability of selecting path x_{1j} in Fig. 13.4 is given by

$$p_{1j} = \frac{\tau_{1j}}{\sum\limits_{p=1}^{7} \tau_{1p}}; \quad j = 1, 2, \ldots, 7$$

where $\tau_{1j} = 0.25$; $j = 1, 2, 4, 5, 6, 7$ and $\tau_{13} = 8.9231$. This gives

$$p_{1j} = \frac{0.25}{10.4231} = 0.0240, \quad j = 1, 2, 4, 5, 6, 7; \quad p_{13} = \frac{8.9231}{10.4231} = 0.8561$$

To determine the discrete value or path selected by an ant using a random number selected in the range (0, 1), cumulative probabilities are associated with different paths as (roulette-wheel selection process):

$x_{11} = (0, 0.0240), \quad x_{12} = (0.0240, 0.0480), \quad x_{13} = (0.0480, 0.9040),$

$x_{14} = (0.9040, 0.9280), \quad x_{15} = (0.9280, 0.9520),$

$x_{16} = (0.9520, 0.9760), \quad x_{17} = (0.9760, 1.0)$

With this information, we go to step 3 and then to step 4. Steps 2, 3, and 4 are repeated until the process converges (until all the ants choose the same best path).

13.6 OPTIMIZATION OF FUZZY SYSTEMS

In traditional designs, the optimization problem is stated in precise mathematical terms. However, in many real-world problems, the design data, objective function, and constraints are stated in vague and linguistic terms. For example, the statement, "This beam carries a load of 1000 lb with a probability of 0.8" is imprecise because of randomness in the material properties of the beam. On the other hand, the statement, "This beam carries a large load" is imprecise because of the fuzzy meaning of "large load." Similarly, in the optimum design of a machine component, the induced stress (σ) is constrained by an upper bound value ($\sigma_{\max}$) as $\sigma \leq \sigma_{\max}$. If $\sigma_{\max} = 30,000$ psi, it implies that a design with $\sigma = 30,000$ psi is acceptable whereas a design with $\sigma = 30,001$ psi is not acceptable. However, there is no substantive difference between designs with $\sigma = 30,000$ psi and $\sigma = 30,001$ psi. It appears that it is more reasonable to have a transition stage from absolute permission to absolute impermission. This implies that the constraint is to be stated in fuzzy terms. Fuzzy theories can be used to model and design systems involving vague and imprecise information [13.22, 13.26, 13.27].

13.6.1 Fuzzy Set Theory

Let X be a classical crisp set of objects, called the *universe,* whose generic elements are denoted by x. Membership in a classical subset A of X can be viewed as a characteristic function μ_A from X to [0, 1] such that

$$\mu_A(x) = \begin{cases} 1 & \text{if } x \in A \\ 0 & \text{if } x \notin A \end{cases} \tag{13.44}$$

The set [0, 1] is called a *valuation set*. A set A is called a *fuzzy set* if the valuation set is allowed to be the whole interval [0, 1]. The fuzzy set A is characterized by the set of all pairs of points denoted as

$$A = \{x, \mu_A(x)\}, \quad x \in X \tag{13.45}$$

where $\mu_A(x)$ is called the *membership function* of x in A. The closer the value of $\mu_A(x)$ is to 1, the more x belongs to A. For example, let $X = \{62 \quad 64 \quad 66 \quad 68 \quad 70 \quad 72 \quad 74 \quad 76 \quad 78 \quad 80\}$ be possible temperature settings of the thermostat ($^\circ$F) in an air-conditioned building. Then the fuzzy set A of "comfortable temperatures for human activity" may be defined as

$$A = \{(62, 0.2) \quad (64, 0.5) \quad (66, 0.8) \quad (68, 0.95) \quad (70, 0.85) \quad (72, 0.75)$$
$$(74, 0.6) \quad (76, 0.4) \quad (78, 0.2) \quad (80, 1.0)\} \tag{13.46}$$

where a grade of membership of 1 implies complete comfort and 0 implies complete discomfort. In general, if X is a finite set, $\{x_1, x_2, \ldots, x_n\}$ the fuzzy set on X can be expressed as

$$A = \mu_A(x_1)|_{x_1} + \mu_A(x_2)|_{x_2} + \cdots + \mu_A(x_n)|_{x_n} = \sum_{i=1}^{n} \mu_A(x_i)|_{x_i} \tag{13.47}$$

or in the limit, we can express A as

$$A = \int_x \mu_A(x)|_x \tag{13.48}$$

Crisp set theory is concerned with membership of precisely defined sets and is suitable for describing objective matters with countable events. Crisp set theory is developed using binary statements and is illustrated in Fig. 13.5a, which shows the support for y_1 with no ambiguity. Since fuzzy set theory is concerned with linguistic statements of support for membership in imprecise sets, a discrete fuzzy set is denoted as in Fig. 13.5b, where the degree of support is shown by the membership values, μ_1, $\mu_2, \ldots, \mu_n$, corresponding to $y_1, y_2, \ldots, y_n$, respectively. The discrete fuzzy set can be generalized to a continuous form as shown in Fig. 13.5c.

The basic crisp set operations of union, intersection, and complement can be represented on Venn diagrams as shown in Fig. 13.6. Similar operations can be defined for fuzzy sets, noting that the sets A and B do not have clear boundaries in this case. The graphs of μ_A and μ_B can be used to define the set-theoretic operations of fuzzy sets. The union of the fuzzy sets A and B is defined as

$$\mu_{A \cup B}(y) = \mu_A(y) \vee \mu_B(y) = \max[\mu_A(y), \mu_B(y)]$$
$$= \begin{cases} \mu_A(y) & \text{if } \mu_A > \mu_B \\ \mu_B(y) & \text{if } \mu_A < \mu_B \end{cases} \tag{13.49}$$

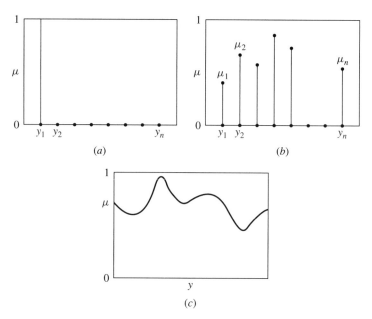

Figure 13.5   Crisp and fuzzy sets: (*a*) crisp set; (*b*) discrete fuzzy set; (*c*) continuous fuzzy set. [13.22], with permission of ASME.

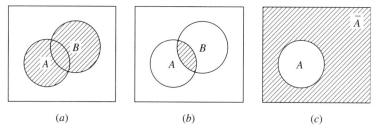

Figure 13.6   Basic set operations in crisp set theory: (*a*) *A* or *B* or both: $A \cup B$; (*b*) *A* and $B : A \cap B$; (*c*) not $A : \overline{A}$. [13.22], with permission of ASME.

The result of this operation is shown in Fig. 13.7*a*. The intersection of the fuzzy sets *A* and *B* is defined as

$$\mu_{A \cap B}(y) = \mu_A(y) \wedge \mu_B(y) = \min[\mu_A(y), \mu_B(y)]$$

$$= \begin{cases} \mu_A(y) & \text{if } \mu_A < \mu_B \\ \mu_B(y) & \text{if } \mu_A > \mu_B \end{cases} \tag{13.50}$$

This operation is shown in Fig. 13.7*b*. The complement of a fuzzy set *A* is shown as $\overline{A}$ in Fig. 13.7*c*, in which for every $\mu_A(y)$, there is a corresponding $\mu_{\overline{A}}(y) = 1 - \mu_A(y)$, which defines the complement of the set $A, \overline{A}$.

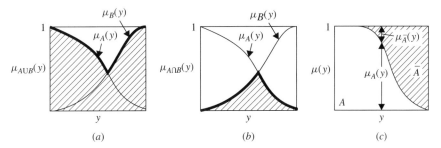

Figure 13.7   Basic set operations in fuzzy set theory: (*a*) union; (*b*) intersection; (*c*) complement. [13.22], with permission of ASME.

13.6.2   Optimization of Fuzzy Systems

The conventional optimization methods deal with selection of the design variables that optimizes an objective function subject to the satisfaction of the stated constraints. For a fuzzy system, this notion of optimization has to be revised. Since the objective and constraint functions are characterized by the membership functions in a fuzzy system, a design (decision) can be viewed as the intersection of the fuzzy objective and constraint functions. For illustration, consider the objective function: "The depth of the crane girder (x) should be substantially greater than 80 in." This can be represented by a membership function, such as

$$\mu_f(x) = \begin{cases} 0 & \text{if } x < 80 \text{ in.} \\ [1 + (x - 80)^{-2}]^{-1} & \text{if } x \geq 80 \text{ in.} \end{cases} \tag{13.51}$$

Let the constraint be "The depth of the crane girder (x) should be in the vicinity of 83 in." This can be described by a membership function of the type

$$\mu_g(x) = [1 + (x - 83)^4]^{-1} \tag{13.52}$$

Then the design (decision) is described by the membership function, $\mu_D(x)$, as

$$\begin{aligned} \mu_D(x) &= \mu_f(x) \wedge \mu_g(x) \\ &= \begin{cases} 0 & x < 80 \text{ in.} \\ \min\{[1 + (x - 80)^{-2}]^{-1}, [1 + (x - 83)^4]^{-1}\} \\ \quad \text{if } x \geq 80 \text{ in.} \end{cases} \end{aligned} \tag{13.53}$$

This relationship is shown in Fig. 13.8.

The conventional optimization problem is usually stated as follows:

$$\text{Find } \mathbf{X} \text{ which minimizes } f(\mathbf{X})$$

subject to

$$g_j^{(l)} \leq g_j(\mathbf{X}) \leq g_j^{(u)}, \quad j = 1, 2, \ldots, m \tag{13.54}$$

where the superscripts l and u denote the lower and upper bound values, respectively. The optimization problem of a fuzzy system is stated as follows:

$$\text{Find } \mathbf{X} \text{ which minimizes } f(\mathbf{X})$$

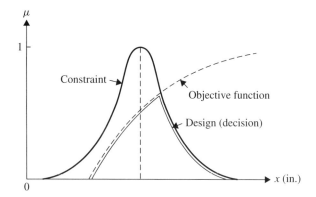

Figure 13.8 Concept of fuzzy decision. [13.22], with permission of ASME.

subject to

$$g_j(\mathbf{X}) \in G_j, \quad j = 1, 2, \ldots, m \tag{13.55}$$

where G_j denotes the fuzzy interval to which the function $g_j(\mathbf{X})$ should belong. Thus the fuzzy feasible region, S, which denotes the intersection of all G_j, is defined by the membership function

$$\mu_S(\mathbf{X}) = \min_{j=1,2,\ldots,m} \{\mu_{G_j}[g_j(\mathbf{X})]\} \tag{13.56}$$

Since a design vector $\mathbf{X}$ is considered feasible when $\mu_S(\mathbf{X}) > 0$, the optimum design is characterized by the maximum value of the intersection of the objective function and the feasible domain:

$$\mu_D(\mathbf{X}^*) = \max \mu_D(\mathbf{X}), \quad \mathbf{X} \in D \tag{13.57}$$

where

$$\mu_D(\mathbf{X}) = \min \left\{ \mu_f(\mathbf{X}), \min_{j=1,2,\ldots,m} \mu_{G_j}[g_j(\mathbf{X})] \right\} \tag{13.58}$$

13.6.3 Computational Procedure

The solution of a fuzzy optimization problem can be determined once the membership functions of f and g_j are known. In practical situations, the constructions of the membership functions is accomplished with the cooperation and assistance of experienced engineers in specific cases. In the absence of other information, linear membership functions are commonly used, based on the expected variations of the objective and constraint functions. Once the membership functions are known, the problem can be posed as a crisp optimization problem as

Find $\mathbf{X}$ and λ which maximize λ

subject to

$$\lambda \leq \mu_f(\mathbf{X})$$

$$\lambda \leq \mu_{g_j^{(l)}(\mathbf{X})}, \quad j = 1, 2, \ldots, m$$

$$\lambda \leq \mu_{g_j^{(u)}(\mathbf{X})}, \quad j = 1, 2, \ldots, m \tag{13.59}$$

13.6.4   Numerical Results

The minimization of the error between the generated and specified outputs of the four-bar mechanism shown in Fig. 13.9 is considered. The design vector is taken as $\mathbf{X} = \{a \quad b \quad c \quad \Omega \quad \beta\}^T$. The mechanism is constrained to be a crank-rocker mechanism so that

$$a - b \leq 0, \quad a - c \leq 0, \quad a \leq 1$$
$$d = [(a + c) - (b + 1)][(c - a)^2 - (b - 1)^2] \leq 0$$

The maximum deviation of the transmission angle (μ) from $90°$ is restricted to be less than a specified value, $t_{max} = 35°$. The specified output angle is

$$\theta_s(\phi) = \begin{cases} 20° + \frac{\phi}{3}, & 0° \leq \phi \leq 240° \\ \text{unspecified}, & 240° \leq \phi < 360° \end{cases}$$

Linear membership functions are assumed for the response characteristics [13.22]. The optimum solution is found to be $\mathbf{X} = \{0.2537 \quad 0.8901 \quad 0.8865 \quad -0.7858 \quad -1.0\}^T$ with $f^* = 1.6562$ and $\lambda^* = 0.4681$. This indicates that the maximum level of satisfaction that can be achieved in the presence of fuzziness in the problem is 0.4681. The transmission angle constraint is found to be active at the optimum solution [13.22].

13.7   NEURAL-NETWORK-BASED OPTIMIZATION

The immense computational power of nervous system to solve perceptional problems in the presence of massive amount of sensory data has been associated with its parallel

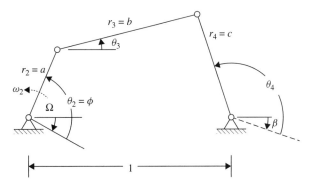

Figure 13.9   Four-bar function generating mechanism.

processing capability. The neural computing strategies have been adopted to solve optimization problems in recent years [13.23, 13.24]. A *neural network* is a massively parallel network of interconnected simple processors (neurons) in which each neuron accepts a set of inputs from other neurons and computes an output that is propagated to the output nodes. Thus a neural network can be described in terms of the individual neurons, the network connectivity, the weights associated with the interconnections between neurons, and the activation function of each neuron. The network maps an input vector from one space to another. The mapping is not specified but is learned.

Consider a single neuron as shown in Fig. 13.10. The neuron receives a set of n inputs, x_i, $i = 1, 2, \ldots, n$, from its neighboring neurons and a bias whose value is equal to 1. Each input has a weight (gain) w_i associated with it. The weighted sum of the inputs determines the state or activity of a neuron, and is given by $a = \sum_{i=1}^{n+1} w_i x_i = \mathbf{W}^T \mathbf{X}$, where $\mathbf{X} = \{x_1 x_2 \cdots x_n 1\}^T$. A simple function is now used to provide a mapping from the n-dimensional space of inputs into a one-dimensional space of the output, which the neuron sends to its neighbors. The output of a neuron is a function of its state and can be denoted as $f(a)$. Usually, no output will be produced unless the activation level of the node exceeds a threshold value. The output of a neuron is commonly described by a sigmoid function as

$$f(a) = \frac{1}{1 + e^{-a}} \tag{13.60}$$

which is shown graphically in Fig. 13.10. The sigmoid function can handle large as well as small input signals. The slope of the function $f(a)$ represents the available gain. Since the output of the neuron depends only on its inputs and the threshold value, each neuron can be considered as a separate processor operating in parallel with other neurons. The learning process consists of determining values for the weights w_i that lead to an optimal association of the inputs and outputs of the neural network.

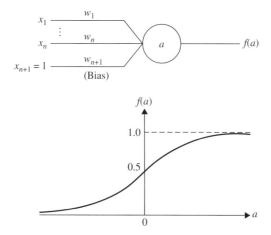

Figure 13.10   Single neuron and its output. [12.23], reprinted with permission of Gordon & Breach Science Publishers.

Several neural network architectures, such as the Hopfield and Kohonen networks, have been proposed to reflect the basic characteristics of a single neuron. These architectures differ one from the other in terms of the number of neurons in the network, the nature of the threshold functions, the connectivities of the various neurons, and the learning procedures. A typical architecture, known as the *multilayer feedforward network,* is shown in Fig. 13.11. In this figure the arcs represent the unidirectional feedforward communication links between the neurons. A weight or gain associated with each of these connections controls the output passing through a connection. The weight can be positive or negative, depending on the excitatory or inhibitory nature of the particular neuron. The strengths of the various interconnections (weights) act as repositories for knowledge representation contained in the network.

The network is trained by minimizing the mean-squared error between the actual output of the output layer and the target output for all the input patterns. The error is minimized by adjusting the weights associated with various interconnections. A number of learning schemes, including a variation of the steepest descent method, have been used in the literature. These schemes govern how the weights are to be varied to minimize the error at the output nodes. For illustration, consider the network shown in Fig. 13.12. This network is to be trained to map the angular displacement and angular velocity relationships, transmission angle, and the mechanical advantage of a four-bar function-generating mechanism (Fig. 13.9). The inputs to the five neurons in the input layer include the three link lengths of the mechanism (r_2, r_3, and r_4) and the angular displacement and velocities of the input link (θ_2 and ω_2). The outputs of the six neurons in the output layer include the angular positions and velocities of the coupler and the output links (θ_3, ω_3, θ_4, and ω_4), the transmission angle (γ), and the mechanical

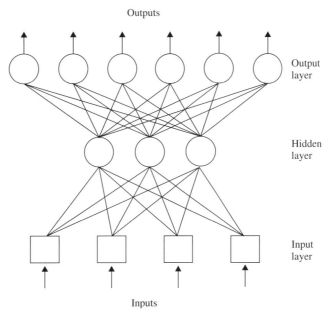

Figure 13.11 Multilayer feedforward network. [13.23], reprinted with permission of Gordon and Breach Science Publishers.

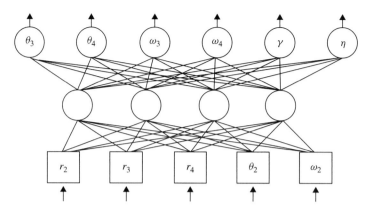

Figure 13.12   Network used to train relationships for a four-bar mechanism. [12.23], reprinted with permission of Gordon & Breach Science Publishers.

advantage (η) of the mechanism. The network is trained by inputting several possible combinations of the values of r_2, r_3, r_4, θ_2, and ω_2 and supplying the corresponding values of $\theta_3, \theta_4, \omega_3, \omega_4, \gamma$, and η. The difference between the values predicted by the network and the actual output is used to adjust the various interconnection weights such that the mean-squared error at the output nodes is minimized. Once trained, the network provides a rapid and efficient scheme that maps the input into the desired output of the four-bar mechanism. It is to be noted that the explicit equations relating r_2, r_3, r_4, θ_2, and ω_2 and the output quantities $\theta_3, \theta_4, \omega_3, \omega_4, \gamma$, and η have not been programmed into the network; rather, the network learns these relationships during the training process by adjusting the weights associated with the various interconnections. The same approach can be used for other mechanical and structural analyses that might require a finite-element-based computations.

Numerical Results. The minimization of the structural weight of the three-bar truss described in Section 7.22.1 (Fig. 7.21) was considered with constraints on the cross-sectional areas and stresses in the members. Two load conditions were considered with $P = 20,000$ lb, $E = 10 \times 10^6$ psi, $\rho = 0.1$ lb/in^3, $H = 100$ in., $\sigma_{\min} = -15,000$ psi, $\sigma_{\max} = 20,000$ psi, $A_i^{(l)} = 0.1$ in^2 ($i = 1, 2$), and $A_i^{(u)} = 5.0$ in^2 ($i = 1, 2$). The solution obtained using neural-network-based optimization is [12.23]: $x_1^* = 0.788$ in^2, $x_2^* = 0.4079$ in^2, and $f^* = 26.3716$ lb. This can be compared with the solution given by nonlinear programming: $x_1^* = 0.7745$ in^2, $x_2^* = 0.4499$ in^2, and $f^* = 26.4051$ lb.

REFERENCES AND BIBLIOGRAPHY

13.1   J. H. Holland, *Adaptation in Natural and Artificial Systems*, University of Michigan Press, Ann Arbor, MI, 1975.

13.2   I. Rechenberg, *Cybernetic Solution Path of an Experimental Problem*, Library Translation 1122, Royal Aircraft Establishment, Farnborough, Hampshire, UK, 1965.

13.3   D. E. Goldberg, *Genetic Algorithms in Search, Optimization, and Machine Linearning*, Addison-Wesley, Reading, MA, 1989.

13.4   S. S. Rao, T. S. Pan, A. K. Dhingra, V. B. Venkayya, and V. Kumar, Genetic-evolution-based optimization methods for engineering design, pp. 318–323 in *Proceedings of the 3rd Air Force/NASA Symposium on Recent Advances in Multidisciplinary Analysis and Optimization,* San Francisco, Sept. 24–26, 1990.

13.5   S. S. Rao, T. S. Pan, and V. B. Venkayya, Optimal placement of actuators in actively controlled structures using genetic algorithms, *AIAA Journal*, Vol. 29, No. 6, pp. 942–943, 1991.

13.6   P. Hajela, Genetic search: an approach to the nonconvex optimization problem, *AIAA Journal*, Vol. 26, No. 7, pp. 1205–1210, 1990.

13.7   P. Hajela and C. Y. Lin, Genetic search strategies in multicriterion optimal design, *Structural Optimization*, Vol. 4, pp. 99–107, 1992.

13.8   D. E. Goldberg, Computer-aided pipeline operation using genetic algorithms and rule learning, Part I: Genetic algorithms in pipeline optimization, *Engineering with Computers*, Vol. 3, pp. 35–45, 1987.

13.9   D. E. Goldberg and C. H. Kuo, Genetic algorithms in pipeline optimization, *ASCE Journal of Computing in Civil Engineering*, Vol. 1, No. 2, pp. 128–141, 1987.

13.10   C. Y. Lin and P. Hajela, Genetic algorithms in optimization problems with discrete and integer design variables, *Engineering Optimization*, Vol. 19, pp. 309–327, 1992.

13.11   Z. Michalewicz, *Genetic Algorithms + Data Structures = Evolution Programs*, 2nd ed., Springer-Verlag, Berlin, 1994.

13.12   B. Hajek, Cooling schedules for optimal annealing, *Mathematics of Operations Research*, Vol. 13, No. 4, pp. 563–571, 1988.

13.13   S. Kirkpatrick, C. D. Gelatt, Jr., and M. P. Vecchi, Optimization by simulated annealing, *Science*, Vol. 220, pp. 671–680, 1983.

13.14   Y. Kim and H. Kim, A stepwise-overlapped parallel simulated annealing algorithm, *Integration, The VLSI Journal*, Vol. 10, pp. 39–54, 1990.

13.15   P. van Laarhoven and E. Aarts, *Simulated Annealing: Theory and Applications*, D. Reidel, Boston, 1987.

13.16   A. Corana, M. Marchesi, C. Martini, and S. Ridella, Minimizing multimodal functions of continuous variables with the simulated annealing algorithm, *ACM Transactions on Mathematical Software*, Vol. 13, No. 3, pp. 262–280, 1987.

13.17   G.-S. Chen, R. J. Bruno, and M. Salama, Optimal placement of active/passive members in truss structures using simulated annealing, *AIAA Journal*, Vol. 29, pp. 1327–1334, 1991.

13.18   M. Lundy and A. Mees, Convergence of an annealing algorithm, *Mathematical Programming*, Vol. 34, pp. 111–124, 1986.

13.19   M. Atiqullah and S. S. Rao, Parallel processing in optimal structural design using simulated annealing, *AIAA Journal*, Vol. 33, pp. 2386–2392, 1995.

13.20   K. Deb, Optimal design of a class of welded structures via genetic algorithms, pp. 444–453 in *Proceedings of the AIAA/ASME/ASCE/AHS/ASC 31st Structures, Structural Dynamics and Materials Conference,* Long Beach, CA, Apr. 2–4, 1990.

13.21   K. M. Ragsdell and D. T. Phillips, Optimal design of a class of welded structure using geometric programming, *ASME Journal of Engineering for Industry*, Vol. 98, No. 3, pp. 1021–1025, 1976.

13.22   S. S. Rao, Description and optimum design of fuzzy mechanical systems, *ASME Journal of Mechanisms, Transmissions, and Automation in Design*, Vol. 109, pp. 126–132, 1987.

13.23 A. K. Dhingra and S. S. Rao, A neural network based approach to mechanical design optimization, *Engineering Optimization*, Vol. 20, pp. 187–203, 1992.

13.24 L. Berke and P. Hajela, Applications of artificial neural nets in structural mechanics, *Structural Optimization*, Vol. 4, pp. 90–98, 1992.

13.25 S. S. Rao, Multiobjective optimization of fuzzy structural systems, *International Journal for Numerical Methods in Engineering*, Vol. 24, pp. 1157–1171, 1987.

13.26 S. S. Rao, K. Sundararaju, B. G. Prakash, and C. Balakrishna, A fuzzy goal programming approach for structural optimization, *AIAA Journal*, Vol. 30, No. 5, pp. 1425–1432, 1992.

13.27 A. K. Dhingra, S. S. Rao, and V. Kumar, Nonlinear membership functions in the fuzzy optimization of mechanical and structural systems, *AIAA Journal*, Vol. 30, No. 1, pp. 251–260, 1992.

13.28 A. K. Dhingra and S. S. Rao, An integrated kinematic–kinetostatic optimal design of planar mechanisms using fuzzy theories, *ASME Journal of Mechanical Design*, Vol. 113, pp. 306–311, 1991.

13.29 R. J. Balling and S. A. May, Large-scale discrete structural optimization: simulated annealing, branch-and-bound, other techniques, *Proceedings of the AIAA/ASME/ASCE/ AHS/ASC 32nd Structures, Structural Dynamics, and Materials Conference,* Long Beach, CA, 1990.

13.30 The Rand Corporation, *A Million Random Digits with 100,000 Normal Deviates*, The Free Press, Glencoe, IL, 1955.

13.31 A. Colorni, M. Dorigo, and V. Maniezzo, Distributed optimization by ant colonies, in *Proceedings of the First European Conference on Artificial Life*, F. J. Varela and P. Bourgine, Eds., MIT Press, Cambridge, MA, pp. 134–142, 1992.

13.32 M. Dorigo, V. Maniezzo, and A. Colorni, The ant system optimization by a colony of cooperating agents, *IEEE Transactions on Systems, Man, and Cybernetics—Part B*, Vol. 26, No. 1, pp. 29–41, 1996.

13.33 J. Kennedy and R. C. Eberhart, *Swarm Intelligence*, Morgan Kaufmann, San Francisco, 2001.

13.34 J. Kennedy and R. C. Eberhart, Particle swarm optimization, *Proceedings of the 1995 IEEE International Conference on Neural Networks*, IEEE Service Center, Piscataway, NJ, 1995.

13.35 J-L. Liu and J-Horng Lin, Evolutionary computation of unconstrained and constrained problems using a novel momentum-type particle swarm optimization, *Engineering Optimization*, Vol. 39, No. 3, pp. 287–305, 2007.

13.36 Y. Shi and R. C. Eberhart, Parameter selection in particle swarm optimization, *Proceedings of the Seventh Annual Conference on Evolutionary Programming*, V. W. Porto, N. Saravanan, D. Waagen, and A. Eibe, Eds., Springer-Verlag, pp. 591–600, Berlin, Germany, 1998.

13.37 N. Metropolis, A. Rosenbluth, M. Rosenbluth, A. Teller, and E. Teller, Equation of state calculations by fast computing machines, *Journal of Chemical Physics*, Vol. 21, No. 6, pp. 1087–1092, 1953.

REVIEW QUESTIONS

13.1 Define the following terms:

(a) Fuzzy parameter

(b) Annealing

(c) Roulette wheel selection process

(d) Pheromone evaporation rate

(e) Neural network

(f) Fuzzy feasible domain

(g) Membership function

(h) Multilayer feedforward network

13.2 Match the following terms:

(a) Fuzzy optimization	Based on shortest path
(b) Genetic algorithms	Analysis equations not programmed
(c) Neural network method	Linguistic data can be used
(d) Simulated annealing	Based on the behavior of a flock of birds
(e) Particle swarm optimization	Based on principle of survival of the fittest
(f) Ant colony optimization	Based on cooling of heated solids

13.3 Answer true or false:

(a) GAs can be used to solve problems with continuous design variables.

(b) GAs do not require derivatives of the objective function.

(c) Crossover involves swapping of the binary digits between two strings.

(d) Mutation operator is used to produce offsprings.

(e) No new strings are formed in the reproduction stage in GAs.

(f) Simulated annealing can be used to solve only discrete optimization problems.

(g) Particle swarm optimization is based on cognitive and social learning rates of groups of birds.

(h) Particle swarm optimization method uses the positions and velocities of particles.

(i) Genetic algorithms basically maximize an unconstrained function.

(j) Simulated annealing basically solves an unconstrained optimization problem.

(k) GAs seek to find a better design point from a trial design point.

(l) GAs can solve a discrete optimization problem with no additional effort.

(m) SA is a type of random search technique.

(n) GAs and SA can find the global minimum with high probability.

(o) GAs are zeroth-order methods.

(p) Discrete variables need not be represented as binary strings in GAs.

(q) SA will find a local minimum if the feasible space is nonconvex.

(r) The expressions relating the input and output are to be programmed in neural-network-based methods.

(s) Several networks architectures can be used in neural-network-based optimization.

(t) A fuzzy quantity is same as a random quantity.

(u) Ant colony optimization solves only discrete optimization problems.

(v) Fuzzy optimization involves the maximization of the intersection of the objective function and feasible domain.

13.4 Give brief answers:

(a) What is Boltzmann's probability distribution?

Modern Methods of Optimization

(b) How is an inequality constrained optimization problem converted into an unconstrained problem for use in GAs?

(c) What is the difference between a crisp set and a fuzzy set?

(d) How is the output of a neuron described commonly?

(e) What are the basic operations used in GAs?

(f) What is a fitness function in GAs?

(g) Can you consider SA as a zeroth-order search method?

(h) How do you select the length of the binary string to represent a design variable?

(i) Construct the objective function to be used in GAs for a minimization problem with mixed equality and inequality constraints.

(j) How is the crossover operation performed in GAs?

(k) What is the purpose of mutation? How is it implemented in GAs?

(l) What is the physical basis of SA?

(m) What is metropolis criterion and where is it used?

(n) What is a neural network?

(o) How is a neuron modeled in neural-network-based models?

(p) What is a sigmoid function?

(q) How is the error in the output minimized during network training?

(r) What is the difference between a random quantity and a fuzzy quantity?

(s) Give two examples of design parameters that can be considered as fuzzy.

(t) What is a valuation set?

(u) What is the significance of membership function?

(v) Define the union of two fuzzy sets A and B?

(w) How is the intersection of two fuzzy sets A and B defined?

(x) Show the complement of a fuzzy set in a Venn diagram.

(y) How is the optimum solution defined in a fuzzy environment?

(z) How is the fuzzy feasible domain defined for a problem with inequality constraints?

PROBLEMS

13.1 Consider the following two strings denoting the vectors X_1 and X_2:

$$X_1 : \{1\ 0\ 0\ 0\ 1\ 0\ 1\ 1\ 0\ 1\}$$
$$X_2 : \{0\ 1\ 1\ 1\ 1\ 1\ 0\ 1\ 1\ 0\}$$

Find the result of crossover at location 2. Also, determine the decimal values of the variables before and after crossover if each string denotes a vector of two variables.

13.2 Two discrete fuzzy sets, A and B are defined as follows:

$$A = \{(60, 0.1)\ \ (62, 0.5)\ \ (64, 0.7)\ \ (66, 0.9)\ \ (68, 1.0)\ \ (70, 0.8)\}$$
$$B = \{(60, 0.0)\ \ (62, 0.2)\ \ (64, 0.4)\ \ (66, 0.8)\ \ (68, 0.9)\ \ (70, 1.0)\}$$

Determine the union and intersection of these sets.

13.3 Determine the size of the binary string to be used to achieve an accuracy of 0.01 for a design variable with the following bounds:

(a) $x^{(l)} = 0, x^{(u)} = 5$

(b) $x^{(l)} = 0, x^{(u)} = 10$

(c) $x^{(l)} = 0, x^{(u)} = 20$

13.4 A design variable, with lower and upper bounds 2 and 13, respectively, is to be represented with an accuracy of 0.02. Determine the size of the binary string to be used.

13.5 Find the minimum of $f = x^5 - 5 x^3 - 20 x + 5$ in the range $(0, 3)$ using the ant colony optimization method. Show detailed calculations for 2 iterations with 4 ants.

13.6 In the ACO method, the amounts of pheromone along the various arcs from node i are given by $\tau_{ij} = 1, 2, 4, 3, 5, 2$ for $j = 1, 2, 3, 4, 5, 6$, respectively. Find the arc (ij) chosen by an ant based on the roulette-wheel selection process based on the random number $r = 0.4921$.

13.7 Solve Example 13.5 by neglecting pheromone evaporation. Show the calculations for 2 iterations.

13.8 Find the maximum of the function $f = -x^5 + 5 x^3 + 20x - 5$ in the range $-4 \le x \le 4$ using the PSO method. Use 4 particles with the initial positions $x_1 = -2, x_2 = 0, x_3 = 1$, and $x_4 = 3$. Show detailed calculations for 2 iterations.

13.9 Solve Example 13.4 using the inertia term when θ varies linearly from 0.9 to 0.4 in Eq. (13.23).

13.10 Find the minimum of the following function using simulated annealing:

$$f(\mathbf{X}) = 6x_1^2 - 6x_1x_2 + 2x_2^2 - x_1 - 2x_2$$

Assume suitable parameters and show detailed calculations for 2 iterations.

13.11 Consider the following function for maximization using simulated annealing: $f(x) = x(1.5 - x)$ in the range $(0, 5)$. If the initial point is $x^{(0)} = 2.0$, generate a neighboring point using a uniformly distributed random number in the range $(0, 1)$. If the temperature is 400, find the pbobability of accepting the neighboring point.

13.12 The population of binary strings in a maximization problem is given below:

String						Fitness
0	0	1	1	0	0	8
0	1	0	1	0	1	12
1	0	1	0	1	1	6
1	1	0	0	0	1	2
0	0	0	1	0	0	18
1	0	0	0	0	0	9
0	1	0	1	0	0	10

Determine the expected number of copies of the best string in the above population in the mating pool using the roulette-wheel selection process.

13.13 Consider the following constrained optimization problem:

$$\text{Minimize}\quad f = x_1^3 - 6x_1^2 + 11x_1 + x_3$$

subject to

$$x_1^2 + x_2^2 - x_3^2 \leq 0$$
$$4 - x_1^2 - x_2^2 - x_3^2 \leq 0$$
$$x_3 - 5 \leq 0$$
$$-x_i \leq 0; \quad i = 1, 2, 3$$

Define the fitness function to be used in GA for this problem.

13.14 The bounds on the design variables in an optimization problem are given by
$$-10 \leq x_1 \leq 10, \quad 0 \leq x_2 \leq 8, \quad 150 \leq x_3 \leq 750$$
Find the minimum binary string length of a design vector $\mathbf{X} = \{x_1, x_2, x_3\}^T$ to achieve an accuracy of 0.01.

14

Practical Aspects of Optimization

14.1 INTRODUCTION

Although the mathematical techniques described in Chapters 3 to 13 can be used to solve all engineering optimization problems, the use of engineering judgment and approximations help in reducing the computational effort involved. In this chapter we consider several types of approximation techniques that can speed up the analysis time without introducing too much error [14.1].

These techniques are especially useful in finite element analysis-based optimization procedures. The practical computation of the derivatives of static displacements, stresses, eigenvalues, eigenvectors, and transient response of mechanical and structural systems is presented. The concept of decomposition, which permits the solution of a large optimization problem through a set of smaller, coordinated subproblems is presented. The use of parallel processing and computation in the solution of large-scale optimization problems is discussed. Many real-life engineering systems involve simultaneous optimization of multiple-objective functions under a specified set of constraints. Several multiobjective optimization techniques are summarized in this chapter.

14.2 REDUCTION OF SIZE OF AN OPTIMIZATION PROBLEM

14.2.1 Reduced Basis Technique

In the optimum design of certain practical systems involving a large number of (n) design variables, some feasible design vectors $\mathbf{X}_1, \mathbf{X}_2, \ldots, \mathbf{X}_r$ may be available to start with. These design vectors may have been suggested by experienced designers or may be available from the design of similar systems in the past. We can reduce the size of the optimization problem by expressing the design vector X as a linear combination of the available feasible design vectors as

$$\mathbf{X} = c_1\mathbf{X}_1 + c_2\mathbf{X}_2 + \cdots + c_r\mathbf{X}_r \qquad (14.1)$$

where $c_1, c_2, \ldots, c_r$ are the unknown constants. Then the optimization problem can be solved using $c_1, c_2, \ldots, c_r$ as design variables. This problem will have a much smaller number of unknowns since $r \ll n$. In Eq. (14.1), the feasible design vectors $\mathbf{X}_1, \mathbf{X}_2, \ldots, \mathbf{X}_r$ serve as the basis vectors. It can be seen that if $c_1 = c_2 = \cdots = c_r = 1/r$, then $\mathbf{X}$ denotes the average of the basis vectors.

14.2.2   Design Variable Linking Technique

When the number of elements or members in a structure is large, it is possible to reduce the number of design variables by using a technique known as *design variable linking* [14.25]. To see this procedure, consider the 12-member truss structure shown in Fig. 14.1. If the area of cross section of each member is varied independently, we will have 12 design variables. On the other hand, if symmetry of members about the vertical (Y) axis is required, the areas of cross section of members 4, 5, 6, 8, and 10 can be assumed to be the same as those of members 1, 2, 3, 7, and 9, respectively. This reduces the number of independent design variables from 12 to 7. In addition, if the cross-sectional area of member 12 is required to be three times that of member 11, we will have six independent design variables only:

$$\mathbf{X} = \begin{Bmatrix} x_1 \\ x_2 \\ x_3 \\ x_4 \\ x_5 \\ x_6 \end{Bmatrix} \equiv \begin{Bmatrix} A_1 \\ A_2 \\ A_3 \\ A_7 \\ A_9 \\ A_{11} \end{Bmatrix} \tag{14.2}$$

Once the vector $\mathbf{X}$ is known, the dependent variables can be determined as $A_4 = A_1$, $A_5 = A_2$, $A_6 = A_3$, $A_8 = A_7$, $A_{10} = A_9$, and $A_{12} = 3A_{11}$. This procedure of treating certain variables as dependent variables is known as *design variable linking*. By defining the vector of all variables as

$$\mathbf{Z}^{\mathrm{T}} = \{z_1 \; z_2 \; \ldots \; z_{12}\}^{\mathrm{T}} \equiv \{A_1 \; A_2 \; \ldots A_{12}\}^{\mathrm{T}}$$

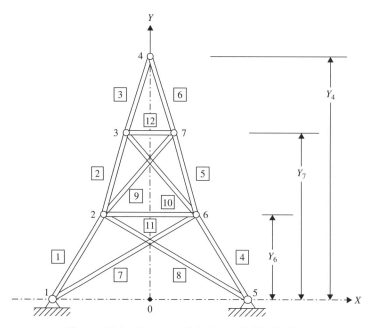

Figure 14.1   Concept of design variable linking.

the relationship between $\mathbf{Z}$ and $\mathbf{X}$ can be expressed as

$$\mathbf{Z}_{12\times1} = [T]_{12\times6} \; \mathbf{X}_{6\times1} \tag{14.3}$$

where the matrix $[T]$ is given by

$$[T] = \begin{bmatrix} 1 & 0 & 0 & 0 & 0 & 0 \\ 0 & 1 & 0 & 0 & 0 & 0 \\ 0 & 0 & 1 & 0 & 0 & 0 \\ 1 & 0 & 0 & 0 & 0 & 0 \\ 0 & 1 & 0 & 0 & 0 & 0 \\ 0 & 0 & 1 & 0 & 0 & 0 \\ 0 & 0 & 0 & 1 & 0 & 0 \\ 0 & 0 & 0 & 1 & 0 & 0 \\ 0 & 0 & 0 & 0 & 1 & 0 \\ 0 & 0 & 0 & 0 & 1 & 0 \\ 0 & 0 & 0 & 0 & 0 & 1 \\ 0 & 0 & 0 & 0 & 0 & 3 \end{bmatrix} \tag{14.4}$$

The concept can be extended to many other situations. For example, if the geometry of the structure is to be varied during optimization (configuration optimization) while maintaining (1) symmetry about the Y axis and (2) alignment of the three nodes 2, 3, and 4 (and 6, 7, and 4), we can define the following independent and dependent design variables:

Independent variables: X_5, X_6, Y_6, Y_7, Y_4

Dependent variables:

$$X_1 = -X_5, \quad X_2 = -X_6, \quad Y_2 = Y_6, \quad Y_3 = Y_7, \quad X_7 = \frac{Y_4 - Y_7}{Y_4 - Y_6}X_6,$$

$$X_3 = -X_7, \quad X_4 = 0, \quad Y_1 = 0, \quad Y_5 = 0$$

Thus the design vector $\mathbf{X}$ is

$$\mathbf{X} = \begin{Bmatrix} x_1 \\ x_2 \\ x_3 \\ x_4 \\ x_5 \end{Bmatrix} \equiv \begin{Bmatrix} X_5 \\ X_6 \\ Y_6 \\ Y_7 \\ Y_4 \end{Bmatrix} \tag{14.5}$$

The relationship between the dependent and independent variables can be defined more systematically, by defining a vector of all geometry variables, $\mathbf{Z}$, as

$$\mathbf{Z} = \{z_1 \, z_2 \ldots z_{14}\}^{\mathrm{T}}$$

$$\equiv \{X_1 \, Y_1 \, X_2 \, Y_2 \, X_3 \, Y_3 \, X_4 \, Y_4 \, X_5 \, Y_5 \, X_6 \, Y_6 \, X_7 \, Y_7\}^{\mathrm{T}}$$

which is related to $\mathbf{X}$ through the relations

$$z_i = f_i(\mathbf{X}), \quad i = 1, 2, \ldots, 14 \tag{14.6}$$

where f_i denotes a function of $\mathbf{X}$.

14.3 FAST REANALYSIS TECHNIQUES

14.3.1 Incremental Response Approach

Let the displacement vector of the structure or machine, $\mathbf{Y}_0$, corresponding to the load vector, $\mathbf{P}_0$, be given by the solution of the equilibrium equations

$$[K_0]\mathbf{Y}_0 = \mathbf{P}_0 \tag{14.7}$$

or

$$\mathbf{Y}_0 = [K_0]^{-1}\mathbf{P}_0 \tag{14.8}$$

where $[K_0]$ is the stiffness matrix corresponding to the design vector, $\mathbf{X}_0$. When the design vector is changed to $\mathbf{X}_0 + \Delta\mathbf{X}$, let the stiffness matrix of the system change to $[K_0] + [\Delta K]$, the displacement vector to $\mathbf{Y}_0 + \Delta\mathbf{Y}$, and the load vector to $\mathbf{P}_0 + \Delta\mathbf{P}$. The equilibrium equations at the new design vector, $\mathbf{X}_0 + \Delta\mathbf{X}$, can be expressed as

$$([K_0] + [\Delta K])(\mathbf{Y}_0 + \Delta\mathbf{Y}) = \mathbf{P}_0 + \Delta\mathbf{P} \tag{14.9}$$

or

$$[K_0]\mathbf{Y}_0 + [\Delta K]\mathbf{Y}_0 + [K_0]\Delta\mathbf{Y} + [\Delta K]\Delta\mathbf{Y} = \mathbf{P}_0 + \Delta\mathbf{P} \tag{14.10}$$

Subtracting Eq. (14.7) from Eq. (14.10), we obtain

$$([K_0] + [\Delta K])\Delta\mathbf{Y} = \Delta\mathbf{P} - [\Delta K]\mathbf{Y}_0 \tag{14.11}$$

By neglecting the term $[\Delta K]\Delta\mathbf{Y}$, Eq. (14.11) can be reduced to

$$[K_0]\Delta\mathbf{Y} \approx \Delta\mathbf{P} - [\Delta K]\mathbf{Y}_0 \tag{14.12}$$

which yields the first approximation to the increment in displacement vector $\Delta\mathbf{Y}$ as

$$\Delta\mathbf{Y}_1 = [K_0]^{-1}(\Delta\mathbf{P} - [\Delta K]\mathbf{Y}_0) \tag{14.13}$$

where $[K_0]^{-1}$ is available from the solution in Eq. (14.8). We can find a better approximation of $\Delta\mathbf{Y}$ by subtracting Eq. (14.12) from Eq. (14.11):

$$([K_0] + [\Delta K])\Delta\mathbf{Y} - [K_0]\Delta\mathbf{Y}_1 = \Delta\mathbf{P} - [\Delta K]\mathbf{Y}_0 - (\Delta\mathbf{P} - [\Delta K]\mathbf{Y}_0) \tag{14.14}$$

or

$$([K_0] + [\Delta K])(\Delta\mathbf{Y} - \Delta\mathbf{Y}_1) = -[\Delta K]\Delta\mathbf{Y}_1 \tag{14.15}$$

By defining

$$\Delta\mathbf{Y}_2 = \Delta\mathbf{Y} - \Delta\mathbf{Y}_1 \tag{14.16}$$

Eq. (14.15) can be expressed as

$$([K_0] + [\Delta K])\Delta\mathbf{Y}_2 = -[\Delta K]\Delta\mathbf{Y}_1 \tag{14.17}$$

Neglecting the term $[\Delta K]\Delta \mathbf{Y}_2$, Eq. (14.17) can be used to obtain the second approximation to $\Delta \mathbf{Y}$, $\Delta \mathbf{Y}_2$, as

$$\Delta \mathbf{Y}_2 = -[K_0]^{-1}([\Delta K]\Delta \mathbf{Y}_1) \tag{14.18}$$

From Eq. (14.16), $\Delta \mathbf{Y}$ can be written as

$$\Delta \mathbf{Y} = \sum_{i=1}^{2} \Delta \mathbf{Y}_i \tag{14.19}$$

This process can be continued and $\Delta \mathbf{Y}$ can be expressed, in general, as

$$\Delta \mathbf{Y} = \sum_{i=1}^{\infty} \Delta \mathbf{Y}_i \tag{14.20}$$

where $\Delta \mathbf{Y}_i$ is found by solving the equations

$$[K_0]\Delta \mathbf{Y}_i = -[\Delta K]\Delta \mathbf{Y}_{i-1} \tag{14.21}$$

Note that the series given by Eq. (14.20) may not converge if the change in the design vector, $\Delta \mathbf{X}$, is not small. Hence it is important to establish the validity of the procedure for each problem, by determining the step sizes for which the series will converge, before using it. The iterative process is usually stopped either by specifying a maximum number of iterations and/or by prescribing a convergence criterion such as

$$\frac{\|\Delta \mathbf{Y}_i\|}{\left\| \sum_{j=1}^{i} \Delta \mathbf{Y}_j \right\|} \leq \varepsilon \tag{14.22}$$

where $\|\Delta \mathbf{Y}_i\|$ is the Euclidean norm of the vector $\Delta \mathbf{Y}_i$ and ε is a small number on the order of 0.01.

Example 14.1 Consider the crane (planar truss) shown in Fig. 14.2. Young's modulus of member e is equal to $E_e = 30 \times 10^6$ psi ($e = 1, 2, 3, 4$), and the other data are shown in Table 14.1. Assuming the base design to be $A_1 = A_2 = 2$ in.^{2} and $A_3 = A_4 = 1$ in.^{2}, and perturbations to be $\Delta A_1 = \Delta A_2 = 0.4$ in.^{2} and $\Delta A_3 = \Delta A_4 = 0.2$ in.^{2}, determine (a) the exact displacements of nodes 3 and 4 at the base design, (b) the displacements of nodes 3 and 4 at the perturbed design using the exact procedure, and, (c) the displacements of nodes 3 and 4 at the perturbed design using the approximation method.

SOLUTION   The stiffness matrix of a typical element e is given by

$$[K^{(e)}] = \frac{A_e E_e}{l_e} \begin{bmatrix} l_{ij}^2 & l_{ij}m_{ij} & -l_{ij}^2 & -l_{ij}m_{ij} \\ l_{ij}m_{ij} & m_{ij}^2 & -l_{ij}m_{ij} & -m_{ij}^2 \\ -l_{ij}^2 & -l_{ij}m_{ij} & l_{ij}^2 & l_{ij}m_{ij} \\ -l_{ij}m_{ij} & -m_{ij}^2 & l_{ij}m_{ij} & m_{ij}^2 \end{bmatrix} \tag{E$_1$}$$

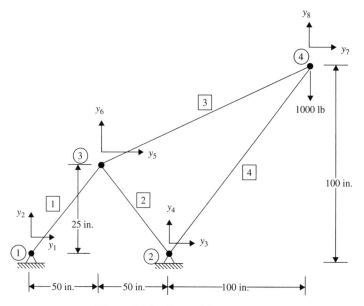

Figure 14.2   Crane (planar truss).

Table 14.1

Member, e	Area of cross section, A_e	Length, l_e (in.)	Global node of: Corner 1, i	Corner 2, j	Direction cosines of member $l_{ij} = \dfrac{X_j - X_i}{l_e}$	$m_{ij} = \dfrac{Y_j - Y_i}{l_e}$
1	A_1	55.9017	1	3	0.8944	0.4472
2	A_2	55.9017	3	2	0.8944	−0.4472
3	A_3	167.7051	3	4	0.8944	0.4472
4	A_4	141.4214	2	4	0.7071	0.7071

where A_e is the cross-sectional area, E_e is Young's modulus, l_e is the length, and (l_{ij}, m_{ij}) are the direction cosines of member e. Equation (E$_1$) can be used to compute the stiffness matrices of the various members using the data of Table 14.1. When the member stiffness matrices are assembled and the boundary conditions ($y_1 = y_2 = y_3 = y_4 = 0$) are applied, the overall stiffness matrix becomes

$$[K] = (30 \times 10^6)$$

$$
\begin{bmatrix}
\left(\dfrac{0.8A_1}{55.9017} + \dfrac{0.8A_2}{55.9017} + \dfrac{0.8A_3}{167.7051}\right) & \left(\dfrac{0.4A_1}{55.9017} - \dfrac{0.4A_2}{55.9017} + \dfrac{0.4A_3}{167.7051}\right) & \left(\dfrac{-0.8A_3}{167.7051}\right) & \left(\dfrac{-0.4A_3}{167.7051}\right) \\[2ex]
 & \left(\dfrac{0.2A_1}{55.9017} + \dfrac{0.2A_2}{55.9017} + \dfrac{0.2A_3}{167.7051}\right) & \left(\dfrac{-0.4A_3}{167.70501}\right) & \left(\dfrac{-0.2A_3}{167.7051}\right) \\[2ex]
\text{symmetric} & & \left(\dfrac{0.8A_3}{167.7051} + \dfrac{0.5A_4}{141.4214}\right) & \left(\dfrac{0.4A_3}{167.7051} + \dfrac{0.5A_4}{141.4214}\right) \\[2ex]
 & & & \left(\dfrac{0.2A_3}{167.7051} + \dfrac{0.5A_4}{141.4214}\right)
\end{bmatrix}
$$

$$\text{(E}_2)$$

Thus the equilibrium equations of the structure can be expressed as

$$[K]\mathbf{Y} = \mathbf{P} \tag{E_3}$$

where

$$\mathbf{Y} = \begin{Bmatrix} y_5 \\ y_6 \\ y_7 \\ y_8 \end{Bmatrix} \quad \text{and} \quad \mathbf{P} = \begin{Bmatrix} p_5 \\ p_6 \\ p_7 \\ p_8 \end{Bmatrix} = \begin{Bmatrix} 0 \\ 0 \\ 0 \\ -1000 \end{Bmatrix}$$

(a) At the base design, $A_1 = A_2 = 2$ in.^{2}, $A_3 = A_4 = 1$ in.^{2}, and the exact solution of Eqs. (E$_3$) gives the displacements of nodes 3 and 4 as

$$\mathbf{Y}_{\text{base}} = \begin{Bmatrix} y_5 \\ y_6 \\ y_7 \\ y_8 \end{Bmatrix}_{\text{base}} = \begin{Bmatrix} 0.001165 \\ 0.002329 \\ 0.05147 \\ -0.07032 \end{Bmatrix} \text{ in.}$$

(b) At the perturbed design, $A_1 = A_2 = 2.4$ in.^{2}, $A_3 = A_4 = 1.2$ in.^{2}, and the exact solution of Eq. (E_3) gives the displacements of nodes 3 and 4 as

$$\mathbf{Y}_{\text{perturb}} = \begin{Bmatrix} y_5 \\ y_6 \\ y_7 \\ y_8 \end{Bmatrix}_{\text{perturb}} = \begin{Bmatrix} 0.0009705 \\ 0.001941 \\ 0.04289 \\ -0.05860 \end{Bmatrix} \text{ in.}$$

(c) The values of $A_1 = A_2 = 2.4$ in.^{2} and $A_3 = A_4 = 1.2$ in.^{2} at the perturbed design are used to compute the new stiffness matrix as $[K]_{\text{perturb}} = [K] + [\Delta K]$, which is then used to compute $\Delta\mathbf{Y}_1, \Delta\mathbf{Y}_2, \ldots$ using the approximation procedure, Eqs. (14.13) and (14.21). The results are shown in Table 14.2. It can be seen that the solution given by Eq. (14.20) converged very fast.

14.3.2   Basis Vector Approach

In structural optimization involving static response, it is possible to conduct an approximate analysis at modified designs based on a limited number of exact analysis results. This results in a substantial saving in computer time since, in most problems, the number of design variables is far smaller than the number of degrees of freedom of the system. Consider the equilibrium equations of the structure in the form

$$\underset{m \times m}{[K]} \underset{m \times 1}{\mathbf{Y}} = \underset{m \times 1}{\mathbf{P}} \tag{14.23}$$

where $[K]$ is the stiffness matrix, $\mathbf{Y}$ the vector of displacements, and $\mathbf{P}$ the load vector. Let the structure have n design variables denoted by the design vector

$$\mathbf{X} = \begin{Bmatrix} x_1 \\ x_2 \\ \vdots \\ x_n \end{Bmatrix}$$

Table 14.2

Exact $\mathbf{Y}_0 = \left\{ \begin{array}{c} 0.116462E-02 \\ 0.232923E-02 \\ 0.514654E-01 \\ -0.703216E-01 \end{array} \right\}$		Exact $(\mathbf{Y}_0 + \Delta\mathbf{Y}) = \left\{ \begin{array}{c} 0.970515E-03 \\ 0.194103E-02 \\ 0.428879E-01 \\ -0.586014E-01 \end{array} \right\}$	

Value of i	$\Delta\mathbf{Y}_i$	$\mathbf{Y}_i = \mathbf{Y}_0 + \sum_{k=1}^{i} \Delta\mathbf{Y}_k$
1	$\left\{ \begin{array}{c} -0.232922E-03 \\ -0.465844E-03 \\ -0.102930E-01 \\ 0.140642E-01 \end{array} \right\}$	$\left\{ \begin{array}{c} 0.931695E-03 \\ 0.186339E-02 \\ 0.411724E-01 \\ -0.562573E-01 \end{array} \right\}$
2	$\left\{ \begin{array}{c} 0.465842E-04 \\ 0.931683E-04 \\ 0.205859E-02 \\ -0.281283E-02 \end{array} \right\}$	$\left\{ \begin{array}{c} 0.978279E-03 \\ 0.195656E-02 \\ 0.432310E-01 \\ -0.590702E-01 \end{array} \right\}$
3	$\left\{ \begin{array}{c} -0.931678E-05 \\ -0.186335E-04 \\ -0.411716E-03 \\ 0.562563E-03 \end{array} \right\}$	$\left\{ \begin{array}{c} 0.968962E-03 \\ 0.193792E-02 \\ 0.428193E-01 \\ -0.585076E-01 \end{array} \right\}$
4	$\left\{ \begin{array}{c} 0.186335E-05 \\ 0.372669E-05 \\ 0.823429E-04 \\ -0.112512E-03 \end{array} \right\}$	$\left\{ \begin{array}{c} 0.970825E-03 \\ 0.194165E-02 \\ 0.429016E-01 \\ -0.586201E-01 \end{array} \right\}$

If we find the exact solution at r *basic* design vectors $\mathbf{X}_1, \mathbf{X}_2, \ldots, \mathbf{X}_r$, the corresponding solutions, $\mathbf{Y}_i$, are found by solving the equations

$$[K_i]\mathbf{Y}_i = \mathbf{P}, \quad i = 1, 2, \ldots, r \tag{14.24}$$

where the stiffness matrix, $[K_i]$, is determined at the design vector $\mathbf{X}_i$. If we consider a new design vector, $\mathbf{X}_N$, in the neighborhood of the basic design vectors, the equilibrium equations at $\mathbf{X}_N$ can be expressed as

$$[K_N]\mathbf{Y}_N = \mathbf{P} \tag{14.25}$$

where $[K_N]$ is the stiffness matrix evaluated at $\mathbf{X}_N$. By approximating $\mathbf{Y}_N$ as a linear combination of the basic displacement vectors $\mathbf{Y}_i$, $i = 1, 2, \ldots, r$, we have

$$\mathbf{Y}_N \approx c_1\mathbf{Y}_1 + c_2\mathbf{Y}_2 + \cdots + c_r\mathbf{Y}_r = [Y]\mathbf{c} \tag{14.26}$$

where $[Y] = [\mathbf{Y}_1, \mathbf{Y}_2, \cdots, \mathbf{Y}_r]$ is an $n \times r$ matrix and $\mathbf{c} = \{c_1, c_2, \cdots, c_r\}^T$ is an r-component column vector. Substitution of Eq. (14.26) into Eq. (14.25) gives

$$[K_N][Y]\mathbf{c} = \mathbf{P} \tag{14.27}$$

By premultiplying Eq. (14.27) by $[Y]^T$ we obtain

$$\underset{r \times r}{[\tilde{K}]} \; \underset{r \times 1}{\mathbf{c}} = \underset{r \times 1}{\tilde{\mathbf{P}}} \tag{14.28}$$

where

$$[\tilde{K}] = [Y]^T [K_N][Y] \tag{14.29}$$

$$\tilde{\mathbf{P}} = [Y]^T \mathbf{P} \tag{14.30}$$

It can be seen that an approximate displacement vector $\mathbf{Y}_N$ can be obtained by solving a smaller (r) system of equations, Eq. (14.28), instead of computing the exact solution $\mathbf{Y}_N$ by solving a larger (n) system of equations, Eq. (14.25). The foregoing method is equivalent to applying the Ritz–Galerkin principle in the subspace spanned by the set of vectors $\mathbf{Y}_1, \mathbf{Y}_2, \ldots, \mathbf{Y}_r$. The assumed modes $\mathbf{Y}_i, i = 1, 2, \ldots, r$, can be considered to be good basis vectors since they are the solutions of similar sets of equations.

Fox and Miura 14.3 applied this method for the analysis of a 124-member, 96-degree-of-freedom space truss (shown in Fig. 14.3). By using a 5-degree-of-freedom approximation, they observed that the solution of Eq. (14.28) required 0.653 s while the solution of Eq. (14.25) required 5.454 s without exceeding 1% error in the maximum displacement components of the structure.

14.4   DERIVATIVES OF STATIC DISPLACEMENTS AND STRESSES

The gradient-based optimization methods require the gradients of the objective and constraint functions. Thus the partial derivatives of the response quantities with respect to the design variables are required. Many practical applications require a finite-element

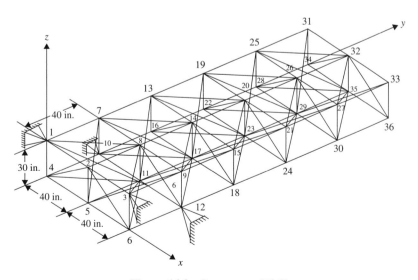

Figure 14.3   Space truss [13.3].

analysis for computing the values of the objective function and/or constraint functions at any design vector. Since the objective and/or constraint functions are to be evaluated at a large number of trial design vectors during optimization, the computation of the derivatives of the response quantities requires substantial computational effort. It is possible to derive approximate expressions for the response quantities. The derivatives of static displacements, stresses, eigenvalues, eigenvectors, and transient response of structural and mechanical systems are presented in this and the following two sections. The equilibrium equations of a machine or structure can be expressed as

$$[K]\mathbf{Y} = \mathbf{P} \tag{14.31}$$

where $[K]$ is the stiffness matrix, $\mathbf{Y}$ the displacement vector, and $\mathbf{P}$ the load vector. By differentiating Eq. (14.31) with respect to the design variable x_i, we obtain

$$\frac{\partial[K]}{\partial x_i}\mathbf{Y} + [K]\frac{\partial\mathbf{Y}}{\partial x_i} = \frac{\partial\mathbf{P}}{\partial x_i} \tag{14.32}$$

where $\partial[K]/\partial x_i$ denotes the matrix formed by differentiating the elements of $[K]$ with respect to x_i. Usually, the matrix is computed using a finite-difference scheme as

$$\frac{\partial[K]}{\partial x_i} \approx \frac{\Delta[K]}{\Delta x_i} = \frac{[K]_{\text{new}} - [K]}{\Delta x_i} \tag{14.33}$$

where $[K]_{\text{new}}$ is the stiffness matrix evaluated at the perturbed design vector $\mathbf{X} + \Delta\mathbf{X}_i$, where the vector $\Delta\mathbf{X}_i$ contains Δx_i in the ith location and zero everywhere else:

$$\Delta\mathbf{X}_i = \{0 \ 0 \ \dots \ 0 \ \Delta x_i \ 0 \ \dots \ 0\}^{\text{T}} \tag{14.34}$$

In most cases the load vector $\mathbf{P}$ is either independent of the design variables or a known function of the design variables, and hence the derivatives, $\partial\mathbf{P}/\partial x_i$, can be evaluated with no difficulty. Equations (14.32) can be solved to find the derivatives of the displacements as

$$\frac{\partial\mathbf{Y}}{\partial x_i} = [K]^{-1}\left(\frac{\partial\mathbf{P}}{\partial x_i} - \frac{\partial[K]}{\partial x_i}\mathbf{Y}\right) \tag{14.35}$$

Since $[K]^{-1}$ or its equivalent is available from the solution of Eqs. (14.31), Eqs. (14.35) can readily be solved to find the derivatives of static displacements with respect to the design variables.

The stresses in a machine or structure (in a particular finite element) can be determined using the relation

$$\boldsymbol{\sigma} = [R]\mathbf{Y} \tag{14.36}$$

where $[R]$ denotes the matrix that relates stresses to nodal displacements. The derivatives of stresses can then be computed as

$$\frac{\partial\boldsymbol{\sigma}}{\partial x_i} = [R]\frac{\partial\mathbf{Y}}{\partial x_i} \tag{14.37}$$

where the matrix $[R]$ is usually independent of the design variables and the vector $\partial\mathbf{Y}/\partial x_i$ is given by Eq. (14.35).

14.5   DERIVATIVES OF EIGENVALUES AND EIGENVECTORS

Let the eigenvalue problem be given by [14.4, 14.6, 14.10]

$$\underset{m \times m}{[K]} \underset{m \times 1}{\mathbf{Y}} = \lambda \underset{m \times m}{[M]} \underset{m \times 1}{\mathbf{Y}} \tag{14.38}$$

where λ is the eigenvalue, $\mathbf{Y}$ the eigenvector, $[K]$ the stiffness matrix, and $[M]$ the mass matrix corresponding to the design vector $\mathbf{X} = \{x_1, x_2, \cdots, x_n\}^{\text{T}}$. Let the solution of Eq. (14.38) be given by the eigenvalues λ_i and the eigenvectors $\mathbf{Y}_i, i = 1, 2, \ldots, m$:

$$[P_i]\mathbf{Y}_i = \mathbf{0} \tag{14.39}$$

where $[P_i]$ is a symmetric matrix given by

$$[P_i] = [K] - \lambda_i[M] \tag{14.40}$$

14.5.1   Derivatives of λ_i

Premultiplication of Eq. (14.39) by $\mathbf{Y}_i^{\text{T}}$ gives

$$\mathbf{Y}_i^{\text{T}}[P_i]\mathbf{Y}_i = 0 \tag{14.41}$$

Differentiation of Eq. (14.41) with respect to the design variable x_j gives

$$\mathbf{Y}_{i,j}^{\text{T}}[P_i]\mathbf{Y}_i + \mathbf{Y}_i^{\text{T}}\frac{\partial[P_i]}{\partial x_j}\mathbf{Y}_i + \mathbf{Y}_i^{\text{T}}[P_i]\mathbf{Y}_{i,j} = 0 \tag{14.42}$$

where $\mathbf{Y}_{i,j} = \partial\mathbf{Y}_i/\partial x_j$. In view of Eq. (14.39), Eq. (14.42) reduces to

$$\mathbf{Y}_i^T\frac{\partial[P_i]}{\partial x_j}\mathbf{Y}_i = 0 \tag{14.43}$$

Differentiation of Eq. (14.40) gives

$$\frac{\partial[P_i]}{\partial x_j} = \frac{\partial[K]}{\partial x_j} - \lambda_i\frac{\partial[M]}{\partial x_j} - \frac{\partial\lambda_i}{\partial x_j}[M] \tag{14.44}$$

where $\partial[K]/\partial x_j$ and $\partial[M]/\partial x_j$ denote the matrices formed by differentiating the elements of $[K]$ and $[M]$ matrices, respectively, with respect to x_j. If the eigenvalues are normalized with respect to the mass matrix, we have [14.10]

$$\mathbf{Y}_i^{\text{T}}[M]\mathbf{Y}_i = 1 \tag{14.45}$$

Substituting Eq. (14.44) into Eq. (14.43) and using Eq. (14.45) gives the derivative of λ_i with respect to x_j as

$$\frac{\partial\lambda_i}{\partial x_j} = \mathbf{Y}_i^{\text{T}}\left[\frac{\partial[K]}{\partial x_j} - \lambda_i\frac{\partial[M]}{\partial x_j}\right]\mathbf{Y}_i \tag{14.46}$$

It can be noted that Eq. (14.46) involves only the eigenvalue and eigenvector under consideration and hence the complete solution of the eigenvalue problem is not required to find the value of $\partial\lambda_i/\partial x_j$.

14.5.2 Derivatives of $\mathbf{Y}_i$

The differentiation of Eqs. (14.39) and (14.45) with respect to x_j results in

$$[P_i]\frac{\partial \mathbf{Y}_i}{\partial x_j} = -\frac{\partial [P_i]}{\partial x_j}\mathbf{Y}_i \qquad (14.47)$$

$$2\mathbf{Y}_i^T[M]\frac{\partial \mathbf{Y}_i}{\partial x_j} = -\mathbf{Y}_i^T\frac{\partial [M]}{\partial x_j}\mathbf{Y}_i \qquad (14.48)$$

where $\partial [P_i]/\partial x_j$ is given by Eq. (14.44). Equations (14.47) and (14.48) can be shown to be linearly independent and can be written together as

$$\underset{(m+1)\times m}{\begin{bmatrix} [P_i] \\ 2\mathbf{Y}_i^T[M] \end{bmatrix}}\underset{m \times 1}{\frac{\partial \mathbf{Y}_i}{\partial x_j}} = -\underset{(m+1)\times m}{\begin{bmatrix} \dfrac{\partial [P_i]}{\partial x_j} \\ \mathbf{Y}_i^T\dfrac{\partial [M]}{\partial x_j} \end{bmatrix}}\underset{m \times 1}{\mathbf{Y}_i} \qquad (14.49)$$

By premultiplying Eq. (14.49) by

$$\begin{bmatrix} [P_i] \\ \mathbf{Y}_i^T[M] \end{bmatrix}^T = \begin{bmatrix} [P_i] & [M]\mathbf{Y}_i \end{bmatrix}$$

we obtain

$$\underset{m \times m}{[[P_i][P_i] + 2[M]\mathbf{Y}_i\mathbf{Y}_i^T[M]]}\underset{m \times 1}{\frac{\partial \mathbf{Y}_i}{\partial x_j}} = -\underset{m \times m}{\left[[P_i]\frac{\partial [P_i]}{\partial x_j} + [M]\mathbf{Y}_i\mathbf{Y}_i^T\frac{\partial [M]}{\partial x_j} \right]}\underset{m \times 1}{\mathbf{Y}_i}$$

$$(14.50)$$

The solution of Eq. (14.50) gives the desired expression for the derivative of the eigenvector, $\partial \mathbf{Y}_i/\partial x_j$, as

$$\frac{\partial \mathbf{Y}_i}{\partial x_j} = -[[P_i][P_i] + 2[M]\mathbf{Y}_i\mathbf{Y}_i^T[M]]^{-1}$$

$$\times \left[[P_i]\frac{\partial [P_i]}{\partial x_j} + [M]\mathbf{Y}_i\mathbf{Y}_i^T\frac{\partial [M]}{\partial x_j} \right]\mathbf{Y}_i \qquad (14.51)$$

Again it can be seen that only the eigenvalue and eigenvector under consideration are involved in the evaluation of the derivatives of eigenvectors.

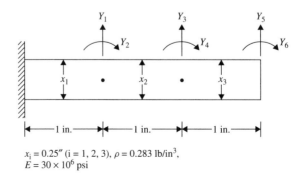

$x_i = 0.25''$ (i = 1, 2, 3), $\rho = 0.283$ lb/in³,
$E = 30 \times 10^6$ psi

Figure 14.4 Cylindrical cantilever beam.

Table 14.3   Derivatives of Eigenvalues [14.4]

i	Eigenvalue, λ_i	$10^{-9}\dfrac{\partial \lambda_i}{\partial x_1}$	$10^{-9}\dfrac{\partial \lambda_i}{\partial x_3}$	$10^{-2}\dfrac{\partial Y_{5i}}{\partial x_1}$	$10^{-2}\dfrac{\partial Y_{6i}}{\partial x_1}$
1	24.66	0.3209	−0.1582	1.478	−2.298
2	974.7	3.86	−0.4144	0.057	−3.046
3	7782.0	23.5	21.67	0.335	−5.307

For illustration, a cylindrical cantilever beam is considered [14.4]. The beam is modeled with three finite elements with six degrees of freedom as indicated in Fig. 14.4. The diameters of the beam are considered as the design variables, x_i, $i = 1, 2, 3$. The first three eigenvalues and their derivatives are shown in Table 14.3 [14.4].

14.6   DERIVATIVES OF TRANSIENT RESPONSE

The equations of motion of an n-degree-of-freedom mechanical/structural system with viscous damping can be expressed as [14.10]

$$[M]\ddot{\mathbf{Y}} + [C]\dot{\mathbf{Y}} + [K]\mathbf{Y} = \mathbf{F}(t) \qquad (14.52)$$

where $[M]$, $[C]$, and $[K]$ are the $n \times n$ mass, damping, and stiffness matrices, respectively, $\mathbf{F}(t)$ is the n-component force vector, $\mathbf{Y}$ is the n-component displacement vector, and a dot over a symbol indicates differentiation with respect to time. Equations (14.52) denote a set of n coupled second-order differential equations. In most practical cases, n will be very large and Eqs. (14.52) are stiff; hence the numerical solution of Eqs. (14.52) will be tedious and produces an accurate solution only for low-frequency components. To reduce the size of the problem, the displacement solution, $\mathbf{Y}$, is expressed in terms of r basis functions $\boldsymbol{\Phi}_1, \boldsymbol{\Phi}_2, \ldots,$ and $\boldsymbol{\Phi}_r$ (with $r \ll n$) as

$$\mathbf{Y} = [\Phi]\mathbf{q} \quad \text{or} \quad y_j = \sum_{k=1}^{r} \Phi_{jk} q_k(t), \quad j = 1, 2, \ldots, n \qquad (14.53)$$

where

$$[\Phi] = [\boldsymbol{\Phi}_1 \ \boldsymbol{\Phi}_2 \ \cdots \ \boldsymbol{\Phi}_r]$$

is the matrix of basis functions, Φ_{jk} the element in row j and column k of the matrix $[\Phi]$, $\mathbf{q}$ an r-component vector of reduced coordinates, and $q_k(t)$ the kth component of the vector $\mathbf{q}$. By substituting Eq. (14.53) into Eq. (14.52) and premultiplying the resulting equation by $[\Phi]^{\mathrm{T}}$, we obtain a system of r differential equations:

$$[\overline{M}]\ddot{\mathbf{q}} + [\overline{C}]\dot{\mathbf{q}} + [\overline{K}]\mathbf{q} = \overline{\mathbf{F}}(t) \qquad (14.54)$$

where

$$[\overline{M}] = [\Phi]^{\mathrm{T}}[M][\Phi] \qquad (14.55)$$

$$[\overline{C}] = [\Phi]^{\mathrm{T}}[C][\Phi] \qquad (14.56)$$

$$[\overline{K}] = [\Phi]^{\mathrm{T}}[K][\Phi] \qquad (14.57)$$

$$\overline{\mathbf{F}}(t) = [\Phi]^{\mathrm{T}}\mathbf{F}(t) \qquad (14.58)$$

Note that if the undamped natural modes of vibration are used as basis functions and if $[C]$ is assumed to be a linear combination of $[M]$ and $[K]$ (called *proportional damping*), Eqs. (14.54) represent a set of r uncoupled second-order differential equations which can be solved independently [14.10]. Once $\mathbf{q}(t)$ is found, the displacement solution $\mathbf{Y}(t)$ can be determined from Eq. (14.53).

In the formulation of optimization problems with restrictions on the dynamic response, the constraints are placed on selected displacement components as

$$|y_j(\mathbf{X}, t)| \leq y_{max}, \quad j = 1, 2, \ldots \tag{14.59}$$

where y_j is the displacement at location j on the machine/structure and y_{max} is the maximum permissible value of the displacement. Constraints on dynamic stresses are also stated in a similar manner. Since Eq. (14.59) is a parametric constraint in terms of the parameter time (t), it is satisfied only at a set of peak or critical values of y_j for computational simplicity. Once Eq. (14.59) is satisfied at the critical points, it will be satisfied (most likely) at all other values of t as well [14.11, 14.12]. The values of y_i at which $dy_j/dt = 0$ or the values of y_i at the end of the time interval denote local maxima and hence are to be considered as candidate critical points. Among the several candidate critical points, only a select number are considered for simplifying the computations. For example, in the response shown in Fig. 14.5, peaks $a, b, c, \ldots, j$ qualify as candidate critical points. However, peaks $a, b, f,$ and j can be discarded as their magnitudes are considerably smaller (less than, for example, 25%) than those of other peaks. Noting that peaks d and e (or g and h) represent essentially a single large peak with high-frequency undulations, we can discard peak e (or g), which has a slightly smaller magnitude than d (or h). Thus finally, only peaks $c, d, h,$ and i need to be considered to satisfy the constraint, Eq. (14.59).

Once the critical points are identified at a reference design $\mathbf{X}$, the sensitivity of the response, $y_j(\mathbf{X}, t)$ with respect to the design variable x_i at the critical point $t = t_c$ can be found using the total derivative of y_j as

$$\frac{dy_j(\mathbf{X}, t)}{dx_i} = \frac{\partial y_j}{\partial x_i} + \frac{\partial y_j}{\partial t}\frac{dt_c}{dx_i}, \quad i = 1, 2, \ldots, n \tag{14.60}$$

The second term on the right-hand side of Eq. (14.60) is always zero since $\partial y_j/\partial t = 0$ at an interior peak $(0 < t_c < t_{max})$ and $dt_c/dx_i = 0$ at the boundary $(t_c = t_{max})$. The

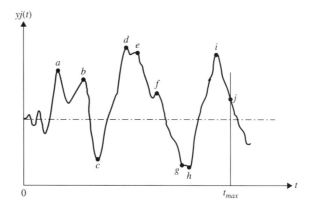

Figure 14.5 Critical points in a typical transient response.

derivative, $\partial y_j / \partial x_i$, can be computed using Eq. (14.53) as

$$\frac{\partial y_j}{\partial x_i} = \sum_{k=1}^{r} \Phi_{jk} \frac{\partial q_k(t)}{\partial x_i}, \quad i = 1, 2, \ldots, n \tag{14.61}$$

where, for simplicity, the elements of the matrix $[\Phi]$ are assumed to be constants (independent of the design vector $\mathbf{X}$). Note that for higher accuracy, the derivatives of Φ_{jk} with respect to x_i (sensitivity of eigenvectors, if the mode shapes are used as the basis vectors) obtained from an equation similar to Eq. (14.51) can be included in finding $\partial y_j / \partial x_i$.

   To find the values of $\partial q_k / \partial x_i$ required in Eq. (14.61), Eq. (14.54) is differentiated with respect to x_i to obtain

$$[\overline{M}] \frac{\partial \ddot{\mathbf{q}}}{\partial x_i} + [\overline{C}] \frac{\partial \dot{\mathbf{q}}}{\partial x_i} + [\overline{K}] \frac{\partial \mathbf{q}}{\partial x_i}$$

$$= \frac{\partial \overline{\mathbf{F}}}{\partial x_i} - \frac{\partial [\overline{M}]}{\partial x_i} \ddot{\mathbf{q}} - \frac{\partial [\overline{C}]}{\partial x_i} \dot{\mathbf{q}} - \frac{\partial [\overline{K}]}{\partial x_i} \mathbf{q}, \quad i = 1, 2, \ldots, n \tag{14.62}$$

The derivatives of the matrices appearing on the right-hand side of Eq. (14.62) can be computed using formulas such as

$$\frac{\partial [\overline{M}]}{\partial x_i} = [\Phi]^{\mathrm{T}} \frac{\partial [M]}{\partial x_i} [\Phi] \tag{14.63}$$

where, for simplicity, $[\Phi]$ is assumed to be constant and $\partial [M]/\partial x_i$ is computed using a finite-difference scheme. In most cases the forcing function $\mathbf{F}(t)$ will be known to be independent of $\mathbf{X}$ or an explicit function of $\mathbf{X}$. Hence the quantity $\partial \overline{\mathbf{F}}/\partial x_i$ can be evaluated without much difficulty. Once the right-hand side is known, Eqs. (14.62) can be integrated numerically in time to find the values of $\partial \ddot{\mathbf{q}}/\partial x_i$, $\partial \dot{\mathbf{q}}/\partial x_i$, and $\partial \mathbf{q}/\partial x_i$. Using the values of $\partial \mathbf{q}/\partial x_i = \{\partial q_k/\partial x_i\}$ at the critical point t_c, the required sensitivity of transient response can be found from Eq. (14.61).

14.7 SENSITIVITY OF OPTIMUM SOLUTION TO PROBLEM PARAMETERS

Any optimum design problem involves a design vector and a set of problem parameters (or preassigned parameters). In many cases, we would be interested in knowing the sensitivities or derivatives of the optimum design (design variables and objective function) with respect to the problem parameters [14.25, 14.26]. As an example, consider the minimum weight design of a machine component or structure subject to a constraint on the induced stress. After solving the problem, we may like to find the effect of changing the material. This means that we would like to know the changes in the optimal dimensions and the minimum weight of the component or structure due to a change in the value of the permissible stress. Usually, the sensitivity derivatives are found by using a finite-difference method. But this requires a costly reoptimization of the problem using incremented values of the parameters. Hence, it is desirable to derive expressions for the sensitivity derivatives from appropriate equations. In this section we discuss two approaches: one based on the Kuhn–Tucker conditions and the other based on the concept of feasible direction.

14.7.1 Sensitivity Equations Using Kuhn–Tucker Conditions

The Kuhn–Tucker conditions satisfied at the constrained optimum design $\mathbf{X}^*$ are given by [see Eqs. (2.73) and (2.74)]

$$\frac{\partial f(\mathbf{X})}{\partial x_i} + \sum_{j \in J_1} \lambda_j \frac{\partial g_j(\mathbf{X})}{\partial x_i} = 0, \qquad i = 1, 2, \ldots, n \tag{14.64}$$

$$g_j(\mathbf{X}) = 0, \qquad j \in J_1 \tag{14.65}$$

$$\lambda_j > 0, \qquad j \in J_1 \tag{14.66}$$

where J_1 is the set of active constraints and Eqs. (14.64) to (14.66) are valid with $\mathbf{X} = \mathbf{X}^*$ and $\lambda_j = \lambda_j^*$. When a problem parameter changes by a small amount, we assume that Eqs. (14.64) to (14.66) remain valid. Treating f, g_j, $\mathbf{X}$, and λ_j as functions of a typical problem parameter p, differentiation of Eqs. (14.64) and (14.65) with respect to p leads to

$$\sum_{k=1}^{n} \left[\frac{\partial^2 f(\mathbf{X})}{\partial x_i \partial x_k} + \sum_{j \in J_1} \lambda_j \frac{\partial^2 g_j(\mathbf{X})}{\partial x_i \partial x_k} \right] \frac{\partial x_k}{\partial p} + \sum_{j \in J_1} \frac{\partial \lambda_j}{\partial p} \frac{\partial g_j(\mathbf{X})}{\partial x_i} + \frac{\partial^2 f(\mathbf{X})}{\partial x_i \partial p}$$

$$+ \sum_{j \in J_1} \lambda_j \frac{\partial^2 g_j(\mathbf{X})}{\partial x_i \partial p} = 0, \qquad i = 1, 2, \ldots, n \tag{14.67}$$

$$\frac{\partial g_j(\mathbf{X})}{\partial p} + \sum_{i=1}^{n} \frac{\partial g_j(\mathbf{X})}{\partial x_i} \frac{\partial x_i}{\partial p} = 0, \qquad j \in J_1 \tag{14.68}$$

Equations (14.67) and (14.68) can be expressed in matrix form as

$$\begin{bmatrix} [P]_{n \times n} & [Q]_{n \times q} \\ [Q]_{q \times n}^T & [0]_{q \times q} \end{bmatrix} \begin{Bmatrix} \dfrac{\partial \mathbf{X}}{\partial p}_{n \times 1} \\ \dfrac{\partial \lambda}{\partial p}_{q \times 1} \end{Bmatrix} + \begin{Bmatrix} \mathbf{a}_{n \times 1} \\ \mathbf{b}_{q \times 1} \end{Bmatrix} = \begin{Bmatrix} \mathbf{0}_{n \times 1} \\ \mathbf{0}_{q \times 1} \end{Bmatrix} \tag{14.69}$$

where q denotes the number of active constraints and the elements of the matrices and vectors in Eq. (14.69) are given by

$$P_{ik} = \frac{\partial^2 f(\mathbf{X})}{\partial x_i \partial x_k} + \sum_{j \in J_1} \lambda_j \frac{\partial^2 g_j(\mathbf{X})}{\partial x_i \partial x_k} \tag{14.70}$$

$$Q_{ij} = \frac{\partial g_j(\mathbf{X})}{\partial x_i}, \qquad j \in J_1 \tag{14.71}$$

$$a_i = \frac{\partial^2 f(\mathbf{X})}{\partial x_i \partial p} + \sum_{j \in J_1} \lambda_j \frac{\partial g_j(\mathbf{X})}{\partial x_i \partial p} \tag{14.72}$$

$$b_j = \frac{\partial g_j(\mathbf{X})}{\partial p}, \qquad j \in J_1 \tag{14.73}$$

$$\frac{\partial X}{\partial p} = \left\{ \begin{array}{c} \frac{\partial x_1}{\partial p} \\ \vdots \\ \frac{\partial x_n}{\partial p} \end{array} \right\}, \quad \frac{\partial \lambda}{\partial p} = \left\{ \begin{array}{c} \frac{\partial \lambda_1}{\partial p} \\ \vdots \\ \frac{\partial \lambda_q}{\partial p} \end{array} \right\} \tag{14.74}$$

The following can be noted in Eqs. (14.69):

1. Equations (14.69) denote $(n + q)$ simultaneous equations in terms of the required sensitivity derivatives, $\partial x_i / \partial p$ $(i = 1, 2, \ldots, n)$ and $\partial \lambda_j / \partial p$ $(j = 1, 2, \ldots, q)$. Both $\mathbf{X}^*$ and λ^* are assumed to be known in Eqs. (14.69). If λ^* are not computed during the optimization process, they can be computed using Eq. (7.263).

2. Equations (14.69) can be solved only if the system is nonsingular. One of the requirements for this is that the active constraints be independent.

3. Second derivatives of f and g_j are required in computing the elements of $[P]$ and $\mathbf{a}$.

4. If sensitivity derivatives are required with respect to several problem parameters $p_1, p_2, \ldots$, only the vectors $\mathbf{a}$ and $\mathbf{b}$ need to be computed for each case and the system of Eqs. (14.69) can be solved efficiently using the techniques of solving simultaneous equations with different right-hand-side vectors.

Once Eqs. (14.69) are solved, the sensitivity of optimum objective value with respect to p can be computed as

$$\frac{df(\mathbf{X})}{dp} = \frac{\partial f(\mathbf{X})}{\partial p} + \sum_{i=1}^{n} \frac{\partial f(\mathbf{X})}{\partial x_i} \frac{\partial x_i}{\partial p} \tag{14.75}$$

The changes in the optimum values of x_i and f necessary to satisfy the Kuhn–Tucker conditions due to a change Δp in the problem parameter can be estimated as

$$\Delta x_i = \frac{\partial x_i}{\partial p} \Delta p, \quad \Delta f = \frac{df}{dp} \Delta p \tag{14.76}$$

The changes in the values of Lagrange multiplier λ_j due to Δp can be estimated as

$$\Delta \lambda_j = \frac{\partial \lambda_j}{\partial p} \Delta p \tag{14.77}$$

Equation (14.77) can be used to determine whether an originally active constraint becomes inactive due to the change, Δp. Since the value of λ_j is zero for an inactive constraint, we have

$$\lambda_j + \Delta \lambda_j = \lambda_j + \frac{\partial \lambda_j}{\partial p} \Delta p = 0 \tag{14.78}$$

from which the value of Δp necessary to make the jth constraint inactive can be found as

$$\Delta p = -\frac{\lambda_j}{\partial \lambda_j / \partial p}, \quad j \in J_1 \tag{14.79}$$

Similarly, a currently inactive constraint will become critical due to Δp if the new value of g_j becomes zero:

$$g_j(\mathbf{X}) + \frac{dg_j}{dp}\Delta p = g_j(\mathbf{X}) + \left(\sum_{i=1}^{n}\frac{\partial g_j}{\partial x_i}\frac{\partial x_i}{\partial p}\right)\Delta p \tag{14.80}$$

Thus the change Δp necessary to make an inactive constraint active can be found as

$$\Delta p = -\frac{g_j(\mathbf{X})}{\displaystyle\sum_{i=1}^{n}\frac{\partial g_j}{\partial x_i}\frac{\partial x_i}{\partial p}} \tag{14.81}$$

14.7.2   Sensitivity Equations Using the Concept of Feasible Direction

Here we treat the problem parameter p as a design variable so that the new design vector becomes

$$\mathbf{X} = \{x_1\ x_2\ \cdots\ x_n\ p\}^{\mathrm{T}} \tag{14.82}$$

As in the case of the method of feasible directions (see Section 7.7), we formulate the direction finding problem as

$$\text{Find } \mathbf{X} \text{ which minimizes } -\mathbf{S}^{\mathrm{T}}\nabla f(\mathbf{X})$$

subject to

$$\mathbf{S}^{\mathrm{T}}\nabla g_j \leq 0, \quad j \in J_1$$
$$\mathbf{S}^{\mathrm{T}}\mathbf{S} \leq 1 \tag{14.83}$$

where the gradients of f and g_j ($j \in J_1$) can be evaluated in the usual manner. The set J_1 can include nearly active constraints also (along with the active constraints) so that we do not violate any constraint due to the change, Δp. The solution of the problem stated in Eqs. (14.83) gives a usable feasible search direction, $\mathbf{S}$. A new design vector along $\mathbf{S}$ can be expressed as

$$\mathbf{X}_{\text{new}} = \mathbf{X}_{\text{current}} + \lambda\mathbf{S} = \mathbf{X}_{\text{current}} + \Delta\mathbf{X} \tag{14.84}$$

where λ is the step length and the components of $\mathbf{S}$ can be considered as

$$s_i = \begin{cases} \dfrac{\partial x_i}{\partial \lambda}, & i = 1, 2, \ldots, n \\[2ex] \dfrac{\partial p}{\partial \lambda}, & i = n + 1 \end{cases} \tag{14.85}$$

so that

$$\Delta p = \lambda s_{n+1} \quad \text{or} \quad \lambda = \frac{\Delta p}{s_{n+1}} \tag{14.86}$$

If the vector $\mathbf{S}$ is normalized by dividing its components by s_{n+1}, Eq. (14.86) gives $\lambda = \Delta p$ and hence Eq. (14.85) gives the desired sensitivity derivatives as

$$
\left\{ \begin{array}{c} \frac{\partial x_1}{\partial p} \\ \vdots \\ \frac{\partial x_n}{\partial p} \end{array} \right\} = \frac{1}{s_{n+1}} \mathbf{S} \tag{14.87}
$$

Thus the sensitivity of the objective function with respect to p can be computed as

$$
\frac{df(\mathbf{X})}{dp} = \nabla f(\mathbf{X})^{\mathrm{T}} \frac{\mathbf{S}}{s_{n+1}} \tag{14.88}
$$

Note that unlike the previous method, this method does not require the values of λ^* and the second derivatives of f and g_j to find the sensitivity derivatives. Also, if sensitivities with respect to several problem parameters $p_1, p_2, \ldots$ are required, all we need to do is to add them to the design vector $\mathbf{X}$ in Eq. (14.82).

14.8 MULTILEVEL OPTIMIZATION

14.8.1 Basic Idea

The design of practical systems involving a large number of elements or subsystems with multiple-load conditions involves excessive number of design variables and constraints. The optimization problem becomes unmanageably large, and the solution process becomes too costly and can easily saturate even the largest computers available. In such cases the optimization problem can be broken into a series of smaller problems using different strategies. The multilevel optimization is a decomposition technique in which the problem is reformulated as several smaller subproblems (one for each subsystem) and a coordination problem (at system level) to preserve the coupling among the subproblems (subsystems). Such approaches have been used in linear and dynamic programming also. In linear programming, the decomposition method (see Section 4.4) involves a number of independent linear subproblems coupled by limitations on the shared resources. When an individual subsystem is solved, the cost of the shared resources is added to its objective function. By a proper variation of the costs of the shared resources, the proposed optimal strategies of the various subproblems are sent to the master program, which, in turn, is optimized so that the overall cost is minimized. In dynamic programming, the problem is treated in stages with an optimal policy determined in each stage (see Chapter 9). This approach is particularly useful when the problem has a serial structure.

For nonlinear design optimization problems, several decomposition methods have been proposed [14.14–14.16]. In the following section we consider a two-level approach in which the system is decomposed into a number of smaller subproblems, each with its own goals and constraints. The individual subsystem optimization problems are solved independently in the first level and the coordinated problem is solved in the second level. The approach is known as the *model-coordination method*.

14.8.2   Method

Let the optimization problem be stated as follows:

$$\text{Find } \mathbf{X} = \{x_1 \ x_2 \ \cdots \ x_n\}^{\mathrm{T}} \text{ which minimizes } f(\mathbf{X}) \tag{14.89}$$

subject to

$$g_j(\mathbf{X}) \leq 0, \qquad j = 1, 2, \ldots, m \tag{14.90}$$

$$h_k(\mathbf{X}) = 0, \qquad k = 1, 2, \ldots, p \tag{14.91}$$

$$x_i^{(l)} \leq x_i \leq x_i^{(u)}, \qquad i = 1, 2, \ldots, n \tag{14.92}$$

where $x_i^{(l)}$ and $x_i^{(u)}$ denote the lower and upper bounds on x_i. Most systems permit the partitioning of the vector $\mathbf{X}$ into two subvectors $\mathbf{Y}$ and $\mathbf{Z}$:

$$\mathbf{X} = \begin{Bmatrix} \mathbf{Y} \\ \mathbf{Z} \end{Bmatrix} \tag{14.93}$$

where the subvector $\mathbf{Y}$ denotes the coordination or interaction variables between the subsystems and the subvector $\mathbf{Z}$ indicates the variables confined to subsystems. The vector $\mathbf{Z}$, in turn, can be partitioned as

$$\mathbf{Z} = \begin{Bmatrix} \mathbf{Z}_1 \\ \vdots \\ \mathbf{Z}_k \\ \vdots \\ \mathbf{Z}_K \end{Bmatrix} \tag{14.94}$$

where $\mathbf{Z}_k$ represents the variables associated with the kth subsystem only and K denotes the number of subsystems. The partitioning of variables, Eq. (14.94), permits us to regroup the constraints as

$$\begin{Bmatrix} g_1(\mathbf{X}) \\ g_2(\mathbf{X}) \\ \vdots \\ g_m(\mathbf{X}) \end{Bmatrix} = \begin{Bmatrix} \mathbf{g}^{(1)}(\mathbf{Y}, \mathbf{Z}_1) \\ \mathbf{g}^{(2)}(\mathbf{Y}, \mathbf{Z}_2) \\ \vdots \\ \mathbf{g}^{(K)}(\mathbf{Y}, \mathbf{Z}_K) \end{Bmatrix} \leq \mathbf{0} \tag{14.95}$$

$$\begin{Bmatrix} l_1(\mathbf{X}) \\ l_2(\mathbf{X}) \\ \vdots \\ l_p(\mathbf{X}) \end{Bmatrix} = \begin{Bmatrix} l^{(1)}(\mathbf{Y}, \mathbf{Z}_1) \\ l^{(2)}(\mathbf{Y}, \mathbf{Z}_2) \\ \vdots \\ l^{(K)}(\mathbf{Y}, \mathbf{Z}_K) \end{Bmatrix} = \mathbf{0} \tag{14.96}$$

where the variables $\mathbf{Y}$ may appear in all the functions while the variables $\mathbf{Z}_k$ appear only in the constraint sets $\mathbf{g}^{(k)} \leq \mathbf{0}$ and $\mathbf{h}^{(k)} = 0$. The bounds on the variables, Eq. (14.92), can be expressed as

$$\mathbf{Y}^{(l)} \leq \mathbf{Y} \leq \mathbf{Y}^{(u)}$$

$$\mathbf{Z}_k^{(l)} \leq \mathbf{Z}_k \leq \mathbf{Z}_k^{(u)}, \qquad k = 1, 2, \ldots, K \tag{14.97}$$

Similarly, the objective function $f(\mathbf{X})$ can be expressed as

$$f(\mathbf{X}) = \sum_{k=1}^{K} f^{(k)}(\mathbf{Y}, \mathbf{Z}_k) \tag{14.98}$$

where $f^{(k)}(\mathbf{Y}, \mathbf{Z}_k)$ denotes the contribution of the kth subsystem to the overall objective function. Using Eqs. (14.95) to (14.98), the two-level approach can be stated as follows.

First-level Problem. Tentatively fix the values of $\mathbf{Y}$ at $\mathbf{Y}^*$ so that the problem of Eqs. (14.89) to (14.92) [or Eqs. (14.95) to (14.98)] can be restated (decomposed) as K independent optimization problems as follows:

$$\text{Find } \mathbf{Z}_k \text{ which minimizes } f^{(k)}(\mathbf{Y}, \mathbf{Z}_k)$$

subject to

$$g^{(k)}(\mathbf{Y}, \mathbf{Z}_k) \leq \mathbf{0}$$

$$\mathbf{h}^{(k)}(\mathbf{Y}, \mathbf{Z}_k) = \mathbf{0} \tag{14.99}$$

$$\mathbf{Z}_k^{(l)} \leq \mathbf{Z}_k \leq \mathbf{Z}_k^{(u)}; \qquad k = 1, 2, \ldots, K$$

It can be seen that the first-level problem seeks to find the minimum of the function

$$f(\mathbf{Y}, \mathbf{Z}) = \sum_{k=1}^{K} f^{(k)}(\mathbf{Y}, \mathbf{Z}_k) \tag{14.100}$$

for the (tentatively) fixed vector $\mathbf{Y}^*$.

Second-level Problem. The following problem is solved in this stage:

$$\text{Find a new } \mathbf{Y}^* \text{ which minimizes } f(\mathbf{Y}) = \sum_{k=1}^{K} f^{(k)}(\mathbf{Y}, \mathbf{Z}_k^*)$$

subject to

$$\mathbf{Y}^{(l)} \leq \mathbf{Y} \leq \mathbf{Y}^{(u)} \tag{14.101}$$

where $\mathbf{Z}_k^*, k = 1, 2, \ldots, K$, are the optimal solutions of the first-level problems. An additional constraint to ensure a finite value of $f(\mathbf{Y}^*)$ is also to be included while solving the problem of Eqs. (14.101). Once the problem is solved and a new $\mathbf{Y}^*$ found, we proceed to solve the first-level problems. This process is to be continued until convergence is achieved. The iterative process can be summarized as follows:

1. Start with an initial coordination vector, $\mathbf{Y}^*$.
2. Solve the K first-level optimization problems, stated in Eqs. (14.99), and find the optimal vectors $\mathbf{Z}_k^*(k = 1, 2, \ldots, K)$.

3. Solve the second-level optimization problem stated in Eqs. (14.101) and find a new vector $\mathbf{Y}^*$.

4. Check for the convergence of f^* and $\mathbf{Y}^*$ (compared to the value $\mathbf{Y}^*$ used earlier).

5. If the process has not converged, go to step 2 and repeat the process until convergence.

The following example illustrates the procedure.

Example 14.2 Find the minimum-weight design of the two-bar truss shown in Fig. 14.6 with constraints on the depth of the truss ($y = h$), cross-sectional areas of the members ($z_1 = A_1$) and ($z_2 = A_2$), and the stresses induced in the bars. Treat the depth of the truss (y) and the cross-sectional areas of bars 1 and 2 (z_1 and z_2) as design variables. The permissible stress in each bar is $\sigma_0 = 10^5$ Pa, unit weight is $76,500\,\text{N/m}^3$, h is constrained as $1\,\text{m} \le h \le 6\,\text{m}$, and the cross-sectional area of each bar is restricted to lie between 0 and $0.1\,\text{m}^2$.

SOLUTION   The stresses induced in the bars can be expressed as

$$\sigma_1 = \frac{P\sqrt{y^2 + 36}}{7yz_1}, \qquad \sigma_2 = \frac{6P\sqrt{y^2 + 1}}{7yz_2}$$

and hence the optimization problem can be stated as follows:

Find $\mathbf{X} = \{y\ z_1\ z_2\}^{\mathrm{T}}$ which minimizes

$$f(\mathbf{X}) = 76,500z_1 \sqrt{y^2 + 36} + 76,500z_2 \sqrt{y^2 + 1}$$

subject to

$$\frac{P\sqrt{y^2 + 36}}{7\sigma_0 yz_1} - 1 \le 0, \qquad \frac{6P\sqrt{y^2 + 1}}{7\sigma_0 yz_2} - 1 \le 0$$

$$1 \le y \le 6, \ 0 \le z_1 \le 0.1, \ 0 \le z_2 \le 0.1$$

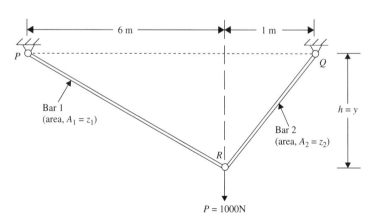

Figure 14.6   Two-bar truss.

We treat the bars 1 and 2 as subsystems 1 and 2, respectively, with y as the coordination variable ($\mathbf{Y} = \{y\}$) and z_1 and z_2 as the subsystem variables ($\mathbf{Z}_1 = \{z_1\}$ and $\mathbf{Z}_2 = \{z_2\}$). By fixing the value of y at y^*, we formulate the first-level problems as follows.

Subproblem 1.
Find z_1 which minimizes

$$f^{(1)}(y^*, z_1) = 76{,}500z_1 \sqrt{(y^*)^2 + 36} \tag{E_1}$$

subject to

$$g_1(y^*, z_1) = \frac{(1428.5714 \times 10^{-6})\sqrt{(y^*)^2 + 36}}{y^* z_1} - 1 \leq 0 \tag{E_2}$$

$$0 \leq z_1 \leq 0.1 \tag{E_3}$$

Subproblem 2.
Find z_2 which minimizes

$$f^{(2)}(y^*, z_2) = 76{,}500z_2 \sqrt{(y^*)^2 + 1} \tag{E_4}$$

subject to

$$g_2(y^*, z_2) = \frac{(8571.4285 \times 10^{-6})\sqrt{(y^*)^2 + 1}}{y^* z_2} - 1 \leq 0 \tag{E_5}$$

$$0 \leq z_2 \leq 0.1 \tag{E_6}$$

We can see that to minimize $f^{(1)}$ we need to make z_1 as small as possible without violating the constraints of Eqs. (E_2) and (E_3). This gives the solution of subproblem 1, z_1^* (which makes g_1 active) as

$$z_1^* = \frac{(1428.5714 \times 10^{-6})\sqrt{(y^*)^2 + 36}}{y^*} \tag{E_7}$$

Similarly, the solution of subproblem 2, z_2^* (which makes g_2 active) can be expressed as

$$z_2^* = \frac{(8571.4285 \times 10^{-6})\sqrt{(y^*)^2 + 1}}{y^*} \tag{E_8}$$

Now we state the second-level problem as follows:

Find y which minimizes $f = f^{(1)}(y, z_1^*) + f^{(2)}(y, z_2^*)$

subject to

$$1 \leq y \leq 6 \tag{E_9}$$

Using Eqs. (E_7) and (E_8), this problem can be restated as (using y for y^*):

Find y which minimizes

$$f = 76{,}500z_1^* \sqrt{y^2 + 36} + 76{,}500z_2^* \sqrt{y^2 + 1}$$

$$= 109.2857\frac{y^2 + 36}{y} + 655.7143\frac{y^2 + 1}{y} \qquad (E_{10})$$

subject to

$$1 \le y \le 6 \text{ and } f \text{ must be defined}$$

The graph of f, given by Eq. (E_{10}), is shown in Fig. 14.7 over the range $1 \le y \le 6$ from which the solution can be determined as $f^* = 3747.7\,\mathrm{N}$, $y^* = h^* = 2.45\,\mathrm{m}$, $z_1^* = A_1^* = 3.7790 \times 10^{-3}\,\mathrm{m}^2$, and $z_2^* = A_2^* = 9.2579 \times 10^{-3}\,\mathrm{m}^2$.

14.9 PARALLEL PROCESSING

Large-scale optimization problems can be solved efficiently using parallel computers. Parallel computers are simply multiple processing units combined in an organized fashion such that multiple independent computations for the same problem could be performed simultaneously or concurrently, thereby increasing the overall computational speed. Optimization problems involving extensive analysis, such as a finite-element analysis, can be solved on parallel computers using the following schemes:

1. A multilevel (decomposition) approach with the subproblems solved in parallel
2. A substructures approach with substructure analyses performed in parallel
3. By implementing the optimization computations in parallel

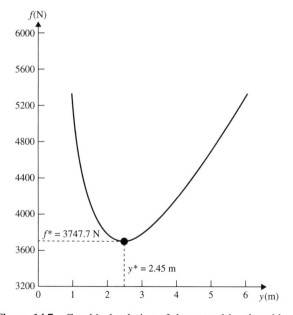

Figure 14.7 Graphical solution of the second-level problem.

If a multilevel (decomposition) approach is used, the optimization of various subsystems (at different levels) can be performed on parallel processors while the solution of the coordinating optimization problem can be accomplished on the main processor. If the optimization problem involves an extensive analysis, such as a finite-element analysis, the problem can be decomposed into subsystems (substructures) and the analyses of subsystems can be conducted on parallel processors with a main processor performing the system-level computations. Such an approach was used by El-Sayed and Hsiung [14.17, 14.20]. The procedure can be summarized as follows:

1. Initialize the optimization process. The current (related) design variables are sent to the various processors.

2. The finite-element analyses of the substructures are performed on different (associated) processors.

3. The main processor collects the stiffness and force contribution matrices from the various processors, solves for the displacements at the shared (common) boundary nodes of substructures, and sends the data to various processors.

4. The associated processors perform the detailed calculations to find the displacements and stresses needed for the evaluation of the constraints.

5. The main processor collects the constraint-related data from the associate processors and checks the convergence of the optimization process. If convergence is not achieved, it performs the computations of the optimization algorithm and the procedure is repeated from step 1 onward.

Numerical examples were solved on a Cray X-MP four-processor supercomputer [14.17]. For a 200-member planar truss, the weight was minimized with constraints on stresses using four substructures. It was reported [14.17] that the parallel computations required 10.585 s of CPU time, while the sequential computations required a CPU time of 13.518 s (with a speedup factor of 1.28)

For most mechanical and structural problems, parallel computers with MIMD (multiple instruction multiple data) architecture are better suited. Atiqullah and Rao [14.21] presented a procedure for the parallel implementation of the simulated annealing algorithm. In this method, certain design variables assigned to each processor perform the variable specific optimization. This information is later combined to complete one cycle of optimization. Since the entire (variable-specific) optimization process is repeated on each processor, all processors will be equally busy most of the time, except for any input/output done by the specific processors. Thus the "divide and conquer" strategy of optimization needs a "communicate and combine" process, which should be kept to a minimum. The detailed procedure is shown as a flow diagram in Fig. 14.8.

The minimum-weight design of a 128-bar planar truss was considered with stress and buckling constraints. A speedup factor of 10.2569 was achieved using the eight-node configuration of an iPSC/860 computer.

14.10 MULTIOBJECTIVE OPTIMIZATION

A multiobjective optimization problem with inequality constraints can be stated as (equality constraints, if they exist, can also be included in the formulation of the problem)

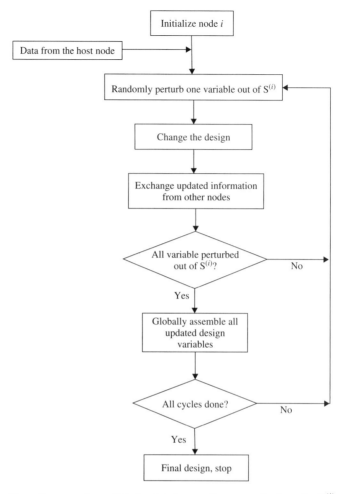

Figure 14.8   Flow diagram of parallel simulated annealing on a single node. $\mathbf{S}^{(i)}$, set of design variables assigned to node i; node i = processor i.

$$\text{Find } \mathbf{X} = \begin{Bmatrix} x_1 \\ x_2 \\ \vdots \\ x_n \end{Bmatrix} \tag{14.102}$$

$$\text{which minimizes } f_1(\mathbf{X}), f_2(\mathbf{X}), \ldots, f_k(\mathbf{X}) \tag{14.103}$$

subject to

$$g_j(\mathbf{X}) \le 0, \quad j = 1, 2, \ldots, m \tag{14.104}$$

where k denotes the number of objective functions to be minimized. Any or all of the functions $f_i(\mathbf{X})$ and $g_j(\mathbf{X})$ may be nonlinear. The multiobjective optimization problem is also known as a *vector minimization problem*.

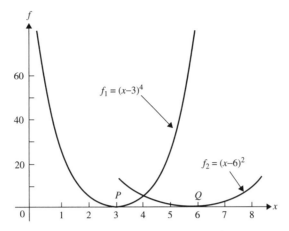

Figure 14.9   Pareto optimal solutions.

In general, no solution vector $\mathbf{X}$ exists that minimizes all the k objective functions simultaneously. Hence, a new concept, known as the *Pareto optimum solution*, is used in multiobjective optimization problems. A feasible solution $\mathbf{X}$ is called *Pareto optimal* if there exists no other feasible solution $\mathbf{Y}$ such that $f_i(\mathbf{Y}) \leq f_i(\mathbf{X})$ for $i = 1, 2, \ldots, k$ with $f_j(\mathbf{Y}) < f_i(\mathbf{X})$ for at least one j. In other words, a feasible vector $\mathbf{X}$ is called Pareto optimal if there is no other feasible solution $\mathbf{Y}$ that would reduce some objective function without causing a simultaneous increase in at least one other objective function. For example, if the objective functions are given by $f_1 = (x - 3)^4$ and $f_2 = (x - 6)^2$, their graphs are shown in Fig. 14.9. For this problem, all the values of x between 3 and 6 (points on the line segment PQ) denote Pareto optimal solutions.

Several methods have been developed for solving a multiobjective optimization problem. Some of these methods are briefly described in the following paragraphs. Most of these methods basically generate a set of Pareto optimal solutions and use some additional criterion or rule to select one particular Pareto optimal solution as the solution of the multiobjective optimization problem.

14.10.1   Utility Function Method

In the utility function method, a utility function $U_i(f_i)$ is defined for each objective depending on the importance of f_i compared to the other objective functions. Then a total or overall utility function U is defined, for example, as

$$U = \sum_{i=1}^{k} U_i(f_i) \tag{14.105}$$

The solution vector $\mathbf{X}^*$ is then found by maximizing the total utility U subjected to the constraints $g_j(\mathbf{X}) \leq 0, j = 1, 2, \ldots, m$. A simple form of Eq. (14.105) is given by

$$U = \sum_{i=1}^{k} U_i = -\sum_{i=1}^{k} w_i f_i(\mathbf{X}) \tag{14.106}$$

where w_i is a scalar weighting factor associated with the ith objective function. This method [Eq.(14.106)] is also known as the *weighting function method*.

14.10.2   Inverted Utility Function Method

In the inverted utility function method, we invert each utility and try to minimize or reduce the total undesirability. Thus if $U_i(f_i)$ denotes the utility function corresponding to the ith objective function, the total undesirability is obtained as

$$U^{-1} = \sum_{i=1}^{k} U_i^{-1} = \sum_{i=1}^{k} \frac{1}{U_i} \qquad (14.107)$$

The solution of the problem is found by minimizing U^{-1} subject to the constraints $g_j(\mathbf{X}) \leq 0, j = 1, 2, \ldots, m$.

14.10.3   Global Criterion Method

In the global criterion method the optimum solution $\mathbf{X}^*$ is found by minimizing a preselected global criterion, $F(\mathbf{X})$, such as the sum of the squares of the relative deviations of the individual objective functions from the feasible ideal solutions. Thus $\mathbf{X}^*$ is found by minimizing

$$F(\mathbf{X}) = \sum_{i=1}^{k} \left\{ \frac{f_i(\mathbf{X}_i^*) - f_i(\mathbf{X})}{f_i(\mathbf{X}_i^*)} \right\}^p$$

subject to $\hfill (14.108)$

$$g_j(\mathbf{X}) \leq 0, \quad j = 1, 2, \ldots, m$$

where p is a constant (an usual value of p is 2) and $\mathbf{X}_i^*$ is the ideal solution for the ith objective function. The solution $\mathbf{X}_i^*$ is obtained by minimizing $f_i(\mathbf{X})$ subject to the constraints $g_j(\mathbf{X}) \leq 0, j = 1, 2, \ldots, m$.

14.10.4   Bounded Objective Function Method

In the bounded objective function method, the minimum and the maximum acceptable achievement levels for each objective function f_i are specified as $L^{(i)}$ and $U^{(i)}$, respectively, for $i = 1, 2, \ldots, k$. Then the optimum solution $\mathbf{X}^*$ is found by minimizing the most important objective function, say, the rth one, as follows:

$$\text{Minimize } f_r(\mathbf{X})$$

subject to

$$g_j(\mathbf{X}) \leq 0, \qquad j = 1, 2, \ldots, m$$

$$L^{(i)} \leq f_i \leq U^{(i)}, \quad i = 1, 2, \ldots, k, \; i \neq r \qquad (14.109)$$

14.10.5 Lexicographic Method

In the lexicographic method, the objectives are ranked in order of importance by the designer. The optimum solutoin $\mathbf{X}^*$ is then found by minimizing the objective functions starting with the most important and proceeding according to the order of importance of the objectives. Let the subscripts of the objectives indicate not only the objective function number, but also the priorities of the objectives. Thus $f_1(\mathbf{X})$ and $f_k(\mathbf{X})$ denote the most and least important objective functions, respectively. The first problem is formulated as

$$\text{Minimize } f_1(\mathbf{X})$$

subject to (14.110)

$$g_j(\mathbf{X}) \le 0, \quad j = 1, 2, \ldots, m$$

and its solution $\mathbf{X}_1^*$ and $f_1^* = f_1(\mathbf{X}_1^*)$ is obtained. Then the second problem is formulated as

$$\text{Minimize } f_2(\mathbf{X})$$

subject to

$$g_j(\mathbf{X}) \le 0, \quad j = 1, 2, \ldots, m$$

$$f_1(\mathbf{X}) = f_1^* \tag{14.111}$$

The solution of this problem is obtained as $\mathbf{X}_2^*$ and $f_2^* = f_2(\mathbf{X}_2^*)$. This procedure is repeated until all the k objectives have been considered. The ith problem is given by

$$\text{Minimize } f_i(\mathbf{X})$$

subject to

$$g_j(\mathbf{X}) \le 0, \quad j = 1, 2, \ldots, m$$

$$f_l(\mathbf{X}) = f_l^*, \quad l = 1, 2, \ldots, i - 1 \tag{14.112}$$

and its solution is found as $\mathbf{X}_i^*$ and $f_i^* = f_i(\mathbf{X}_i^*)$. Finally, the solution obtained at the end (i.e., $\mathbf{X}_k^*$) is taken as the desired solution $\mathbf{X}^*$ of the original multiobjective optimization problem.

14.10.6 Goal Programming Method

In the simplest version of goal programming, the designer sets goals for each objective that he or she wishes to attain. The optimum solution $\mathbf{X}^*$ is then defined as the one that minimizes the deviations from the set goals. Thus the goal programming formulation of the multiobjective optimization problem leads to

$$\text{Minimize } \left[\sum_{j=1}^{k} (d_j^+ + d_j^-)^p \right]^{1/p}, \quad p \ge 1$$

subject to

$$g_j(\mathbf{X}) \le 0, \quad j = 1, 2, \ldots, m$$

$$f_j(\mathbf{X}) + d_j^+ - d_j^- = b_j, \quad j = 1, 2, \ldots, k$$

$$d_j^+ \ge 0, \quad j = 1, 2, \ldots, k \qquad (14.113)$$

$$d_j^- \ge 0, \quad j = 1, 2, \ldots, k$$

$$d_j^+ d_j^- = 0, \quad j = 1, 2, \ldots, k$$

where b_j is the goal set by the designer for the jth objective and d_j^+ and d_j^- are, respectively, the underachievement and overachievement of the jth goal. The value of p is based on the utility function chosen by the designer. Often the goal for the jth objective, b_j, is found by first solving the following problem:

$$\text{Minimize} f_j(\mathbf{X})$$

subject to $\qquad (14.114)$

$$g_j(\mathbf{X}) \le 0, \quad j = 1, 2, \ldots, m$$

If the solution of the problem stated in Eq. (14.114) is denoted by $\mathbf{X}_j^*$, then b_j is taken as $b_j = f_j(\mathbf{X}_j^*)$.

14.10.7 Goal Attainment Method

In the goal attainment method, goals are set as b_i for the objective function $f_i(\mathbf{X}), i = 1, 2, \ldots, k$. In addition, a weight $w_i > 0$ is defined for the objective function $f_i(\mathbf{X})$ to denote the importance of the ith objective function relative to other objective functions in meeting the goal $b_i, i = 1, 2, \ldots, k$. Often the goal b_i is found by first solving the single objective optimization problem:

$$\text{Minimize} f_i(\mathbf{X})$$

subject to $\qquad (14.115)$

$$g_j(\mathbf{X}) \le 0; j = 1, 2, \ldots, m$$

If the solution of the problem stated in Eq. (14.115) is denoted $\mathbf{X}_j^*$ then b_i can be taken as the optimum value of the objective f_i, $f_i^* = f(\mathbf{X}_i^*)$. A scalar γ is introduced as a design variable in addition to the n design variables $x_i, i = 1, 2, \ldots, n$. Then the following problem is solved:

Find $x_1, x_2, \ldots, x_n$ and γ

to minimize $F(x_1, x_2, \ldots, x_n, \gamma) = \gamma$

subject to $\qquad (14.116)$

$$g_j(\mathbf{X}) \le 0; j = 1, 2, \ldots, m$$

$$f_i(\mathbf{X}) - \gamma w_i \le b_i; i = 1, 2, \ldots, k$$

with the weights satisfying the normalization condition

$$\sum_{i=1}^{k} w_i = 1$$

14.11   SOLUTION OF MULTIOBJECTIVE PROBLEMS USING MATLAB

The MATLAB function `fgoalattain` can be used to solve a multiobjective optimization problem using the goal attainment method. The following example illustrates the procedure.

Example 14.3   Find the solution of the following three-objective optimization problem using goal attainment method using the MATLAB function `fgoalattain`.
Minimize

$$f_1 = \tfrac{1}{2}(x_1 - 2)^2 + \tfrac{1}{13}(x_2 + 1)^2 + 3$$

$$f_2 = \tfrac{1}{175}(x_1 + x_2 - 3)^2 + \tfrac{1}{17}(2x_2 - x_1)^2 - 13$$

$$f_3 = \tfrac{1}{8}(3x_1 - 2x_2 + 4)^2 + \tfrac{1}{27}(x_1 - x_2 + 1)^2 + 15$$

subject to

$$-4 \le x_i \le 4; i = 1, 2$$

$$4x_1 + x_2 - 4 \le 0$$

$$-x_1 - 1 \le 0$$

$$x_1 - x_2 - 2 \le 0$$

Assume the initial design variables to be $x_1 = x_2 = 0.1$, the weights to be $w_1 = 0.2$, $w_2 = 0.5$, and $w_3 = 0.3$, and the goals to be $b_1 = 5$, $b_2 = -8$, and $b_3 = 20$.

SOLUTION

Step 1: Create an m-file for the objective functions and save it as `fgoalattain_obj.m`

```
function f = fgoalattainobj(x)
f(1)  =  (x(1)-2)^2/2+(x(2)+1)^2/13+3
f(2)  =  (x(1)+x(2)-3)^2/175+(2*x(2)-x(1))^2/17-13
f(3)  =  (3*x(1)-2*x(2)+4)^2/8+(x(1)-x(2)+1)^2/27+15
```

Step 2: Create an m-file for the constraints and save it as `fgoalattain_con.m`

```
function [c ceq] = fgoalattaincon(x)
c= [- 4- x(1); ...
   x(1)- 4; ...
```

```
       - 4- x(2); ...
       x(2)- 4; ...
       x(2)+4*x(1)- 4; ...
       - 1- x(1); ...
       x(1)- 2- x(2)]
```

```
ceq = [];
```

Step 3: Ctreate an m-file for the main program and save it as `fgoalat-tain_main.m`

```
clc; clear all;
x0 = [0.1 0.1]
weight = [0.2 0.5 0.3]
goal = [5 -8 20]
x,fval,attainfactor,exitflag] = fgoalattain (@fgoalattainobj,
   x0,goal,weight,[],[],[],[],[],[],@fgoalattaincon)
```

Step 4: Run the program `fgoalattain_main.m` to obtain the following result:

Initial design vector:	0.1,0.1
Initial objective values:	4.8981 -12.9546 17.1383
Constraints at initial design:	-4.1000
	-3.9000
	-4.1000
	-3.9000
	-3.5000
	-1.1000
	-2.0000
Optimum design vector:	0.8308 0.6769
Optimum objective values:	3.8999 -12.9712 18.3498
Constraints at optimum design:	-4.8308
	-3.1692
	-4.6769
	-3.3231
	-0.0000
	-1.8308
	-1.8462

REFERENCES AND BIBLIOGRAPHY

14.1 L. A. Schmidt, Jr., and B. Farshi, Some approximation concepts for structural synthesis, *AIAA Journal*, Vol. 12, No. 5, pp. 692–699, 1974.

14.2 E. J. Haug, K. K. Choi, and V. Komkov, *Design Sensitivity Analysis of Structural Systems*, Academic Press, New York, 1986.

14.3   R. L. Fox and H. Miura, An approximate analysis technique for design calculations, *AIAA Journal*, Vol. 9, No. 1, pp. 177–179, 1971.

14.4   R. L. Fox and M. P. Kapoor, Rates of change of eigenvalues and eigenvectors, *AIAA Journal*, Vol. 6, No. 12, pp. 2426–2429, 1968.

14.5   D. V. Murthy and R. T. Haftka, Derivatives of eigenvalues and eigenvectors of general complex matrix, *International Journal for Numerical Methods in Engineering*, Vol. 26, pp. 293–311, 1988.

14.6   R. B. Nelson, Simplified calculation of eigenvector derivatives, *AIAA Journal*, Vol. 14, pp. 1201–1205, 1976.

14.7   S. S. Rao, Rates of change of flutter Mach number and flutter frequency, *AIAA Journal*, Vol. 10, pp. 1526–1528, 1972.

14.8   T. R. Sutter, C. J. Camarda, J. L. Walsh, and H. M. Adelman, Comparison of several methods for the calculation of vibration mode shape derivatives, *AIAA Journal*, Vol. 26, No. 12, pp. 1506–1511, 1988.

14.9   S. S. Rao, *The Finite Element Method in Engineering*, 4th ed., Elsevier Butterworth Heinemann, Burlington, MA, 2005.

14.10   S. S. Rao, *Mechanical Vibrations*, 4th ed., Pearson Prentice Hall, Upper Saddle River, NJ, 2004.

14.11   R. V. Grandhi, R. T. Haftka, and L. T. Watson, Efficient identification of critical stresses in structures subjected to dynamic loads, *Computers and Structures*, Vol. 22, pp. 373–386, 1986.

14.12   W. H. Greene and R. T. Haftka, Computational aspects of sensitivity calculations in transient structural analysis, *Computers and Structures*, Vol. 32, No. 2, pp. 433–443, 1989.

14.13   U. Kirsch, M. Reiss, and U. Shamir, Optimum design by partitioning into substructures, *ASCE Journal of the Structural Division*, Vol. 98, No. ST1, pp. 249–267, 1972.

14.14   U. Kirsch, Multilevel approach to optimum structural design, *ASCE Journal of the Structural Division*, Vol. 101, No. ST4, pp. 957–974, 1975.

14.15   J. Sobieszczanski-Sobieski, B. James, and A. Dovi, Structural optimization by multilevel decomposition, *AIAA Journal*, Vol. 23, No. 11, pp. 1775–1782, 1985.

14.16   J. Sobieszczanski-Sobieski, B. B. James, and M. F. Riley, Structural sizing by generalized, multilevel optimization, *AIAA Journal*, Vol. 25, No. 1, pp. 139–145, 1987.

14.17   M.E.M. El-Sayed and C.-K. Hsiung, Parallel structural optimization with parallel analysis interfaces, pp. 398–403 in *Proceedings of the 3rd Air Force/NASA Symposium on Recent Advances in Multidisciplinary Analysis and Optimization*, San Francisco, Sept. 24–26, 1990.

14.18   E. S. Sikiotis and V. E. Saouma, Parallel structural optimization on a network of computer workstations, *Computers and Structures*, Vol. 29, No. 1, pp. 141–150, 1988.

14.19   H. Adeli and O. Kamat, Concurrent optimization of large structures; Part I: Algorithms, Part II: Applications, *ASCE Journal of Aerospace Engineering*, Vol. 5, No. 1, pp. 79–110, 1992.

14.20   M.E.M. El-Sayed and C.-K. Hsiung, Optimum structural design with parallel finite element analysis. *Computers and Structures*, Vol. 40, No. 6, pp. 1469–1474, 1991.

14.21   M. M. Atiqullah and S. S. Rao, Parallel processing in optimal structural design using simulated annealing, *AIAA Journal*, Vol. 33, pp. 2386–2392, 1995.

14.22   L. A. Schmit, Jr., and H. Miura, *Approximation Concepts for Efficient Structural Synthesis*, NASA CR-2552, 1976.

14.23 L. A. Schmit and C. Fleury, Structural synthesis by combining approximation concepts and dual methods, *AIAA Journal*, Vol. 18, pp. 1252–1260, 1980.

14.24 T. S. Pan, S. S. Rao, and V. B. Venkayya, Rates of change of closed-loop eigenvalues and eigenvectors of actively controlled structures, *International Journal for Numerical Methods in Engineering*, Vol. 30, No. 5, pp. 1013–1028, 1990.

14.25 G. N. Vanderplaats, *Numerical Optimization Techniques for Engineering Design with Applications*, McGraw-Hill, New York, 1984.

14.26 J. Sobieszczanski-Sobieski, J. F. Barthelemy, and K. M. Riley, "Sensitivity of Optimum Solutions to Problem Parameters," *AIAA Journal*, Vol. 20, pp. 1291–1299, 1982.

14.27 U. Kirsch, *Optimum Structural Design. Concepts, Methods, and Applications*, McGraw-Hill, New York, 1981.

14.28 R. T. Haftka and Z. Gürdal, *Elements of Structural Optimization*, 3rd ed., Kluwer Academic, Dordrecht, The Netherlands, 1992.

14.29 E. E. Rosinger, Interactive algorithm for multiobjective optimization, *Journal of Optimization Theory and Applications*, Vol. 35, pp. 339–365, 1981; Errata in Vol. 38, pp. 147–148, 1982.

14.30 T. L. Vincent and W. J. Grantham, *Optimality in Parametric Systems*, Wiley, New York, 1981.

14.31 W. Stadler, A survey of multicriteria optimization of the vector maximum problem, *Journal of Optimization Theory and Applications*, Vol. 29, pp. 1–52, 1979.

14.32 D. Koo, *Elements of Optimization*, Springer-Verlag, New York, 1977.

14.33 J. P. Ignizio (Ed.), *Linear Programming in Single- and Multiple-objective Systems*, Prentice-Hall, Englewood Cliffs, NJ, 1982.

14.34 S. S. Rao, Game theory approach for multiobjective structural optimization, *Computers and Structures*, Vol. 25, No. 1, pp. 119–127, 1987.

14.35 C. L. Hwang and A.S.M. Masud, *Multiple Objective Decision Making: Methods and Applications*, Springer-Verlag, Berlin, 1979.

14.36 S. S. Rao, V. B. Venkayya, and N. S. Khot, Game theory approach for the integrated design of structures and controls, *AIAA Journal*, Vol. 26, No. 4, pp. 463–469, 1988.

14.37 S. S. Rao and T. I. Freiheit, A modified game theory approach to multiobjective optimization, *ASME Journal of Mechanical Design*, Vol. 113, pp. 286–291, 1991.

14.38 S. S. Rao and R. L. Kaplan, Optimal balancing of high-speed linkages using multiobjective programming techniques, *ASME Journal of Mechanisms, Transmissions, and Automation in Design*, Vol. 108, pp. 454–460, 1986.

14.39 S. S. Rao and H. R. Eslampour, Multistage multiobjective optimization of gearboxes, *ASME Journal of Mechanisms, Transmissions, and Automation in Design*, Vol. 108, pp. 461–468, 1986.

14.40 S. K. Hati and S. S. Rao, Cooperative solution in the synthesis of multi-degree of freedom shock isolation systems, *ASME Journal of Vibration, Acoustics, Stress and Reliability in Design*, Vol. 105, pp. 101–103, 1983.

14.41 S. S. Rao and S. K. Hati, "Game theory approach in multicriteria optimization of function generating mechanisms, *ASME Journal of Mechanical Design*, Vol: 101, pp. 398–406, 1979.

14.42 S. S. Rao, A. K. Dhingra, and H. Miura, Pareto-optimal solutions in helicopter design problems, *Engineering Optimization*, Vol. 15, No. 3, pp. 211–231, 1990.

14.43 H. Eschenauer, J. Koski, and A. Osyczka, *Multicriteria Design Optimization: Procedures and Applications*, Springer-Verlag, New York, 1990.

14.44   W. Stadler, Ed., *Multicriteria Optimization in Engineering and in the Sciences*, Plenum Press, New York, 1988.

14.45   The Rand Corporation, *A Million Random Digits with 100,000 Normal Deviates*, The Free Press, Glencoe, IL, 1955.

14.46   C. A. Coello Coello, D. A. Van Veldhuizen, and G. B. Lamont, *Evolutionary Algorithms for Solving Multi-objective Problems*, Kluwer Academic/Plenum, New York, 2002.

REVIEW QUESTIONS

14.1   What is a reduced basis technique?

14.2   State two methods of reducing the size of an optimization problem.

14.3   What is design variable linking? Can it always be used?

14.4   Under what condition(s) is the convergence of the quantity $\Sigma_i \Delta \mathbf{Y}_i$ in the fast reanalysis method ensured?

14.5   How do you compute the derivatives of the stiffness matrix with respect to a design variable, $\partial[K]/\partial x_i$?

14.6   What is a MIMD computer?

14.7   Indicate various ways by which parallel computations can be performed in a large-scale optimization problem.

14.8   How are the goals determined in the goal programming method?

14.9   Answer *true* or *false*:

(a) The computation of the derivatives of a particular λ_i requires other eigenvalues besides λ_i.

(b) The derivatives of the ith eigenvector can be found without knowledge of the eigenvectors other than $\mathbf{Y}_i$.

(c) There is only one way to derive expressions for the sensitivity of optimal objective function with respect to problem parameters.

(d) Multilevel optimization is same as decomposition.

(e) In multilevel optimization, the suboptimization problems are to be solved iteratively.

(f) All multiobjective optimization methods find only a Pareto optimum solution.

(g) All multiobjective optimization techniques convert the problem into a single objective problem.

(h) A vector optimization problem is same as a multiobjective optimization problem.

(i) Only one Pareto optimal solution exists for a multiobjective optimization problem.

(j) The weighting function method can be considered as the utility function method.

(k) It is possible to achieve the optimum value of each objective function simultaneously in a multiobjective optimization problem.

14.10   Define the following terms:

(a) Pareto optimal point

(b) Utility function method

(c) Weighting function method

(d) Global criterion function method

(e) Bounded objective function method

(f) Lexicographic method

PROBLEMS

14.1 Consider the minimum-volume design of the four-bar truss shown in Fig. 14.2 subject to a constraint on the vertical displacement of node 4. Let $\mathbf{X}_1 = \{1, 1, 0.5, 0.5\}^T$ and $\mathbf{X}_2 = \{0.5, 0.5, 1, 1\}^T$ be two design vectors, with x_i denoting the area of cross section of bar $i (i = 1, 2, 3, 4)$. By expressing the optimum design vectors as $\mathbf{X} = c_1 \mathbf{X}_1 + c_2 \mathbf{X}_2$, determine the values of c_1 and c_2 through graphical optimization when the maximum permissible vertical deflection of node 4 is restricted to a magnitude of 0.1 in.

14.2 Consider the configuration (shape) optimization of the 10-bar truss shown in Fig. 14.10. The (X, Y) coordinates of the nodes are to be varied while maintaining **(a)** symmetry of the structure about the X axis, and **(b)** alignment of nodes 1, 2, and 3 (4, 5, and 6). Identify the independent and dependent design variables and derive the relevant design variable linking relationships.

14.3 For the four-bar truss considered in Example 14.1 (shown in Fig. 14.2), a base design vector is given by $\mathbf{X}_0 = \{A_1, A_2, A_3, A_4\}^T = \{2.0, 1.0, 2.0, 1.0\}^T$ in^2. If $\Delta \mathbf{X}$ is given by $\Delta \mathbf{X} = \{0.4, 0.4, -0.4, -0.4\}^T$ in^2, determine

(a) The exact displacement vector $\mathbf{Y}_0 = \{y_5, y_6, y_7, y_8\}^T$ at $\mathbf{X}_0$

(b) The exact displacement vector $(\mathbf{Y}_0 + \Delta \mathbf{Y})$ at $(\mathbf{X}_0 + \Delta \mathbf{X})$

(c) The displacement vector $(\mathbf{Y}_0 + \Delta \mathbf{Y})$ where $\Delta \mathbf{Y}$ is given by Eq. (14.20) with five terms

14.4 Consider the 11-member truss shown in Fig. 5.1 with loads $Q = -1000$ lb, $R = 1000$ lb, and $S = 2000$ lb. If $A_i = x_i$ denotes the area of cross section of member i, and $u_1, u_2, \ldots, u_{10}$ indicate the displacement components of the nodes, the equilibrium equations can be expressed as shown in Eqs. (E$_1$) to (E$_{10}$) of Example 5.1. Assuming that $E = 30 \times 10^6$ psi, $l = 50$ in., $x_i = 1$ in$^2 (i = 1, 2, \ldots, 11)$, $\Delta x_i = 0.1$ in$^2 (i = 1, 2, \ldots, 5)$, and $\Delta x_i = -0.1$ in$^2 (i = 6, 7, \ldots, 11)$, determine

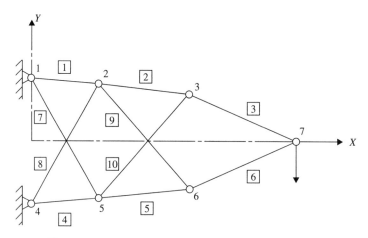

Figure 14.10 Design variable linking of a 10-bar truss.

(a) Exact displacement solution $\mathbf{U}_0$ at $\mathbf{X}_0$

(b) Exact displacement solution $(\mathbf{U}_0 + \Delta\mathbf{U})$ at the perturbed design, $(\mathbf{X}_0 + \Delta\mathbf{X})$

(c) Approximate displacement solution, $(\mathbf{U}_0 + \Delta\mathbf{U})$, at $(\mathbf{X}_0 + \Delta\mathbf{X})$ using Eq. (14.20) with four terms for $\Delta\mathbf{U}$

14.5 Consider the four-bar truss shown in Fig. 14.2 whose stiffness matrix is given by Eq. (E$_2$) of Example 14.1. Determine the values of the derivatives of y_i with respect to the area A_1, $\partial y_i / \partial x_1 (i = 5, 6, 7, 8)$ at the reference design $\mathbf{X}_0 = \{A_1\ A_2\ A_3\ A_4\}^{\mathrm{T}} = \{2.0, 2.0, 1.0, 1.0\}^{\mathrm{T}}$ in^2.

14.6 Find the values of $\partial y_i / \partial x_2$ $(i = 5, 6, 7, 8)$ in Problem 14.5.

14.7 Find the values of $\partial y_i / \partial x_3$ $(i = 5, 6, 7, 8)$ in Problem 14.5.

14.8 Find the values of $\partial y_i / \partial x_4$ $(i = 5, 6, 7, 8)$ in Problem 14.5.

14.9 The equilibrium equations of the stepped bar shown in Fig. 14.11 are given by

$$[K]\mathbf{Y} = \mathbf{P} \tag{1}$$

with

$$[K] = \begin{bmatrix} \dfrac{A_1 E_1}{l_1} + \dfrac{A_2 E_2}{l_2} & -\dfrac{A_2 E_2}{l_2} \\ -\dfrac{A_2 E_2}{l_2} & \dfrac{A_2 E_2}{l_2} \end{bmatrix} \tag{2}$$

$$\mathbf{Y} = \begin{Bmatrix} Y_1 \\ Y_2 \end{Bmatrix}, \quad \mathbf{P} = \begin{Bmatrix} P_1 \\ P_2 \end{Bmatrix} \tag{3}$$

If $A_1 = 2$ in.^{2}, $A_2 = 1$ in.^{2}, $E_1 = E_2 = 30 \times 10^6$ psi, $2l_1 = l_2 = 50$ in., $P_1 = 100$ lb, and $P_2 = 200$ lb, determine

(a) Displacements, $\mathbf{Y}$

(b) Values of $\partial\mathbf{Y}/\partial A_1$ and $\partial\mathbf{Y}/\partial A_2$ using the method of Section 14.4

(c) Values of $\partial\boldsymbol{\sigma}/\partial A_1$ and $\partial\boldsymbol{\sigma}/\partial A_2$, where $\boldsymbol{\sigma} = \{\sigma_1, \sigma_2\}^{\mathrm{T}}$ denotes the vector of stresses in the bars and $\sigma_1 = E_1 Y_1 / l_1$ and $\sigma_2 = E_2(Y_2 - Y_1)/l_2$

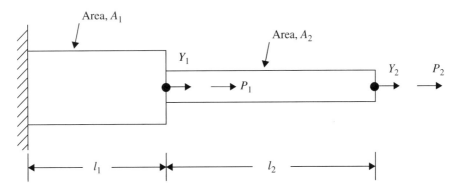

Figure 14.11   Stepped bar.

14.10   The eigenvalue problem for the stepped bar shown in Fig. 14.11 can be expressed as $[K]\mathbf{Y} = \lambda[M]\mathbf{Y}$ with the mass matrix, $[M]$, given by

$$[M] = \begin{bmatrix} (2\rho_1 A_1 l_1 + \rho_2 A_2 l_2) & \rho_2 A_2 l_2 \\ \rho_2 A_2 l_2 & \rho_2 A_2 l_2 \end{bmatrix}$$

where ρ_i, A_i, and l_i denote the mass density, area of cross section, and length of the segment i, and the stiffness matrix, $[K]$, is given by Eq. (2) of Problem 14.9. If $A_1 = 2$ in^2, $A_2 = 1$ in^2, $E_1 = E_2 = 30 \times 10^6$ psi, $2l_1 = l_2 = 50$ in., and $\rho_1 g = \rho_2 g = 0.283$ lb/in^3, determine

(a) Eigenvalues λ_i and the eigenvectors $\mathbf{Y}_i$, $i = 1, 2$
(b) Values of $\partial\lambda_i/\partial A_1$, $i = 1, 2$, using the method of Section 14.5
(c) Values of $\partial\mathbf{Y}_i/\partial\mathbf{Y}_1$, $i = 1, 2$, using the method of Section 14.5

14.11   For the stepped bar considered in Problem 14.10, determine the following using the method of Section 14.5.

(a) Values of $\partial\lambda_i/\partial A_2$, $i = 1, 2$
(b) Values of $\partial\mathbf{Y}_i/\partial A_2$, $i = 1, 2$

14.12   A cantilever beam with a hollow circular section with outside diameter d and wall thickness t (Fig. 14.12) is modeled with one beam finite element. The resulting static equilibrium equations can be expressed as

$$\frac{2EI}{l^3} \begin{bmatrix} 6 & -3l \\ -3l & 2l^2 \end{bmatrix} \begin{Bmatrix} Y_1 \\ Y_2 \end{Bmatrix} = \begin{Bmatrix} P_1 \\ P_2 \end{Bmatrix}$$

where I is the area moment of intertia of the cross section, E is Young's modulus, and l the length. Determine the displacements, Y_i, and the sensitivities of the deflections, $\partial Y_i/\partial d$ and $\partial Y_i/\partial t$ $(i = 1, 2)$, for the following data: $E = 30 \times 10^6$ psi, $l = 20$ in., $d = 2$ in., $t = 0.1$ in., $P_1 = 100$ lb, and $P_2 = 0$.

14.13   The eigenvalues of the cantilever beam shown in Fig. 14.12 are governed by the equation

$$\frac{2EI}{l^3} \begin{bmatrix} 6 & -3l \\ -3l & 2l^2 \end{bmatrix} \begin{Bmatrix} Y_1 \\ Y_2 \end{Bmatrix} = \frac{\lambda\rho Al}{420} \begin{bmatrix} 156 & -22l \\ -22l & 4l^2 \end{bmatrix} \begin{Bmatrix} Y_1 \\ Y_2 \end{Bmatrix}$$

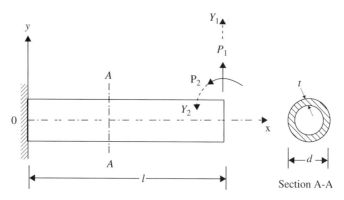

Figure 14.12   Hollow circular cantilever beam.

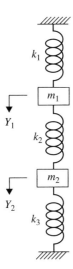

Figure 14.13 Two-degree-of-freedom spring–mass system.

where E is Young's modulus, I the area moment of inertia, l the length, ρ the mass density, A the cross-sectional area, λ the eigenvalue, and $\mathbf{Y} = \{Y_1, Y_2\}^{\mathrm{T}}$ = eigenvector. If $E = 30 \times 10^6$ psi, $d = 2$ in., $t = 0.1$ in., $l = 20$ in., and $\rho g = 0.283 \, \mathrm{lb/in}^3$, determine

(a) Eigenvalues λ_i and eigenvectors $\mathbf{Y}_i$ $(i = 1, 2)$

(b) Values of $\partial \lambda_i / \partial d$ and $\partial \lambda_i / \partial t$ $(i = 1, 2)$

14.14 In Problem 14.13, determine the derivatives of the eigenvectors $\partial \mathbf{Y}_i / \partial d$ and $\partial \mathbf{Y}_i / \partial t$ $(i = 1, 2)$.

14.15 The natural frequencies of the spring–mass system shown in Fig. 14.13 are given by (for $k_i = k$, $i = 1, 2, 3$ and $m_i = m$, $i = 1, 2$)

$$\lambda_1 = \frac{k}{m} = \omega_1^2, \qquad \lambda_2 = \frac{3k}{m} = \omega_2^2$$

$$\mathbf{Y}_1 = c_1 \begin{Bmatrix} 1 \\ 1 \end{Bmatrix}, \qquad \mathbf{Y}_2 = c_2 \begin{Bmatrix} 1 \\ -1 \end{Bmatrix}$$

where ω_1 and ω_2 are the natural frequencies of vibration of the system and c_1 and c_2 are constants. The stiffness of each helical spring is given by

$$k = \frac{d^4 G}{8 D^3 n}$$

where d is the wire diameter, D the coil diameter, G the shear modulus, and n the number of turns of the spring. Determine the values of $\partial \omega_i / \partial D$ and $\partial \mathbf{Y}_i / \partial D$ for the following data: $d = 0.04$ in., $G = 11.5 \times 10^6$ psi, $D = 0.4$ in., $n = 10$, and $m = 32.2 \, \mathrm{lb}$ $\text{-}s^2/\mathrm{in}$. The stiffness and mass matrices of the system are given by

$$[K] = k \begin{bmatrix} 2 & -1 \\ -1 & 2 \end{bmatrix}, \quad [M] = m \begin{bmatrix} 1 & 0 \\ 0 & 1 \end{bmatrix}$$

14.16 Find the minimum volume design of the truss shown in Fig. 14.14 with constraints on the depth of the truss (y), cross-sectional areas of the bars (A_1 and A_2), and the stresses

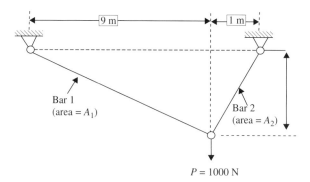

Figure 14.14   Two-bar truss.

induced in the bars (σ_1 and σ_2). Treat y, A_1, and A_2 as design variables with $\sigma_i \leq 10^5$ Pa ($i = 1, 2$), 1 m $\leq y \leq 4$ m, and $0 \leq A_i \leq 0.2$ m^2 ($i = 1, 2$). Use multilevel optimization approach for the solution.

14.17   Find the sensitivities of x_1^*, x_2^*, and f^* with respect to Young's modulus of the tubular column considered in Example 1.1.

14.18   Consider the two-bar truss shown in Fig. 1.15. The problem of design of the truss for minimum weight subject to stress constraints can be stated as follows:

Find x_1, A_1, and A_2 which minimize

$$f = 28.30A_1 \sqrt{1 + x^2} + 14.15A_2 \sqrt{1 + x^2}$$

subject to

$$g_1 = \frac{0.1768(1 + x)\sqrt{1 + x^2}}{A_1 x} - 1 \leq 0$$

$$g_2 = \left| \frac{0.1768(x - 1)\sqrt{1 + x^2}}{A_2 x} \right| - 1 \leq 0$$

$$0.1 \leq x \leq 2.5, \quad 1.0 \leq A_i \leq 2.5 \ (i = 1, 2)$$

where the members are assumed to be made up of different materials. Solve this optimization problem using the multilevel approach.

14.19   Consider the design of the two-bar truss shown in Fig. 14.15 with the location of nodes 1 and 2(x) and the area of cross section of bars (A) as design variables. If the weight and the displacement of node 3 are to be minimized with constraints on the stresses induced in the bars along with bounds on the design variables, the problem can be stated as follows [14.34]:

Find $\mathbf{X} = \{x_1 x_2\}^\mathrm{T}$ which minimizes

$$f_1(\mathbf{X}) = 2\rho h x_2 \sqrt{1 + x_1^2}$$

$$f_2 = \frac{Ph(1 + x_1^2)^{1.5}\sqrt{1 + x_1^4}}{2\sqrt{2}Ex_1^2 x_2}$$

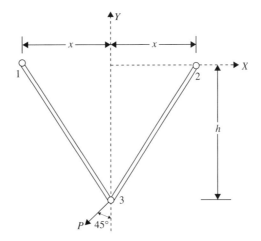

Figure 14.15 Two-bar truss.

subject to

$$g_1(\mathbf{X}) = \frac{P(1 + x_1)\sqrt{1 + x_1^2}}{2\sqrt{2}x_1 x_2} - \sigma_0 \leq 0$$

$$g_2(\mathbf{X}) = \frac{P(x_1 - 1)\sqrt{1 + x_1^2}}{2\sqrt{2}x_1 x_2} - \sigma_0 \leq 0$$

$$x_i \geq x_i^{(l)}, \quad i = 1, 2$$

where $x_1 = x/h$, $x_2 = A/A_{\text{ref}}$, h the depth, E is Young's modulus, ρ the weight density, σ_0 the permissible stress, and $x_i^{(l)}$ the lower bound on x_i. Find the optimum solutions of the individual objective functions subject to the stated constraints using a graphical procedure. Data: $P = 10,000\,\text{lb}$, $\rho = 0.283\,\text{lb/in}^3$, $E = 30 \times 10^6$ psi, $h = 100$ in., $A_{\text{ref}} = 1$ in.^{2}, $\sigma_0 = 20,000$ psi, $x_1^{(l)} = 0.1$, and $x_2^{(l)} = 1.0$.

14.20 Solve the two-objective optimization problem stated in Problem 14.19 using the weighting method with equal weights to the two objective functions. Use a graphical method of solution.

14.21 Solve the two-objective optimization problem stated in Problem 14.19 using the global criterion method with $p = 2$. Use a graphical method of solution.

14.22 Formulate the two-objective optimization problem stated in Problem 14.19 as a goal programming problem using the goals of 30 lb and 0.015 in. for the objectives f_1 and f_2, respectively. Solve the problem using a graphical procedure.

14.23 Consider the following two-objective optimization problem:

Find $\mathbf{X} = \{x_1 \ x_2 \ x_3 \ x_4 \ x_5 \ x_6\}^{\text{T}}$
to minimize

$$f_1(\mathbf{X}) = -25(x_1 - 2)^2 - (x_2 - 2)^2 - (x_3 - 1)^2 - (x_4 - 4)^2 - (x_5 - 1)^2$$
$$f_2(\mathbf{X}) = x_1^2 + x_2^2 + x_3^2 + x_4^2 + x_5^2 + x_6^2$$

subject to

$$-x_1 - x_2 + 2 \le 0; \ x_1 + x_2 - 6 \le 0; \ -x_1 + x_2 - 2 \le 0; \ x_1 - 3x_2 - 2 \le 0;$$

$$(x_3 - 3)^2 + x_4^2 - 4 \le 0; \ -(x_5 - 3)^2 - x_6 + 4 \le 0; \ 0 \le x_i \le 10, i = 1, 2, 6$$

$$1 \le x_i \le 5, \ i = 3, 5; \ 0 \le x_4 \le 6$$

Find the minima of the individual objective functions under the stated constraints using the MATLAB function `fmincon`.

14.24 Find the solution of the two-objective optimization problem stated in Problem 14.23 using the weighting function method with the weights $w_1 = w_2 = 1$. Use the MATLAB function `fmincon` for the solution.

14.25 Find the solution of the two-objective optimization problem stated in Problem 14.23 using the global criterion method with $p = 2$. Use the MATLAB function `fmincon` for the solution.

14.26 Find the solution of the two-objective optimization problem stated in Problem 14.23 using the bounded objective function method. Take the lower and upper bounds on f_2 as 80 and 120% of the optimum value f_2^* found in Problem 13.23. Use the MATLAB function `fmincon` for the solution.

14.27 Find the solution of the two-objective optimization problem stated in Problem 14.23 using the goal attainment method. Use the MATLAB function `fgoalattain` for the solution. Use suitable goals for the objectives.

14.28 Consider the following three-objective optimization problem:
Find $\mathbf{X} = \{x_1 \ x_2\}^{\mathrm{T}}$ to minimize

$$f_1(\mathbf{X}) = 1.5 - x_1(1 - x_2)$$
$$f_2(\mathbf{X}) = 2.25 - x_1(1 - x_2^2)$$
$$f_3(\mathbf{X}) = 2.625 - x_1(1 - x_2^3)$$

subject to

$$-x_1^2 - (x_2 - 0.5)^2 + 9 \le 0$$
$$(x_1 - 1)^2 + (x_2 - 0.5)^2 - 6.25 \le 0$$
$$-10 \le x_i \le 10; i = 1, 2$$

Find the minima of the individual objectives under the stated constraints using the MATLAB function `fmincon`.

14.29 Find the solution of the 3-objective problem stated in Problem 14.28 using the weighting function method with the weights $w_1 = w_2 = w_3 = 1$. Use the MATLAB function `fmincon` for the solution.

14.30 Find the solution of the multiobjective problem stated in Problem 14.28 using the goal attainment method. Use the MATLAB function `fgoalattain` for the solution. Use suitable goals for the objectives.

A

Convex and Concave Functions

Convex Function. A function $f(\mathbf{X})$ is said to be convex if for any pair of points

$$
\mathbf{X}_1 = \begin{Bmatrix} x_1^{(1)} \\ x_2^{(1)} \\ \vdots \\ x_n^{(1)} \end{Bmatrix} \quad \text{and} \quad \mathbf{X}_2 = \begin{Bmatrix} x_1^{(2)} \\ x_2^{(2)} \\ \vdots \\ x_n^{(2)} \end{Bmatrix}
$$

and all λ, $0 \le \lambda \le 1$,

$$
f[\lambda \mathbf{X}_2 + (1 - \lambda)\mathbf{X}_1] \le \lambda f(\mathbf{X}_2) + (1 - \lambda) f(\mathbf{X}_1) \tag{A.1}
$$

that is, if the segment joining the two points lies entirely above or on the graph of $f(\mathbf{X})$. Figures A.1a and A.2a illustrate a convex function in one and two dimensions, respectively. It can be seen that a convex function is always bending upward and hence it is apparent that the local minimum of a convex function is also a global minimum.

Concave Function. A function $f(\mathbf{X})$ is called a concave function if for any two points $\mathbf{X}_1$ and $\mathbf{X}_2$, and for all $0 \le \lambda \le 1$,

$$
f[\lambda \mathbf{X}_2 + (1 - \lambda)\mathbf{X}_1] \ge \lambda f(\mathbf{X}_2) + (1 - \lambda) f(\mathbf{X}_1) \tag{A.2}
$$

that is, if the line segment joining the two points lies entirely below or on the graph of $f(\mathbf{X})$.

Figures A.1b and A.2b give a concave function in one and two dimensions, respectively. It can be seen that a concave function bends downard and hence the local maximum will also be its global maximum. It can be seen that the negative of a convex function is a concave function, and vice versa. Also note that the sum of convex functions is a convex function and the sum of the concave functions is a concave function. A function $f(\mathbf{X})$ is strictly convex or concave if the strict inequality holds in Eqs. (A.1) or (A.2) for any $\mathbf{X}_1 \ne \mathbf{X}_2$. A linear function will be both convex and concave since it satisfies both inequalities (A.1) and (A.2). A function may be convex within a region and concave elsewhere. An example of such a function is shown in Fig. A.3.

Testing for Convexity or Concavity. In addition to the definition given, the following equivalent relations can be used to identify a convex function.

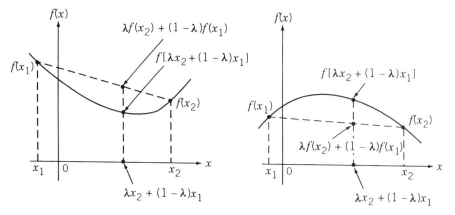

Figure A.1   Functions of one variable: (*a*) convex function in one variable; (*b*) concave function in one variable.

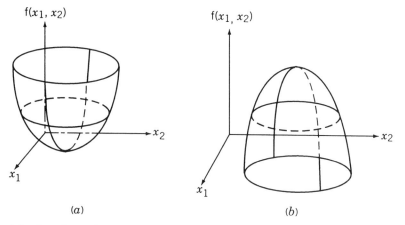

Figure A.2   Functions of two variables: (*a*) convex function in two variables; (*b*) concave function in two variables.

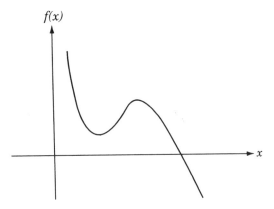

Figure A.3   Function that is convex over certain region and concave over certain other region.

Theorem A.1 A function $f(\mathbf{X})$ is convex if for any two points $\mathbf{X}_1$ and $\mathbf{X}_2$, we have

$$f(\mathbf{X}_2) \geq f(\mathbf{X}_1) + \nabla f^{\mathrm{T}}(\mathbf{X}_1)(\mathbf{X}_2 - \mathbf{X}_1)$$

Proof: If $f(\mathbf{X})$ is convex, we have by definition

$$f[\lambda \mathbf{X}_2 + (1 - \lambda)\mathbf{X}_1] \leq \lambda f(\mathbf{X}_2) + (1 - \lambda) f(\mathbf{X}_1)$$

that is,

$$f[\mathbf{X}_1 + \lambda(\mathbf{X}_2 - \mathbf{X}_1)] \leq f(\mathbf{X}_1) + \lambda[f(\mathbf{X}_2) - f(\mathbf{X}_1)] \tag{A.3}$$

This inequality can be rewritten as

$$f(\mathbf{X}_2) - f(\mathbf{X}_1) \geq \left\{ \frac{f[\mathbf{X}_1 + \lambda(\mathbf{X}_2 - \mathbf{X}_1)] - f(\mathbf{X}_1)}{\lambda(\mathbf{X}_2 - \mathbf{X}_1)} \right\} (\mathbf{X}_2 - \mathbf{X}_1) \tag{A.4}$$

By defining $\Delta \mathbf{X} = \lambda(\mathbf{X}_2 - \mathbf{X}_1)$, the inequality (A.4) can be written as

$$f(\mathbf{X}_2) - f(\mathbf{X}_1) \geq \frac{f[\mathbf{X}_1 + \Delta \mathbf{X}] - f(\mathbf{X}_1)}{\Delta \mathbf{X}} (\mathbf{X}_2 - \mathbf{X}_1) \tag{A.5}$$

By taking the limit as $\Delta \mathbf{X} \to \mathbf{0}$, inequality (A.5) becomes

$$f(\mathbf{X}_2) - f(\mathbf{X}_1) \geq \nabla f^{T}(\mathbf{X}_1)(\mathbf{X}_2 - \mathbf{X}_1) \tag{A.6}$$

which can be seen to be the desired result. If $f(\mathbf{X})$ is concave, the opposite type of inequality holds true in (A.6).

Theorem A.2 A function $f(\mathbf{X})$ is convex if the Hessian matrix $\mathbf{H}(\mathbf{X}) = [\partial^2 f(\mathbf{X})/\partial x_i \, \partial x_j]$ is positive semidefinite.

Proof: From Taylor's theorem we have

$$f(\mathbf{X}^* + \mathbf{h}) = f(\mathbf{X}^*) + \sum_{i=1}^{n} h_i \frac{\partial f}{\partial x_i}(\mathbf{X}^*)$$

$$+ \frac{1}{2!} \sum_{i=1}^{n} \sum_{j=1}^{n} h_i h_j \left. \frac{\partial^2 f}{\partial x_i \, \partial x_j} \right|_{\mathbf{X} = \mathbf{X}^* + \theta \mathbf{h}} \tag{A.7}$$

where $0 < \theta < 1$. By letting $\mathbf{X}^* = \mathbf{X}_1$, $\mathbf{X}^* + \mathbf{h} = \mathbf{X}_2$ and $\mathbf{h} = \mathbf{X}_2 - \mathbf{X}_1$, Eq. (A.7) can be rewritten as

$$f(\mathbf{X}_2) = f(\mathbf{X}_1) + \nabla f^{\mathrm{T}}(\mathbf{X}_1)(\mathbf{X}_2 - X_1) + \tfrac{1}{2}(\mathbf{X}_2 - \mathbf{X}_1)^{\mathrm{T}}$$

$$\times \mathbf{H}\{\mathbf{X}_1 + \theta(\mathbf{X}_2 - \mathbf{X}_1)\}(\mathbf{X}_2 - \mathbf{X}_1) \tag{A.8}$$

It can be seen that inequality (A.6) is satisfied [and hence $f(\mathbf{X})$ will be convex] if $\mathbf{H}(\mathbf{X})$ is positive semidefinite. Further, if $\mathbf{H}(\mathbf{X})$ is positive definite, the function $f(\mathbf{X})$ will be strictly convex. It can also be proved that if $f(\mathbf{X})$ is concave, the Hessian matrix is negative semidefinite.

The following theorem establishes a very important relation, namely, that any local minimum is a global minimum for a convex function.

Theorem A.3 Any local minimum of a convex function $f(\mathbf{X})$ is a global minimum.

Proof: Let us prove this theorem by contradiction. Suppose that there exist two different local minima, say, $\mathbf{X}_1$ and $\mathbf{X}_2$, for the function $f(\mathbf{X})$. Let $f(\mathbf{X}_2) < f(\mathbf{X}_1)$. Since $f(\mathbf{X})$ is convex, $\mathbf{X}_1$ and $\mathbf{X}_2$ have to satisfy the relation (A.6), that is,

$$f(\mathbf{X}_2) - f(\mathbf{X}_1) \geq \nabla f^T(\mathbf{X}_1)(\mathbf{X}_2 - \mathbf{X}_1) \tag{A.6}$$

or

$$\nabla f^T(\mathbf{X}_1)\mathbf{S} \leq \mathbf{0} \tag{A.9}$$

where $\mathbf{S} = (\mathbf{X}_2 - \mathbf{X}_1)$ is a vector joining the points $\mathbf{X}_1$ to $\mathbf{X}_2$. Equation (A.9) indicates that the value of the function $f(\mathbf{X})$ can be decreased further by moving in the direction $\mathbf{S} = (\mathbf{X}_2 - \mathbf{X}_1)$ from point $\mathbf{X}_1$. This conclusion contradicts the original assumption that $\mathbf{X}_1$ is a local minimum. Thus there cannot exist more than one minimum for a convex function.

Example A.1   Determine whether the following functions are convex or concave.

(a) $f(x) = e^x$
(b) $f(x) = -8x^2$
(c) $f(x_1, x_2) = 3x_1^3 - 6x_2^2$
(d) $f(x_1, x_1, x_3) = 4x_1^2 + 3x_2^2 + 5x_3^2 + 6x_1x_2 + x_1x_3 - 3x_1 - 2x_2 + 15$

SOLUTION

(a) $f(x) = e^x$: $H(x) = d^2f/dx^2 = e^x > 0$ for all real values of x. Hence $f(x)$ is strictly convex.
(b) $f(x) = -8x^2$: $H(x) = d^2f/dx^2 = -16 < 0$ for all real values of x. Hence $f(x)$ is strictly concave.
(c) $f = 2x_1^3 - 6x_2^2$:

$$\mathbf{H}(\mathbf{X}) = \begin{bmatrix} \partial^2 f/\partial x_1^2 & \partial^2 f/\partial x_1\,\partial x_2 \\ \partial^2 f/\partial x_1\,\partial x_2 & \partial^2 f/\partial x_2^2 \end{bmatrix} = \begin{bmatrix} 12x_1 & 0 \\ 0 & -12 \end{bmatrix}$$

Here $\partial^2 f/\partial x_1^2 = 12x_1 \leq 0$ for $x_1 \leq 0$ and ≥ 0 for $x_1 \geq 0$, and

$$\left| \mathbf{H}(\mathbf{X}) \right| = -144x_1 \geq 0 \quad \text{for} \quad x_1 \leq 0 \quad \text{and} \quad \leq 0 \quad \text{for } x_1 \geq 0$$

Hence $\mathbf{H}(\mathbf{X})$ will be negative semidefinite and $f(\mathbf{X})$ is concave for $x_1 \leq 0$.

(d) $f = 4x_1^2 + 3x_2^2 + 5x_3^2 + 6x_1x_2 + x_1x_3 - 3x_1 - 2x_2 + 15$:

$$\mathbf{H(X)} = \begin{bmatrix} \partial^2 f/\partial x_1^2 & \partial^2 f/\partial x_1\,\partial x_2 & \partial^2 f/\partial x_1\,\partial x_3 \\ \partial^2 f/\partial x_1\,\partial x_2 & \partial^2 f/\partial x_2^2 & \partial^2 f/\partial x_2\,\partial x_3 \\ \partial^2 f/\partial x_1\partial x_3 & \partial^2 f/\partial x_2\,\partial x_3 & \partial^2 f/\partial x_3^2 \end{bmatrix}$$

$$= \begin{bmatrix} 8 & 6 & 1 \\ 6 & 6 & 0 \\ 1 & 0 & 10 \end{bmatrix}$$

Here the principal minors are given by

$$|8| = 8 > 0$$

$$\begin{vmatrix} 8 & 6 \\ 6 & 6 \end{vmatrix} = 12 > 0$$

$$\begin{vmatrix} 8 & 6 & 1 \\ 6 & 6 & 0 \\ 1 & 0 & 10 \end{vmatrix} = 114 > 0$$

and hence the matrix $\mathbf{H(X)}$ is positive definite for all real values of x_1, x_2, and x_3. Therefore, $f(\mathbf{X})$ is a strictly convex function.

B

Some Computational Aspects
of Optimization

Several methods were presented for solving different types of optimization problems in Chapters 3 to 14. This appendix is intended to give some guidance to the reader in choosing a suitable method for solving a particular problem along with some computational details. Most of the discussion is aimed at the solution of nonlinear programming problems.

B.1 CHOICE OF METHOD

Several factors are to be considered in deciding a particular method to solve a given optimization problem. Some of them are

1. The type of problem to be solved (general nonlinear programming problem, geometric programming problem, etc.)
2. The availability of a ready-made computer program
3. The calender time required for the development of a program
4. The necessity of derivatives of the functions f and g_j, $j = 1, 2, \ldots, m$
5. The available knowledge about the efficiency of the method
6. The accuracy of the solution desired
7. The programming language and quality of coding desired
8. The robustness and dependability of the method in finding the true optimum solution
9. The generality of the program for solving other problems
10. The ease with which the program can be used and its output interpreted

B.2 COMPARISON OF UNCONSTRAINED METHODS

A number of studies have been made to evaluate the various unconstrained minimization methods. Moré, Garbow, and Hillstrom [B.1] provided a collection of 35 test functions for testing the reliability and robustness of unconstrained minimization software. The performance of eight unconstrained minimization methods was evaluated by Box [B.2] using a set of test problems with up to 20 variables. Straeter and Hogge [B.3] compared four gradient-based unconstrained optimization techniques using two test problems.

A comparison of several variable metric algorithms was made by Shanno and Phua [B.4]. Sargent and Sebastian presented numerical experiences with unconstrained minimization algorithms [B.5]. On the basis of these studies, the following general conclusions can be drawn.

If the first and second derivatives of the objective function (f) can be evaluated easily (either in closed form or by a finite-difference scheme), and if the number of design variables is not large ($n \leq 50$), Newton's method can be used effectively. For n greater than about 50, the storage and inversion of the Hessian matrix at each stage becomes quite tedious and the variable metric methods might prove to be more useful. As the problem size increases (beyond $n = 100$ or so), the conjugate gradient method becomes more powerful.

In many practical problems, the first derivatives of f can be computed more accurately than the second derivatives. In such cases, the BFGS and DFP methods become an obvious choice of minimization. Of these two, the BFGS method is more stable and efficient. If the evaluation of the derivatives of f is extremely difficult or if the function does not possess continuous derivatives, Powell's method can be used to solve the problem efficiently.

With regard to the one-dimensional minimization required in all the unconstrained methods, the Newton and cubic interpolation methods are most efficient when the derivatives of f are available. Otherwise, the Fibonacci or the golden section method has to be used.

B.3   COMPARISON OF CONSTRAINED METHODS

The comparative evaluation of nonlinear programming techniques was conducted by several investigators. Colville [B.6] compared the efficiencies of 30 codes using eight test problems that involve 3 to 16 design variables and 0 to 14 constraints. However, the codes were tested at different sites on different computers and hence the study was not considered reliable. Eason and Fenton [B.7] conducted a comparative study of 20 codes using 13 problems that also included the problems used by Colville. However, their study was confined primarily to penalty function-type methods. Sandgren and Ragsdell [B.8] studied the relative efficiencies of the leading nonlinear programming methods of the day more systematically. They studied 24 codes using 35 problems, including some of those used by Colville and Eason and Fenton.

The number of design variables varied from 2 to 48 and the number of constraints ranged from 0 to 19; some problems involved equality constraints, too. They found the GRG method to be most robust and efficient followed by the exterior and interior penalty function methods.

Schittkowski published the results of his study of nonlinear programming codes in 1980 [B.9]. He experimented with 20 codes on 180 randomly generated test problems using multiple starting points. Based on his study, the sequential quadratic programming was found to be most efficient, followed by the GRG, method of multipliers, and penalty function methods, in that order. Similar comparative studies of geometric programming codes were also conducted [B.10–B.12]. Although the studies above were quite extensive, the conclusion may not be of much use in practice since the studies were limited to relatively few methods and further they are limited to specially formulated test problems that are not related to real-life problems. Thus each new practical problem has to be tackled almost independently based on past experience. The following guidelines are applicable for a general problem.

The sequential quadratic programming approach can be used for solving a variety of problems efficiently. The GRG method and Zoutendijk's method of feasible directions, although slightly less efficient, can also be used for the efficient solution of constrained problems. The ALM and penalty function methods are less efficient but are robust and reliable in finding the solution of constrained problems.

B.4   AVAILABILITY OF COMPUTER PROGRAMS

Many computer programs are available to solve nonlinear programming problems. Notable among these is the book by Kuester and Mize [B.13], which gives Fortran programs for solving linear, quadratic, geometric, dynamic, and nonlinear programming problems. During practical computations, it is important to note that a method that works well for a given class of problems may work poorly for others. Hence it is usually necessary to try more than one method to solve a particular problem efficiently. Further, the efficiency of any nonlinear programming method depends largely on the values of adjustable parameters such as starting point, step length, and convergence requirements. Hence a proper set of values to these adjustable parameters can be given only by using a trial-and-error procedure or through experience gained in working with the method for similar problems. It is also desirable to run the program with different starting points to avoid local and false optima. It is advisable to test the two convergence criteria stated in Section 7.21 before accepting a point as a local minimum.

Moré and Wright present information on the current state of numerical optimization software in [B.16]. Several software systems such as IMSL, MATLAB, and ACM contain programs to solve optimization problems. The relevant addresses are

IMSL
7500 Bellaire Boulevard
Houston, TX 77036

MATLAB
The MathWorks, Inc.
24 Prime Park Way
Natick, MA 01760

ACM Distribution Service
c/o International Mathematics and Statistics Service
7500 Bellaire Boulevard
Houston, TX 77036

In addition, the commercial structural optimization packages listed in Table B.1 are available in the market [B.14, B.15]. Most of these softwares are based on a finite-element-based analysis for objective and constraint function evaluations and use several types of approximation strategies.

Table B.1   Summary of Some Structural Optimization Packages

Software system (program)	Source (developer)	Capabilities and characteristics
ASTROS (Automated STRuctural Optimization System)	Air Force Wright Laboratories FIBRA Wright-Patterson Air Force Base, OH 45433-6553	Structural optimization with static, eigenvalue, modal analysis, and flutter constraints; approximation concepts; compatibility with NASTRAN; sensitivity analysis
ANSYS	Swanson Analysis Systems, Inc. P.O. Box 65 Johnson Road Houston, PA 15342-0065	Optimum design based on curve-fitting technique to approximate the response using several trial design vectors
MSC/NASTRAN MacNeal Schwendler Corporation/NAsa STRuctural ANalysis)	MacNeal-Schwendler Corporation 15 Colorado Boulevard Los Angeles, CA 90041	Structural optimization capability based on static, natural frequency, and buckling analysis; approximation concepts and sensitivity analysis
NISAOPT	Engineering Mechanics Research Corporation P.O. Box 696 Troy, MI 48099	Minimum-weight design subject to displacement, stress, natural frequency and buckling constraints; shape optimization
GENESIS	VMA Exngineering Inc. Manderin Avenue, Suite F Goleta, CA 93117	Structural optimization; approximation concepts used to tightly couple the analysis and redesign tasks

B.5   SCALING OF DESIGN VARIABLES AND CONSTRAINTS

In some problems there may be an enormous difference in scale between variables due to difference in dimensions. For example, if the speed of the engine (n) and the cylinder wall thickness (t) are taken as design variables in the design of an IC engine, n will be of the order of 10^3 (revolutions per minute) and t will be of the order of 1 (cm). These differences in scale of the variables may cause some difficulties while selecting increments for step lengths or calculating numerical derivatives. Sometimes the objective function contours will be distorted due to these scale disparities. Hence it is a good practice to scale the variables so that all the variables will be dimensionless and vary between 0 and 1 approximately. For scaling the variables, it is necessary to establish an approximate range for each variable. For this we can take some estimates (based on judgment and experience) for the lower and upper limits on $x_i (x_i^{\min}$ and

$x_i^{\max}$), $i = 1, 2, \ldots, n$. The values of these bounds are not critical and there will not be any harm even if they span partially the infeasible domain. Another aspect of scaling is encountered with constraint functions. This becomes necessary whenever the values of the constraint functions differ by large magnitudes. This aspect of scaling (normalization) of constraints was discussed in Section 7.13.

B.6   COMPUTER PROGRAMS FOR MODERN METHODS OF OPTIMIZATION

Fuzzy Logic Toolbox.   Matlab has a fuzzy logic toolbox for designing systems based on fuggy logic. Graphical user interfaces (GUI) are available to guide the user through the steps of fuzzy interface system design. The toolbox can be used to model complex system behaviors using simple logic rules and then implement the rules in a fuzzy interface system. Fuzzy optimization can be implemented using fuzzy logic toolbox in conjunction with an optimization program such as fmincon.

Genetic Algorithm and Direct Search Toolbox.   The genetic algorithm and direct search toolbox, which can be used to solve problems that are difficult to solve with traditional optimization techniques, is available with Matlab. The genetic algorithm of the toolbox can be used when the function, such as the objective or constraint function, is discontinuous, highly nonlinear, stochastic, or has unreliable or undefined derivatives. In this toolbox also, graphical user interfaces (GUI) are available for quick setting up of problems, selecting algorithmic options, and monitoring progress. Naturally, the options of creating initial population, fitness scaling, parent selection, crossover and mutation are available in the toolbox. The Matlab optimization programs (using direct search methods) can be integrated with the genetic algorithm.

Neural Network Toolbox.   The neural network toolbox is available with Matlab for designing, implementing, visualizing and simulating neural networks. The GUI available with the toolbox helps in creating, training and simulating neural networks. It permits modular network representation to have any number of input-setting layers and network interconnection and a graphical view of the network architecture. Optimization programs can be used in conjunction with the functions of the neural network toolbox to accomplish neural network-based optimization. The neural network toolbox can also be used to apply neural networks for the identification and control of nonlinear systems.

Simulated Annealing Algorithm.   An m-file to implement the simulated annealing algorithm to solve function minimization problems in the Matlab environment was created by Joachim Vandekerckhove. The link is given below:

http://www.mathworks.com/matlabcentral/fileexchange/10548

Particle Swarm Optimization.   An m-file to implement the particle swarm optimization method in the Matlab environment was created by Wael Korani. The link is given below:

http://www.mathworks.com/matlabcentral/fileexchange/20205

Ant Colony Optimization.   An m-file to implement the ant colony optimization method in the Matlab environment for the solution of symmetrical and unsymmetrical traveling salesman problem was created by H. Wang. The link is given below:

http://www.mathworks.com/matlabcentral/fileexchange/14543

Multiobjective Optimization.   An m-file to implement multiobjective optimization using evolutionary algorithms (based on nondominated sorting genetic algorithm, abbreviated NSGA) in the Matlab environment was created by Arvind Seshadri. The link is given below:

http://www.mathworks.com/matlabcentral/fileexchange/10429

REFERENCES AND BIBLIOGRAPHY

B.1   J. J. Moré, B. S. Garbow, and K. E. Hillstrom, Testing unconstrained optimization software, *ACM Transactions on Mathematical Software*, Vol. 7, No. 1, pp. 17–41, 1981.

B.2   M. J. Box, A comparison of several current optimization methods, and the use of transformations in constrained problems, *Computer Journal*, Vol. 9, No. 1, pp. 67–77, 1966.

B.3   T. A. Straeter and J. E. Hogge, A comparison of gradient dependent techniques for the minimization of an unconstrained function of several variables, *AIAA Journal*, Vol. 8, No. 12, pp. 2226–2229, 1970.

B.4   D. F. Shanno and K. H. Phua, Numerical comparison of several variable-metric algorithms, *Journal of Optimization Theory and Applications*, Vol. 25, No. 4, pp. 507–518, 1978.

B.5   R.W.H. Sargent and D. J. Sebastian, Numerical exprience with algorithms for unconstrained minimization, pp. 45–113 in *Numerical Methods for Nonlinear Optimization*, F. A. Lootsma, Ed., Academic Press, London, 1972.

B.6   A. R. Colville, *A Comparative Study of Nonlinear Programming Codes*, Technical Report 320–2949, IBM New York Scientific Center, June 1968.

B.7   E. D. Eason and R. G. Fenton, A comparison of numerical optimization methods for engineering design, *ASME Journal of Engineering for Industry*, Vol. 96, No. 1, pp. 196–200, 1974.

B.8   E. Sandgren and K. M. Ragsdell, The utility of nonlinear programming algorithms: A comparative study, Parts I and II, *ASME Journal of Mechanical Design*, Vol. 102, No. 3, pp. 540–551, 1980.

B.9   K. Schittkowski, *Nonlinear Programming Codes: Information, Tests, Performance*, Lecture Notes in Economics and Mathematical Systems, Vol. 183, Springer-Verlag, New York, 1980.

B.10 P.V.L.N. Sarma, X. M. Martens, G. V. Reklaitis, and M. J. Rijckaert, A comparison of computational strategies for geometric programs, *Journal of Optimization Theory and Applications*, Vol. 26, No. 2, pp. 185–203, 1978.

B.11 J. E. Fattler, Y. T. Sin, R. R. Root, K. M. Ragsdell, and G. V. Reklaitis, On the computational utility of posynomial geometric programming solution methods, *Mathematical Programming*, Vol. 22, pp. 163–201, 1982.

B.12 R. S. Dembo, The current state-of-the-art of algorithms and computer software for geometric programming, *Journal of Optimization Theory and Applications*, Vol. 26, p. 149, 1978.

B.13   J. L. Kuester and J. H. Mize, *Optimization Techniques with Fortran*, McGraw-Hill, New York, 1973.

B.14   E. H. Johnson, Tools for structural optimization, Chapter 29, pp. 851–863, in *Structural Optimization: Status and Promise*, M. P. Kamat, Ed., AIAA, Washington, DC, 1993.

B.15   H.R.E.M. Hörnlein and K. Schittkowski, *Software Systems for Structural Optimization*, Birkhauser, Basel, 1993.

B.16   J. J. Moré and S. J. Wright, *Optimization Software Guide*, Society of Industrial and Applied Mathematics, Philadelphia, PA, 1993.

C

Introduction to MATLAB®

MATLAB, derived from MATrix LABoratory, is a software package that was originally developed in the late 1970s for the solution of scientific and engineering problems. The software can be used to execute a single statement or a list of statements, called a script or m-file. MATLAB family includes the Optimization Toolbox, which is a library of programs or m-files to solve different types of optimization problems. Some basic features of MATLAB are summarized in this appendix.

C.1 FEATURES AND SPECIAL CHARACTERS

Some of the important features and special characters used in MATLAB are indicated below:

1. Symbol ≫	This is the default *prompt* symbol in MATLAB
2. Symbol ;	A semicolon at the end of a line *avoids the echoing* the information entered before the semicolon
3. Symbol ...	Three periods at the end of a line indicates the *continuation* of the code in the next line
4. help command_name	This displays information on different ways the command can be used
5. Symbol %	Any text after this symbol is considered a *comment* and will not be operational

6. MATLAB is case sensitive. Uppercase and lowercase letters are treated separately.

7. MATLAB assumes all variables to be arrays. As such, separate dimension statements are not needed. Scalar quantities need not be given as arrays.

8. Names of variables: variable names should start with a letter and can have a length of up to 31 characters in any combination of letters, digits, and underscores.

9. The symbols for the basic arithmetic operations of addition, subtraction, multiplication, division, and exponentiation are $+$, $-$, $*$, $/$, and $^\wedge$, respectively.

10. MATLAB has some built-in variable names and, as such, we should avoid using those names for variables in writing a MATLAB program or m-file. Examples of built-in names: pi (for π), sin (for sine of an angle), etc.

C.2 DEFINING MATRICES IN MATLAB

Before performing arithmetic operations or using them in developing MATLAB programs or m-files, the relevant matrices need to be defined using statements such as the following.

1. A row vector or $1 \times n$ matrix, denoted A, can be defined by enclosing its elements in brackets and separated by either spaces or commas.

    Example: $A = [1 \quad 2 \quad 3]$

2. A column vector or $n \times 1$ matrix, denoted A, can be defined by entering its elements in different lines or in a single line using a semicolon to separate them or in a single line using a row vector with a prime on the right-side bracket (to denote the transpose).

    Example: $A = \begin{bmatrix} 1 \\ 2 \\ 3 \end{bmatrix}$, $A = [1; \quad 2; \quad 3]$, or $A = [1 \quad 2 \quad 3]'$.

3. A matrix of size $m \times n$, denoted A, can be defined as follows (similar to the procedure used for a column vector).

    Example: $A = \begin{bmatrix} 1 & 2 & 3 \\ 4 & 5 & 6 \\ 7 & 8 & 9 \end{bmatrix}$, or $A = [1 \quad 2 \quad 3; \quad 4 \quad 5 \quad 6; \quad 7 \quad 8 \quad 9]$.

4. Definitions of some special matrices:

    $A = $ eye (3)
    implies an identity matrix of order 3: $A = \begin{bmatrix} 1 & 0 & 0 \\ 0 & 1 & 0 \\ 0 & 0 & 1 \end{bmatrix}$.

    $A = $ ones (3)
    implies a square matrix of order 3 with all elements equal to one: $A = \begin{bmatrix} 1 & 1 & 1 \\ 1 & 1 & 1 \\ 1 & 1 & 1 \end{bmatrix}$.

    $A = $ zeros (2, 3)
    implies a 2×3 matrix with all elements equal to zero: $A = \begin{bmatrix} 0 & 0 & 0 \\ 0 & 0 & 0 \end{bmatrix}$.

5. Some uses of the colon operator (:):

    (i) To generate all numbers between 100 and 50 in increments of -7

    $$>> \quad 100 : -7 : 50$$

    This command generates the numbers 100   93   86   79   65   58   51

(ii) To generate all numbers between 0 and π in increments of $\pi/6$

$$>> 0 : pi/6 : pi$$

This command generates the numbers
0   0.5236   1.0472   1.5708   2.0944   2.6180   3.1416

C.3 CREATING m-FILES

MATLAB can be used in an interactive mode by typing each command from the keyboard. In this mode, MATLAB performs the operations much like an extended calculator. However, there are situations in which this mode of operation is inefficient. For example, if the same set of commands is to be repeated a number of times with different values of the input parameters, developing a MATLAB program will be quicker and efficient.

A MATLAB program consists of a sequence of MATLAB instructions written outside MATLAB and then executed in MATLAB as a single block of commands. Such a program is called a script file, or m-file. It is necessary to give a name to the script file. The name should end with .m (a dot followed by the letter m). A typical m-file (called `fibo.m`) is

```
file "fibo.m"

% m-file to compute Fibonacci numbers
f=[1 1];
i=1;
while f(i)+f(i+1)<1000
    f(i+2)=f(i)+f(i+1);
    i=i+1;
end
```

C.4 OPTIMIZATION TOOLBOX

The Optimization Toolbox includes programs or m-files that can be used to solve different types of optimization problems. The following publication gives information on the optimization toolbox, including algorithms and examples for different programs:

T. F. Coleman, M. A. Branch, and A. Grace, *Optimization Toolbox—for Use with MATLAB®*, User's Guide, Version 2, Math Works, Inc., Natick, MA, 1999.

The use of any program or m-file in the optimization toolbox requires the following:

- Selecting the appropriate program or m-file to solve the specific problem at hand.
- Formulation of the optimization problem in the format expected by MATLAB. In general, this involves stating the objective function in a specific form such as a "minimization" type and the constraints in a specific form such as "less than or equal to zero" type.

- Distinction between linear and nonlinear constraints.
- Identification of lower and upper bounds on design variables.
- Setting/changing the parameters of the optimization algorithm (based on the available options).

Using MATLAB Programs. Each program or m-file in MATLAB can be implemented in several ways. The details can be found either in the reference given above or online using the help command. For illustration, the help command and the response for the program *fmincon* are shown below.

The function *fmincon* can be used in 12 different ways as indicated below (by the help command). The differences depend on the available data in the mathematical model of the problem and the information required from the solution of the problem. In using the different function calls, any data missing in the mathematical model of the optimization problem need to be indicated using a null vector as [ ]. Note that the response is edited for brevity.

```
>> help fmincon

  FMINCON Finds the constrained minimum of a function of
several variables.
  FMINCON solves problems of the form:
    min F(X) subject to:
  A*X <= B, Aeq*X = Beq (linear constraints)
  C(X) <= 0, Ceq(X) = 0 (nonlinear constraints)
  LB <= X <= UB

X=FMINCON (FUN,X0,A,B)
X=FMINCON (FUN,X0,A,B,Aeq,Beq)
X=FMINCON (FUN,X0,A,B,Aeq,Beq,LB,UB)
X=FMINCON (FUN,X0,A,B,Aeq,Beq,LB,UB,NONLCON)
X=FMINCON (FUN,X0,A,B,Aeq,Beq,LB,UB,NONLCON,OPTIONS)
X=FMINCON (FUN,X0,A,B,Aeq,Beq,LB,UB,NONLCON,OPTIONS,...
 P1,P2,...)
[X,FVAL] = FMINCON (FUN,X0,...)
[X,FVAL,EXITFLAG] = FMINCON (FUN,X0,...)
[X,FVAL,EXITFLAG,OUTPUT]=FMINCON (FUN,X0,...)
[X,FVAL,EXITFLAG,OUTPUT,LAMBDA] =FMINCON (FUN,X0,...)
[X,FVAL,EXITFLAG,OUTPUT,LAMBDA,GRAD]=FMINCON (FUN,X0,...)
[X,FVAL,EXITFLAG,OUTPUT,LAMBDA,GRAD,HESSIAN]=FMINCON
 (FUN,X0,...).
```

The solution of representative constrained nonlinear programming problems using the function fmincon is illustrated in Chapters 1 and 7.

Answers to Selected Problems

CHAPTER 1

1.1 Min. $f = 5x_A - 80x_B + 160x_C + 15x_D$, $0.05x_A + 0.05x_B + 0.1x_C + 0.15x_D \leq 1000$, $0.1x_A + 0.15x_B + 0.2x_C + 0.05x_D \leq 2000$, $0.05x_A + 0.1x_B + 0.1x_C + 0.15x_D \leq 1500$, $x_A \geq 5000$, $x_B \geq 0$, $x_C \geq 0$, $x_D \geq 4000$

1.2(a) $\mathbf{X}^* = \{0.65, 0.53521\}$ **(b)** $\mathbf{X}^* = \{0.9, 2.5\}$

(c) $\mathbf{X}^* = \{0.65, 0.53521\}$ **1.5** $x_1^* = x_2^* = 300$

1.9(a) $R_1^* = 4.472$, $R_2^* = 2.236$ **(b)** $R_1^* = 3.536$, $R_2^* = 3.536$

(c) $R_1^* = 6.67$, $R_2^* = 3.33$

1.11(a) $y_1 = \ln x_1$, $y_2 = \ln x_2$, $\ln f = 2y_1 + 3y_2$

(b) $f = 10^{y_2 x_2}$, $x_1 = 10^{y_2}$, $\ln(\log_{10} f) = \ln(\log_{10} x_1) + \ln x_2$

1.14 $x_{ij} = 1$ if city j is visited immediately after city i, and $= 0$ otherwise.
Find $\{x_{ij}\}$ to minimize $f = \sum\limits_{i=1}^{n} \sum\limits_{j=1}^{n} d_{ij}x_{ij}$ subject to $\sum\limits_{i=1}^{n} x_{ij} = 1$ $(i = 1, 2, \ldots, n)$,

$i \neq j$ and $\sum\limits_{j=1}^{n} x_{ij} = 1$ $(i = 1, 2, \ldots, n)$, $j \neq i$

1.19 Min. $f = \rho l b d$, $\dfrac{P_y}{bd} + \dfrac{6P_x l}{bd^2} \leq \sigma_y$, $\dfrac{P_y}{bd} + \dfrac{6P_x l}{bd^2} \leq \dfrac{\pi^2 E d^2}{48 l^2}$, $b \geq 0.5$, $b \leq 2d$.

1.25 Max. $f = \frac{2}{3}t_m + \frac{3}{5}t_d$, $t_m + t_d \leq 40$, $t_d \geq 1.25t_m$, $0 \leq t_m \leq 24$, $0 \leq t_d \leq 20$.

1.29 Min. $f = \pi x_3 [x_1^2 - (x_1 - x_2)^2] + \frac{4}{3}\pi[x_1^3 - (x_1 - x_4)^3]$, $\pi x_3 (x_1 - x_2)^2 + \frac{4}{3}\pi(x_1 - x_4)^3 - 4{,}619{,}606 \leq 0$, $x_2 - \dfrac{pR_0}{S.e + 0.4p} \leq 0$, $x_4 - \dfrac{pR_0}{S.e + 0.8p} \leq 0$

CHAPTER 2

2.1 $r^* = R$ **2.3** $x^* = 1.5$ (inflection point)

2.5 $x = -1$ (not min, not max), $x = 2$ (min)   **2.9** $d = \left(\dfrac{D^5}{8fl}\right)^{1/4}$

2.10 35.36 m   **2.11(a)** $79.28°$ **(b)** 0.911 from end of stroke

2.13 positive semidefinite   **2.15** positive definite

2.17 negative definite   **2.19** indefinite

2.21 $x_1^* = 0.2507\,\text{m}, \ x_2^* = 5.0879 \times 10^{-3}\,\text{m}$

2.23 $a = 328, \ b = -376$ **2.26** $x^* = 27, \ y^* = 21$

2.27 $x^* = 100$ **2.28(a)** minimum   **(b)** minimum

(c) saddle point   **(d)** none   **2.30** saddle point at $(0, 0)$

2.33 $dx_1 = $ arbitrary, $dx_2 = 0$ **2.36** radius $= 2r/3$, length $= h/3$

2.38 length $= (a^{2/3} + b^{2/3})^{3/2}$ **2.40** $h^* = \left(\dfrac{4V}{\pi}\right)^{1/3}$, $r^* = \dfrac{h*}{2}$

2.41 $x_1^* = x_3^* = (S/3)^{1/2}, \ x_2^* = (S/12)^{1/2}$

2.43 $d^* = \frac{1}{6}\{(a + b) - \sqrt{a^2 - ab + b^2}\}$ **2.47** $200\,\text{mm} \times 250\,\text{mm}$

2.50 $\mathbf{X}^* = \{4, 2, 2\}$ **2.53** $198.43\,\text{ft} \times 113.39\,\text{ft}$

2.55(a) $f_{\text{new}}^* = 15\pi$ **(b)** $f_{\text{new}}^* = 18\pi$ **2.57(a)** $f^* = 1/3$

(b) $f^* = -1/9$ **2.61** $\mathbf{X}_2$ is local minimum

2.63(a) Kuhn–Tucker conditions satisfied

(b) $\lambda_1 = 0.4, \ \lambda_2 = 0.2, \ \lambda_3 = 0$ **2.65(a)** $\mathbf{S} = \{1, -3\}$ **(b)** none

2.67 optimum   **2.69** $x_1^* = \frac{3}{4}, \ x_2^* = 4\frac{9}{16}$ **2.73** convex   **2.75** none optimum

CHAPTER 3

3.3 $x_1 = 1, \ x_2 = 2, \ x_3 = 3$ **3.5** $x_1 = 2, \ x_2 = 4, \ x_3 = 6$

3.7 $x_1^* = 1/3, \ x_2^* = 4/3$ **3.9** $x_1^* = 2\frac{2}{5}, \ x_2^* = 1\frac{1}{5}$

3.12 $x^* = 3\frac{3}{11}, \ y^* = 3\frac{2}{11}$ **3.15** $x^* = 5\frac{1}{13}, \ y^* = 1\frac{1}{13}$

3.17 all points on line joining $(2, 10)$ and $(7.4286, 15.4286)$

3.18 $x^* = 10, \ y^* = 18$ **3.20** $x^* = 9/7, \ y^* = 40/7$

3.23 $x^* = 6, \ y^* = 1$ **3.25** $x^* = 6, \ y^* = 0$

3.27 $x^* = 75/8, \ y^* = 27/8$ **3.29** $x^* = 3, \ y^* = -2.5$

3.31 $x^* = 4, \ y^* = 0$ **3.33** unbounded   **3.35** $x^* = 4/7, \ y^* = 30/7$

3.37 $x^* = 36/7, \ y^* = 15/7$ **3.39** $x^* = 16/5, \ y^* = 1/5$

3.41 infeasible   **3.43** unbounded

3.48 $x_1^* = 3000.0, \ x_2^* = 416.7, \ x_3^* = 1200.0$

3.50 x_1^* (barley) $= 40, \ x_2^* = x_3^* = x_4^* = 0, \ x_5^*$ (leased) $= 160$

3.55 $x_A^* = 1.5$, $x_B^* = 0$ **3.57** $x_m^* = 16$, $x_d^* = 20$

3.60 $x^* = 36/11$, $y^* = 35/11$

3.66 all points on the line joining (7.4286, 15.4286) and (10, 18)

3.71 $x^* = 3.6207$, $y^* = 8.4483$ **3.75** $x^* = 2/7$, $y^* = 30/7$

3.79 $x^* = 56/23$, $y^* = 45/23$ **3.85** $x^* = -4/3$, $y^* = 7$

3.89 $x^* = 0$, $y^* = 3$

3.92 (x_1, x_2) = amounts of mixed nuts (A, B) used, lb. $x_1^* = 80/7$, $x_2^* = 120/7$

3.94 $x_A^* = 62.5$, $x_B^* = 31.25$

3.96 x_i = number of units of P_i produced per week. $x_1^* = 100/3$, $x_2^* = 250/3$

3.99 (x_1, x_2) = number of units of (A, B) sold per month. $x_1^* = 19.17$, $x_2^* = 45$

3.102 x_i = number of days used in a month for process type i $(i = 1, 2, 3, 4)$. $x_1^* = 30$, $x_2^* = x_3^* = x_4^* = 0$

CHAPTER 4

4.1 $\mathbf{X}^* = \{2.333, 1.333, 0, 0\}$

4.3 $x_i^* = 0$, $i = 1, 2, 3$, $x_4^* = 2/5$, $x_5^* = 4/5$ **4.5** solution unbounded

4.9 $x_i^* = 0$, $i = 1, 2, 5, 6, 7$, $x_3^* = 0.5$, $x_4^* = 1.5$

4.12 $x_1^* = 2.35$, $x_2^* = 0.1$, $x_3^* = 2.7$, $x_4^* = 1.2$

4.15 $x_1^* = x_2^* = x_3^* = x_6^* = 0$, $x_4^* = 120$, $x_5^* = 100$

4.17 optimum solution remains same, $f_{new}^* = -27,600/3$

4.19 (x_1, x_2, x_3, x_4) = number of units of products (A, B, C, D) produced. $x_1^* = 4000/3$, $x_2^* = x_3^* = 0$, $x_4^* = 200/3$

4.23 $x_1^* = 1000/3$, $x_2^* = x_3^* = 0$, $x_4^* = 800/3$

4.29 $x_1^* = 0$, $x_2^* = 0.5$ **4.31** $x_1^* = 0$, $x_2^* = 0.5$

4.33 infinite solutions   **4.35** $x_1^* = 0$, $x_2^* = 0.5$

4.37 $\mathbf{X}^{(2)} = \{0.3367, 0.3112, 0.3250\}$

4.40 $x_1^* = 0.9815$, $x_2^* = 1.2323$, $x_3^* = 0.4471$

CHAPTER 5

5.2 0.484   **5.3** 0.481   **5.4** 0.49   **5.6** 0.8   **5.9** 0.7817

5.11(a) 0.786151   **(b)** 0.786142   **(c)** 0.786192   **5.14(a)** 999

(b) 20 **(c)** 19 **(d)** 14 **(e)** 14 **5.17(a)** 2.7814

(b) 2.7183 **5.18(a)** 2.7183 **(b)** 2.7289 **(c)** 2.7183

5.20 0.25 **5.21** 0.001257 **5.22** 0.00126 **5.24** 0.00125631

CHAPTER 6

6.1 Min. $f = P_0(0.5u_1^2 + 0.5u_2^2 - u_1u_2 - u_2)$

6.2 $\tilde{f}_1 = 7.0751$, $\tilde{f}_2 = 74.8087$ where $\tilde{f} = \dfrac{3f\rho l^4}{Eh^2}$

6.4 $x_1 = 65.567$, $x_2 = 52.974$ **6.5** $x_1^* = 4.5454$, $x_2^* = 5.4545$

6.7 $f = 4250x_1^2 - 1000x_1x_2 - 2500x_1x_3 + 1500x_2^2 - 500x_2x_3 + 5750x_3^2 - 1000x_1 - 2000x_2 - 3000x_3$, $\mathbf{X}^* = \{0.3241, 0.8360, 0.3677\}$

6.9 $\mathbf{X}^* \approx \{1, 1\}$ **6.12** $\mathbf{X}^* = \{0.9465, 2.0615, 2.9671\}$

6.14 $f(z_1, z_2) = -5 + 1.0429z_1 - 0.7244z_2 + 0.5z_1^2 + 0.5z_2^2$

6.16(a) yes **(b)** no **6.19(a)** 60,002.0 **(b)** 241.3729

6.30 $\mathbf{X}_1 = \{2, -1, -8\}$ $\mathbf{X}_2 = \{2, -0.7, -8\}$ $\mathbf{X}_3 = \{2.26, -0.85, -8\}$ $\mathbf{X}_4 = \{2.15, -0.74, -7.755\}$ **6.35** $\mathbf{X}_2 = \{5.57, 0\}$, $f_2 > f_1$

6.38 $x_1^* = 1$, $x_2^* = 1$ **6.45** $\mathbf{X}_5 = \{2.0869, 1.7390\}$, $f_5 = -8.3477$

6.47 $\mathbf{X}^* = \{-2, 1, 4\}$ **6.48** $x^* = 1.1423$, $y^* = 0.8337$

6.50 $x_1^* = 1.698105$, $x_2^* = 0.883407$ **6.52** $\mathbf{X}^* = \{5, -8\}$

6.55(a) no **(b)** yes

CHAPTER 7

7.1 $\mathbf{X}^* = \{2, 3\}$, $f^* = -50$

7.6(a) Min. $f = 12x_1^2 + 30x_2^2 - 8x_1x_2 - 22x_1 + 60x_2 - 78$, $x_2 + 2 = 0$, $x_1 + x_2 \leq 0$

(b) Min. $f = 18x_1 - 68x_2 - 70$, $x_2 + 2 = 0$, $x_1 + x_2 \leq 0$

7.8 $\mathbf{X}^* = \{1.74558, 1.95265\}$, $f^* = -9.23478$

7.11 Max. $f = 3.5483d^4w$, $2.2227 \times 10^{-6}d^4 - 1 \leq 0$, $0.2223d^2w - 150 \leq 0$, $d \leq 25$ **7.13** $-8s_1 + 4s_2 < 0$, $s_1 + 2s_2 \leq 0$, $-s_1 \leq 0$

7.15 $\mathbf{X}^* = \{0.75, 4.56249\}$, $f^* = 0.25391$

7.18 $\mathbf{X}^* = \{3, 3\}$, $f^* = 18$ **7.21** $x_1^* = 24$ cm, $x_2^* = x_3^* = 12$ cm

7.23(a) $\phi_k = 2x - r_k \left(\dfrac{1}{2 - x} + \dfrac{1}{x - 10} \right)$,

(b) $\phi_k = 2x + r_k(\langle 2 - x \rangle^2 + \langle x - 10 \rangle^2)$

7.27 $x_1^* = 0.989637$, $x_2^* = 1.979274$

7.29 $\frac{1}{4}x_1^2 + \frac{1}{16}x_2^2 - 1 \leq 0$, $x_1/5 + x_2/3 - 1 \leq 0$, $r_1 = 1.5$

7.31 $x_1^* = 4.1$, $x_2^* = 5.9$ **7.34** $\mathbf{X}^* \approx \{0.8984, 0\}$, $f^* \approx 2.2079$

7.36 $\mathbf{X}^* \approx \{1.671, 17.6\}$ **7.39** $x_1 = 0.4028$, $x_2 = 0.8056$

7.42 optimum, $\lambda_1 = \lambda_2 = \dfrac{1}{4\sqrt{2}}$, $\lambda_3 = 11$

7.45 $\mathbf{X}^* \approx \{1.3480, 0.7722, 0.4299\}$, $f^* \approx 0.1154$

CHAPTER 8

8.1 $f \geq 2.268866$ **8.2** $f \geq 3.464102$ **8.3** $f \geq 3$

8.5 radius $= 0.4174$ m, height $= 1.6695$ m

8.6 radius $= 0.3633$ m, height $= 2.9067$ m

8.7 $x_1^* = 1.5 \times 10^6$, $x_2^* = 1.0 \times 10^6$

8.9 $x_1^* = 5.7224$, $x_2^* = 0.8737$, $x_3^* = 7.2813$

8.10 $x_1^* = 1.0845$, $x_2^* = 1.1761$

8.11 $x_1^* = 8.6365$, $x_2^* = 0.9397$, $x_3^* = 6.8219$, $x_4^* = 0.9609$

8.12 $x_1^* = 1.1262$, $x_2^* = 1.1945$, $x_3^* = 1.6575$

8.13 $x_1^* = 2.2629$, $x_2^* = 7.1689$, $x_3^* = 4.5850$

8.14 $x_1^* = 0.3780$, $x_2^* = 0.5345$, $x_3^* = 0.5714$

8.17 $d^* = 0.002808$ m, $D^* = 0.02935$ m

8.18 $V^* = 323.3201$ ft/min, $F^* = 0.005$ in/rev   **8.20** 2

8.22 $R^* = 0.2118$, $L^* = 0.2907$

8.23 $R^* = 1.2821$, $L^* = 0.5266$, $f^* = 16.2056$

CHAPTER 9

9.1 $x_1^* = 2$, $x_2^* = x_3^* = 0$, $x_4^* = 3$ **9.2** *A-B-F-J-K-L-P*

9.3 $n_1 = 2$, $n_2 = 3$, $n_3 = 1$ **9.4** 24,000 ft at *B, C, D*, and *E*

9.5 *D-H-L-K-J-I-M*

9.6 stage 1 $(0, n)$, stage 2 $(0, 2n/3)$, stage 3 $(4n/9, 0)$

9.7 *A* B_1 C_1 D_1 *E*   **9.9** units invested in stations 1, 2, 3: $(0, 2, 1)$

9.10 $x_1^* = 7.5$, $x_2^* = 10.0$ **9.11** $x_1^* = 60$, $x_2^* = 70$, $x_3^* = 80$

9.13 $x_1^* = 5$, $x_2^* = 0$, $x_3^* = 5$, $x_4^* = 0$

CHAPTER 10

10.1 $\mathbf{X}^* = \{2, 1\}$, $f^* = 13$ **10.3** $\mathbf{X}^* = \{0, 9\}$, $f^* = 27$

10.4 $\mathbf{X}^* = \{1, 0\}$, $f^* = 3$ **10.5** $\mathbf{X}^* = \{0, 3\}$, $f^* = 3$

10.6 $\mathbf{X}^* = \{3, 3\}$, $f^* = 39$ **10.7** $\mathbf{X}^* = \{4, 3\}$, $f^* = 10$

10.8 $187 = 1\ 0\ 1\ 1\ 1\ 0\ 1\ 1$ **10.9** $\mathbf{X}^* = \{1, 2, 0\}$, $f^* = 3$

10.12 $\mathbf{X}^* = \{1, 1, 1\}$, $f^* = 18$ **10.13** $\mathbf{X}^* = \{1, 1, 1, 1, 0\}$, $f^* = 9$

10.15 $\mathbf{X}^* = \{4, 0\}$, $f^* = 4$ **10.16** $\mathbf{X}^* = \{2, 2.5\}$, $f^* = 20.5$

CHAPTER 11

11.2 $\overline{V} = \dfrac{2}{\sqrt{\pi h}}$, $\sigma_v = \dfrac{2}{h}\sqrt{\dfrac{3}{8} - \dfrac{1}{\pi}}$ **11.3** $\overline{X} = 3.2$, $\sigma_X = 0.8$

11.4 $a = 769.2308$, $\mu_X = 1$, $\sigma_X = 0.048038$

11.7 $f_X(x) = x + 1.5x^2$, $f_Y(y) = y + 1.5y^2$

11.8 $\sigma_X = 0.006079$ cm, rejects $= 1.32\%$ **11.9** independent

11.10 dependent   **11.11(a)** 0.99904,   **(b)** 0.0475,

(c) $3616\,\text{kg}_\text{f}/\text{cm}^2$ **11.12** 0.6767

11.13 $\overline{R} = 268.9520$ ft, $\sigma_R = 56.1941$ ft, $\overline{R}_{\text{second order}} = 270.1673$ ft

11.15 $\mathbf{X}^* = \{0, 0, 0, 12\}$, $f^* = 12$

11.17(a) $\mathbf{X}^* = \{0.0, 36.93, 174.40\}$, $f^* = 1{,}891.72$

(b) $\mathbf{X}^* =$ same as in (a), $\sigma_f^* = 524.50$

(c) $\mathbf{X}^* =$ same as in (a), $(\overline{f} + \sigma_f)^* = 2{,}416.22$

CHAPTER 12

12.3 $x(t) = c_1 e^t + (2 - c_1)e^{-t} - t$ where c_1 is a constant

12.4 circle of radius $L/(2\pi)$

12.6 $\mathbf{X}^* = \begin{Bmatrix} 0.2 \\ 0.2 \end{Bmatrix}$ in^2 **12.7** $\mathbf{X}^* = \begin{Bmatrix} 1.2169 \\ 0.3805 \end{Bmatrix}$

CHAPTER 13

13.1 Before: $\mathbf{X}_1 = \begin{Bmatrix} 17 \\ 13 \end{Bmatrix}$, $\mathbf{X}_2 = \begin{Bmatrix} 15 \\ 22 \end{Bmatrix}$; After: $\mathbf{X}_1 = \begin{Bmatrix} 23 \\ 22 \end{Bmatrix}$, $\mathbf{X}_2 = \begin{Bmatrix} 9 \\ 13 \end{Bmatrix}$

13.3 (a) 9, (b) 10, (c) 11   **13.4** 10   **13.6** $(i \ j) = (i \ 4)$ **13.8** $x^* = 2$

13.9 $x_1(2) = 2.8297$, $x_2(2) = 1.9345$, $x_3(2) = 1.6362$, $x_4(2) = 1.1887$

13.12 Number of copies of strings 1, 2, 3, 4, 5, 6, 7 are 0, 0, 1, 2, 5, 2, 2, respectively

13.14 String length $= 37$

CHAPTER 14

14.1 $c_1^* = 0.04$, $c_2^* = 0.81$

14.3(a) $\{0.001165, 0.002329, 0.03949, -0.05635\}$,

(b) $\{0.0009705, 0.001941, 0.05273, -0.084102\}$,

(c) $\{0.0009704, 0.001941, 0.05265, -0.08395\}$

14.5 $\left\{ \dfrac{\partial y_i}{\partial x_1} \right\} = \{-0.000582, -0.001165, -0.002329, 0.002329\}$

14.7 $\left\{ \dfrac{\partial y_i}{\partial x_3} \right\} = \{0.4693 \times 10^{-7}, 0.9477 \times 10^{-7}, -0.027948, 0.027947\}$

14.9(a) $\begin{Bmatrix} 0.000125 \\ 0.000458 \end{Bmatrix}$ **(b)** $\begin{Bmatrix} -0.000229 \\ -0.000229 \end{Bmatrix}$, $\begin{Bmatrix} 0.0 \\ 0.000333 \end{Bmatrix}$

(c) $\begin{Bmatrix} -275 \\ 0 \end{Bmatrix}$, $\begin{Bmatrix} 0 \\ 200 \end{Bmatrix}$

14.11 $\dfrac{\partial \lambda_1}{\partial A_2} = 2.28840$, $\dfrac{\partial \lambda_2}{\partial A_2} = 46.8649$, $\dfrac{\partial \mathbf{Y}_1}{\partial A_2} = \begin{Bmatrix} -0.312639 \times 10^{-12} \\ 0.391666 \times 10^{-6} \end{Bmatrix}$,

$\dfrac{\partial \mathbf{Y}_2}{\partial A_2} = \begin{Bmatrix} 0.698492 \times 10^{-8} \\ 0.883790 \times 10^{-2} \end{Bmatrix}$

14.15 $\dfrac{\partial \omega_1}{\partial D} = -1.584664$, $\dfrac{\partial \omega_2}{\partial D} = -2.744719$

14.16 $y^* = 3$, $A_1^* = 0.316228 \times 10^{-7}$, $A_2^* = 0.948683 \times 10^{-7}$, $f^* = 0.6 \times 10^{-6}$

14.18 $y^* = 0.25$, $A_1^* = 1.0$, $A_2^* = 1.0$, $f^* = 43.7565$

14.20 $\mathbf{X}^* = \{0.7635, 1.0540\}$, $f^* = 187.5670$ with $f = 0.625 f_1 + 1061.0 f_2$

14.21 $\mathbf{X}^* = \{0.8, 1.1\}$, $F^* = 3.1267$

14.22 $\mathbf{X}^* = \{0.75, 1.25\}$

Index